Square Foot Costs with RSMeans data

Wafaa Hamitou, Senior Editor

2018
39th annual edition

Chief Data Officer
Noam Reininger

Engineering Director
Bob Mewis, CCP

Contributing Editors
Christopher Babbitt
Sam Babbitt
Michelle Curran
Matthew Doheny (8)
Cheryl Elsmore
Linval Gentles
John Gomes (13, 41)
Derrick Hale, PE (2, 31, 32, 33, 34, 35, 44, 46)
Wafaa Hamitou (11, 12)

Joseph Kelble (14, 21, 22, 23, 25, 26, 27, 28, 48)
Charles Kibbee (1, 4)
Gerard Lafond, PE
Thomas Lane (6, 7)
Genevieve Medeiros
Elisa Mello
Ken Monty
Marilyn Phelan, AIA (9, 10)
Stephen C. Plotner (3, 5)
Callum Riley
Stephen Rosenberg
Jeff Sessions
Gabe Sirota

Matthew Sorrentino
Kevin Souza
Keegan Spraker
Tim Tonello
Jen Walsh
David Yazbek

Product Manager
Andrea Sillah

Production Manager
Debbie Panarelli

Production
Jonathan Forgit
Mary Lou Geary

Sharon Larsen
Sheryl Rose

Technical Support
Judy Abbruzzese
Gary L. Hoitt

Cover Design
Blaire Collins

Data Analytics
Tim Duggan
Todd Glowac
Matthew Kelliher-Gibson

Numbers in italics are the divisional responsibilities for each editor. Please contact the designated editor directly with any questions.

Gordian RSMeans data
Construction Publishers & Consultants
1099 Hingham Street, Suite 201
Rockland, MA 02370
United States of America
1-800-448-8182
www.RSMeans.com

Copyright 2017 by The Gordian Group Inc.
All rights reserved.
Cover photo © iStock.com/ClaudioVentrella

Printed in the United States of America
ISSN 1540-6326
ISBN 978-1-946872-18-0

0058 $362.99 per copy (in United States)
Price is subject to change without prior notice.

Related Data and Services

The residential section of *2018 Square Foot Costs with RSMeans data* is aimed at new residential construction projects valued at up to $1,000,000. The labor rate used in the calculations is for residential construction. The commercial/industrial/institutional section is aimed primarily at new construction projects valued at $1,500,000 and up. The labor rates in the calculations are union.

Our engineers recommend the following products and services to complement *Square Foot Costs with RSMeans data*:

Annual Cost Data Books

2018 Assemblies Costs with RSMeans data
2018 Building Construction Costs with RSMeans data
2018 Residential Costs with RSMeans data
2018 Light Commercial Costs with RSMeans data
2018 Square Foot Renovation Costs with RSMeans data
Life Cycle Cost Package

Reference Books

Estimating Building Costs
RSMeans Estimating Handbook
Green Building: Project Planning & Estimating
How to Estimate with RSMeans data
Plan Reading & Material Takeoff
Project Scheduling & Management for Construction
Universal Design for Style, Comfort & Safety

Seminars and In-House Training

Training for Conceptual Estimating on CD
Training for our online estimating solution
Plan Reading & Material Take Off

RSMeans data Online

For access to the latest cost data, an intuitive search, and an easy-to-use estimate builder, take advantage of the time savings available from our online application. To learn more visit: www.RSMeans.com/2018online.

Enterprise Solutions

Building owners, facility managers, building product manufacturers, and attorneys across the public and private sectors engage with RSMeans data Enterprise to solve unique challenges where trusted construction cost data is critical. To learn more visit: www.RSMeans.com/Enterprise.

Custom Built Data Sets

Building and Space Models: Quickly plan construction costs across multiple locations based on geography, project size, building system component, product options, and other variables for precise budgeting and cost control.

Predictive Analytics: Accurately plan future builds with custom graphical interactive dashboards, negotiate future costs of tenant build-outs, and identify and compare national account pricing.

Consulting

Building Product Manufacturing Analytics: Validate your claims and assist with new product launches.

Third-Party Legal Resources: Used in cases of construction cost or estimate disputes, construction product failure vs. installation failure, eminent domain, class action construction product liability, and more.

API

For resellers or internal application integration, RSMeans data is offered via API. Deliver Unit, Assembly, and Square Foot Model data within your interface. To learn more about how you can provide your customers with the latest in localized construction cost data visit: www.RSMeans.com/API.

Table of Contents

Foreword

The Value of RSMeans data from Gordian

Since 1942, RSMeans data has been the industry-standard materials, labor, and equipment cost information database for contractors, facility owners and managers, architects, engineers, and anyone else that requires the latest localized construction cost information. Over 75 years later, the objective remains the same: to provide facility and construction professionals with the most current and comprehensive construction cost database possible.

With the constant influx of new construction methods and materials, in addition to ever-changing labor and material costs, last year's cost data is not reliable for today's designs, estimates, or budgets. The RSMeans data engineers invest over 22,000 hours in cost research annually and apply real-world construction experience to identify and quantify new building products and methodologies, adjust productivity rates, and adjust costs to local market conditions across the nation. This unparalleled construction cost expertise is why so many facility and construction professionals rely on RSMeans data year over year.

About Gordian

Gordian originated in the spirit of innovation and a strong commitment to helping clients reach and exceed their construction goals. In 1982, Gordian's Chairman and Founder, Harry H. Mellon, created Job Order Contracting while serving as Chief Engineer at the Supreme Headquarters Allied Powers Europe. Job Order Contracting is a unique indefinite delivery/indefinite quantity (IDIQ) process, which enables facility owners to complete a substantial number of repair, maintenance, and construction projects with a single, competitively awarded contract. Realizing facility and infrastructure owners across various industries could greatly benefit from the time and cost saving advantages of this innovative construction procurement solution, he established Gordian in 1990.

Continuing the commitment to providing the most relevant and accurate facility and construction data, software, and expertise in the industry, Gordian enhanced the fortitude of its data with the acquisition of RSMeans in 2014. And in an effort to expand its facility management capabilities, Gordian acquired Sightlines, the leading provider of facilities benchmarking data and analysis, in 2015.

Our Offerings

Gordian is the leader in facility and construction cost data, software, and expertise for all phases of the building life cycle. From planning to design, procurement, construction, and operations, Gordian's solutions help clients maximize efficiency, optimize cost savings, and increase building quality with its highly specialized data engineers, software, and unique proprietary data sets.

Our Commitment

At Gordian, we do more than talk about the quality of our data and the usefulness of its application. We stand behind all of our RSMeans data — from historical cost indexes to construction materials and techniques — to craft current costs and predict future trends. If you have any questions about our products or services, please call us toll-free at 800-448-8182 or visit our website at www.gordian.com.

How the Cost Data Is Built: An Overview

Unit Prices*

All cost data have been divided into 50 divisions according to the MasterFormat® system of classification and numbering.

Assemblies*

The cost data in this section have been organized in an "Assemblies" format. These assemblies are the functional elements of a building and are arranged according to the 7 elements of the UNIFORMAT II classification system. For a complete explanation of a typical "Assembly", see "RSMeans data: Assemblies—How They Work."

Residential Models*

Model buildings for four classes of construction—economy, average, custom, and luxury—are developed and shown with complete costs per square foot.

Commercial/Industrial/Institutional Models*

This section contains complete costs for 77 typical model buildings expressed as costs per square foot.

Green Commercial/Industrial/Institutional Models*

This section contains complete costs for 25 green model buildings expressed as costs per square foot.

References*

This section includes information on Equipment Rental Costs, Crew Listings, Historical Cost Indexes, City Cost Indexes, Location Factors, Reference Tables, and Change Orders, as well as a listing of abbreviations.

- **Equipment Rental Costs:** Included are the average costs to rent and operate hundreds of pieces of construction equipment.
- **Crew Listings:** This section lists all the crews referenced in the cost data. A crew is composed of more than one trade classification and/or the addition of power equipment to any trade classification. Power equipment is included in the cost of the crew. Costs are shown both with bare labor rates and with the installing contractor's overhead and profit added. For each, the total crew cost per eight-hour day and the composite cost per labor-hour are listed.

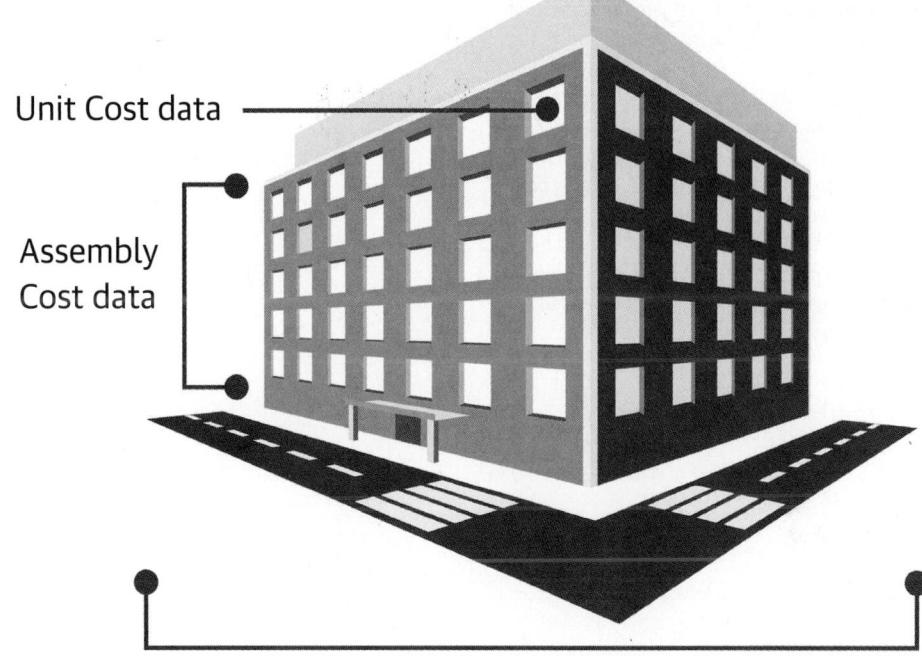

Unit Cost data

Assembly Cost data

Square Foot Models

- **Historical Cost Indexes:** These indexes provide you with data to adjust construction costs over time.
- **City Cost Indexes:** All costs in this data set are U.S. national averages. Costs vary by region. You can adjust for this by CSI Division to over 730 cities in 900+ 3-digit zip codes throughout the U.S. and Canada by using this data.
- **Location Factors:** You can adjust total project costs to over 730 cities in 900+ 3-digit zip codes throughout the U.S. and Canada by using the weighted number, which applies across all divisions.
- **Reference Tables:** At the beginning of selected major classifications in the Unit Prices are reference numbers indicators. These numbers refer you to related information in the Reference Section. In this section, you'll find reference tables, explanations, and estimating information that support how we develop the unit price data, technical data, and estimating procedures.
- **Change Orders:** This section includes information on the factors that influence the pricing of change orders.

- **Abbreviations:** A listing of abbreviations used throughout this information, along with the terms they represent, is included.

Index (printed versions only)

A comprehensive listing of all terms and subjects will help you quickly find what you need when you are not sure where it occurs in MasterFormat®.

Conclusion

This information is designed to be as comprehensive and easy to use as possible.

The Construction Specifications Institute (CSI) and Construction Specifications Canada (CSC) have produced the 2016 edition of MasterFormat®, a system of titles and numbers used extensively to organize construction information.

All unit prices in the RSMeans cost data are now arranged in the 50-division MasterFormat® 2016 system.

---

* Not all information is available in all data sets

Note: The material prices in RSMeans cost data are "contractor's prices." They are the prices that contractors can expect to pay at the lumberyards, suppliers'/distributors' warehouses, etc. Small orders of specialty items would be higher than the costs shown, while very large orders, such as truckload lots, would be less. The variation would depend on the size, timing, and negotiating power of the contractor. The labor costs are primarily for new construction or major renovation rather than repairs or minor alterations. With reasonable exercise of judgment, the figures can be used for any building work.

Residential Section

Table of Contents

Introduction to the Residential Square Foot Cost Section

The Residential Square Foot Cost Section contains costs per square foot for four classes of construction in seven building types. Costs are listed for various exterior wall materials which are typical of the class and building type. There are cost tables for wings and ells with modification tables to adjust the base cost of each class of building. Non-standard items can easily be added to the standard structures.

Cost estimating for a residence is a three-step process:

1. Identification
2. Listing dimensions
3. Calculations

Guidelines and a sample cost estimating procedure are shown on the following pages.

Identification

To properly identify a residential building, the class of construction, type, and exterior wall material must be determined. The "Building Classes" information has drawings and guidelines for determining the class of construction. There are also detailed specifications and additional drawings at the beginning of each set of tables to further aid in proper building class and type identification.

Profiles for eight types of residential buildings and their configurations follow. Definitions of living area are next to each profile. Sketches and definitions of garage types follow.

Selection of Appropriate Costs

To select the appropriate cost from the base tables, the following information is needed:

1. Class of construction
   - Economy
   - Average
   - Custom
   - Luxury
2. Type of residence
   - 1 story
   - 1-1/2 story
   - 2 story
   - 2-1/2 story
   - 3 story
   - Bi-level
   - Tri-level
3. Occupancy
   - One family
   - Two family
   - Three family
4. Building configuration
   - Detached
   - Town/Rowhouse
   - Semi-detached
5. Exterior wall construction
   - Wood frame
   - Brick veneer
   - Solid masonry
6. Living areas

Modifications are classified by class, type, and size.

Computations

The computation process should take the following sequence:

1. Multiply the base cost by the area.
2. Add or subtract the modifications, adjustments, and alternatives.
3. Apply the location modifier.

When selecting costs, interpolate or use the cost that most nearly matches the structure under study. This applies to size, exterior wall construction, and class.

Base Tables and Modifications

Base cost tables show the base cost per square foot without a basement, with one full bath and one full kitchen for economy and average homes, and an additional half bath for custom and luxury models. Adjustments for finished and unfinished basements are part of the base cost tables. Adjustments for multi-family residences, additional bathrooms, townhouses, alternative roofs, and air conditioning and heating systems are listed in Modifications, Adjustments, and Alternatives tables below the base cost tables.

Costs for other modifications, adjustments, and alternatives, including garages, breezeways, and site improvements, follow the base tables.

Listing of Dimensions

To use this section, only the dimensions used to calculate the horizontal area of the building and additions, modifications, adjustments, and alternatives are needed. The dimensions, normally the length and width, can come from drawings or field measurements. For ease in calculation, consider measuring in tenths of feet, i.e., 9 ft. 6 in. = 9.5 ft. and 9 ft. 4 in. = 9.3 ft.

In all cases, make a sketch of the building. Any protrusions or other variations in shape should be noted on the sketch with dimensions.

Calculations

The calculations portion of the estimate is a two-step activity:

1. The selection of appropriate costs from the tables
2. Computations

Living Area

Base cost tables are prepared as costs per square foot of living area. The living area of a residence is the area which is suitable and normally designed for full time living. It does not include basement recreation rooms or finished attics, although these areas are often considered full time living areas by the owners.

Living area is calculated from the exterior dimensions without the need to adjust for exterior wall thickness. When calculating the living area of a 1-1/2 story, two story, three story, or tri-level residence, overhangs and other differences in size and shape between floors must be considered.

Only the floor area with a ceiling height of seven feet or more in a 1-1/2 story residence is considered living area. In bi-levels and tri-levels, the areas that are below grade are considered living areas, even when these areas may not be completely finished.

How to Use the Residential Square Foot Cost Section

The following is a detailed explanation of a sample entry in the Residential Square Foot Cost Section. Each bold number below corresponds to the item being described in the following list with the appropriate component of the sample entry in parentheses. Prices listed are costs that include overhead and profit of the installing contractor. Total model costs include an additional markup for the general contractor's overhead and profit and fees specific to the class of construction.

| RESIDENTIAL | Average ❶ | 2 Story ❷ |

- Simple design from standard plans
- Single family — 1 full bath, 1 kitchen
- No basement ❸
- Asphalt shingles on roof
- Hot air heat
- Gypsum wallboard interior finishes
- Materials and workmanship are average
- Detail specifications on p. 27

Note: The illustration shown may contain some optional components (for example: garages and/or fireplaces) whose costs are shown in the modifications, adjustments, & alternatives below or at the end of the square foot section.

Base cost per square foot of living area

Exterior Wall ❹	Living Area ❺										
	1000	1200	1400	1600	1800	2000	2200	2600	3000	3400	3800
Wood Siding - Wood Frame	149.75	135.30	128.30	123.35	118.45 ❻	113.25	109.75	103.15	96.85	93.90	91.20
Brick Veneer - Wood Frame	156.15	141.25	133.80	128.60	123.40	118.05	114.25	107.25	100.60	97.50	94.65
Stucco on Wood Frame	144.50	130.50	123.75	119.00	114.40	109.35	106.00	99.80	93.65	90.90	88.40
Solid Masonry	170.60	154.65	146.35	140.60	134.75	128.90	124.65	116.60	109.25	105.65	102.45
Finished Basement, Add ❼	23.25	22.90	22.10	21.65	21.05	20.70	20.30 ❽	19.55	18.95	18.60	18.30
Unfinished Basement, Add	9.20	8.55	8.05	7.75	7.40	7.15	6.95	6.45	6.15	5.90	5.70

Modifications

Add to the total cost

Upgrade Kitchen Cabinets ❾	$ + 5950
Solid Surface Countertops (Included)	
Full Bath - including plumbing, wall and floor finishes	+ 8050
Half Bath - including plumbing, wall and floor finishes	+ 4767
One Car Attached Garage	+ 15,086
One Car Detached Garage	+ 19,847
Fireplace & Chimney	+ 7671

Adjustments

For multi family - add to total cost

Additional Kitchen	$ + 9963
Additional Bath	+ 8050
Additional Entry & Exit	+ 1804
Separate Heating	+ 1650
Separate Electric	+ 1817

*For Townhouse/Rowhouse -
Multiply cost per square foot by*

Inner Unit ❿	.90
End Unit	.95

Alternatives

Add to or deduct from the cost per square foot of living area

Cedar Shake Roof	+ 1.75
Clay Tile Roof	+ 3.45
Slate Roof	+ 3.85
Upgrade Walls to Skim Coat Plaster ⓫	+ .59
Upgrade Ceilings to Textured Finish	+ .60
Air Conditioning, in Heating Ductwork	+ 3
In Separate Ductwork	+ 5.81
Heating Systems, Hot Water	+ 1.52
Heat Pump	+ 1.59
Electric Heat	− .62
Not Heated	− 3.36

Additional upgrades or components

Kitchen Cabinets & Countertops	Page 58
Bathroom Vanities	59
Fireplaces & Chimneys	59
Windows, Skylights & Dormers	59
Appliances	60
Breezeways & Porches	60
Finished Attic	60
Garages	61
Site Improvements ⓬	61
Wings & Ells	37

1 Class of Construction (Average)

The class of construction depends upon the design and specifications of the plan. The four classes are economy, average, custom, and luxury.

2 Type of Residence (2 Story)

The building type describes the number of stories or levels in the model. The seven building types are 1 story, 1-1/2 story, 2 story, 2-1/2 story, 3 story, bi-level, and tri-level.

3 Specification Highlights (Hot Air Heat)

These specifications include information concerning the components of the model, including the number of baths, roofing types, HVAC systems, materials, and workmanship. If the components listed are not appropriate, modifications can be made by consulting the information shown below or in the Assemblies Section.

4 Exterior Wall System (Wood Siding-Wood Frame)

This section includes the types of exterior wall systems and the structural frames used. The exterior wall systems shown are typical of the class of construction and the building type shown.

5 Living Areas (2,000 S.F.)

The living area is that area of the residence which is suitable and normally designed for full time living. It does not include basement recreation rooms or finished attics. Living area is calculated from the exterior dimensions without the need to adjust for exterior wall thickness. When calculating the living area of a 1-1/2 story, 2 story, 3 story, or tri-level residence, overhangs and other differences in size and shape between floors must be considered. Only the floor area with a ceiling height of seven feet or more in a 1-1/2 story residence is considered living area. In bi-levels and tri-levels, the areas that are below grade are considered living areas, even when these areas may not be completely finished. A range of various living areas for the residential model is shown to aid in the selection of values from the matrix.

6 Base Costs per Square Foot of Living Area ($113.25)

Base cost tables show the cost per square foot of living area without a basement, with one full bath and one full kitchen for economy and average homes, and an additional half bath for custom and luxury models. When selecting costs, interpolate or use the cost that most nearly matches the residence under consideration for size, exterior wall system, and class of construction. Prices listed are costs that include overhead and profit of the installing contractor, a general contractor markup, and an allowance for plans that vary by class of construction. For additional information on contractor overhead and architectural fees, see the Reference Section.

7 Basement Types (Finished)

The two types of basements are finished and unfinished. The specifications and components for both are shown under Building Classes in the Introduction to this section.

8 Additional Costs for Basements ($20.70 or $7.15)

These values indicate the additional cost per square foot of living area for either a finished or an unfinished basement.

9 Modifications and Adjustments (Upgrade Kitchen Cabinets $5,950)

Modifications and Adjustments are costs added to or subtracted from the total cost of the residence. The total cost of the residence is equal to the cost per square foot of living area times the living area. Typical modifications and adjustments include kitchens, baths, garages, and fireplaces.

10 Multiplier for Townhouse/Rowhouse (Inner Unit 0.90)

The multipliers shown adjust the base costs per square foot of living area for the common wall condition encountered in townhouses or rowhouses.

11 Alternatives (Skim Coat Plaster $0.59)

Alternatives are costs added to or subtracted from the base cost per square foot of living area. Typical alternatives include variations in kitchens, baths, roofing, and air conditioning and heating systems.

12 Additional Upgrades or Components (Wings & Ells)

Costs for additional upgrades or components, including wings or ells, breezeways, porches, finished attics, and site improvements, are shown at the end of each quality section and at the end of the Residential Square Foot Section.

Building Classes

Economy Class

An economy class residence is usually built from stock plans. The materials and workmanship are sufficient to satisfy building codes. Low construction cost is more important than distinctive features. The overall shape of the foundation and structure is seldom other than square or rectangular.

An unfinished basement includes a 7' high, 8" thick foundation wall composed of either concrete block or cast-in-place concrete.

Included in the finished basement cost are inexpensive paneling or drywall as the interior finish on the foundation walls, a low cost sponge backed carpeting adhered to the concrete floor, a drywall ceiling, and overhead lighting.

Custom Class

A custom class residence is usually built from plans and specifications with enough features to give the building a distinct design. Materials and workmanship are generally above average with obvious attention given to construction details. Construction normally exceeds building code requirements.

An unfinished basement includes a 7'-6" high, 10" thick cast-in-place concrete foundation wall or a 7'-6" high, 12" thick concrete block foundation wall.

A finished basement includes painted drywall on insulated 2" × 4" wood furring as the interior finish to the concrete walls, a suspended ceiling, carpeting adhered to the concrete floor, overhead lighting, and heating.

Average Class

An average class residence is a simple design and built from standard plans. Materials and workmanship are average but often exceed minimum building codes. There are frequently special features that give the residence some distinctive characteristics.

An unfinished basement includes a 7'-6" high, 8" thick foundation wall composed of either cast-in-place concrete or concrete block.

Included in the finished basement are plywood paneling or drywall on furring that is fastened to the foundation walls, sponge backed carpeting adhered to the concrete floor, a suspended ceiling, overhead lighting, and heating.

Luxury Class

A luxury class residence is built from an architect's plan for a specific owner. It is unique both in design and workmanship. There are many special features, and construction usually exceeds all building codes. It is obvious that primary attention is placed on the owner's comfort and pleasure. Construction is supervised by an architect.

An unfinished basement includes an 8' high, 12" thick foundation wall that is composed of cast-in-place concrete or concrete block.

A finished basement includes painted drywall on 2" × 4" wood furring as the interior finish, a suspended ceiling, tackless carpet on wood subfloor with sleepers, overhead lighting, and heating.

Configurations

Detached House

This category of residence is a free-standing separate building with or without an attached garage. It has four complete walls.

Semi-Detached House

This category of residence has two side-by-side living units. The common wall is fireproof. Semi-detached residences can be treated as a rowhouse with two end units. Semi-detached residences can be any of the building types.

Town/Rowhouse

This category of residence has a number of attached units made up of inner units and end units. The units are joined by common walls. The inner units have only two exterior walls. The common walls are fireproof. The end units have three walls and a common wall. Town/rowhouses can be any of the building types.

Building Types

One Story

This is an example of a one-story dwelling. The living area of this type of residence is confined to the ground floor. The headroom in the attic is usually too low for use as a living area.

One-and-one-half Story

The living area on the upper level of this type of residence is 50% to 90% of the ground floor. This is made possible by a combination of this design's high-peaked roof and/or dormers. Only the upper level area with a ceiling height of seven feet or more is considered living area. The living area of this residence is the sum of the ground floor area plus the area on the second level with a ceiling height of seven feet or more.

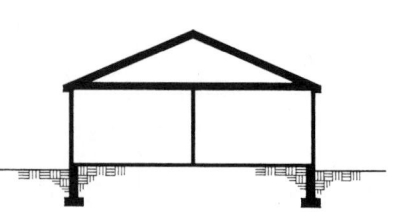

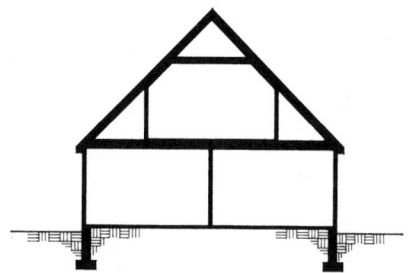

One Story with Finished Attic

The main living area in this type of residence is the ground floor. The upper level or attic area has sufficient headroom for use as a living area. This is made possible by a high-peaked roof. The living area in the attic is less than 50% of the ground floor. The living area of this type of residence is the ground floor area only. The finished attic is considered an adjustment.

Two Story

This type of residence has a second floor or upper level area which is equal or nearly equal to the ground floor area. The upper level of this type of residence can range from 90% to 110% of the ground floor area, depending on setbacks or overhangs. The living area is the sum of the ground floor area and the upper level floor area.

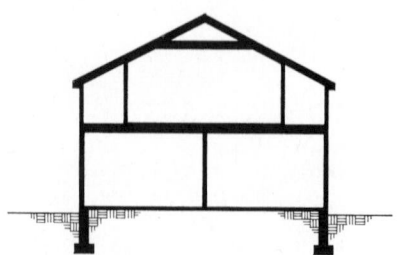

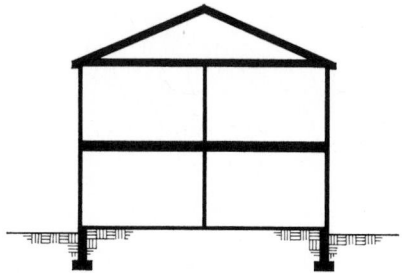

Two-and-one-half Story

This type of residence has two levels of equal or nearly equal area and a third level which has a living area that is 50% to 90% of the ground floor. This is made possible by a high-peaked roof, extended wall heights, and/or dormers. Only the upper level area with a ceiling height of seven feet or more is considered living area. The living area of this residence is the sum of the ground floor area, the second floor area, and the area on the third level with a ceiling height of seven feet or more.

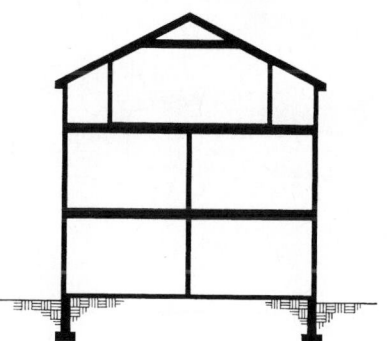

Bi-level

This type of residence has two living areas, one above the other. One area is about four feet below grade and the second is about four feet above grade. Both areas are equal in size. The lower level in this type of residence is designed and built to serve as a living area and not as a basement. Both levels have full ceiling heights. The living area is the sum of the lower level area and the upper level area.

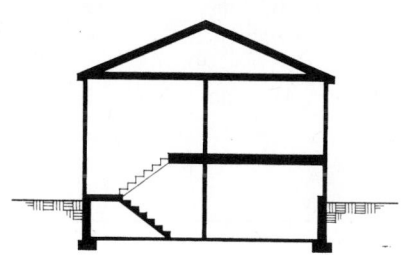

Three Story

This type of residence has three levels which are equal or nearly equal. As in the two story residence, the second and third floor areas may vary slightly depending on setbacks or overhangs. The living area is the sum of the ground floor area and the two upper level floor areas.

Tri-level

This type of residence has three levels of living area: one at grade level, one about four feet below grade, and one about four feet above grade. All levels are designed to serve as living areas. All levels have full ceiling heights. The living area is a sum of the areas of each of the three levels.

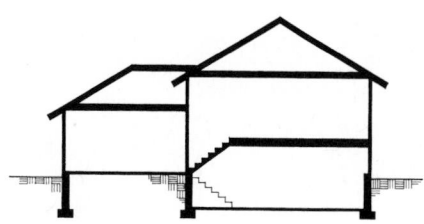

Garage Types

Attached Garage

Shares a common wall with the dwelling. Access is typically through a door between the dwelling and garage.

Basement Garage

Constructed under the roof of the dwelling but below the living area.

Built-In Garage

Constructed under the second floor living space and above the basement level of the dwelling. Reduces gross square feet of the living area.

Detached Garage

Constructed apart from the main dwelling. Shares no common area or wall with the dwelling.

Building Components

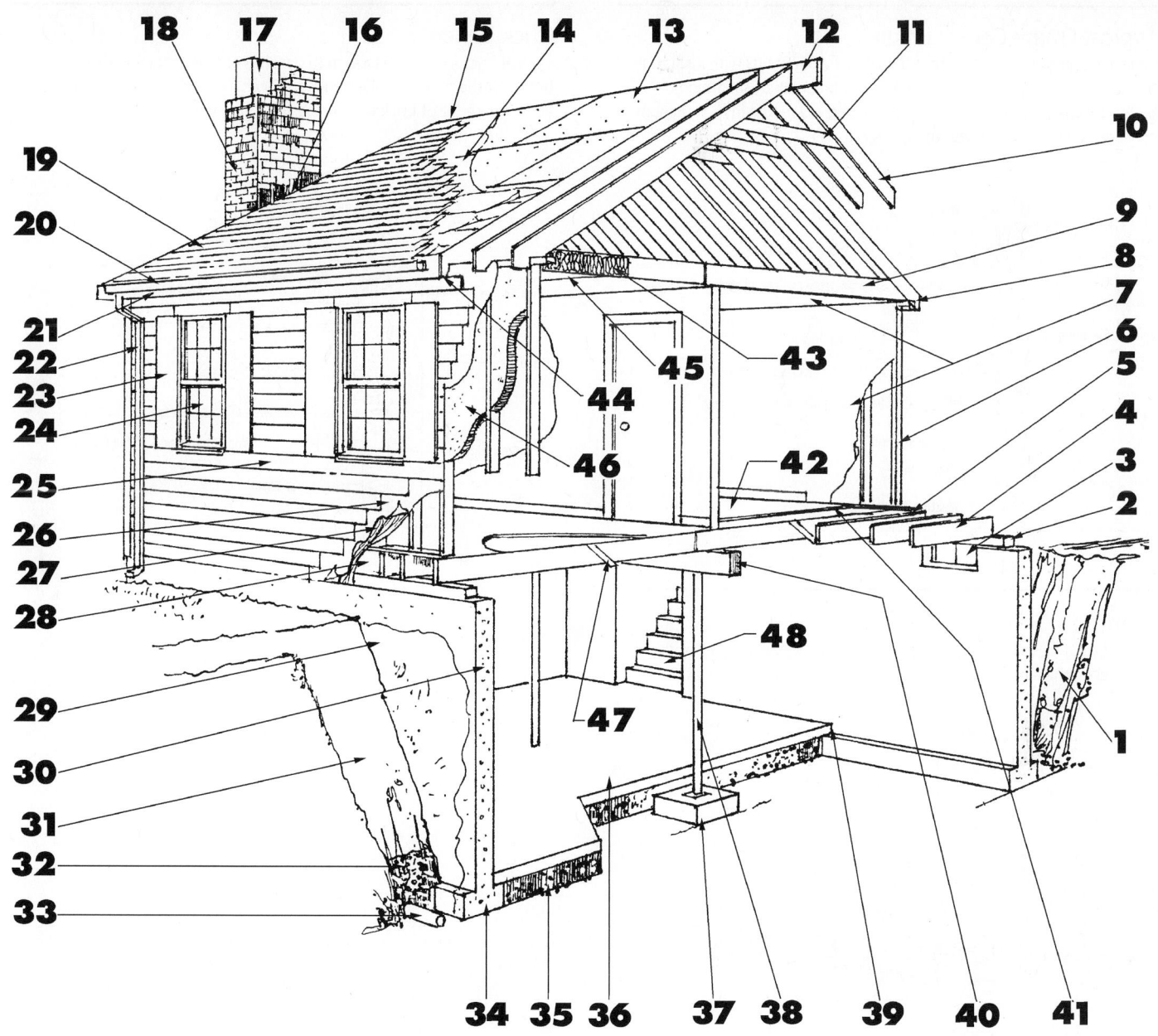

1. Excavation
2. Sill Plate
3. Basement Window
4. Floor Joist
5. Shoe Plate
6. Studs
7. Drywall
8. Plate
9. Ceiling Joists
10. Rafters
11. Collar Ties
12. Ridge Board
13. Roof Sheathing
14. Roof Felt
15. Roof Shingles
16. Flashing
17. Flue Lining
18. Chimney
19. Roof Shingles
20. Gutter
21. Fascia
22. Downspout
23. Shutter
24. Window
25. Wall Shingles
26. Weather Barrier
27. Wall Sheathing
28. Fire Stop
29. Dampproofing
30. Foundation Wall
31. Backfill
32. Drainage Stone
33. Drainage Tile
34. Wall Footing
35. Gravel
36. Concrete Slab
37. Column Footing
38. Pipe Column
39. Expansion Joint
40. Girder
41. Sub-floor
42. Finish Floor
43. Attic Insulation
44. Soffit
45. Ceiling Strapping
46. Wall Insulation
47. Cross Bridging
48. Bulkhead Stairs

Exterior Wall Construction

Typical Frame Construction

Typical wood frame construction consists of wood studs with insulation between them. A typical exterior surface is made up of sheathing, building paper, and exterior siding consisting of wood, vinyl, aluminum, or stucco over the wood sheathing.

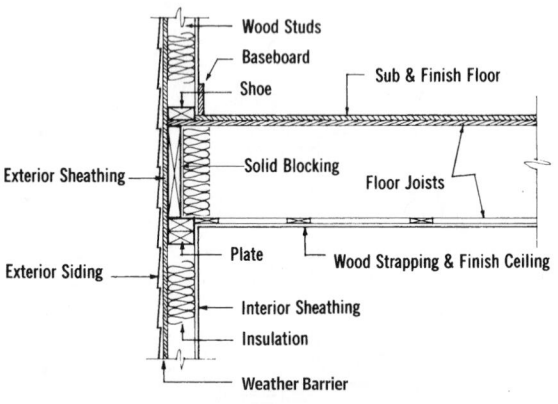

Brick Veneer

Typical brick veneer construction consists of wood studs with insulation between them. A typical exterior surface is sheathing, building paper, and an exterior of brick tied to the sheathing with metal strips.

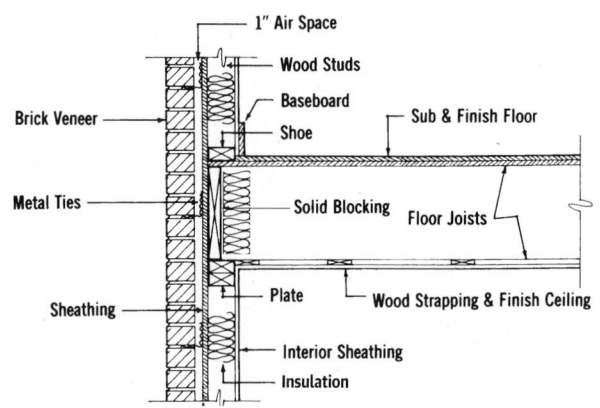

Stone

Typical solid masonry construction consists of a stone or block wall covered on the exterior with brick, stone, or other masonry.

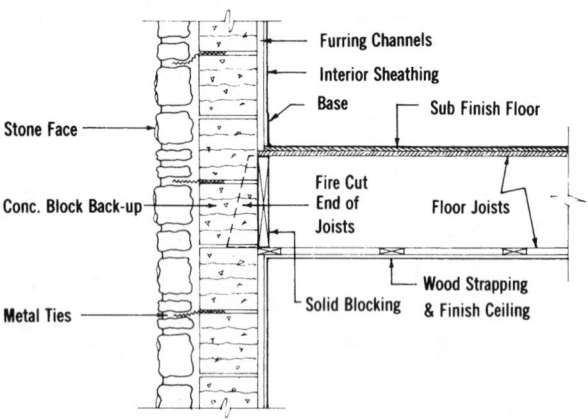

Residential Cost Estimate Worksheet

Worksheet Instructions

The residential cost estimate worksheet can be used as an outline for developing a residential construction or replacement cost. It is also useful for insurance appraisals. The design of the worksheet helps eliminate errors and omissions. To use the worksheet, follow the example below.

1. Fill out the owner's name, residence address, the estimator or appraiser's name, some type of project identifying number or code, and the date.

2. Determine from the plans, specifications, owner's description, photographs, or any other means possible the class of construction. The models in this data set use economy, average, custom, and luxury as classes. Fill in the appropriate box.

3. Fill in the appropriate box for the residence type, configuration, occupancy, and exterior wall. If you require clarification, the pages preceding this worksheet describe each of these.

4. Next, the living area of the residence must be established. The heated or air conditioned space of the residence, not including the basement, should be measured. It is easiest to break the structure up into separate components as shown in the example: the main house (A), a one-and-one-half story wing (B), and a one story wing (C). The breezeway (D), garage (E), and open covered porch (F) will be treated differently. Data entry blocks for the living area are included on the worksheet for your use. Keep each level of each component separate, and fill out the blocks as shown.

5. By using the information on the worksheet, find the model, wing, or ell in the following square foot cost pages that best matches the class, type, exterior finish, and size of the residence being estimated. Use the Modifications, Adjustments, and Alternatives to determine the adjusted cost per square foot of living area for each component.

6. For each component, multiply the cost per square foot by the living area square footage. If the residence is a town/rowhouse, a multiplier should be applied based upon the configuration.

7. The second page of the residential cost estimate worksheet has space for the additional components of a house. The cost for additional bathrooms, finished attic space, breezeways, porches, fireplaces, appliance or cabinet upgrades, and garages should be added on this page. The information for each of these components is found with the model being used or in the Modifications, Adjustments, and Alternatives.

8. Add the total from page one of the estimate worksheet and the items listed on page two. The sum is the adjusted total building cost.

9. Depending on the use of the final estimated cost, one of the remaining two boxes should be filled out. Any additional items or exclusions should be added or subtracted at this time. The data contained in this data set are a national average. Construction costs are different throughout the country. To allow for this difference, a location factor based upon the first three digits of the residence's zip code must be applied. The location factor is a multiplier that increases or decreases the adjusted total building cost. Find the appropriate location factor and calculate the local cost. If depreciation is a concern, a dollar figure should be subtracted at this point.

10. No residence will match a model exactly. Many differences will be found. At this level of estimating, a variation of +/- 10% should be expected.

Adjustments Instructions

No residence matches a model exactly in shape, material, or specifications. The common differences are:

1. Two or more exterior wall systems:
   - Partial basement
   - Partly finished basement
2. Specifications or features that are between two classes
3. Crawl space instead of a basement

Examples

Below are quick examples. See pages 15–17 for complete examples of cost adjustments for these differences:

1. Residence "A" is an average one-story structure with 1,600 S.F. of living area and no basement. Three walls are wood siding on wood frame, and the fourth wall is brick veneer on wood frame. The brick veneer wall is 35% of the exterior wall area.

   Use page 28 to calculate the Base Cost per S.F. of Living Area. Wood Siding for 1,600 S.F. = $116.75 per S.F. and Brick Veneer for 1,600 S.F. = $120.60 per S.F.

   0.65 ($116.75) + 0.35 ($120.60) = $118.10 per S.F. of Living Area.

2a. Residence "B" is the same as Residence "A" but it has an unfinished basement under 50% of the building. To adjust the $118.10 per S.F. of living area for this partial basement, use page 28.

   $118.10 + 0.5 ($11.65) = $123.93 per S.F. of Living Area.

2b. Residence "C" is the same as Residence "A" but it has a full basement under the entire building. 640 S.F. or 40% of the basement area is finished.

   Using Page 28:

   $118.10 + 0.40 ($33.30) + 0.60 ($11.65) = $138.41 per S.F. of Living Area.

3. When specifications or features of a building are between classes, estimate the percent deviation, and use two tables to calculate the cost per S.F.

   A two-story residence with wood siding and 1,800 S.F. of living area has features 30% better than Average, but 70% less than Custom.

   From pages 30 and 42:

Custom 1,800 S.F. Base Cost	=	**$152.10 per S.F.**
Average 1,800 S.F. Base Cost	=	**$118.45 per S.F.**
DIFFERENCE	=	**$33.65 per S.F.**
Cost is $118.45 + 0.30 ($33.65)	=	**$128.55 per S.F. of Living Area.**

4. To add the cost of a crawl space, use the cost of an unfinished basement as a maximum. For specific costs of components to be added or deducted, such as vapor barrier, underdrain, and floor, see the "Assemblies" section (pages 285 to 518).

Model Residence Example

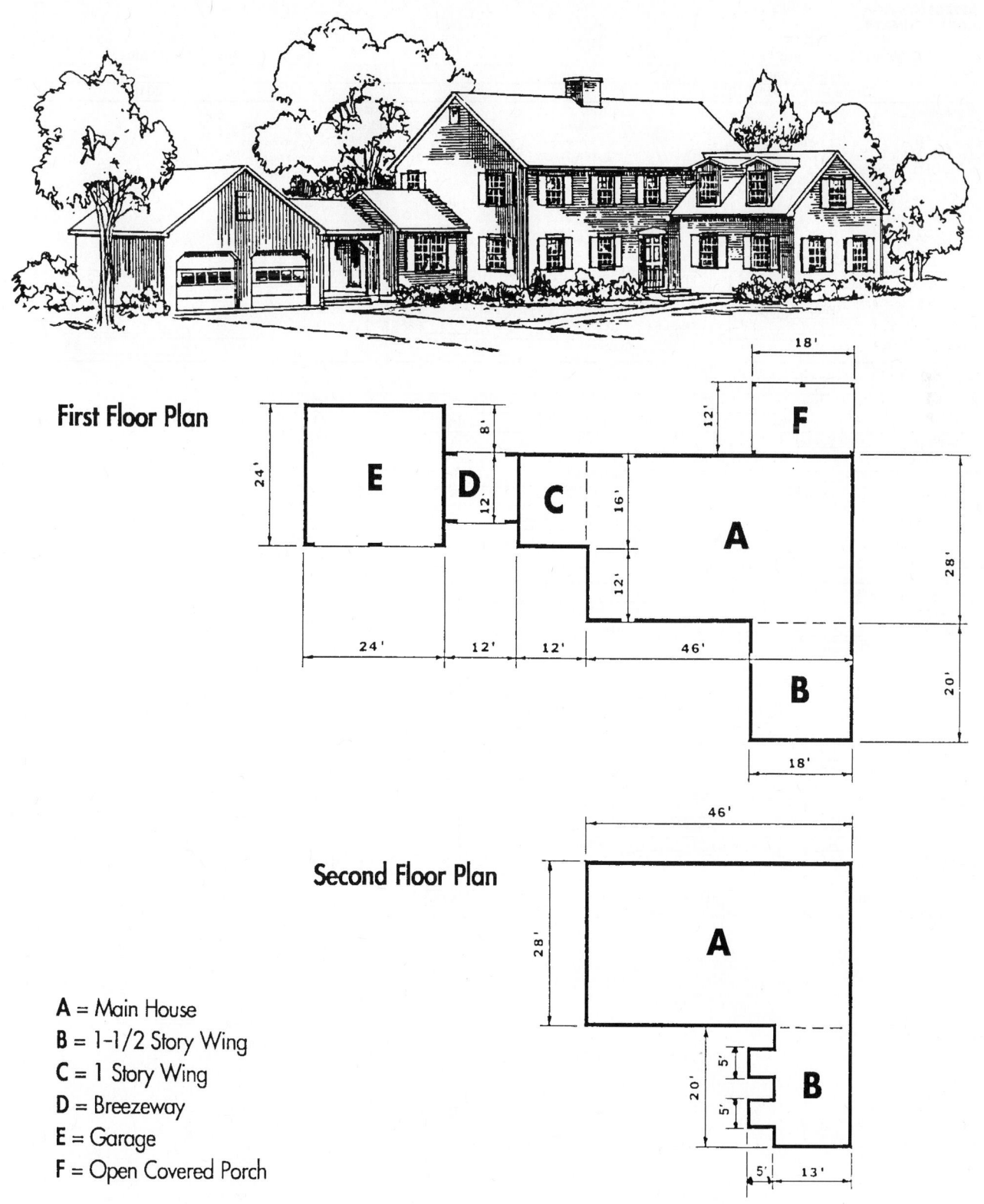

First Floor Plan

E

D

C

B

A

F

24'

8'

12'

12'

16'

12'

28'

20'

24'

12'

12'

46'

18'

18'

Second Floor Plan

A

B

46'

28'

20'

5'

5'

5'

13'

A = Main House
B = 1-1/2 Story Wing
C = 1 Story Wing
D = Breezeway
E = Garage
F = Open Covered Porch

RESIDENTIAL COST ESTIMATE

OWNERS NAME:	**Albert Westenberg**	APPRAISER:	**Nicole Wojtowicz**
RESIDENCE ADDRESS:	**300 Sygiel Road**	PROJECT:	**# 55**
CITY, STATE, ZIP CODE:	**Three Rivers, MA 01080**	DATE:	**Jan. 1, 2018**

CLASS OF CONSTRUCTION
- ☐ ECONOMY
- ☑ AVERAGE
- ☐ CUSTOM
- ☐ LUXURY

RESIDENCE TYPE
- ☐ 1 STORY
- ☐ 1 1/2 STORY
- ☑ 2 STORY
- ☐ 2 1/2 STORY
- ☐ 3 STORY
- ☐ BI-LEVEL
- ☐ TRI-LEVEL

CONFIGURATION
- ☑ DETACHED
- ☐ TOWN/ROW HOUSE
- ☐ SEMI-DETACHED

OCCUPANCY
- ☑ ONE FAMILY
- ☐ TWO FAMILY
- ☐ THREE FAMILY
- ☐ OTHER

EXTERIOR WALL SYSTEM
- ☑ WOOD SIDING - WOOD FRAME
- ☐ BRICK VENEER - WOOD FRAME
- ☐ STUCCO ON WOOD FRAME
- ☐ PAINTED CONCRETE BLOCK
- ☐ SOLID MASONRY (AVERAGE & CUSTOM)
- ☐ STONE VENEER - WOOD FRAME
- ☐ SOLID BRICK (LUXURY)
- ☐ SOLID STONE (LUXURY)

* LIVING AREA (Main Building)		
First Level	1288	S.F.
Second level	1288	S.F.
Third Level		S.F.
Total	2576	S.F.

* LIVING AREA (Wing or Ell) (**B**)		
First Level	360	S.F.
Second level	310	S.F.
Third Level		S.F.
Total	670	S.F.

* LIVING AREA (Wing or Ell) (**C**)		
First Level	192	S.F.
Second level		S.F.
Third Level		S.F.
Total	192	S.F.

* Basement Area is not part of living area.

MAIN BUILDING			COSTS PER S.F. LIVING AREA	
Cost per Square Foot of Living Area, from Page		30	$	103.15
Basement Addition:	% Finished, 100	% Unfinished	+	6.45
Roof Cover Adjustment: **Cedar Shake**	Type, Page 30	(Add or Deduct)	( )	1.65
Central Air Conditioning: ☐ Separate Ducts	☑ Heating Ducts, Page	30	+	2.96
Heating System Adjustment:	Type, Page	(Add or Deduct)	( )	
Main Building: Adjusted Cost per S.F. of Living Area			$	# 114.21

MAIN BUILDING TOTAL COST	$ 114.21 /S.F.	x	2,576 S.F.	x		=	$ 294,205
	Cost per S.F. Living Area		Living Area		Town/Row House Multiplier (Use 1 for Detached)		TOTAL COST

WING OR ELL (B) 1 - 1/2 STORY			COSTS PER S.F. LIVING AREA	
Cost per Square Foot of Living Area, from Page		36	$	98.20
Basement Addition: 100	% Finished,	% Unfinished	+	29.95
Roof Cover Adjustment:	Type, Page	(Add or Deduct)	( )	—
Central Air Conditioning: ☐ Separate Ducts	☑ Heating Ducts, Page	30	+	2.96
Heating System Adjustment:	Type, Page	(Add or Deduct)	( )	—
Wing or Ell (**B**): Adjusted Cost per S.F. of Living Area			$	# 131.11

WING OR ELL (B) TOTAL COST	$ 131.11 /S.F.	x	670 S.F.		=	$ 87,844
	Cost per S.F. Living Area		Living Area			TOTAL COST

WING OR ELL (C) 1 STORY			COSTS PER S.F. LIVING AREA	
Cost per Square Foot of Living Area, from Page		# 36 (WOOD SIDING)	$	# 150.10
Basement Addition:	% Finished,	% Unfinished	+	—
Roof Cover Adjustment:	Type, Page #	(Add or Deduct)	( )	—
Central Air Conditioning: ☐ Separate Ducts	☐ Heating Ducts, Page		+	—
Heating System Adjustment:	Type, Page	(Add or Deduct)	( )	—
Wing or Ell (C) Adjusted Cost per S.F. of Living Area			$	# 150.1

WING OR ELL (C) TOTAL COST	$ 150.10 /S.F.	x	192 S.F.		=	$ 28,819
	Cost per S.F. Living Area		Living Area			TOTAL COST

	TOTAL THIS PAGE	410,868

Page 1 of 2

RESIDENTIAL
COST ESTIMATE

						QUANTITY	UNIT COST		
Total Page 1								$	410,868
Additional Bathrooms:	2	Full,	1	Half	2 @ 7405 1 @ 4331			+	19,141
Finished Attic:	N/A	Ft. x		Ft.			S.F.		
Breezeway: ☑ Open ☐ closed	12	Ft. x	12	Ft.		144	S.F.	40.07	+ 5,770
Covered Porch: ☑ Open ☐ Enclosed	18	Ft. x	12	Ft.		216	S.F.	37.81	+ 8,167
Fireplace: ☑ Interior Chimney ☐ Exterior Chimney								+	
☑ No. of Flues (2) ☑ Additional Fireplaces 1 - 2nd Story								+	12,746
Appliances:								+	—
Kitchen Cabinets Adjustments:				(±)					—
☑ Garage ☐ Carport: 2 Car(s) Description **Wood, Attached** (±)									25,350
Miscellaneous:								+	

	ADJUSTED TOTAL BUILDING COST	$	482,042

REPLACEMENT COST				INSURANCE COST			
ADJUSTED TOTAL BUILDING COST	$	482,042		ADJUSTED TOTAL BUILDING COST	$		
Site Improvements				Insurance Exclusions			
(A) Paving & Sidewalks	$			(A) Footings, sitework, Underground Piping	-$		
(B) Landscaping	$			(B) Architects Fees	-$		
(C) Fences	$			Total Building Cost Less Exclusion	$		
(D) Swimming Pools	$			Location Factor	x		
(E) Miscellaneous	$						
TOTAL	$	482,042		LOCAL INSURABLE REPLACEMENT COST	$		
Location Factor	x	1.060					
Location Replacement Cost	$	510,964					
Depreciation	-$	52,592					
LOCAL DEPRECIATED COST	$	458,372					

SKETCH AND ADDITIONAL CALCULATIONS

1 Story

1-1/2 Story

2 Story

Bi-Level

Tri-Level

	Components
1 Site Work	Site preparation for slab or excavation for lower level; 4' deep trench excavation for foundation wall.
2 Foundations	Continuous reinforced concrete footing, 8" deep x 18" wide; dampproofed and insulated 8" thick reinforced concrete block foundation wall, 4' deep; 4" concrete slab on 4" crushed stone base and polyethylene vapor barrier, trowel finish.
3 Framing	Exterior walls—2 x 4 wood studs, 16" O.C.; 1/2" insulation board sheathing; wood truss roof 24" O.C. or 2 x 6 rafters 16" O.C. with 1/2" plywood sheathing, 4 in 12 or 8 in 12 roof pitch.
4 Exterior Walls	Beveled wood siding and #15 felt building paper on insulated wood frame walls; sliding sash or double hung wood windows; 2 flush solid core wood exterior doors. **Alternates:** • Brick veneer on wood frame, has 4" veneer of common brick. • Stucco on wood frame has 1" thick colored stucco finish. • Painted concrete block has 8" concrete block sealed and painted on exterior and furring on the interior for the drywall.
5 Roofing	20 year asphalt shingles; #15 felt building paper; aluminum gutters, downspouts, drip edge and flashings.
6 Interiors	Walls and ceilings—1/2" taped and finished drywall, primed and painted with 2 coats; painted baseboard and trim; rubber backed carpeting 80%, asphalt tile 20%; hollow core wood interior doors.
7 Specialties	Economy grade kitchen cabinets—6 L.F. wall and base with plastic laminate counter top and kitchen sink; 30 gallon electric water heater.
8 Mechanical	1 lavatory, white, wall hung; 1 water closet, white; 1 bathtub, enameled steel, white; gas fired warm air heat.
9 Electrical	100 Amp. service; romex wiring; incandescent lighting fixtures, switches, receptacles.
10 Overhead and Profit	General Contractor overhead and profit.

Adjustments

Unfinished Basement:
7' high 8" concrete block or cast-in-place concrete walls.

Finished Basement:
Includes inexpensive paneling or drywall on foundation walls. Inexpensive sponge backed carpeting on concrete floor, drywall ceiling, and lighting.

- **Mass produced from stock plans**
- **Single family — 1 full bath, 1 kitchen**
- **No basement**
- **Asphalt shingles on roof**
- **Hot air heat**
- **Gypsum wallboard interior finishes**
- **Materials and workmanship are sufficient to meet codes**
- **Detail specifications on page 19**

Note: The illustration shown may contain some optional components (for example: garages and/or fireplaces) whose costs are shown in the modifications, adjustments, & alternatives below or at the end of the square foot section.

©Home Planners, Inc.

Base cost per square foot of living area

Exterior Wall	Living Area										
	600	800	1000	1200	1400	1600	1800	2000	2400	2800	3200
Wood Siding - Wood Frame	147.60	133.20	122.20	113.55	105.85	100.95	98.60	95.25	88.80	84.10	80.75
Brick Veneer - Wood Frame	153.35	138.30	126.90	117.70	109.70	104.60	102.05	98.50	91.70	86.80	83.30
Stucco on Wood Frame	138.95	125.45	115.20	107.20	100.10	95.60	93.40	90.35	84.30	79.90	76.95
Painted Concrete Block	143.95	129.85	119.25	110.90	103.45	98.70	96.40	93.25	86.85	82.30	79.20
Finished Basement, Add	33.15	31.20	29.80	28.50	27.50	26.80	26.45	25.90	25.15	24.55	24.05
Unfinished Basement, Add	14.85	13.25	12.15	11.10	10.35	9.80	9.55	9.10	8.45	8.00	7.60

Modifications

Add to the total cost

Upgrade Kitchen Cabinets	$ + 1216
Solid Surface Countertops	+ 870
Full Bath - including plumbing, wall and floor finishes	+ 6440
Half Bath - including plumbing, wall and floor finishes	+ 3814
One Car Attached Garage	+ 13,994
One Car Detached Garage	+ 18,103
Fireplace & Chimney	+ 6624

Adjustments

For multi family - add to total cost

Additional Kitchen	$ + 5931
Additional Bath	+ 6440
Additional Entry & Exit	+ 1804
Separate Heating	+ 1650
Separate Electric	+ 1093

For Townhouse/Rowhouse -
Multiply cost per square foot by

Inner Unit	.95
End Unit	.97

Alternatives

Add to or deduct from the cost per square foot of living area

Composition Roll Roofing	– 1.05
Cedar Shake Roof	+ 4.20
Upgrade Walls and Ceilings to Skim Coat Plaster	+ .80
Upgrade Ceilings to Textured Finish	+ .60
Air Conditioning, in Heating Ductwork	+ 4.79
In Separate Ductwork	+ 7.05
Heating Systems, Hot Water	+ 1.60
Heat Pump	+ 1.32
Electric Heat	– 1.44
Not Heated	– 4.31

Additional upgrades or components

Kitchen Cabinets & Countertops	Page 58
Bathroom Vanities	59
Fireplaces & Chimneys	59
Windows, Skylights & Dormers	59
Appliances	60
Breezeways & Porches	60
Finished Attic	60
Garages	61
Site Improvements	61
Wings & Ells	25

- **Mass produced from stock plans**
- **Single family — 1 full bath, 1 kitchen**
- **No basement**
- **Asphalt shingles on roof**
- **Hot air heat**
- **Gypsum wallboard interior finishes**
- **Materials and workmanship are sufficient to meet codes**
- **Detail specifications on page 19**

Note: The illustration shown may contain some optional components (for example: garages and/or fireplaces) whose costs are shown in the modifications, adjustments, & alternatives below or at the end of the square foot section.

Base cost per square foot of living area

	Living Area										
Exterior Wall	**600**	**800**	**1000**	**1200**	**1400**	**1600**	**1800**	**2000**	**2400**	**2800**	**3200**
Wood Siding - Wood Frame	173.65	144.00	129.25	122.20	117.10	109.25	105.45	101.55	93.35	90.25	86.80
Brick Veneer - Wood Frame	181.90	150.00	134.80	127.40	122.00	113.75	109.70	105.60	96.95	93.60	89.90
Stucco on Wood Frame	161.20	135.00	120.90	114.35	109.65	102.45	99.05	95.45	87.95	85.10	82.10
Painted Concrete Block	168.15	140.05	125.55	118.75	113.85	106.30	102.60	98.85	90.95	88.00	84.70
Finished Basement, Add	25.50	21.60	20.65	19.90	19.40	18.65	18.20	17.80	16.95	16.60	16.20
Unfinished Basement, Add	13.10	10.00	9.25	8.65	8.20	7.60	7.25	6.95	6.30	6.05	5.70

Modifications

Add to the total cost

Upgrade Kitchen Cabinets	$ + 1216
Solid Surface Countertops	+ 870
Full Bath - including plumbing, wall and floor finishes	+ 6440
Half Bath - including plumbing, wall and floor finishes	+ 3814
One Car Attached Garage	+ 13,994
One Car Detached Garage	+ 18,103
Fireplace & Chimney	+ 6624

Adjustments

For multi family - add to total cost

Additional Kitchen	$ + 5931
Additional Bath	+ 6440
Additional Entry & Exit	+ 1804
Separate Heating	+ 1650
Separate Electric	+ 1093

For Townhouse/Rowhouse -
Multiply cost per square foot by

Inner Unit	.95
End Unit	.97

Alternatives

Add to or deduct from the cost per square foot of living area

Composition Roll Roofing	– .75
Cedar Shake Roof	+ 3.05
Upgrade Walls and Ceilings to Skim Coat Plaster	+ .81
Upgrade Ceilings to Textured Finish	+ .60
Air Conditioning, in Heating Ductwork	+ 3.57
In Separate Ductwork	+ 6.19
Heating Systems, Hot Water	+ 1.52
Heat Pump	+ 1.45
Electric Heat	– 1.14
Not Heated	– 3.96

Additional upgrades or components

Kitchen Cabinets & Countertops	Page 58
Bathroom Vanities	59
Fireplaces & Chimneys	59
Windows, Skylights & Dormers	59
Appliances	60
Breezeways & Porches	60
Finished Attic	60
Garages	61
Site Improvements	61
Wings & Ells	25

- Mass produced from stock plans
- Single family — 1 full bath, 1 kitchen
- No basement
- Asphalt shingles on roof
- Hot air heat
- Gypsum wallboard interior finishes
- Materials and workmanship are sufficient to meet codes
- Detail specifications on page 19

Note: The illustration shown may contain some optional components (for example: garages and/or fireplaces) whose costs are shown in the modifications, adjustments, & alternatives below or at the end of the square foot section.

Base cost per square foot of living area

Exterior Wall	Living Area										
	1000	1200	1400	1600	1800	2000	2200	2600	3000	3400	3800
Wood Siding - Wood Frame	132.05	119.65	113.60	109.45	105.30	100.70	97.65	91.75	86.10	83.55	81.25
Brick Veneer - Wood Frame	138.00	125.15	118.70	114.35	109.90	105.15	101.85	95.60	89.70	86.90	84.45
Stucco on Wood Frame	123.05	111.35	105.80	102.00	98.35	93.95	91.20	86.00	80.80	78.45	76.40
Painted Concrete Block	128.55	116.40	110.55	106.55	102.60	98.05	95.10	89.50	84.05	81.55	79.35
Finished Basement, Add	17.40	16.60	16.00	15.65	15.20	14.95	14.60	14.05	13.65	13.40	13.10
Unfinished Basement, Add	8.05	7.45	6.95	6.65	6.30	6.10	5.85	5.35	5.10	4.85	4.65

Modifications

Add to the total cost

Upgrade Kitchen Cabinets	$ + 1216
Solid Surface Countertops	+ 870
Full Bath - including plumbing, wall and floor finishes	+ 6440
Half Bath - including plumbing, wall and floor finishes	+ 3814
One Car Attached Garage	+ 13,994
One Car Detached Garage	+ 18,103
Fireplace & Chimney	+ 7318

Adjustments

For multi family - add to total cost

Additional Kitchen	$ + 5931
Additional Bath	+ 6440
Additional Entry & Exit	+ 1804
Separate Heating	+ 1650
Separate Electric	+ 1093

For Townhouse/Rowhouse -
Multiply cost per square foot by

Inner Unit	.93
End Unit	.96

Alternatives

Add to or deduct from the cost per square foot of living area

Composition Roll Roofing	– .55
Cedar Shake Roof	+ 2.10
Upgrade Walls and Ceilings to Skim Coat Plaster	+ .82
Upgrade Ceilings to Textured Finish	+ .60
Air Conditioning, in Heating Ductwork	+ 2.91
In Separate Ductwork	+ 5.68
Heating Systems, Hot Water	+ 1.48
Heat Pump	+ 1.53
Electric Heat	– 1
Not Heated	– 3.75

Additional upgrades or components

Kitchen Cabinets & Countertops	Page 58
Bathroom Vanities	59
Fireplaces & Chimneys	59
Windows, Skylights & Dormers	59
Appliances	60
Breezeways & Porches	60
Finished Attic	60
Garages	61
Site Improvements	61
Wings & Ells	25

- **Mass produced from stock plans**
- **Single family — 1 full bath, 1 kitchen**
- **No basement**
- **Asphalt shingles on roof**
- **Hot air heat**
- **Gypsum wallboard interior finishes**
- **Materials and workmanship are sufficient to meet codes**
- **Detail specifications on page 19**

Note: The illustration shown may contain some optional components (for example: garages and/or fireplaces) whose costs are shown in the modifications, adjustments, & alternatives below or at the end of the square foot section.

Base cost per square foot of living area

Exterior Wall	Living Area										
	1000	1200	1400	1600	1800	2000	2200	2600	3000	3400	3800
Wood Siding - Wood Frame	122.25	110.55	105.10	101.30	97.70	93.35	90.65	85.45	80.30	78.00	76.00
Brick Veneer - Wood Frame	126.75	114.75	109.05	105.05	101.20	96.70	93.85	88.45	83.00	80.55	78.40
Stucco on Wood Frame	115.40	104.20	99.20	95.65	92.35	88.20	85.75	81.05	76.25	74.10	72.30
Painted Concrete Block	119.55	108.00	102.75	99.05	95.50	91.30	88.70	83.70	78.65	76.45	74.50
Finished Basement, Add	17.40	16.60	16.00	15.65	15.20	14.95	14.60	14.05	13.65	13.40	13.10
Unfinished Basement, Add	8.05	7.45	6.95	6.65	6.30	6.10	5.85	5.35	5.10	4.85	4.65

Modifications

Add to the total cost

Upgrade Kitchen Cabinets	$ + 1216
Solid Surface Countertops	+ 870
Full Bath - including plumbing, wall and floor finishes	+ 6440
Half Bath - including plumbing, wall and floor finishes	+ 3814
One Car Attached Garage	+ 13,994
One Car Detached Garage	+ 18,103
Fireplace & Chimney	+ 6624

Adjustments

For multi family - add to total cost

Additional Kitchen	$ + 5931
Additional Bath	+ 6440
Additional Entry & Exit	+ 1804
Separate Heating	+ 1650
Separate Electric	+ 1093

For Townhouse/Rowhouse -
Multiply cost per square foot by

Inner Unit	.94
End Unit	.97

Alternatives

Add to or deduct from the cost per square foot of living area

Composition Roll Roofing	– .55
Cedar Shake Roof	+ 2.10
Upgrade Walls and Ceilings to Skim Coat Plaster	+ .77
Upgrade Ceilings to Textured Finish	+ .60
Air Conditioning, in Heating Ductwork	+ 2.91
In Separate Ductwork	+ 5.68
Heating Systems, Hot Water	+ 1.48
Heat Pump	+ 1.53
Electric Heat	– 1
Not Heated	– 3.75

Additional upgrades or components

Kitchen Cabinets & Countertops	Page 58
Bathroom Vanities	59
Fireplaces & Chimneys	59
Windows, Skylights & Dormers	59
Appliances	60
Breezeways & Porches	60
Finished Attic	60
Garages	61
Site Improvements	61
Wings & Ells	25

Important: See the Reference Section for Location Factors (to adjust for your city) and Estimating Forms.

- **Mass produced from stock plans**
- **Single family — 1 full bath, 1 kitchen**
- **No basement**
- **Asphalt shingles on roof**
- **Hot air heat**
- **Gypsum wallboard interior finishes**
- **Materials and workmanship are sufficient to meet codes**
- **Detail specifications on page 19**

Note: The illustration shown may contain some optional components (for example: garages and/or fireplaces) whose costs are shown in the modifications, adjustments, & alternatives below or at the end of the square foot section.

©Design Basics, Inc.

Base cost per square foot of living area

Exterior Wall	Living Area										
	1200	1500	1800	2000	2200	2400	2800	3200	3600	4000	4400
Wood Siding - Wood Frame	113.85	104.35	97.40	94.55	90.50	87.15	84.60	81.00	77.05	75.60	72.40
Brick Veneer - Wood Frame	118.00	108.15	100.80	97.85	93.60	90.10	87.40	83.60	79.45	77.95	74.55
Stucco on Wood Frame	107.45	98.55	92.20	89.60	85.80	82.75	80.35	77.10	73.40	72.00	69.05
Solid Masonry	111.20	102.00	95.25	92.50	88.55	85.35	82.80	79.40	75.50	74.15	71.05
Finished Basement, Add*	20.80	19.85	19.05	18.65	18.30	17.90	17.60	17.15	16.75	16.60	16.25
Unfinished Basement, Add*	8.85	8.10	7.45	7.10	6.85	6.55	6.30	5.95	5.65	5.50	5.25

*Basement under middle level only.

Modifications

Add to the total cost

Upgrade Kitchen Cabinets	$ + 1216
Solid Surface Countertops	+ 870
Full Bath - including plumbing, wall and floor finishes	+ 6440
Half Bath - including plumbing, wall and floor finishes	+ 3814
One Car Attached Garage	+ 13,994
One Car Detached Garage	+ 18,103
Fireplace & Chimney	+ 6624

Adjustments

For multi family - add to total cost

Additional Kitchen	$ + 5931
Additional Bath	+ 6440
Additional Entry & Exit	+ 1804
Separate Heating	+ 1650
Separate Electric	+ 1093

For Townhouse/Rowhouse -
Multiply cost per square foot by

Inner Unit	.93
End Unit	.96

Alternatives

Add to or deduct from the cost per square foot of living area

Composition Roll Roofing	– .75
Cedar Shake Roof	+ 3.05
Upgrade Walls and Ceilings to Skim Coat Plaster	+ .70
Upgrade Ceilings to Textured Finish	+ .60
Air Conditioning, in Heating Ductwork	+ 2.48
In Separate Ductwork	+ 5.28
Heating Systems, Hot Water	+ 1.42
Heat Pump	+ 1.59
Electric Heat	– .86
Not Heated	– 3.64

Additional upgrades or components

Kitchen Cabinets & Countertops	Page 58
Bathroom Vanities	59
Fireplaces & Chimneys	59
Windows, Skylights & Dormers	59
Appliances	60
Breezeways & Porches	60
Finished Attic	60
Garages	61
Site Improvements	61
Wings & Ells	25

1 Story — Base cost per square foot of living area

Exterior Wall	Living Area							
	50	100	200	300	400	500	600	700
Wood Siding - Wood Frame	196.40	149.15	128.70	105.90	99.25	95.20	92.60	93.10
Brick Veneer - Wood Frame	209.55	158.55	136.55	111.20	104.00	99.65	96.80	97.15
Stucco on Wood Frame	176.35	134.85	116.80	98.00	92.10	88.55	86.20	86.95
Painted Concrete Block	189.15	144.05	124.45	103.10	96.70	92.90	90.30	90.85
Finished Basement, Add	50.85	41.40	37.50	30.95	29.60	28.85	28.35	27.95
Unfinished Basement, Add	28.40	21.05	18.05	12.95	11.90	11.35	10.90	10.60

1-1/2 Story — Base cost per square foot of living area

Exterior Wall	Living Area							
	100	200	300	400	500	600	700	800
Wood Siding - Wood Frame	155.70	125.15	106.35	94.30	88.65	85.80	82.30	81.40
Brick Veneer - Wood Frame	167.50	134.60	114.20	100.45	94.30	91.15	87.35	86.40
Stucco on Wood Frame	137.85	110.85	94.45	85.05	80.05	77.70	74.70	73.80
Painted Concrete Block	149.30	120.00	102.05	90.95	85.55	82.90	79.60	78.65
Finished Basement, Add	33.95	30.00	27.35	24.50	23.70	23.20	22.70	22.65
Unfinished Basement, Add	17.45	14.45	12.35	10.15	9.50	9.10	8.75	8.70

2 Story — Base cost per square foot of living area

Exterior Wall	Living Area							
	100	200	400	600	800	1000	1200	1400
Wood Siding - Wood Frame	159.60	118.30	100.30	81.60	75.75	72.25	69.90	70.65
Brick Veneer - Wood Frame	172.85	127.70	108.20	86.90	80.45	76.65	74.10	74.70
Stucco on Wood Frame	139.65	104.00	88.40	73.65	68.65	65.55	63.55	64.50
Painted Concrete Block	152.45	113.20	96.05	78.80	73.20	69.85	67.60	68.45
Finished Basement, Add	25.45	20.75	18.80	15.55	14.85	14.45	14.20	14.05
Unfinished Basement, Add	14.25	10.55	9.05	6.50	5.95	5.70	5.45	5.30

Base costs do not include bathroom or kitchen facilities. Use Modifications/Adjustments/Alternatives on pages 58-61 where appropriate.

Important: See the Reference Section for Location Factors (to adjust for your city) and Estimating Forms.

1 Story

1-1/2 Story

2 Story

2-1/2 Story

Bi-Level

Tri-Level

Components

1 Site Work

Site preparation for slab or excavation for lower level; 4' deep trench excavation for foundation wall.

2 Foundations

Continuous reinforced concrete footing, 8" deep x 18" wide; dampproofed and insulated 8" thick reinforced concrete foundation wall, 4' deep; 4" concrete slab on 4" crushed stone base and polyethylene vapor barrier, trowel finish.

3 Framing

Exterior walls—2 x 4 wood studs, 16" O.C.; 1/2" plywood sheathing; 2 x 6 rafters 16" O.C. with 1/2" plywood sheathing, 4 in 12 or 8 in 12 roof pitch, 2 x 6 ceiling joists; 1/2" plywood subfloor on 1 x 2 wood sleepers 16" O.C.; 2 x 8 floor joists with 5/8" plywood subfloor on models with more than one level.

4 Exterior Walls

Beveled wood siding and #15 felt building paper on insulated wood frame walls; double hung windows; 3 flush solid core wood exterior doors with storms.
  Alternates:
  • Brick veneer on wood frame, has 4" veneer of common brick.
  • Stucco on wood frame has 1" thick colored stucco finish.
  • Solid masonry has a 6" concrete block load bearing wall with insulation and a brick or stone exterior. Also solid brick or stone walls.

5 Roofing

25 year asphalt shingles; #15 felt building paper; aluminum gutters, downspouts, drip edge and flashings.

6 Interiors

Walls and ceilings—1/2" taped and finished drywall, primed and painted with 2 coats; painted baseboard and trim, finished hardwood floor 40%, carpet with 1/2" underlayment 40%, vinyl tile with 1/2" underlayment 15%, ceramic tile with 1/2" underlayment 5%; hollow core and louvered interior doors.

7 Specialties

Average grade kitchen cabinets—14 L.F. wall and base with plastic laminate counter top and kitchen sink; 40 gallon electric water heater.

8 Mechanical

1 lavatory, white, wall hung; 1 water closet, white; 1 bathtub, enameled steel, white; gas fired warm air heat.

9 Electrical

100 Amp. service; romex wiring; incandescent lighting fixtures, switches, receptacles.

10 Overhead and Profit

General Contractor overhead and profit.

Adjustments

Unfinished Basement:
7'-6" high cast-in-place concrete walls 8" thick or 8" concrete block.

Finished Basement:
Includes plywood paneling or drywall on furring on foundation walls, sponge backed carpeting on concrete floor, suspended ceiling, lighting and heating.

- **Simple design from standard plans**
- **Single family — 1 full bath, 1 kitchen**
- **No basement**
- **Asphalt shingles on roof**
- **Hot air heat**
- **Gypsum wallboard interior finishes**
- **Materials and workmanship are average**
- **Detail specifications on p. 27**

Note: The illustration shown may contain some optional components (for example: garages and/or fireplaces) whose costs are shown in the modifications, adjustments, & alternatives below or at the end of the square foot section.

eHome Planners, Inc.

Base cost per square foot of living area

Exterior Wall	Living Area										
	600	800	1000	1200	1400	1600	1800	2000	2400	2800	3200
Wood Siding - Wood Frame	172.50	154.65	141.65	131.25	122.50	116.75	113.60	109.85	102.45	97.05	93.30
Brick Veneer - Wood Frame	178.70	160.15	146.60	135.70	126.60	120.60	117.35	113.30	105.65	100.00	96.00
Stucco on Wood Frame	167.35	150.00	137.45	127.45	119.00	113.50	110.50	106.85	99.80	94.60	91.05
Solid Masonry	191.80	171.85	157.20	145.30	135.30	128.75	125.20	120.70	112.40	106.25	101.80
Finished Basement, Add	40.50	39.05	37.25	35.55	34.15	33.30	32.80	32.05	31.05	30.25	29.55
Unfinished Basement, Add	16.80	15.15	14.00	13.00	12.15	11.65	11.25	10.85	10.25	9.75	9.35

Modifications

Add to the total cost

Upgrade Kitchen Cabinets	$ + 5950
Solid Surface Countertops (Included)	
Full Bath - including plumbing, wall and floor finishes	+ 8050
Half Bath - including plumbing, wall and floor finishes	+ 4767
One Car Attached Garage	+ 15,086
One Car Detached Garage	+ 19,847
Fireplace & Chimney	+ 6942

Adjustments

For multi family - add to total cost

Additional Kitchen	$ + 9963
Additional Bath	+ 8050
Additional Entry & Exit	+ 1804
Separate Heating	+ 1650
Separate Electric	+ 1817

For Townhouse/Rowhouse -
Multiply cost per square foot by

Inner Unit	.92
End Unit	.96

Alternatives

Add to or deduct from the cost per square foot of living area

Cedar Shake Roof	+ 3.50
Clay Tile Roof	+ 6.90
Slate Roof	+ 7.70
Upgrade Walls to Skim Coat Plaster	+ .49
Upgrade Ceilings to Textured Finish	+ .60
Air Conditioning, in Heating Ductwork	+ 4.95
In Separate Ductwork	+ 7.32
Heating Systems, Hot Water	+ 1.63
Heat Pump	+ 1.37
Electric Heat	– .79
Not Heated	– 3.61

Additional upgrades or components

Kitchen Cabinets & Countertops	Page 58
Bathroom Vanities	59
Fireplaces & Chimneys	59
Windows, Skylights & Dormers	59
Appliances	60
Breezeways & Porches	60
Finished Attic	60
Garages	61
Site Improvements	61
Wings & Ells	37

- **Simple design from standard plans**
- **Single family — 1 full bath, 1 kitchen**
- **No basement**
- **Asphalt shingles on roof**
- **Hot air heat**
- **Gypsum wallboard interior finishes**
- **Materials and workmanship are average**
- **Detail specifications on p. 27**

Note: The illustration shown may contain some optional components (for example: garages and/or fireplaces) whose costs are shown in the modifications, adjustments, & alternatives below or at the end of the square foot section.

©By Designer

Base cost per square foot of living area

Exterior Wall	Living Area										
	600	800	1000	1200	1400	1600	1800	2000	2400	2800	3200
Wood Siding - Wood Frame	197.00	163.20	145.95	137.45	131.35	122.40	118.05	113.50	104.30	100.65	96.70
Brick Veneer - Wood Frame	205.80	169.65	151.85	143.00	136.60	127.20	122.55	117.85	108.15	104.30	100.05
Stucco on Wood Frame	189.40	157.75	140.85	132.65	126.85	118.30	114.15	109.75	101.00	97.50	93.85
Solid Masonry	223.70	182.60	163.90	154.25	147.35	136.95	131.80	126.55	115.85	111.60	106.80
Finished Basement, Add	33.10	28.75	27.55	26.55	25.90	24.80	24.25	23.75	22.60	22.20	21.60
Unfinished Basement, Add	14.55	11.40	10.55	9.95	9.50	8.90	8.50	8.20	7.55	7.25	6.90

Modifications

Add to the total cost

Upgrade Kitchen Cabinets	$ + 5950
Solid Surface Countertops (Included)	
Full Bath - including plumbing, wall and floor finishes	+ 8050
Half Bath - including plumbing, wall and floor finishes	+ 4767
One Car Attached Garage	+ 15,086
One Car Detached Garage	+ 19,847
Fireplace & Chimney	+ 6942

Adjustments

For multi family - add to total cost

Additional Kitchen	$ + 9963
Additional Bath	+ 8050
Additional Entry & Exit	+ 1804
Separate Heating	+ 1650
Separate Electric	+ 1817

For Townhouse/Rowhouse -
Multiply cost per square foot by

Inner Unit	.92
End Unit	.96

Alternatives

Add to or deduct from the cost per square foot of living area

Cedar Shake Roof	+ 2.55
Clay Tile Roof	+ 5
Slate Roof	+ 5.60
Upgrade Walls to Skim Coat Plaster	+ .57
Upgrade Ceilings to Textured Finish	+ .60
Air Conditioning, in Heating Ductwork	+ 3.76
In Separate Ductwork	+ 6.40
Heating Systems, Hot Water	+ 1.55
Heat Pump	+ 1.51
Electric Heat	- .72
Not Heated	- 3.46

Additional upgrades or components

Kitchen Cabinets & Countertops	Page 58
Bathroom Vanities	59
Fireplaces & Chimneys	59
Windows, Skylights & Dormers	59
Appliances	60
Breezeways & Porches	60
Finished Attic	60
Garages	61
Site Improvements	61
Wings & Ells	37

Important: See the Reference Section for Location Factors (to adjust for your city) and Estimating Forms.

- Simple design from standard plans
- Single family — 1 full bath, 1 kitchen
- No basement
- Asphalt shingles on roof
- Hot air heat
- Gypsum wallboard interior finishes
- Materials and workmanship are average
- Detail specifications on p. 27

Note: The illustration shown may contain some optional components (for example: garages and/or fireplaces) whose costs are shown in the modifications, adjustments, & alternatives below or at the end of the square foot section.

Base cost per square foot of living area

Exterior Wall	Living Area										
	1000	1200	1400	1600	1800	2000	2200	2600	3000	3400	3800
Wood Siding - Wood Frame	149.75	135.30	128.30	123.35	118.45	113.25	109.75	103.15	96.85	93.90	91.20
Brick Veneer - Wood Frame	156.15	141.25	133.80	128.60	123.40	118.05	114.25	107.25	100.60	97.50	94.65
Stucco on Wood Frame	144.50	130.50	123.75	119.00	114.40	109.35	106.00	99.80	93.65	90.90	88.40
Solid Masonry	170.60	154.65	146.35	140.60	134.75	128.90	124.65	116.60	109.25	105.65	102.45
Finished Basement, Add	23.25	22.90	22.10	21.65	21.05	20.70	20.30	19.55	18.95	18.60	18.30
Unfinished Basement, Add	9.20	8.55	8.05	7.75	7.40	7.15	6.95	6.45	6.15	5.90	5.70

Modifications

Add to the total cost

Upgrade Kitchen Cabinets	$ + 5950
Solid Surface Countertops (Included)	
Full Bath - including plumbing, wall and floor finishes	+ 8050
Half Bath - including plumbing, wall and floor finishes	+ 4767
One Car Attached Garage	+ 15,086
One Car Detached Garage	+ 19,847
Fireplace & Chimney	+ 7671

Adjustments

For multi family - add to total cost

Additional Kitchen	$ + 9963
Additional Bath	+ 8050
Additional Entry & Exit	+ 1804
Separate Heating	+ 1650
Separate Electric	+ 1817

For Townhouse/Rowhouse -
Multiply cost per square foot by

Inner Unit	.90
End Unit	.95

Alternatives

Add to or deduct from the cost per square foot of living area

Cedar Shake Roof	+ 1.75
Clay Tile Roof	+ 3.45
Slate Roof	+ 3.85
Upgrade Walls to Skim Coat Plaster	+ .59
Upgrade Ceilings to Textured Finish	+ .60
Air Conditioning, in Heating Ductwork	+ 3
In Separate Ductwork	+ 5.81
Heating Systems, Hot Water	+ 1.52
Heat Pump	+ 1.59
Electric Heat	– .62
Not Heated	– 3.36

Additional upgrades or components

Kitchen Cabinets & Countertops	Page 58
Bathroom Vanities	59
Fireplaces & Chimneys	59
Windows, Skylights & Dormers	59
Appliances	60
Breezeways & Porches	60
Finished Attic	60
Garages	61
Site Improvements	61
Wings & Ells	37

- **Simple design from standard plans**
- **Single family — 1 full bath, 1 kitchen**
- **No basement**
- **Asphalt shingles on roof**
- **Hot air heat**
- **Gypsum wallboard interior finishes**
- **Materials and workmanship are average**
- **Detail specifications on p. 27**

Note: The illustration shown may contain some optional components (for example: garages and/or fireplaces) whose costs are shown in the modifications, adjustments, & alternatives below or at the end of the square foot section.

Base cost per square foot of living area

Exterior Wall	Living Area										
	1200	1400	1600	1800	2000	2400	2800	3200	3600	4000	4400
Wood Siding - Wood Frame	149.55	139.90	127.50	125.05	120.30	112.75	106.90	100.90	97.90	92.55	90.70
Brick Veneer - Wood Frame	156.50	146.20	133.30	130.85	125.70	117.65	111.60	105.20	102.00	96.30	94.40
Stucco on Wood Frame	143.70	134.60	122.65	120.20	115.80	108.60	103.00	97.35	94.55	89.40	87.70
Solid Masonry	171.65	159.90	145.95	143.50	137.65	128.40	121.95	114.50	110.75	104.50	102.35
Finished Basement, Add	19.70	19.30	18.55	18.45	17.95	17.20	16.85	16.25	15.90	15.55	15.40
Unfinished Basement, Add	7.65	7.05	6.65	6.50	6.25	5.75	5.55	5.20	5.00	4.75	4.65

Modifications

Add to the total cost

Upgrade Kitchen Cabinets	$ + 5950
Solid Surface Countertops (Included)	
Full Bath - including plumbing, wall and floor finishes	+ 8050
Half Bath - including plumbing, wall and floor finishes	+ 4767
One Car Attached Garage	+ 15,086
One Car Detached Garage	+ 19,847
Fireplace & Chimney	+ 8415

Adjustments

For multi family - add to total cost

Additional Kitchen	$ + 9963
Additional Bath	+ 8050
Additional Entry & Exit	+ 1804
Separate Heating	+ 1650
Separate Electric	+ 1817

For Townhouse/Rowhouse -
Multiply cost per square foot by

Inner Unit	.90
End Unit	.95

Alternatives

Add to or deduct from the cost per square foot of living area

Cedar Shake Roof	+ 1.55
Clay Tile Roof	+ 3
Slate Roof	+ 3.35
Upgrade Walls to Skim Coat Plaster	+ .56
Upgrade Ceilings to Textured Finish	+ .60
Air Conditioning, in Heating Ductwork	+ 2.73
In Separate Ductwork	+ 5.62
Heating Systems, Hot Water	+ 1.38
Heat Pump	+ 1.63
Electric Heat	– 1.11
Not Heated	– 3.92

Additional upgrades or components

Kitchen Cabinets & Countertops	Page 58
Bathroom Vanities	59
Fireplaces & Chimneys	59
Windows, Skylights & Dormers	59
Appliances	60
Breezeways & Porches	60
Finished Attic	60
Garages	61
Site Improvements	61
Wings & Ells	37

- **Simple design from standard plans**
- **Single family — 1 full bath, 1 kitchen**
- **No basement**
- **Asphalt shingles on roof**
- **Hot air heat**
- **Gypsum wallboard interior finishes**
- **Materials and workmanship are average**
- **Detail specifications on p. 27**

Note: The illustration shown may contain some optional components (for example: garages and/or fireplaces) whose costs are shown in the modifications, adjustments, & alternatives below or at the end of the square foot section.

Base cost per square foot of living area

Exterior Wall	Living Area										
	1500	1800	2100	2500	3000	3500	4000	4500	5000	5500	6000
Wood Siding - Wood Frame	135.70	122.80	116.95	112.30	103.80	100.00	94.85	89.25	87.45	85.40	83.25
Brick Veneer - Wood Frame	141.95	128.65	122.35	117.45	108.55	104.40	98.90	93.00	91.15	88.90	86.50
Stucco on Wood Frame	130.55	118.10	112.55	108.00	99.95	96.35	91.55	86.20	84.50	82.55	80.55
Solid Masonry	156.60	142.20	135.05	129.55	119.50	114.85	108.35	101.70	99.55	97.05	94.15
Finished Basement, Add	16.85	16.65	16.15	15.75	15.15	14.85	14.40	14.00	13.90	13.70	13.45
Unfinished Basement, Add	6.20	5.80	5.45	5.25	4.90	4.70	4.40	4.15	4.05	3.95	3.80

Modifications

Add to the total cost

Upgrade Kitchen Cabinets	$ + 5950
Solid Surface Countertops (Included)	
Full Bath - including plumbing, wall and floor finishes	+ 8050
Half Bath - including plumbing, wall and floor finishes	+ 4767
One Car Attached Garage	+ 15,086
One Car Detached Garage	+ 19,847
Fireplace & Chimney	+ 8415

Adjustments

For multi family - add to total cost

Additional Kitchen	$ + 9963
Additional Bath	+ 8050
Additional Entry & Exit	+ 1804
Separate Heating	+ 1650
Separate Electric	+ 1817

For Townhouse/Rowhouse -
Multiply cost per square foot by

Inner Unit	.88
End Unit	.94

Alternatives

Add to or deduct from the cost per square foot of living area

Cedar Shake Roof	+ 1.15
Clay Tile Roof	+ 2.30
Slate Roof	+ 2.55
Upgrade Walls to Skim Coat Plaster	+ .59
Upgrade Ceilings to Textured Finish	+ .60
Air Conditioning, in Heating Ductwork	+ 2.73
In Separate Ductwork	+ 5.62
Heating Systems, Hot Water	+ 1.38
Heat Pump	+ 1.63
Electric Heat	– .86
Not Heated	– 3.66

Additional upgrades or components

Kitchen Cabinets & Countertops	Page 58
Bathroom Vanities	59
Fireplaces & Chimneys	59
Windows, Skylights & Dormers	59
Appliances	60
Breezeways & Porches	60
Finished Attic	60
Garages	61
Site Improvements	61
Wings & Ells	37

- **Simple design from standard plans**
- **Single family — 1 full bath, 1 kitchen**
- **No basement**
- **Asphalt shingles on roof**
- **Hot air heat**
- **Gypsum wallboard interior finishes**
- **Materials and workmanship are average**
- **Detail specifications on p. 27**

Note: The illustration shown may contain some optional components (for example: garages and/or fireplaces) whose costs are shown in the modifications, adjustments, & alternatives below or at the end of the square foot section.

Base cost per square foot of living area

Exterior Wall	Living Area										
	1000	1200	1400	1600	1800	2000	2200	2600	3000	3400	3800
Wood Siding - Wood Frame	139.50	125.85	119.40	114.85	110.45	105.55	102.45	96.55	90.70	88.05	85.70
Brick Veneer - Wood Frame	144.35	130.35	123.60	118.90	114.25	109.20	105.90	99.70	93.60	90.85	88.30
Stucco on Wood Frame	135.45	122.10	115.95	111.55	107.35	102.50	99.55	93.95	88.30	85.80	83.50
Solid Masonry	155.15	140.35	132.95	127.80	122.60	117.30	113.55	106.60	100.05	96.90	94.15
Finished Basement, Add	23.25	22.90	22.10	21.65	21.05	20.70	20.30	19.55	18.95	18.60	18.30
Unfinished Basement, Add	9.20	8.55	8.05	7.75	7.40	7.15	6.95	6.45	6.15	5.90	5.70

Modifications

Add to the total cost

Upgrade Kitchen Cabinets	$ + 5950
Solid Surface Countertops (Included)	
Full Bath - including plumbing, wall and floor finishes	+ 8050
Half Bath - including plumbing, wall and floor finishes	+ 4767
One Car Attached Garage	+ 15,086
One Car Detached Garage	+ 19,847
Fireplace & Chimney	+ 6942

Adjustments

For multi family - add to total cost

Additional Kitchen	$ + 9963
Additional Bath	+ 8050
Additional Entry & Exit	+ 1804
Separate Heating	+ 1650
Separate Electric	+ 1817

For Townhouse/Rowhouse -
Multiply cost per square foot by

Inner Unit	.91
End Unit	.96

Alternatives

Add to or deduct from the cost per square foot of living area

Cedar Shake Roof	+ 1.75
Clay Tile Roof	+ 3.45
Slate Roof	+ 3.85
Upgrade Walls to Skim Coat Plaster	+ .54
Upgrade Ceilings to Textured Finish	+ .60
Air Conditioning, in Heating Ductwork	+ 3
In Separate Ductwork	+ 5.81
Heating Systems, Hot Water	+ 1.52
Heat Pump	+ 1.59
Electric Heat	– .62
Not Heated	– 3.36

Additional upgrades or components

Kitchen Cabinets & Countertops	Page 58
Bathroom Vanities	59
Fireplaces & Chimneys	59
Windows, Skylights & Dormers	59
Appliances	60
Breezeways & Porches	60
Finished Attic	60
Garages	61
Site Improvements	61
Wings & Ells	37

- **Simple design from standard plans**
- **Single family — 1 full bath, 1 kitchen**
- **No basement**
- **Asphalt shingles on roof**
- **Hot air heat**
- **Gypsum wallboard interior finishes**
- **Materials and workmanship are average**
- **Detail specifications on p. 27**

Note: The illustration shown may contain some optional components (for example: garages and/or fireplaces) whose costs are shown in the modifications, adjustments, & alternatives below or at the end of the square foot section.

©Design Basics, Inc.

Base cost per square foot of living area

Exterior Wall	Living Area										
	1200	1500	1800	2100	2400	2700	3000	3400	3800	4200	4600
Wood Siding - Wood Frame	129.80	118.85	110.75	103.65	99.15	96.60	93.65	90.95	86.65	83.10	81.35
Brick Veneer - Wood Frame	134.30	122.95	114.45	107.00	102.30	99.65	96.50	93.70	89.20	85.45	83.70
Stucco on Wood Frame	126.05	115.45	107.60	100.85	96.50	94.05	91.25	88.65	84.45	81.10	79.40
Solid Masonry	144.20	131.85	122.50	114.30	109.15	106.25	102.65	99.70	94.80	90.70	88.70
Finished Basement, Add*	26.70	26.20	25.05	24.20	23.55	23.25	22.80	22.45	21.95	21.60	21.35
Unfinished Basement, Add*	10.25	9.50	8.75	8.25	7.85	7.65	7.35	7.15	6.85	6.60	6.50

*Basement under middle level only.

Modifications

Add to the total cost

Upgrade Kitchen Cabinets	$ + 5950
Solid Surface Countertops (Included)	
Full Bath - including plumbing, wall and floor finishes	+ 8050
Half Bath - including plumbing, wall and floor finishes	+ 4767
One Car Attached Garage	+ 15,086
One Car Detached Garage	+ 19,847
Fireplace & Chimney	+ 6942

Adjustments

For multi family - add to total cost

Additional Kitchen	$ + 9963
Additional Bath	+ 8050
Additional Entry & Exit	+ 1804
Separate Heating	+ 1650
Separate Electric	+ 1817

For Townhouse/Rowhouse -
Multiply cost per square foot by

Inner Unit	.90
End Unit	.95

Alternatives

Add to or deduct from the cost per square foot of living area

Cedar Shake Roof	+ 2.55
Clay Tile Roof	+ 5
Slate Roof	+ 5.60
Upgrade Walls to Skim Coat Plaster	+ .47
Upgrade Ceilings to Textured Finish	+ .60
Air Conditioning, in Heating Ductwork	+ 2.52
In Separate Ductwork	+ 5.47
Heating Systems, Hot Water	+ 1.47
Heat Pump	+ 1.66
Electric Heat	– .53
Not Heated	– 3.24

Additional upgrades or components

Kitchen Cabinets & Countertops	Page 58
Bathroom Vanities	59
Fireplaces & Chimneys	59
Windows, Skylights & Dormers	59
Appliances	60
Breezeways & Porches	60
Finished Attic	60
Garages	61
Site Improvements	61
Wings & Ells	37

- Post and beam frame
- Log exterior walls
- Simple design from standard plans
- Single family — 1 full bath, 1 kitchen
- No basement
- Asphalt shingles on roof
- Hot air heat
- Gypsum wallboard interior finishes
- Materials and workmanship are average
- Detail specification on page 27

Note: The illustration shown may contain some optional components (for example: garages and/or fireplaces) whose costs are shown in the modifications, adjustments, & alternatives below or at the end of the square foot section.

Base cost per square foot of living area

Exterior Wall	Living Area										
	600	800	1000	1200	1400	1600	1800	2000	2400	2800	3200
6" Log - Solid Wall	191.45	172.75	159.00	147.95	138.70	132.65	129.40	125.30	117.50	111.80	107.70
8" Log - Solid Wall	179.10	161.70	149.00	138.95	130.50	124.95	121.95	118.35	111.15	105.90	102.30
Finished Basement, Add	40.50	39.05	37.25	35.55	34.15	33.30	32.80	32.05	31.05	30.25	29.55
Unfinished Basement, Add	16.80	15.15	14.00	13.00	12.15	11.65	11.25	10.85	10.25	9.75	9.35

Modifications

Add to the total cost

Upgrade Kitchen Cabinets	$ + 5950
Solid Surface Countertops (Included)	
Full Bath - including plumbing, wall and floor finishes	+ 8050
Half Bath - including plumbing, wall and floor finishes	+ 4767
One Car Attached Garage	+ 15,086
One Car Detached Garage	+ 19,847
Fireplace & Chimney	+ 6942

Adjustments

For multi family - add to total cost

Additional Kitchen	$ + 9963
Additional Bath	+ 8050
Additional Entry & Exit	+ 1804
Separate Heating	+ 1650
Separate Electric	+ 1817

For Townhouse/Rowhouse - Multiply cost per square foot by

Inner Unit	.92
End Unit	.96

Alternatives

Add to or deduct from the cost per square foot of living area

Cedar Shake Roof	+ 3.50
Air Conditioning, in Heating Ductwork	+ 4.95
In Separate Ductwork	+ 7.31
Heating Systems, Hot Water	+ 1.63
Heat Pump	+ 1.36
Electric Heat	– .83
Not Heated	– 3.61

Additional upgrades or components

Kitchen Cabinets & Countertops	Page 58
Bathroom Vanities	59
Fireplaces & Chimneys	59
Windows, Skylights & Dormers	59
Appliances	60
Breezeways & Porches	60
Finished Attic	60
Garages	61
Site Improvements	61
Wings & Ells	37

Important: See the Reference Section for Location Factors (to adjust for your city) and Estimating Forms.

- Post and beam frame
- Log exterior walls
- Simple design from standard plans
- Single family — 1 full bath, 1 kitchen
- No basement
- Asphalt shingles on roof
- Hot air heat
- Gypsum wallboard interior finishes
- Materials and workmanship are average
- Detail specification on page 27

Note: The illustration shown may contain some optional components (for example: garages and/or fireplaces) whose costs are shown in the modifications, adjustments, & alternatives below or at the end of the square foot section.

Base cost per square foot of living area

Exterior Wall	Living Area										
	1000	1200	1400	1600	1800	2000	2200	2600	3000	3400	3800
6" Log - Solid Wall	163.30	148.30	140.75	135.50	130.25	124.80	121.00	113.85	107.15	103.90	101.10
8" Log - Solid Wall	150.50	136.45	129.70	124.95	120.30	115.25	111.85	105.60	99.55	96.70	94.20
Finished Basement, Add	23.25	22.90	22.10	21.65	21.05	20.70	20.30	19.55	18.95	18.60	18.30
Unfinished Basement, Add	9.20	8.55	8.05	7.75	7.40	7.15	6.95	6.45	6.15	5.90	5.70

Modifications

Add to the total cost

Upgrade Kitchen Cabinets	$ + 5950
Solid Surface Countertops (Included)	
Full Bath - including plumbing, wall and floor finishes	+ 8050
Half Bath - including plumbing, wall and floor finishes	+ 4767
One Car Attached Garage	+ 15,086
One Car Detached Garage	+ 19,847
Fireplace & Chimney	+ 7671

Adjustments

For multi family - add to total cost

Additional Kitchen	$ + 9963
Additional Bath	+ 8050
Additional Entry & Exit	+ 1804
Separate Heating	+ 1650
Separate Electric	+ 1817

For Townhouse/Rowhouse -
Multiply cost per square foot by

Inner Unit	.92
End Unit	.96

Alternatives

Add to or deduct from the cost per square foot of living area

Cedar Shake Roof	+ 1.75
Air Conditioning, in Heating Ductwork	+ 3
In Separate Ductwork	+ 5.81
Heating Systems, Hot Water	+ 1.52
Heat Pump	+ 1.59
Electric Heat	– .62
Not Heated	– 3.36

Additional upgrades or components

Kitchen Cabinets & Countertops	Page 58
Bathroom Vanities	59
Fireplaces & Chimneys	59
Windows, Skylights & Dormers	59
Appliances	60
Breezeways & Porches	60
Finished Attic	60
Garages	61
Site Improvements	61
Wings & Ells	37

1 Story — Base cost per square foot of living area

Exterior Wall	Living Area							
	50	100	200	300	400	500	600	700
Wood Siding - Wood Frame	223.85	172.40	150.10	125.90	118.60	114.15	111.30	112.05
Brick Veneer - Wood Frame	224.60	169.10	145.10	118.10	110.20	105.55	102.40	103.00
Stucco on Wood Frame	212.20	164.00	143.05	121.10	114.25	110.10	107.45	108.35
Solid Masonry	275.45	209.35	180.85	146.40	137.05	131.40	129.55	129.45
Finished Basement, Add	65.20	53.90	48.45	39.35	37.50	36.40	35.65	35.20
Unfinished Basement, Add	30.75	23.20	20.05	14.80	13.75	13.15	12.70	12.40

1-1/2 Story — Base cost per square foot of living area

Exterior Wall	Living Area							
	100	200	300	400	500	600	700	800
Wood Siding - Wood Frame	181.10	145.85	124.70	111.65	105.30	102.10	98.20	97.10
Brick Veneer - Wood Frame	235.90	177.00	147.15	128.80	119.75	114.80	109.65	107.70
Stucco on Wood Frame	213.10	158.75	131.90	116.90	108.80	104.50	99.90	98.00
Solid Masonry	269.30	203.80	169.40	146.15	135.80	130.05	124.00	121.90
Finished Basement, Add	43.85	39.60	35.95	31.90	30.80	30.10	29.45	29.30
Unfinished Basement, Add	19.25	16.10	14.00	11.70	11.10	10.65	10.30	10.25

2 Story — Base cost per square foot of living area

Exterior Wall	Living Area							
	100	200	400	600	800	1000	1200	1400
Wood Siding - Wood Frame	180.70	135.45	115.75	95.80	89.40	85.45	82.90	83.95
Brick Veneer - Wood Frame	240.30	168.30	135.55	109.00	100.05	94.75	91.20	91.50
Stucco on Wood Frame	214.75	150.00	120.35	98.85	91.00	86.25	83.10	83.70
Solid Masonry	277.75	195.10	157.80	123.85	113.50	107.25	103.05	102.95
Finished Basement, Add	34.70	29.05	26.35	21.75	20.90	20.30	19.95	19.70
Unfinished Basement, Add	15.50	11.75	10.20	7.55	7.05	6.75	6.50	6.35

Base costs do not include bathroom or kitchen facilities. Use Modifications/Adjustments/Alternatives on pages 58-61 where appropriate.

Important: See the Reference Section for Location Factors (to adjust for your city) and Estimating Forms.

1 Story

1 - 1/2 Story

2 Story

2 - 1/2 Story

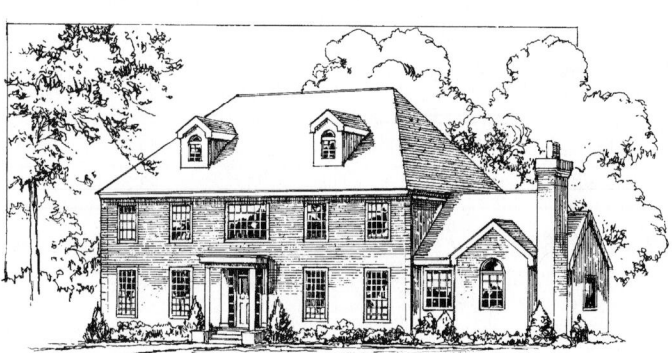

Bi-Level

Tri-Level

	Components
1 Site Work	Site preparation for slab or excavation for lower level; 4' deep trench excavation for foundation wall.
2 Foundations	Continuous reinforced concrete footing, 8" deep x 18" wide; dampproofed and insulated 8" thick reinforced concrete foundation wall, 4' deep; 4" concrete slab on 4" crushed stone base and polyethylene vapor barrier, trowel finish.
3 Framing	Exterior walls—2 x 6 wood studs, 16" O.C.; 1/2" plywood sheathing; 2 x 8 rafters 16" O.C. with 1/2" plywood sheathing, 4 in 12, 6 in 12 or 8 in 12 roof pitch, 2 x 6 or 2 x 8 ceiling joists; 1/2" plywood subfloor on 1 x 3 wood sleepers 16" O.C.; 2 x 10 floor joists with 5/8" plywood subfloor on models with more than one level.
4 Exterior Walls	Horizontal beveled wood siding and #15 felt building paper on insulated wood frame walls; double hung windows; 3 flush solid core wood exterior doors with storms. **Alternates:** • Brick veneer on wood frame, has 4" veneer high quality face brick or select common brick. • Stone veneer on wood frame has exterior veneer of field stone or 2" thick limestone. • Solid masonry has an 8" concrete block wall with insulation and a brick or stone facing. It may be a solid brick or stone structure.
5 Roofing	30 year asphalt shingles; #15 felt building paper; aluminum gutters, downspouts and drip edge; copper flashings.
6 Interiors	Walls and ceilings—5/8" drywall, skim coat plaster, primed and painted with 2 coats; hardwood baseboard and trim; sanded and finished, hardwood floor 70%, ceramic tile with 1/2" underlayment 20%, vinyl tile with 1/2" underlayment 10%; wood panel interior doors, primed and painted with 2 coats.
7 Specialties	Custom grade kitchen cabinets—20 L.F. wall and base with plastic laminate counter top and kitchen sink; 4 L.F. bathroom vanity; 75 gallon electric water heater; medicine cabinet.
8 Mechanical	Gas fired warm air heat/air conditioning; one full bath including bathtub, corner shower, built in lavatory and water closet; one 1/2 bath including built in lavatory and water closet.
9 Electrical	100 Amp. service; romex wiring; incandescent lighting fixtures, switches, receptacles.
10 Overhead and Profit	General Contractor overhead and profit.

Adjustments

Unfinished Basement:
7'-6" high cast-in-place concrete walls 10" thick or 12" concrete block.

Finished Basement:
Includes painted drywall on 2 x 4 wood furring with insulation, suspended ceiling, carpeting on concrete floor, heating and lighting.

- A distinct residence from designer's plans
- Single family — 1 full bath, 1 half bath, 1 kitchen
- No basement
- Asphalt shingles on roof
- Forced hot air heat/air conditioning
- Gypsum wallboard interior finishes
- Materials and workmanship are above average
- Detail specifications on page 39

Note: The illustration shown may contain some optional components (for example: garages and/or fireplaces) whose costs are shown in the modifications, adjustments, & alternatives below or at the end of the square foot section.

©Design Basics, Inc.

Base cost per square foot of living area

Exterior Wall	Living Area										
	800	1000	1200	1400	1600	1800	2000	2400	2800	3200	3600
Wood Siding - Wood Frame	216.45	195.05	178.40	165.10	155.95	150.70	144.80	134.00	126.05	120.45	114.80
Brick Veneer - Wood Frame	227.10	204.80	187.15	173.00	163.40	157.90	151.55	140.10	131.70	125.75	119.70
Stone Veneer - Wood Frame	236.95	213.70	195.20	180.35	170.30	164.55	157.80	145.80	137.00	130.60	124.25
Solid Masonry	237.50	214.20	195.70	180.80	170.75	164.90	158.15	146.15	137.25	130.90	124.50
Finished Basement, Add	60.80	60.55	57.95	55.85	54.50	53.65	52.55	51.00	49.75	48.70	47.75
Unfinished Basement, Add	26.70	25.20	23.90	22.80	22.10	21.65	21.10	20.25	19.60	19.05	18.55

Modifications

Add to the total cost

Upgrade Kitchen Cabinets	$ + 1850
Solid Surface Countertops (Included)	
Full Bath - including plumbing, wall and floor finishes	+ 9660
Half Bath - including plumbing, wall and floor finishes	+ 5720
Two Car Attached Garage	+ 29,653
Two Car Detached Garage	+ 33,939
Fireplace & Chimney	+ 7266

Adjustments

For multi family - add to total cost

Additional Kitchen	$ + 22,193
Additional Full Bath & Half Bath	+ 15,380
Additional Entry & Exit	+ 1804
Separate Heating & Air Conditioning	+ 7787
Separate Electric	+ 1817

For Townhouse/Rowhouse -
Multiply cost per square foot by

Inner Unit	.90
End Unit	.95

Alternatives

Add to or deduct from the cost per square foot of living area

Cedar Shake Roof	+ 2.75
Clay Tile Roof	+ 6.10
Slate Roof	+ 6.90
Upgrade Ceilings to Textured Finish	+ .60
Air Conditioning, in Heating Ductwork	Base System
Heating Systems, Hot Water	+ 1.67
Heat Pump	+ 1.36
Electric Heat	– 2.33
Not Heated	– 4.52

Additional upgrades or components

Kitchen Cabinets & Countertops	Page 58
Bathroom Vanities	59
Fireplaces & Chimneys	59
Windows, Skylights & Dormers	59
Appliances	60
Breezeways & Porches	60
Finished Attic	60
Garages	61
Site Improvements	61
Wings & Ells	47

Important: See the Reference Section for Location Factors (to adjust for your city) and Estimating Forms.

- A distinct residence from designer's plans
- Single family — 1 full bath, 1 half bath, 1 kitchen
- No basement
- Asphalt shingles on roof
- Forced hot air heat/air conditioning
- Gypsum wallboard interior finishes
- Materials and workmanship are above average
- Detail specifications on page 39

Note: The illustration shown may contain some optional components (for example: garages and/or fireplaces) whose costs are shown in the modifications, adjustments, & alternatives below or at the end of the square foot section.

©Donald A. Gardner Architects, Inc.

Base cost per square foot of living area

Exterior Wall	Living Area										
	1000	1200	1400	1600	1800	2000	2400	2800	3200	3600	4000
Wood Siding - Wood Frame	195.55	181.35	171.25	159.45	152.45	145.80	133.30	127.50	122.10	118.05	112.45
Brick Veneer - Wood Frame	207.05	192.15	181.50	168.75	161.35	154.20	140.70	134.50	128.70	124.35	118.30
Stone Veneer - Wood Frame	217.70	202.15	191.00	177.35	169.55	161.95	147.55	141.05	134.65	130.10	123.70
Solid Masonry	218.20	202.55	191.50	177.75	169.85	162.35	147.85	141.35	134.90	130.40	123.90
Finished Basement, Add	40.35	40.65	39.50	37.95	37.00	36.15	34.50	33.75	32.90	32.45	31.75
Unfinished Basement, Add	17.95	17.15	16.65	15.75	15.30	14.90	14.05	13.65	13.15	12.95	12.60

Modifications

Add to the total cost

Upgrade Kitchen Cabinets	$ + 1850
Solid Surface Countertops (Included)	
Full Bath - including plumbing, wall and floor finishes	+ 9660
Half Bath - including plumbing, wall and floor finishes	+ 5720
Two Car Attached Garage	+ 29,653
Two Car Detached Garage	+ 33,939
Fireplace & Chimney	+ 7266

Adjustments

For multi family - add to total cost

Additional Kitchen	$ + 22,193
Additional Full Bath & Half Bath	+ 15,380
Additional Entry & Exit	+ 1804
Separate Heating & Air Conditioning	+ 7787
Separate Electric	+ 1817

For Townhouse/Rowhouse -
Multiply cost per square foot by

Inner Unit	.90
End Unit	.95

Alternatives

Add to or deduct from the cost per square foot of living area

Cedar Shake Roof	+ 2
Clay Tile Roof	+ 4.40
Slate Roof	+ 5
Upgrade Ceilings to Textured Finish	+ .60
Air Conditioning, in Heating Ductwork	Base System
Heating Systems, Hot Water	+ 1.60
Heat Pump	+ 1.42
Electric Heat	− 2.05
Not Heated	− 4.17

Additional upgrades or components

- A distinct residence from designer's plans
- Single family — 1 full bath, 1 half bath, 1 kitchen
- No basement
- Asphalt shingles on roof
- Forced hot air heat/air conditioning
- Gypsum wallboard interior finishes
- Materials and workmanship are above average
- Detail specifications on page 39

Note: The illustration shown may contain some optional components (for example: garages and/or fireplaces) whose costs are shown in the modifications, adjustments, & alternatives below or at the end of the square foot section.

Base cost per square foot of living area

Exterior Wall	Living Area										
	1200	1400	1600	1800	2000	2400	2800	3200	3600	4000	4400
Wood Siding - Wood Frame	178.40	167.15	159.15	152.10	144.70	134.05	125.05	119.15	115.60	111.90	108.65
Brick Veneer - Wood Frame	189.80	177.85	169.40	161.70	154.00	142.40	132.65	126.30	122.50	118.30	114.90
Stone Veneer - Wood Frame	200.40	187.75	178.85	170.60	162.55	150.10	139.70	132.85	128.85	124.25	120.65
Solid Masonry	201.35	188.60	179.55	171.40	163.25	150.80	140.30	133.40	129.40	124.75	121.10
Finished Basement, Add	32.45	32.60	31.75	30.90	30.30	28.95	27.90	27.20	26.85	26.25	25.90
Unfinished Basement, Add	14.40	13.80	13.40	12.90	12.65	11.95	11.40	11.05	10.80	10.55	10.40

Modifications

Add to the total cost

Upgrade Kitchen Cabinets	$ + 1850
Solid Surface Countertops (Included)	
Full Bath - including plumbing, wall and floor finishes	+ 9660
Half Bath - including plumbing, wall and floor finishes	+ 5720
Two Car Attached Garage	+ 29,653
Two Car Detached Garage	+ 33,939
Fireplace & Chimney	+ 8202

Adjustments

For multi family - add to total cost

Additional Kitchen	$ + 22,193
Additional Full Bath & Half Bath	+ 15,380
Additional Entry & Exit	+ 1804
Separate Heating & Air Conditioning	+ 7787
Separate Electric	+ 1817

For Townhouse/Rowhouse -
Multiply cost per square foot by

Inner Unit	.87
End Unit	.93

Alternatives

Add to or deduct from the cost per square foot of living area

Cedar Shake Roof	+ 1.35
Clay Tile Roof	+ 3.05
Slate Roof	+ 3.45
Upgrade Ceilings to Textured Finish	+ .60
Air Conditioning, in Heating Ductwork	Base System
Heating Systems, Hot Water	+ 1.55
Heat Pump	+ 1.58
Electric Heat	− 2.05
Not Heated	− 3.93

Additional upgrades or components

Kitchen Cabinets & Countertops	Page 58
Bathroom Vanities	59
Fireplaces & Chimneys	59
Windows, Skylights & Dormers	59
Appliances	60
Breezeways & Porches	60
Finished Attic	60
Garages	61
Site Improvements	61
Wings & Ells	47

- **A distinct residence from designer's plans**
- **Single family — 1 full bath, 1 half bath, 1 kitchen**
- **No basement**
- **Asphalt shingles on roof**
- **Forced hot air heat/air conditioning**
- **Gypsum wallboard interior finishes**
- **Materials and workmanship are above average**
- **Detail specifications on page 39**

Note: The illustration shown may contain some optional components (for example: garages and/or fireplaces) whose costs are shown in the modifications, adjustments, & alternatives below or at the end of the square foot section.

Base cost per square foot of living area

Exterior Wall	Living Area										
	1500	1800	2100	2400	2800	3200	3600	4000	4500	5000	5500
Wood Siding - Wood Frame	175.50	157.75	147.20	140.75	132.60	124.75	120.50	113.70	110.25	106.95	103.70
Brick Veneer - Wood Frame	187.70	169.00	157.35	150.30	141.70	133.05	128.35	121.05	117.10	113.50	109.90
Stone Veneer - Wood Frame	199.00	179.40	166.60	159.10	150.15	140.70	135.55	127.75	123.45	119.55	115.75
Solid Masonry	199.80	180.15	167.30	159.75	150.75	141.25	136.10	128.15	123.90	120.00	116.15
Finished Basement, Add	25.75	25.75	24.40	23.75	23.20	22.30	21.75	21.20	20.80	20.40	20.10
Unfinished Basement, Add	11.55	10.95	10.25	9.95	9.70	9.15	8.90	8.60	8.35	8.15	8.00

Modifications

Add to the total cost

Upgrade Kitchen Cabinets	$ + 1850
Solid Surface Countertops (Included)	
Full Bath - including plumbing, wall and floor finishes	+ 9660
Half Bath - including plumbing, wall and floor finishes	+ 5720
Two Car Attached Garage	+ 29,653
Two Car Detached Garage	+ 33,939
Fireplace & Chimney	+ 8202

Adjustments

For multi family - add to total cost

Additional Kitchen	$ + 22,193
Additional Full Bath & Half Bath	+ 15,380
Additional Entry & Exit	+ 1804
Separate Heating & Air Conditioning	+ 7787
Separate Electric	+ 1817

For Townhouse/Rowhouse -
Multiply cost per square foot by

Inner Unit	.87
End Unit	.94

Alternatives

Add to or deduct from the cost per square foot of living area

Cedar Shake Roof	+ 1.20
Clay Tile Roof	+ 2.65
Slate Roof	+ 3
Upgrade Ceilings to Textured Finish	+ .60
Air Conditioning, in Heating Ductwork	Base System
Heating Systems, Hot Water	+ 1.40
Heat Pump	+ 1.63
Electric Heat	– 3.61
Not Heated	– 3.93

Additional upgrades or components

Kitchen Cabinets & Countertops	Page 58
Bathroom Vanities	59
Fireplaces & Chimneys	59
Windows, Skylights & Dormers	59
Appliances	60
Breezeways & Porches	60
Finished Attic	60
Garages	61
Site Improvements	61
Wings & Ells	47

- A distinct residence from designer's plans
- Single family — 1 full bath, 1 half bath, 1 kitchen
- No basement
- Asphalt shingles on roof
- Forced hot air heat/air conditioning
- Gypsum wallboard interior finishes
- Materials and workmanship are above average
- Detail specifications on page 39

Note: The illustration shown may contain some optional components (for example: garages and/or fireplaces) whose costs are shown in the modifications, adjustments, & alternatives below or at the end of the square foot section.

Base cost per square foot of living area

Exterior Wall	Living Area										
	1500	1800	2100	2500	3000	3500	4000	4500	5000	5500	6000
Wood Siding - Wood Frame	172.25	154.80	146.20	138.90	127.90	122.40	115.70	108.90	106.25	103.60	100.85
Brick Veneer - Wood Frame	184.40	166.05	156.75	148.95	137.05	131.10	123.50	116.10	113.25	110.35	107.15
Stone Veneer - Wood Frame	195.65	176.45	166.50	158.25	145.50	139.05	130.80	122.85	119.75	116.60	113.00
Solid Masonry	196.70	177.45	167.45	159.05	146.25	139.75	131.45	123.50	120.30	117.15	113.60
Finished Basement, Add	22.55	22.55	21.65	21.10	20.20	19.65	18.90	18.30	18.10	17.85	17.50
Unfinished Basement, Add	10.10	9.60	9.15	8.85	8.40	8.15	7.75	7.45	7.35	7.20	7.00

Modifications

Add to the total cost

Upgrade Kitchen Cabinets	$ + 1850
Solid Surface Countertops (Included)	
Full Bath - including plumbing, wall and floor finishes	+ 9660
Half Bath - including plumbing, wall and floor finishes	+ 5720
Two Car Attached Garage	+ 29,653
Two Car Detached Garage	+ 33,939
Fireplace & Chimney	+ 9262

Adjustments

For multi family - add to total cost

Additional Kitchen	$ + 22,193
Additional Full Bath & Half Bath	+ 15,380
Additional Entry & Exit	+ 1804
Separate Heating & Air Conditioning	+ 7787
Separate Electric	+ 1817

For Townhouse/Rowhouse -
Multiply cost per square foot by

Inner Unit	.85
End Unit	.93

Alternatives

Add to or deduct from the cost per square foot of living area

Cedar Shake Roof	+ .90
Clay Tile Roof	+ 2.05
Slate Roof	+ 2.30
Upgrade Ceilings to Textured Finish	+ .60
Air Conditioning, in Heating Ductwork	Base System
Heating Systems, Hot Water	+ 1.40
Heat Pump	+ 1.63
Electric Heat	– 3.61
Not Heated	– 3.81

Additional upgrades or components

Kitchen Cabinets & Countertops	Page 58
Bathroom Vanities	59
Fireplaces & Chimneys	59
Windows, Skylights & Dormers	59
Appliances	60
Breezeways & Porches	60
Finished Attic	60
Garages	61
Site Improvements	61
Wings & Ells	47

- A distinct residence from designer's plans
- Single family — 1 full bath, 1 half bath, 1 kitchen
- No basement
- Asphalt shingles on roof
- Forced hot air heat/air conditioning
- Gypsum wallboard interior finishes
- Materials and workmanship are above average
- Detail specifications on page 39

Note: The illustration shown may contain some optional components (for example: garages and/or fireplaces) whose costs are shown in the modifications, adjustments, & alternatives below or at the end of the square foot section.

Base cost per square foot of living area

Exterior Wall	Living Area										
	1200	1400	1600	1800	2000	2400	2800	3200	3600	4000	4400
Wood Siding - Wood Frame	168.95	158.35	150.70	144.15	137.05	127.15	118.75	113.30	109.90	106.55	103.45
Brick Veneer - Wood Frame	177.60	166.45	158.50	151.45	144.10	133.50	124.55	118.70	115.15	111.45	108.20
Stone Veneer - Wood Frame	185.65	174.00	165.70	158.25	150.60	139.35	129.85	123.70	120.00	116.00	112.60
Solid Masonry	186.30	174.55	166.25	158.75	151.10	139.80	130.30	124.10	120.35	116.30	112.95
Finished Basement, Add	32.45	32.60	31.75	30.90	30.30	28.95	27.90	27.20	26.85	26.25	25.90
Unfinished Basement, Add	14.40	13.80	13.40	12.90	12.65	11.95	11.40	11.05	10.80	10.55	10.40

Modifications

Add to the total cost

Upgrade Kitchen Cabinets	$ + 1850
Solid Surface Countertops (Included)	
Full Bath - including plumbing, wall and floor finishes	+ 9660
Half Bath - including plumbing, wall and floor finishes	+ 5720
Two Car Attached Garage	+ 29,653
Two Car Detached Garage	+ 33,939
Fireplace & Chimney	+ 7266

Adjustments

For multi family - add to total cost

Additional Kitchen	$ + 22,193
Additional Full Bath & Half Bath	+ 15,380
Additional Entry & Exit	+ 1804
Separate Heating & Air Conditioning	+ 7787
Separate Electric	+ 1817

For Townhouse/Rowhouse -
Multiply cost per square foot by

Inner Unit	.89
End Unit	.95

Alternatives

Add to or deduct from the cost per square foot of living area

Cedar Shake Roof	+ 1.35
Clay Tile Roof	+ 3.05
Slate Roof	+ 3.45
Upgrade Ceilings to Textured Finish	+ .60
Air Conditioning, in Heating Ductwork	Base System
Heating Systems, Hot Water	+ 1.55
Heat Pump	+ 1.58
Electric Heat	– 2.05
Not Heated	– 3.81

Additional upgrades or components

Kitchen Cabinets & Countertops	Page 58
Bathroom Vanities	59
Fireplaces & Chimneys	59
Windows, Skylights & Dormers	59
Appliances	60
Breezeways & Porches	60
Finished Attic	60
Garages	61
Site Improvements	61
Wings & Ells	47

Important: See the Reference Section for Location Factors (to adjust for your city) and Estimating Forms.

- **A distinct residence from designer's plans**
- **Single family — 1 full bath, 1 half bath, 1 kitchen**
- **No basement**
- **Asphalt shingles on roof**
- **Forced hot air heat/air conditioning**
- **Gypsum wallboard interior finishes**
- **Materials and workmanship are above average**
- **Detail specifications on page 39**

Note: The illustration shown may contain some optional components (for example: garages and/or fireplaces) whose costs are shown in the modifications, adjustments, & alternatives below or at the end of the square foot section.

eDesign Basics, Inc.

Base cost per square foot of living area

Exterior Wall	Living Area										
	1200	1500	1800	2100	2400	2800	3200	3600	4000	4500	5000
Wood Siding - Wood Frame	174.65	157.95	145.55	135.35	128.45	123.35	117.50	111.45	108.75	103.10	99.90
Brick Veneer - Wood Frame	183.35	165.90	152.70	141.80	134.55	129.25	122.95	116.55	113.60	107.65	104.10
Stone Veneer - Wood Frame	191.45	173.20	159.25	147.85	140.20	134.60	127.95	121.20	118.15	111.75	108.00
Solid Masonry	192.05	173.75	159.75	148.25	140.60	135.00	128.30	121.50	118.45	112.10	108.30
Finished Basement, Add*	40.55	40.40	38.70	37.25	36.35	35.75	34.80	34.00	33.65	32.85	32.30
Unfinished Basement, Add*	17.85	16.85	15.95	15.20	14.75	14.40	13.95	13.50	13.30	12.90	12.60

*Basement under middle level only.

Modifications

Add to the total cost

Upgrade Kitchen Cabinets	$ + 1850
Solid Surface Countertops (Included)	
Full Bath - including plumbing, wall and floor finishes	+ 9660
Half Bath - including plumbing, wall and floor finishes	+ 5720
Two Car Attached Garage	+ 29,653
Two Car Detached Garage	+ 33,939
Fireplace & Chimney	+ 7266

Adjustments

For multi family - add to total cost

Additional Kitchen	$ + 22,193
Additional Full Bath & Half Bath	+ 15,380
Additional Entry & Exit	+ 1804
Separate Heating & Air Conditioning	+ 7787
Separate Electric	+ 1817

*For Townhouse/Rowhouse -
Multiply cost per square foot by*

Inner Unit	.87
End Unit	.94

Alternatives

Add to or deduct from the cost per square foot of living area

Cedar Shake Roof	+ 2
Clay Tile Roof	+ 4.40
Slate Roof	+ 5
Upgrade Ceilings to Textured Finish	+ .60
Air Conditioning, in Heating Ductwork	Base System
Heating Systems, Hot Water	+ 1.50
Heat Pump	+ 1.65
Electric Heat	− 1.83
Not Heated	− 3.81

Additional upgrades or components

Kitchen Cabinets & Countertops	Page 58
Bathroom Vanities	59
Fireplaces & Chimneys	59
Windows, Skylights & Dormers	59
Appliances	60
Breezeways & Porches	60
Finished Attic	60
Garages	61
Site Improvements	61
Wings & Ells	47

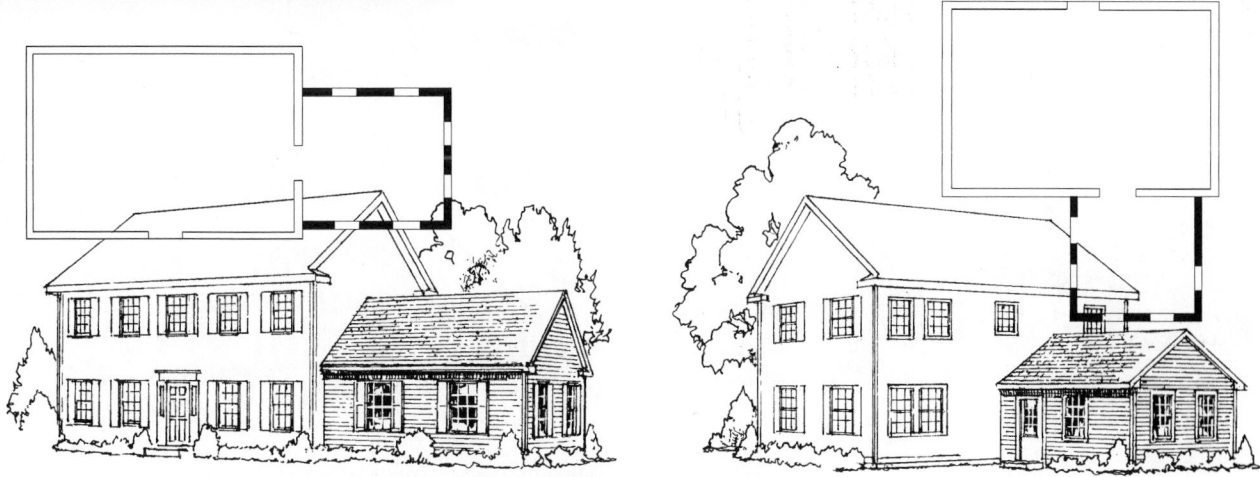

1 Story — Base cost per square foot of living area

Exterior Wall	Living Area							
	50	100	200	300	400	500	600	700
Wood Siding - Wood Frame	257.45	199.95	175.05	148.30	140.05	135.15	131.90	132.85
Brick Veneer - Wood Frame	284.90	219.60	191.45	159.15	149.90	144.35	140.60	141.25
Stone Veneer - Wood Frame	310.35	237.75	206.55	169.25	158.95	152.80	148.70	149.05
Solid Masonry	316.10	241.85	209.95	171.50	161.00	154.75	150.45	150.80
Finished Basement, Add	98.20	83.60	75.70	62.50	59.85	58.30	57.25	56.50
Unfinished Basement, Add	72.90	50.40	40.10	30.80	28.35	26.90	25.90	25.20

1-1/2 Story — Base cost per square foot of living area

Exterior Wall	Living Area							
	100	200	300	400	500	600	700	800
Wood Siding - Wood Frame	204.80	167.90	145.45	131.90	125.10	121.85	117.70	116.50
Brick Veneer - Wood Frame	229.35	187.55	161.80	144.70	136.85	133.00	128.20	126.95
Stone Veneer - Wood Frame	252.05	205.70	176.95	156.50	147.80	143.30	137.95	136.60
Solid Masonry	257.15	209.75	180.35	159.15	150.20	145.60	140.15	138.70
Finished Basement, Add	65.50	60.55	55.30	49.50	47.85	46.85	45.85	45.70
Unfinished Basement, Add	43.60	33.25	28.40	24.15	22.65	21.70	20.90	20.60

2 Story — Base cost per square foot of living area

Exterior Wall	Living Area							
	100	200	400	600	800	1000	1200	1400
Wood Siding - Wood Frame	204.10	155.60	134.40	113.80	106.80	102.60	99.90	101.15
Brick Veneer - Wood Frame	231.65	175.25	150.75	124.65	116.65	111.80	108.60	109.60
Stone Veneer - Wood Frame	257.05	193.40	165.90	134.75	125.75	120.30	116.65	117.35
Solid Masonry	262.75	197.50	169.30	137.00	127.75	122.20	118.45	119.10
Finished Basement, Add	49.15	41.85	37.85	31.30	29.90	29.15	28.65	28.25
Unfinished Basement, Add	36.45	25.25	20.05	15.40	14.15	13.45	12.90	12.55

Base costs do not include bathroom or kitchen facilities. Use Modifications/Adjustments/Alternatives on pages 58-61 where appropriate.

Important: See the Reference Section for Location Factors (to adjust for your city) and Estimating Forms.

1 Story

1-1/2 Story

2 Story

2-1/2 Story

Bi-Level

Tri-Level

Components

1 Site Work		Site preparation for slab or excavation for lower level; 4' deep trench excavation for foundation wall.
2 Foundations		Continuous reinforced concrete footing, 8" deep x 18" wide; dampproofed and insulated 12" thick reinforced concrete foundation wall, 4' deep; 4" concrete slab on 4" crushed stone base and polyethylene vapor barrier, trowel finish.
3 Framing		Exterior walls—2 x 6 wood studs, 16" O.C.; 1/2" plywood sheathing; 2 x 8 rafters 16" O.C. with 1/2" plywood sheathing, 4 in 12, 6 in 12 or 8 in 12 roof pitch, 2 x 6 or 2 x 8 ceiling joists; 1/2" plywood subfloor on 1 x 3 wood sleepers 16" O.C.; 2 x 10 floor joists with 5/8" plywood subfloor on models with more than one level.
4 Exterior Walls		Face brick veneer and #15 felt building paper on insulated wood frame walls; double hung windows; 3 flush solid core wood exterior doors with storms. **Alternates:** • Wood siding on wood frame, has top quality cedar or redwood siding or hand split cedar shingles or shakes. • Solid brick may have solid brick exterior wall or brick on concrete block. • Solid stone concrete block with selected fieldstone or limestone exterior.
5 Roofing		Red cedar shingles; #15 felt building paper; aluminum gutters, downspouts and drip edge; copper flashings.
6 Interiors		Walls and ceilings—5/8" drywall, skim coat plaster, primed and painted with 2 coats; hardwood baseboard and trim; sanded and finished, hardwood floor 70%, ceramic tile with 1/2" underlayment 20%, vinyl tile with 1/2" underlayment 10%; wood panel interior doors, primed and painted with 2 coats.
7 Specialties		Luxury grade kitchen cabinets—25 L.F. wall and base with plastic laminate counter top and kitchen sink; 6 L.F. bathroom vanity; 75 gallon electric water heater; medicine cabinet.
8 Mechanical		Gas fired warm air heat/air conditioning; one full bath including bathtub, corner shower, built in lavatory and water closet; one 1/2 bath including built in lavatory and water closet.
9 Electrical		100 Amp. service; romex wiring; incandescent lighting fixtures, switches, receptacles.
10 Overhead and Profit		General Contractor overhead and profit.

Adjustments

Unfinished Basement:
8'-0" high cast-in-place concrete walls 12" thick or 12" concrete block.

Finished Basement:
Includes painted drywall on 2 x 4 wood furring with insulation, suspended ceiling, carpeting on subfloor with sleepers, heating and lighting.

- **Unique residence built from an architect's plan**
- **Single family — 1 full bath, 1 half bath, 1 kitchen**
- **No basement**
- **Cedar shakes on roof**
- **Forced hot air heat/air conditioning**
- **Gypsum wallboard interior finishes**
- **Many special features**
- **Extraordinary materials and workmanship**
- **Detail specifications on page 49**

Note: The illustration shown may contain some optional components (for example: garages and/or fireplaces) whose costs are shown in the modifications, adjustments, & alternatives below or at the end of the square foot section.

©Home Planners, Inc.

Base cost per square foot of living area

Exterior Wall	Living Area										
	1000	1200	1400	1600	1800	2000	2400	2800	3200	3600	4000
Wood Siding - Wood Frame	230.40	210.10	194.05	183.05	176.65	169.35	156.45	147.05	140.35	133.60	128.10
Brick Veneer - Wood Frame	241.20	219.85	202.95	191.35	184.65	176.90	163.30	153.40	146.25	139.10	133.20
Solid Brick	254.05	231.45	213.45	201.25	194.15	185.85	171.45	160.95	153.25	145.60	139.30
Solid Stone	261.80	238.50	219.90	207.25	199.90	191.25	176.40	165.45	157.45	149.55	143.00
Finished Basement, Add	59.80	64.35	61.70	60.05	59.00	57.60	55.65	54.15	52.80	51.70	50.70
Unfinished Basement, Add	27.00	25.30	24.00	23.10	22.65	21.95	20.90	20.10	19.45	18.90	18.40

Modifications

Add to the total cost

Upgrade Kitchen Cabinets	$ + 2604
Solid Surface Countertops (Included)	
Full Bath - including plumbing, wall and floor finishes	+ 11,593
Half Bath - including plumbing, wall and floor finishes	+ 6864
Two Car Attached Garage	+ 33,913
Two Car Detached Garage	+ 38,535
Fireplace & Chimney	+ 10,406

Adjustments

For multi family - add to total cost

Additional Kitchen	$ + 30,297
Additional Full Bath & Half Bath	+ 18,457
Additional Entry & Exit	+ 2418
Separate Heating & Air Conditioning	+ 7787
Separate Electric	+ 1817

For Townhouse/Rowhouse -
Multiply cost per square foot by

Inner Unit	.90
End Unit	.95

Alternatives

Add to or deduct from the cost per square foot of living area

Heavyweight Asphalt Shingles	– 2.75
Clay Tile Roof	+ 3.40
Slate Roof	+ 4.20
Upgrade Ceilings to Textured Finish	+ .60
Air Conditioning, in Heating Ductwork	Base System
Heating Systems, Hot Water	+ 1.80
Heat Pump	+ 1.46
Electric Heat	– 2.05
Not Heated	– 4.92

Additional upgrades or components

Kitchen Cabinets & Countertops	Page 58
Bathroom Vanities	59
Fireplaces & Chimneys	59
Windows, Skylights & Dormers	59
Appliances	60
Breezeways & Porches	60
Finished Attic	60
Garages	61
Site Improvements	61
Wings & Ells	57

Important: See the Reference Section for Location Factors (to adjust for your city) and Estimating Forms.

- **Unique residence built from an architect's plan**
- **Single family — 1 full bath, 1 half bath, 1 kitchen**
- **No basement**
- **Cedar shakes on roof**
- **Forced hot air heat/air conditioning**
- **Gypsum wallboard interior finishes**
- **Many special features**
- **Extraordinary materials and workmanship**
- **Detail specifications on page 49**

Note: The illustration shown may contain some optional components (for example: garages and/or fireplaces) whose costs are shown in the modifications, adjustments, & alternatives below or at the end of the square foot section.

eLarry E. Belk Designs

Base cost per square foot of living area

Exterior Wall	Living Area										
	1000	1200	1400	1600	1800	2000	2400	2800	3200	3600	4000
Wood Siding - Wood Frame	231.70	214.10	201.60	187.40	179.00	170.90	156.00	148.85	142.40	137.50	130.85
Brick Veneer - Wood Frame	244.60	226.15	213.05	197.85	188.90	180.30	164.30	156.75	149.65	144.55	137.40
Solid Brick	259.20	239.85	226.10	209.70	200.10	190.95	173.75	165.65	157.90	152.45	144.80
Solid Stone	268.90	249.00	234.75	217.55	207.60	198.05	180.00	171.55	163.40	157.80	149.70
Finished Basement, Add	42.00	45.65	44.25	42.25	41.15	40.15	38.05	37.10	36.05	35.45	34.60
Unfinished Basement, Add	19.45	18.55	17.90	16.90	16.35	15.80	14.75	14.30	13.70	13.40	12.95

Modifications

Add to the total cost

Upgrade Kitchen Cabinets	$ + 2604
Solid Surface Countertops (Included)	
Full Bath - including plumbing, wall and floor finishes	+ 11,593
Half Bath - including plumbing, wall and floor finishes	+ 6864
Two Car Attached Garage	+ 33,913
Two Car Detached Garage	+ 38,535
Fireplace & Chimney	+ 10,406

Adjustments

For multi family - add to total cost

Additional Kitchen	$ + 30,297
Additional Full Bath & Half Bath	+ 18,457
Additional Entry & Exit	+ 2418
Separate Heating & Air Conditioning	+ 7787
Separate Electric	+ 1817

For Townhouse/Rowhouse -
Multiply cost per square foot by

Inner Unit	.90
End Unit	.95

Alternatives

Add to or deduct from the cost per square foot of living area

Heavyweight Asphalt Shingles	– 2
Clay Tile Roof	+ 2.45
Slate Roof	+ 3.05
Upgrade Ceilings to Textured Finish	+ .60
Air Conditioning, in Heating Ductwork	Base System
Heating Systems, Hot Water	+ 1.72
Heat Pump	+ 1.62
Electric Heat	– 2.05
Not Heated	– 4.54

Additional upgrades or components

Kitchen Cabinets & Countertops	Page 58
Bathroom Vanities	59
Fireplaces & Chimneys	59
Windows, Skylights & Dormers	59
Appliances	60
Breezeways & Porches	60
Finished Attic	60
Garages	61
Site Improvements	61
Wings & Ells	57

- Unique residence built from an architect's plan
- Single family — 1 full bath, 1 half bath, 1 kitchen
- No basement
- Cedar shakes on roof
- Forced hot air heat/air conditioning
- Gypsum wallboard interior finishes
- Many special features
- Extraordinary materials and workmanship
- Detail specifications on page 49

Note: The illustration shown may contain some optional components (for example: garages and/or fireplaces) whose costs are shown in the modifications, adjustments, & alternatives below or at the end of the square foot section.

Base cost per square foot of living area

Exterior Wall	Living Area										
	1200	1400	1600	1800	2000	2400	2800	3200	3600	4000	4400
Wood Siding - Wood Frame	210.20	196.60	186.80	178.25	169.45	156.60	145.95	138.90	134.60	130.10	126.35
Brick Veneer - Wood Frame	223.05	208.55	198.15	189.05	179.75	165.95	154.40	146.85	142.30	137.25	133.35
Solid Brick	239.30	223.70	212.65	202.70	192.95	177.80	165.20	156.95	152.10	146.45	142.20
Solid Stone	247.65	231.50	220.15	209.75	199.70	183.95	170.80	162.20	157.15	151.15	146.70
Finished Basement, Add	33.75	36.70	35.60	34.55	33.85	32.20	30.90	30.05	29.55	28.85	28.45
Unfinished Basement, Add	15.60	14.85	14.40	13.85	13.55	12.65	12.00	11.60	11.30	10.95	10.75

Modifications

Add to the total cost

Upgrade Kitchen Cabinets	$ + 2604
Solid Surface Countertops (Included)	
Full Bath - including plumbing, wall and floor finishes	+ 11,593
Half Bath - including plumbing, wall and floor finishes	+ 6864
Two Car Attached Garage	+ 33,913
Two Car Detached Garage	+ 38,535
Fireplace & Chimney	+ 11,407

Adjustments

For multi family - add to total cost

Additional Kitchen	$ + 30,297
Additional Full Bath & Half Bath	+ 18,457
Additional Entry & Exit	+ 2418
Separate Heating & Air Conditioning	+ 7787
Separate Electric	+ 1817

For Townhouse/Rowhouse - Multiply cost per square foot by

Inner Unit	.86
End Unit	.93

Alternatives

Add to or deduct from the cost per square foot of living area

Heavyweight Asphalt Shingles	– 1.35
Clay Tile Roof	+ 1.70
Slate Roof	+ 2.10
Upgrade Ceilings to Textured Finish	+ .60
Air Conditioning, in Heating Ductwork	Base System
Heating Systems, Hot Water	+ 1.67
Heat Pump	+ 1.71
Electric Heat	– 1.86
Not Heated	– 4.29

Additional upgrades or components

Kitchen Cabinets & Countertops	Page 58
Bathroom Vanities	59
Fireplaces & Chimneys	59
Windows, Skylights & Dormers	59
Appliances	60
Breezeways & Porches	60
Finished Attic	60
Garages	61
Site Improvements	61
Wings & Ells	57

- **Unique residence built from an architect's plan**
- **Single family — 1 full bath, 1 half bath, 1 kitchen**
- **No basement**
- **Cedar shakes on roof**
- **Forced hot air heat/air conditioning**
- **Gypsum wallboard interior finishes**
- **Many special features**
- **Extraordinary materials and workmanship**
- **Detail specifications on page 49**

Note: The illustration shown may contain some optional components (for example: garages and/or fireplaces) whose costs are shown in the modifications, adjustments, & alternatives below or at the end of the square foot section.

•Larry W. Garnett & Associates, Inc

Base cost per square foot of living area

Exterior Wall	Living Area										
	1500	1800	2100	2500	3000	3500	4000	4500	5000	5500	6000
Wood Siding - Wood Frame	205.60	184.70	172.00	162.55	149.90	140.90	132.25	128.10	124.20	120.30	116.40
Brick Veneer - Wood Frame	219.25	197.30	183.30	173.20	159.55	149.70	140.40	135.75	131.50	127.30	123.05
Solid Brick	235.90	212.70	197.05	186.15	171.30	160.40	150.30	145.15	140.50	135.80	131.25
Solid Stone	245.35	221.35	204.85	193.45	178.05	166.50	155.90	150.45	145.55	140.70	135.90
Finished Basement, Add	26.95	29.05	27.35	26.45	25.20	24.15	23.40	22.85	22.40	21.95	21.65
Unfinished Basement, Add	12.65	11.90	11.05	10.60	10.00	9.45	9.05	8.80	8.55	8.30	8.20

Modifications

Add to the total cost

Upgrade Kitchen Cabinets	$ + 2604
Solid Surface Countertops (Included)	
Full Bath - including plumbing, wall and floor finishes	+ 11,593
Half Bath - including plumbing, wall and floor finishes	+ 6864
Two Car Attached Garage	+ 33,913
Two Car Detached Garage	+ 38,535
Fireplace & Chimney	+ 12,482

Adjustments

For multi family - add to total cost

Additional Kitchen	$ + 30,297
Additional Full Bath & Half Bath	+ 18,457
Additional Entry & Exit	+ 2418
Separate Heating & Air Conditioning	+ 7787
Separate Electric	+ 1817

For Townhouse/Rowhouse -
Multiply cost per square foot by

Inner Unit	.86
End Unit	.93

Alternatives

Add to or deduct from the cost per square foot of living area

Heavyweight Asphalt Shingles	– 1.20
Clay Tile Roof	+ 1.45
Slate Roof	+ 1.80
Upgrade Ceilings to Textured Finish	+ .60
Air Conditioning, in Heating Ductwork	Base System
Heating Systems, Hot Water	+ 1.51
Heat Pump	+ 1.76
Electric Heat	– 3.65
Not Heated	– 4.29

Additional upgrades or components

Kitchen Cabinets & Countertops	Page 58
Bathroom Vanities	59
Fireplaces & Chimneys	59
Windows, Skylights & Dormers	59
Appliances	60
Breezeways & Porches	60
Finished Attic	60
Garages	61
Site Improvements	61
Wings & Ells	57

- **Unique residence built from an architect's plan**
- **Single family — 1 full bath, 1 half bath, 1 kitchen**
- **No basement**
- **Cedar shakes on roof**
- **Forced hot air heat/air conditioning**
- **Gypsum wallboard interior finishes**
- **Many special features**
- **Extraordinary materials and workmanship**
- **Detail specifications on page 49**

Note: The illustration shown may contain some optional components (for example: garages and/or fireplaces) whose costs are shown in the modifications, adjustments, & alternatives below or at the end of the square foot section.

Base cost per square foot of living area

Exterior Wall	Living Area										
	1500	1800	2100	2500	3000	3500	4000	4500	5000	5500	6000
Wood Siding - Wood Frame	202.00	181.45	171.00	162.05	149.15	142.45	134.60	126.60	123.50	120.30	117.00
Brick Veneer - Wood Frame	215.60	194.05	182.80	173.30	159.35	152.10	143.30	134.70	131.30	127.80	124.05
Solid Brick	233.30	210.45	198.05	187.90	172.60	164.70	154.75	145.25	141.55	137.65	133.35
Solid Stone	241.85	218.40	205.50	195.00	179.05	170.75	160.30	150.40	146.50	142.40	137.80
Finished Basement, Add	23.65	25.50	24.45	23.70	22.55	21.85	21.00	20.25	20.00	19.65	19.20
Unfinished Basement, Add	11.05	10.40	9.95	9.55	9.00	8.70	8.20	7.85	7.70	7.55	7.30

Modifications

Add to the total cost

Upgrade Kitchen Cabinets	$ + 2604
Solid Surface Countertops (Included)	
Full Bath - including plumbing, wall and floor finishes	+ 11,593
Half Bath - including plumbing, wall and floor finishes	+ 6864
Two Car Attached Garage	+ 33,913
Two Car Detached Garage	+ 38,535
Fireplace & Chimney	+ 12,482

Adjustments

For multi family - add to total cost

Additional Kitchen	$ + 30,297
Additional Full Bath & Half Bath	+ 18,457
Additional Entry & Exit	+ 2418
Separate Heating & Air Conditioning	+ 7787
Separate Electric	+ 1817

For Townhouse/Rowhouse -
Multiply cost per square foot by

Inner Unit	.84
End Unit	.92

Alternatives

Add to or deduct from the cost per square foot of living area

Heavyweight Asphalt Shingles	– .90
Clay Tile Roof	+ 1.15
Slate Roof	+ 1.40
Upgrade Ceilings to Textured Finish	+ .60
Air Conditioning, in Heating Ductwork	Base System
Heating Systems, Hot Water	+ 1.51
Heat Pump	+ 1.76
Electric Heat	– 3.65
Not Heated	– 4.17

Additional upgrades or components

Kitchen Cabinets & Countertops	Page 58
Bathroom Vanities	59
Fireplaces & Chimneys	59
Windows, Skylights & Dormers	59
Appliances	60
Breezeways & Porches	60
Finished Attic	60
Garages	61
Site Improvements	61
Wings & Ells	57

- **Unique residence built from an architect's plan**
- **Single family — 1 full bath, 1 half bath, 1 kitchen**
- **No basement**
- **Cedar shakes on roof**
- **Forced hot air heat/air conditioning**
- **Gypsum wallboard interior finishes**
- **Many special features**
- **Extraordinary materials and workmanship**
- **Detail specifications on page 49**

Note: The illustration shown may contain some optional components (for example: garages and/or fireplaces) whose costs are shown in the modifications, adjustments, & alternatives below or at the end of the square foot section.

Base cost per square foot of living area

Exterior Wall	Living Area										
	1200	1400	1600	1800	2000	2400	2800	3200	3600	4000	4400
Wood Siding - Wood Frame	199.15	186.25	176.90	169.00	160.50	148.60	138.60	132.00	128.05	123.95	120.30
Brick Veneer - Wood Frame	208.90	195.35	185.65	177.20	168.40	155.75	145.05	138.10	133.90	129.35	125.65
Solid Brick	221.00	206.60	196.45	187.35	178.20	164.50	153.15	145.60	141.15	136.20	132.25
Solid Stone	227.55	212.80	202.30	192.90	183.50	169.35	157.50	149.65	145.15	139.90	135.80
Finished Basement, Add	33.75	36.70	35.60	34.55	33.85	32.20	30.90	30.05	29.55	28.85	28.45
Unfinished Basement, Add	15.60	14.85	14.40	13.85	13.55	12.65	12.00	11.60	11.30	10.95	10.75

Modifications

Add to the total cost

Upgrade Kitchen Cabinets	$ + 2604
Solid Surface Countertops (Included)	
Full Bath - including plumbing, wall and floor finishes	+ 11,593
Half Bath - including plumbing, wall and floor finishes	+ 6864
Two Car Attached Garage	+ 33,913
Two Car Detached Garage	+ 38,535
Fireplace & Chimney	+ 10,406

Adjustments

For multi family - add to total cost

Additional Kitchen	$ + 30,297
Additional Full Bath & Half Bath	+ 18,457
Additional Entry & Exit	+ 2418
Separate Heating & Air Conditioning	+ 7787
Separate Electric	+ 1817

For Townhouse/Rowhouse -
Multiply cost per square foot by

Inner Unit	.89
End Unit	.94

Alternatives

Add to or deduct from the cost per square foot of living area

Heavyweight Asphalt Shingles	– 1.35
Clay Tile Roof	+ 1.70
Slate Roof	+ 2.10
Upgrade Ceilings to Textured Finish	+ .60
Air Conditioning, in Heating Ductwork	Base System
Heating Systems, Hot Water	+ 1.67
Heat Pump	+ 1.71
Electric Heat	– 1.86
Not Heated	– 4.29

Additional upgrades or components

Kitchen Cabinets & Countertops	Page 58
Bathroom Vanities	59
Fireplaces & Chimneys	59
Windows, Skylights & Dormers	59
Appliances	60
Breezeways & Porches	60
Finished Attic	60
Garages	61
Site Improvements	61
Wings & Ells	57

Important: See the Reference Section for Location Factors (to adjust for your city) and Estimating Forms.

- **Unique residence built from an architect's plan**
- **Single family — 1 full bath, 1 half bath, 1 kitchen**
- **No basement**
- **Cedar shakes on roof**
- **Forced hot air heat/air conditioning**
- **Gypsum wallboard interior finishes**
- **Many special features**
- **Extraordinary materials and workmanship**
- **Detail specifications on page 49**

Note: The illustration shown may contain some optional components (for example: garages and/or fireplaces) whose costs are shown in the modifications, adjustments, & alternatives below or at the end of the square foot section.

©Home Planners, Inc.

Base cost per square foot of living area

Exterior Wall	Living Area										
	1500	1800	2100	2400	2800	3200	3600	4000	4500	5000	5500
Wood Siding - Wood Frame	185.55	170.65	158.40	150.15	143.90	136.95	129.85	126.45	119.90	116.10	111.90
Brick Veneer - Wood Frame	194.40	178.65	165.70	156.90	150.45	143.05	135.50	131.90	124.90	120.80	116.35
Solid Brick	205.20	188.40	174.55	165.20	158.45	150.45	142.30	138.60	131.05	126.60	121.85
Solid Stone	211.35	193.90	179.60	170.00	162.95	154.60	146.25	142.40	134.60	129.85	124.95
Finished Basement, Add*	39.85	42.90	41.10	40.00	39.20	38.10	37.05	36.65	35.70	35.05	34.40
Unfinished Basement, Add*	18.00	16.90	16.05	15.40	15.05	14.50	13.95	13.70	13.20	12.90	12.60

*Basement under middle level only.

Modifications

Add to the total cost

Upgrade Kitchen Cabinets	$ + 2604
Solid Surface Countertops (Included)	
Full Bath - including plumbing, wall and floor finishes	+ 11,593
Half Bath - including plumbing, wall and floor finishes	+ 6864
Two Car Attached Garage	+ 33,913
Two Car Detached Garage	+ 38,535
Fireplace & Chimney	+ 10,406

Adjustments

For multi family - add to total cost

Additional Kitchen	$ + 30,297
Additional Full Bath & Half Bath	+ 18,457
Additional Entry & Exit	+ 2418
Separate Heating & Air Conditioning	+ 7787
Separate Electric	+ 1817

For Townhouse/Rowhouse -
Multiply cost per square foot by

Inner Unit	.86
End Unit	.93

Alternatives

Add to or deduct from the cost per square foot of living area

Heavyweight Asphalt Shingles	– 2
Clay Tile Roof	+ 2.45
Slate Roof	+ 3.05
Upgrade Ceilings to Textured Finish	+ .60
Air Conditioning, in Heating Ductwork	Base System
Heating Systems, Hot Water	+ 1.62
Heat Pump	+ 1.78
Electric Heat	– 1.63
Not Heated	– 4.17

Additional upgrades or components

Kitchen Cabinets & Countertops	Page 58
Bathroom Vanities	59
Fireplaces & Chimneys	59
Windows, Skylights & Dormers	59
Appliances	60
Breezeways & Porches	60
Finished Attic	60
Garages	61
Site Improvements	61
Wings & Ells	57

1 Story — Base cost per square foot of living area

Exterior Wall	Living Area							
	50	100	200	300	400	500	600	700
Wood Siding - Wood Frame	281.20	217.20	189.35	159.55	150.40	145.00	141.35	142.40
Brick Veneer - Wood Frame	312.70	239.65	208.10	172.05	161.65	155.50	151.35	152.05
Solid Brick	358.90	272.70	235.65	190.40	178.20	170.90	166.00	166.20
Solid Stone	376.90	285.50	246.35	197.55	184.60	176.90	171.70	171.70
Finished Basement, Add	107.55	97.40	87.30	70.55	67.20	65.15	63.75	62.85
Unfinished Basement, Add	56.35	43.45	38.05	29.05	27.30	26.20	25.50	24.95

1-1/2 Story — Base cost per square foot of living area

Exterior Wall	Living Area							
	100	200	300	400	500	600	700	800
Wood Siding - Wood Frame	223.00	181.80	156.75	141.90	134.20	130.55	125.95	124.60
Brick Veneer - Wood Frame	251.15	204.30	175.50	156.50	147.70	143.30	137.95	136.55
Solid Brick	292.35	237.35	203.05	177.95	167.55	162.05	155.75	154.10
Solid Stone	308.40	250.20	213.75	186.25	175.20	169.35	162.60	160.90
Finished Basement, Add	70.70	69.85	63.15	55.75	53.75	52.40	51.15	51.00
Unfinished Basement, Add	35.80	30.40	26.80	22.90	21.80	21.10	20.45	20.35

2 Story — Base cost per square foot of living area

Exterior Wall	Living Area							
	100	200	400	600	800	1000	1200	1400
Wood Siding - Wood Frame	218.85	164.85	141.15	118.30	110.55	105.90	102.85	104.30
Brick Veneer - Wood Frame	250.40	187.35	159.95	130.75	121.75	116.40	112.85	113.90
Solid Brick	296.65	220.35	187.45	149.10	138.30	131.85	127.50	128.10
Solid Stone	314.60	233.20	198.15	156.30	144.70	137.80	133.15	133.60
Finished Basement, Add	53.80	48.75	43.70	35.25	33.60	32.60	31.95	31.45
Unfinished Basement, Add	28.15	21.70	19.05	14.55	13.65	13.10	12.75	12.50

Base costs do not include bathroom or kitchen facilities. Use Modifications/Adjustments/Alternatives on pages 58-61 where appropriate.

Important: See the Reference Section for Location Factors (to adjust for your city) and Estimating Forms.

Kitchen cabinets - Base units, hardwood *(Cost per Unit)*

	Economy	Average	Custom	Luxury
24″ deep, 35″ high,				
One top drawer,				
One door below				
12″ wide	$ 278	$ 370	$ 492	$ 648
15″ wide	289	385	512	674
18″ wide	311	415	552	726
21″ wide	323	430	572	753
24″ wide	371	495	658	866
Four drawers				
12″ wide	293	390	519	683
15″ wide	296	395	525	691
18″ wide	323	430	572	753
24″ wide	353	470	625	823
Two top drawers,				
Two doors below				
27″ wide	401	535	712	936
30″ wide	435	580	771	1015
33″ wide	450	600	798	1050
36″ wide	465	620	825	1085
42″ wide	491	655	871	1146
48″ wide	525	700	931	1225
Range or sink base				
(Cost per unit)				
Two doors below				
30″ wide	368	490	652	858
33″ wide	390	520	692	910
36″ wide	409	545	725	954
42″ wide	428	570	758	998
48″ wide	450	600	798	1050
Corner Base Cabinet				
(Cost per unit)				
36″ wide	653	870	1157	1523
Lazy Susan *(Cost per unit)*				
With revolving door	825	1100	1463	1925

Kitchen cabinets - Wall cabinets, hardwood *(Cost per Unit)*

	Economy	Average	Custom	Luxury
12″ deep, 2 doors				
12″ high				
30″ wide	$ 255	$ 340	$ 452	$ 595
36″ wide	92	122	162	214
15″ high				
30″ wide	259	345	459	604
33″ wide	311	415	552	726
36″ wide	304	405	539	709
24″ high				
30″ wide	338	450	599	788
36″ wide	368	490	652	858
42″ wide	337	449	597	786
30″ high, 1 door				
12″ wide	248	330	439	578
15″ wide	259	345	459	604
18″ wide	281	375	499	656
24″ wide	323	430	572	753
30″ high, 2 doors				
27″ wide	353	470	625	823
30″ wide	375	500	665	875
36″ wide	424	565	751	989
42″ wide	454	605	805	1059
48″ wide	506	675	898	1181
Corner wall, 30″ high				
24″ wide	368	490	652	858
30″ wide	390	520	692	910
36″ wide	443	590	785	1033
Broom closet				
84″ high, 24″ deep				
18″ wide	686	915	1217	1601
Oven Cabinet				
84″ high, 24″ deep				
27″ wide	1031	1375	1829	2406

Kitchen countertops *(Cost per L.F.)*

	Economy	Average	Custom	Luxury
Solid Surface				
24″ wide, no backsplash	$ 110	$ 146	$ 194	$ 256
with backsplash	123	164	218	287
Stock plastic laminate, 24″ wide				
with backsplash	26	34	46	60
Custom plastic laminate, no splash				
7/8″ thick, alum. molding	40	53	71	94
1-1/4″ thick, no splash	44	59	78	103
Marble				
1/2″ - 3/4″ thick w/splash	61	81	108	143
Maple, laminated				
1-1/2″ thick w/splash	100	133	177	233
Stainless steel				
(per S.F.)	162	216	287	378
Cutting blocks, recessed				
16″ x 20″ x 1″ (each)	132	176	234	308

    For customer support on your Square Foot Costs with RSMeans data, call 800.448.8182.

Vanity bases *(Cost per Unit)*

2 door, 30" high, 21" deep	Economy	Average	Custom	Luxury
24" wide	$ 315	$ 420	$ 559	$ 735
30" wide	379	505	672	884
36" wide	379	505	672	884
48" wide	488	650	865	1138

Solid surface vanity tops *(Cost Each)*

Center bowl	Economy	Average	Custom	Luxury
22" x 25"	$ 410	$ 443	$ 478	$ 516
22" x 31"	470	508	549	593
22" x 37"	540	583	630	680
22" x 49"	665	718	775	837

Fireplaces & Chimneys *(Cost per Unit)*

	1-1/2 Story	2 Story	3 Story
Economy (prefab metal)			
Exterior chimney & 1 fireplace	$ 6623	$ 7318	$ 8027
Interior chimney & 1 fireplace	6346	7056	7382
Average (masonry)			
Exterior chimney & 1 fireplace	6942	7671	8414
Interior chimney & 1 fireplace	6652	7396	7738
For more than 1 flue, add	472	804	1348
For more than 1 fireplace, add	4658	4658	4658
Custom (masonry)			
Exterior chimney & 1 fireplace	7266	8202	9262
Interior chimney & 1 fireplace	6813	7712	8311
For more than 1 flue, add	570	986	1651
For more than 1 fireplace, add	5219	5219	5219
Luxury (masonry)			
Exterior chimney & 1 fireplace	10406	11407	12482
Interior chimney & 1 fireplace	9927	10854	11489
For more than 1 flue, add	859	1434	2002
For more than 1 fireplace, add	8202	8202	8202

Windows and Skylights *(Cost Each)*

	Economy	Average	Custom	Luxury
Fixed Picture Windows				
3'-6" x 4'-0"	$ 375	$ 500	$ 665	$ 875
4'-0" x 6'-0"	648	864	1150	1513
5'-0" x 6'-0"	733	977	1300	1711
6'-0" x 6'-0"	761	1015	1350	1776
Bay/Bow Windows				
8'-0" x 5'-0"	874	1165	1550	2039
10'-0" x 5'-0"	930	1240	1650	2171
10'-0" x 6'-0"	1523	2030	2700	3553
12'-0" x 6'-0"	1988	2650	3525	4638
Palladian Windows				
3'-2" x 6'-4"		1522	2025	2664
4'-0" x 6'-0"		1672	2225	2928
5'-5" x 6'-10"		2293	3050	4013
8'-0" x 6'-0"		2763	3675	4836
Skylights				
46" x 21-1/2"	614	945	1145	1260
46" x 28"	631	970	1200	1320
57" x 44"	696	1070	1185	1304

Dormers *(Cost/S.F. of plan area)*

	Economy	Average	Custom	Luxury
Framing and Roofing Only				
Gable dormer, 2" x 6" roof frame	$ 33	$ 37	$ 41	$ 67
2" x 8" roof frame	34	38	42	70
Shed dormer, 2" x 6" roof frame	21	25	28	43
2" x 8" roof frame	23	26	29	44
2" x 10" roof frame	25	28	30	45

Appliances *(Cost per Unit)*

	Economy	Average	Custom	Luxury
Range				
30" free standing, 1 oven	$ 585	$ 1643	$ 2172	$ 2700
2 oven	1175	2438	3069	3700
30" built-in, 1 oven	970	1685	2043	2400
2 oven	1500	1725	1838	1950
21" free standing				
1 oven	590	748	827	905
Counter Top Ranges				
4 burner standard	450	1275	1688	2100
As above with griddle	1675	2913	3532	4150
Microwave Oven	238	504	637	770
Compactor				
4 to 1 compaction	855	1140	1283	1425
Deep Freeze				
15 to 23 C.F.	725	875	950	1025
30 C.F.	970	1073	1124	1175
Dehumidifier, portable, auto.				
15 pint	355	409	436	462
30 pint	415	478	509	540
Washing Machine, automatic	885	1443	1722	2000
Water Heater				
Electric, glass lined				
30 gal.	1125	1363	1482	1600
80 gal.	2150	2675	2938	3200
Water Heater, Gas, glass lined				
30 gal.	1975	2388	2594	2800
50 gal.	2325	2838	3094	3350
Dishwasher, built-in				
2 cycles	490	660	745	830
4 or more cycles	605	750	1188	1625
Dryer, automatic	720	1360	1680	2000
Garage Door Opener	545	650	703	755
Garbage Disposal	182	236	263	290
Heater, Electric, built-in				
1250 watt ceiling type	259	335	373	410
1250 watt wall type	350	370	380	390
Wall type w/blower				
1500 watt	340	391	417	442
3000 watt	705	811	864	917
Hood For Range, 2 speed				
30" wide	222	711	956	1200
42" wide	305	1353	1877	2400
Humidifier, portable				
7 gal. per day	162	187	199	211
15 gal. per day	210	242	258	273
Ice Maker, automatic				
13 lb. per day	1450	1668	1777	1885
51 lb. per day	1750	2013	2144	2275
Refrigerator, no frost				
10-12 C.F.	555	633	672	710
14-16 C.F.	690	835	908	980
18-20 C.F.	865	1433	1717	2000
21-29 C.F.	1275	2038	2419	2800
Sump Pump, 1/3 H.P.	330	428	477	525

Breezeway *(Cost per S.F.)*

Class	Type	Area (S.F.)			
		50	100	150	200
Economy	Open	$ 48.00	$ 39.84	$ 34.94	$ 32.49
	Enclosed	177.53	126.09	103.62	89.06
Average	Open	56.47	46.00	40.33	37.50
	Enclosed	181.46	131.09	108.53	97.01
Custom	Open	68.39	56.09	49.33	45.95
	Enclosed	244.30	175.68	144.85	129.18
Luxury	Open	76.43	62.94	55.78	52.20
	Enclosed	288.16	204.69	167.38	148.45

Porches *(Cost per S.F.)*

Class	Type	Area (S.F.)				
		25	50	100	200	300
Economy	Open	$ 85.19	$ 63.00	$ 47.28	$ 36.07	$ 32.28
	Enclosed	187.30	148.18	97.76	73.91	64.25
Average	Open	88.78	67.97	49.77	37.31	33.11
	Enclosed	205.26	162.13	104.14	76.50	65.59
Custom	Open	150.62	108.97	83.56	62.40	60.71
	Enclosed	249.90	199.63	131.69	99.36	105.77
Luxury	Open	163.59	118.89	90.71	66.13	60.04
	Enclosed	287.23	260.12	167.60	121.82	100.45

Finished attic *(Cost per S.F.)*

Class	Area (S.F.)				
	400	500	600	800	1000
Economy	$ 23.20	$ 22.42	$ 21.49	$ 21.13	$ 20.34
Average	35.76	34.98	34.12	33.69	32.75
Custom	45.75	44.73	43.71	43.04	42.16
Luxury	58.36	56.94	55.61	54.29	53.40

Alarm system *(Cost per System)*

	Burglar Alarm	Smoke Detector
Economy	$ 440	$ 73
Average	500	84
Custom	879	217
Luxury	1525	268

Sauna, prefabricated
(Cost per unit, including heater and controls—7' high)

Size	Cost
6' x 4'	$ 5400
6' x 5'	6350
6' x 6'	6675
6' x 9'	8425
8' x 10'	9875
8' x 12'	11,400
10' x 12'	13,200

Garages *

(Costs include exterior wall systems comparable with the quality of the residence. Included in the cost is an allowance for one personnel door, manual overhead door(s) and electrical fixture.)

Class	Type									
	Detached			Attached			Built-in		Basement	
	One Car	Two Car	Three Car	One Car	Two Car	Three Car	One Car	Two Car	One Car	Two Car
Economy										
Wood	$18,103	$27,607	$37,111	$13,994	$24,039	$33,543	$-1971	$-3941	$1832	$2524
Masonry	24,566	35,696	46,826	18,039	29,709	40,839	-2764	-5529		
Average										
Wood	19,847	29,790	39,733	15,086	25,569	35,512	-2185	-4370	2084	3028
Masonry	24,628	35,773	46,919	18,078	29,763	40,908	-2772	-4415		
Custom										
Wood	22,141	33,939	45,737	17,160	29,653	41,451	-3112	-2939	3046	4952
Masonry	26,881	39,871	52,861	20,126	33,811	46,801	-3694	-4103		
Luxury										
Wood	24,775	38,535	52,296	19,459	33,913	47,674	-3223	-3159	4121	6583
Masonry	32,126	47,735	63,344	24,059	40,362	55,971	-4125	-4965		

*See the Introduction to this section for definitions of garage types.

Swimming pools *(Cost per S.F.)*

Residential	
In-ground	$ 40.50 - 97.50
Deck equipment	1.30
Paint pool, preparation & 3 coats (epoxy)	5.31
Rubber base paint	4.75
Pool Cover	1.91
Swimming Pool Heaters	
(not including wiring, external piping, base or pad)	
Gas	
155 MBH	$ 3100.00
190 MBH	3700.00
500 MBH	13,100.
Electric	
15 KW 7200 gallon pool	2875.00
24 KW 9600 gallon pool	3350.00
54 KW 24,000 gallon pool	5250.00

Wood and coal stoves

Wood Only	
Free Standing (minimum)	$ 2200
Fireplace Insert (minimum)	1871
Coal Only	
Free Standing	$ 2114
Fireplace Insert	2314
Wood and Coal	
Free Standing	$ 4348
Fireplace Insert	4456

Sidewalks *(Cost per S.F.)*

Concrete, 3000 psi with wire mesh	4" thick	$ 4.33
	5" thick	5.28
	6" thick	5.94
Precast concrete patio blocks (natural)	2" thick	6.95
Precast concrete patio blocks (colors)	2" thick	7.25
Flagstone, bluestone	1" thick	21.95
Flagstone, bluestone	1-1/2" thick	30.95
Slate (natural, irregular)	3/4" thick	19.60
Slate (random rectangular)	1/2" thick	30.55
Seeding		
Fine grading & seeding includes lime, fertilizer & seed	per S.Y.	3.11
Lawn Sprinkler System	per S.F.	1.14

Fencing *(Cost per L.F.)*

Chain Link, 4' high, galvanized	$ 17.75
Gate, 4' high (each)	238.00
Cedar Picket, 3' high, 2 rail	15.75
Gate (each)	235.00
3 Rail, 4' high	18.95
Gate (each)	253.00
Cedar Stockade, 3 Rail, 6' high	18.40
Gate (each)	256.00
Board & Battens, 2 sides 6' high, pine	35.50
6' high, cedar	39.00
No. 1 Cedar, basketweave, 6' high	37.00
Gate, 6' high (each)	315.00

Carport *(Cost per S.F.)*

Economy	$ 10.61
Average	15.79
Custom	23.17
Luxury	26.91

Insurance Exclusions

Insurance exclusions are a matter of policy coverage and are not standard or universal. When making an appraisal for insurance purposes, it is recommended that some time be taken in studying the insurance policy to determine the specific items to be excluded. Most homeowners insurance policies have, as part of their provisions, statements that read "the policy permits the insured to exclude from value, in determining whether the amount of the insurance equals or exceeds 80% of its replacement cost, such items as building excavation, basement footings, underground piping and wiring, piers and other supports which are below the undersurface of the lowest floor or where there is no basement below the ground." Loss to any of these items, however, is covered.

Costs for Excavation, Spread and Strip Footings and Underground Piping

This chart shows excluded items expressed as a percentage of total building cost.

Class	Number of Stories			
	1	1-1/2	2	3
Luxury	3.4%	2.7%	2.4%	1.9%
Custom	3.5%	2.7%	2.4%	1.9%
Average	5.0%	4.3%	3.8%	3.3%
Economy	5.0%	4.4%	4.1%	3.5%

Architect/Designer Fees

Architect/designer fees as presented in the following chart are typical ranges for 4 classes of residences. Factors affecting these ranges include economic conditions, size and scope of project, site selection, and standardization. Where superior quality and detail is required or where closer supervision is required, fees may run higher than listed. Lower quality or simplicity may dictate lower fees.

The editors have included architect/designer fees from the "mean" column in calculating costs.

Listed below are average costs expressed as a range and mean of total building cost for 4 classes of residences.

Class			Range	Mean
Luxury	—	Architecturally designed and supervised	5%–15%	10%
Custom	—	Architecturally modified designer plans	$2000–$3500	$2500
Average	—	Designer plans	$800–$1500	$1100
Economy	—	Stock plans	$100–$500	$300

Depreciation as generally defined is "the loss of value due to any cause." Specifically, depreciation can be broken down into three categories: **Physical Depreciation, Functional Obsolescence,** and **Economic Obsolescence.**

Physical Depreciation is the loss of value from the wearing out of the building's components. Such causes may be decay, dry rot, cracks, or structural defects.

Functional Obsolescence is the loss of value from features which render the residence less useful or desirable. Such features may be higher than normal ceilings, design and/or style changes, or outdated mechanical systems.

Economic Obsolescence is the loss of value from external features which render the residence less useful or desirable. These features may be changes in neighborhood socio-economic grouping, zoning changes, legislation, etc.

Depreciation as it applies to the residential portion of this manual deals with the observed physical condition of the residence being appraised. It is in essence *"the cost to cure."*

The ultimate method to arrive at *"the cost to cure"* is to analyze each component of the residence.

For example:

> Component, Roof Covering – Cost = $2400
> Average Life = 15 years
> Actual Life = 5 years
> Depreciation = 5 ÷ 15 = .33 or 33%
>
> $2400 × 33% = $792 = Amount of Depreciation

The following table, however, can be used as a guide in estimating the % depreciation using the actual age and general condition of the residence.

Depreciation Table — Residential

Age in Years	Good	Average	Poor
2	2%	3%	10%
5	4	6	20
10	7	10	25
15	10	15	30
20	15	20	35
25	18	25	40
30	24	30	45
35	28	35	50
40	32	40	55
45	36	45	60
50	40	50	65

Commercial/Industrial/ Institutional Models

Table of Contents

Introduction to the Commercial/ Industrial/Institutional Section

General

The Commercial/Industrial/Institutional section contains base building costs per square foot of floor area for 77 model buildings. Each model has a table of square foot costs for combinations of exterior wall and framing systems. This table is supplemented by a list of common additives and their unit costs. In printed versions, a breakdown of all component costs used to develop the base cost for the model is included. In electronic products, the cost breakdown is available as a printable report. Modifications to the standard models can be performed manually in printed products and automatically using the "swapper" feature in electronic products.

This data may be used directly to estimate the construction cost of most types of buildings when only the floor area, exterior wall construction, and framing systems are known. To adjust the base cost for components that are different than the model, use the tables from the Assemblies Section.

Building Identification & Model Selection

The building models in this section represent structures by use. Occupancy, however, does not necessarily identify the building, i.e., a restaurant could be a converted warehouse. In all instances, the building should be described and identified by its own physical characteristics. The model selection should also be guided by comparing specifications with the model. In the case of converted use, data from one model may be used to supplement data from another.

Green Models

Consistent with expanding green trends in the design and construction industry, 25 green building models are currently included. Although similar to our standard models in building type and structural system, the green models align with Energy Star requirements and address many of the items necessary to obtain LEED certification. Although our models do not include site-specific information, our design assumption is that they are located in climate zone 5. DOE's eQuest software was used to perform an energy analysis of the model buildings. By reducing energy use, we were able to reduce the size of the service entrances, switchgear, power feeds, and generators. Each of the following building systems was researched and analyzed: building envelope, HVAC, plumbing fixtures, lighting, and electrical service. These systems were targeted because of their impact on energy usage and green building.

Wall and roof insulation was increased to reduce heat loss and a recycled vapor barrier was added to the foundation. White roofs were specified for models with flat roofs. Interior finishes include materials with recycled content and low VOC paint. Stainless steel toilet partitions were selected because the recycled material is easy to maintain. Plumbing fixtures were selected to conserve water. Water closets are low-flow and equipped with auto-sensor valves. Waterless urinals are specified throughout. Faucets use auto-sensor flush valves and are powered by a hydroelectric power unit inside the faucet. Energy–efficient, low–flow water coolers were specified to reduce energy consumption and water usage throughout each building model.

Lighting efficiency was achieved by specifying LED fixtures in place of standard fluorescents. Daylight on/off lighting control systems were incorporated into the models. These controls are equipped with sensors that automatically turn off the lights when sufficient daylight is available. Energy monitoring systems were also included in all green models.

Adjustments

The base cost tables represent the base cost per square foot of floor area for buildings without a basement and without unusual special features. Basement costs and other common additives are available. Cost adjustments can also be made to the model by using the tables from the Assemblies Section.

Dimensions

All base cost tables are developed so that measurements can be readily made during the inspection process. Areas are calculated from exterior dimensions and story heights are measured from the top surface of one floor to the top surface of the floor above. Roof areas are measured by horizontal area covered and costs related to inclines are converted with appropriate factors. The precision of measurement is a matter of the user's choice and discretion. For ease in calculation, consideration should be given to measuring in tenths of a foot, i.e., 9 ft. 6 in. = 9.5 ft., 9 ft. 4 in. = 9.3 ft.

Floor Area

The term "Floor Area" as used in this section includes the sum of floor plate at grade level and above. This dimension is measured from the outside face of the foundation wall. Basement costs are calculated separately. The user must exercise his/her own judgment, where the lowest level floor is slightly below grade, whether to consider it at grade level or make the basement adjustment.

How to Use the Commercial/ Industrial/Institutional Section

The following is a detailed explanation of a sample entry in the Commercial/Industrial/Institutional Square Foot Cost Section. Each bold number below corresponds to the described item on the following page with the appropriate component or cost of the sample entry following in parentheses.

Prices listed are costs that include overhead and profit of the installing contractor and additional markups for General Conditions and Architects' Fees.

COMMERCIAL/INDUSTRIAL/ INSTITUTIONAL	M.010 ❶	Apartment, 1-3 Story ❷

Costs per square foot of floor area

Exterior Wall ❸	S.F. Area	8000	12000	15000	19000	22500 ❹	25000	29000	32000	36000
	L.F. Perimeter	213	280	330	350	400	433	442	480	520
Fiber Cement	Wood Frame	183.45	175.45	172.30	166.50	164.95	163.90	160.90	160.25	159.25
Stone Veneer	Wood Frame	204.05	193.45	189.25	180.75	178.60 ❺	177.35	172.65	171.80	170.40
Brick Veneer	Rigid Steel	198.80	189.15	185.25	177.75	175.80	174.60	170.50	169.75	168.45
E.I.F.S. and Metal Studs	Rigid Steel	193.20	184.60	181.15	174.75	173.05	172.00	168.55	167.90	166.75
Brick Veneer	Reinforced Concrete	220.40	208.95	204.35	194.80	192.60	191.10	185.85	185.00	183.40
Stucco and Concrete Block	Reinforced Concrete	208.10	198.15	194.20	186.30	184.35	183.10	178.85	178.10	176.70
Perimeter Adj., Add or Deduct ❻	Per 100 L.F.	15.60	10.40	8.25	6.55	5.55 ❼	5.05	4.25	3.90	3.50
Story Hgt. Adj., Add or Deduct	Per 1 Ft.	2.45	2.15	2.00	1.70	1.55	1.65	1.35	1.40	1.30
❽ For *Basement*, add $38.70 per square foot of basement area										

The above costs were calculated using the basic specifications shown on the facing page. These costs should be adjusted where necessary for design alternatives and owner's requirements.

Common additives ❾

Description	Unit	$ Cost
Appliances		
Cooking range, 30" free standing		
1 oven	Ea.	610 - 2775
2 oven	Ea.	1200 - 3725
30" built-in		
1 oven	Ea.	995 - 2550
2 oven	Ea.	1575 - 3125
Counter top cook tops, 4 burner	Ea.	475 - 2150
Microwave oven	Ea.	281 - 855
Combination range, refrig. & sink, 30" wide	Ea.	2075 - 2625
72" wide	Ea.	3075
Combination range, refrigerator, sink,		
microwave oven & icemaker	Ea.	7100
Compactor, residential, 4-1 compaction	Ea.	885 - 1475
Dishwasher, built-in, 2 cycles	Ea.	685 - 1225
4 cycles	Ea.	800 - 2025
Garbage disposer, sink type	Ea.	260 - 370
Hood for range, 2 speed, vented, 30" wide	Ea.	370 - 1450
42" wide	Ea.	450 - 2650
Refrigerator, no frost 10-12 C.F.	Ea.	580 - 750
18-20 C.F.	Ea.	895 - 2075

Description	Unit	$ Cost
Closed Circuit Surveillance, One station		
Camera and monitor	Ea.	1475
For additional camera stations, add	Ea.	665
Elevators, Hydraulic passenger, 2 stops		
2000# capacity	Ea.	74,100
2500# capacity	Ea.	76,600
3500# capacity	Ea.	81,600
Additional stop, add	Ea.	8375
Emergency Lighting, 25 watt, battery operated		
Lead battery	Ea.	330
Nickel cadmium	Ea.	550
Laundry Equipment		
Dryer, gas, 16 lb. capacity	Ea.	1025
30 lb. capacity	Ea.	4100
Washer, 4 cycle	Ea.	1325
Commercial	Ea.	1700
Smoke Detectors		
Ceiling type	Ea.	244
Duct type	Ea.	585

1 Model Number (M.010)

"M" distinguishes this section of the data and stands for model. The number designation is a sequential number. Our newest models are green and are designated with the letter "G" followed by the building type number.

2 Type of Building (Apartment, 1-3 Story)

There are 50 different types of commercial/industrial/institutional buildings highlighted in this section.

3 Exterior Wall Construction and Building Framing Options (Stone Veneer with Wood Frame)

Three or more commonly used exterior walls and, in most cases, two typical building framing systems are presented for each type of building. The model selected should be based on the actual characteristics of the building being estimated.

4 Total Square Foot of Floor Area and Perimeter Used to Compute Base Costs (22,500 Square Feet and 400 Linear Feet)

Square foot of floor area is the total gross area of all floors at grade and above and does not include a basement. The perimeter in linear feet used for the base cost is generally for a rectangular, economical building shape.

5 Cost per Square Foot of Floor Area ($178.60)

The highlighted cost is for a building of the selected exterior wall and framing system and floor area. Costs for buildings with floor areas other than those calculated may be interpolated between the costs shown.

6 Building Perimeter and Story Height Adjustments

Square foot costs for a building with a perimeter or floor-to-floor story height significantly different from the model used to calculate the base cost may be adjusted (add or deduct) to reflect the actual building geometry.

7 Cost per Square Foot of Floor Area for the Perimeter and/or Height Adjustment ($5.55 for Perimeter Difference and $1.55 for Story Height Difference)

Add (or deduct) $5.55 to the base square foot cost for each 100 feet of perimeter difference between the model and the actual building. Add (or deduct) $1.55 to the base square foot cost for each 1 foot of story height difference between the model and the actual building.

8 Optional Cost per Square Foot of Basement Floor Area ($38.70)

The cost of an unfinished basement for the building being estimated is $38.70 times the gross floor area of the basement.

9 Common Additives

Common components and/or systems used in this type of building are listed. These costs should be added to the total building cost. Additional selections may be found in the Assemblies Section.

The following is a detailed explanation of the specification and costs for a model building in the Commercial/Industrial/Institutional Square Foot Cost Section. Each bold number below corresponds to the described item on the following page with the appropriate component of the sample entry following in parentheses.

Prices listed in the specification are costs that include overhead and profit of the installing contractor but not the general contractor's markup and architect's fees.

Model costs calculated for a 3 story building with 10' story height and 22,500 square feet of floor area ❶ ❷ Apartment, 1-3 Story

				Unit	Unit Cost	Cost Per S.F.	% Of Sub-Total
A. SUBSTRUCTURE							
1010	Standard Foundations	Poured concrete; strip and spread footings; 4' foundation wall		S.F. Ground	10.26	3.42	
1020	Special Foundations	N/A		—	—	—	
1030	Slab on Grade	4" reinforced concrete		S.F. Slab	5.80	1.93	4.1%
2010	Basement Excavation	Site preparation for slab and trench for foundation wall and footing		S.F. Ground	.33	.11	
2020	Basement Walls	N/A		—	—	—	
B. SHELL							
	B10 Superstructure						
1010	Floor Construction	Wood Beam and joist on wood columns, fireproofed		S.F. Floor	17.27	11.51	10.6%
1020	Roof Construction	Wood roof, truss, 4/12 slope		S.F. Roof	7.77	2.59	
	B20 Exterior Enclosure						
2010	Exterior Walls	Ashlar stone veneer on wood stud back up, insulated	80% of wall	S.F. Wall	41.34	17.64	
2020	Exterior Windows	Aluminum horizontal sliding	20% of wall	Each	529	3.76	16.4%
2030	Exterior Doors	Steel door , hollow metal with frame		Each	3015	.53	
	B30 Roofing						
3010	Roof Coverings	Asphalt roofing, strip shingles, flashing		S.F. Roof	2.16	.72	0.5%
3020	Roof Openings	N/A		—	—	—	
C. INTERIORS ❸			❹		❻ ❼ ❽		❾
1010	Partitions	Gypsum board on wood studs	10 S.F. Floor/L.F. Partition	S.F. Partition	7.73	6.87	
1020	Interior Doors	Interior single leaf doors, frames and hardware	130 S.F. Floor/Door	Each	1369	10.37	
1030	Fittings	Residential wall and base wood cabinets, laminate counters	❺	S.F. Floor	4.74	4.74	
2010	Stair Construction	Wood Stairs with wood rails		Flight	2915	.78	26.3%
3010	Wall Finishes	95% paint, 5% ceramic wall tile		S.F. Surface	1.47	2.61	
3020	Floor Finishes	80% Carpet tile, 10% vinyl composition tile, 10% ceramic tile		S.F. Floor	5.28	5.28	
3030	Ceiling Finishes	Painted gypsum board ceiling on resilient channels		S.F. Ceiling	4.54	4.54	
D. SERVICES							
	D10 Conveying						
1010	Elevators & Lifts	Hydraulic passenger elevator		Each	117,900	5.24	3.9%
1020	Escalators & Moving Walks	N/A		—	—	—	
	D20 Plumbing						
2010	Plumbing Fixtures	Kitchen, bath, laundry and service fixtures, supply and drainage	1 Fixture/285 S.F. Floor	Each	2029	7.12	
2020	Domestic Water Distribution	Electric water heater		S.F. Floor	8.38	8.38	12.0%
2040	Rain Water Drainage	Roof drains		S.F. Roof	1.59	.53	
	D30 HVAC						
3010	Energy Supply	Oil fired hot water, baseboard radiation		S.F. Floor	8.47	8.47	
3020	Heat Generating Systems	N/A		—	—	—	
3030	Cooling Generating Systems	Chilled water, air cooled condenser system		S.F. Floor	8.99	8.99	13.1%
3050	Terminal & Package Units	N/A		—	—	—	
3090	Other HVAC Sys. & Equipment	N/A		—	—	—	
	D40 Fire Protection						
4010	Sprinklers	Wet pipe spinkler system, light hazard		S.F. Floor	3.71	3.71	2.8%
4020	Standpipes	N/A		—	—	—	
	D50 Electrical						
5010	Electrical Service/Distribution	800 ampere service, panel board and feeders		S.F. Floor	2.67	2.67	
5020	Lighting & Branch Wiring	Incandescent fixtures, receptacles, switches, A.C. and misc. power		S.F. Floor	7.66	7.66	9.1%
5030	Communications & Security	Addressable alarm systems, emergency lighting, internet and phone wiring		S.F. Floor	1.80	1.80	
5090	Other Electrical Systems	N/A		—	—	—	
E. EQUIPMENT & FURNISHINGS							
1010	Commercial Equipment	N/A		—	—	—	
1020	Institutional Equipment	N/A		—	—	—	
1030	Vehicular Equipment	N/A		—	—	—	1.2 %
1090	Other Equipment	Residential freestanding gas ranges, dishwashers		S.F. Floor	1.60	1.60	
F. SPECIAL CONSTRUCTION							
1020	Integrated Construction	N/A		—	—	—	0.0 %
1040	Special Facilities	N/A		—	—	—	
G. BUILDING SITEWORK	**N/A**						
				❿ **Sub-Total**		133.57	**100%**
	CONTRACTOR FEES (General Requirements: 10%, Overhead: 5%, Profit: 10%) ⓫				25%	33.34	
	ARCHITECT FEES				7%	11.69	
				⓬ **Total Building Cost**		**178.60**	

1 Building Description (Model costs are calculated for a 3-story apartment building with a 10' story height and 22,500 square feet of floor area)
The model highlighted is described in terms of building type, number of stories, typical story height, and square footage.

2 Type of Building
(Apartment, 1–3 Story)

3 Division C Interiors
(C1020 Interior Doors)
System costs are presented in divisions according to the 7-element UNIFORMAT II classifications. Each of the component systems is listed.

4 Specification Highlights
(Interior single leaf doors, frames and hardware)
All systems in each subdivision are described with the material and proportions used.

5 Quantity Criteria (130 S.F. Floor/Door)
The criteria used in determining quantities for the calculations are shown.

6 Unit (Each)
The unit of measure shown in this column is the unit of measure of the particular system shown that corresponds to the unit cost.

7 Unit Cost ($1,369)
The cost per unit of measure of each system subdivision.

8 Cost per Square Foot ($10.37)
The cost per square foot for each system is the unit cost of the system times the total number of units, divided by the total square feet of building area.

9 % of Sub-Total (26.3%)
The percent of sub-total is the total cost per square foot of all systems in the division divided by the sub-total cost per square foot of the building.

10 Sub-Total ($133.57)
The sub-total is the total of all the system costs per square foot.

11 Project Fees
(Contractor Fees) (25%);
(Architects' Fee) (7%)
Contractor fees to cover the general requirements, overhead, and profit of the General Contractor are added as a percentage of the sub-total. An Architect's Fee, also as a percentage of the sub-total, is also added. These values may vary significantly with the building type or project.

12 Total Building Cost ($178.60)
The total building cost per square foot of building area is the sum of the square foot costs of all the systems, the General Contractor's general requirements, overhead and profit, and the Architect's fee. The total building cost is the amount that appears shaded in the Cost per Square Foot of Floor Area table shown previously.

Examples

Example 1

This example illustrates the use of the base cost tables. The base cost is adjusted for different exterior wall systems, different story height and a partial basement.

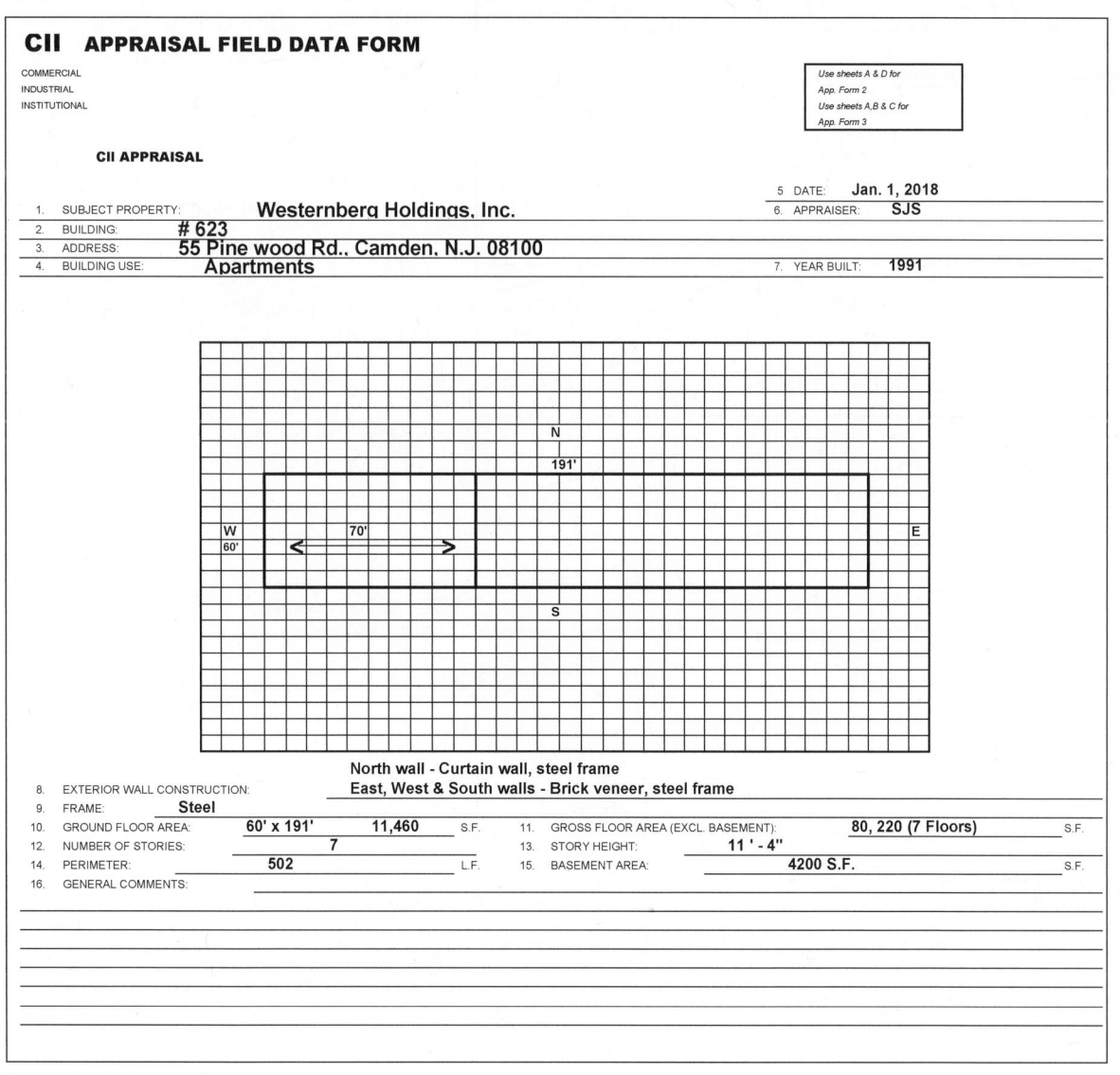

CII APPRAISAL FIELD DATA FORM

COMMERCIAL
INDUSTRIAL
INSTITUTIONAL

Use sheets A & D for
App. Form 2
Use sheets A,B & C for
App. Form 3

CII APPRAISAL

			5. DATE:	Jan. 1, 2018
1.	SUBJECT PROPERTY:	**Westernberg Holdings, Inc.**	6. APPRAISER:	**SJS**
2.	BUILDING:	**# 623**		
3.	ADDRESS:	**55 Pine wood Rd., Camden, N.J. 08100**		
4.	BUILDING USE:	**Apartments**	7. YEAR BUILT:	**1991**

North wall - Curtain wall, steel frame
East, West & South walls - Brick veneer, steel frame

8. EXTERIOR WALL CONSTRUCTION:
9. FRAME: **Steel**
10. GROUND FLOOR AREA: **60' x 191'** **11,460** S.F.
11. GROSS FLOOR AREA (EXCL. BASEMENT): **80, 220 (7 Floors)** S.F.
12. NUMBER OF STORIES: **7**
13. STORY HEIGHT: **11 ' - 4"**
14. PERIMETER: **502** L.F.
15. BASEMENT AREA: **4200 S.F.** S.F.
16. GENERAL COMMENTS:

Example 1 (continued)

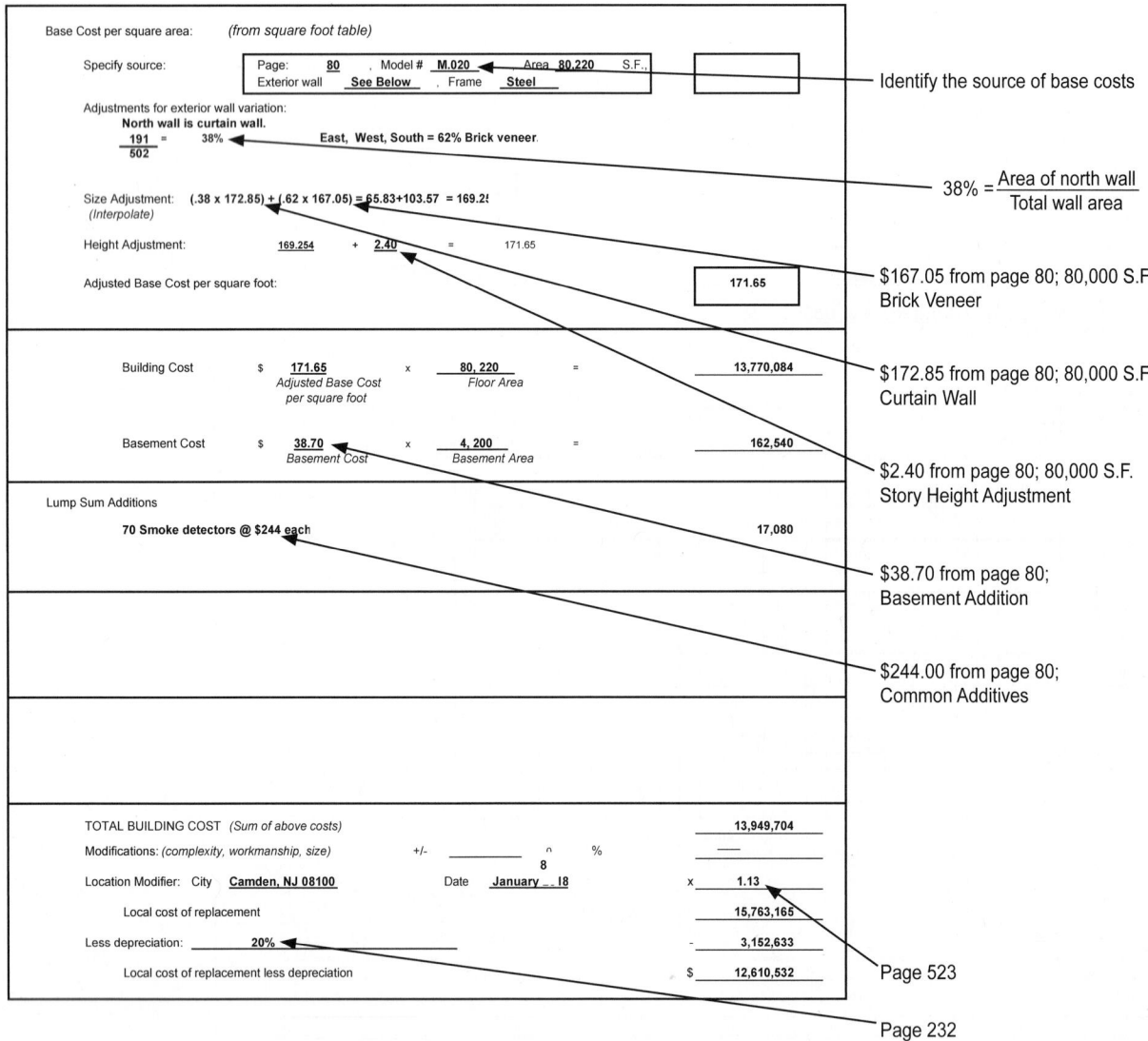

Base Cost per square area: *(from square foot table)*

Specify source:

| Page: __80__ , Model # __M.020__ Area __80,220__ S.F., | |
| Exterior wall __See Below__ , Frame __Steel__ | |

— Identify the source of base costs

Adjustments for exterior wall variation:
North wall is curtain wall.

$\frac{191}{502}$ = **38%** **East, West, South = 62% Brick veneer**.

$38\% = \dfrac{\text{Area of north wall}}{\text{Total wall area}}$

Size Adjustment: (.38 x 172.85) + (.62 x 167.05) = 65.83+103.57 = 169.2!
(Interpolate)

$167.05 from page 80; 80,000 S.F. Brick Veneer

Height Adjustment: __169.254__ + __2.40__ = 171.65

Adjusted Base Cost per square foot: __171.65__

$172.85 from page 80; 80,000 S.F. Curtain Wall

Building Cost $ __171.65__ x __80, 220__ = __13,770,084__
      Adjusted Base Cost   *Floor Area*
      per square foot

$2.40 from page 80; 80,000 S.F. Story Height Adjustment

Basement Cost $ __38.70__ x __4, 200__ = __162,540__
      Basement Cost   *Basement Area*

Lump Sum Additions

    70 Smoke detectors @ $244 each   17,080

$38.70 from page 80; Basement Addition

$244.00 from page 80; Common Additives

TOTAL BUILDING COST *(Sum of above costs)*   __13,949,704__

Modifications: *(complexity, workmanship, size)* +/- _____ ^ % ——
                                               8

Location Modifier: City __Camden, NJ 08100__ Date __January __ l8__ x __1.13__

    Local cost of replacement   __15,763,165__

Less depreciation: __20%__ _____ - __3,152,633__

    Local cost of replacement less depreciation $ __12,610,532__

Page 523

Page 232

74

Example 2

This example shows how to modify a model building.

Model #M.020, 4–7 Story Apartment, page 80, matches the example quite closely. The model specifies a 6-story building with a 10' story height.

The following adjustments must be made:

- add stone ashlar wall
- adjust model to a 5-story building
- change partitions
- change floor finish

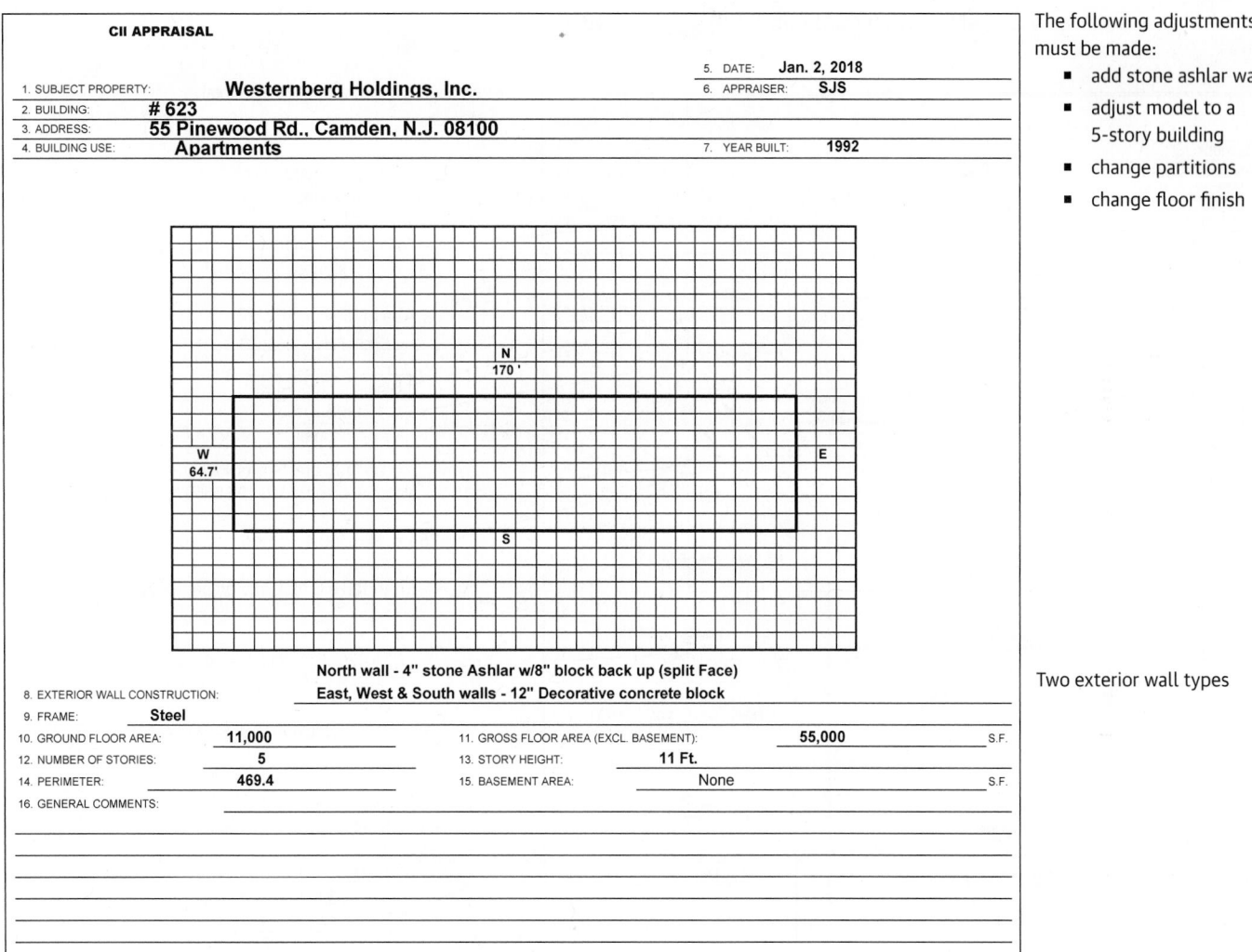

Two exterior wall types

CII APPRAISAL		
	5. DATE: **Jan. 2, 2018**	
1. SUBJECT PROPERTY: **Westernberg Holdings, Inc.**	6. APPRAISER: **SJS**	
2. BUILDING: **# 623**		
3. ADDRESS: **55 Pinewood Rd., Camden, N.J. 08100**		
4. BUILDING USE: **Apartments**	7. YEAR BUILT: **1992**	

8. EXTERIOR WALL CONSTRUCTION:
North wall - 4" stone Ashlar w/8" block back up (split Face)
East, West & South walls - 12" Decorative concrete block

9. FRAME: **Steel**

10. GROUND FLOOR AREA: **11,000**   11. GROSS FLOOR AREA (EXCL. BASEMENT): **55,000** S.F.

12. NUMBER OF STORIES: **5**   13. STORY HEIGHT: **11 Ft.**

14. PERIMETER: **469.4**   15. BASEMENT AREA: None S.F.

16. GENERAL COMMENTS:

(building diagram labels: N 170', W 64.7', E, S)

Example 2 (continued)

Square foot area (excluding basement) from item 11 _____ 55,000 _____ S.F.
Perimeter from item 14 _____ 469.4 _____ Lin. Ft.
Item 17-Model square foot costs (from model sub-total) _____ $135.48 _____

From page 81, Model #M.020 Sub-total

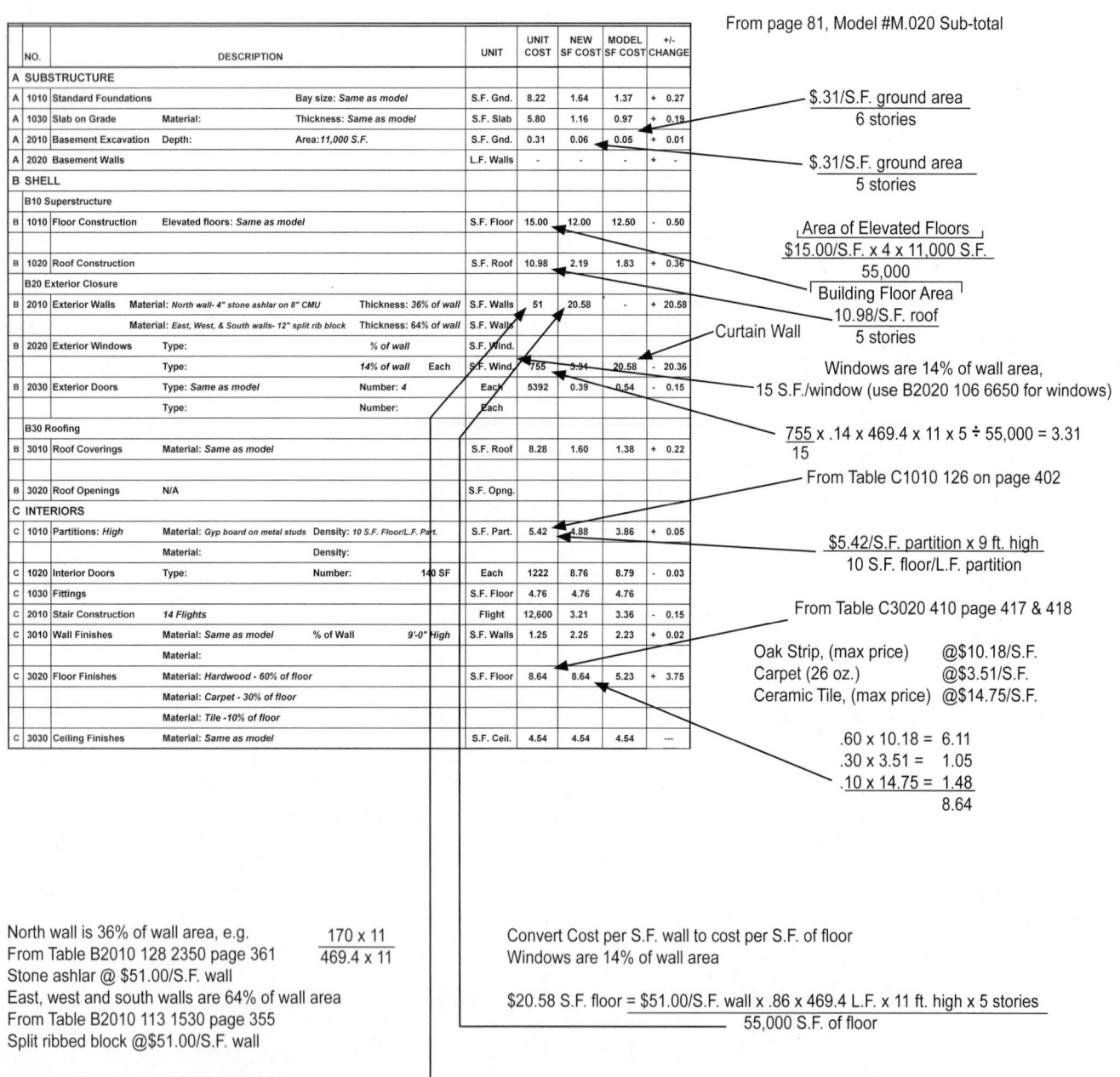

NO.	DESCRIPTION			UNIT	UNIT COST	NEW SF COST	MODEL SF COST	+/- CHANGE
A SUBSTRUCTURE								
A 1010 Standard Foundations		Bay size: Same as model		S.F. Gnd.	8.22	1.64	1.37	+ 0.27
A 1030 Slab on Grade	Material:	Thickness: Same as model		S.F. Slab	5.80	1.16	0.97	+ 0.19
A 2010 Basement Excavation	Depth:	Area:11,000 S.F.		S.F. Gnd.	0.31	0.06	0.05	+ 0.01
A 2020 Basement Walls				L.F. Walls	-	-	-	+ -
B SHELL								
B10 Superstructure								
B 1010 Floor Construction	Elevated floors: Same as model			S.F. Floor	15.00	12.00	12.50	- 0.50
B 1020 Roof Construction				S.F. Roof	10.98	2.19	1.83	+ 0.36
B20 Exterior Closure								
B 2010 Exterior Walls	Material: North wall- 4" stone ashlar on 8" CMU	Thickness: 36% of wall		S.F. Walls	51	20.58	-	+ 20.58
	Material: East, West, & South walls- 12" split rib block	Thickness: 64% of wall		S.F. Walls				
B 2020 Exterior Windows	Type:	% of wall		S.F. Wind.				
	Type:	14% of wall	Each	S.F. Wind.	755	3.34	20.58	- 20.36
B 2030 Exterior Doors	Type: Same as model	Number: 4		Each	5392	0.39	0.54	- 0.15
	Type:	Number:		Each				
B30 Roofing								
B 3010 Roof Coverings	Material: Same as model			S.F. Roof	8.28	1.60	1.38	+ 0.22
B 3020 Roof Openings	N/A			S.F. Opng.				
C INTERIORS								
C 1010 Partitions: High	Material: Gyp board on metal studs	Density: 10 S.F. Floor/L.F. Part.		S.F. Part.	5.42	4.88	3.86	+ 0.05
	Material:	Density:						
C 1020 Interior Doors	Type:	Number:	140 SF	Each	1222	8.76	8.79	- 0.03
C 1030 Fittings				S.F. Floor	4.76	4.76	4.76	
C 2010 Stair Construction	14 Flights			Flight	12,600	3.21	3.36	- 0.15
C 3010 Wall Finishes	Material: Same as model	% of Wall	9'-0" High	S.F. Walls	1.25	2.25	2.23	+ 0.02
	Material:							
C 3020 Floor Finishes	Material: Hardwood - 60% of floor			S.F. Floor	8.64	8.64	5.23	+ 3.75
	Material: Carpet - 30% of floor							
	Material: Tile -10% of floor							
C 3030 Ceiling Finishes	Material: Same as model			S.F. Ceil.	4.54	4.54	4.54	---

$.31/S.F. ground area
_____
6 stories

$.31/S.F. ground area
_____
5 stories

Area of Elevated Floors
$15.00/S.F. x 4 x 11,000 S.F.
_____
55,000

Building Floor Area
10.98/S.F. roof
_____
5 stories

Curtain Wall

Windows are 14% of wall area,
15 S.F./window (use B2020 106 6650 for windows)

755 x .14 x 469.4 x 11 x 5 ÷ 55,000 = 3.31
_____
15

From Table C1010 126 on page 402

$5.42/S.F. partition x 9 ft. high
_____
10 S.F. floor/L.F. partition

From Table C3020 410 page 417 & 418

Oak Strip, (max price)   @$10.18/S.F.
Carpet (26 oz.)   @$3.51/S.F.
Ceramic Tile, (max price)   @$14.75/S.F.

.60 x 10.18 = 6.11
.30 x 3.51 = 1.05
.10 x 14.75 = 1.48
              8.64

North wall is 36% of wall area, e.g.   170 x 11
From Table B2010 128 2350 page 361   _____
Stone ashlar @ $51.00/S.F. wall   469.4 x 11
East, west and south walls are 64% of wall area
From Table B2010 113 1530 page 355
Split ribbed block @$51.00/S.F. wall

New wall cost
(.36 x $51.00/S.F.) + (.64 x $51.00/S.F.) = $51.00/S.F. wall

Convert Cost per S.F. wall to cost per S.F. of floor
Windows are 14% of wall area

$20.58 S.F. floor = $51.00/S.F. wall x .86 x 469.4 L.F. x 11 ft. high x 5 stories
_____
55,000 S.F. of floor

Example 2 (continued)

	NO.	SYSTEM/COMPONENT	DESCRIPTION	UNIT	UNIT COST	NEW S.F. COST	MODEL S.F. COST	+/- CHANGE
D		**SERVICES**						
		D10 Conveying						
D	1010	Elevators & Lifts 2	Type: *Same as model* Capacity: Stops:	Each	25,200	9.11	8.40	+ 0.71
D	1020	Escalators & Moving Walks	Type:	Each				
		D20 Plumbing						
D	2010	Plumbing *Kitchen, bath, + service fixtures*	1 Fixture/285 S.F. Floor	Each	1,919	6.73	6.71	- 0.21
D	2020	Domestic Water Distribution		S.F. Floor	8.44	8.44	8.44	---
D	2040	Rain Water Drainage		S.F. Roof	1.74	0.34	0.29	+ 0.05
		D30 HVAC						
D	3010	Energy Supply		S.F. Floor	8.24	8.24	8.24	---
D	3020	Heat Generating Systems	Type: *Same as model*	S.F. Floor				
D	3030	Cooling Generating Systems	Type: *Same as model*	S.F. Floor	9.49	9.49	9.49	---
D	3090	Other HVAC Sys. & Equipment		Each				
		D40 Fire Protection						
D	4010	Sprinklers	*Same as model*	S.F. Floor	2.95	2.95	2.95	---
D	4020	Standpipes	*Same as model*	S.F. Floor	0.98	0.98	0.98	---
		D50 Electrical						
D	5010	Electrical Service/Distribution	*Same as model*	S.F. Floor	2.76	2.76	2.76	---
D	5020	Lighting & Branch Wiring		S.F. Floor	7.62	7.62	7.62	---
D	5030	Communications & Security		S.F. Floor	1.71	1.71	1.71	---
D	5090	Other Electrical Systems		S.F. Floor				
E		**EQUIPMENT & FURNISHINGS**						
E	1010	Commercial Equipment		Each				
E	1020	Institutional Equipment		Each				
E	1030	Vehicular Equipment		Each				
E	1090	Other Equipment		Each	1.62	1.62	1.62	
F		**SPECIAL CONSTRUCTION**						
F	1020	Integrated Construction	N/A	S.F.				
F	1040	Special Facilities	N/A	S.F.				
G		**BUILDING SITEWORK**						

ITEM						Total Change	- 4.13
17	Model 020 total $ 135.48						
18	Adjusted S.F. cost $ **131.35** item 17 +/- changes						
19	Building area - from item 11 **55,000** S.F. x adjusted S.F. cost..........................				$		6,908,550
20	Basement area - from item 15 S.F. x S.F. cost $				$		
21	Base building sub-total - item 20 + item 19				$		6,908,550
22	Miscellaneous addition (quality, etc.)				$		
23	Sub-total - item 22 + 21				$		6,908,550
24	General conditions -25 % of item 23				$		1,727,138
25	Sub-total - item 24 + item 23..				$		8,635,688
26	Architects fees **7** % of item 25				$		604,498
27	Sub-total - item 26 + item 25				$		9,240,186
28	Location modifier .... **zip code 08100**				x		1.13
29	Local replacement cost - item 28 x item 27				$		10,441,410
30	Depreciation **20** % of item 29				$		2,088,282
31	Depreciated local replacement cost - item 29 less item 30				$		8,353,128
32	Exclusions				$		
33	Net depreciated replacement cost - item 31 less item 32				$		8,353,128

From page 523

Use table on page 232

Costs per square foot of floor area

Exterior Wall	S.F. Area	8000	12000	15000	19000	22500	25000	29000	32000	36000
	L.F. Perimeter	213	280	330	350	400	433	442	480	520
Fiber Cement	Wood Frame	183.45	175.45	172.30	166.50	164.95	163.90	160.90	160.25	159.25
Stone Veneer	Wood Frame	204.05	193.45	189.25	180.75	178.60	177.35	172.65	171.80	170.40
Brick Veneer	Rigid Steel	198.80	189.15	185.25	177.75	175.80	174.60	170.50	169.75	168.45
E.I.F.S. and Metal Studs	Rigid Steel	193.20	184.60	181.15	174.75	173.05	172.00	168.55	167.90	166.75
Brick Veneer	Reinforced Concrete	220.40	208.95	204.35	194.80	192.60	191.10	185.85	185.00	183.40
Stucco and Concrete Block	Reinforced Concrete	208.10	198.15	194.20	186.30	184.35	183.10	178.85	178.10	176.70
Perimeter Adj., Add or Deduct	Per 100 L.F.	15.60	10.40	8.25	6.55	5.55	5.05	4.25	3.90	3.50
Story Hgt. Adj., Add or Deduct	Per 1 Ft.	2.45	2.15	2.00	1.70	1.55	1.65	1.35	1.40	1.30

For Basement, add $ 38.70 per square foot of basement area

The above costs were calculated using the basic specifications shown on the facing page. These costs should be adjusted where necessary for design alternatives and owner's requirements.

Common additives

Description	Unit	$ Cost
Appliances		
Cooking range, 30" free standing		
1 oven	Ea.	610 - 2775
2 oven	Ea.	1200 - 3725
30" built-in		
1 oven	Ea.	995 - 2550
2 oven	Ea.	1575 - 3125
Counter top cook tops, 4 burner	Ea.	475 - 2150
Microwave oven	Ea.	281 - 855
Combination range, refrig. & sink, 30" wide	Ea.	2075 - 2625
72" wide	Ea.	3075
Combination range, refrigerator, sink,		
microwave oven & icemaker	Ea.	7100
Compactor, residential, 4-1 compaction	Ea.	885 - 1475
Dishwasher, built-in, 2 cycles	Ea.	685 - 1225
4 cycles	Ea.	800 - 2025
Garbage disposer, sink type	Ea.	260 - 370
Hood for range, 2 speed, vented, 30" wide	Ea.	370 - 1450
42" wide	Ea.	450 - 2650
Refrigerator, no frost 10-12 C.F.	Ea.	580 - 750
18-20 C.F.	Ea.	895 - 2075

Description	Unit	$ Cost
Closed Circuit Surveillance, One station		
Camera and monitor	Ea.	1475
For additional camera stations, add	Ea.	665
Elevators, Hydraulic passenger, 2 stops		
2000# capacity	Ea.	74,100
2500# capacity	Ea.	76,600
3500# capacity	Ea.	81,600
Additional stop, add	Ea.	8375
Emergency Lighting, 25 watt, battery operated		
Lead battery	Ea.	330
Nickel cadmium	Ea.	550
Laundry Equipment		
Dryer, gas, 16 lb. capacity	Ea.	1025
30 lb. capacity	Ea.	4100
Washer, 4 cycle	Ea.	1325
Commercial	Ea.	1700
Smoke Detectors		
Ceiling type	Ea.	244
Duct type	Ea.	585

Important: See the Reference Section for Location Factors.

Model costs calculated for a 3 story building with 10' story height and 22,500 square feet of floor area

				Unit	Unit Cost	Cost Per S.F.	% Of Sub-Total
A. SUBSTRUCTURE							
1010	Standard Foundations	Poured concrete; strip and spread footings; 4' foundation wall		S.F. Ground	10.26	3.42	
1020	Special Foundations	N/A		—	—	—	
1030	Slab on Grade	4" reinforced concrete		S.F. Slab	5.80	1.93	4.1%
2010	Basement Excavation	Site preparation for slab and trench for foundation wall and footing		S.F. Ground	.33	.11	
2020	Basement Walls	N/A		—	—	—	
B. SHELL							
	B10 Superstructure						
1010	Floor Construction	Wood Beam and joist on wood columns, fireproofed		S.F. Floor	17.27	11.51	10.6%
1020	Roof Construction	Wood roof, truss, 4/12 slope		S.F. Roof	7.77	2.59	
	B20 Exterior Enclosure						
2010	Exterior Walls	Ashlar stone veneer on wood stud back up, insulated	80% of wall	S.F. Wall	41.34	17.64	
2020	Exterior Windows	Aluminum horizontal sliding	20% of wall	Each	529	3.76	16.4%
2030	Exterior Doors	Steel door , hollow metal with frame		Each	3015	.53	
	B30 Roofing						
3010	Roof Coverings	Asphalt roofing, strip shingles, flashing		S.F. Roof	2.16	.72	0.5%
3020	Roof Openings	N/A		—	—	—	
C. INTERIORS							
1010	Partitions	Gypsum board on wood studs	10 S.F. Floor/L.F. Partition	S.F. Partition	7.73	6.87	
1020	Interior Doors	Interior single leaf doors, frames and hardware	130 S.F. Floor/Door	Each	1369	10.37	
1030	Fittings	Residential wall and base wood cabinets, laminate counters		S.F. Floor	4.74	4.74	
2010	Stair Construction	Wood Stairs with wood rails		Flight	2915	.78	26.3%
3010	Wall Finishes	95% paint, 5% ceramic wall tile		S.F. Surface	1.47	2.61	
3020	Floor Finishes	80% Carpet tile, 10% vinyl composition tile, 10% ceramic tile		S.F. Floor	5.28	5.28	
3030	Ceiling Finishes	Painted gypsum board ceiling on resilient channels		S.F. Ceiling	4.54	4.54	
D. SERVICES							
	D10 Conveying						
1010	Elevators & Lifts	Hydraulic passenger elevator		Each	117,900	5.24	3.9%
1020	Escalators & Moving Walks	N/A		—	—	—	
	D20 Plumbing						
2010	Plumbing Fixtures	Kitchen, bath, laundry and service fixtures, supply and drainage	1 Fixture/285 S.F. Floor	Each	2029	7.12	
2020	Domestic Water Distribution	Electric water heater		S.F. Floor	8.38	8.38	12.0%
2040	Rain Water Drainage	Roof drains		S.F. Roof	1.59	.53	
	D30 HVAC						
3010	Energy Supply	Oil fired hot water, baseboard radiation		S.F. Floor	8.47	8.47	
3020	Heat Generating Systems	N/A		—	—	—	
3030	Cooling Generating Systems	Chilled water, air cooled condenser system		S.F. Floor	8.99	8.99	13.1%
3050	Terminal & Package Units	N/A		—	—	—	
3090	Other HVAC Sys. & Equipment	N/A		—	—	—	
	D40 Fire Protection						
4010	Sprinklers	Wet pipe spinkler system, light hazard		S.F. Floor	3.71	3.71	2.8%
4020	Standpipes	N/A		—	—	—	
	D50 Electrical						
5010	Electrical Service/Distribution	800 ampere service, panel board and feeders		S.F. Floor	2.67	2.67	
5020	Lighting & Branch Wiring	Incandescent fixtures, receptacles, switches, A.C. and misc. power		S.F. Floor	7.66	7.66	9.1%
5030	Communications & Security	Addressable alarm systems, emergency lighting, internet and phone wiring		S.F. Floor	1.80	1.80	
5090	Other Electrical Systems	N/A		—	—	—	
E. EQUIPMENT & FURNISHINGS							
1010	Commercial Equipment	N/A		—	—	—	
1020	Institutional Equipment	N/A		—	—	—	
1030	Vehicular Equipment	N/A		—	—	—	1.2 %
1090	Other Equipment	Residential freestanding gas ranges, dishwashers		S.F. Floor	1.60	1.60	
F. SPECIAL CONSTRUCTION							
1020	Integrated Construction	N/A		—	—	—	0.0 %
1040	Special Facilities	N/A		—	—	—	
G. BUILDING SITEWORK	**N/A**						

		Sub-Total	133.57	100%
CONTRACTOR FEES (General Requirements: 10%, Overhead: 5%, Profit: 10%)		25%	33.34	
ARCHITECT FEES		7%	11.69	
	Total Building Cost		**178.60**	

For customer support on your Square Foot Costs with RSMeans data, call 800.448.8182.

Costs per square foot of floor area

Exterior Wall	S.F. Area	40000	45000	50000	55000	60000	70000	80000	90000	100000
	L.F. Perimeter	366	400	433	466	500	510	530	550	594
Curtain Wall	Rigid Steel	187.25	185.20	183.60	182.25	181.20	176.15	172.85	170.30	169.25
Brick Veneer	Rigid Steel	179.15	177.40	175.95	174.75	173.85	169.70	167.05	164.95	164.05
E.I.F.S. and Metal Studs	Rigid Steel	173.20	171.55	170.25	169.25	168.40	164.95	162.65	160.90	160.15
Precast Concrete	Reinforced Concrete	208.50	206.00	204.00	202.35	201.05	194.30	189.95	186.50	185.20
Brick Veneer	Reinforced Concrete	197.00	194.90	193.15	191.70	190.60	185.15	181.60	178.80	177.70
Stucco	Reinforced Concrete	185.75	183.90	182.45	181.20	180.30	175.90	173.10	170.90	169.95
Perimeter Adj., Add or Deduct	Per 100 L.F.	10.20	9.15	8.15	7.45	6.80	5.80	5.10	4.55	4.05
Story Hgt. Adj., Add or Deduct	Per 1 Ft.	3.35	3.25	3.15	3.10	3.05	2.60	2.40	2.20	2.20

For Basement, add $38.70 per square foot of basement area

The above costs were calculated using the basic specifications shown on the facing page. These costs should be adjusted where necessary for design alternatives and owner's requirements.

Common additives

Description	Unit	$ Cost
Appliances		
Cooking range, 30" free standing		
1 oven	Ea.	610 - 2775
2 oven	Ea.	1200 - 3725
30" built-in		
1 oven	Ea.	995 - 2550
2 oven	Ea.	1575 - 3125
Counter top cook tops, 4 burner	Ea.	475 - 2150
Microwave oven	Ea.	281 - 855
Combination range, refrig. & sink, 30" wide	Ea.	2075 - 2625
72" wide	Ea.	3075
Combination range, refrigerator, sink,		
microwave oven & icemaker	Ea.	7100
Compactor, residential, 4-1 compaction	Ea.	885 - 1475
Dishwasher, built-in, 2 cycles	Ea.	685 - 1225
4 cycles	Ea.	800 - 2025
Garbage disposer, sink type	Ea.	260 - 370
Hood for range, 2 speed, vented, 30" wide	Ea.	370 - 1450
42" wide	Ea.	450 - 2650
Refrigerator, no frost 10-12 C.F.	Ea.	580 - 750
18-20 C.F.	Ea.	895 - 2075

Description	Unit	$ Cost
Closed Circuit Surveillance, One station		
Camera and monitor	Ea.	1475
For additional camera stations, add	Ea.	665
Elevators, Electric passenger, 5 stops		
2000# capacity	Ea.	186,000
3500# capacity	Ea.	193,000
5000# capacity	Ea.	200,000
Additional stop, add	Ea.	10,900
Emergency Lighting, 25 watt, battery operated		
Lead battery	Ea.	330
Nickel cadmium	Ea.	550
Laundry Equipment		
Dryer, gas, 16 lb. capacity	Ea.	1025
30 lb. capacity	Ea.	4100
Washer, 4 cycle	Ea.	1325
Commercial	Ea.	1700
Smoke Detectors		
Ceiling type	Ea.	244
Duct type	Ea.	585

Important: See the Reference Section for Location Factors.

Model costs calculated for a 6 story building with 10'-4" story height and 60,000 square feet of floor area

				Unit	Unit Cost	Cost Per S.F.	% Of Sub-Total
A. SUBSTRUCTURE							
1010	Standard Foundations	Poured concrete: strip and spread footings; 4' foundation wall		S.F. Ground	8.22	1.37	
1020	Special Foundations	N/A		—	—	—	
1030	Slab on Grade	4" reinforced slab on grade		S.F. Slab	5.80	.97	1.8%
2010	Basement Excavation	Site preparation for slab and trench for foundation wall and footing		S.F. Ground	.31	.05	
2020	Basement Walls	N/A		—	—	—	
B. SHELL							
B10 Superstructure							
1010	Floor Construction	Open web steel joists, slab form, concrete, fireproofed steel columns		S.F. Floor	15	12.50	10.6%
1020	Roof Construction	Metal deck, open web steel joists, beams, columns		S.F. Roof	10.98	1.83	
B20 Exterior Enclosure							
2010	Exterior Walls	N/A		—	—	—	
2020	Exterior Windows	Curtain wall glazing system, thermo-break frame	100% of wall	S.F. Wall	51	20.58	15.6 %
2030	Exterior Doors	Aluminum and glass single and double doors		Each	5392	.54	
B30 Roofing							
3010	Roof Coverings	Single ply membrane, stone ballast, rigid insulation		S.F. Roof	8.28	1.38	1.0%
3020	Roof Openings	N/A		—	—	—	
C. INTERIORS							
1010	Partitions	Gypsum board on metal studs	10 S.F. Floor/L.F. Partition	S.F. Partition	9.10	8.09	
1020	Interior Doors	Interior single leaf doors, frames and hardware	140 S.F. Floor/Door	Each	1222	8.79	
1030	Fittings	Residential wall and base wood cabinets, laminate counters		S.F. Floor	4.76	4.76	
2010	Stair Construction	Concrete filled metal pan		Flight	12,600	3.36	27.3%
3010	Wall Finishes	95% paint, 5% ceramic wall tile		S.F. Surface	1.25	2.23	
3020	Floor Finishes	80% Carpet tile, 10% vinyl composition tile, 10% ceramic tile		S.F. Floor	5.28	5.28	
3030	Ceiling Finishes	Painted gypsum board ceiling on resilient channels		S.F. Ceiling	4.54	4.54	
D. SERVICES							
D10 Conveying							
1010	Elevators & Lifts	Two geared passenger elevators		Each	252,000	8.40	6.2%
1020	Escalators & Moving Walks	N/A		—	—	—	
D20 Plumbing							
2010	Plumbing Fixtures	Kitchen, bath, laundry and service fixtures, supply and drainage	1 Fixture/285 S.F. Floor	Each	1919	6.71	
2020	Domestic Water Distribution	Electric water heater		S.F. Floor	8.44	8.44	11.4%
2040	Rain Water Drainage	Roof drains		S.F. Roof	1.74	.29	
D30 HVAC							
3010	Energy Supply	Oil fired hot water, baseboard radiation		S.F. Floor	8.24	8.24	
3020	Heat Generating Systems	N/A		—	—	—	
3030	Cooling Generating Systems	Chilled water, air cooled condenser system		S.F. Floor	9.49	9.49	13.1%
3050	Terminal & Package Units	N/A		—	—	—	
3090	Other HVAC Sys. & Equipment	N/A		—	—	—	
D40 Fire Protection							
4010	Sprinklers	Wet pipe spinkler system, light hazard		S.F. Floor	2.95	2.95	2.9%
4020	Standpipes	Standpipes and hose systems, with pumps		S.F. Floor	.98	.98	
D50 Electrical							
5010	Electrical Service/Distribution	1600 Ampere service, panel boards and feeders		S.F. Floor	2.76	2.76	
5020	Lighting & Branch Wiring	Incandescent fixtures, receptacles, switches, A.C. and misc. power		S.F. Floor	7.62	7.62	8.9%
5030	Communications & Security	Addressable alarm systems, emergency lighting, internet and phone wiring		S.F. Floor	1.71	1.71	
5090	Other Electrical Systems	N/A		—	—	—	
E. EQUIPMENT & FURNISHINGS							
1010	Commercial Equipment	N/A		—	—	—	
1020	Institutional Equipment	N/A		—	—	—	
1030	Vehicular Equipment	N/A		—	—	—	1.2 %
1090	Other Equipment	Residential freestanding gas ranges, dishwashers		S.F. Floor	1.62	1.62	
F. SPECIAL CONSTRUCTION							
1020	Integrated Construction	N/A		—	—	—	0.0 %
1040	Special Facilities	N/A		—	—	—	
G. BUILDING SITEWORK	**N/A**						

		Sub-Total	135.48	100%
	CONTRACTOR FEES (General Requirements: 10%, Overhead: 5%, Profit: 10%)	25%	33.87	
	ARCHITECT FEES	7%	11.85	
	Total Building Cost		**181.20**	

For customer support on your Square Foot Costs with RSMeans data, call 800.448.8182.

81

Costs per square foot of floor area

Exterior Wall	S.F. Area	95000	112000	129000	145000	170000	200000	275000	400000	600000
	L.F. Perimeter	345	386	406	442	480	510	530	660	800
Curtain Wall	Rigid Steel	228.40	225.30	221.30	219.65	216.55	213.15	205.75	202.10	198.25
Brick Veneer	Rigid Steel	220.40	217.70	214.35	212.90	210.35	207.50	201.50	198.45	195.30
E.I.F.S.	Rigid Steel	214.60	212.20	209.30	208.05	205.85	203.40	198.40	195.80	193.20
Precast Concrete	Reinforced Concrete	243.60	239.85	234.70	232.65	228.80	224.35	214.70	210.05	205.10
Brick Veneer	Reinforced Concrete	232.45	229.25	225.05	223.30	220.15	216.60	208.85	205.00	201.00
Stucco	Reinforced Concrete	224.15	221.40	217.85	216.35	213.70	210.70	204.40	201.20	197.95
Perimeter Adj., Add or Deduct	Per 100 L.F.	10.65	9.05	7.85	6.95	5.90	5.05	3.70	2.50	1.65
Story Hgt. Adj., Add or Deduct	Per 1 Ft.	3.25	3.10	2.80	2.70	2.55	2.25	1.70	1.50	1.20
For Basement, add $39.70 per square foot of basement area										

The above costs were calculated using the basic specifications shown on the facing page. These costs should be adjusted where necessary for design alternatives and owner's requirements.

Common additives

Description	Unit	$ Cost
Appliances		
Cooking range, 30" free standing		
1 oven	Ea.	610 - 2775
2 oven	Ea.	1200 - 3725
30" built-in		
1 oven	Ea.	995 - 2550
2 oven	Ea.	1575 - 3125
Counter top cook tops, 4 burner	Ea.	475 - 2150
Microwave oven	Ea.	281 - 855
Combination range, refrig. & sink, 30" wide	Ea.	2075 - 2625
72" wide	Ea.	3075
Combination range, refrigerator, sink, microwave oven & icemaker	Ea.	7100
Compactor, residential, 4-1 compaction	Ea.	885 - 1475
Dishwasher, built-in, 2 cycles	Ea.	685 - 1225
4 cycles	Ea.	800 - 2025
Garbage disposer, sink type	Ea.	260 - 370
Hood for range, 2 speed, vented, 30" wide	Ea.	370 - 1450
42" wide	Ea.	450 - 2650
Refrigerator, no frost 10-12 C.F.	Ea.	580 - 750
18-20 C.F.	Ea.	895 - 2075

Description	Unit	$ Cost
Closed Circuit Surveillance, One station		
Camera and monitor	Ea.	1475
For additional camera stations, add	Ea.	665
Elevators, Electric passenger, 10 stops		
3000# capacity	Ea.	423,500
4000# capacity	Ea.	426,500
5000# capacity	Ea.	432,500
Additional stop, add	Ea.	10,900
Emergency Lighting, 25 watt, battery operated		
Lead battery	Ea.	330
Nickel cadmium	Ea.	550
Laundry Equipment		
Dryer, gas, 16 lb. capacity	Ea.	1025
30 lb. capacity	Ea.	4100
Washer, 4 cycle	Ea.	1325
Commercial	Ea.	1700
Smoke Detectors		
Ceiling type	Ea.	244
Duct type	Ea.	585

Important: See the Reference Section for Location Factors.

Model costs calculated for a 15 story building with 10'-6" story height and 145,000 square feet of floor area

				Unit	Unit Cost	Cost Per S.F.	% Of Sub-Total
A. SUBSTRUCTURE							
1010	Standard Foundations	CIP concrete pile caps; 4' foundation wall		S.F. Ground	22.95	1.53	
1020	Special Foundations	Steel H piles, concrete grade beams		S.F. Ground	289	19.29	
1030	Slab on Grade	4" thick reinforced		S.F. Slab	5.80	.38	12.8%
2010	Basement Excavation	Site preparation for slab, piles and grade beams		S.F. Ground	.31	.02	
2020	Basement Walls	N/A		—	—	—	
B. SHELL							
	B10 Superstructure						
1010	Floor Construction	Open web steel joists, slab form, concrete, fireproofed steel columns		S.F. Floor	15.14	14.13	9.0%
1020	Roof Construction	Metal deck, open web steel joists, beams, columns		S.F. Roof	10.95	.73	
	B20 Exterior Enclosure						
2010	Exterior Walls	N/A		—	—	—	
2020	Exterior Windows	Curtain wall glazing system, thermo-break frame	100% of wall	S.F. Wall	51	18.82	13.1%
2030	Exterior Doors	Aluminum and glass single and double doors		Each	3340	2.96	
	B30 Roofing						
3010	Roof Coverings	Single ply membrane, stone ballast, rigid insulation		S.F. Roof	7.95	.53	0.3%
3020	Roof Openings	N/A		—	—	—	
C. INTERIORS							
1010	Partitions	Gypsum board on metal studs	10 S.F. Floor/L.F. Partition	S.F. Partition	12.56	11.16	
1020	Interior Doors	Interior single leaf doors, frames and hardware	140 S.F. Floor/Door	Each	1221	8.54	
1030	Fittings	Residential wall and base wood cabinets, laminate counters		S.F. Floor	4.76	4.76	
2010	Stair Construction	Cement filled metal pan		Flight	12,600	3.73	24.3%
3010	Wall Finishes	95% paint, 5% ceramic wall tile		S.F. Surface	1.25	2.23	
3020	Floor Finishes	80% Carpet tile, 10% vinyl composition tile, 10% ceramic tile		S.F. Floor	5.28	5.28	
3030	Ceiling Finishes	Painted gypsum board ceiling on resilient channels		S.F. Ceiling	4.54	4.54	
D. SERVICES							
	D10 Conveying						
1010	Elevators & Lifts	Four geared passenger elevators		Each	504,963	13.93	8.4%
1020	Escalators & Moving Walks	N/A		—	—	—	
	D20 Plumbing						
2010	Plumbing Fixtures	Kitchen, bath, laundry and service fixtures, supply and drainage	1 Fixture/280 S.F. Floor	Each	1868	6.67	
2020	Domestic Water Distribution	Electric water heater		S.F. Floor	8.42	8.42	9.3%
2040	Rain Water Drainage	Roof drains		S.F. Roof	3.75	.25	
	D30 HVAC						
3010	Energy Supply	Oil fired hot water, baseboard radiation		S.F. Floor	8.24	8.24	
3020	Heat Generating Systems	N/A		—	—	—	
3030	Cooling Generating Systems	Chilled water, air cooled condenser system		S.F. Floor	9.49	9.49	10.7%
3050	Terminal & Package Units	N/A		—	—	—	
3090	Other HVAC Sys. & Equipment	N/A		—	—	—	
	D40 Fire Protection						
4010	Sprinklers	Wet pipe spinkler system, light hazard and accessory package		S.F. Floor	3.02	3.02	3.0%
4020	Standpipes	Standpipes and hose systems, with pumps		S.F. Floor	1.90	1.90	
	D50 Electrical						
5010	Electrical Service/Distribution	2000 Ampere service, panel boards and feeders		S.F. Floor	2	2	
5020	Lighting & Branch Wiring	Incandescent fixtures, receptacles, switches, A.C. and misc. power		S.F. Floor	8.28	8.28	8.2%
5030	Communications & Security	Addressable alarm systems, emergency lighting, internet and phone wiring		S.F. Floor	3.30	3.30	
5090	Other Electrical Systems	N/A		—	—	—	
E. EQUIPMENT & FURNISHINGS							
1010	Commercial Equipment	N/A		—	—	—	
1020	Institutional Equipment	N/A		—	—	—	
1030	Vehicular Equipment	N/A		—	—	—	1.0%
1090	Other Equipment	Residential freestanding gas ranges, dishwashers		S.F. Floor	1.62	1.62	
F. SPECIAL CONSTRUCTION							
1020	Integrated Construction	N/A		—	—	—	0.0%
1040	Special Facilities	N/A		—	—	—	
G. BUILDING SITEWORK	**N/A**						

			Sub-Total	165.75	100%
CONTRACTOR FEES (General Requirements: 10%, Overhead: 5%, Profit: 10%)			25%	41.47	
ARCHITECT FEES			6%	12.43	
		Total Building Cost		**219.65**	

For customer support on your Square Foot Costs with RSMeans data, call 800.448.8182.

83

Costs per square foot of floor area

Exterior Wall	S.F. Area	2000	3000	4000	5000	7500	10000	15000	25000	50000
	L.F. Perimeter	188	224	260	300	356	400	504	700	1200
Brick Veneer	Wood Frame	257.85	229.45	215.30	207.40	192.60	184.45	177.00	170.70	166.05
	Rigid Steel	286.05	257.75	243.75	235.80	221.15	213.05	205.60	199.35	194.80
Brick Veneer and Concrete Block	Wood Truss	261.45	231.55	216.70	208.40	192.70	183.95	176.00	169.30	164.45
	Bearing Walls	300.15	268.95	253.45	244.80	228.25	219.05	210.70	203.55	198.45
Wood Clapboard	Wood Frame	245.20	219.35	206.50	199.25	186.20	179.05	172.40	166.90	162.85
Vinyl Clapboard	Wood Frame	247.10	220.90	207.85	200.50	187.20	179.85	173.10	167.45	163.35
Perimeter Adj., Add or Deduct	Per 100 L.F.	35.10	23.40	17.55	14.00	9.40	7.00	4.65	2.75	1.45
Story Hgt. Adj., Add or Deduct	Per 1 Ft.	4.10	3.25	2.85	2.55	2.10	1.75	1.40	1.15	1.05

For Basement, add $34.90 per square foot of basement area

The above costs were calculated using the basic specifications shown on the facing page. These costs should be adjusted where necessary for design alternatives and owner's requirements.

Common additives

Description	Unit	$ Cost
Appliances		
Cooking range, 30" free standing		
1 oven	Ea.	610 - 2775
2 oven	Ea.	1200 - 3725
30" built-in		
1 oven	Ea.	995 - 2550
2 oven	Ea.	1575 - 3125
Counter top cook tops, 4 burner	Ea.	475 - 2150
Microwave oven	Ea.	281 - 855
Combination range, refrig. & sink, 30" wide	Ea.	2075 - 2625
60" wide	Ea.	2425
72" wide	Ea.	3075
Combination range, refrigerator, sink, microwave oven & icemaker	Ea.	7100
Compactor, residential, 4-1 compaction	Ea.	885 - 1475
Dishwasher, built-in, 2 cycles	Ea.	685 - 1225
4 cycles	Ea.	800 - 2025
Garbage disposer, sink type	Ea.	260 - 370
Hood for range, 2 speed, vented, 30" wide	Ea.	370 - 1450
42" wide	Ea.	450 - 2650

Description	Unit	$ Cost
Appliances, cont.		
Refrigerator, no frost 10-12 C.F.	Ea.	580 - 750
14-16 C.F.	Ea.	720 - 1025
18-20 C.F.	Ea.	895 - 2075
Laundry Equipment		
Dryer, gas, 16 lb. capacity	Ea.	1025
30 lb. capacity	Ea.	4100
Washer, 4 cycle	Ea.	1325
Commercial	Ea.	1700
Sound System		
Amplifier, 250 watts	Ea.	2025
Speaker, ceiling or wall	Ea.	240
Trumpet	Ea.	460

Important: See the Reference Section for Location Factors.

Model costs calculated for a 1 story building with 10' story height and 10,000 square feet of floor area

Assisted - Senior Living

				Unit	Unit Cost	Cost Per S.F.	% Of Sub-Total
A.	**SUBSTRUCTURE**						
1010	Standard Foundations	Poured concrete strip footings; 4' foundation wall		S.F. Ground	7.45	7.45	
1020	Special Foundations	N/A		—	—	—	
1030	Slab on Grade	4" reinforced concrete with vapor barrier and granular base		S.F. Slab	5.80	5.80	10.4%
2010	Basement Excavation	Site preparation for slab and trench for foundation wall and footing		S.F. Ground	.57	.57	
2020	Basement Walls	N/A		—	—	—	
B.	**SHELL**						
	B10 Superstructure						
1010	Floor Construction	Steel column		S.F. Floor	.62	.62	6.3%
1020	Roof Construction	Plywood on wood trusses (pitched); fiberglass batt insulation		S.F. Roof	7.72	7.72	
	B20 Exterior Enclosure						
2010	Exterior Walls	Brick veneer on wood studs, insulated	80% of wall	S.F. Wall	26.25	8.40	
2020	Exterior Windows	Double hung wood	20% of wall	Each	665	3.33	10.4%
2030	Exterior Doors	Aluminum and glass; 18 ga. steel flush		Each	3538	2.13	
	B30 Roofing						
3010	Roof Coverings	Asphalt shingles with flashing (pitched); al. gutters & downspouts		S.F. Roof	3.04	3.04	2.3%
3020	Roof Openings	N/A		—	—	—	
C.	**INTERIORS**						
1010	Partitions	Gypsum board on wood studs	6 S.F. Floor/L.F. Partition	S.F. Partition	5.72	7.62	
1020	Interior Doors	Single leaf wood	80 S.F. Floor/Door	Each	720	8.47	
1030	Fittings	N/A		—	—	—	
2010	Stair Construction	N/A		—	—	—	26.5%
3010	Wall Finishes	Paint on drywall		S.F. Surface	.89	2.38	
3020	Floor Finishes	80% carpet, 20% ceramic tile		S.F. Floor	7.61	7.61	
3030	Ceiling Finishes	Gypsum board on wood furring, painted		S.F. Ceiling	9.08	9.08	
D.	**SERVICES**						
	D10 Conveying						
1010	Elevators & Lifts	N/A		—	—	—	0.0%
1020	Escalators & Moving Walks	N/A		—	—	—	
	D20 Plumbing						
2010	Plumbing Fixtures	Kitchen, toilet and service fixtures, supply and drainage	1 Fixture/200 S.F. Floor	Each	3258	16.29	
2020	Domestic Water Distribution	Gas water heater		S.F. Floor	2.52	2.52	14.2%
2040	Rain Water Drainage	N/A		—	—	—	
	D30 HVAC						
3010	Energy Supply	N/A		—	—	—	
3020	Heat Generating Systems	N/A		—	—	—	
3030	Cooling Generating Systems	N/A		—	—	—	3.2%
3050	Terminal & Package Units	Gas fired hot water boiler; terminal package heat/A/C units		S.F. Floor	4.27	4.27	
3090	Other HVAC Sys. & Equipment	N/A		—	—	—	
	D40 Fire Protection						
4010	Sprinklers	Wet pipe sprinkler system		S.F. Floor	3.79	3.79	3.6%
4020	Standpipes	Standpipes		S.F. Floor	.95	.95	
	D50 Electrical						
5010	Electrical Service/Distribution	600 ampere service, panel board, feeders and branch wiring		S.F. Floor	8.67	8.67	
5020	Lighting & Branch Wiring	Fuorescent & incandescent fixtures, receptacles, switches, A.C. and misc. power		S.F. Floor	11.62	11.62	22.0%
5030	Communications & Security	Addressable alarm, fire detection and command center, communication system		S.F. Floor	8.68	8.68	
5090	Other Electrical Systems	Emergency generator, 7.5 kW		S.F. Floor	.22	.22	
E.	**EQUIPMENT & FURNISHINGS**						
1010	Commercial Equipment	Residential washers and dryers		S.F. Floor	.48	.48	
1020	Institutional Equipment	N/A		—	—	—	1.3%
1030	Vehicular Equipment	N/A		—	—	—	
1090	Other Equipment	Residential ranges & refrigerators		S.F. Floor	1.21	1.21	
F.	**SPECIAL CONSTRUCTION**						
1020	Integrated Construction	N/A		—	—	—	0.0%
1040	Special Facilities	N/A		—	—	—	
G.	**BUILDING SITEWORK**	**N/A**					

		Sub-Total	132.92	100%
	CONTRACTOR FEES (General Requirements: 10%, Overhead: 5%, Profit: 10%)	25%	33.25	
	ARCHITECT FEES	11%	18.28	
	Total Building Cost		**184.45**	

For customer support on your Square Foot Costs with RSMeans data, call 800.448.8182.

85

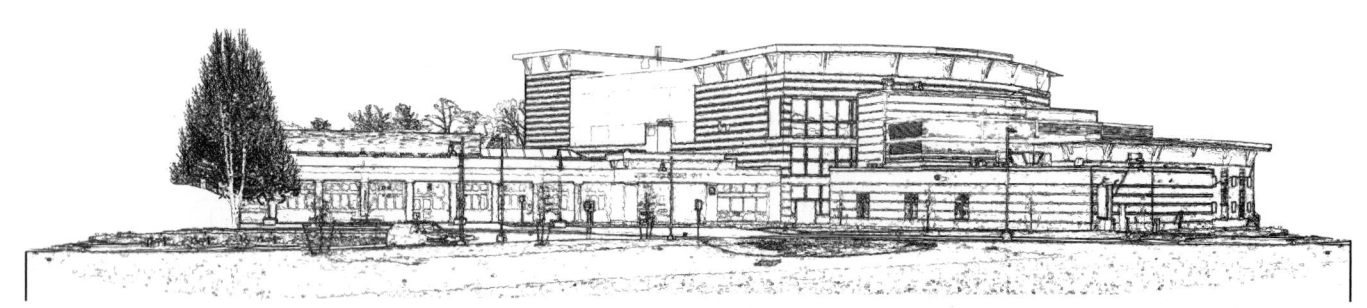

Costs per square foot of floor area

Exterior Wall	S.F. Area	12000	15000	18000	21000	24000	27000	30000	33000	36000
	L.F. Perimeter	440	500	540	590	640	665	700	732	770
Face Brick and Concrete Block	Rigid Steel	217.10	211.05	205.15	201.65	199.10	195.55	193.25	191.25	189.80
	Bearing Walls	209.10	203.05	197.15	193.70	191.15	187.60	185.30	183.25	181.85
Precast Concrete	Rigid Steel	218.30	210.95	203.75	199.50	196.45	192.05	189.25	186.80	185.00
Decorative Concrete Block	Bearing Walls	191.85	187.45	183.05	180.50	178.60	176.00	174.35	172.80	171.75
Concrete Block	Rigid Steel	176.80	174.45	172.20	170.80	169.85	168.55	167.70	166.85	166.35
	Bearing Walls	186.80	182.80	178.95	176.60	174.90	172.60	171.10	169.80	168.80
Perimeter Adj., Add or Deduct	Per 100 L.F.	14.30	11.45	9.50	8.25	7.20	6.35	5.70	5.20	4.75
Story Hgt. Adj., Add or Deduct	Per 1 Ft.	2.20	2.05	1.75	1.75	1.65	1.50	1.40	1.35	1.30

For Basement, add $31.70 per square foot of basement area

The above costs were calculated using the basic specifications shown on the facing page. These costs should be adjusted where necessary for design alternatives and owner's requirements.

Common additives

Description	Unit	$ Cost
Closed Circuit Surveillance, One station		
Camera and monitor	Ea.	1475
For additional camera stations, add	Ea.	665
Emergency Lighting, 25 watt, battery operated		
Lead battery	Ea.	330
Nickel cadmium	Ea.	550
Seating		
Auditorium chair, all veneer	Ea.	355
Veneer back, padded seat	Ea.	380
Upholstered, spring seat	Ea.	335
Classroom, movable chair & desk	Set	81 - 171
Lecture hall, pedestal type	Ea.	350 - 650
Smoke Detectors		
Ceiling type	Ea.	244
Duct type	Ea.	585
Sound System		
Amplifier, 250 watts	Ea.	2025
Speaker, ceiling or wall	Ea.	240
Trumpet	Ea.	460

Important: See the Reference Section for Location Factors.

Model costs calculated for a 1 story building with 24' story height and 24,000 square feet of floor area

				Unit	Unit Cost	Cost Per S.F.	% Of Sub-Total
A.	**SUBSTRUCTURE**						
1010	Standard Foundations	Poured concrete; strip and spread footings; 4' foundation wall		S.F. Ground	4.25	4.25	
1020	Special Foundations	N/A		—	—	—	
1030	Slab on Grade	6" reinforced concrete with vapor barrier and granular base		S.F. Slab	7.01	7.01	7.8%
2010	Basement Excavation	Site preparation for slab and trench for foundation wall and footing		S.F. Ground	.18	.18	
2020	Basement Walls	N/A		—	—	—	
B.	**SHELL**						
	B10 Superstructure						
1010	Floor Construction	Open web steel joists, slab form, concrete	(balcony)	S.F. Floor	23.04	2.88	3.0%
1020	Roof Construction	Metal deck on steel truss		S.F. Roof	1.51	1.51	
	B20 Exterior Enclosure						
2010	Exterior Walls	Precast concrete panel	80% of wall (adjusted for end walls)	S.F. Wall	58	29.70	
2020	Exterior Windows	Glass curtain wall	20% of wall	Each	42.95	5.50	26.0%
2030	Exterior Doors	Double aluminum and glass and hollow metal		Each	5883	2.93	
	B30 Roofing						
3010	Roof Coverings	Built-up tar and gravel with flashing; perlite/EPS composite insulation		S.F. Roof	6.55	6.55	4.6%
3020	Roof Openings	Roof hatches		S.F. Roof	.20	.20	
C.	**INTERIORS**						
1010	Partitions	Concrete Block and toilet partitions	40 S.F. Floor/L.F. Partition	S.F. Partition	10.40	4.16	
1020	Interior Doors	Single leaf hollow metal	400 S.F. Floor/Door	Each	1222	3.06	
1030	Fittings	N/A		—	—	—	
2010	Stair Construction	Concrete filled metal pan		Flight	18,900	2.37	19.0%
3010	Wall Finishes	70% paint, 30% epoxy coating		S.F. Surface	5.64	4.51	
3020	Floor Finishes	70% vinyl tile, 30% carpet		S.F. Floor	7.51	7.51	
3030	Ceiling Finishes	Fiberglass board, suspended		S.F. Ceiling	6.30	6.30	
D.	**SERVICES**						
	D10 Conveying						
1010	Elevators & Lifts	One hydraulic passenger elevator		Each	101,280	4.22	2.9%
1020	Escalators & Moving Walks	N/A		—	—	—	
	D20 Plumbing						
2010	Plumbing Fixtures	Toilet and service fixtures, supply and drainage	1 Fixture/800 S.F. Floor	Each	7328	9.16	
2020	Domestic Water Distribution	Gas fired water heater		S.F. Floor	4.25	4.25	10.0%
2040	Rain Water Drainage	Roof drains		S.F. Roof	1.34	1.34	
	D30 HVAC						
3010	Energy Supply	N/A		—	—	—	
3020	Heat Generating Systems	Included in D3050		—	—	—	
3030	Cooling Generating Systems	N/A		—	—	—	9.9%
3050	Terminal & Package Units	Single zone rooftop unit, gas heating, electric cooling		S.F. Floor	14.50	14.50	
3090	Other HVAC Sys. & Equipment	N/A		—	—	—	
	D40 Fire Protection						
4010	Sprinklers	Wet pipe sprinkler system		S.F. Floor	3.79	3.79	2.8%
4020	Standpipes	Standpipe		S.F. Floor	.39	.39	
	D50 Electrical						
5010	Electrical Service/Distribution	800 ampere service, panel board and feeders		S.F. Floor	1.89	1.89	
5020	Lighting & Branch Wiring	Fluorescent fixtures, receptacles, switches, A.C. and misc. power		S.F. Floor	12.88	12.88	14.0%
5030	Communications & Security	Addressable alarm systems, emergency lighting and public address system		S.F. Floor	4.22	4.22	
5090	Other Electrical Systems	Emergency generator, 100 kW		S.F. Floor	1.60	1.60	
E.	**EQUIPMENT & FURNISHINGS**						
1010	Commercial Equipment	N/A		—	—	—	
1020	Institutional Equipment	N/A		—	—	—	0.0%
1030	Vehicular Equipment	N/A		—	—	—	
1090	Other Equipment	N/A		—	—	—	
F.	**SPECIAL CONSTRUCTION**						
1020	Integrated Construction	N/A		—	—	—	0.0%
1040	Special Facilities	N/A		—	—	—	
G.	**BUILDING SITEWORK**	**N/A**					

			Sub-Total	146.86	**100%**
CONTRACTOR FEES (General Requirements: 10%, Overhead: 5%, Profit: 10%)			25%	36.74	
ARCHITECT FEES			7%	12.85	
		Total Building Cost		**196.45**	

For customer support on your Square Foot Costs with RSMeans data, call 800.448.8182.

87

Costs per square foot of floor area

Exterior Wall	S.F. Area	2000	2700	3400	4100	4800	5500	6200	6900	7600
	L.F. Perimeter	180	208	236	256	280	303	317	337	357
Face Brick and Concrete Block	Rigid Steel	276.90	260.45	250.75	242.30	237.20	233.20	228.65	225.90	223.60
	Reinforced Concrete	294.10	277.60	267.95	259.50	254.40	250.40	245.80	243.10	240.80
Precast Concrete	Rigid Steel	293.45	274.50	263.35	253.60	247.75	243.05	237.80	234.60	231.95
	Reinforced Concrete	311.25	292.25	281.10	271.35	265.50	260.85	255.55	252.35	249.70
Limestone and Concrete Block	Rigid Steel	305.35	284.70	272.70	262.00	255.65	250.60	244.75	241.30	238.45
	Reinforced Concrete	322.55	301.90	289.85	279.20	272.85	267.80	262.00	258.50	255.60
Perimeter Adj., Add or Deduct	Per 100 L.F.	51.85	38.40	30.45	25.30	21.60	18.80	16.70	15.00	13.65
Story Hgt. Adj., Add or Deduct	Per 1 Ft.	4.90	4.10	3.75	3.35	3.20	3.00	2.80	2.65	2.55

For Basement, add $36.60 per square foot of basement area

The above costs were calculated using the basic specifications shown on the facing page. These costs should be adjusted where necessary for design alternatives and owner's requirements.

Common additives

Description	Unit	$ Cost
Bulletproof Teller Window, 44" x 60"	Ea.	5850
60" x 48"	Ea.	7700
Closed Circuit Surveillance, One station		
Camera and monitor	Ea.	1475
For additional camera stations, add	Ea.	665
Counters, Complete	Station	6475
Door & Frame, 3' x 6'-8", bullet resistant steel		
with vision panel	Ea.	7375 - 9525
Drive-up Window, Drawer & micr., not incl. glass	Ea.	9525 - 13,300
Emergency Lighting, 25 watt, battery operated		
Lead battery	Ea.	330
Nickel cadmium	Ea.	550
Night Depository	Ea.	10,000 - 15,200
Package Receiver, painted	Ea.	1975
stainless steel	Ea.	3050
Partitions, Bullet resistant to 8' high	L.F.	345 - 545
Pneumatic Tube Systems, 2 station	Ea.	34,200
With TV viewer	Ea.	62,500

Description	Unit	$ Cost
Service Windows, Pass thru, steel		
24" x 36"	Ea.	4100
48" x 48"	Ea.	4175
72" x 40"	Ea.	6650
Smoke Detectors		
Ceiling type	Ea.	244
Duct type	Ea.	585
Twenty-four Hour Teller		
Automatic deposit cash & memo	Ea.	54,500
Vault Front, Door & frame		
1 hour test, 32"x 78"	Opng.	8000
2 hour test, 32" door	Opng.	9800
40" door	Opng.	11,300
4 hour test, 32" door	Opng.	12,100
40" door	Opng.	13,800
Time lock, two movement, add	Ea.	2400

Important: See the Reference Section for Location Factors.

Model costs calculated for a 1 story building with 14' story height and 4,100 square feet of floor area

				Unit	Unit Cost	Cost Per S.F.	% Of Sub-Total
A. SUBSTRUCTURE							
1010	Standard Foundations	Poured concrete; strip and spread footings; 4' foundation wall		S.F. Ground	11.37	11.37	
1020	Special Foundations	N/A		—	—	—	
1030	Slab on Grade	4" reinforced concrete with vapor barrier and granular base		S.F. Slab	5.80	5.80	9.1%
2010	Basement Excavation	Site preparation for slab and trench for foundation wall and footing		S.F. Ground	.33	.33	
2020	Basement Walls	N/A		—	—	—	
B. SHELL							
	B10 Superstructure						
1010	Floor Construction	Cast-in-place columns		L.F. Column	131	6.39	12.3%
1020	Roof Construction	Cast-in-place concrete flat plate		S.F. Roof	17.25	17.25	
	B20 Exterior Enclosure						
2010	Exterior Walls	Face brick with concrete block backup	80% of wall	S.F. Wall	35.25	24.65	
2020	Exterior Windows	Horizontal aluminum sliding	20% of wall	Each	589	6.86	17.6%
2030	Exterior Doors	Double aluminum and glass and hollow metal		Each	4585	2.24	
	B30 Roofing						
3010	Roof Coverings	Built-up tar and gravel with flashing; perlite/EPS composite insulation		S.F. Roof	8.21	8.21	4.3%
3020	Roof Openings	N/A		—	—	—	
C. INTERIORS							
1010	Partitions	Gypsum board on metal studs	20 S.F. of Floor/L.F. Partition	S.F. Partition	11.74	5.87	
1020	Interior Doors	Single leaf hollow metal	200 S.F. Floor/Door	Each	1222	6.11	
1030	Fittings	N/A		—	—	—	
2010	Stair Construction	N/A		—	—	—	14.0%
3010	Wall Finishes	50% vinyl wall covering, 50% paint		S.F. Surface	1.49	1.49	
3020	Floor Finishes	50% carpet tile, 40% vinyl composition tile, 10% quarry tile		S.F. Floor	5.79	5.79	
3030	Ceiling Finishes	Mineral fiber tile on concealed zee bars		S.F. Ceiling	7.60	7.60	
D. SERVICES							
	D10 Conveying						
1010	Elevators & Lifts	N/A		—	—	—	0.0 %
1020	Escalators & Moving Walks	N/A		—	—	—	
	D20 Plumbing						
2010	Plumbing Fixtures	Toilet and service fixtures, supply and drainage	1 Fixture/580 S.F. Floor	Each	5980	10.31	
2020	Domestic Water Distribution	Gas fired water heater		S.F. Floor	1.74	1.74	7.1%
2040	Rain Water Drainage	Roof drains		S.F. Roof	1.67	1.67	
	D30 HVAC						
3010	Energy Supply	N/A		—	—	—	
3020	Heat Generating Systems	Included in D3050		—	—	—	
3030	Cooling Generating Systems	N/A		—	—	—	7.4 %
3050	Terminal & Package Units	Single zone rooftop unit, gas heating, electric cooling		S.F. Floor	14.30	14.30	
3090	Other HVAC Sys. & Equipment	N/A		—	—	—	
	D40 Fire Protection						
4010	Sprinklers	Wet pipe sprinkler system		S.F. Floor	5.13	5.13	4.4%
4020	Standpipes	Standpipe		S.F. Floor	3.25	3.25	
	D50 Electrical						
5010	Electrical Service/Distribution	200 ampere service, panel board and feeders		S.F. Floor	2.74	2.74	
5020	Lighting & Branch Wiring	Fluorescent fixtures, receptacles, switches, A.C. and misc. power		S.F. Floor	8.06	8.06	11.5%
5030	Communications & Security	Alarm systems, internet and phone wiring, emergency lighting, and security television		S.F. Floor	11.14	11.14	
5090	Other Electrical Systems	Emergency generator, 15 kW, Uninterruptible power supply		S.F. Floor	.23	.23	
E. EQUIPMENT & FURNISHINGS							
1010	Commercial Equipment	Automatic teller, drive up window, night depository		S.F. Floor	8.37	8.37	
1020	Institutional Equipment	Closed circuit TV monitoring system		S.F. Floor	2.95	2.95	5.9%
1030	Vehicular Equipment	N/A		—	—	—	
1090	Other Equipment	N/A		—	—	—	
F. SPECIAL CONSTRUCTION							
1020	Integrated Construction	N/A		—	—	—	6.4 %
1040	Special Facilities	Security vault door		S.F. Floor	12.36	12.36	
G. BUILDING SITEWORK	**N/A**						

			Sub-Total	192.21	100%
CONTRACTOR FEES (General Requirements: 10%, Overhead: 5%, Profit: 10%)			25%	48.07	
ARCHITECT FEES			8%	19.22	
		Total Building Cost		**259.50**	

For customer support on your Square Foot Costs with RSMeans data, call 800.448.8182.

89

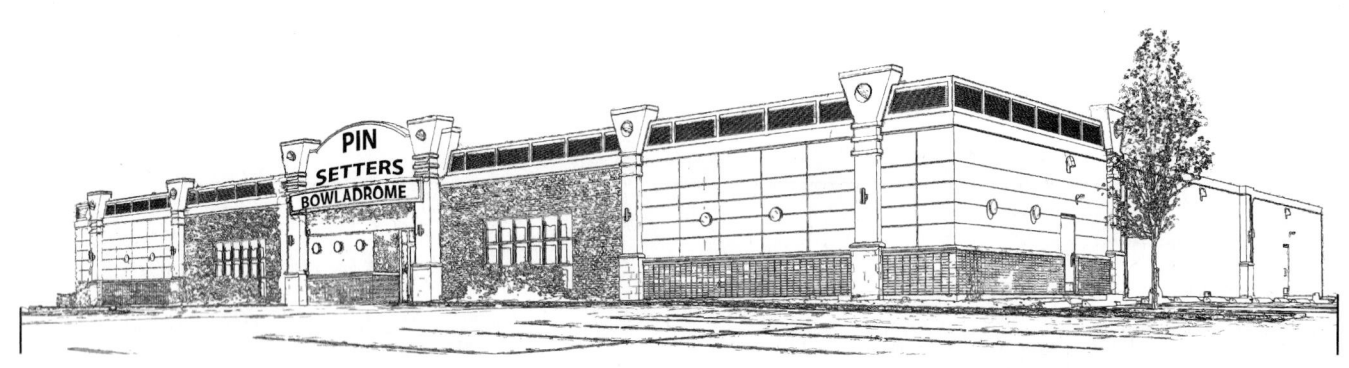

Costs per square foot of floor area

Exterior Wall	S.F. Area	12000	14000	16000	18000	20000	22000	24000	26000	28000
	L.F. Perimeter	460	491	520	540	566	593	620	645	670
Concrete Block	Steel Roof Deck	184.25	180.85	178.35	176.00	174.30	173.00	171.85	170.90	170.05
Decorative Concrete Block	Steel Roof Deck	188.90	185.20	182.30	179.65	177.75	176.30	175.05	173.95	173.00
Face Brick and Concrete Block	Steel Roof Deck	197.60	193.10	189.65	186.45	184.15	182.40	180.90	179.55	178.40
Jumbo Brick and Concrete Block	Steel Roof Deck	193.75	189.60	186.45	183.45	181.35	179.70	178.30	177.10	176.00
Stucco and Concrete Block	Steel Roof Deck	188.00	184.30	181.55	178.95	177.10	175.60	174.40	173.30	172.40
Precast Concrete	Steel Roof Deck	196.15	191.80	188.50	185.35	183.15	181.40	179.95	178.65	177.55
Perimeter Adj., Add or Deduct	Per 100 L.F.	5.55	4.90	4.25	3.75	3.45	3.10	2.90	2.60	2.40
Story Hgt. Adj., Add or Deduct	Per 1 Ft.	1.20	1.20	1.05	0.95	0.95	0.90	0.90	0.80	0.80
For Basement, add $36.10 per square foot of basement area										

The above costs were calculated using the basic specifications shown on the facing page. These costs should be adjusted where necessary for design alternatives and owner's requirements.

Common additives

Description	Unit	$ Cost
Bowling Alleys, Incl. alley, pinsetter,		
scorer, counter & misc. supplies, average	Lane	67,500
For automatic scorer, add (maximum)	Lane	11,200
Emergency Lighting, 25 watt, battery operated		
Lead battery	Ea.	330
Nickel cadmium	Ea.	550
Lockers, Steel, Single tier, 60" or 72"	Opng.	232 - 395
2 tier, 60" or 72" total	Opng.	139 - 172
5 tier, box lockers	Opng.	73 - 87
Locker bench, lam., maple top only	L.F.	36.50
Pedestals, steel pipe	Ea.	79
Seating		
Auditorium chair, all veneer	Ea.	355
Veneer back, padded seat	Ea.	380
Upholstered, spring seat	Ea.	335
Sound System		
Amplifier, 250 watts	Ea.	2025
Speaker, ceiling or wall	Ea.	240
Trumpet	Ea.	460

Important: See the Reference Section for Location Factors.

Model costs calculated for a 1 story building with 14' story height and 20,000 square feet of floor area

Bowling Alley

				Unit	Unit Cost	Cost Per S.F.	% Of Sub-Total
A. SUBSTRUCTURE							
1010	Standard Foundations	Poured concrete; strip and spread footings; 4' foundation wall		S.F. Ground	3.89	3.89	
1020	Special Foundations	N/A		—	—	—	
1030	Slab on Grade	4" reinforced concrete with vapor barrier and granular base		S.F. Slab	5.80	5.80	7.8%
2010	Basement Excavation	Site preparation for slab and trench for foundation wall and footing		S.F. Ground	.57	.57	
2020	Basement Walls	N/A		—	—	—	
B. SHELL							
	B10 Superstructure						
1010	Floor Construction	N/A		—	—	—	8.0 %
1020	Roof Construction	Metal deck on open web steel joists with columns and beams		S.F. Roof	10.52	10.52	
	B20 Exterior Enclosure						
2010	Exterior Walls	Concrete block	90% of wall	S.F. Wall	14.39	5.13	
2020	Exterior Windows	Horizontal pivoted	10% of wall	Each	2055	3.39	7.1%
2030	Exterior Doors	Double aluminum and glass and hollow metal		Each	3526	.88	
	B30 Roofing						
3010	Roof Coverings	Built-up tar and gravel with flashing; perlite/EPS composite insulation		S.F. Roof	6.29	6.29	4.9%
3020	Roof Openings	Roof hatches		S.F. Roof	.20	.20	
C. INTERIORS							
1010	Partitions	Concrete block	50 S.F. Floor/L.F. Partition	S.F. Partition	11.55	2.31	
1020	Interior Doors	Hollow metal single leaf	1000 S.F. Floor/Door	Each	1222	1.22	
1030	Fittings	N/A		—	—	—	
2010	Stair Construction	N/A		—	—	—	6.4%
3010	Wall Finishes	Paint and block filler		S.F. Surface	6.53	2.61	
3020	Floor Finishes	Vinyl tile	25% of floor	S.F. Floor	2.76	.69	
3030	Ceiling Finishes	Suspended fiberglass board	25% of area	S.F. Ceiling	6.30	1.57	
D. SERVICES							
	D10 Conveying						
1010	Elevators & Lifts	N/A		—	—	—	0.0 %
1020	Escalators & Moving Walks	N/A		—	—	—	
	D20 Plumbing						
2010	Plumbing Fixtures	Toilet and service fixtures, supply and drainage	1 Fixture/2200 S.F. Floor	Each	3740	1.70	
2020	Domestic Water Distribution	Gas fired water heater		S.F. Floor	.32	.32	2.1%
2040	Rain Water Drainage	Roof drains		S.F. Roof	.72	.72	
	D30 HVAC						
3010	Energy Supply	N/A		—	—	—	
3020	Heat Generating Systems	Included in D3050		—	—	—	
3030	Cooling Generating Systems	N/A		—	—	—	12.5 %
3050	Terminal & Package Units	Single zone rooftop unit, gas heating, electric cooling		S.F. Floor	16.45	16.45	
3090	Other HVAC Sys. & Equipment	N/A		—	—	—	
	D40 Fire Protection						
4010	Sprinklers	Sprinklers, light hazard		S.F. Floor	3.79	3.79	3.4%
4020	Standpipes	Standpipe, wet, Class III		S.F. Floor	.67	.67	
	D50 Electrical						
5010	Electrical Service/Distribution	800 ampere service, panel board and feeders		S.F. Floor	2.27	2.27	
5020	Lighting & Branch Wiring	High efficiency fluorescent fixtures, receptacles, switches, A.C. and misc. power		S.F. Floor	11.43	11.43	11.5%
5030	Communications & Security	Addressable alarm systems and emergency lighting		S.F. Floor	1.38	1.38	
5090	Other Electrical Systems	Emergency generator, 7.5 kW		S.F. Floor	.09	.09	
E. EQUIPMENT & FURNISHINGS							
1010	Commercial Equipment	N/A		—	—	—	
1020	Institutional Equipment	N/A		—	—	—	
1030	Vehicular Equipment	N/A		—	—	—	36.2 %
1090	Other Equipment	Bowling Alley, ballrack, automatic scorers		S.F. Floor	47.66	47.66	
F. SPECIAL CONSTRUCTION							
1020	Integrated Construction	N/A		—	—	—	0.0 %
1040	Special Facilities	N/A		—	—	—	
G. BUILDING SITEWORK	**N/A**						

			Sub-Total	131.55	100%
	CONTRACTOR FEES (General Requirements: 10%, Overhead: 5%, Profit: 10%)		25%	32.88	
	ARCHITECT FEES		6%	9.87	
		Total Building Cost		**174.30**	

Costs per square foot of floor area

Exterior Wall	S.F. Area	6000	8000	10000	12000	14000	16000	18000	20000	22000
	L.F. Perimeter	320	386	453	475	505	530	570	600	630
Face Brick and Concrete Block	Bearing Walls	181.25	174.65	170.85	164.85	161.10	158.05	156.35	154.60	153.10
	Rigid Steel	184.35	177.75	174.05	168.15	164.45	161.45	159.80	158.05	156.60
Decorative Concrete Block	Bearing Walls	171.85	166.15	162.85	157.90	154.75	152.20	150.80	149.30	148.05
	Rigid Steel	174.95	169.30	166.10	161.25	158.10	155.60	154.25	152.75	151.55
Precast Concrete	Bearing Walls	184.75	177.85	173.85	167.50	163.45	160.15	158.40	156.55	155.00
	Rigid Steel	187.85	181.00	177.00	170.80	166.80	163.60	161.85	160.05	158.50
Perimeter Adj., Add or Deduct	Per 100 L.F.	15.20	11.40	9.15	7.60	6.50	5.60	5.10	4.55	4.20
Story Hgt. Adj., Add or Deduct	Per 1 Ft.	2.45	2.25	2.10	1.90	1.70	1.50	1.50	1.45	1.35
Basement—Not Applicable										

The above costs were calculated using the basic specifications shown on the facing page. These costs should be adjusted where necessary for design alternatives and owner's requirements.

Common additives

Description	Unit	$ Cost
Directory Boards, Plastic, glass covered		
30" x 20"	Ea.	640
36" x 48"	Ea.	1625
Aluminum, 24" x 18"	Ea.	630
36" x 24"	Ea.	790
48" x 32"	Ea.	1025
48" x 60"	Ea.	2200
Emergency Lighting, 25 watt, battery operated		
Lead battery	Ea.	330
Nickel cadmium	Ea.	550
Benches, Hardwood	L.F.	141 - 233
Ticket Printer	Ea.	8025
Turnstiles, One way		
4 arm, 46" dia., manual	Ea.	2300
Electric	Ea.	3425
High security, 3 arm		
65" dia., manual	Ea.	7775
Electric	Ea.	10,800
Gate with horizontal bars		
65" dia., 7' high transit type	Ea.	7450

Important: See the Reference Section for Location Factors.

Model costs calculated for a 1 story building with 14' story height and 12,000 square feet of floor area

				Unit	Unit Cost	Cost Per S.F.	% Of Sub-Total
A. SUBSTRUCTURE							
1010	Standard Foundations	Poured concrete; strip and spread footings; 4' foundation wall		S.F. Ground	6.30	6.30	
1020	Special Foundations	N/A		—	—	—	
1030	Slab on Grade	6" reinforced concrete with vapor barrier and granular base		S.F. Slab	8.15	8.15	11.9%
2010	Basement Excavation	Site preparation for slab and trench for foundation wall and footing		S.F. Ground	.33	.33	
2020	Basement Walls	N/A		—	—	—	
B. SHELL							
B10 Superstructure							
1010	Floor Construction	N/A		—	—	—	5.8 %
1020	Roof Construction	Metal deck on open web steel joists with columns and beams		S.F. Roof	7.19	7.19	
B20 Exterior Enclosure							
2010	Exterior Walls	Face brick with concrete block backup	70% of wall	S.F. Wall	35.24	13.67	
2020	Exterior Windows	Store front	30% of wall	Each	33.85	5.63	17.4%
2030	Exterior Doors	Double aluminum and glass		Each	7200	2.40	
B30 Roofing							
3010	Roof Coverings	Built-up tar and gravel with flashing; perlite/EPS composite insulation		S.F. Roof	7.16	7.16	5.8%
3020	Roof Openings	N/A		—	—	—	
C. INTERIORS							
1010	Partitions	Lightweight concrete block	15 S.F. Floor/L.F. Partition	S.F. Partition	11.58	7.72	
1020	Interior Doors	Hollow metal	150 S.F. Floor/Door	Each	1222	.81	
1030	Fittings	N/A		—	—	—	
2010	Stair Construction	N/A		—	—	—	22.5%
3010	Wall Finishes	Glazed coating		S.F. Surface	2.12	2.83	
3020	Floor Finishes	Quarry tile and vinyl composition tile		S.F. Floor	10.34	10.34	
3030	Ceiling Finishes	Mineral fiber tile on concealed zee bars		S.F. Ceiling	6.30	6.30	
D. SERVICES							
D10 Conveying							
1010	Elevators & Lifts	N/A		—	—	—	0.0 %
1020	Escalators & Moving Walks	N/A		—	—	—	
D20 Plumbing							
2010	Plumbing Fixtures	Toilet and service fixtures, supply and drainage	1 Fixture/850 S.F. Floor	Each	7812	9.19	
2020	Domestic Water Distribution	Electric water heater		S.F. Floor	2.36	2.36	10.3%
2040	Rain Water Drainage	Roof drains		S.F. Roof	1.29	1.29	
D30 HVAC							
3010	Energy Supply	N/A		—	—	—	
3020	Heat Generating Systems	Included in D3050		—	—	—	
3030	Cooling Generating Systems	N/A		—	—	—	10.1 %
3050	Terminal & Package Units	Single zone rooftop unit, gas heating, electric cooling		S.F. Floor	12.55	12.55	
3090	Other HVAC Sys. & Equipment	N/A		—	—	—	
D40 Fire Protection							
4010	Sprinklers	Wet pipe sprinkler system		S.F. Floor	4.96	4.96	4.9%
4020	Standpipes	Standpipe		S.F. Floor	1.11	1.11	
D50 Electrical							
5010	Electrical Service/Distribution	400 ampere service, panel board and feeders		S.F. Floor	1.89	1.89	
5020	Lighting & Branch Wiring	Fluorescent fixtures, receptacles, switches, A.C. and misc. power		S.F. Floor	9.69	9.69	11.4%
5030	Communications & Security	Alarm systems and emergency lighting		S.F. Floor	2.11	2.11	
5090	Other Electrical Systems	Emergency generator, 7.5 kW		S.F. Floor	.45	.45	
E. EQUIPMENT & FURNISHINGS							
1010	Commercial Equipment	N/A		—	—	—	
1020	Institutional Equipment	N/A		—	—	—	
1030	Vehicular Equipment	N/A		—	—	—	0.0 %
1090	Other Equipment	N/A		—	—	—	
F. SPECIAL CONSTRUCTION							
1020	Integrated Construction	N/A		—	—	—	0.0 %
1040	Special Facilities	N/A		—	—	—	
G. BUILDING SITEWORK	**N/A**						

		Sub-Total	124.43	100%
CONTRACTOR FEES (General Requirements: 10%, Overhead: 5%, Profit: 10%)		25%	31.09	
ARCHITECT FEES		6%	9.33	
	Total Building Cost		**164.85**	

For customer support on your Square Foot Costs with RSMeans data, call 800.448.8182.

93

Costs per square foot of floor area

Exterior Wall	S.F. Area	600	800	1000	1200	1600	2000	2400	3000	4000
	L.F. Perimeter	100	114	128	139	164	189	210	235	280
Face Brick and Concrete Block	Rigid Steel	369.60	354.10	344.80	337.05	328.45	323.35	318.85	313.00	307.70
	Bearing Walls	362.05	346.60	337.25	329.55	320.95	315.80	311.35	305.45	300.15
Concrete Block	Rigid Steel	324.00	315.10	309.80	305.40	300.45	297.50	294.95	291.55	288.50
	Bearing Walls	316.50	307.60	302.25	297.85	292.90	289.95	287.40	284.05	280.95
Metal Panel	Rigid Steel	327.70	318.30	312.70	307.90	302.75	299.65	296.95	293.35	290.15
Metal Sandwich Panel	Rigid Steel	328.60	318.95	313.10	308.20	302.80	299.60	296.80	293.10	289.75
Perimeter Adj., Add or Deduct	Per 100 L.F.	106.75	80.05	64.05	53.35	40.05	32.05	26.75	21.35	16.00
Story Hgt. Adj., Add or Deduct	Per 1 Ft.	6.50	5.60	5.05	4.55	4.05	3.70	3.50	3.05	2.75
Basement—Not Applicable										

The above costs were calculated using the basic specifications shown on the facing page. These costs should be adjusted where necessary for design alternatives and owner's requirements.

Common additives

Description	Unit	$ Cost
Air Compressors, Electric		
1-1/2 H.P., standard controls	Ea.	1700
Dual controls	Ea.	2300
5 H.P., 115/230 volt, standard control	Ea.	4000
Dual controls	Ea.	5300
Emergency Lighting, 25 watt, battery operated		
Lead battery	Ea.	330
Nickel cadmium	Ea.	550
Fence, Chain link, 6' high		
9 ga. wire, galvanized	L.F.	29
6 ga. wire	L.F.	34.50
Gate	Ea.	400
Product Dispenser with vapor recovery		
for 6 nozzles	Ea.	27,900
Laundry Equipment		
Dryer, gas, 16 lb. capacity	Ea.	1025
30 lb. capacity	Ea.	4100
Washer, 4 cycle	Ea.	1325
Commercial	Ea.	1700

Description	Unit	$ Cost
Lockers, Steel, single tier, 60" or 72"	Opng.	232 - 395
2 tier, 60" or 72" total	Opng.	139 - 172
5 tier, box lockers	Opng.	73 - 87
Locker bench, lam. maple top only	L.F.	36.50
Pedestals, steel pipe	Ea.	79
Paving, Bituminous		
Wearing course plus base course	S.Y.	7.40
Safe, Office type, 1 hour rating		
30" x 18" x 18"	Ea.	2475
60" x 36" x 18", double door	Ea.	10,200
Sidewalks, Concrete 4" thick	S.F.	4.99
Yard Lighting, 20' aluminum pole with 400 watt high pressure sodium fixture	Ea.	3600

Important: See the Reference Section for Location Factors.

Model costs calculated for a 1 story building with 12' story height and 800 square feet of floor area

			Unit	Unit Cost	Cost Per S.F.	% Of Sub-Total
A. SUBSTRUCTURE						
1010	Standard Foundations	Poured concrete; strip and spread footings; 4' foundation wall	S.F. Ground	15.16	15.16	
1020	Special Foundations	N/A	—	—	—	
1030	Slab on Grade	5" reinforced concrete with vapor barrier and granular base	S.F. Slab	7.47	7.47	8.9%
2010	Basement Excavation	Site preparation for slab and trench for foundation wall and footing	S.F. Ground	1.06	1.06	
2020	Basement Walls	N/A	—	—	—	
B. SHELL						
B10 Superstructure						
1010	Floor Construction	N/A	—	—	—	3.6 %
1020	Roof Construction	Metal deck, open web steel joists, beams, columns	S.F. Roof	9.58	9.58	
B20 Exterior Enclosure						
2010	Exterior Walls	Face brick with concrete block backup 70% of wall	S.F. Wall	35.25	42.20	
2020	Exterior Windows	Horizontal pivoted steel 5% of wall	Each	826	7.85	27.4%
2030	Exterior Doors	Steel overhead hollow metal	Each	4520	22.60	
B30 Roofing						
3010	Roof Coverings	Built-up tar and gravel with flashing; perlite/EPS composite insulation	S.F. Roof	9.35	9.35	3.5%
3020	Roof Openings	N/A	—	—	—	
C. INTERIORS						
1010	Partitions	Concrete block 20 S.F. Floor/S.F. Partition	S.F. Partition	10.90	5.45	
1020	Interior Doors	Hollow metal 600 S.F. Floor/Door	Each	1222	2.03	
1030	Fittings	N/A	—	—	—	
2010	Stair Construction	N/A	—	—	—	3.4%
3010	Wall Finishes	Paint	S.F. Surface	1.39	1.39	
3020	Floor Finishes	N/A	—	—	—	
3030	Ceiling Finishes	N/A	—	—	—	
D. SERVICES						
D10 Conveying						
1010	Elevators & Lifts	N/A	—	—	—	0.0 %
1020	Escalators & Moving Walks	N/A	—	—	—	
D20 Plumbing						
2010	Plumbing Fixtures	Toilet and service fixtures, supply and drainage 1 Fixture/160 S.F. Floor	Each	2962	18.51	
2020	Domestic Water Distribution	Gas fired water heater	S.F. Floor	39.58	39.58	23.7%
2040	Rain Water Drainage	Roof drains	S.F. Roof	4.72	4.72	
D30 HVAC						
3010	Energy Supply	N/A	—	—	—	
3020	Heat Generating Systems	N/A	—	—	—	
3030	Cooling Generating Systems	N/A	—	—	—	3.7 %
3050	Terminal & Package Units	Single zone rooftop unit, gas heating, electric cooling	S.F. Floor	9.75	9.75	
3090	Other HVAC Sys. & Equipment	N/A	—	—	—	
D40 Fire Protection						
4010	Sprinklers	N/A	—	—	—	0.0 %
4020	Standpipes	N/A	—	—	—	
D50 Electrical						
5010	Electrical Service/Distribution	400 ampere service, panel board and feeders	S.F. Floor	25.23	25.23	
5020	Lighting & Branch Wiring	Fluorescent fixtures, receptacles, switches and misc. power	S.F. Floor	40.90	40.90	25.7%
5030	Communications & Security	N/A	—	—	—	
5090	Other Electrical Systems	Emergency generator, 7.5 kW	S.F. Floor	1.94	1.94	
E. EQUIPMENT & FURNISHINGS						
1010	Commercial Equipment	N/A	—	—	—	
1020	Institutional Equipment	N/A	—	—	—	0.0 %
1030	Vehicular Equipment	N/A	—	—	—	
1090	Other Equipment	N/A	—	—	—	
F. SPECIAL CONSTRUCTION						
1020	Integrated Construction	N/A	—	—	—	0.0 %
1040	Special Facilities	N/A	—	—	—	
G. BUILDING SITEWORK	**N/A**					

		Sub-Total	264.77	100%
CONTRACTOR FEES (General Requirements: 10%, Overhead: 5%, Profit: 10%)		25%	66.16	
ARCHITECT FEES		7%	23.17	
		Total Building Cost	354.10	

Costs per square foot of floor area

Exterior Wall	S.F. Area	2000	7000	12000	17000	22000	27000	32000	37000	42000
	L.F. Perimeter	180	340	470	540	640	740	762	793	860
Decorative Concrete Block	Wood Arch	278.45	205.50	189.70	178.15	173.85	171.10	165.70	162.10	160.60
	Steel Truss	280.05	207.10	191.30	179.75	175.45	172.75	167.30	163.70	162.20
Stone and Concrete Block	Wood Arch	345.20	241.50	218.75	201.65	195.40	191.40	183.30	177.95	175.80
	Steel Truss	346.75	243.15	220.40	203.30	197.05	193.05	184.95	179.60	177.40
Face Brick and Concrete Block	Wood Arch	310.15	222.60	203.50	189.35	184.10	180.75	174.05	169.65	167.85
	Steel Truss	311.80	224.25	205.15	190.95	185.75	182.40	175.70	171.25	169.45
Perimeter Adj., Add or Deduct	Per 100 L.F.	72.00	20.65	12.00	8.45	6.45	5.35	4.50	3.90	3.50
Story Hgt. Adj., Add or Deduct	Per 1 Ft.	4.20	2.30	1.85	1.45	1.35	1.25	1.10	1.00	0.95

For Basement, add $37.70 per square foot of basement area

The above costs were calculated using the basic specifications shown on the facing page. These costs should be adjusted where necessary for design alternatives and owner's requirements.

Common additives

Description	Unit	$ Cost
Altar, Wood, custom design, plain	Ea.	3300
Deluxe	Ea.	16,800
Granite or marble, average	Ea.	17,100
Deluxe	Ea.	46,100
Ark, Prefabricated, plain	Ea.	11,800
Deluxe	Ea.	151,000
Baptistry, Fiberglass, incl. plumbing	Ea.	7700 - 12,600
Bells & Carillons, 48 bells	Ea.	1,168,500
24 bells	Ea.	550,000
Confessional, Prefabricated wood		
Single, plain	Ea.	4550
Deluxe	Ea.	11,000
Double, plain	Ea.	8475
Deluxe	Ea.	23,800
Emergency Lighting, 25 watt, battery operated		
Lead battery	Ea.	330
Nickel cadmium	Ea.	550
Lecterns, Wood, plain	Ea.	940
Deluxe	Ea.	6775

Description	Unit	$ Cost
Pews/Benches, Hardwood	L.F.	141 - 233
Pulpits, Prefabricated, hardwood	Ea.	2000 - 11,600
Railing, Hardwood	L.F.	257
Steeples, translucent fiberglass		
30" square, 15' high	Ea.	12,000
25' high	Ea.	13,700
Painted fiberglass, 24" square, 14' high	Ea.	10,200
28' high	Ea.	9000
Aluminum		
20' high, 3'- 6" base	Ea.	11,800
35' high, 8'- 0" base	Ea.	42,200
60' high, 14'- 0" base	Ea.	84,000

Important: See the Reference Section for Location Factors.

Model costs calculated for a 1 story building with 24' story height and 17,000 square feet of floor area

				Unit	Unit Cost	Cost Per S.F.	% Of Sub-Total
A.	**SUBSTRUCTURE**						
1010	Standard Foundations	Poured concrete; strip and spread footings; 4' foundation wall		S.F. Ground	8.65	8.65	
1020	Special Foundations	N/A		—	—	—	
1030	Slab on Grade	4" reinforced concrete with vapor barrier and granular base		S.F. Slab	5.80	5.80	10.6%
2010	Basement Excavation	Site preparation for slab and trench for foundation wall and footing		S.F. Ground	.18	.18	
2020	Basement Walls	N/A		—	—	—	
B.	**SHELL**						
	B10 Superstructure						
1010	Floor Construction	N/A		—	—	—	16.1 %
1020	Roof Construction	Wood deck on laminated wood arches		S.F. Roof	20.78	22.17	
	B20 Exterior Enclosure						
2010	Exterior Walls	Face brick with concrete block backup	80% of wall (adjusted for end walls)	S.F. Wall	41.86	25.53	
2020	Exterior Windows	Aluminum, top hinged, in-swinging and curtain wall panels	20% of wall	S.F. Window	33.45	5.10	22.9%
2030	Exterior Doors	Double hollow metal swinging, single hollow metal		Each	2385	.84	
	B30 Roofing						
3010	Roof Coverings	Asphalt shingles with flashing; polystyrene insulation		S.F. Roof	6.52	6.52	4.7%
3020	Roof Openings	N/A		—	—	—	
C.	**INTERIORS**						
1010	Partitions	Plaster on metal studs	40 S.F. Floor/L.F. Partitions	S.F. Partition	16	9.60	
1020	Interior Doors	Hollow metal	400 S.F. Floor/Door	Each	1222	3.06	
1030	Fittings	N/A		—	—	—	
2010	Stair Construction	N/A		—	—	—	14.1%
3010	Wall Finishes	Paint		S.F. Surface	1.27	1.52	
3020	Floor Finishes	Carpet		S.F. Floor	5.24	5.24	
3030	Ceiling Finishes	N/A		—	—	—	
D.	**SERVICES**						
	D10 Conveying						
1010	Elevators & Lifts	N/A		—	—	—	0.0 %
1020	Escalators & Moving Walks	N/A		—	—	—	
	D20 Plumbing						
2010	Plumbing Fixtures	Kitchen, toilet and service fixtures, supply and drainage	1 Fixture/2430 S.F. Floor	Each	8578	3.53	
2020	Domestic Water Distribution	Gas fired hot water heater		S.F. Floor	.41	.41	2.9%
2040	Rain Water Drainage	N/A		—	—	—	
	D30 HVAC						
3010	Energy Supply	Oil fired hot water, wall fin radiation		S.F. Floor	10.89	10.89	
3020	Heat Generating Systems	N/A		—	—	—	
3030	Cooling Generating Systems	N/A		—	—	—	15.7%
3050	Terminal & Package Units	Split systems with air cooled condensing units		S.F. Floor	10.79	10.79	
3090	Other HVAC Sys. & Equipment	N/A		—	—	—	
	D40 Fire Protection						
4010	Sprinklers	Wet pipe sprinkler system		S.F. Floor	3.79	3.79	3.7%
4020	Standpipes	Standpipe		S.F. Floor	1.34	1.34	
	D50 Electrical						
5010	Electrical Service/Distribution	400 ampere service, panel board and feeders		S.F. Floor	.81	.81	
5020	Lighting & Branch Wiring	Fluorescent fixtures, receptacles, switches, A.C. and misc. power		S.F. Floor	8.71	8.71	9.3%
5030	Communications & Security	Alarm systems, sound system and emergency lighting		S.F. Floor	3.09	3.09	
5090	Other Electrical Systems	Emergency generator, 7.5 kW		S.F. Floor	.13	.13	
E.	**EQUIPMENT & FURNISHINGS**						
1010	Commercial Equipment	N/A		—	—	—	
1020	Institutional Equipment	N/A		—	—	—	0.0 %
1030	Vehicular Equipment	N/A		—	—	—	
1090	Other Equipment	N/A		—	—	—	
F.	**SPECIAL CONSTRUCTION**						
1020	Integrated Construction	N/A		—	—	—	0.0 %
1040	Special Facilities	N/A		—	—	—	
G.	**BUILDING SITEWORK**	**N/A**					

		Sub-Total	137.70	100%
CONTRACTOR FEES (General Requirements: 10%, Overhead: 5%, Profit: 10%)		25%	34.44	
ARCHITECT FEES		10%	17.21	
	Total Building Cost		**189.35**	

For customer support on your Square Foot Costs with RSMeans data, call 800.448.8182.

97

Costs per square foot of floor area

Exterior Wall	S.F. Area	2000	4000	6000	8000	12000	15000	18000	20000	22000
	L.F. Perimeter	180	280	340	386	460	535	560	600	640
Stone Ashlar and Concrete Block	Wood Truss	275.40	246.40	230.20	220.35	209.05	205.85	200.95	199.60	198.55
	Steel Joists	280.65	251.00	234.35	224.25	212.60	209.30	204.20	202.95	201.80
Stucco and Concrete Block	Wood Truss	245.55	223.25	211.40	204.35	196.35	194.00	190.60	189.70	188.95
	Steel Joists	250.80	227.85	215.55	208.20	199.85	197.45	193.90	192.95	192.15
Brick Veneer	Wood Frame	255.55	231.00	217.75	209.70	200.60	197.95	194.10	193.05	192.20
Wood Shingles	Wood Frame	242.80	221.05	209.65	202.85	195.20	192.95	189.65	188.75	188.00
Perimeter Adj., Add or Deduct	Per 100 L.F.	49.20	24.65	16.40	12.30	8.20	6.50	5.45	4.90	4.50
Story Hgt. Adj., Add or Deduct	Per 1 Ft.	5.65	4.45	3.55	3.05	2.40	2.20	1.95	1.90	1.85

For Basement, add $33.10 per square foot of basement area

The above costs were calculated using the basic specifications shown on the facing page. These costs should be adjusted where necessary for design alternatives and owner's requirements.

Common additives

Description	Unit	$ Cost	Description	Unit	$ Cost
Bar, Front bar	L.F.	450	Sauna, Prefabricated, complete, 6' x 4'	Ea.	6600
Back bar	L.F.	360	6' x 9'	Ea.	10,000
Booth, Upholstered, custom, straight	L.F.	258 - 475	8' x 8'	Ea.	10,600
"L" or "U" shaped	L.F.	267 - 450	10' x 12'	Ea.	15,500
Fireplaces, Brick not incl. chimney			Smoke Detectors		
or foundation, 30" x 24" opening	Ea.	3400	Ceiling type	Ea.	244
Chimney, standard brick			Duct type	Ea.	585
Single flue 16" x 20"	V.L.F.	113	Sound System		
20" x 20"	V.L.F.	133	Amplifier, 250 watts	Ea.	2025
2 flue, 20" x 32"	V.L.F.	203	Speaker, ceiling or wall	Ea.	240
Lockers, Steel, single tier, 60" or 72"	Opng.	232 - 395	Trumpet	Ea.	460
2 tier, 60" or 72" total	Opng.	139 - 172	Steam Bath, Complete, to 140 C.F.	Ea.	3175
5 tier, box lockers	Opng.	73 - 87	To 300 C.F.	Ea.	3525
Locker bench, lam. maple top only	L.F.	36.50	To 800 C.F.	Ea.	7125
Pedestals, steel pipe	Ea.	79	To 2500 C.F.	Ea.	9725
Refrigerators, Prefabricated, walk-in			Swimming Pool Complete, gunite	S.F.	106 - 132
7'-6" High, 6' x 6'	S.F.	145	Tennis Court, Complete with fence		
10' x 10'	S.F.	114	Bituminous	Ea.	52,500 - 94,000
12' x 14'	S.F.	99.50	Clay	Ea.	56,500 - 97,500
12' x 20'	S.F.	137			

Important: See the Reference Section for Location Factors.

Model costs calculated for a 1 story building with 12' story height and 6,000 square feet of floor area

				Unit	Unit Cost	Cost Per S.F.	% Of Sub-Total
A. SUBSTRUCTURE							
1010	Standard Foundations	Poured concrete; strip and spread footings; 4' foundation wall		S.F. Ground	9.31	9.31	
1020	Special Foundations	N/A		—	—	—	
1030	Slab on Grade	4" reinforced concrete with vapor barrier and granular base		S.F. Slab	5.80	5.80	9.1%
2010	Basement Excavation	Site preparation for slab and trench for foundation wall and footing		S.F. Ground	.33	.33	
2020	Basement Walls	N/A		—	—	—	
B. SHELL							
	B10 Superstructure						
1010	Floor Construction	6 x 6 wood columns		S.F. Floor	.25	.25	4.7%
1020	Roof Construction	Wood truss with plywood sheathing		S.F. Roof	7.76	7.76	
	B20 Exterior Enclosure						
2010	Exterior Walls	Stone ashlar veneer on concrete block	65% of wall	S.F. Wall	47.65	21.06	
2020	Exterior Windows	Aluminum horizontal sliding	35% of wall	Each	529	8.39	19.6%
2030	Exterior Doors	Double aluminum and glass, hollow metal		Each	3997	4	
	B30 Roofing						
3010	Roof Coverings	Asphalt shingles		S.F. Roof	18.82	3.22	1.9%
3020	Roof Openings	N/A		—	—	—	
C. INTERIORS							
1010	Partitions	Gypsum board on metal studs, load bearing	14 S.F. Floor/L.F. Partition	S.F. Partition	8.88	6.34	
1020	Interior Doors	Single leaf wood	140 S.F. Floor/Door	Each	720	5.14	
1030	Fittings	N/A		—	—	—	
2010	Stair Construction	N/A		—	—	—	16.9%
3010	Wall Finishes	40% vinyl wall covering, 40% paint, 20% ceramic tile		S.F. Surface	2.81	4.01	
3020	Floor Finishes	50% carpet, 30% hardwood tile, 20% ceramic tile		S.F. Floor	8.54	8.54	
3030	Ceiling Finishes	Gypsum plaster on wood furring		S.F. Ceiling	4.85	4.85	
D. SERVICES							
	D10 Conveying						
1010	Elevators & Lifts	N/A		—	—	—	0.0 %
1020	Escalators & Moving Walks	N/A		—	—	—	
	D20 Plumbing						
2010	Plumbing Fixtures	Kitchen, toilet and service fixtures, supply and drainage	1 Fixture/125 S.F. Floor	Each	2320	18.56	
2020	Domestic Water Distribution	Gas fired hot water		S.F. Floor	9.04	9.04	16.2%
2040	Rain Water Drainage	N/A		—	—	—	
	D30 HVAC						
3010	Energy Supply	N/A		—	—	—	
3020	Heat Generating Systems	Included in D3050		—	—	—	
3030	Cooling Generating Systems	N/A		—	—	—	18.2 %
3050	Terminal & Package Units	Multizone rooftop unit, gas heating, electric cooling		S.F. Floor	31.05	31.05	
3090	Other HVAC Sys. & Equipment	N/A		—	—	—	
	D40 Fire Protection						
4010	Sprinklers	Wet pipe sprinkler system		S.F. Floor	6.09	6.09	4.7%
4020	Standpipes	Standpipe		S.F. Floor	1.91	1.91	
	D50 Electrical						
5010	Electrical Service/Distribution	400 ampere service, panel board and feeders		S.F. Floor	3.56	3.56	
5020	Lighting & Branch Wiring	High efficiency fluorescent fixtures, receptacles, switches, A.C. and misc. power		S.F. Floor	9.03	9.03	8.7%
5030	Communications & Security	Addressable alarm systems and emergency lighting		S.F. Floor	2.03	2.03	
5090	Other Electrical Systems	Emergency generator, 11.5 kW		S.F. Floor	.24	.24	
E. EQUIPMENT & FURNISHINGS							
1010	Commercial Equipment	N/A		—	—	—	
1020	Institutional Equipment	N/A		—	—	—	
1030	Vehicular Equipment	N/A		—	—	—	0.0 %
1090	Other Equipment	N/A		—	—	—	
F. SPECIAL CONSTRUCTION							
1020	Integrated Construction	N/A		—	—	—	0.0 %
1040	Special Facilities	N/A		—	—	—	
G. BUILDING SITEWORK	**N/A**						

		Sub-Total	170.51	100%
CONTRACTOR FEES (General Requirements: 10%, Overhead: 5%, Profit: 10%)		25%	42.64	
ARCHITECT FEES		8%	17.05	
	Total Building Cost		**230.20**	

For customer support on your Square Foot Costs with RSMeans data, call 800.448.8182.

99

Costs per square foot of floor area

Exterior Wall	S.F. Area	4000	8000	12000	17000	22000	27000	32000	37000	42000
	L.F. Perimeter	280	386	520	585	640	740	840	940	940
Stone Ashlar and Concrete Block	Steel Joists	230.10	199.30	191.65	180.45	173.90	171.55	170.00	168.80	165.25
	Wood Joists	223.10	193.40	185.95	175.20	168.90	166.60	165.10	164.00	160.65
Face Brick and Concrete Block	Steel Joists	221.00	193.05	186.05	176.00	170.10	168.00	166.60	165.55	162.40
	Wood Joists	214.35	187.40	180.60	170.95	165.30	163.25	161.85	160.85	157.85
Decorative Concrete Block	Steel Joists	211.10	186.25	179.90	171.10	166.05	164.15	162.85	161.95	159.20
	Wood Joists	204.15	180.35	174.30	165.90	161.00	159.20	158.00	157.10	154.55
Perimeter Adj., Add or Deduct	Per 100 L.F.	27.60	13.85	9.25	6.50	5.00	4.10	3.45	3.05	2.70
Story Hgt. Adj., Add or Deduct	Per 1 Ft.	5.00	3.45	3.15	2.45	2.10	2.00	1.90	1.85	1.65
For Basement, add $33.70 per square foot of basement area										

The above costs were calculated using the basic specifications shown on the facing page. These costs should be adjusted where necessary for design alternatives and owner's requirements.

Common additives

Description	Unit	$ Cost	Description	Unit	$ Cost
Bar, Front bar	L.F.	450	Kitchen Equipment, cont.		
Back bar	L.F.	360	Freezer, 44 C.F., reach-in	Ea.	5525
Booth, Upholstered, custom, straight	L.F.	258 - 475	Ice cube maker, 50 lb. per day	Ea.	2100
"L" or "U" shaped	L.F.	267 - 450	Lockers, Steel, single tier, 60" or 72"	Opng.	232 - 395
Emergency Lighting, 25 watt, battery operated			2 tier, 60" or 72" total	Opng.	139 - 172
Lead battery	Ea.	330	5 tier, box lockers	Opng.	73 - 87
Nickel cadmium	Ea.	550	Locker bench, lam. maple top only	L.F.	36.50
Flagpoles, Complete			Pedestals, steel pipe	Ea.	79
Aluminum, 20' high	Ea.	1900	Refrigerators, Prefabricated, walk-in		
40' High	Ea.	4500	7'-6" High, 6' x 6'	S.F.	145
70' High	Ea.	11,600	10' x 10'	S.F.	114
Fiberglass, 23' High	Ea.	1450	12' x 14'	S.F.	99.50
39'-5" High	Ea.	3325	12' x 20'	S.F.	137
59' High	Ea.	7300	Smoke Detectors		
Kitchen Equipment			Ceiling type	Ea.	244
Broiler	Ea.	4425	Duct type	Ea.	585
Coffee urn, twin 6 gallon	Ea.	2900	Sound System		
Cooler, 6 ft. long, reach-in	Ea.	3750	Amplifier, 250 watts	Ea.	2025
Dishwasher, 10-12 racks per hr.	Ea.	4150	Speaker, ceiling or wall	Ea.	240
Food warmer, counter 1.2 kw	Ea.	790	Trumpet	Ea.	460

Important: See the Reference Section for Location Factors.

Model costs calculated for a 1 story building with 12' story height and 22,000 square feet of floor area

				Unit	Unit Cost	Cost Per S.F.	% Of Sub-Total
A. SUBSTRUCTURE							
1010	Standard Foundations	Poured concrete; strip and spread footings; 4' foundation wall		S.F. Ground	4.13	4.13	
1020	Special Foundations	N/A		—	—	—	
1030	Slab on Grade	4" reinforced concrete with vapor barrier and granular base		S.F. Slab	5.80	5.80	7.8%
2010	Basement Excavation	Site preparation for slab and trench for foundation wall and footing		S.F. Ground	.18	.18	
2020	Basement Walls	N/A		—	—	—	
B. SHELL							
B10 Superstructure							
1010	Floor Construction	Steel column		S.F. Floor	.65	.65	6.3%
1020	Roof Construction	Metal deck on open web steel joists		S.F. Roof	7.55	7.55	
B20 Exterior Enclosure							
2010	Exterior Walls	Stone, ashlar veneer on concrete block	65% of wall	S.F. Wall	47.64	10.81	
2020	Exterior Windows	Window wall	35% of wall	Each	56	6.83	14.4%
2030	Exterior Doors	Double aluminum and glass doors		Each	3997	1.10	
B30 Roofing							
3010	Roof Coverings	Built-up tar and gravel with flashing; perlite/EPS composite insulation		S.F. Roof	6.67	6.67	5.1%
3020	Roof Openings	N/A		—	—	—	
C. INTERIORS							
1010	Partitions	Lightweight concrete block	14 S.F. Floor/L.F. Partition	S.F. Partition	12.63	9.02	
1020	Interior Doors	Single leaf wood	140 S.F. Floor/Door	Each	720	5.14	
1030	Fittings	N/A		—	—	—	
2010	Stair Construction	N/A		—	—	—	23.9%
3010	Wall Finishes	65% paint, 25% vinyl wall covering, 10% ceramic tile		S.F. Surface	1.90	2.72	
3020	Floor Finishes	60% carpet, 35% hardwood, 15% ceramic tile		S.F. Floor	7.86	7.86	
3030	Ceiling Finishes	Mineral fiber tile on concealed zee bars		S.F. Ceiling	6.30	6.30	
D. SERVICES							
D10 Conveying							
1010	Elevators & Lifts	N/A		—	—	—	0.0 %
1020	Escalators & Moving Walks	N/A		—	—	—	
D20 Plumbing							
2010	Plumbing Fixtures	Kitchen, toilet and service fixtures, supply and drainage	1 Fixture/1050 S.F. Floor	Each	9555	9.10	
2020	Domestic Water Distribution	Gas fired hot water heater		S.F. Floor	.37	.37	7.9%
2040	Rain Water Drainage	Roof drains		S.F. Roof	.85	.85	
D30 HVAC							
3010	Energy Supply	N/A		—	—	—	
3020	Heat Generating Systems	Included in D3050		—	—	—	
3030	Cooling Generating Systems	N/A		—	—	—	21.1 %
3050	Terminal & Package Units	Multizone rooftop unit, gas heating, electric cooling		S.F. Floor	27.40	27.40	
3090	Other HVAC Sys. & Equipment	N/A		—	—	—	
D40 Fire Protection							
4010	Sprinklers	Wet pipe sprinkler system		S.F. Floor	5.23	5.23	4.4%
4020	Standpipes	Standpipe		S.F. Floor	.52	.52	
D50 Electrical							
5010	Electrical Service/Distribution	400 ampere service, panel board and feeders		S.F. Floor	1.25	1.25	
5020	Lighting & Branch Wiring	High efficiency fluorescent fixtures, receptacles, switches, A.C. and misc. power		S.F. Floor	8.86	8.86	9.1%
5030	Communications & Security	Addressable alarm systems and emergency lighting		S.F. Floor	1.54	1.54	
5090	Other Electrical Systems	Emergency generator, 11.5 kW		S.F. Floor	.13	.13	
E. EQUIPMENT & FURNISHINGS							
1010	Commercial Equipment	N/A		—	—	—	
1020	Institutional Equipment	N/A		—	—	—	0.0 %
1030	Vehicular Equipment	N/A		—	—	—	
1090	Other Equipment	N/A		—	—	—	
F. SPECIAL CONSTRUCTION							
1020	Integrated Construction	N/A		—	—	—	0.0 %
1040	Special Facilities	N/A		—	—	—	
G. BUILDING SITEWORK	**N/A**						

		Sub-Total	130.01	100%
	CONTRACTOR FEES (General Requirements: 10%, Overhead: 5%, Profit: 10%)	25%	32.51	
	ARCHITECT FEES	7%	11.38	
	Total Building Cost		**173.90**	

For customer support on your Square Foot Costs with RSMeans data, call 800.448.8182.

101

Costs per square foot of floor area

Exterior Wall	S.F. Area	15000	20000	28000	38000	50000	65000	85000	100000	150000
	L.F. Perimeter	350	400	480	550	630	660	750	825	1035
Fiber Cement	Wood Frame	182.05	173.45	166.05	160.25	156.30	152.00	149.35	148.10	145.55
Stone Veneer	Wood Frame	199.55	188.40	178.80	170.95	165.50	159.30	155.65	154.05	150.40
Brick Veneer	Rigid Steel	197.60	187.35	178.60	171.50	166.55	161.05	157.75	156.35	153.10
Curtain Wall	Rigid Steel	201.55	190.35	180.75	172.85	167.35	161.10	157.45	155.85	152.20
Brick Veneer	Reinforced Concrete	216.05	204.55	194.80	186.75	181.10	174.75	171.00	169.30	165.55
Stone Veneer	Reinforced Concrete	231.45	217.80	206.15	196.30	189.50	181.50	176.85	174.80	170.15
Perimeter Adj., Add or Deduct	Per 100 L.F.	8.40	6.30	4.55	3.35	2.50	1.95	1.50	1.30	0.90
Story Hgt. Adj., Add or Deduct	Per 1 Ft.	1.20	1.05	0.95	0.80	0.65	0.45	0.40	0.50	0.40
For Basement, add $39.30 per square foot of basement area										

The above costs were calculated using the basic specifications shown on the facing page. These costs should be adjusted where necessary for design alternatives and owner's requirements.

Common additives

Description	Unit	$ Cost	Description	Unit	$ Cost
Carrels Hardwood	Ea.	740 - 1925	Lockers, Steel, single tier, 60" or 72"	Opng.	232 - 395
Clock System			2 tier, 60" or 72" total	Opng.	139 - 172
20 Room	Ea.	20,200	5 tier, box lockers	Opng.	73 - 87
50 Room	Ea.	48,200	Locker bench, lam. maple top only	L.F.	36.50
Elevators, Hydraulic passenger, 2 stops			Pedestals, steel pipe	Ea.	79
1500# capacity	Ea.	73,100	Seating		
2500# capacity	Ea.	76,600	Auditorium chair, all veneer	Ea.	355
3500# capacity	Ea.	81,600	Veneer back, padded seat	Ea.	380
Additional stop, add	Ea.	8375	Upholstered, spring seat	Ea.	335
Emergency Lighting, 25 watt, battery operated			Classroom, movable chair & desk	Set	81 - 171
Lead battery	Ea.	330	Lecture hall, pedestal type	Ea.	350 - 650
Nickel cadmium	Ea.	550	Smoke Detectors		
Flagpoles, Complete			Ceiling type	Ea.	244
Aluminum, 20' high	Ea.	1900	Duct type	Ea.	585
40' High	Ea.	4500	Sound System		
70' High	Ea.	11,600	Amplifier, 250 watts	Ea.	2025
Fiberglass, 23' High	Ea.	1450	Speaker, ceiling or wall	Ea.	240
39'-5" High	Ea.	3325	Trumpet	Ea.	460
59' High	Ea.	7300	TV Antenna, Master system, 12 outlet	Outlet	242
			30 outlet	Outlet	335
			100 outlet	Outlet	370

Important: See the Reference Section for Location Factors.

Model costs calculated for a 2 story building with 12' story height and 50,000 square feet of floor area

College, Classroom, 2-3 Story

				Unit	Unit Cost	Cost Per S.F.	% Of Sub-Total
A. SUBSTRUCTURE							
1010	Standard Foundations	Poured concrete; strip and spread footings; 4' foundation wall		S.F. Ground	4.42	2.21	
1020	Special Foundations	N/A		—	—	—	
1030	Slab on Grade	4" reinforced slab on grade		S.F. Slab	5.80	2.90	4.2%
2010	Basement Excavation	Site preparation for slab and trench for foundation wall and footing		S.F. Ground	.31	.16	
2020	Basement Walls	N/A		—	—	—	
B. SHELL							
B10 Superstructure							
1010	Floor Construction	Open web steel joists, slab form, concrete, fireproofed steel columns		S.F. Floor	16.38	8.19	10.9%
1020	Roof Construction	Metal deck on open web steel joists, columns		S.F. Roof	10.92	5.46	
B20 Exterior Enclosure							
2010	Exterior Walls	N/A		—	—	—	
2020	Exterior Windows	Curtain wall glazing system, thermo-break frame	100% of wall	S.F. Wall	51	12.45	10.6 %
2030	Exterior Doors	Aluminum and glass double doors		Each	7200	.87	
B30 Roofing							
3010	Roof Coverings	Single ply membrane, stone ballast, rigid insulation		S.F. Roof	7.08	3.54	2.9%
3020	Roof Openings	Roof and smoke hatches		S.F. Roof	2587	.11	
C. INTERIORS							
1010	Partitions	Gypsum board on Metal studs	20 S.F. Floor/L.F. Partition	S.F. Partition	6.70	3.35	
1020	Interior Doors	Single leaf hollow metal doors	420 S.F. Floor/Door	Each	1222	2.90	
1030	Fittings	Toilet partitions, chalkboards, classroom cabinetry		S.F. Floor	5.35	5.35	
2010	Stair Construction	Cement filled metal pan		Flight	18,900	1.89	20.5%
3010	Wall Finishes	95% paint, 5% ceramic wall tile		S.F. Surface	1.46	1.46	
3020	Floor Finishes	25% Carpet tile, 35% vinyl composition tile, 5% ceramic tile		S.F. Floor	4.42	4.42	
3030	Ceiling Finishes	Acoustic ceiling tiles on suspended channel grid		S.F. Ceiling	6.30	6.30	
D. SERVICES							
D10 Conveying							
1010	Elevators & Lifts	Two hydraulic passenger elevators		Each	91,250	3.65	2.9%
1020	Escalators & Moving Walks	N/A		—	—	—	
D20 Plumbing							
2010	Plumbing Fixtures	Restroom and service fixtures, supply and drainage	1 Fixture/395 S.F. Floor	Each	2435	6.18	
2020	Domestic Water Distribution	Gas fired water heater		S.F. Floor	3.12	3.12	7.8%
2040	Rain Water Drainage	Roof drains		S.F. Roof	1.04	.52	
D30 HVAC							
3010	Energy Supply	N/A		—	—	—	
3020	Heat Generating Systems	Included in D3050		—	—	—	
3030	Cooling Generating Systems	N/A		—	—	—	17.1 %
3050	Terminal & Package Units	Multizone gas heating, electric cooling unit		S.F. Floor	21.45	21.45	
3090	Other HVAC Sys. & Equipment	N/A		—	—	—	
D40 Fire Protection							
4010	Sprinklers	Wet pipe spinkler system, light hazard		S.F. Floor	3.30	3.30	2.9%
4020	Standpipes	Standpipes		S.F. Floor	.39	.39	
D50 Electrical							
5010	Electrical Service/Distribution	2000 ampere service, panel board and feeders		S.F. Floor	3.77	3.77	
5020	Lighting & Branch Wiring	Incandescent fixtures, receptacles, switches, A.C. and misc. power		S.F. Floor	13.75	13.75	19.9%
5030	Communications & Security	Addressable alarm systems, emergency lighting, internet and phone wiring		S.F. Floor	7.43	7.43	
5090	Other Electrical Systems	N/A		—	—	—	
E. EQUIPMENT & FURNISHINGS							
1010	Commercial Equipment	N/A		—	—	—	
1020	Institutional Equipment	N/A		—	—	—	
1030	Vehicular Equipment	N/A		—	—	—	0.0 %
1090	Other Equipment	N/A		—	—	—	
F. SPECIAL CONSTRUCTION							
1020	Integrated Construction	N/A		—	—	—	0.0 %
1040	Special Facilities	N/A		—	—	—	
G. BUILDING SITEWORK	**N/A**						

	Unit	Cost Per S.F.	% Of Sub-Total
Sub-Total		125.12	100%
CONTRACTOR FEES (General Requirements: 10%, Overhead: 5%, Profit: 10%)	25%	31.28	
ARCHITECT FEES	7%	10.95	
Total Building Cost		**167.35**	

For customer support on your Square Foot Costs with RSMeans data, call 800.448.8182.

Costs per square foot of floor area

Exterior Wall	S.F. Area	10000	15000	25000	40000	55000	70000	80000	90000	100000
	L.F. Perimeter	260	320	400	476	575	628	684	721	772
Fiber Cement	Wood Frame	215.65	198.10	181.85	171.15	166.85	163.50	162.25	161.05	160.25
Stone Veneer	Wood Frame	244.25	221.25	199.00	183.80	177.90	172.95	171.25	169.45	168.40
Brick Veneer	Rigid Steel	230.70	209.70	189.70	176.25	170.95	166.60	165.15	163.55	162.60
Curtain Wall	Rigid Steel	245.60	222.50	200.00	184.70	178.75	173.75	172.05	170.25	169.15
Brick Veneer	Reinforced Concrete	249.55	226.60	204.35	189.20	183.30	178.30	176.65	174.85	173.80
Stone Veneer	Reinforced Concrete	279.65	251.35	222.90	203.00	195.45	188.70	186.60	184.15	182.75
Perimeter Adj., Add or Deduct	Per 100 L.F.	14.15	9.45	5.65	3.55	2.55	2.05	1.75	1.60	1.45
Story Hgt. Adj., Add or Deduct	Per 1 Ft.	2.20	1.85	1.30	1.00	0.90	0.70	0.80	0.70	0.65

For Basement, add $ 38.30 per square foot of basement area

The above costs were calculated using the basic specifications shown on the facing page. These costs should be adjusted where necessary for design alternatives and owner's requirements.

Common additives

Description	Unit	$ Cost		Description	Unit	$ Cost
Carrels Hardwood	Ea.	740 - 1925		Kitchen Equipment		
Closed Circuit Surveillance, One station				Broiler	Ea.	4425
Camera and monitor	Ea.	1475		Coffee urn, twin 6 gallon	Ea.	2900
For additional camera stations, add	Ea.	665		Cooler, 6 ft. long	Ea.	3750
Elevators, Hydraulic passenger, 2 stops				Dishwasher, 10-12 racks per hr.	Ea.	4150
2000# capacity	Ea.	74,100		Food warmer	Ea.	790
2500# capacity	Ea.	76,600		Freezer, 44 C.F., reach-in	Ea.	5525
3500# capacity	Ea.	81,600		Ice cube maker, 50 lb. per day	Ea.	2100
Additional stop, add	Ea.	8375		Range with 1 oven	Ea.	3575
Emergency Lighting, 25 watt, battery operated				Laundry Equipment		
Lead battery	Ea.	330		Dryer, gas, 16 lb. capacity	Ea.	1025
Nickel cadmium	Ea.	550		30 lb. capacity	Ea.	4100
Furniture	Student	2800 - 5350		Washer, 4 cycle	Ea.	1325
Intercom System, 25 station capacity				Commercial	Ea.	1700
Master station	Ea.	3225		Smoke Detectors		
Intercom outlets	Ea.	212		Ceiling type	Ea.	244
Handset	Ea.	590		Duct type	Ea.	585
				TV Antenna, Master system, 12 outlet	Outlet	242
				30 outlet	Outlet	335
				100 outlet	Outlet	370

Important: See the Reference Section for Location Factors.

Model costs calculated for a 3 story building with 12' story height and 25,000 square feet of floor area

College, Dormitory, 2-3 Story

				Unit	Unit Cost	Cost Per S.F.	% Of Sub-Total
A. SUBSTRUCTURE							
1010	Standard Foundations	Poured concrete; strip and spread footings; 4' foundation wall		S.F. Ground	14.31	4.77	
1020	Special Foundations	N/A		—	—	—	
1030	Slab on Grade	4" reinforced slab on grade		S.F. Slab	5.80	1.93	5.0%
2010	Basement Excavation	Site preparation for slab and trench for foundation wall and footing		S.F. Ground	.18	.06	
2020	Basement Walls	N/A		—	—	—	
B. SHELL							
	B10 Superstructure						
1010	Floor Construction	Wood Beam and Joist on wood columns, fireproofed		S.F. Floor	17.84	11.89	10.7%
1020	Roof Construction	Wood roof, truss, 4/12 slope		S.F. Roof	7.77	2.59	
	B20 Exterior Enclosure						
2010	Exterior Walls	Vinyl siding on wood studs, insulated	80% of wall	S.F. Wall	16.84	7.76	
2020	Exterior Windows	Aluminum awning windows	20% of wall	Each	755	3.78	9.9%
2030	Exterior Doors	Aluminum and glass double doors		Each	7800	1.88	
	B30 Roofing						
3010	Roof Coverings	Asphalt roofing, strip shingles, flashing, gutters and downspouts		S.F. Roof	4.23	1.41	1.0%
3020	Roof Openings	N/A		—	—	—	
C. INTERIORS							
1010	Partitions	Sound deadening gypsum board on wood studs	10 S.F. Floor/L.F. Partition	S.F. Partition	9.47	10.52	
1020	Interior Doors	Single leaf solid core wood doors	100 S.F. Floor/Door	Each	1222	5.09	
1030	Fittings	Toilet partitions and bathroom accessories		S.F. Floor	2.62	2.62	
2010	Stair Construction	Prefabricated wood stairs with rails		Flight	12,600	5.54	24.7%
3010	Wall Finishes	95% paint, 5% ceramic wall tile		S.F. Surface	1.80	4	
3020	Floor Finishes	80% Carpet tile, 10% vinyl composition tile, 10% ceramic tile		S.F. Floor	5.16	5.16	
3030	Ceiling Finishes	Acoustic ceiling tiles on suspended channel grid		S.F. Ceiling	.63	.63	
D. SERVICES							
	D10 Conveying						
1010	Elevators & Lifts	Hydraulic passenger elevator		Each	124,000	4.96	3.6%
1020	Escalators & Moving Walks	N/A		—	—	—	
	D20 Plumbing						
2010	Plumbing Fixtures	Restroom and service fixtures, supply and drainage	1 Fixture/300 S.F. Floor	Each	4620	15.35	
2020	Domestic Water Distribution	Electric water heater		S.F. Floor	6.46	6.46	16.0%
2040	Rain Water Drainage	Roof drains		—	—	—	
	D30 HVAC						
3010	Energy Supply	N/A		—	—	—	
3020	Heat Generating Systems	Included in D3050		—	—	—	
3030	Cooling Generating Systems	N/A		—	—	—	10.0 %
3050	Terminal & Package Units	Multizone gas heating, electric cooling unit		S.F. Floor	13.55	13.55	
3090	Other HVAC Sys. & Equipment	N/A		—	—	—	
	D40 Fire Protection						
4010	Sprinklers	Wet pipe spinkler system, light hazard		S.F. Floor	3.12	3.12	3.0%
4020	Standpipes	Standpipes		S.F. Floor	1	1	
	D50 Electrical						
5010	Electrical Service/Distribution	600 ampere service, panel board and feeders		S.F. Floor	2.30	2.30	
5020	Lighting & Branch Wiring	Fluorescent fixtures, receptacles, switches, A.C. and misc. power		S.F. Floor	10.27	10.27	14.9%
5030	Communications & Security	Addressable alarm systems, emergency lighting, internet and phone wiring		S.F. Floor	7.63	7.63	
5090	Other Electrical Systems	N/A		—	—	—	
E. EQUIPMENT & FURNISHINGS							
1010	Commercial Equipment	N/A		—	—	—	
1020	Institutional Equipment	N/A		—	—	—	
1030	Vehicular Equipment	N/A		—	—	—	1.2 %
2020	Moveable Furnishings	Dormitory furniture		S.F. Floor	1.67	1.67	
F. SPECIAL CONSTRUCTION							
1020	Integrated Construction	N/A		—	—	—	0.0 %
1040	Special Facilities	N/A		—	—	—	
G. BUILDING SITEWORK	**N/A**						

		Sub-Total	135.94	**100%**
CONTRACTOR FEES (General Requirements: 10%, Overhead: 5%, Profit: 10%)		25%	34.01	
ARCHITECT FEES		7%	11.90	
	Total Building Cost		**181.85**	

For customer support on your Square Foot Costs with RSMeans data, call 800.448.8182.

Costs per square foot of floor area

Exterior Wall	S.F. Area	20000	35000	45000	65000	85000	110000	135000	160000	200000
	L.F. Perimeter	260	340	400	440	500	540	560	590	640
Brick Veneer	Rigid Steel	192.95	178.15	174.40	165.75	162.00	158.05	155.00	153.15	151.20
Curtain Wall	Rigid Steel	217.00	196.25	190.95	178.40	173.10	167.40	162.95	160.30	157.45
E.I.F.S.	Rigid Steel	184.45	171.95	168.70	161.55	158.40	155.10	152.55	151.10	149.45
Precast Concrete	Precast Concrete	230.40	208.00	202.30	188.75	182.95	176.75	171.95	169.05	165.95
Brick Veneer	Reinforced Concrete	241.55	216.90	210.60	195.55	189.15	182.35	176.95	173.75	170.35
Stone Veneer	Reinforced Concrete	268.00	235.85	227.70	207.85	199.50	190.45	183.30	179.10	174.55
Perimeter Adj., Add or Deduct	Per 100 L.F.	18.60	10.75	8.30	5.65	4.35	3.40	2.75	2.35	1.85
Story Hgt. Adj., Add or Deduct	Per 1 Ft.	3.40	2.60	2.30	1.70	1.50	1.25	1.05	0.95	0.85
For Basement, add $38.30 per square foot of basement area										

The above costs were calculated using the basic specifications shown on the facing page. These costs should be adjusted where necessary for design alternatives and owner's requirements.

Common additives

Description	Unit	$ Cost
Carrels Hardwood	Ea.	740 - 1925
Closed Circuit Surveillance, One station		
Camera and monitor	Ea.	1475
For additional camera stations, add	Ea.	665
Elevators, Electric passenger, 5 stops		
2000# capacity	Ea.	186,000
2500# capacity	Ea.	190,500
3500# capacity	Ea.	193,000
Additional stop, add	Ea.	10,900
Emergency Lighting, 25 watt, battery operated		
Lead battery	Ea.	330
Nickel cadmium	Ea.	550
Furniture	Student	2800 - 5350
Intercom System, 25 station capacity		
Master station	Ea.	3225
Intercom outlets	Ea.	212
Handset	Ea.	590

Description	Unit	$ Cost
Kitchen Equipment		
Broiler	Ea.	4425
Coffee urn, twin, 6 gallon	Ea.	2900
Cooler, 6 ft. long	Ea.	3750
Dishwasher, 10-12 racks per hr.	Ea.	4150
Food warmer	Ea.	790
Freezer, 44 C.F., reach-in	Ea.	5525
Ice cube maker, 50 lb. per day	Ea.	2100
Range with 1 oven	Ea.	3575
Laundry Equipment		
Dryer, gas, 16 lb. capacity	Ea.	1025
30 lb. capacity	Ea.	4100
Washer, 4 cycle	Ea.	1325
Commercial	Ea.	1700
Smoke Detectors		
Ceiling type	Ea.	244
Duct type	Ea.	585
TV Antenna, Master system, 12 outlet	Outlet	242
30 outlet	Outlet	335
100 outlet	Outlet	370

Important: See the Reference Section for Location Factors.

Model costs calculated for a 6 story building with 12′ story height and 85,000 square feet of floor area

				Unit	Unit Cost	Cost Per S.F.	% Of Sub-Total
A. SUBSTRUCTURE							
1010	Standard Foundations	Poured concrete; strip and spread footings; 4′ foundation wall		S.F. Ground	9.84	1.64	
1020	Special Foundations	N/A		—	—	—	
1030	Slab on Grade	4″ reinforced slab on grade		S.F. Slab	5.80	.97	1.9%
2010	Basement Excavation	Site preparation for slab and trench for foundation wall and footing		S.F. Ground	.31	.05	
2020	Basement Walls	N/A		—	—	—	
B. SHELL							
	B10 Superstructure						
1010	Floor Construction	Cast-in-place column, I beam, precast double T beam & topping		S.F. Floor	19.94	16.62	14.6%
1020	Roof Construction	Precast double T beams with 2″ topping		S.F. Roof	20.22	3.37	
	B20 Exterior Enclosure						
2010	Exterior Walls	Precast concrete panels, insulated	80% of wall	S.F. Wall	60	20.49	
2020	Exterior Windows	Aluminum sliding windows	20% of wall	Each	529	2.99	17.5%
2030	Exterior Doors	Aluminum and glass double doors and single doors		Each	4663	.44	
	B30 Roofing						
3010	Roof Coverings	Single ply membrane, stone ballast, rigid insulation		S.F. Roof	7.62	1.27	1.0%
3020	Roof Openings	Roof and smoke hatches		S.F. Roof	2587	.11	
C. INTERIORS							
1010	Partitions	Sound deadening gypsum board on metal studs, CMU partitions	10 S.F. Floor/L.F. Partition	S.F. Partition	6.49	7.21	
1020	Interior Doors	Single leaf solid core wood doors	100 S.F. Floor/Door	Each	1222	5.09	
1030	Fittings	Toilet partitions and bathroom accessories		S.F. Floor	1.15	1.15	
2010	Stair Construction	Cement filled metal pan with rails		Flight	18,900	4.45	21.8%
3010	Wall Finishes	95% paint, 5% ceramic wall tile		S.F. Surface	2.73	6.06	
3020	Floor Finishes	80% Carpet tile, 10% vinyl composition tile, 10% ceramic tile		S.F. Floor	5.16	5.16	
3030	Ceiling Finishes	Acoustic ceiling tiles on suspended channel grid		S.F. Ceiling	.63	.63	
D. SERVICES							
	D10 Conveying						
1010	Elevators & Lifts	Traction geared passenger elevators		Each	267,113	12.57	9.2%
1020	Escalators & Moving Walks	N/A		—	—	—	
	D20 Plumbing						
2010	Plumbing Fixtures	Restroom and service fixtures, supply and drainage	1 Fixture/300 S.F. Floor	Each	2116	7.03	
2020	Domestic Water Distribution	Electric water heater		S.F. Floor	2.57	2.57	7.2%
2040	Rain Water Drainage	Roof drains		S.F. Roof	1.62	.27	
	D30 HVAC						
3010	Energy Supply	Oil fired hot water, wall fin radiation		S.F. Floor	4.94	4.94	
3020	Heat Generating Systems	N/A		—	—	—	
3030	Cooling Generating Systems	Chilled water, air cooled condenser system		S.F. Floor	9.49	9.49	10.6%
3050	Terminal & Package Units	N/A		—	—	—	
3090	Other HVAC Sys. & Equipment	N/A		—	—	—	
	D40 Fire Protection						
4010	Sprinklers	Wet pipe spinkler system, light hazard		S.F. Floor	3.06	3.06	2.8%
4020	Standpipes	Standpipes		S.F. Floor	.71	.71	
	D50 Electrical						
5010	Electrical Service/Distribution	800 ampere service, panel board and feeders		S.F. Floor	.84	.84	
5020	Lighting & Branch Wiring	Fluorescent fixtures, receptacles, switches, A.C. and misc. power		S.F. Floor	10.19	10.19	12.9%
5030	Communications & Security	Addressable alarm systems, emergency lighting, internet and phone wiring		S.F. Floor	6.66	6.66	
5090	Other Electrical Systems	N/A		—	—	—	
E. EQUIPMENT & FURNISHINGS							
1010	Commercial Equipment	N/A		—	—	—	
1020	Institutional Equipment	N/A		—	—	—	
1030	Vehicular Equipment	N/A		—	—	—	0.5 %
2020	Moveable Furnishings	Dormitory furniture		S.F. Floor	.74	.74	
F. SPECIAL CONSTRUCTION							
1020	Integrated Construction	N/A		—	—	—	0.0 %
1040	Special Facilities	N/A		—	—	—	
G. BUILDING SITEWORK	**N/A**						

				Sub-Total	136.77	**100%**
	CONTRACTOR FEES (General Requirements: 10%, Overhead: 5%, Profit: 10%)			25%	34.21	
	ARCHITECT FEES			7%	11.97	

	Total Building Cost	182.95

For customer support on your Square Foot Costs with RSMeans data, call 800.448.8182.

Costs per square foot of floor area

Exterior Wall	S.F. Area	12000	20000	28000	37000	45000	57000	68000	80000	92000
	L.F. Perimeter	470	600	698	793	900	1000	1075	1180	1275
Brick Veneer	Rigid Steel	229.85	199.65	185.05	175.60	171.10	165.20	161.15	158.55	156.50
Curtain Wall	Rigid Steel	243.75	209.90	193.25	182.50	177.40	170.45	165.80	162.75	160.25
E.I.F.S.	Rigid Steel	227.45	198.55	184.60	175.65	171.35	165.75	162.00	159.50	157.50
Precast Concrete	Reinforced Concrete	255.75	221.30	204.35	193.30	188.15	181.10	176.25	173.15	170.65
Brick Veneer	Reinforced Concrete	247.35	214.90	199.00	188.70	183.80	177.30	172.90	170.00	167.70
Stone Veneer	Reinforced Concrete	261.35	225.65	207.90	196.35	191.00	183.60	178.50	175.30	172.65
Perimeter Adj., Add or Deduct	Per 100 L.F.	11.90	7.15	5.10	3.85	3.15	2.50	2.20	1.85	1.50
Story Hgt. Adj., Add or Deduct	Per 1 Ft.	1.35	1.10	0.85	0.80	0.70	0.60	0.60	0.55	0.45
For Basement, add $24.70 per square foot of basement area										

The above costs were calculated using the basic specifications shown on the facing page. These costs should be adjusted where necessary for design alternatives and owner's requirements.

Common additives

Description	Unit	$ Cost
Cabinets, Base, door units, metal	L.F.	350
Drawer units	L.F.	690
Tall storage cabinets, open	L.F.	640
With doors	L.F.	1025
Wall, metal 12-1/2" deep, open	L.F.	294
With doors	L.F.	520
Carrels Hardwood	Ea.	740 - 1925
Countertops, not incl. base cabinets, acid proof	S.F.	67 - 70
Stainless steel	S.F.	196
Fume Hood, Not incl. ductwork	L.F.	845 - 2000
Ductwork	Hood	6450 - 10,200
Glassware Washer, Distilled water rinse	Ea.	8150 - 17,500
Seating		
Auditorium chair, all veneer	Ea.	355
Veneer back, padded seat	Ea.	380
Upholstered, spring seat	Ea.	335
Classroom, movable chair & desk	Set	81 - 171
Lecture hall, pedestal type	Ea.	350 - 650

Description	Unit	$ Cost
Safety Equipment, Eye wash, hand held	Ea.	455
Deluge shower	Ea.	900
Sink, One piece plastic		
Flask wash, freestanding	Ea.	2700
Tables, acid resist. top, drawers	L.F.	229
Titration Unit, Four 2000 ml reservoirs	Ea.	6650

Important: See the Reference Section for Location Factors.

Model costs calculated for a 1 story building with 12' story height and 45,000 square feet of floor area

				Unit	Unit Cost	Cost Per S.F.	% Of Sub-Total
A.	**SUBSTRUCTURE**						
1010	Standard Foundations	Poured concrete; strip and spread footings; 4' foundation wall		S.F. Ground	3.83	3.83	
1020	Special Foundations	N/A		—	—	—	
1030	Slab on Grade	4" reinforced slab on grade		S.F. Slab	5.80	5.80	7.5%
2010	Basement Excavation	Site preparation for slab and trench for foundation wall and footing		S.F. Ground	.18	.18	
2020	Basement Walls	N/A		—	—	—	
B.	**SHELL**						
	B10 Superstructure						
1010	Floor Construction	Structural fireproofing		S.F. Floor	7.23	.94	9.1%
1020	Roof Construction	Metal deck on open web steel joists, columns		S.F. Roof	10.90	10.90	
	B20 Exterior Enclosure						
2010	Exterior Walls	N/A		—	—	—	
2020	Exterior Windows	Curtain wall glazing system, thermo-break frame	100% of wall	S.F. Wall	51	12.23	11.5 %
2030	Exterior Doors	Aluminum and glass single and double doors		Each	6168	2.74	
	B30 Roofing						
3010	Roof Coverings	Single ply membrane, stone ballast, rigid insulation		S.F. Roof	6.80	6.80	5.5%
3020	Roof Openings	Roof and smoke hatches		S.F. Roof	.35	.35	
C.	**INTERIORS**						
1010	Partitions	Gypsum board on metal studs	10 S.F. Floor/L.F. Partition	S.F. Partition	5.87	5.87	
1020	Interior Doors	Single leaf steel doors	820 S.F. Floor/Door	Each	1452	1.77	
1030	Fittings	Toilet partitions and lockers		S.F. Floor	.65	.65	
2010	Stair Construction	N/A		—	—	—	17.5%
3010	Wall Finishes	80% paint, 20% ceramic wall tile		S.F. Surface	.89	1.78	
3020	Floor Finishes	80% Carpet tile, 10% vinyl composition tile, 10% ceramic tile		S.F. Floor	6.41	6.41	
3030	Ceiling Finishes	Acoustic ceiling tiles on suspended channel grid		S.F. Ceiling	6.30	6.30	
D.	**SERVICES**						
	D10 Conveying						
1010	Elevators & Lifts	N/A		—	—	—	0.0 %
1020	Escalators & Moving Walks	N/A		—	—	—	
	D20 Plumbing						
2010	Plumbing Fixtures	Toilet and service fixtures, supply and drainage	1 Fixture/260 S.F. Floor	Each	1461	5.62	
2020	Domestic Water Distribution	Gas fired water heater		S.F. Floor	3.48	3.48	7.3%
2040	Rain Water Drainage	Roof drains		S.F. Roof	.45	.45	
	D30 HVAC						
3010	Energy Supply	N/A		—	—	—	
3020	Heat Generating Systems	Included in D3050		—	—	—	
3030	Cooling Generating Systems	N/A		—	—	—	16.5 %
3050	Terminal & Package Units	Multizone gas heating, electric cooling unit		S.F. Floor	21.45	21.45	
3090	Other HVAC Sys. & Equipment	N/A		—	—	—	
	D40 Fire Protection						
4010	Sprinklers	Wet pipe spinkler system, light hazard		S.F. Floor	2.93	2.93	2.5%
4020	Standpipes	Standpipes		S.F. Floor	.35	.35	
	D50 Electrical						
5010	Electrical Service/Distribution	1000 ampere service, panel board and feeders		S.F. Floor	1.61	1.61	
5020	Lighting & Branch Wiring	Fluorescent fixtures, receptacles, switches, A.C. and misc. power		S.F. Floor	11.81	11.81	13.1%
5030	Communications & Security	Addressable alarm systems, emergency lighting, internet and phone wiring		S.F. Floor	3.58	3.58	
5090	Other Electrical Systems	Emergency generator, 11.5 kW, Uninterruptible power supply		S.F. Floor	.11	.11	
E.	**EQUIPMENT & FURNISHINGS**						
1010	Commercial Equipment	N/A		—	—	—	
1020	Institutional Equipment	Laboratory equipment		S.F. Floor	12.24	12.24	9.4 %
1030	Vehicular Equipment	N/A		—	—	—	
1090	Other Equipment	N/A		—	—	—	
F.	**SPECIAL CONSTRUCTION**						
1020	Integrated Construction	N/A		—	—	—	0.0 %
1040	Special Facilities	N/A		—	—	—	
G.	**BUILDING SITEWORK**	**N/A**					

	Sub-Total	130.18	100%
CONTRACTOR FEES (General Requirements: 10%, Overhead: 5%, Profit: 10%)	25%	32.57	
ARCHITECT FEES	9%	14.65	
Total Building Cost		**177.40**	

For customer support on your Square Foot Costs with RSMeans data, call 800.448.8182.

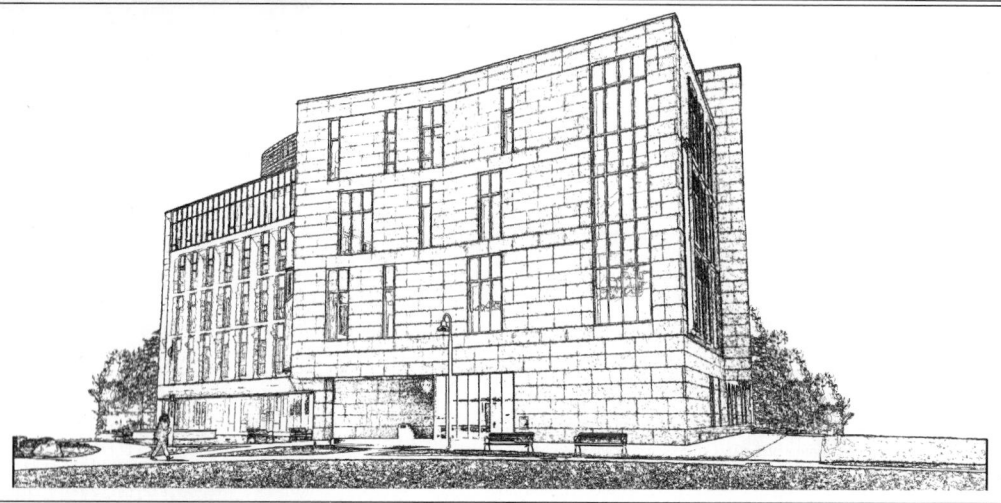

Costs per square foot of floor area

Exterior Wall	S.F. Area	15000	20000	25000	30000	35000	40000	45000	50000	55000
	L.F. Perimeter	354	425	457	513	568	583	629	644	683
Face Brick and Concrete Block	Rigid Steel	193.85	187.65	181.50	178.65	176.60	173.40	172.05	170.00	169.05
	Reinforced Concrete	183.55	177.35	171.15	168.30	166.25	163.10	161.70	159.65	158.70
Precast Concrete	Rigid Steel	206.25	198.95	191.30	187.85	185.40	181.40	179.75	177.15	175.90
	Reinforced Concrete	196.20	188.75	181.00	177.50	174.95	170.90	169.20	166.60	165.35
Limestone and Concrete Block	Rigid Steel	205.75	198.40	190.70	187.30	184.75	180.75	179.10	176.50	175.30
	Reinforced Concrete	195.45	188.05	180.35	176.90	174.40	170.45	168.80	166.15	164.90
Perimeter Adj., Add or Deduct	Per 100 L.F.	10.55	7.95	6.35	5.25	4.55	4.00	3.55	3.15	2.85
Story Hgt. Adj., Add or Deduct	Per 1 Ft.	2.65	2.40	2.05	1.90	1.80	1.65	1.60	1.45	1.35
For Basement, add $41.00 per square foot of basement area										

The above costs were calculated using the basic specifications shown on the facing page. These costs should be adjusted where necessary for design alternatives and owner's requirements.

Common additives

Description	Unit	$ Cost
Carrels Hardwood	Ea.	740 - 1925
Elevators, Hydraulic passenger, 2 stops		
2000# capacity	Ea.	74,100
2500# capacity	Ea.	76,600
3500# capacity	Ea.	81,600
Emergency Lighting, 25 watt, battery operated		
Lead battery	Ea.	330
Nickel cadmium	Ea.	550
Escalators, Metal		
32" wide, 10' story height	Ea.	162,000
20' story height	Ea.	199,000
48" wide, 10' Story height	Ea.	171,000
20' story height	Ea.	209,000
Glass		
32" wide, 10' story height	Ea.	153,500
20' story height	Ea.	188,500
48" wide, 10' story height	Ea.	161,000
20' story height	Ea.	198,000

Description	Unit	$ Cost
Lockers, Steel, Single tier, 60" or 72"	Opng.	232 - 395
2 tier, 60" or 72" total	Opng.	139 - 172
5 tier, box lockers	Opng.	73 - 87
Locker bench, lam. maple top only	L.F.	36.50
Pedestals, steel pipe	Ea.	79
Sound System		
Amplifier, 250 watts	Ea.	2025
Speaker, ceiling or wall	Ea.	240
Trumpet	Ea.	460

Important: See the Reference Section for Location Factors.

College, Student Union

Model costs calculated for a 2 story building with 12' story height and 25,000 square feet of floor area

				Unit	Unit Cost	Cost Per S.F.	% Of Sub-Total
A. SUBSTRUCTURE							
1010	Standard Foundations	Poured concrete; strip and spread footings; 4' foundation wall		S.F. Ground	8.40	4.20	
1020	Special Foundations	N/A		—	—	—	
1030	Slab on Grade	4" reinforced concrete with vapor barrier and granular base		S.F. Slab	5.80	2.90	5.6%
2010	Basement Excavation	Site preparation for slab and trench for foundation wall and footing		S.F. Ground	.18	.09	
2020	Basement Walls	N/A		—	—	—	
B. SHELL							
	B10 Superstructure						
1010	Floor Construction	Concrete flat plate		S.F. Floor	29.10	14.55	17.9%
1020	Roof Construction	Concrete flat plate		S.F. Roof	16.62	8.31	
	B20 Exterior Enclosure						
2010	Exterior Walls	Face brick with concrete block backup	75% of wall	S.F. Wall	35.25	11.60	
2020	Exterior Windows	Window wall	25% of wall	Each	48.25	5.29	13.6%
2030	Exterior Doors	Double aluminum and glass		Each	3550	.57	
	B30 Roofing						
3010	Roof Coverings	Built-up tar and gravel with flashing; perlite/EPS composite insulation		S.F. Roof	7	3.50	2.8%
3020	Roof Openings	Roof hatches		S.F. Roof	.14	.07	
C. INTERIORS							
1010	Partitions	Gypsum board on metal studs	14 S.F. Floor/L.F. Partition	S.F. Partition	6.76	4.83	
1020	Interior Doors	Single leaf hollow metal	140 S.F. Floor/Door	Each	1222	2.44	
1030	Fittings	N/A		—	—	—	
2010	Stair Construction	Cast in place concrete		Flight	8950	1.43	18.4%
3010	Wall Finishes	50% paint, 50% vinyl wall covering		S.F. Surface	2.54	3.63	
3020	Floor Finishes	50% carpet, 50% vinyl composition tile		S.F. Floor	4.86	4.86	
3030	Ceiling Finishes	Suspended fiberglass board		S.F. Ceiling	6.30	6.30	
D. SERVICES							
	D10 Conveying						
1010	Elevators & Lifts	One hydraulic passenger elevator		Each	91,250	3.65	2.9%
1020	Escalators & Moving Walks	N/A		—	—	—	
	D20 Plumbing						
2010	Plumbing Fixtures	Toilet and service fixtures, supply and drainage	1 Fixture/1040 S.F. Floor	Each	3328	3.20	
2020	Domestic Water Distribution	Gas fired water heater		S.F. Floor	.99	.99	3.9%
2040	Rain Water Drainage	Roof drains		S.F. Roof	1.58	.79	
	D30 HVAC						
3010	Energy Supply	N/A		—	—	—	
3020	Heat Generating Systems	Included in D3050		—	—	—	
3030	Cooling Generating Systems	N/A		—	—	—	16.8 %
3050	Terminal & Package Units	Multizone unit, gas heating, electric cooling		S.F. Floor	21.45	21.45	
3090	Other HVAC Sys. & Equipment	N/A		—	—	—	
	D40 Fire Protection						
4010	Sprinklers	Wet pipe sprinkler system		S.F. Floor	3.30	3.30	3.3%
4020	Standpipes	Standpipe, wet, Class III		S.F. Floor	.98	.98	
	D50 Electrical						
5010	Electrical Service/Distribution	600 ampere service, panel board and feeders		S.F. Floor	1.79	1.79	
5020	Lighting & Branch Wiring	High efficiency fluorescent fixtures, receptacles, switches, A.C. and misc. power		S.F. Floor	13.43	13.43	14.9%
5030	Communications & Security	Addressable alarm systems, internet wiring, communications systems and emergency lighting		S.F. Floor	3.65	3.65	
5090	Other Electrical Systems	Emergency generator, 11.5 kW		S.F. Floor	.17	.17	
E. EQUIPMENT & FURNISHINGS							
1010	Commercial Equipment	N/A		—	—	—	
1020	Institutional Equipment	N/A		—	—	—	
1030	Vehicular Equipment	N/A		—	—	—	0.0 %
1090	Other Equipment	N/A		—	—	—	
F. SPECIAL CONSTRUCTION							
1020	Integrated Construction	N/A		—	—	—	
1040	Special Facilities	N/A		—	—	—	0.0 %
G. BUILDING SITEWORK	**N/A**						

		Sub-Total	127.97	100%
CONTRACTOR FEES (General Requirements: 10%, Overhead: 5%, Profit: 10%)		25%	31.98	
ARCHITECT FEES		7%	11.20	
	Total Building Cost		**171.15**	

For customer support on your Square Foot Costs with RSMeans data, call 800.448.8182.

111

Costs per square foot of floor area

Exterior Wall	S.F. Area	4000	6000	8000	10000	12000	14000	16000	18000	20000
	L.F. Perimeter	260	340	420	453	460	510	560	610	600
Face Brick and Concrete Block	Bearing Walls	171.90	163.60	159.45	153.00	146.90	145.10	143.70	142.65	139.40
	Rigid Steel	175.10	166.75	162.60	156.10	150.05	148.20	146.90	145.80	142.55
Decorative Concrete Block	Bearing Walls	157.10	150.65	147.45	142.60	138.15	136.75	135.75	134.90	132.50
	Rigid Steel	160.20	153.80	150.55	145.80	141.25	139.90	138.90	138.10	135.65
Tilt-up Concrete Panels	Bearing Walls	152.05	146.30	143.40	139.15	135.15	133.95	133.05	132.30	130.20
	Rigid Steel	155.20	149.40	146.50	142.30	138.35	137.10	136.20	135.45	133.35
Perimeter Adj., Add or Deduct	Per 100 L.F.	20.85	13.90	10.45	8.35	6.90	6.00	5.25	4.70	4.15
Story Hgt. Adj., Add or Deduct	Per 1 Ft.	3.10	2.70	2.45	2.10	1.80	1.70	1.70	1.65	1.40

For Basement, add $35.90 per square foot of basement area

The above costs were calculated using the basic specifications shown on the facing page. These costs should be adjusted where necessary for design alternatives and owner's requirements.

Common additives

Description	Unit	$ Cost
Bar, Front bar	L.F.	450
Back bar	L.F.	360
Booth, Upholstered, custom straight	L.F.	258 - 475
"L" or "U" shaped	L.F.	267 - 450
Bowling Alleys, incl. alley, pinsetter		
Scorer, counter & misc. supplies, average	Lane	67,500
For automatic scorer, add	Lane	11,200
Emergency Lighting, 25 watt, battery operated		
Lead battery	Ea.	330
Nickel cadmium	Ea.	550
Kitchen Equipment		
Broiler	Ea.	4425
Coffee urn, twin 6 gallon	Ea.	2900
Cooler, 6 ft. long	Ea.	3750
Dishwasher, 10-12 racks per hr.	Ea.	4150
Food warmer	Ea.	790
Freezer, 44 C.F., reach-in	Ea.	5525
Ice cube maker, 50 lb. per day	Ea.	2100
Range with 1 oven	Ea.	3575

Description	Unit	$ Cost
Movie Equipment		
Projector, 35mm	Ea.	13,600 - 18,600
Screen, wall or ceiling hung	S.F.	9.50 - 13.85
Partitions, Folding leaf, wood		
Acoustic type	S.F.	85 - 141
Seating		
Auditorium chair, all veneer	Ea.	355
Veneer back, padded seat	Ea.	380
Upholstered, spring seat	Ea.	335
Classroom, movable chair & desk	Set	81 - 171
Lecture hall, pedestal type	Ea.	350 - 650
Sound System		
Amplifier, 250 watts	Ea.	2025
Speaker, ceiling or wall	Ea.	240
Trumpet	Ea.	460
Stage Curtains, Medium weight	S.F.	11.50 - 42
Curtain Track, Light duty	L.F.	95.50
Swimming Pools, Complete, gunite	S.F.	106 - 132

Important: See the Reference Section for Location Factors.

Community Center

Model costs calculated for a 1 story building with 12' story height and 10,000 square feet of floor area

			Unit	Unit Cost	Cost Per S.F.	% Of Sub-Total
A. SUBSTRUCTURE						
1010	Standard Foundations	Poured concrete; strip and spread footings; 4' foundation wall	S.F. Ground	7.03	7.03	
1020	Special Foundations	N/A	—	—	—	
1030	Slab on Grade	4" reinforced concrete with vapor barrier and granular base	S.F. Slab	5.80	5.80	11.6%
2010	Basement Excavation	Site preparation for slab and trench for foundation wall and footing	S.F. Ground	.33	.33	
2020	Basement Walls	N/A	—	—	—	
B. SHELL						
B10 Superstructure						
1010	Floor Construction	N/A	—	—	—	7.4 %
1020	Roof Construction	Metal deck on open web steel joists	S.F. Roof	8.40	8.40	
B20 Exterior Enclosure						
2010	Exterior Walls	Face brick with concrete block backup 80% of wall	S.F. Wall	35.23	15.32	
2020	Exterior Windows	Aluminum sliding 20% of wall	Each	633	2.15	16.7%
2030	Exterior Doors	Double aluminum and glass and hollow metal	Each	3703	1.49	
B30 Roofing						
3010	Roof Coverings	Built-up tar and gravel with flashing; perlite/EPS composite insulation	S.F. Roof	7.42	7.42	6.6%
3020	Roof Openings	Roof hatches	S.F. Roof	.11	.11	
C. INTERIORS						
1010	Partitions	Gypsum board on metal studs 14 S.F. Floor/L.F. Partition	S.F. Partition	9.20	6.57	
1020	Interior Doors	Single leaf hollow metal 140 S.F. Floor/Door	Each	1222	1.22	
1030	Fittings	Toilet partitions, directory board, mailboxes	S.F. Floor	1.78	1.78	
2010	Stair Construction	N/A	—	—	—	20.6%
3010	Wall Finishes	Paint	S.F. Surface	1.95	2.78	
3020	Floor Finishes	50% carpet, 50% vinyl tile	S.F. Floor	4.74	4.74	
3030	Ceiling Finishes	Mineral fiber tile on concealed zee bars	S.F. Ceiling	6.30	6.30	
D. SERVICES						
D10 Conveying						
1010	Elevators & Lifts	N/A	—	—	—	0.0 %
1020	Escalators & Moving Walks	N/A	—	—	—	
D20 Plumbing						
2010	Plumbing Fixtures	Kitchen, toilet and service fixtures, supply and drainage 1 Fixture/910 S.F. Floor	Each	3413	3.75	
2020	Domestic Water Distribution	Electric water heater	S.F. Floor	10.73	10.73	13.4%
2040	Rain Water Drainage	Roof drains	S.F. Roof	.65	.65	
D30 HVAC						
3010	Energy Supply	N/A	—	—	—	
3020	Heat Generating Systems	Included in D3050	—	—	—	
3030	Cooling Generating Systems	N/A	—	—	—	10.1 %
3050	Terminal & Package Units	Single zone rooftop unit, gas heating, electric cooling	S.F. Floor	11.50	11.50	
3090	Other HVAC Sys. & Equipment	N/A	—	—	—	
D40 Fire Protection						
4010	Sprinklers	Wet pipe sprinkler system	S.F. Floor	3.79	3.79	3.3%
4020	Standpipes	N/A	—	—	—	
D50 Electrical						
5010	Electrical Service/Distribution	200 ampere service, panel board and feeders	S.F. Floor	1.14	1.14	
5020	Lighting & Branch Wiring	High efficiency fluorescent fixtures, receptacles, switches, A.C. and misc. power	S.F. Floor	6.27	6.27	8.1%
5030	Communications & Security	Addressable alarm systems and emergency lighting	S.F. Floor	1.56	1.56	
5090	Other Electrical Systems	Emergency generator, 15 kW	S.F. Floor	.19	.19	
E. EQUIPMENT & FURNISHINGS						
1010	Commercial Equipment	Freezer, chest type	S.F. Floor	.90	.90	
1020	Institutional Equipment	N/A	—	—	—	
1030	Vehicular Equipment	N/A	—	—	—	2.0%
1090	Other Equipment	Kitchen equipment, directory board, mailboxes, built-in coat racks	S.F. Floor	1.39	1.39	
F. SPECIAL CONSTRUCTION						
1020	Integrated Construction	N/A	—	—	—	0.0 %
1040	Special Facilities	N/A	—	—	—	
G. BUILDING SITEWORK	**N/A**					

		Sub-Total	113.31	100%
CONTRACTOR FEES (General Requirements: 10%, Overhead: 5%, Profit: 10%)		25%	28.36	
ARCHITECT FEES		8%	11.33	
	Total Building Cost		**153**	

For customer support on your Square Foot Costs with RSMeans data, call 800.448.8182.

113

Costs per square foot of floor area

Exterior Wall	S.F. Area	10000	12500	15000	17500	20000	22500	25000	30000	40000
	L.F. Perimeter	400	450	500	550	600	625	650	700	800
Fiber Cement	Wood Frame	339.90	336.75	334.65	333.15	332.05	330.35	328.95	326.90	324.30
Stone Veneer	Wood Frame	360.90	355.50	351.85	349.25	347.35	344.35	341.90	338.40	333.95
Curtain Wall	Rigid Steel	366.80	361.35	357.65	355.05	353.10	350.10	347.60	344.00	339.45
E.I.F.S.	Rigid Steel	346.50	343.35	341.25	339.80	338.60	336.95	335.50	333.45	330.85
Brick Veneer	Reinforced Concrete	373.40	367.70	363.95	361.20	359.20	356.05	353.50	349.80	345.10
Metal Panel	Reinforced Concrete	370.75	365.60	362.20	359.75	358.00	355.20	352.85	349.55	345.25
Perimeter Adj., Add or Deduct	Per 100 L.F.	7.55	6.05	5.10	4.35	3.80	3.35	3.05	2.55	1.95
Story Hgt. Adj., Add or Deduct	Per 1 Ft.	1.00	0.90	0.80	0.80	0.80	0.70	0.65	0.65	0.55

For Basement, add $39.40 per square foot of basement area

The above costs were calculated using the basic specifications shown on the facing page. These costs should be adjusted where necessary for design alternatives and owner's requirements.

Common additives

Description	Unit	$ Cost
Clock System		
20 room	Ea.	20,200
50 room	Ea.	48,200
Closed Circuit Surveillance, one station		
Camera and monitor	Total	1475
For additional camera stations, add	Ea.	665
For zoom lens - remote control, add	Ea.	2775 - 6950
For automatic iris for low light, add	Ea.	1675
Directory Boards, plastic, glass covered		
30" x 20"	Ea.	640
36" x 48"	Ea.	1625
Aluminum, 24" x 18"	Ea.	630
36" x 24"	Ea.	790
48" x 32"	Ea.	1025
48" x 60"	Ea.	2200
Emergency Lighting, 25 watt, battery operated		
Lead battery	Ea.	330
Nickel cadmium	Ea.	550

Description	Unit	$ Cost
Smoke Detectors		
Ceiling type	Ea.	244
Duct type	Ea.	585
Sound System		
Amplifier, 250 watts	Ea.	2025
Speakers, ceiling or wall	Ea.	240
Trumpets	Ea.	460

Important: See the Reference Section for Location Factors.

Model costs calculated for a 1 story building with 16'-6" story height and 22,500 square feet of floor area

Computer Data Center

				Unit	Unit Cost	Cost Per S.F.	% Of Sub-Total
A. SUBSTRUCTURE							
1010	Standard Foundations	Poured concrete; strip and spread footings; 4' foundation wall		S.F. Ground	4.07	4.07	
1020	Special Foundations	N/A		—	—	—	
1030	Slab on Grade	4" reinforced concrete		S.F. Slab	5.80	5.80	4.1%
2010	Basement Excavation	Site preparation for slab and trench for foundation wall and footing		S.F. Ground	.18	.20	
2020	Basement Walls	N/A		—	—	—	
B. SHELL							
	B10 Superstructure						
1010	Floor Construction	Wood columns, fireproofed		S.F. Floor	3.30	3.30	4.6%
1020	Roof Construction	Wood roof truss 4:12 pitch		S.F. Roof	8.10	8.10	
	B20 Exterior Enclosure						
2010	Exterior Walls	Fiber cement siding on wood studs, insulated	95% of wall	S.F. Wall	19.34	8.42	
2020	Exterior Windows	Aluminum horizontal sliding	5% of wall	Each	589	.90	4.3%
2030	Exterior Doors	Aluminum and glass single and double doors		Each	4933	1.31	
	B30 Roofing						
3010	Roof Coverings	Asphalt roofing, strip shingles, gutters and downspouts		S.F. Roof	2.31	2.31	1.0%
3020	Roof Openings	Roof hatch		S.F. Roof	.25	.25	
C. INTERIORS							
1010	Partitions	Sound deadening gypsum board on wood studs	15 S.F. Floor/L.F. Partition	S.F. Partition	9.56	8.82	
1020	Interior Doors	Solid core wood doors in metal frames	370 S.F. Floor/Door	Each	926	2.51	
1030	Fittings	Plastic laminate toilet partitions		S.F. Floor	.29	.29	
2010	Stair Construction	N/A		—	—	—	11.8%
3010	Wall Finishes	90% Paint, 10% ceramic wall tile		S.F. Surface	1.83	3.37	
3020	Floor Finishes	65% Carpet, 10% porcelain tile, 10% quarry tile		S.F. Floor	6.61	6.61	
3030	Ceiling Finishes	Acoustic ceiling tiles on suspended channel grid		S.F. Ceiling	7.60	7.60	
D. SERVICES							
	D10 Conveying						
1010	Elevators & Lifts	N/A		—	—	—	0.0 %
1020	Escalators & Moving Walks	N/A		—	—	—	
	D20 Plumbing						
2010	Plumbing Fixtures	Restroom and service fixtures, supply and drainage	1 Fixture/1180 S.F. Floor	Each	2218	1.88	
2020	Domestic Water Distribution	Gas fired water heater		S.F. Floor	.90	.90	1.1%
2040	Rain Water Drainage	N/A		—	—	—	
	D30 HVAC						
3010	Energy Supply	Hot water reheat system		S.F. Floor	6.91	6.91	
3020	Heat Generating Systems	Oil fired hot water boiler and pumps		Each	54,925	10.72	
3030	Cooling Generating Systems	Cooling tower and chiller		S.F. Floor	9.90	9.90	41.9%
3050	Terminal & Package Units	N/A		—	—	—	
3090	Other HVAC Sys. & Equipment	Plate heat exchanger, ductwork, AHUs and VAV terminals		S.F. Floor	76	75.86	
	D40 Fire Protection						
4010	Sprinklers	85% light hazard wet pipe sprinkler system, 15% preaction system		S.F. Floor	4.55	4.55	2.1%
4020	Standpipes	Standpipes and hose systems		S.F. Floor	.72	.72	
	D50 Electrical						
5010	Electrical Service/Distribution	1200 ampere service, panel boards and feeders		S.F. Floor	3.68	3.68	
5020	Lighting & Branch Wiring	Fluorescent fixtures, receptacles, switches, A.C. and misc. power		S.F. Floor	17.46	17.46	27.1%
5030	Communications & Security	Telephone systems, internet wiring, and addressable alarm systems		S.F. Floor	31.96	31.96	
5090	Other Electrical Systems	Emergency generator, UPS system with 15 minute pack		S.F. Floor	13.73	13.73	
E. EQUIPMENT & FURNISHINGS							
1010	Commercial Equipment	N/A		—	—	—	
1020	Institutional Equipment	N/A		—	—	—	
1030	Vehicular Equipment	N/A		—	—	—	0.0 %
1090	Other Equipment	N/A		—	—	—	
F. SPECIAL CONSTRUCTION & DEMOLITION							
1020	Integrated Construction	Pedestal access floor		S.F. Floor	4.84	4.84	2.0%
1040	Special Facilities	N/A		—	—	—	
G. BUILDING SITEWORK	**N/A**						

		Sub-Total	246.97	100%
CONTRACTOR FEES (General Requirements: 10%, Overhead: 5%, Profit: 10%)		25%	61.77	
ARCHITECT FEES		7%	21.61	
	Total Building Cost		**330.35**	

For customer support on your Square Foot Costs with RSMeans data, call 800.448.8182.

Costs per square foot of floor area

Exterior Wall	S.F. Area	16000	23000	30000	37000	44000	51000	58000	65000	72000
	L.F. Perimeter	597	763	821	954	968	1066	1090	1132	1220
Limestone and Concrete Block	Reinforced Concrete	247.90	237.10	226.30	222.45	215.90	213.55	210.00	207.60	206.50
	Rigid Steel	245.70	234.95	224.15	220.25	213.80	211.40	207.80	205.45	204.30
Face Brick and Concrete Block	Reinforced Concrete	231.25	222.30	214.10	210.95	206.10	204.25	201.60	199.85	198.95
	Rigid Steel	229.10	220.15	211.95	208.80	204.00	202.05	199.45	197.65	196.80
Stone and Concrete Block	Reinforced Concrete	237.75	228.10	218.90	215.45	209.95	207.90	204.85	202.85	201.85
	Rigid Steel	235.60	225.90	216.70	213.25	207.75	205.70	202.70	200.70	199.70
Perimeter Adj., Add or Deduct	Per 100 L.F.	8.80	6.15	4.65	3.75	3.25	2.75	2.40	2.15	1.95
Story Hgt. Adj., Add or Deduct	Per 1 Ft.	3.15	2.80	2.30	2.15	1.90	1.75	1.55	1.45	1.45

For Basement, add $33.90 per square foot of basement area

The above costs were calculated using the basic specifications shown on the facing page. These costs should be adjusted where necessary for design alternatives and owner's requirements.

Common additives

Description	Unit	$ Cost
Benches, Hardwood	L.F.	141 - 233
Clock System		
20 room	Ea.	20,200
50 room	Ea.	48,200
Closed Circuit Surveillance, One station		
Camera and monitor	Ea.	1475
For additional camera stations, add	Ea.	665
Directory Boards, Plastic, glass covered		
30" x 20"	Ea.	640
36" x 48"	Ea.	1625
Aluminum, 24" x 18"	Ea.	630
36" x 24"	Ea.	790
48" x 32"	Ea.	1025
48" x 60"	Ea.	2200
Emergency Lighting, 25 watt, battery operated		
Lead battery	Ea.	330
Nickel cadmium	Ea.	550

Description	Unit	$ Cost
Flagpoles, Complete		
Aluminum, 20' high	Ea.	1900
40' high	Ea.	4500
70' high	Ea.	11,600
Fiberglass, 23' high	Ea.	1450
39'-5" high	Ea.	3325
59' high	Ea.	7300
Intercom System, 25 station capacity		
Master station	Ea.	3225
Intercom outlets	Ea.	212
Handset	Ea.	590
Safe, Office type, 1 hour rating		
30" x 18" x 18"	Ea.	2475
60" x 36" x 18", double door	Ea.	10,200
Smoke Detectors		
Ceiling type	Ea.	244
Duct type	Ea.	585

Important: See the Reference Section for Location Factors.

Model costs calculated for a 1 story building with 14' story height and 30,000 square feet of floor area

				Unit	Unit Cost	Cost Per S.F.	% Of Sub-Total

A. SUBSTRUCTURE

				Unit	Unit Cost	Cost Per S.F.	% Of Sub-Total
1010	Standard Foundations	Poured concrete; strip and spread footings; 4' foundation wall		S.F. Ground	4.64	4.64	
1020	Special Foundations	N/A		—	—	—	
1030	Slab on Grade	4" reinforced concrete with vapor barrier and granular base		S.F. Slab	5.80	5.80	6.3%
2010	Basement Excavation	Site preparation for slab and trench for foundation wall and footing		S.F. Ground	.18	.18	
2020	Basement Walls	N/A		—	—	—	

B. SHELL

B10 Superstructure

				Unit	Unit Cost	Cost Per S.F.	% Of Sub-Total
1010	Floor Construction	Cast-in-place columns		L.F. Column	78	1.75	14.8%
1020	Roof Construction	Cast-in-place concrete waffle slab		S.F. Roof	23.30	23.30	

B20 Exterior Enclosure

				Unit	Unit Cost	Cost Per S.F.	% Of Sub-Total
2010	Exterior Walls	Limestone panels with concrete block backup	75% of wall	S.F. Wall	67	19.26	
2020	Exterior Windows	Aluminum with insulated glass	25% of wall	Each	755	3.15	13.7%
2030	Exterior Doors	Double wood		Each	3221	.74	

B30 Roofing

				Unit	Unit Cost	Cost Per S.F.	% Of Sub-Total
3010	Roof Coverings	Built-up tar and gravel with flashing; perlite/EPS composite insulation		S.F. Roof	6.25	6.25	3.8%
3020	Roof Openings	Roof hatches		S.F. Roof	.11	.11	

C. INTERIORS

				Unit	Unit Cost	Cost Per S.F.	% Of Sub-Total
1010	Partitions	Plaster on metal studs	10 S.F. Floor/L.F. Partition	S.F. Partition	14.54	17.45	
1020	Interior Doors	Single leaf wood	100 S.F. Floor/Door	Each	720	7.20	
1030	Fittings	Toilet partitions		S.F. Floor	.52	.52	
2010	Stair Construction	N/A		—	—	—	33.1%
3010	Wall Finishes	70% paint, 20% wood paneling, 10% vinyl wall covering		S.F. Surface	2.29	5.49	
3020	Floor Finishes	60% hardwood, 20% carpet, 20% terrazzo		S.F. Floor	14.32	14.32	
3030	Ceiling Finishes	Gypsum plaster on metal lath, suspended		S.F. Ceiling	11.10	11.10	

D. SERVICES

D10 Conveying

				Unit	Unit Cost	Cost Per S.F.	% Of Sub-Total
1010	Elevators & Lifts	N/A		—	—	—	0.0%
1020	Escalators & Moving Walks	N/A		—	—	—	

D20 Plumbing

				Unit	Unit Cost	Cost Per S.F.	% Of Sub-Total
2010	Plumbing Fixtures	Toilet and service fixtures, supply and drainage	1 Fixture/1110 S.F. Floor	Each	4851	4.37	
2020	Domestic Water Distribution	Electric hot water heater		S.F. Floor	3.86	3.86	5.8%
2040	Rain Water Drainage	Roof drains		S.F. Roof	1.53	1.53	

D30 HVAC

				Unit	Unit Cost	Cost Per S.F.	% Of Sub-Total
3010	Energy Supply	N/A		—	—	—	
3020	Heat Generating Systems	Included in D3050		—	—	—	
3030	Cooling Generating Systems	N/A		—	—	—	12.7%
3050	Terminal & Package Units	Multizone unit, gas heating, electric cooling		S.F. Floor	21.45	21.45	
3090	Other HVAC Sys. & Equipment	N/A		—	—	—	

D40 Fire Protection

				Unit	Unit Cost	Cost Per S.F.	% Of Sub-Total
4010	Sprinklers	Wet pipe sprinkler system		S.F. Floor	2.93	2.93	2.1%
4020	Standpipes	Standpipe, wet, Class III		S.F. Floor	.54	.54	

D50 Electrical

				Unit	Unit Cost	Cost Per S.F.	% Of Sub-Total
5010	Electrical Service/Distribution	400 ampere service, panel board and feeders		S.F. Floor	1.05	1.05	
5020	Lighting & Branch Wiring	High efficiency fluorescent fixtures, receptacles, switches, A.C. and misc. power		S.F. Floor	10.33	10.33	7.8%
5030	Communications & Security	Addressable alarm systems, internet wiring, and emergency lighting		S.F. Floor	1.74	1.74	
5090	Other Electrical Systems	Emergency generator, 11.5 kW		S.F. Floor	.14	.14	

E. EQUIPMENT & FURNISHINGS

				Unit	Unit Cost	Cost Per S.F.	% Of Sub-Total
1010	Commercial Equipment	N/A		—	—	—	
1020	Institutional Equipment	N/A		—	—	—	0.0%
1030	Vehicular Equipment	N/A		—	—	—	
1090	Other Equipment	N/A		—	—	—	

F. SPECIAL CONSTRUCTION

				Unit	Unit Cost	Cost Per S.F.	% Of Sub-Total
1020	Integrated Construction	N/A		—	—	—	0.0%
1040	Special Facilities	N/A		—	—	—	

G. BUILDING SITEWORK   N/A

	Sub-Total	169.20	**100%**
CONTRACTOR FEES (General Requirements: 10%, Overhead: 5%, Profit: 10%)	25%	42.29	
ARCHITECT FEES	7%	14.81	
	Total Building Cost	**226.30**	

For customer support on your Square Foot Costs with RSMeans data, call 800.448.8182.

117

Costs per square foot of floor area

Exterior Wall	S.F. Area	30000	40000	45000	50000	60000	70000	80000	90000	100000
	L.F. Perimeter	400	466	500	533	600	666	733	800	867
Limestone and Concrete Block	Reinforced Concrete	265.55	252.15	247.80	244.25	238.90	235.10	232.25	230.05	228.25
	Rigid Steel	267.25	253.85	249.50	245.90	240.60	236.75	234.00	231.70	229.95
Face Brick and Concrete Block	Reinforced Concrete	250.40	238.95	235.15	232.15	227.55	224.25	221.85	219.95	218.45
	Rigid Steel	252.10	240.60	236.90	233.80	229.25	225.95	223.55	221.60	220.10
Stone and Concrete Block	Reinforced Concrete	256.35	244.05	240.10	236.85	231.95	228.50	225.90	223.85	222.30
	Rigid Steel	258.00	245.80	241.80	238.55	233.70	230.20	227.65	225.55	223.95
Perimeter Adj., Add or Deduct	Per 100 L.F.	12.80	9.60	8.55	7.65	6.45	5.45	4.80	4.25	3.90
Story Hgt. Adj., Add or Deduct	Per 1 Ft.	4.00	3.50	3.35	3.15	3.00	2.85	2.75	2.65	2.60

For Basement, add $34.60 per square foot of basement area

The above costs were calculated using the basic specifications shown on the facing page. These costs should be adjusted where necessary for design alternatives and owner's requirements.

Common additives

Description	Unit	$ Cost
Benches, Hardwood	L.F.	141 - 233
Clock System		
20 room	Ea.	20,200
50 room	Ea.	48,200
Closed Circuit Surveillance, One station		
Camera and monitor	Ea.	1475
For additional camera stations, add	Ea.	665
Directory Boards, Plastic, glass covered		
30" x 20"	Ea.	640
36" x 48"	Ea.	1625
Aluminum, 24" x 18"	Ea.	630
36" x 24"	Ea.	790
48" x 32"	Ea.	1025
48" x 60"	Ea.	2200
Elevators, Hydraulic passenger, 2 stops		
1500# capacity	Ea.	73,100
2500# capacity	Ea.	76,600
3500# capacity	Ea.	81,600
Additional stop, add	Ea.	8375

Description	Unit	$ Cost
Emergency Lighting, 25 watt, battery operated		
Lead battery	Ea.	330
Nickel cadmium	Ea.	550
Flagpoles, Complete		
Aluminum, 20' high	Ea.	1900
40' high	Ea.	4500
70' high	Ea.	11,600
Fiberglass, 23' high	Ea.	1450
39'-5" high	Ea.	3325
59' high	Ea.	7300
Intercom System, 25 station capacity		
Master station	Ea.	3225
Intercom outlets	Ea.	212
Handset	Ea.	590
Safe, Office type, 1 hour rating		
30" x 18" x 18"	Ea.	2475
60" x 36" x 18", double door	Ea.	10,200
Smoke Detectors		
Ceiling type	Ea.	244
Duct type	Ea.	585

Important: See the Reference Section for Location Factors.

Model costs calculated for a 3 story building with 12' story height and 60,000 square feet of floor area

				Unit	Unit Cost	Cost Per S.F.	% Of Sub-Total
A.	**SUBSTRUCTURE**						
1010	Standard Foundations	Poured concrete; strip and spread footings; 4' foundation wall		S.F. Ground	5.28	1.76	
1020	Special Foundations	N/A		—	—	—	
1030	Slab on Grade	4" reinforced concrete with vapor barrier and granular base		S.F. Slab	5.80	1.93	2.2%
2010	Basement Excavation	Site preparation for slab and trench for foundation wall and footing		S.F. Ground	.18	.06	
2020	Basement Walls	N/A		—	—	—	
B.	**SHELL**						
	B10 Superstructure						
1010	Floor Construction	Concrete slab with metal deck and beams		S.F. Floor	36.48	24.32	18.1%
1020	Roof Construction	Concrete slab with metal deck and beams		S.F. Roof	20.79	6.93	
	B20 Exterior Enclosure						
2010	Exterior Walls	Face brick with concrete block backup	75% of wall	S.F. Wall	38.63	10.43	
2020	Exterior Windows	Horizontal pivoted steel	25% of wall	Each	768	7.68	10.7%
2030	Exterior Doors	Double aluminum and glass and hollow metal		Each	4722	.47	
	B30 Roofing						
3010	Roof Coverings	Built-up tar and gravel with flashing; perlite/EPS composite insulation		S.F. Roof	6.72	2.24	1.3%
3020	Roof Openings	N/A		—	—	—	
C.	**INTERIORS**						
1010	Partitions	Plaster on metal studs	10 S.F. Floor/L.F. Partition	S.F. Partition	14.70	14.70	
1020	Interior Doors	Single leaf wood	100 S.F. Floor/Door	Each	720	7.20	
1030	Fittings	Toilet partitions		S.F. Floor	.26	.26	
2010	Stair Construction	Concrete filled metal pan		Flight	22,075	1.84	31.2%
3010	Wall Finishes	70% paint, 20% wood paneling, 10% vinyl wall covering		S.F. Surface	2.28	4.56	
3020	Floor Finishes	60% hardwood, 20% terrazzo, 20% carpet		S.F. Floor	14.32	14.32	
3030	Ceiling Finishes	Gypsum plaster on metal lath, suspended		S.F. Ceiling	11.10	11.10	
D.	**SERVICES**						
	D10 Conveying						
1010	Elevators & Lifts	Five hydraulic passenger elevators		Each	145,080	12.09	7.0%
1020	Escalators & Moving Walks	N/A		—	—	—	
	D20 Plumbing						
2010	Plumbing Fixtures	Toilet and service fixtures, supply and drainage	1 Fixture/665 S.F. Floor	Each	3571	5.37	
2020	Domestic Water Distribution	Electric water heater		S.F. Floor	6.43	6.43	7.4%
2040	Rain Water Drainage	Roof drains		S.F. Roof	2.79	.93	
	D30 HVAC						
3010	Energy Supply	N/A		—	—	—	
3020	Heat Generating Systems	Included in D3050		—	—	—	
3030	Cooling Generating Systems	N/A		—	—	—	12.4 %
3050	Terminal & Package Units	Multizone unit, gas heating, electric cooling		S.F. Floor	21.45	21.45	
3090	Other HVAC Sys. & Equipment	N/A		—	—	—	
	D40 Fire Protection						
4010	Sprinklers	Wet pipe sprinkler system		S.F. Floor	3.21	3.21	2.1%
4020	Standpipes	Standpipe, wet, Class III		S.F. Floor	.49	.49	
	D50 Electrical						
5010	Electrical Service/Distribution	800 ampere service, panel board and feeders		S.F. Floor	.96	.96	
5020	Lighting & Branch Wiring	High efficiency fluorescent fixtures, receptacles, switches, A.C. and misc. power		S.F. Floor	10.46	10.46	7.7%
5030	Communications & Security	Addressable alarm systems, internet wiring, and emergency lighting		S.F. Floor	1.59	1.59	
5090	Other Electrical Systems	Emergency generator, 15 kW		S.F. Floor	.24	.24	
E.	**EQUIPMENT & FURNISHINGS**						
1010	Commercial Equipment	N/A		—	—	—	
1020	Institutional Equipment	N/A		—	—	—	
1030	Vehicular Equipment	N/A		—	—	—	0.0 %
1090	Other Equipment	N/A		—	—	—	
F.	**SPECIAL CONSTRUCTION**						
1020	Integrated Construction	N/A		—	—	—	0.0 %
1040	Special Facilities	N/A		—	—	—	
G.	**BUILDING SITEWORK**	**N/A**					

		Sub-Total	173.02	100%
CONTRACTOR FEES (General Requirements: 10%, Overhead: 5%, Profit: 10%)		25%	43.25	
ARCHITECT FEES		6%	12.98	
	Total Building Cost		**229.25**	

For customer support on your Square Foot Costs with RSMeans data, call 800.448.8182.

119

Costs per square foot of floor area

Exterior Wall	S.F. Area	2000	5000	7000	10000	12000	15000	18000	21000	25000
	L.F. Perimeter	200	310	360	440	480	520	560	600	660
Tilt-up Concrete Panels	Steel Joists	227.25	194.95	186.80	181.05	178.10	174.45	171.95	170.15	168.65
Decorative Concrete Block	Bearing Walls	233.75	198.25	189.25	182.80	179.55	175.45	172.70	170.70	169.05
Face Brick and Concrete Block	Steel Joists	270.25	220.80	208.15	199.10	194.55	188.75	184.80	182.00	179.60
Stucco and Concrete Block	Wood Truss	226.30	195.50	187.60	181.95	179.05	175.35	172.90	171.10	169.55
Brick Veneer	Wood Frame	237.55	203.05	194.05	187.65	184.35	180.10	177.30	175.25	173.55
Wood Clapboard	Wood Frame	221.00	192.75	185.55	180.35	177.75	174.40	172.15	170.50	169.15
Perimeter Adj., Add or Deduct	Per 100 L.F.	29.45	11.75	8.40	5.90	4.85	3.85	3.25	2.80	2.35
Story Hgt. Adj., Add or Deduct	Per 1 Ft.	2.90	1.80	1.45	1.25	1.15	0.95	0.90	0.85	0.80

For Basement, add $33.70 per square foot of basement area

The above costs were calculated using the basic specifications shown on the facing page. These costs should be adjusted where necessary for design alternatives and owner's requirements.

Common additives

Description	Unit	$ Cost
Emergency Lighting, 25 watt, battery operated		
Lead battery	Ea.	330
Nickel cadmium	Ea.	550
Flagpoles, Complete		
Aluminum, 20' high	Ea.	1900
40' high	Ea.	4500
70' high	Ea.	11,600
Fiberglass, 23' high	Ea.	1450
39'-5" high	Ea.	3325
59' high	Ea.	7300
Gym Floor, Incl. sleepers and finish, maple	S.F.	16.70
Intercom System, 25 Station capacity		
Master station	Ea.	3225
Intercom outlets	Ea.	212
Handset	Ea.	590

Description	Unit	$ Cost
Lockers, Steel, single tier, 60" to 72"	Opng.	232 - 395
2 tier, 60" to 72" total	Opng.	139 - 172
5 tier, box lockers	Opng.	73 - 87
Locker bench, lam. maple top only	L.F.	36.50
Pedestals, steel pipe	Ea.	79
Smoke Detectors		
Ceiling type	Ea.	244
Duct type	Ea.	585
Sound System		
Amplifier, 250 watts	Ea.	2025
Speaker, ceiling or wall	Ea.	240
Trumpet	Ea.	460

Important: See the Reference Section for Location Factors.

Model costs calculated for a 1 story building with 12' story height and 10,000 square feet of floor area

				Unit	Unit Cost	Cost Per S.F.	% Of Sub-Total
A. SUBSTRUCTURE							
1010	Standard Foundations	Poured concrete; strip and spread footings; 4' foundation wall		S.F. Ground	6.27	6.27	
1020	Special Foundations	N/A		—	—	—	
1030	Slab on Grade	4" concrete with vapor barrier and granular base		S.F. Slab	5.25	5.25	8.6%
2010	Basement Excavation	Site preparation for slab and trench for foundation wall and footing		S.F. Ground	.33	.33	
2020	Basement Walls	N/A		—	—	—	
B. SHELL							
	B10 Superstructure						
1010	Floor Construction	Wood beams on columns		S.F. Floor	.24	.24	6.1%
1020	Roof Construction	Wood trusses		S.F. Roof	8.10	8.10	
	B20 Exterior Enclosure						
2010	Exterior Walls	Brick veneer on wood studs	85% of wall	S.F. Wall	26.76	12.01	
2020	Exterior Windows	Window wall	15% of wall	Each	56	4.42	16.8%
2030	Exterior Doors	Aluminum and glass; steel		Each	3513	6.68	
	B30 Roofing						
3010	Roof Coverings	Asphalt shingles, 9" fiberglass batt insulation, gutters and downspouts		S.F. Roof	4.55	4.55	3.3%
3020	Roof Openings	N/A		—	—	—	
C. INTERIORS							
1010	Partitions	Gypsum board on wood studs	8 S.F. Floor/S.F. Partition	S.F. Partition	8.45	5.63	
1020	Interior Doors	Single leaf hollow metal	700 S.F. Floor/Door	Each	1222	3.26	
1030	Fittings	Toilet partitions		S.F. Floor	.47	.47	
2010	Stair Construction	N/A		—	—	—	15.7%
3010	Wall Finishes	Paint		S.F. Surface	1.70	2.26	
3020	Floor Finishes	5% quarry tile, 95% vinyl composition tile		S.F. Floor	3.72	3.72	
3030	Ceiling Finishes	Fiberglass tile on tee grid		S.F. Ceiling	6.30	6.30	
D. SERVICES							
	D10 Conveying						
1010	Elevators & Lifts	N/A		—	—	—	0.0 %
1020	Escalators & Moving Walks	N/A		—	—	—	
	D20 Plumbing						
2010	Plumbing Fixtures	Toilet and service fixtures, supply and drainage	1 Fixture/455 S.F. Floor	Each	3225	18.12	
2020	Domestic Water Distribution	Electric water heater		S.F. Floor	3.06	3.06	15.4%
2040	Rain Water Drainage	N/A		—	—	—	
	D30 HVAC						
3010	Energy Supply	Oil fired hot water, wall fin radiation		S.F. Floor	11.98	11.98	
3020	Heat Generating Systems	N/A		—	—	—	
3030	Cooling Generating Systems	N/A		—	—	—	18.7%
3050	Terminal & Package Units	Split systems with air cooled condensing units		S.F. Floor	13.81	13.81	
3090	Other HVAC Sys. & Equipment	N/A		—	—	—	
	D40 Fire Protection						
4010	Sprinklers	Sprinkler, light hazard		S.F. Floor	3.79	3.79	3.4%
4020	Standpipes	Standpipe		S.F. Floor	.95	.95	
	D50 Electrical						
5010	Electrical Service/Distribution	200 ampere service, panel board and feeders		S.F. Floor	1.14	1.14	
5020	Lighting & Branch Wiring	High efficiency fluorescent fixtures, receptacles, switches, A.C. and misc. power		S.F. Floor	8.56	8.56	8.3%
5030	Communications & Security	Addressable alarm systems and emergency lighting		S.F. Floor	1.58	1.58	
5090	Other Electrical Systems	Emergency generator, 15 kW		S.F. Floor	.09	.09	
E. EQUIPMENT & FURNISHINGS							
1010	Commercial Equipment	N/A		—	—	—	
1020	Institutional Equipment	Cabinets and countertop		S.F. Floor	5.16	5.16	3.7 %
1030	Vehicular Equipment	N/A		—	—	—	
1090	Other Equipment	N/A		—	—	—	
F. SPECIAL CONSTRUCTION							
1020	Integrated Construction	N/A		—	—	—	0.0 %
1040	Special Facilities	N/A		—	—	—	
G. BUILDING SITEWORK	**N/A**						

	Sub-Total	137.73	**100%**
CONTRACTOR FEES (General Requirements: 10%, Overhead: 5%, Profit: 10%)	25%	34.43	
ARCHITECT FEES	9%	15.49	
	Total Building Cost	**187.65**	

For customer support on your Square Foot Costs with RSMeans data, call 800.448.8182.

Costs per square foot of floor area

Exterior Wall	S.F. Area	12000	18000	24000	30000	36000	42000	48000	54000	60000
	L.F. Perimeter	460	580	713	730	826	880	965	1006	1045
Concrete Block	Rigid Steel	146.00	138.10	134.65	129.75	128.05	126.10	125.15	123.80	122.75
	Bearing Walls	143.55	135.70	132.25	127.35	125.65	123.70	122.70	121.35	120.30
Precast Concrete	Rigid Steel	155.75	146.30	142.20	135.95	133.85	131.45	130.25	128.50	127.15
Metal Panel	Rigid Steel	148.10	139.85	136.30	131.05	129.30	127.30	126.20	124.85	123.65
Face Brick and Common Brick	Rigid Steel	167.45	156.15	151.30	143.35	140.85	137.85	136.40	134.20	132.45
Tilt-up Concrete Panels	Rigid Steel	147.60	139.50	135.85	130.75	129.00	127.00	125.95	124.55	123.45
Perimeter Adj., Add or Deduct	Per 100 L.F.	5.95	4.00	2.95	2.40	2.00	1.80	1.50	1.30	1.15
Story Hgt. Adj., Add or Deduct	Per 1 Ft.	0.80	0.75	0.65	0.55	0.50	0.50	0.45	0.45	0.35

For Basement, add $34.30 per square foot of basement area

The above costs were calculated using the basic specifications shown on the facing page. These costs should be adjusted where necessary for design alternatives and owner's requirements.

Common additives

Description	Unit	$ Cost
Clock System		
20 room	Ea.	20,200
50 room	Ea.	48,200
Dock Bumpers, Rubber blocks		
4-1/2" thick, 10" high, 14" long	Ea.	84
24" long	Ea.	136
36" long	Ea.	335
12" high, 14" long	Ea.	106
24" long	Ea.	113
36" long	Ea.	170
6" thick, 10" high, 14" long	Ea.	105
24" long	Ea.	172
36" long	Ea.	269
20" high, 11" long	Ea.	185
Dock Boards, Heavy		
60" x 60" Aluminum, 5,000# cap.	Ea.	1550
9000# cap.	Ea.	1475
15,000# cap.	Ea.	1400

Description	Unit	$ Cost
Dock Levelers, Hinged 10 ton cap.		
6' x 8'	Ea.	6250
7' x 8'	Ea.	8625
Partitions, Woven wire, 10 ga., 1-1/2" mesh		
4' wide x 7' high	Ea.	208
8' high	Ea.	225
10' High	Ea.	284
Platform Lifter, Portable, 6'x 6'		
3000# cap.	Ea.	10,500
4000# cap.	Ea.	12,900
Fixed, 6' x 8', 5000# cap.	Ea.	12,800

Important: See the Reference Section for Location Factors.

Model costs calculated for a 1 story building with 20' story height and 30,000 square feet of floor area

				Unit	Unit Cost	Cost Per S.F.	% Of Sub-Total
A. SUBSTRUCTURE							
1010	Standard Foundations	Poured concrete; strip and spread footings; 4' foundation wall		S.F. Ground	4.30	4.30	
1020	Special Foundations	N/A		—	—	—	
1030	Slab on Grade	4" reinforced concrete with vapor barrier and granular base		S.F. Slab	7.47	7.47	12.3%
2010	Basement Excavation	Site preparation for slab and trench for foundation wall and footing		S.F. Ground	.18	.18	
2020	Basement Walls	N/A		—	—	—	
B. SHELL							
	B10 Superstructure						
1010	Floor Construction	N/A		—	—	—	
1020	Roof Construction	Metal deck, open web steel joists, beams and columns		S.F. Roof	10.40	10.40	10.7 %
	B20 Exterior Enclosure						
2010	Exterior Walls	Concrete block	75% of wall	S.F. Wall	9.59	3.50	
2020	Exterior Windows	Industrial horizontal pivoted steel	25% of wall	Each	873	3.31	8.7%
2030	Exterior Doors	Double aluminum and glass, hollow metal, steel overhead		Each	3332	1.66	
	B30 Roofing						
3010	Roof Coverings	Built-up tar and gravel with flashing; perlite/EPS composite insulation		S.F. Roof	6.45	6.45	7.1%
3020	Roof Openings	Roof hatches		S.F. Roof	.46	.46	
C. INTERIORS							
1010	Partitions	Concrete block	60 S.F. Floor/L.F. Partition	S.F. Partition	9.70	1.94	
1020	Interior Doors	Single leaf hollow metal and fire doors	600 S.F. Floor/Door	Each	1222	2.03	
1030	Fittings	Toilet partitions		S.F. Floor	1.04	1.04	
2010	Stair Construction	N/A		—	—	—	8.4%
3010	Wall Finishes	Paint		S.F. Surface	5.13	2.05	
3020	Floor Finishes	Vinyl composition tile	10% of floor	S.F. Floor	3.10	.31	
3030	Ceiling Finishes	Fiberglass board on exposed grid system	10% of area	S.F. Ceiling	7.60	.77	
D. SERVICES							
	D10 Conveying						
1010	Elevators & Lifts	N/A		—	—	—	0.0 %
1020	Escalators & Moving Walks	N/A		—	—	—	
	D20 Plumbing						
2010	Plumbing Fixtures	Toilet and service fixtures, supply and drainage	1 Fixture/1000 S.F. Floor	Each	5320	5.32	
2020	Domestic Water Distribution	Gas fired water heater		S.F. Floor	.67	.67	7.5%
2040	Rain Water Drainage	Roof drains		S.F. Roof	1.28	1.28	
	D30 HVAC						
3010	Energy Supply	Oil fired hot water, unit heaters		S.F. Floor	10.21	10.21	
3020	Heat Generating Systems	N/A		—	—	—	
3030	Cooling Generating Systems	Chilled water, air cooled condenser system		S.F. Floor	11.83	11.83	22.7%
3050	Terminal & Package Units	N/A		—	—	—	
3090	Other HVAC Sys. & Equipment	N/A		—	—	—	
	D40 Fire Protection						
4010	Sprinklers	Sprinklers, ordinary hazard		S.F. Floor	4.35	4.35	5.2%
4020	Standpipes	Standpipe, wet, Class III		S.F. Floor	.69	.69	
	D50 Electrical						
5010	Electrical Service/Distribution	600 ampere service, panel board and feeders		S.F. Floor	1.06	1.06	
5020	Lighting & Branch Wiring	High intensity discharge fixtures, receptacles, switches, A.C. and misc. power		S.F. Floor	13.66	13.66	16.7%
5030	Communications & Security	Addressable alarm systems and emergency lighting		S.F. Floor	1.52	1.52	
5090	Other Electrical Systems	N/A		—	—	—	
E. EQUIPMENT & FURNISHINGS							
1010	Commercial Equipment	N/A		—	—	—	
1020	Institutional Equipment	N/A		—	—	—	
1030	Vehicular Equipment	Dock shelters		S.F. Floor	.54	.54	0.6 %
1090	Other Equipment	N/A		—	—	—	
F. SPECIAL CONSTRUCTION							
1020	Integrated Construction	N/A		—	—	—	0.0 %
1040	Special Facilities	N/A		—	—	—	
G. BUILDING SITEWORK	**N/A**						

				Sub-Total	97	100%
	CONTRACTOR FEES (General Requirements: 10%, Overhead: 5%, Profit: 10%)		25%	24.26		
	ARCHITECT FEES		7%	8.49		

Total Building Cost	**129.75**

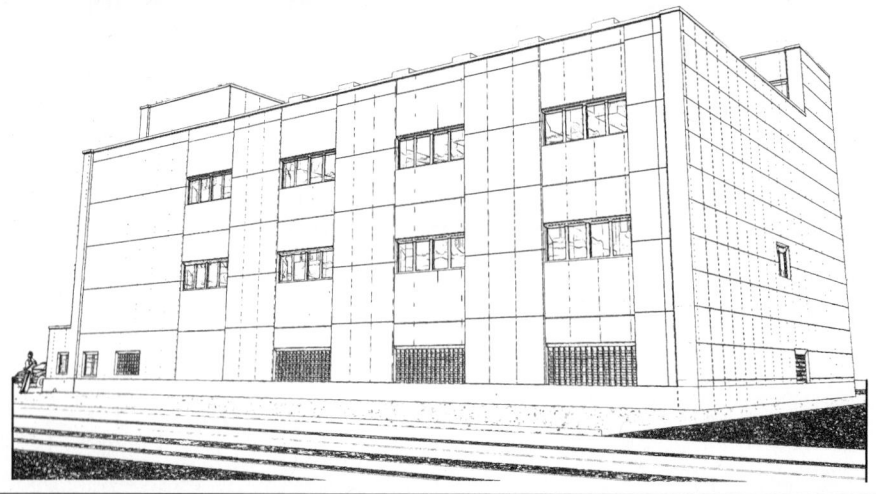

Costs per square foot of floor area

Exterior Wall	S.F. Area	20000	30000	40000	50000	60000	70000	80000	90000	100000
	L.F. Perimeter	625	750	850	950	1015	1095	1150	1275	1300
Face Brick and Common Brick	Rigid Steel	211.65	191.30	179.50	172.35	166.05	162.15	158.35	157.45	154.10
	Reinforced Concrete	203.10	182.75	170.90	163.80	157.50	153.55	149.75	**148.95**	145.60
Face Brick and Concrete Block	Rigid Steel	209.35	189.45	177.95	171.00	164.80	161.00	157.25	156.40	153.15
	Reinforced Concrete	200.80	180.90	169.40	162.40	156.25	152.40	148.70	147.90	144.60
Stucco and Concrete Block	Rigid Steel	189.40	173.55	164.40	158.90	154.05	151.00	148.15	147.40	144.90
	Reinforced Concrete	180.90	164.95	155.85	150.30	145.45	142.45	139.55	138.85	136.30
Perimeter Adj., Add or Deduct	Per 100 L.F.	13.50	9.00	6.75	5.40	4.55	3.85	3.40	3.00	2.70
Story Hgt. Adj., Add or Deduct	Per 1 Ft.	6.45	5.10	4.40	3.95	3.50	3.20	2.95	2.95	2.75

For Basement, add $37.40 per square foot of basement area

The above costs were calculated using the basic specifications shown on the facing page. These costs should be adjusted where necessary for design alternatives and owner's requirements.

Common additives

Description	Unit	$ Cost
Clock System		
20 room	Ea.	20,200
50 room	Ea.	48,200
Dock Bumpers, Rubber blocks		
4-1/2" thick, 10" high, 14" long	Ea.	84
24" long	Ea.	136
36" long	Ea.	335
12" high, 14" long	Ea.	106
24" long	Ea.	113
36" long	Ea.	170
6" thick, 10" high, 14" long	Ea.	105
24" long	Ea.	172
36" long	Ea.	269
20" high, 11" long	Ea.	185
Dock Boards, Heavy		
60" x 60" Aluminum, 5,000# cap.	Ea.	1550
9000# cap.	Ea.	1475
15,000# cap.	Ea.	1400

Description	Unit	$ Cost
Dock Levelers, Hinged 10 ton cap.		
6' x 8'	Ea.	6250
7' x 8'	Ea.	8625
Elevator, Hydraulic freight, 2 stops		
3500# capacity	Ea.	145,100
4000# capacity	Ea.	150,100
Additional stop, add	Ea.	9750
Partitions, Woven wire, 10 ga., 1-1/2" mesh		
4' Wide x 7" high	Ea.	208
8' High	Ea.	225
10' High	Ea.	284
Platform Lifter, Portable, 6'x 6'		
3000# cap.	Ea.	10,500
4000# cap.	Ea.	12,900
Fixed, 6'x 8', 5000# cap.	Ea.	12,800

Important: See the Reference Section for Location Factors.

Model costs calculated for a 3 story building with 12' story height and 90,000 square feet of floor area

Factory, 3 Story

				Unit	Unit Cost	Cost Per S.F.	% Of Sub-Total
A. SUBSTRUCTURE							
1010	Standard Foundations	Poured concrete; strip and spread footings; 4' foundation wall		S.F. Ground	6.24	2.08	
1020	Special Foundations	N/A		—	—	—	
1030	Slab on Grade	4" reinforced concrete with vapor barrier and granular base		S.F. Slab	11.56	3.85	5.3%
2010	Basement Excavation	Site preparation for slab and trench for foundation wall and footing		S.F. Ground	.18	.06	
2020	Basement Walls	N/A		—	—	—	
B. SHELL							
B10 Superstructure							
1010	Floor Construction	Concrete flat slab		S.F. Floor	23.55	15.70	19.1%
1020	Roof Construction	Concrete flat slab		S.F. Roof	17.19	5.73	
B20 Exterior Enclosure							
2010	Exterior Walls	Face brick with common brick backup	70% of wall	S.F. Wall	37.45	13.37	
2020	Exterior Windows	Industrial, horizontal pivoted steel	30% of wall	Each	1371	13.11	24.1%
2030	Exterior Doors	Double aluminum & glass, hollow metal, overhead doors		Each	3222	.61	
B30 Roofing							
3010	Roof Coverings	Built-up tar and gravel with flashing; perlite/EPS composite insulation		S.F. Roof	7.02	2.34	2.2%
3020	Roof Openings	Roof hatches		S.F. Roof	.48	.16	
C. INTERIORS							
1010	Partitions	Gypsum board on metal studs	50 S.F. Floor/L.F. Partition	S.F. Partition	4.50	.90	
1020	Interior Doors	Single leaf fire doors	500 S.F. Floor/Door	Each	1452	2.90	
1030	Fittings	Toilet partitions		S.F. Floor	.52	.52	
2010	Stair Construction	Concrete		Flight	8950	.39	8.2%
3010	Wall Finishes	Paint		S.F. Surface	.90	.36	
3020	Floor Finishes	90% metallic hardener, 10% vinyl composition tile		S.F. Floor	3.48	3.48	
3030	Ceiling Finishes	Fiberglass board on exposed grid systems	10% of area	S.F. Ceiling	6.30	.63	
D. SERVICES							
D10 Conveying							
1010	Elevators & Lifts	Two hydraulic freight elevators		Each	205,650	4.57	4.1%
1020	Escalators & Moving Walks	N/A		—	—	—	
D20 Plumbing							
2010	Plumbing Fixtures	Toilet and service fixtures, supply and drainage	1 Fixture/1345 S.F. Floor	Each	8917	6.63	
2020	Domestic Water Distribution	Gas fired water heater		S.F. Floor	.40	.40	6.8%
2040	Rain Water Drainage	Roof drains		S.F. Roof	1.71	.57	
D30 HVAC							
3010	Energy Supply	Oil fired hot water, unit heaters		S.F. Floor	5.20	5.20	
3020	Heat Generating Systems	N/A		—	—	—	
3030	Cooling Generating Systems	Chilled water, air cooled condenser system		S.F. Floor	11.83	11.83	15.2%
3050	Terminal & Package Units	N/A		—	—	—	
3090	Other HVAC Sys. & Equipment	N/A		—	—	—	
D40 Fire Protection							
4010	Sprinklers	Wet pipe sprinkler system		S.F. Floor	3.81	3.81	4.0%
4020	Standpipes	Standpipe, wet, Class III		S.F. Floor	.67	.67	
D50 Electrical							
5010	Electrical Service/Distribution	800 ampere service, panel board and feeders		S.F. Floor	.63	.63	
5020	Lighting & Branch Wiring	High intensity discharge fixtures, switches, A.C. and misc. power		S.F. Floor	10.25	10.25	11.0%
5030	Communications & Security	Addressable alarm systems and emergency lighting		S.F. Floor	1.25	1.25	
5090	Other Electrical Systems	Emergency generator, 30 kW		S.F. Floor	.21	.21	
E. EQUIPMENT & FURNISHINGS							
1010	Commercial Equipment	N/A		—	—	—	
1020	Institutional Equipment	N/A		—	—	—	
1030	Vehicular Equipment	Dock Shelters		S.F. Floor	.18	.18	0.2 %
1090	Other Equipment	N/A		—	—	—	
F. SPECIAL CONSTRUCTION							
1020	Integrated Construction	N/A		—	—	—	
1040	Special Facilities	N/A		—	—	—	0.0 %
G. BUILDING SITEWORK	**N/A**						

				Sub-Total	112.39	100%
	CONTRACTOR FEES (General Requirements: 10%, Overhead: 5%, Profit: 10%)			25%	28.13	
	ARCHITECT FEES			6%	8.43	
				Total Building Cost	**148.95**	

For customer support on your Square Foot Costs with RSMeans data, call 800.448.8182.

125

Costs per square foot of floor area

Exterior Wall	S.F. Area	4000	4500	5000	5500	6000	6500	7000	7500	8000
	L.F. Perimeter	260	280	300	310	320	336	353	370	386
Face Brick and Concrete Block	Steel Joists	202.85	198.80	195.65	191.40	187.90	185.70	183.95	182.50	181.05
	Bearing Walls	198.75	194.75	191.55	187.30	183.80	181.65	179.85	178.40	177.00
Decorative Concrete Block	Steel Joists	186.60	183.25	180.65	177.25	174.55	172.75	171.35	170.15	169.00
	Bearing Walls	182.55	179.15	176.55	173.20	170.45	168.75	167.30	166.10	164.90
Limestone and Concrete Block	Steel Joists	222.10	217.25	213.40	208.05	203.70	201.05	198.90	197.15	195.40
	Bearing Walls	218.05	213.15	209.35	204.00	199.60	196.95	194.85	193.10	191.30
Perimeter Adj., Add or Deduct	Per 100 L.F.	22.15	19.70	17.75	16.10	14.80	13.65	12.70	11.85	11.15
Story Hgt. Adj., Add or Deduct	Per 1 Ft.	3.00	2.85	2.75	2.55	2.45	2.40	2.30	2.30	2.25

For Basement, add $41.90 per square foot of basement area

The above costs were calculated using the basic specifications shown on the facing page. These costs should be adjusted where necessary for design alternatives and owner's requirements.

Common additives

Description	Unit	$ Cost
Appliances		
Cooking range, 30" free standing		
1 oven	Ea.	610 - 2775
2 oven	Ea.	1200 - 3725
30" built-in		
1 oven	Ea.	995 - 2550
2 oven	Ea.	1575 - 3125
Counter top cook tops, 4 burner	Ea.	475 - 2150
Microwave oven	Ea.	281 - 855
Combination range, refrig. & sink, 30" wide	Ea.	2075 - 2625
60" wide	Ea.	2425
72" wide	Ea.	3075
Combination range refrigerator, sink		
microwave oven & icemaker	Ea.	7100
Compactor, residential, 4-1 compaction	Ea.	885 - 1475
Dishwasher, built-in, 2 cycles	Ea.	685 - 1225
4 cycles	Ea.	800 - 2025
Garbage disposer, sink type	Ea.	260 - 370
Hood for range, 2 speed, vented, 30" wide	Ea.	370 - 1450
42" wide	Ea.	450 - 2650

Description	Unit	$ Cost
Appliances, cont.		
Refrigerator, no frost 10-12 C.F.	Ea.	580 - 750
14-16 C.F.	Ea.	720 - 1025
18-20 C.F.	Ea.	895 - 2075
Lockers, Steel, single tier, 60" or 72"	Opng.	232 - 395
2 tier, 60" or 72" total	Opng.	139 - 172
5 tier, box lockers	Opng.	73 - 87
Locker bench, lam. maple top only	L.F.	36.50
Pedestals, steel pipe	Ea.	79
Sound System		
Amplifier, 250 watts	Ea.	2025
Speaker, ceiling or wall	Ea.	240
Trumpet	Ea.	460

Important: See the Reference Section for Location Factors.

Model costs calculated for a 1 story building with 14' story height and 6,000 square feet of floor area

Fire Station, 1 Story

				Unit	Unit Cost	Cost Per S.F.	% Of Sub-Total
A. SUBSTRUCTURE							
1010	Standard Foundations	Poured concrete; strip and spread footings; 4' foundation wall		S.F. Ground	7.69	7.69	
1020	Special Foundations	N/A		—	—	—	
1030	Slab on Grade	6" reinforced concrete with vapor barrier and granular base		S.F. Slab	7.47	7.47	11.3%
2010	Basement Excavation	Site preparation for slab and trench for foundation wall and footing		S.F. Ground	.57	.57	
2020	Basement Walls	N/A		—	—	—	
B. SHELL							
B10 Superstructure							
1010	Floor Construction	N/A		—	—	—	
1020	Roof Construction	Metal deck, open web steel joists, beams on columns		S.F. Roof	11.34	11.34	8.1 %
B20 Exterior Enclosure							
2010	Exterior Walls	Face brick with concrete block backup	75% of wall	S.F. Wall	35.25	19.74	
2020	Exterior Windows	Aluminum insulated glass	10% of wall	Each	873	2.04	19.7%
2030	Exterior Doors	Single aluminum and glass, overhead, hollow metal	15% of wall	S.F. Door	50	5.62	
B30 Roofing							
3010	Roof Coverings	Built-up tar and gravel with flashing; perlite/EPS composite insulation		S.F. Roof	7.78	7.78	5.7%
3020	Roof Openings	Skylights, roof hatches		S.F. Roof	.19	.19	
C. INTERIORS							
1010	Partitions	Concrete block	17 S.F. Floor/L.F. Partition	S.F. Partition	10.40	6.12	
1020	Interior Doors	Single leaf hollow metal	500 S.F. Floor/Door	Each	1222	2.44	
1030	Fittings	Toilet partitions		S.F. Floor	.52	.52	
2010	Stair Construction	N/A		—	—	—	13.8%
3010	Wall Finishes	Paint		S.F. Surface	3.27	3.85	
3020	Floor Finishes	50% vinyl tile, 50% paint		S.F. Floor	2.52	2.52	
3030	Ceiling Finishes	Fiberglass board on exposed grid, suspended	50% of area	S.F. Ceiling	7.60	3.81	
D. SERVICES							
D10 Conveying							
1010	Elevators & Lifts	N/A		—	—	—	0.0 %
1020	Escalators & Moving Walks	N/A		—	—	—	
D20 Plumbing							
2010	Plumbing Fixtures	Kitchen, toilet and service fixtures, supply and drainage	1 Fixture/375 S.F. Floor	Each	4185	11.16	
2020	Domestic Water Distribution	Gast fired water heater		S.F. Floor	3.40	3.40	11.7%
2040	Rain Water Drainage	Roof drains		S.F. Roof	1.75	1.75	
D30 HVAC							
3010	Energy Supply	N/A		—	—	—	
3020	Heat Generating Systems	Included in D3050		—	—	—	
3030	Cooling Generating Systems	N/A		—	—	—	18.0 %
3050	Terminal & Package Units	Rooftop multizone unit system		S.F. Floor	25.09	25.09	
3090	Other HVAC Sys. & Equipment	N/A		—	—	—	
D40 Fire Protection							
4010	Sprinklers	Wet pipe sprinkler system		S.F. Floor	5.13	5.13	4.8%
4020	Standpipes	Standpipe, wet, Class III		S.F. Floor	1.59	1.59	
D50 Electrical							
5010	Electrical Service/Distribution	200 ampere service, panel board and feeders		S.F. Floor	1.66	1.66	
5020	Lighting & Branch Wiring	High efficiency fluorescent fixtures, receptacles, switches, A.C. and misc. power		S.F. Floor	5.66	5.66	6.7%
5030	Communications & Security	Addressable alarm systems		S.F. Floor	2.03	2.03	
5090	Other Electrical Systems	N/A		—	—	—	
E. EQUIPMENT & FURNISHINGS							
1010	Commercial Equipment	N/A		—	—	—	
1020	Institutional Equipment	N/A		—	—	—	0.0 %
1030	Vehicular Equipment	N/A		—	—	—	
1090	Other Equipment	N/A		—	—	—	
F. SPECIAL CONSTRUCTION							
1020	Integrated Construction	N/A		—	—	—	0.0 %
1040	Special Facilities	N/A		—	—	—	
G. BUILDING SITEWORK	**N/A**						

			Sub-Total	139.17	**100%**
CONTRACTOR FEES (General Requirements: 10%, Overhead: 5%, Profit: 10%)		25%	34.81	
ARCHITECT FEES		8%	13.92	

Total Building Cost 187.90

For customer support on your Square Foot Costs with RSMeans data, call 800.448.8182.

127

Costs per square foot of floor area

Exterior Wall	S.F. Area	6000	7000	8000	9000	10000	11000	12000	13000	14000
	L.F. Perimeter	220	240	260	280	286	303	320	336	353
Face Brick and Concrete Block	Steel Joists	214.40	207.45	202.30	198.25	192.90	190.05	187.60	185.55	183.85
	Precast Concrete	225.95	219.00	213.80	209.80	204.45	201.50	199.10	197.00	195.30
Decorative Concrete Block	Steel Joists	197.80	191.80	187.35	183.80	179.40	176.95	174.85	173.10	171.60
	Precast Concrete	211.80	205.80	201.35	197.85	193.40	190.95	188.85	187.10	185.70
Limestone and Concrete Block	Steel Joists	233.65	225.30	219.10	214.20	207.35	203.80	200.85	198.35	196.30
	Precast Concrete	247.65	239.30	233.10	228.20	221.35	217.85	214.90	212.35	210.30
Perimeter Adj., Add or Deduct	Per 100 L.F.	25.95	22.25	19.40	17.30	15.55	14.15	12.95	11.90	11.10
Story Hgt. Adj., Add or Deduct	Per 1 Ft.	3.40	3.20	3.05	2.90	2.70	2.50	2.45	2.30	2.30
For Basement, add $40.70 per square foot of basement area										

The above costs were calculated using the basic specifications shown on the facing page. These costs should be adjusted where necessary for design alternatives and owner's requirements.

Common additives

Description	Unit	$ Cost
Appliances		
Cooking range, 30" free standing		
1 oven	Ea.	610 - 2775
2 oven	Ea.	1200 - 3725
30" built-in		
1 oven	Ea.	995 - 2550
2 oven	Ea.	1575 - 3125
Counter top cook tops, 4 burner	Ea.	475 - 2150
Microwave oven	Ea.	281 - 855
Combination range, refrig. & sink, 30" wide	Ea.	2075 - 2625
60" wide	Ea.	2425
72" wide	Ea.	3075
Combination range, refrigerator, sink, microwave oven & icemaker	Ea.	7100
Compactor, residential, 4-1 compaction	Ea.	885 - 1475
Dishwasher, built-in, 2 cycles	Ea.	685 - 1225
4 cycles	Ea.	800 - 2025
Garbage disposer, sink type	Ea.	260 - 370
Hood for range, 2 speed, vented, 30" wide	Ea.	370 - 1450
42" wide	Ea.	450 - 2650

Description	Unit	$ Cost
Appliances, cont.		
Refrigerator, no frost 10-12 C.F.	Ea.	580 - 750
14-16 C.F.	Ea.	720 - 1025
18-20 C.F.	Ea.	895 - 2075
Elevators, Hydraulic passenger, 2 stops		
1500# capacity	Ea.	73,100
2500# capacity	Ea.	76,600
3500# capacity	Ea.	81,600
Lockers, Steel, single tier, 60" or 72"	Opng.	232 - 395
2 tier, 60" or 72" total	Opng.	139 - 172
5 tier, box lockers	Opng.	73 - 87
Locker bench, lam. maple top only	L.F.	36.50
Pedestals, steel pipe	Ea.	79
Sound System		
Amplifier, 250 watts	Ea.	2025
Speaker, ceiling or wall	Ea.	240
Trumpet	Ea.	460

Important: See the Reference Section for Location Factors.

Model costs calculated for a 2 story building with 14' story height and 10,000 square feet of floor area

Fire Station, 2 Story

				Unit	Unit Cost	Cost Per S.F.	% Of Sub-Total
A. SUBSTRUCTURE							
1010	Standard Foundations	Poured concrete; strip and spread footings; 4' foundation wall		S.F. Ground	8.28	4.14	
1020	Special Foundations	N/A		—	—	—	
1030	Slab on Grade	6" reinforced concrete with vapor barrier and granular base		S.F. Slab	7.47	3.74	6.1%
2010	Basement Excavation	Site preparation for slab and trench for foundation wall and footing		S.F. Ground	.57	.29	
2020	Basement Walls	N/A		—	—	—	
B. SHELL							
	B10 Superstructure						
1010	Floor Construction	Open web steel joists, slab form, concrete		S.F. Floor	16.98	8.49	8.5%
1020	Roof Construction	Metal deck on open web steel joists		S.F. Roof	5.60	2.80	
	B20 Exterior Enclosure						
2010	Exterior Walls	Decorative concrete block	75% of wall	S.F. Wall	21.69	13.03	
2020	Exterior Windows	Aluminum insulated glass	10% of wall	Each	755	2.63	15.2%
2030	Exterior Doors	Single aluminum and glass, steel overhead, hollow metal	15% of wall	S.F. Door	37.38	4.49	
	B30 Roofing						
3010	Roof Coverings	Built-up tar and gravel with flashing; perlite/EPS composite insulation		S.F. Roof	7.96	3.98	3.0%
3020	Roof Openings	N/A		—	—	—	
C. INTERIORS							
1010	Partitions	Concrete block	10 S.F. Floor/L.F. Partition	S.F. Partition	10.40	6.12	
1020	Interior Doors	Single leaf hollow metal	500 S.F. Floor/Door	Each	1452	2.90	
1030	Fittings	Toilet partitions		S.F. Floor	.47	.47	
2010	Stair Construction	Concrete filled metal pan		Flight	22,075	4.42	18.1%
3010	Wall Finishes	Paint		S.F. Surface	3.81	4.48	
3020	Floor Finishes	50% vinyl tile, 50% paint		S.F. Floor	2.52	2.52	
3030	Ceiling Finishes	Fiberglass board on exposed grid, suspended	50% of area	S.F. Ceiling	6.30	3.16	
D. SERVICES							
	D10 Conveying						
1010	Elevators & Lifts	One hydraulic passenger elevator		Each	91,200	9.12	6.9%
1020	Escalators & Moving Walks	N/A		—	—	—	
	D20 Plumbing						
2010	Plumbing Fixtures	Kitchen toilet and service fixtures, supply and drainage	1 Fixture/400 S.F. Floor	Each	4260	10.65	
2020	Domestic Water Distribution	Gas fired water heater		S.F. Floor	2.68	2.68	10.9%
2040	Rain Water Drainage	Roof drains		S.F. Roof	2.34	1.17	
	D30 HVAC						
3010	Energy Supply	N/A		—	—	—	
3020	Heat Generating Systems	Included in D3050		—	—	—	
3030	Cooling Generating Systems	N/A		—	—	—	18.9 %
3050	Terminal & Package Units	Rooftop multizone unit system		S.F. Floor	25.09	25.09	
3090	Other HVAC Sys. & Equipment	N/A		—	—	—	
	D40 Fire Protection						
4010	Sprinklers	Wet pipe sprinkler system		S.F. Floor	4.08	4.08	4.3%
4020	Standpipes	Standpipe, wet, Class III		S.F. Floor	1.64	1.64	
	D50 Electrical						
5010	Electrical Service/Distribution	400 ampere service, panel board and feeders		S.F. Floor	1.67	1.67	
5020	Lighting & Branch Wiring	High efficiency fluorescent fixtures, receptacles, switches, A.C. and misc. power		S.F. Floor	7.30	7.30	8.1%
5030	Communications & Security	Addressable alarm systems and emergency lighting		S.F. Floor	1.58	1.58	
5090	Other Electrical Systems	Emergency generator, 15 kW		S.F. Floor	.26	.26	
E. EQUIPMENT & FURNISHINGS							
1010	Commercial Equipment	N/A		—	—	—	
1020	Institutional Equipment	N/A		—	—	—	
1030	Vehicular Equipment	N/A		—	—	—	0.0 %
1090	Other Equipment	N/A		—	—	—	
F. SPECIAL CONSTRUCTION							
1020	Integrated Construction	N/A		—	—	—	0.0 %
1040	Special Facilities	N/A		—	—	—	
G. BUILDING SITEWORK	**N/A**						

	Sub-Total	132.90	**100%**
CONTRACTOR FEES (General Requirements: 10%, Overhead: 5%, Profit: 10%)	25%	33.21	
ARCHITECT FEES	8%	13.29	
	Total Building Cost	179.40	

For customer support on your Square Foot Costs with RSMeans data, call 800.448.8182.

Costs per square foot of floor area

Exterior Wall	S.F. Area	4000	5000	6000	8000	10000	12000	14000	16000	18000
	L.F. Perimeter	180	205	230	260	300	340	353	386	420
Wood Clapboard	Wood Frame	226.45	217.70	211.90	202.95	198.30	195.15	191.55	189.75	188.35
Aluminum Clapboard	Wood Frame	223.20	214.80	209.15	200.60	196.15	193.10	189.75	188.00	186.70
Wood Board and Batten	Wood Frame	225.65	217.00	211.25	202.40	197.75	194.65	191.15	189.30	187.95
Face Brick and Concrete Block	Bearing Walls	231.30	220.80	213.75	202.20	196.40	192.50	187.55	185.20	183.50
Stucco and Concrete Block	Bearing Walls	213.10	204.20	198.25	189.05	184.30	181.00	177.35	175.40	174.05
Decorative Concrete Block	Wood Joists	231.70	222.50	216.40	206.75	201.80	198.40	194.55	192.55	191.05
Perimeter Adj., Add or Deduct	Per 100 L.F.	17.60	14.05	11.75	8.75	6.95	5.85	5.00	4.35	3.90
Story Hgt. Adj., Add or Deduct	Per 1 Ft.	2.25	2.15	2.00	1.60	1.55	1.40	1.35	1.20	1.20

For Basement, add $25.90 per square foot of basement area

The above costs were calculated using the basic specifications shown on the facing page. These costs should be adjusted where necessary for design alternatives and owner's requirements.

Common additives

Description	Unit	$ Cost
Appliances		
Cooking range, 30" free standing		
1 oven	Ea.	610 - 2775
2 oven	Ea.	1200 - 3725
30" built-in		
1 oven	Ea.	995 - 2550
2 oven	Ea.	1575 - 3125
Counter top cook tops, 4 burner	Ea.	475 - 2150
Microwave oven	Ea.	281 - 855
Combination range, refrig. & sink, 30" wide	Ea.	2075 - 2625
60" wide	Ea.	2425
72" wide	Ea.	3075
Combination range, refrigerator, sink,		
microwave oven & icemaker	Ea.	7100
Compactor, residential, 4-1 compaction	Ea.	885 - 1475
Dishwasher, built-in, 2 cycles	Ea.	685 - 1225
4 cycles	Ea.	800 - 2025
Garbage disposer, sink type	Ea.	260 - 370
Hood for range, 2 speed, vented, 30" wide	Ea.	370 - 1450
42" wide	Ea.	450 - 2650

Description	Unit	$ Cost
Appliances, cont.		
Refrigerator, no frost 10-12 C.F.	Ea.	580 - 750
14-16 C.F.	Ea.	720 - 1025
18-20 C.F.	Ea.	895 - 2075
Elevators, Hydraulic passenger, 2 stops		
1500# capacity	Ea.	73,100
2500# capacity	Ea.	76,600
3500# capacity	Ea.	81,600
Laundry Equipment		
Dryer, gas, 16 lb. capacity	Ea.	1025
30 lb. capacity	Ea.	4100
Washer, 4 cycle	Ea.	1325
Commercial	Ea.	1700
Sound System		
Amplifier, 250 watts	Ea.	2025
Speaker, ceiling or wall	Ea.	240
Trumpet	Ea.	460

Important: See the Reference Section for Location Factors.

Model costs calculated for a 2 story building with 10' story height and 10,000 square feet of floor area

Fraternity/Sorority House

				Unit	Unit Cost	Cost Per S.F.	% Of Sub-Total
A. SUBSTRUCTURE							
1010	Standard Foundations	Poured concrete; strip and spread footings; 4' foundation wall		S.F. Ground	7.98	3.99	
1020	Special Foundations	N/A		—	—	—	
1030	Slab on Grade	4" reinforced concrete with vapor barrier and granular base		S.F. Slab	5.80	2.90	4.9%
2010	Basement Excavation	Site preparation for slab and trench for foundation wall and footing		S.F. Ground	.57	.29	
2020	Basement Walls	N/A		—	—	—	
B. SHELL							
B10 Superstructure							
1010	Floor Construction	Plywood on wood joists		S.F. Floor	41.46	20.73	15.7%
1020	Roof Construction	Plywood on wood rafters (pitched)		S.F. Roof	3.91	2.19	
B20 Exterior Enclosure							
2010	Exterior Walls	Cedar bevel siding on wood studs, insulated	80% of wall	S.F. Wall	14.10	6.77	
2020	Exterior Windows	Double hung wood	20% of wall	Each	665	3.19	8.4%
2030	Exterior Doors	Solid core wood		Each	3139	2.20	
B30 Roofing							
3010	Roof Coverings	Asphalt shingles with flashing (pitched); rigid fiberglass insulation		S.F. Roof	5.46	2.73	1.9%
3020	Roof Openings	N/A		—	—	—	
C. INTERIORS							
1010	Partitions	Gypsum board on wood studs	25 S.F. Floor/L.F. Partition	S.F. Partition	8.97	2.87	
1020	Interior Doors	Single leaf wood	200 S.F. Floor/Door	Each	720	3.60	
1030	Fittings	N/A		—	—	—	
2010	Stair Construction	Wood		Flight	2915	.87	14.4%
3010	Wall Finishes	Paint		S.F. Surface	.89	.57	
3020	Floor Finishes	10% hardwood, 70% carpet, 20% ceramic tile		S.F. Floor	8.14	8.14	
3030	Ceiling Finishes	Gypsum board on wood furring		S.F. Ceiling	4.85	4.85	
D. SERVICES							
D10 Conveying							
1010	Elevators & Lifts	One hydraulic passenger elevator		Each	91,200	9.12	6.3%
1020	Escalators & Moving Walks	N/A		—	—	—	
D20 Plumbing							
2010	Plumbing Fixtures	Kitchen toilet and service fixtures, supply and drainage	1 Fixture/150 S.F. Floor	Each	2346	15.64	
2020	Domestic Water Distribution	Gas fired water heater		S.F. Floor	3.65	3.65	13.3%
2040	Rain Water Drainage	N/A		—	—	—	
D30 HVAC							
3010	Energy Supply	Oil fired hot water, baseboard radiation		S.F. Floor	8.43	8.43	
3020	Heat Generating Systems	N/A		—	—	—	
3030	Cooling Generating Systems	N/A		—	—	—	11.2%
3050	Terminal & Package Units	Split system with air cooled condensing unit		S.F. Floor	7.87	7.87	
3090	Other HVAC Sys. & Equipment	N/A		—	—	—	
D40 Fire Protection							
4010	Sprinklers	Wet pipe sprinkler system		S.F. Floor	4.08	4.08	3.6%
4020	Standpipes	Standpipe		S.F. Floor	1.17	1.17	
D50 Electrical							
5010	Electrical Service/Distribution	600 ampere service, panel board and feeders		S.F. Floor	3.96	3.96	
5020	Lighting & Branch Wiring	High efficiency fluorescent fixtures, receptacles, switches, A.C. and misc. power		S.F. Floor	10.58	10.58	20.4%
5030	Communications & Security	Addressable alarm, communication system, internet and phone wiring, and generator set		S.F. Floor	14.93	14.93	
5090	Other Electrical Systems	Emergency generator, 7.5 kW		S.F. Floor	.22	.22	
E. EQUIPMENT & FURNISHINGS							
1010	Commercial Equipment	N/A		—	—	—	
1020	Institutional Equipment	N/A		—	—	—	0.0%
1030	Vehicular Equipment	N/A		—	—	—	
1090	Other Equipment	N/A		—	—	—	
F. SPECIAL CONSTRUCTION							
1020	Integrated Construction	N/A		—	—	—	0.0%
1040	Special Facilities	N/A		—	—	—	
G. BUILDING SITEWORK	**N/A**						

			Sub-Total	145.54	**100%**
CONTRACTOR FEES (General Requirements: 10%, Overhead: 5%, Profit: 10%)			25%	36.39	
ARCHITECT FEES			9%	16.37	
		Total Building Cost		**198.30**	

For customer support on your Square Foot Costs with RSMeans data, call 800.448.8182.

131

Costs per square foot of floor area

Exterior Wall	S.F. Area	4000	6000	8000	10000	12000	14000	16000	18000	20000
	L.F. Perimeter	260	320	384	424	460	484	510	540	576
Wood Board and Batten	Wood Frame	218.35	190.10	176.20	166.65	160.00	154.85	151.05	148.25	146.10
Brick Veneer	Wood Frame	227.60	197.70	183.05	172.65	165.45	159.75	155.55	152.50	150.20
Aluminum Clapboard	Wood Frame	211.75	184.65	171.35	162.30	156.10	151.35	147.80	145.20	143.20
Face Brick and Concrete Block	Wood Truss	234.60	203.25	188.00	176.95	169.35	163.20	158.65	155.40	152.95
Limestone and Concrete Block	Wood Truss	256.00	220.90	203.85	191.00	182.00	174.60	169.20	165.30	162.45
Stucco and Concrete Block	Wood Truss	216.80	188.70	174.90	165.35	158.85	153.75	149.90	147.15	145.05
Perimeter Adj., Add or Deduct	Per 100 L.F.	13.40	8.95	6.70	5.25	4.45	3.80	3.35	2.95	2.70
Story Hgt. Adj., Add or Deduct	Per 1 Ft.	1.90	1.55	1.40	1.20	1.15	1.05	0.95	0.85	0.85

For Basement, add $34.50 per square foot of basement area

The above costs were calculated using the basic specifications shown on the facing page. These costs should be adjusted where necessary for design alternatives and owner's requirements.

Common additives

Description	Unit	$ Cost
Autopsy Table, Standard	Ea.	11,600
Deluxe	Ea.	19,400
Directory Boards, Plastic, glass covered		
30" x 20"	Ea.	640
36" x 48"	Ea.	1625
Aluminum, 24" x 18"	Ea.	630
36" x 24"	Ea.	790
48" x 32"	Ea.	1025
48" x 60"	Ea.	2200
Emergency Lighting, 25 watt, battery operated		
Lead battery	Ea.	330
Nickel cadmium	Ea.	550
Mortuary Refrigerator, End operated		
Two capacity	Ea.	9800
Six capacity	Ea.	18,400

Description	Unit	$ Cost
Planters, Precast concrete		
48" diam., 24" high	Ea.	790
7" diam., 36" high	Ea.	1850
Fiberglass, 36" diam., 24" high	Ea.	825
60" diam., 24" high	Ea.	1350
Smoke Detectors		
Ceiling type	Ea.	244
Duct type	Ea.	585

Important: See the Reference Section for Location Factors.

Model costs calculated for a 1 story building with 12' story height and 10,000 square feet of floor area

Funeral Home

				Unit	Unit Cost	Cost Per S.F.	% Of Sub-Total
A. SUBSTRUCTURE							
1010	Standard Foundations	Poured concrete; strip and spread footings; 4' foundation wall		S.F. Ground	6.18	6.18	
1020	Special Foundations	N/A		—	—	—	
1030	Slab on Grade	4" reinforced concrete with vapor barrier and granular base		S.F. Slab	5.80	5.80	10.1%
2010	Basement Excavation	Site preparation for slab and trench for foundation wall and footing		S.F. Ground	.33	.33	
2020	Basement Walls	N/A		—	—	—	
B. SHELL							
B10 Superstructure							
1010	Floor Construction	Wood beams on columns		S.F. Floor	.89	.89	5.2%
1020	Roof Construction	Plywood on wood truss		S.F. Roof	5.53	5.53	
B20 Exterior Enclosure							
2010	Exterior Walls	1" x 4" vertical T & G redwood siding on wood studs	90% of wall	S.F. Wall	17.10	7.83	
2020	Exterior Windows	Double hung wood	10% of wall	Each	533	2.26	9.9%
2030	Exterior Doors	Wood swinging double doors, single leaf hollow metal		Each	3360	2.02	
B30 Roofing							
3010	Roof Coverings	Single ply membrane, fully adhered; polyisocyanurate sheets		S.F. Roof	6.10	6.10	5.0%
3020	Roof Openings	N/A		—	—	—	
C. INTERIORS							
1010	Partitions	Gypsum board on wood studs with sound deadening board	15 S.F. Floor/L.F. Partition	S.F. Partition	10.16	5.42	
1020	Interior Doors	Single leaf wood	150 S.F. Floor/Door	Each	720	4.80	
1030	Fittings	N/A		—	—	—	
2010	Stair Construction	N/A		—	—	—	23.9%
3010	Wall Finishes	50% wallpaper, 25% wood paneling, 25% paint		S.F. Surface	3.42	3.65	
3020	Floor Finishes	70% carpet, 30% ceramic tile		S.F. Floor	9.07	9.07	
3030	Ceiling Finishes	Fiberglass board on exposed grid, suspended		S.F. Ceiling	6.30	6.30	
D. SERVICES							
D10 Conveying							
1010	Elevators & Lifts	N/A		—	—	—	0.0 %
1020	Escalators & Moving Walks	N/A		—	—	—	
D20 Plumbing							
2010	Plumbing Fixtures	Toilet and service fixtures, supply and drainage	1 Fixture/770 S.F. Floor	Each	3065	3.98	
2020	Domestic Water Distribution	Electric water heater		S.F. Floor	17.46	17.46	17.5%
2040	Rain Water Drainage	Roof drain		S.F. Roof			
D30 HVAC							
3010	Energy Supply	N/A		—	—	—	
3020	Heat Generating Systems	Included in D3030		—	—	—	
3030	Cooling Generating Systems	Multizone rooftop unit, gas heating, electric cooling		S.F. Floor	19.05	19.05	15.6 %
3050	Terminal & Package Units	N/A		—	—	—	
3090	Other HVAC Sys. & Equipment	N/A		—	—	—	
D40 Fire Protection							
4010	Sprinklers	Wet pipe sprinkler system		S.F. Floor	3.79	3.79	3.9%
4020	Standpipes	Standpipe, wet, Class III		S.F. Floor	.95	.95	
D50 Electrical							
5010	Electrical Service/Distribution	400 ampere service, panel board and feeders		S.F. Floor	1.98	1.98	
5020	Lighting & Branch Wiring	High efficiency fluorescent fixtures, receptacles, switches, A.C. and misc. power		S.F. Floor	7.23	7.23	8.9%
5030	Communications & Security	Addressable alarm systems and emergency lighting		S.F. Floor	1.58	1.58	
5090	Other Electrical Systems	Emergency generator, 15 kW		S.F. Floor	.09	.09	
E. EQUIPMENT & FURNISHINGS							
1010	Commercial Equipment	N/A		—	—	—	
1020	Institutional Equipment	N/A		—	—	—	0.0 %
1030	Vehicular Equipment	N/A		—	—	—	
1090	Other Equipment	N/A		—	—	—	
F. SPECIAL CONSTRUCTION							
1020	Integrated Construction	N/A		—	—	—	0.0 %
1040	Special Facilities	N/A		—	—	—	
G. BUILDING SITEWORK	**N/A**						

		Sub-Total	122.29	100%
CONTRACTOR FEES (General Requirements: 10%, Overhead: 5%, Profit: 10%)		25%	30.60	
ARCHITECT FEES		9%	13.76	
	Total Building Cost		**166.65**	

For customer support on your Square Foot Costs with RSMeans data, call 800.448.8182.

133

Costs per square foot of floor area

Exterior Wall	S.F. Area	12000	14000	16000	19000	21000	23000	26000	28000	30000
	L.F. Perimeter	440	474	510	556	583	607	648	670	695
E.I.F.S. and Concrete Block	Rigid Steel	138.60	135.75	133.60	131.00	129.45	128.10	126.70	125.65	124.95
Tilt-up Concrete Panels	Rigid Steel	136.10	133.40	131.50	128.95	127.55	126.30	124.95	124.05	123.35
Face Brick and Concrete Block	Bearing Walls	138.85	135.35	132.85	129.70	127.80	126.20	124.45	123.20	122.30
	Rigid Steel	146.40	142.85	140.35	137.20	135.30	133.70	131.95	130.75	129.80
Stucco and Concrete Block	Bearing Walls	129.95	127.10	125.05	122.50	121.05	119.75	118.35	117.35	116.65
	Rigid Steel	137.45	134.60	132.60	130.05	128.55	127.25	125.85	124.85	124.15
Perimeter Adj., Add or Deduct	Per 100 L.F.	7.35	6.30	5.55	4.65	4.25	3.85	3.35	3.20	2.90
Story Hgt. Adj., Add or Deduct	Per 1 Ft.	1.60	1.40	1.40	1.25	1.25	1.10	1.05	1.05	0.95

For Basement, add $39.30 per square foot of basement area

The above costs were calculated using the basic specifications shown on the facing page. These costs should be adjusted where necessary for design alternatives and owner's requirements.

Common additives

Description	Unit	$ Cost
Emergency Lighting, 25 watt, battery operated		
Lead battery	Ea.	330
Nickel cadmium	Ea.	550
Smoke Detectors		
Ceiling type	Ea.	244
Duct type	Ea.	585
Sound System		
Amplifier, 250 watts	Ea.	2025
Speaker, ceiling or wall	Ea.	240
Trumpet	Ea.	460

Important: See the Reference Section for Location Factors.

Model costs calculated for a 1 story building with 14' story height and 21,000 square feet of floor area

Garage, Auto Sales

				Unit	Unit Cost	Cost Per S.F.	% Of Sub-Total
A. SUBSTRUCTURE							
1010	Standard Foundations	Poured concrete; strip and spread footings; 4' foundation wall		S.F. Ground	4.08	4.08	
1020	Special Foundations	N/A		—	—	—	
1030	Slab on Grade	4" reinforced concrete with vapor barrier and granular base		S.F. Slab	8.15	8.15	13.0%
2010	Basement Excavation	Site preparation for slab and trench for foundation wall and footing		S.F. Ground	.33	.33	
2020	Basement Walls	N/A		—	—	—	
B. SHELL							
	B10 Superstructure						
1010	Floor Construction	N/A		—	—	—	13.0 %
1020	Roof Construction	Metal deck, open web steel joists, beams, columns		S.F. Roof	12.57	12.57	
	B20 Exterior Enclosure						
2010	Exterior Walls	E.I.F.S.	70% of wall	S.F. Wall	18.67	5.08	
2020	Exterior Windows	Window wall	30% of wall	Each	59	6.91	15.8%
2030	Exterior Doors	Double aluminum and glass, hollow metal, steel overhead		Each	3842	3.28	
	B30 Roofing						
3010	Roof Coverings	Built-up tar and gravel with flashing; perlite/EPS composite insulation		S.F. Roof	7.32	7.32	7.6%
3020	Roof Openings	N/A		—	—	—	
C. INTERIORS							
1010	Partitions	Gypsum board on metal studs	28 S.F. Floor/L.F. Partition	S.F. Partition	5.34	2.29	
1020	Interior Doors	Hollow metal	280 S.F. Floor/Door	Each	1222	4.36	
1030	Fittings	N/A		—	—	—	
2010	Stair Construction	N/A		—	—	—	13.8%
3010	Wall Finishes	Paint		S.F. Surface	1.17	1	
3020	Floor Finishes	50% vinyl tile, 50% paint		S.F. Floor	2.52	2.52	
3030	Ceiling Finishes	Fiberglass board on exposed grid, suspended	50% of area	S.F. Ceiling	6.30	3.16	
D. SERVICES							
	D10 Conveying						
1010	Elevators & Lifts	N/A		—	—	—	0.0 %
1020	Escalators & Moving Walks	N/A		—	—	—	
	D20 Plumbing						
2010	Plumbing Fixtures	Toilet and service fixtures, supply and drainage	1 Fixture/1500 S.F. Floor	Each	3165	2.11	
2020	Domestic Water Distribution	Gas fired water heater		S.F. Floor	2.25	2.25	7.1%
2040	Rain Water Drainage	Roof drains		S.F. Roof	2.52	2.52	
	D30 HVAC						
3010	Energy Supply	N/A		—	—	—	
3020	Heat Generating Systems	N/A		—	—	—	
3030	Cooling Generating Systems	N/A		—	—	—	11.2 %
3050	Terminal & Package Units	Single zone rooftop unit, gas heating, electric cooling		S.F. Floor	10.11	10.11	
3090	Other HVAC Sys. & Equipment	Underfloor garage exhaust system		S.F. Floor	.72	.72	
	D40 Fire Protection						
4010	Sprinklers	Wet pipe sprinkler system		S.F. Floor	4.96	4.96	5.6%
4020	Standpipes	Standpipe		S.F. Floor	.50	.50	
	D50 Electrical						
5010	Electrical Service/Distribution	200 ampere service, panel board and feeders		S.F. Floor	.49	.49	
5020	Lighting & Branch Wiring	T-8 fluorescent fixtures, receptacles, switches, A.C. and misc. power		S.F. Floor	6.99	6.99	11.4%
5030	Communications & Security	Addressable alarm systems, partial internet wiring and emergency lighting		S.F. Floor	3.44	3.44	
5090	Other Electrical Systems	Emergency generator, 7.5 kW		S.F. Floor	.08	.08	
E. EQUIPMENT & FURNISHINGS							
1010	Commercial Equipment	N/A		—	—	—	
1020	Institutional Equipment	N/A		—	—	—	1.6 %
1030	Vehicular Equipment	Hoists, compressor, fuel pump		S.F. Floor	1.59	1.59	
1090	Other Equipment	N/A		—	—	—	
F. SPECIAL CONSTRUCTION							
1020	Integrated Construction	N/A		—	—	—	0.0 %
1040	Special Facilities	N/A		—	—	—	
G. BUILDING SITEWORK	**N/A**						

		Sub-Total	96.81	100%
CONTRACTOR FEES (General Requirements: 10%, Overhead: 5%, Profit: 10%)		25%	24.17	
ARCHITECT FEES		7%	8.47	
	Total Building Cost		**129.45**	

For customer support on your Square Foot Costs with RSMeans data, call 800.448.8182.

135

Costs per square foot of floor area

Exterior Wall	S.F. Area	85000	115000	145000	175000	205000	235000	265000	295000	325000
	L.F. Perimeter	529	638	723	823	875	910	975	1027	1075
Face Brick and Concrete Block	Rigid Steel	84.60	82.75	81.40	80.70	79.80	79.05	78.60	78.20	77.80
	Precast Concrete	80.25	78.40	77.10	76.35	75.45	74.70	74.25	73.85	73.50
Precast Concrete	Rigid Steel	94.30	91.45	89.25	88.10	86.45	85.10	84.35	83.65	83.05
	Precast Concrete	90.00	87.15	84.95	83.75	82.15	80.80	80.00	79.30	78.70
Cast in Place Concrete	Rigid Steel	81.85	80.35	79.20	78.60	77.90	77.30	76.95	76.65	76.35
	Reinforced Concrete	74.25	72.75	71.65	71.05	70.30	69.70	69.40	69.05	68.80
Perimeter Adj., Add or Deduct	Per 100 L.F.	1.85	1.40	1.10	0.95	0.80	0.65	0.60	0.55	0.55
Story Hgt. Adj., Add or Deduct	Per 1 Ft.	0.85	0.80	0.70	0.65	0.55	0.55	0.50	0.45	0.50
Basement—Not Applicable										

The above costs were calculated using the basic specifications shown on the facing page. These costs should be adjusted where necessary for design alternatives and owner's requirements.

Common additives

Description	Unit	$ Cost
Elevators, Electric passenger, 5 stops		
2000# capacity	Ea.	186,000
3500# capacity	Ea.	193,000
5000# capacity	Ea.	200,000
Barrier gate w/programmable controller	Ea.	3825
Booth for attendant, average	Ea.	14,100
Fee computer	Ea.	13,900
Ticket spitter with time/date stamp	Ea.	8025
Mag strip encoding	Ea.	21,500
Collection station, pay on foot	Ea.	130,000
Parking control software	Ea.	27,800 - 111,500
Painting, Parking stalls	Stall	9.05
Parking Barriers		
Timber with saddles, 4" x 4"	L.F.	8.10
Precast concrete, 6" x 10" x 6'	Ea.	80
Traffic Signs, directional, 12" x 18", high density	Ea.	100

Important: See the Reference Section for Location Factors.

Model costs calculated for a 5 story building with 10' story height and 145,000 square feet of floor area

				Unit	Unit Cost	Cost Per S.F.	% Of Sub-Total
A.	**SUBSTRUCTURE**						
1010	Standard Foundations	Poured concrete; strip and spread footings; 4' foundation wall		S.F. Ground	4.75	.95	
1020	Special Foundations	N/A		—	—	—	
1030	Slab on Grade	6" reinforced concrete with vapor barrier and granular base		S.F. Slab	7.42	1.48	4.2%
2010	Basement Excavation	Site preparation for slab and trench for foundation wall and footing		S.F. Ground	.18	.04	
2020	Basement Walls	N/A		—	—	—	
B.	**SHELL**						
	B10 Superstructure						
1010	Floor Construction	Double tee precast concrete slab, precast concrete columns		S.F. Floor	36.35	29.08	50.0%
1020	Roof Construction	N/A		—	—	—	
	B20 Exterior Enclosure						
2010	Exterior Walls	Face brick with concrete block backup	40% of story height	S.F. Wall	53	5.27	
2020	Exterior Windows	N/A		—	—	—	9.1%
2030	Exterior Doors	N/A		—	—	—	
	B30 Roofing						
3010	Roof Coverings	N/A		—	—	—	0.0%
3020	Roof Openings	N/A		—	—	—	
C.	**INTERIORS**						
1010	Partitions	Concrete block		S.F. Partition	36.14	1.39	
1020	Interior Doors	Hollow metal		Each	24,440	.17	
1030	Fittings	N/A		—	—	—	
2010	Stair Construction	Concrete		Flight	12,600	1.39	13.5%
3010	Wall Finishes	Paint		S.F. Surface	1.95	.15	
3020	Floor Finishes	Parking deck surface coating		S.F. Floor	4.76	4.76	
3030	Ceiling Finishes	N/A		—	—	—	
D.	**SERVICES**						
	D10 Conveying						
1010	Elevators & Lifts	Two hydraulic passenger elevators		Each	176,900	2.44	4.2%
1020	Escalators & Moving Walks	N/A		—	—	—	
	D20 Plumbing						
2010	Plumbing Fixtures	Toilet and service fixtures, supply and drainage	1 Fixture/18,125 S.F. Floor	Each	725	.04	
2020	Domestic Water Distribution	Electric water heater		S.F. Floor	.12	.12	2.9%
2040	Rain Water Drainage	Roof drains		S.F. Roof	7.50	1.50	
	D30 HVAC						
3010	Energy Supply	N/A		—	—	—	
3020	Heat Generating Systems	N/A		—	—	—	
3030	Cooling Generating Systems	N/A		—	—	—	0.0%
3050	Terminal & Package Units	N/A		—	—	—	
3090	Other HVAC Sys. & Equipment	N/A		—	—	—	
	D40 Fire Protection						
4010	Sprinklers	Dry pipe sprinkler system		S.F. Floor	4.48	4.48	7.9%
4020	Standpipes	Standpipes and hose systems		S.F. Floor	.10	.10	
	D50 Electrical						
5010	Electrical Service/Distribution	400 ampere service, panel board and feeders		S.F. Floor	.25	.25	
5020	Lighting & Branch Wiring	T-8 fluorescent fixtures, receptacles, switches and misc. power		S.F. Floor	3.17	3.17	6.2%
5030	Communications & Security	Addressable alarm systems and emergency lighting		S.F. Floor	.12	.12	
5090	Other Electrical Systems	Emergency generator, 7.5 kW		S.F. Floor	.06	.06	
E.	**EQUIPMENT & FURNISHINGS**						
1010	Commercial Equipment	N/A		—	—	—	
1020	Institutional Equipment	N/A		—	—	—	
1030	Vehicular Equipment	Ticket dispensers, booths, automatic gates		S.F. Floor	1.21	1.21	2.1%
1090	Other Equipment	N/A		—	—	—	
F.	**SPECIAL CONSTRUCTION**						
1020	Integrated Construction	N/A		—	—	—	0.0%
1040	Special Facilities	N/A		—	—	—	
G.	**BUILDING SITEWORK**	**N/A**					

	Sub-Total	58.17	100%
CONTRACTOR FEES (General Requirements: 10%, Overhead: 5%, Profit: 10%)	25%	14.57	
ARCHITECT FEES	6%	4.36	
Total Building Cost		**77.10**	

For customer support on your Square Foot Costs with RSMeans data, call 800.448.8182.

137

Costs per square foot of floor area

Exterior Wall	S.F. Area	20000	30000	40000	50000	75000	100000	125000	150000	175000
	L.F. Perimeter	400	500	600	650	775	900	1000	1100	1185
Cast in Place Concrete	Reinforced Concrete	108.50	101.25	97.60	94.20	89.55	87.30	85.65	84.60	83.65
Perimeter Adj., Add or Deduct	Per 100 L.F.	6.40	4.25	3.20	2.55	1.70	1.25	1.05	0.80	0.75
Story Hgt. Adj., Add or Deduct	Per 1 Ft.	2.45	2.05	1.85	1.60	1.25	1.05	0.95	0.85	0.85
Basement—Not Applicable										

The above costs were calculated using the basic specifications shown on the facing page. These costs should be adjusted where necessary for design alternatives and owner's requirements.

Common additives

Description	Unit	$ Cost
Elevators, Hydraulic passenger, 2 stops		
1500# capacity	Ea.	73,100
2500# capacity	Ea.	76,600
3500# capacity	Ea.	81,600
Barrier gate w/programmable controller	Ea.	3825
Booth for attendant, average	Ea.	14,100
Fee computer	Ea.	13,900
Ticket spitter with time/date stamp	Ea.	8025
Mag strip encoding	Ea.	21,500
Collection station, pay on foot	Ea.	130,000
Parking control software	Ea.	27,800 - 111,500
Painting, Parking stalls	Stall	9.05
Parking Barriers		
Timber with saddles, 4" x 4"	L.F.	8.10
Precast concrete, 6" x 10" x 6'	Ea.	80
Traffic Signs, directional, 12" x 18"	Ea.	100

Important: See the Reference Section for Location Factors.

Model costs calculated for a 2 story building with 10' story height and 100,000 square feet of floor area

Garage, Underground Parking

				Unit	Unit Cost	Cost Per S.F.	% Of Sub-Total
A. SUBSTRUCTURE							
1010	Standard Foundations	Poured concrete; strip and spread footings and waterproofing		S.F. Ground	8.46	4.23	
1020	Special Foundations	N/A		—	—	—	
1030	Slab on Grade	5" reinforced concrete with vapor barrier and granular base		S.F. Slab	7.47	3.74	20.4%
2010	Basement Excavation	Excavation 24' deep		S.F. Ground	10.40	5.20	
2020	Basement Walls	N/A		—	—	—	
B. SHELL							
	B10 Superstructure						
1010	Floor Construction	Cast-in-place concrete beam and slab, concrete columns		S.F. Floor	33.12	16.56	47.0%
1020	Roof Construction	Cast-in-place concrete beam and slab, concrete columns		S.F. Roof	27.60	13.80	
	B20 Exterior Enclosure						
2010	Exterior Walls	Cast-in place concrete		S.F. Wall	24.33	4.38	
2020	Exterior Windows	N/A		—	—	—	7.1%
2030	Exterior Doors	Steel overhead, hollow metal		Each	5548	.23	
	B30 Roofing						
3010	Roof Coverings	Neoprene membrane traffic deck		S.F. Roof	5.42	2.71	4.2%
3020	Roof Openings	N/A		—	—	—	
C. INTERIORS							
1010	Partitions	Concrete block		S.F. Partition	47.84	.92	
1020	Interior Doors	Hollow metal		Each	9776	.10	
1030	Fittings	N/A		—	—	—	
2010	Stair Construction	Concrete		Flight	7400	.37	2.3%
3010	Wall Finishes	Paint		S.F. Surface	2.60	.10	
3020	Floor Finishes	N/A		—	—	—	
3030	Ceiling Finishes	N/A		—	—	—	
D. SERVICES							
	D10 Conveying						
1010	Elevators & Lifts	Two hydraulic passenger elevators		Each	91,000	1.82	2.8%
1020	Escalators & Moving Walks	N/A		—	—	—	
	D20 Plumbing						
2010	Plumbing Fixtures	Drainage in parking areas, toilets, & service fixtures	1 Fixture/5000 S.F. Floor	Each	.05	.05	
2020	Domestic Water Distribution	Electric water heater		S.F. Floor	.17	.17	2.3%
2040	Rain Water Drainage	Roof drains		S.F. Roof	2.50	1.25	
	D30 HVAC						
3010	Energy Supply	N/A		—	—	—	
3020	Heat Generating Systems	N/A		—	—	—	
3030	Cooling Generating Systems	N/A		—	—	—	0.3 %
3050	Terminal & Package Units	Exhaust fans		S.F. Floor	.18	.18	
3090	Other HVAC Sys. & Equipment	N/A		—	—	—	
	D40 Fire Protection						
4010	Sprinklers	Dry pipe sprinkler system		S.F. Floor	4.51	4.51	7.2%
4020	Standpipes	Dry standpipe system, class III		S.F. Floor	.17	.17	
	D50 Electrical						
5010	Electrical Service/Distribution	200 ampere service, panel board and feeders		S.F. Floor	.13	.13	
5020	Lighting & Branch Wiring	T-8 fluorescent fixtures, receptacles, switches and misc. power		S.F. Floor	3.41	3.41	5.9%
5030	Communications & Security	Addressable alarm systems and emergency lighting		S.F. Floor	.18	.18	
5090	Other Electrical Systems	Emergency generator, 11.5 kW		S.F. Floor	.07	.07	
E. EQUIPMENT & FURNISHINGS							
1010	Commercial Equipment	N/A		—	—	—	
1020	Institutional Equipment	N/A		—	—	—	
1030	Vehicular Equipment	Ticket dispensers, booths, automatic gates		S.F. Floor	.38	.38	0.6 %
1090	Other Equipment	N/A		—	—	—	
F. SPECIAL CONSTRUCTION							
1020	Integrated Construction	N/A		—	—	—	0.0 %
1040	Special Facilities	N/A		—	—	—	
G. BUILDING SITEWORK	**N/A**						

			Sub-Total	64.66	100%
CONTRACTOR FEES (General Requirements: 10%, Overhead: 5%, Profit: 10%)			25%	16.17	
ARCHITECT FEES			8%	6.47	
		Total Building Cost		**87.30**	

For customer support on your Square Foot Costs with RSMeans data, call 800.448.8182.

139

Costs per square foot of floor area

Exterior Wall	S.F. Area	2000	4000	6000	8000	10000	12000	14000	16000	18000
	L.F. Perimeter	180	260	340	420	450	500	550	575	600
Concrete Block	Wood Joists	168.50	148.40	141.75	138.35	133.70	131.45	129.80	127.85	126.20
	Steel Joists	179.00	156.40	148.95	145.10	139.90	137.45	135.65	133.40	131.55
Cast in Place Concrete	Wood Joists	180.30	157.20	149.55	145.70	139.95	137.30	135.35	132.85	130.85
	Steel Joists	192.30	166.00	157.30	152.90	146.55	143.60	141.45	138.70	136.55
Stucco	Wood Frame	166.30	147.20	140.90	137.65	133.15	131.00	129.40	127.45	125.85
Metal Panel	Rigid Steel	192.05	165.55	156.75	152.30	146.05	143.10	140.95	138.20	136.10
Perimeter Adj., Add or Deduct	Per 100 L.F.	27.20	13.60	9.05	6.80	5.40	4.55	3.90	3.35	3.00
Story Hgt. Adj., Add or Deduct	Per 1 Ft.	2.35	1.65	1.45	1.40	1.10	1.05	1.05	0.85	0.85
For Basement, add $ 36.90 per square foot of basement area										

The above costs were calculated using the basic specifications shown on the facing page. These costs should be adjusted where necessary for design alternatives and owner's requirements.

Common additives

Description	Unit	$ Cost
Air Compressors		
Electric 1-1/2 H.P., standard controls	Ea.	1700
Dual controls	Ea.	2300
5 H.P. 115/230 Volt, standard controls	Ea.	4000
Dual controls	Ea.	5300
Product Dispenser		
with vapor recovery for 6 nozzles	Ea.	27,900
Lifts, Single post		
8000# cap., swivel arm	Ea.	11,700
Two post, adjustable frames, 12,000# cap.	Ea.	5775
24,000# cap.	Ea.	13,400
30,000# cap.	Ea.	62,000
Four post, roll on ramp, 25,000# cap.	Ea.	24,700
Lockers, Steel, single tier, 60" or 72"	Opng.	232 - 395
2 tier, 60" or 72" total	Opng.	139 - 172
5 tier, box lockers	Opng.	73 - 87
Locker bench, lam. maple top only	L.F.	36.50
Pedestals, steel pipe	Ea.	79
Lube Equipment		
3 reel type, with pumps, no piping	Ea.	14,700
Spray Painting Booth, 26' long, complete	Ea.	15,900

Important: See the Reference Section for Location Factors.

Model costs calculated for a 1 story building with 14' story height and 10,000 square feet of floor area

				Unit	Unit Cost	Cost Per S.F.	% Of Sub-Total
A. SUBSTRUCTURE							
1010	Standard Foundations	Poured concrete; strip and spread footings; 4' foundation wall		S.F. Ground	6.16	6.16	
1020	Special Foundations	N/A		—	—	—	
1030	Slab on Grade	6" reinforced concrete with vapor barrier and granular base		S.F. Slab	8.15	8.15	14.1%
2010	Basement Excavation	Site preparation for slab and trench for foundation wall and footing		S.F. Ground	.33	.33	
2020	Basement Walls	N/A		—	—	—	
B. SHELL							
B10 Superstructure							
1010	Floor Construction	N/A		—	—	—	6.0 %
1020	Roof Construction	Metal deck on open web steel joists		S.F. Roof	6.19	6.19	
B20 Exterior Enclosure							
2010	Exterior Walls	Concrete block	80% of wall	S.F. Wall	14.40	7.26	
2020	Exterior Windows	Hopper type commercial steel	5% of wall	Each	529	1.11	10.5%
2030	Exterior Doors	Steel overhead and hollow metal	15% of wall	S.F. Door	26.35	2.49	
B30 Roofing							
3010	Roof Coverings	Built-up tar and gravel; perlite/EPS composite insulation		S.F. Roof	7.14	7.14	6.9%
3020	Roof Openings	Skylight		S.F. Roof	.02	.02	
C. INTERIORS							
1010	Partitions	Concrete block	50 S.F. Floor/L.F. Partition	S.F. Partition	25.15	5.03	
1020	Interior Doors	Single leaf hollow metal	3000 S.F. Floor/Door	Each	1222	.41	
1030	Fittings	Toilet partitions		S.F. Floor	.16	.16	
2010	Stair Construction	N/A		—	—	—	10.0%
3010	Wall Finishes	Paint		S.F. Surface	7.80	3.12	
3020	Floor Finishes	90% metallic floor hardener, 10% vinyl composition tile		S.F. Floor	1.25	1.25	
3030	Ceiling Finishes	Gypsum board on wood joists in office and washrooms	10% of area	S.F. Ceiling	4.35	.44	
D. SERVICES							
D10 Conveying							
1010	Elevators & Lifts	N/A		—	—	—	0.0 %
1020	Escalators & Moving Walks	N/A		—	—	—	
D20 Plumbing							
2010	Plumbing Fixtures	Toilet and service fixtures, supply and drainage	1 Fixture/500 S.F. Floor	Each	1730	3.46	
2020	Domestic Water Distribution	Gas fired water heater		S.F. Floor	.71	.71	6.4%
2040	Rain Water Drainage	Roof drains		S.F. Roof	2.43	2.43	
D30 HVAC							
3010	Energy Supply	N/A		—	—	—	
3020	Heat Generating Systems	N/A		—	—	—	
3030	Cooling Generating Systems	N/A		—	—	—	11.0 %
3050	Terminal & Package Units	Single zone AC unit		S.F. Floor	10.35	10.35	
3090	Other HVAC Sys. & Equipment	Garage exhaust system		S.F. Floor	1.01	1.01	
D40 Fire Protection							
4010	Sprinklers	Sprinklers, ordinary hazard		S.F. Floor	4.96	4.96	5.8%
4020	Standpipes	Standpipe		S.F. Floor	1.04	1.04	
D50 Electrical							
5010	Electrical Service/Distribution	200 ampere service, panel board and feeders		S.F. Floor	.48	.48	
5020	Lighting & Branch Wiring	T-8 fluorescent fixtures, receptacles, switches, A.C. and misc. power		S.F. Floor	8.70	8.70	12.5%
5030	Communications & Security	Addressable alarm systems, partial internet wiring and emergency lighting		S.F. Floor	3.64	3.64	
5090	Other Electrical Systems	Emergency generator, 15 kW		S.F. Floor	.09	.09	
E. EQUIPMENT & FURNISHINGS							
1010	Commercial Equipment	N/A		—	—	—	
1020	Institutional Equipment	N/A		—	—	—	16.9 %
1030	Vehicular Equipment	Hoists		S.F. Floor	17.51	17.51	
1090	Other Equipment	N/A		—	—	—	
F. SPECIAL CONSTRUCTION							
1020	Integrated Construction	N/A		—	—	—	0.0 %
1040	Special Facilities	N/A		—	—	—	
G. BUILDING SITEWORK	**N/A**						

		Sub-Total	103.64	100%
CONTRACTOR FEES (General Requirements: 10%, Overhead: 5%, Profit: 10%)		25%	25.90	
ARCHITECT FEES		8%	10.36	
		Total Building Cost	**139.90**	

For customer support on your Square Foot Costs with RSMeans data, call 800.448.8182.

141

Costs per square foot of floor area

Exterior Wall	S.F. Area	600	800	1000	1200	1400	1600	1800	2000	2200
	L.F. Perimeter	100	120	126	140	153	160	170	180	190
Face Brick and Concrete Block	Wood Truss	304.90	285.95	263.30	253.60	246.05	237.40	231.95	227.70	224.10
	Steel Joists	326.05	306.05	282.20	271.90	263.95	254.85	249.10	244.65	240.90
Metal Sandwich Panel	Rigid Steel	289.75	271.70	250.85	241.70	234.60	226.60	221.65	217.65	214.45
Face Brick and Glazed Block	Steel Joists	345.05	323.20	296.55	285.20	276.45	266.20	259.90	254.90	250.80
Aluminum Clapboard	Wood Frame	260.35	245.05	228.10	220.50	214.60	208.15	204.05	200.80	198.15
Wood Clapboard	Wood Frame	261.75	246.30	229.10	221.40	215.55	208.95	204.85	201.55	198.85
Perimeter Adj., Add or Deduct	Per 100 L.F.	133.55	100.15	80.10	66.75	57.25	50.05	44.60	40.05	36.50
Story Hgt. Adj., Add or Deduct	Per 1 Ft.	8.60	7.80	6.55	6.00	5.65	5.20	4.95	4.70	4.50
Basement—Not Applicable										

The above costs were calculated using the basic specifications shown on the facing page. These costs should be adjusted where necessary for design alternatives and owner's requirements.

Common additives

Description	Unit	$ Cost
Air Compressors		
Electric 1-1/2 H.P., standard controls	Ea.	1700
Dual controls	Ea.	2300
5 H.P. 115/230 volt, standard controls	Ea.	4000
Dual controls	Ea.	5300
Product Dispenser		
with vapor recovery for 6 nozzles	Ea.	27,900
Lifts, Single post		
8000# cap. swivel arm	Ea.	11,700
Two post, adjustable frames, 12,000# cap.	Ea.	5775
24,000# cap.	Ea.	13,400
30,000# cap.	Ea.	62,000
Four post, roll on ramp, 25,000# cap.	Ea.	24,700
Lockers, Steel, single tier, 60" or 72"	Opng.	232 - 395
2 tier, 60" or 72" total	Opng.	139 - 172
5 tier, box lockers	Ea.	73 - 87
Locker bench, lam. maple top only	L.F.	36.50
Pedestals, steel pipe	Ea.	79
Lube Equipment		
3 reel type, with pumps, no piping	Ea.	14,700

Important: See the Reference Section for Location Factors.

Model costs calculated for a 1 story building with 12' story height and 1,400 square feet of floor area

					Unit	Unit Cost	Cost Per S.F.	% Of Sub-Total
A. SUBSTRUCTURE								
1010	Standard Foundations	Poured concrete; strip and spread footings; 4' foundation wall			S.F. Ground	13.40	13.40	
1020	Special Foundations	N/A			—	—	—	
1030	Slab on Grade	6" reinforced concrete with vapor barrier and granular base			S.F. Slab	8.15	8.15	12.4%
2010	Basement Excavation	Site preparation for slab and trench for foundation wall and footing			S.F. Ground	1.06	1.06	
2020	Basement Walls	N/A			—	—	—	
B. SHELL								
	B10 Superstructure							
1010	Floor Construction	N/A			—	—	—	4.5 %
1020	Roof Construction	Plywood on wood trusses			S.F. Roof	8.17	8.17	
	B20 Exterior Enclosure							
2010	Exterior Walls	Face brick with concrete block backup	60% of wall		S.F. Wall	35.25	27.74	
2020	Exterior Windows	Store front and metal top hinged outswinging	20% of wall		Each	48.25	12.66	30.1%
2030	Exterior Doors	Steel overhead, aluminum & glass and hollow metal	20% of wall		Each	55	14.51	
	B30 Roofing							
3010	Roof Coverings	Asphalt shingles with flashing; perlite/EPS composite insulation			S.F. Roof	5.61	5.61	3.1%
3020	Roof Openings	N/A			—	—	—	
C. INTERIORS								
1010	Partitions	Concrete block	25 S.F. Floor/L.F. Partition		S.F. Partition	8.53	2.73	
1020	Interior Doors	Single leaf hollow metal	700 S.F. Floor/Door		Each	1222	1.75	
1030	Fittings	Toilet partitions			S.F. Floor	2.23	2.23	
2010	Stair Construction	N/A			—	—	—	7.4%
3010	Wall Finishes	Paint			S.F. Surface	6.20	3.97	
3020	Floor Finishes	Vinyl composition tile	35% of floor area		S.F. Floor	3.06	1.07	
3030	Ceiling Finishes	Painted gypsum board on furring in sales area & washrooms	35% of floor area		S.F. Ceiling	4.85	1.70	
D. SERVICES								
	D10 Conveying							
1010	Elevators & Lifts	N/A			—	—	—	0.0 %
1020	Escalators & Moving Walks	N/A			—	—	—	
	D20 Plumbing							
2010	Plumbing Fixtures	Toilet and service fixtures, supply and drainage	1 Fixture/235 S.F. Floor		Each	2381	10.13	
2020	Domestic Water Distribution	Gas fired water heater			S.F. Floor	5.10	5.10	8.4%
2040	Rain Water Drainage	N/A			—	—	—	
	D30 HVAC							
3010	Energy Supply	N/A			—	—	—	
3020	Heat Generating Systems	N/A			—	—	—	
3030	Cooling Generating Systems	N/A			—	—	—	14.5 %
3050	Terminal & Package Units	Single zone AC unit			S.F. Floor	20.54	20.54	
3090	Other HVAC Sys. & Equipment	Underfloor exhaust system			S.F. Floor	5.91	5.91	
	D40 Fire Protection							
4010	Sprinklers	Wet pipe sprinkler system			S.F. Floor	10.15	10.15	8.1%
4020	Standpipes	Standpipe			S.F. Floor	4.66	4.66	
	D50 Electrical							
5010	Electrical Service/Distribution	200 ampere service, panel board and feeders			S.F. Floor	4.92	4.92	
5020	Lighting & Branch Wiring	High efficiency fluorescent fixtures, receptacles, switches, A.C. and misc. power			S.F. Floor	6.61	6.61	11.5%
5030	Communications & Security	Addressable alarm systems and emergency lighting			S.F. Floor	8.81	8.81	
5090	Other Electrical Systems	Emergency generator			S.F. Floor	.67	.67	
E. EQUIPMENT & FURNISHINGS								
1010	Commercial Equipment	N/A			—	—	—	
1020	Institutional Equipment	N/A			—	—	—	0.0 %
1030	Vehicular Equipment	N/A			—	—	—	
1090	Other Equipment	N/A			—	—	—	
F. SPECIAL CONSTRUCTION								
1020	Integrated Construction	N/A			—	—	—	0.0 %
1040	Special Facilities	N/A			—	—	—	
G. BUILDING SITEWORK	**N/A**							

		Sub-Total	182.25	100%
CONTRACTOR FEES (General Requirements: 10%, Overhead: 5%, Profit: 10%)		25%	45.58	
ARCHITECT FEES		8%	18.22	
	Total Building Cost		**246.05**	

For customer support on your Square Foot Costs with RSMeans data, call 800.448.8182.

143

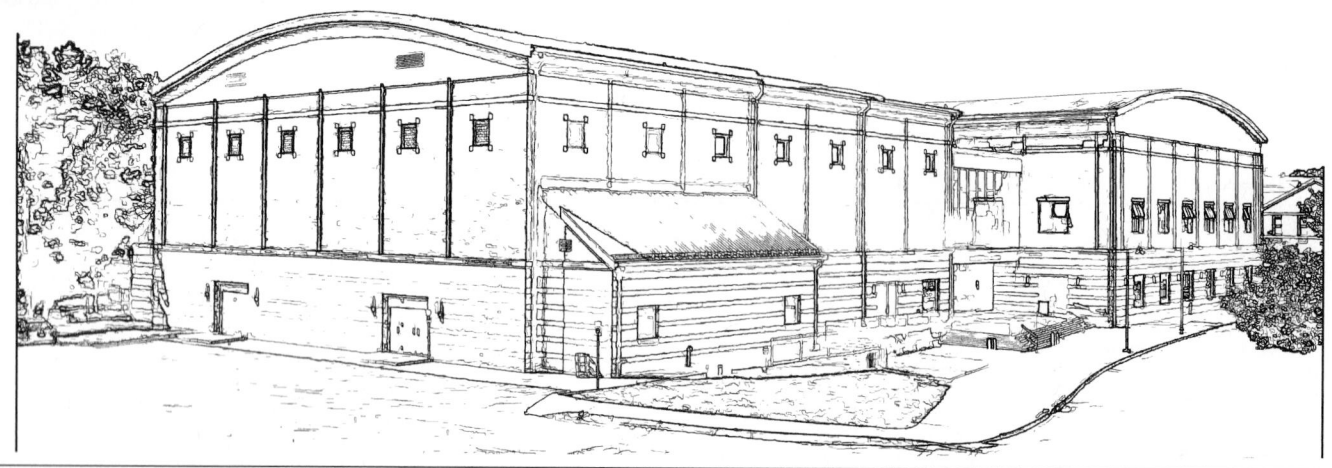

Costs per square foot of floor area

Exterior Wall	S.F. Area	12000	16000	20000	25000	30000	35000	40000	45000	50000
	L.F. Perimeter	440	520	600	700	708	780	841	910	979
Concrete Block	Wood Arch	186.85	182.05	179.20	176.80	172.65	171.20	169.90	169.05	168.35
	Rigid Steel	195.00	190.20	187.30	184.90	180.60	179.20	177.90	177.00	176.30
Face Brick and Concrete Block	Wood Arch	211.65	204.10	199.55	195.80	188.60	186.30	184.15	182.70	181.60
	Rigid Steel	219.40	211.75	207.20	203.50	196.30	193.95	191.85	190.45	189.30
Metal Sandwich Panel	Wood Arch	197.05	191.10	187.55	184.65	179.25	177.45	175.75	174.65	173.80
	Rigid Steel	208.55	202.15	198.35	195.25	189.35	187.40	185.60	184.45	183.55
Perimeter Adj., Add or Deduct	Per 100 L.F.	7.25	5.50	4.35	3.55	2.90	2.55	2.20	1.90	1.75
Story Hgt. Adj., Add or Deduct	Per 1 Ft.	1.00	0.95	0.80	0.85	0.65	0.60	0.60	0.60	0.55
Basement—Not Applicable										

The above costs were calculated using the basic specifications shown on the facing page. These costs should be adjusted where necessary for design alternatives and owner's requirements.

Common additives

Description	Unit	$ Cost
Bleachers, Telescoping, manual		
To 15 tier	Seat	164 - 222
16-20 tier	Seat	340 - 405
21-30 tier	Seat	340 - 460
For power operation, add	Seat	63 - 101
Gym Divider Curtain, Mesh top		
Manual roll-up	S.F.	14.80
Gym Mats		
2" naugahyde covered	S.F.	5.70
2" nylon	S.F.	9.10
1-1/2" wall pads	S.F.	8.85
1" wrestling mats	S.F.	5.50
Scoreboard		
Basketball, one side	Ea.	3625 - 14,600
Basketball Backstop		
Wall mtd., 6' extended, fixed	Ea.	2825 - 3150
Swing up, wall mtd.	Ea.	3000 - 4425

Description	Unit	$ Cost
Lockers, Steel, single tier, 60" or 72"	Opng.	232 - 395
2 tier, 60" or 72" total	Opng.	139 - 172
5 tier, box lockers	Opng.	73 - 87
Locker bench, lam. maple top only	L.F.	36.50
Pedestals, steel pipe	Ea.	79
Sound System		
Amplifier, 250 watts	Ea.	2025
Speaker, ceiling or wall	Ea.	240
Trumpet	Ea.	460
Emergency Lighting, 25 watt, battery operated		
Lead battery	Ea.	330
Nickel cadmium	Ea.	550

Important: See the Reference Section for Location Factors.

Model costs calculated for a 1 story building with 25' story height and 20,000 square feet of floor area

Gymnasium

				Unit	Unit Cost	Cost Per S.F.	% Of Sub-Total
A. SUBSTRUCTURE							
1010	Standard Foundations	Poured concrete; strip and spread footings; 4' foundation wall		S.F. Ground	4.08	4.08	
1020	Special Foundations	N/A		—	—	—	
1030	Slab on Grade	4" reinforced concrete with vapor barrier and granular base		S.F. Slab	5.80	5.80	7.5%
2010	Basement Excavation	Site preparation for slab and trench for foundation wall and footing		S.F. Ground	.18	.18	
2020	Basement Walls	N/A		—	—	—	
B. SHELL							
B10 Superstructure							
1010	Floor Construction	Steel column		S.F. Floor	2.36	2.36	17.1%
1020	Roof Construction	Wood deck on laminated wood arches		S.F. Roof	20.60	20.60	
B20 Exterior Enclosure							
2010	Exterior Walls	Reinforced concrete block (end walls included)	90% of wall	S.F. Wall	12.74	8.60	
2020	Exterior Windows	Metal horizontal pivoted	10% of wall	Each	673	5.05	10.7%
2030	Exterior Doors	Aluminum and glass, hollow metal, steel overhead		Each	2285	.69	
B30 Roofing							
3010	Roof Coverings	EPDM, 60 mils, fully adhered; polyisocyanurate insulation		S.F. Roof	5.36	5.36	4.0%
3020	Roof Openings	N/A		—	—	—	
C. INTERIORS							
1010	Partitions	Concrete block	50 S.F. Floor/L.F. Partition	S.F. Partition	10.40	2.08	
1020	Interior Doors	Single leaf hollow metal	500 S.F. Floor/Door	Each	1222	2.44	
1030	Fittings	Toilet partitions		S.F. Floor	.31	.31	
2010	Stair Construction	N/A		—	—	—	20.0%
3010	Wall Finishes	50% paint, 50% ceramic tile		S.F. Surface	10.88	4.35	
3020	Floor Finishes	90% hardwood, 10% ceramic tile		S.F. Floor	16.43	16.43	
3030	Ceiling Finishes	Mineral fiber tile on concealed zee bars	15% of area	S.F. Ceiling	7.60	1.14	
D. SERVICES							
D10 Conveying							
1010	Elevators & Lifts	N/A		—	—	—	0.0 %
1020	Escalators & Moving Walks	N/A		—	—	—	
D20 Plumbing							
2010	Plumbing Fixtures	Toilet and service fixtures, supply and drainage	1 Fixture/515 S.F. Floor	Each	5006	9.72	
2020	Domestic Water Distribution	Electric water heater		S.F. Floor	5.42	5.42	11.3%
2040	Rain Water Drainage	N/A		—	—	—	
D30 HVAC							
3010	Energy Supply	N/A		—	—	—	
3020	Heat Generating Systems	Included in D3050		—	—	—	
3030	Cooling Generating Systems	N/A		—	—	—	9.4 %
3050	Terminal & Package Units	Single zone rooftop unit, gas heating, electric cooling		S.F. Floor	12.55	12.55	
3090	Other HVAC Sys. & Equipment	N/A		—	—	—	
D40 Fire Protection							
4010	Sprinklers	Wet pipe sprinkler system		S.F. Floor	3.79	3.79	3.7%
4020	Standpipes	Standpipe		S.F. Floor	1.13	1.13	
D50 Electrical							
5010	Electrical Service/Distribution	400 ampere service, panel board and feeders		S.F. Floor	1.14	1.14	
5020	Lighting & Branch Wiring	High efficiency fluorescent fixtures, receptacles, switches, A.C. and misc. power		S.F. Floor	9.73	9.73	10.5%
5030	Communications & Security	Addressable alarm systems, sound system and emergency lighting		S.F. Floor	3	3	
5090	Other Electrical Systems	Emergency generator, 7.5 kW		S.F. Floor	.22	.22	
E. EQUIPMENT & FURNISHINGS							
1010	Commercial Equipment	N/A		—	—	—	
1020	Institutional Equipment	N/A		—	—	—	5.8 %
1030	Vehicular Equipment	N/A		—	—	—	
1090	Other Equipment	Bleachers, sauna, weight room		S.F. Floor	7.80	7.80	
F. SPECIAL CONSTRUCTION							
1020	Integrated Construction	N/A		—	—	—	0.0 %
1040	Special Facilities	N/A		—	—	—	
G. BUILDING SITEWORK	N/A						

		Sub-Total	133.97	**100%**
CONTRACTOR FEES (General Requirements: 10%, Overhead: 5%, Profit: 10%)		25%	33.51	
ARCHITECT FEES		7%	11.72	
		Total Building Cost	179.20	

For customer support on your Square Foot Costs with RSMeans data, call 800.448.8182.

145

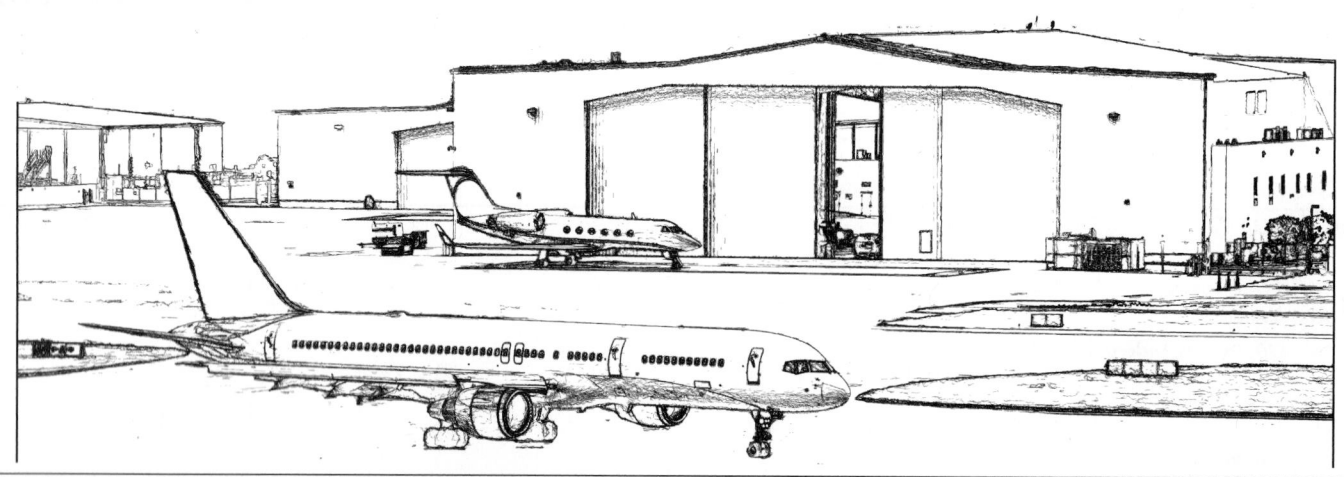

Costs per square foot of floor area

Exterior Wall	S.F. Area	5000	10000	15000	20000	30000	40000	50000	75000	100000
	L.F. Perimeter	300	410	500	580	710	830	930	1150	1300
Concrete Block	Rigid Steel	192.50	156.95	143.70	136.50	128.30	123.90	120.85	116.30	113.30
	Bearing Walls	201.95	166.20	152.85	145.60	137.35	132.85	129.80	125.20	122.20
Precast Concrete	Rigid Steel	256.65	200.80	179.35	167.60	153.65	146.10	140.75	132.70	127.25
	Bearing Walls	260.00	203.85	182.25	170.40	156.40	148.75	143.40	135.30	129.75
Metal Panel	Rigid Steel	186.90	153.15	140.60	133.85	126.10	121.95	119.15	114.90	112.10
Metal Sandwich Panel	Rigid Steel	194.10	158.05	144.55	137.30	128.90	124.45	121.35	116.75	113.65
Perimeter Adj., Add or Deduct	Per 100 L.F.	21.35	10.65	7.10	5.35	3.55	2.70	2.15	1.50	1.10
Story Hgt. Adj., Add or Deduct	Per 1 Ft.	2.05	1.40	1.15	1.00	0.85	0.70	0.65	0.50	0.45
Basement—Not Applicable										

The above costs were calculated using the basic specifications shown on the facing page. These costs should be adjusted where necessary for design alternatives and owner's requirements.

Common additives

Description	Unit	$ Cost
Closed Circuit Surveillance, One station		
Camera and monitor	Ea.	1475
For additional camera stations, add	Ea.	665
Emergency Lighting, 25 watt, battery operated		
Lead battery	Ea.	330
Nickel cadmium	Ea.	550
Lockers, Steel, single tier, 60" or 72"	Opng.	232 - 395
2 tier, 60" or 72" total	Opng.	139 - 172
5 tier, box lockers	Opng.	73 - 87
Locker bench, lam. maple top only	L.F.	36.50
Pedestals, steel pipe	Ea.	79
Safe, Office type, 1 hour rating		
30" x 18" x 18"	Ea.	2475
60" x 36" x 18", double door	Ea.	10,200
Sound System		
Amplifier, 250 watts	Ea.	2025
Speaker, ceiling or wall	Ea.	240
Trumpet	Ea.	460

Important: See the Reference Section for Location Factors.

Model costs calculated for a 1 story building with 24' story height and 20,000 square feet of floor area

Hangar, Aircraft

				Unit	Unit Cost	Cost Per S.F.	% Of Sub-Total
A. SUBSTRUCTURE							
1010	Standard Foundations	Poured concrete; strip and spread footings; 4' foundation wall		S.F. Ground	4.19	4.19	
1020	Special Foundations	N/A		—	—	—	
1030	Slab on Grade	6" reinforced concrete with vapor barrier and granular base		S.F. Slab	8.15	8.15	12.5%
2010	Basement Excavation	Site preparation for slab and trench for foundation wall and footing		S.F. Ground	.18	.18	
2020	Basement Walls	N/A		—	—	—	
B. SHELL							
	B10 Superstructure						
1010	Floor Construction	N/A		—	—	—	13.1 %
1020	Roof Construction	Metal deck, open web steel joists, beams, columns		S.F. Roof	13.08	13.08	
	B20 Exterior Enclosure						
2010	Exterior Walls	Galvanized steel siding		S.F. Wall	12.21	4.25	
2020	Exterior Windows	Industrial horizontal pivoted steel	20% of wall	Each	2055	11.92	25.0%
2030	Exterior Doors	Steel overhead and sliding	30% of wall	Each	42.19	8.81	
	B30 Roofing						
3010	Roof Coverings	Elastomeric membrane; fiberboard insulation		S.F. Roof	5.63	5.63	5.7%
3020	Roof Openings	Roof hatches		S.F. Roof	.12	.12	
C. INTERIORS							
1010	Partitions	Concrete block	200 S.F. Floor/L.F. Partition	S.F. Partition	11	.55	
1020	Interior Doors	Single leaf hollow metal	5000 S.F. Floor/Door	Each	1222	.25	
1030	Fittings	Toilet partitions		S.F. Floor	.31	.31	
2010	Stair Construction	N/A		—	—	—	1.2%
3010	Wall Finishes	Paint		S.F. Surface	1.40	.14	
3020	Floor Finishes	N/A		—	—	—	
3030	Ceiling Finishes	N/A		—	—	—	
D. SERVICES							
	D10 Conveying						
1010	Elevators & Lifts	N/A		—	—	—	0.0 %
1020	Escalators & Moving Walks	N/A		—	—	—	
	D20 Plumbing						
2010	Plumbing Fixtures	Toilet and service fixtures, supply and drainage	1 Fixture/1000 S.F. Floor	Each	2840	2.84	
2020	Domestic Water Distribution	Gas fired water heater		S.F. Floor	2.72	2.72	7.2%
2040	Rain Water Drainage	Roof drains		S.F. Roof	1.61	1.61	
	D30 HVAC						
3010	Energy Supply	N/A		—	—	—	
3020	Heat Generating Systems	Unit heaters		S.F. Floor	8.09	8.09	
3030	Cooling Generating Systems	N/A		—	—	—	8.4 %
3050	Terminal & Package Units	Exhaust fan		S.F. Floor	.30	.30	
3090	Other HVAC Sys. & Equipment	N/A		—	—	—	
	D40 Fire Protection						
4010	Sprinklers	Sprinklers, extra hazard, deluge system		S.F. Floor	8.09	8.09	13.4%
4020	Standpipes	Standpipe and fire pump		S.F. Floor	5.33	5.33	
	D50 Electrical						
5010	Electrical Service/Distribution	200 ampere service, panel board and feeders		S.F. Floor	1.02	1.02	
5020	Lighting & Branch Wiring	High intensity discharge fixtures, receptacles, switches and misc. power		S.F. Floor	10.97	10.97	13.5%
5030	Communications & Security	Addressable alarm systems and emergency lighting		S.F. Floor	1.38	1.38	
5090	Other Electrical Systems	Emergency generator, 7.5 kW		S.F. Floor	.13	.13	
E. EQUIPMENT & FURNISHINGS							
1010	Commercial Equipment	N/A		—	—	—	
1020	Institutional Equipment	N/A		—	—	—	
1030	Vehicular Equipment	N/A		—	—	—	0.0 %
1090	Other Equipment	N/A		—	—	—	
F. SPECIAL CONSTRUCTION							
1020	Integrated Construction	N/A		—	—	—	0.0 %
1040	Special Facilities	N/A		—	—	—	
G. BUILDING SITEWORK	**N/A**						

			Sub-Total	100.06	**100%**
CONTRACTOR FEES (General Requirements: 10%, Overhead: 5%, Profit: 10%)			25%	25.03	
ARCHITECT FEES			7%	8.76	
		Total Building Cost		**133.85**	

For customer support on your Square Foot Costs with RSMeans data, call 800.448.8182.

147

Costs per square foot of floor area

Exterior Wall	S.F. Area	25000	40000	55000	70000	85000	100000	115000	130000	145000
	L.F. Perimeter	388	520	566	666	766	866	878	962	1045
Fiber Cement	Wood Frame	302.85	295.40	289.65	287.55	286.15	285.20	283.35	282.75	282.25
Brick Veneer	Wood Frame	309.60	301.05	294.15	291.65	290.05	288.95	286.65	285.95	285.40
Stone Veneer	Rigid Steel	328.95	317.55	307.55	304.25	302.05	300.50	297.10	296.10	295.35
Curtain Wall	Rigid Steel	328.10	316.55	306.30	302.90	300.70	299.20	295.65	294.65	293.85
E.I.F.S.	Reinforced Concrete	325.55	315.20	306.30	303.35	301.35	300.00	296.95	296.10	295.40
Metal Panel	Reinforced Concrete	331.05	319.90	310.15	306.85	304.70	303.25	299.85	298.95	298.10
Perimeter Adj., Add or Deduct	Per 100 L.F.	6.05	3.80	2.75	2.15	1.75	1.50	1.25	1.15	1.00
Story Hgt. Adj., Add or Deduct	Per 1 Ft.	1.30	1.15	0.90	0.80	0.80	0.80	0.65	0.65	0.60

For Basement, add $38.60 per square foot of basement area

The above costs were calculated using the basic specifications shown on the facing page. These costs should be adjusted where necessary for design alternatives and owner's requirements.

Common additives

Description	Unit	$ Cost		Description	Unit	$ Cost
Cabinet, Base, door units, metal	L.F.	350		Nurses Call Station		
Drawer units	L.F.	690		Single bedside call station	Ea.	266
Tall storage cabinets, 7' high, open	L.F.	640		Ceiling speaker station	Ea.	178
With doors	L.F.	1025		Emergency call station	Ea.	185
Wall, metal 12-1/2" deep, open	L.F.	294		Pillow speaker	Ea.	325
With doors	L.F.	520		Double bedside call station	Ea.	370
Closed Circuit TV (Patient monitoring)				Duty station	Ea.	320
One station camera & monitor	Ea.	1475		Standard call button	Ea.	205
For additional camera, add	Ea.	665		Master control station for 20 stations	Ea.	6025
For automatic iris for low light, add	Ea.	1675		Sound System		
Hubbard Tank, with accessories				Amplifier, 250 watts	Ea.	2025
Stainless steel, 125 GPM 45 psi	Ea.	16,500		Speaker, ceiling or wall	Ea.	240
For electric hoist, add	Ea.	3575		Trumpet	Ea.	460
Mortuary Refrigerator, End operated				Station, Dietary with ice	Ea.	23,300
2 capacity	Ea.	9800		Sterilizers		
6 capacity	Ea.	18,400		Single door, steam	Ea.	148,500
				Double door, steam	Ea.	246,000
				Portable, countertop, steam	Ea.	3000 - 5450
				Gas	Ea.	47,000
				Automatic washer/sterilizer	Ea.	65,000

Important: See the Reference Section for Location Factors.

Model costs calculated for a 3 story building with 12' story height and 55,000 square feet of floor area

				Unit	Unit Cost	Cost Per S.F.	% Of Sub-Total
A. SUBSTRUCTURE							
1010	Standard Foundations	Poured concrete; strip and spread footings; 4' foundation wall		S.F. Ground	5.91	1.97	
1020	Special Foundations	N/A		—	—	—	
1030	Slab on Grade	4" reinforced slab on grade		S.F. Slab	5.80	1.93	1.8%
2010	Basement Excavation	Site preparation for slab and trench for foundation wall and footing		S.F. Ground	.18	.06	
2020	Basement Walls	N/A		—	—	—	
B. SHELL							
	B10 Superstructure						
1010	Floor Construction	Wood Beam and Joist on wood columns, fireproofed		S.F. Floor	15.78	10.52	6.1%
1020	Roof Construction	Wood roof, truss, 4/12 slope		S.F. Roof	8.10	2.70	
	B20 Exterior Enclosure						
2010	Exterior Walls	Brick veneer on wood stud back up, insulated	80% of wall	S.F. Wall	28.54	8.46	
2020	Exterior Windows	Aluminum awning, insulated	20% of wall	Each	755	2.43	5.3%
2030	Exterior Doors	Aluminum and glass double doors, hollow metal egress doors		Each	4410	.48	
	B30 Roofing						
3010	Roof Coverings	Asphalt roofing, strip shingles, flashing, gutters and downspouts		S.F. Roof	2.49	.83	0.5%
3020	Roof Openings	Roof hatches		S.F. Roof	.57	.19	
C. INTERIORS							
1010	Partitions	Concrete block partition, gypsum board on wood studs	10 S.F. Floor/L.F. Partition	S.F. Partition	10.16	11.29	
1020	Interior Doors	Hollow metal fire doors	90 S.F. Floor/Door	Each	1268	9.39	
1030	Fittings	Hospital curtains		S.F. Floor	1.20	1.20	
2010	Stair Construction	Cast in place concrete stairs, with landings and rails		Flight	8950	1.30	21.7%
3010	Wall Finishes	65% paint, 35% ceramic tile		S.F. Surface	3.44	7.64	
3020	Floor Finishes	60% vinyl tile, 20% ceramic , 15% terrazzo, 5% epoxy		S.F. Floor	9.70	9.70	
3030	Ceiling Finishes	Acoustic ceiling tiles on suspended channel grid		S.F. Ceiling	6.30	6.30	
D. SERVICES							
	D10 Conveying						
1010	Elevators & Lifts	Two hydraulic passenger elevators		Each	136,950	4.98	2.3%
1020	Escalators & Moving Walks	N/A		—	—	—	
	D20 Plumbing						
2010	Plumbing Fixtures	Medical, patient & specialty, supply and drainage	1 Fixture/225 S.F. Floor	Each	1812	8.09	
2020	Domestic Water Distribution	Electric water heater		S.F. Floor	18.99	18.99	12.5%
2040	Rain Water Drainage	Roof drains		S.F. Roof			
	D30 HVAC						
3010	Energy Supply	Conditioned air with hot water reheat system		S.F. Floor	5.16	5.16	
3020	Heat Generating Systems	Boiler		Each	128,350	3.73	
3030	Cooling Generating Systems	Conditioned air, chillers, cooling towers		S.F. Floor	6.67	6.67	27.6%
3050	Terminal & Package Units	N/A		—	—	—	
3090	Other HVAC Sys. & Equipment	Conditioned air, ductwork, surgey air curtain		S.F. Floor	43.98	43.98	
	D40 Fire Protection						
4010	Sprinklers	Wet pipe spinkler system		S.F. Floor	3.23	3.23	1.9%
4020	Standpipes	Standpipe		S.F. Floor	.89	.89	
	D50 Electrical						
5010	Electrical Service/Distribution	1200 ampere service, panel board and feeders		S.F. Floor	2.64	2.64	
5020	Lighting & Branch Wiring	Fluorescent fixtures, receptacles, switches, A.C. and misc. power		S.F. Floor	17.85	17.85	12.8%
5030	Communications & Security	Addressable alarm systems, emergency lighting, internet and phone wiring		S.F. Floor	5.69	5.69	
5090	Other Electrical Systems	Emergency generator, 200 Kw with fuel tank, UPS system		S.F. Floor	1.40	1.40	
E. EQUIPMENT & FURNISHINGS							
1010	Commercial Equipment	N/A		—	—	—	
1020	Institutional Equipment	Laboratory and commercial kitchen equipment		S.F. Floor	12.77	12.77	7.5 %
1030	Vehicular Equipment	N/A		—	—	—	
2020	Moveable Furnishings	Patient wall systems		S.F. Floor	3.40	3.40	
F. SPECIAL CONSTRUCTION							
1020	Integrated Construction	N/A		—	—	—	0.0 %
1040	Special Facilities	N/A		—	—	—	
G. BUILDING SITEWORK	**N/A**						

		Sub-Total	215.86	100%
CONTRACTOR FEES (General Requirements: 10%, Overhead: 5%, Profit: 10%)		25%	54.01	
ARCHITECT FEES		9%	24.28	
	Total Building Cost		**294.15**	

For customer support on your Square Foot Costs with RSMeans data, call 800.448.8182.

Costs per square foot of floor area

Exterior Wall	S.F. Area	100000	125000	150000	175000	200000	225000	250000	275000	300000
	L.F. Perimeter	594	705	816	783	866	950	1033	1116	1200
Fiber Cement	Rigid Steel	282.50	280.25	278.65	274.75	273.80	273.10	272.55	272.00	271.60
Curtain Wall	Rigid Steel	301.60	298.50	296.40	289.95	288.55	287.55	286.75	286.00	285.50
Metal Panel	Rigid Steel	286.10	283.80	282.20	278.25	277.25	276.50	275.95	275.45	275.05
Brick Veneer	Reinforced Concrete	302.50	299.55	297.50	291.40	290.10	289.10	288.30	287.65	287.15
Stone Veneer	Reinforced Concrete	314.50	310.95	308.50	300.45	298.85	297.70	296.70	295.85	295.25
E.I.F.S.	Reinforced Concrete	295.45	292.80	291.05	286.05	285.00	284.10	283.40	282.85	282.40
Perimeter Adj., Add or Deduct	Per 100 L.F.	3.25	2.60	2.20	1.90	1.60	1.50	1.25	1.20	1.05
Story Hgt. Adj., Add or Deduct	Per 1 Ft.	1.10	1.05	1.05	0.90	0.85	0.85	0.75	0.80	0.75

For Basement, add $39.80 per square foot of basement area

The above costs were calculated using the basic specifications shown on the facing page. These costs should be adjusted where necessary for design alternatives and owner's requirements.

Common additives

Description	Unit	$ Cost	Description	Unit	$ Cost
Cabinets, Base, door units, metal	L.F.	350	Nurses Call Station		
Drawer units	L.F.	690	Single bedside call station	Ea.	266
Tall storage cabinets, 7' high, open	L.F.	640	Ceiling speaker station	Ea.	178
With doors	L.F.	1025	Emergency call station	Ea.	185
Wall, metal 12-1/2" deep, open	L.F.	294	Pillow speaker	Ea.	325
With doors	L.F.	520	Double bedside call station	Ea.	370
Closed Circuit TV (Patient monitoring)			Duty station	Ea.	320
One station camera & monitor	Ea.	1475	Standard call button	Ea.	205
For additional camera add	Ea.	665	Master control station for 20 stations	Ea.	6025
For automatic iris for low light add	Ea.	1675	Sound System		
Hubbard Tank, with accessories			Amplifier, 250 watts	Ea.	2025
Stainless steel, 125 GPM 45 psi	Ea.	16,500	Speaker, ceiling or wall	Ea.	240
For electric hoist, add	Ea.	3575	Trumpet	Ea.	460
Mortuary Refrigerator, End operated			Station, Dietary with ice	Ea.	23,300
2 capacity	Ea.	9800	Sterilizers		
6 capacity	Ea.	18,400	Single door, steam	Ea.	148,500
			Double door, steam	Ea.	246,000
			Portable, counter top, steam	Ea.	3000 - 5450
			Gas	Ea.	47,000
			Automatic washer/sterilizer	Ea.	65,000

Important: See the Reference Section for Location Factors.

Model costs calculated for a 6 story building with 12' story height and 200,000 square feet of floor area

				Unit	Unit Cost	Cost Per S.F.	% Of Sub-Total
A. SUBSTRUCTURE							
1010	Standard Foundations	Poured concrete; strip and spread footings; 4' foundation wall		S.F. Ground	4.92	.82	
1020	Special Foundations	N/A		—	—	—	
1030	Slab on Grade	4" reinforced slab on grade		S.F. Slab	14.85	2.48	1.7%
2010	Basement Excavation	Site preparation for slab and trench for foundation wall and footing		S.F. Ground	.18	.03	
2020	Basement Walls	N/A		—	—	—	
B. SHELL							
B10 Superstructure							
1010	Floor Construction	Open web steel joists, slab form, concrete, fireproofed steel columns		S.F. Floor	15.16	12.63	7.2%
1020	Roof Construction	Metal deck on open web steel joints, columns		S.F. Roof	10.98	1.83	
B20 Exterior Enclosure							
2010	Exterior Walls	Fiber cement siding on metal studs, insulated	70% of wall	S.F. Wall	19.25	4.20	
2020	Exterior Windows	Aluminum sliding windows	30% of wall	Each	589	2.45	3.8%
2030	Exterior Doors	Aluminum and glass double doors, hollow metal egress doors		Each	7340	1.02	
B30 Roofing							
3010	Roof Coverings	Single ply membrane, stone ballast		S.F. Roof	7.02	1.17	0.6%
3020	Roof Openings	Roof hatches		S.F. Roof	.24	.04	
C. INTERIORS							
1010	Partitions	Concrete block partitions, gypsum board on metal studs	10 S.F. Floor/L.F. Partition	S.F. Partition	7.08	7.87	
1020	Interior Doors	Single leaf hollow metal doors, fire doors	90 S.F. Floor/Door	Each	1268	9.39	
1030	Fittings	Hospital curtains		S.F. Floor	1.32	1.32	
2010	Stair Construction	Concrete filled metal pan		Flight	12,600	1.63	21.5%
3010	Wall Finishes	65% Paint, 35 % ceramic wall tile		S.F. Surface	3.42	7.59	
3020	Floor Finishes	60% vinyl tile, 20% ceramic , 20% terrazzo		S.F. Floor	9.20	9.20	
3030	Ceiling Finishes	Acoustic ceiling tiles on suspended channel grid		S.F. Ceiling	6.30	6.30	
D. SERVICES							
D10 Conveying							
1010	Elevators & Lifts	Six traction geared hospital elevators		Each	235,000	7.05	3.5%
1020	Escalators & Moving Walks	N/A		—	—	—	
D20 Plumbing							
2010	Plumbing Fixtures	Medical, patient & specialty, supply and drainage	1 Fixture/415 S.F. Floor	Each	3149	7.57	
2020	Domestic Water Distribution	Electric water heater		S.F. Floor	11.60	11.60	9.9%
2040	Rain Water Drainage	Roof drains		S.F. Roof	4.56	.76	
D30 HVAC							
3010	Energy Supply	Conditioned air with hot water reheat system		S.F. Floor	4.48	4.48	
3020	Heat Generating Systems	Steam boiler for services		Each	43,750	.55	
3030	Cooling Generating Systems	Chillers		S.F. Floor	3.19	3.19	24.2%
3050	Terminal & Package Units	N/A		—	—	—	
3090	Other HVAC Sys. & Equipment	Hot water boilers, ductwork, VAV terminals, ventilation system		S.F. Floor	40.40	40.40	
D40 Fire Protection							
4010	Sprinklers	Wet pipe spinkler system		S.F. Floor	3.02	3.02	1.8%
4020	Standpipes	Standpipe		S.F. Floor	.57	.57	
D50 Electrical							
5010	Electrical Service/Distribution	2000 ampere service, panel board and feeders		S.F. Floor	2.25	2.25	
5020	Lighting & Branch Wiring	Fluorescent fixtures, receptacles, switches, A.C. and misc. power		S.F. Floor	20.87	20.87	13.3%
5030	Communications & Security	Addressable alarms, emergency lighting, internet and phone wiring		S.F. Floor	2.54	2.54	
5090	Other Electrical Systems	Emergency generator, 400 Kw with fuel tank, UPS system		S.F. Floor	.97	.97	
E. EQUIPMENT & FURNISHINGS							
1010	Commercial Equipment	N/A		—	—	—	
1020	Institutional Equipment	Laboratory and kitchen equipment		S.F. Floor	21.76	21.76	12.5 %
1030	Vehicular Equipment	N/A		—	—	—	
2020	Moveable Furnishings	Patient wall systems		S.F. Floor	3.41	3.41	
F. SPECIAL CONSTRUCTION							
1020	Integrated Construction	N/A		—	—	—	0.0 %
1040	Special Facilities	N/A		—	—	—	
G. BUILDING SITEWORK	**N/A**						
				Sub-Total		200.96	**100%**
	CONTRACTOR FEES (General Requirements: 10%, Overhead: 5%, Profit: 10%)				25%	50.23	
	ARCHITECT FEES				9%	22.61	
				Total Building Cost		**273.80**	

For customer support on your Square Foot Costs with RSMeans data, call 800.448.8182.

151

Costs per square foot of floor area

Exterior Wall	S.F. Area	35000	55000	75000	95000	115000	135000	155000	175000	195000
	L.F. Perimeter	314	401	497	555	639	722	754	783	850
Brick Veneer	Reinforced Concrete	222.75	208.60	202.70	196.85	194.45	192.75	189.50	186.90	185.95
Stone Veneer	Rigid Steel	210.15	198.25	193.20	188.40	186.35	184.90	182.30	180.20	179.45
E.I.F.S.	Reinforced Concrete	196.50	187.65	183.85	180.50	179.00	177.90	176.15	174.85	174.25
Curtain Wall	Rigid Steel	210.75	198.70	193.65	188.75	186.70	185.25	182.60	180.45	179.70
Precast Concrete	Reinforced Concrete	209.45	197.10	191.85	186.85	184.80	183.25	180.50	178.30	177.50
Fiber Cement	Rigid Steel	186.50	178.45	174.95	172.00	170.60	169.60	168.10	166.90	166.40
Perimeter Adj., Add or Deduct	Per 100 L.F.	16.60	10.60	7.75	6.20	5.10	4.30	3.75	3.35	3.00
Story Hgt. Adj., Add or Deduct	Per 1 Ft.	2.90	2.40	2.15	1.95	1.80	1.75	1.65	1.45	1.45
For Basement, add $37.90 per square foot of basement area										

The above costs were calculated using the basic specifications shown on the facing page. These costs should be adjusted where necessary for design alternatives and owner's requirements.

Common additives

Description	Unit	$ Cost	Description	Unit	$ Cost
Bar, Front bar	L.F.	450	Laundry Equipment		
Back bar	L.F.	360	Folders, blankets & sheets, king size	Ea.	77,000
Booth, Upholstered, custom, straight	L.F.	258 - 475	Ironers, 110" single roll	Ea.	40,900
"L" or "U" shaped	L.F.	267 - 450	Combination washer extractor 50#	Ea.	15,600
Closed Circuit Surveillance, One station			125#	Ea.	39,000
Camera and monitor	Ea.	1475	Sauna, Prefabricated, complete		
For additional camera stations, add	Ea.	665	6' x 4'	Ea.	6600
Directory Boards, Plastic, glass covered			6' x 6'	Ea.	8850
30" x 20"	Ea.	640	6' x 9'	Ea.	10,000
36" x 48"	Ea.	1625	8' x 8'	Ea.	10,600
Aluminum, 24" x 18"	Ea.	630	10' x 12'	Ea.	15,500
48" x 32"	Ea.	1025	Smoke Detectors		
48" x 60"	Ea.	2200	Ceiling type	Ea.	244
Elevators, Electric passenger, 5 stops			Duct type	Ea.	585
3500# capacity	Ea.	193,000	Sound System		
5000# capacity	Ea.	200,000	Amplifier, 250 watts	Ea.	2025
Additional stop, add	Ea.	10,900	Speaker, ceiling or wall	Ea.	240
Emergency Lighting, 25 watt, battery operated			Trumpet	Ea.	460
Lead battery	Ea.	330	TV Antenna, Master system, 12 outlet	Outlet	242
Nickel cadmium	Ea.	550	30 outlet	Outlet	335
			100 outlet	Outlet	370

Important: See the Reference Section for Location Factors.

Model costs calculated for a 6 story building with 10' story height and 135,000 square feet of floor area

				Unit	Unit Cost	Cost Per S.F.	% Of Sub-Total
A. SUBSTRUCTURE							
1010	Standard Foundations	Poured concrete; strip and spread footings; 4' foundation wall		S.F. Ground	6	1	
1020	Special Foundations	N/A		—	—	—	
1030	Slab on Grade	4" reinforced slab on grade		S.F. Slab	5.80	.97	1.4%
2010	Basement Excavation	Site preparation for slab and trench for foundation wall and footing		S.F. Ground	.18	.03	
2020	Basement Walls	N/A		—	—	—	
B. SHELL							
	B10 Superstructure						
1010	Floor Construction	Cast in place concrete columns, beams, and slab, fireproofed		S.F. Floor	30.66	25.55	17.6%
1020	Roof Construction	Included in B1010		—	—	—	
	B20 Exterior Enclosure						
2010	Exterior Walls	Face brick with concrete block backup. Insulated	80% of wall	S.F. Wall	41.21	10.58	
2020	Exterior Windows	Aluminum awning windows	20% of wall	Each	755	2.11	9.0%
2030	Exterior Doors	Aluminum and glass double doors, hollow metal egress doors		Each	4938	.36	
	B30 Roofing						
3010	Roof Coverings	Single ply membrane, stone ballast, rigid insulation		S.F. Roof	7.50	1.25	0.9%
3020	Roof Openings	Roof hatches		S.F. Roof	.24	.04	
C. INTERIORS							
1010	Partitions	Fire rated gypsum board on metal studs, CMU partitions	10 S.F. Floor/L.F. Partition	S.F. Partition	6.80	6.04	
1020	Interior Doors	Single leaf hollow metal	90 S.F. Floor/Door	Each	1222	13.58	
1030	Fittings	N/A		—	—	—	
2010	Stair Construction	Concrete filled metal pan		Flight	15,800	3.04	25.4%
3010	Wall Finishes	80% paint, 20% ceramic tile		S.F. Surface	2.46	4.38	
3020	Floor Finishes	80% carpet, 10% vinyl composition tile, 10% ceramic tile		S.F. Floor	5.69	5.69	
3030	Ceiling Finishes	Painted gypsum board ceiling on metal furring		S.F. Ceiling	4.19	4.19	
D. SERVICES							
	D10 Conveying						
1010	Elevators & Lifts	Four geared passenger elevators		Each	252,113	7.47	5.1%
1020	Escalators & Moving Walks	N/A		—	—	—	
	D20 Plumbing						
2010	Plumbing Fixtures	Kitchen, bath, laundry and service fixtures, supply and drainage	1 Fixture/150 S.F. Floor	Each	3060	19.74	
2020	Domestic Water Distribution	Gas fired water heater		S.F. Floor	.70	.70	14.3%
2040	Rain Water Drainage	Roof drains		S.F. Roof	2.34	.39	
	D30 HVAC						
3010	Energy Supply	Gas fired hot water, wall fin radiation		S.F. Floor	5.44	5.44	
3020	Heat Generating Systems	N/A		—	—	—	
3030	Cooling Generating Systems	Chilled water, fan coil units		S.F. Floor	14	14	13.4%
3050	Terminal & Package Units	N/A		—	—	—	
3090	Other HVAC Sys. & Equipment	N/A		—	—	—	
	D40 Fire Protection						
4010	Sprinklers	Wet pipe spinkler system, light hazard and accessory package		S.F. Floor	3.03	3.03	2.4%
4020	Standpipes	Standpipes and hose systems, with pumps		S.F. Floor	.41	.41	
	D50 Electrical						
5010	Electrical Service/Distribution	2000 Ampere service, panel boards and feeders		S.F. Floor	1.44	1.44	
5020	Lighting & Branch Wiring	Fluorescent fixtures, receptacles, switches, A.C. and misc. power		S.F. Floor	9.13	9.13	10.6%
5030	Communications & Security	Addressable alarm systems, emergency lighting, internet and phone wiring		S.F. Floor	4.45	4.45	
5090	Other Electrical Systems	Emergency generator muffler and transfer switch, 250 kW		S.F. Floor	.46	.46	
E. EQUIPMENT & FURNISHINGS							
1010	Commercial Equipment	N/A		—	—	—	
1020	Institutional Equipment	N/A		—	—	—	0.0 %
1030	Vehicular Equipment	N/A		—	—	—	
1090	Other Equipment	N/A		—	—	—	
F. SPECIAL CONSTRUCTION							
1020	Integrated Construction	N/A		—	—	—	0.0 %
1040	Special Facilities	N/A		—	—	—	
G. BUILDING SITEWORK	**N/A**						

			Sub-Total	145.47	100%
CONTRACTOR FEES (General Requirements: 10%, Overhead: 5%, Profit: 10%)			25%	36.37	
ARCHITECT FEES			6%	10.91	
		Total Building Cost		**192.75**	

For customer support on your Square Foot Costs with RSMeans data, call 800.448.8182.

153

Costs per square foot of floor area

Exterior Wall	S.F. Area	140000	243000	346000	450000	552000	655000	760000	860000	965000
	L.F. Perimeter	403	587	672	800	936	1073	1213	1195	1312
Brick Veneer	Reinforced Concrete	228.15	218.45	209.80	206.65	205.05	203.85	203.10	199.55	198.95
Stone Veneer	Rigid Steel	211.90	204.75	198.70	196.50	195.35	194.45	193.85	191.45	191.10
E.I.F.S.	Reinforced Concrete	201.40	196.35	192.35	190.90	190.10	189.50	189.05	187.65	187.40
Curtain Wall	Rigid Steel	211.55	204.50	198.50	196.30	195.10	194.25	193.65	191.35	190.95
Precast Concrete	Reinforced Concrete	221.55	213.30	206.10	203.50	202.15	201.15	200.50	197.65	197.20
Fiber Cement	Rigid Steel	192.95	188.35	184.90	183.70	182.90	182.35	182.00	180.85	180.55
Perimeter Adj., Add or Deduct	Per 100 L.F.	11.80	6.80	4.80	3.75	2.95	2.55	2.10	1.90	1.75
Story Hgt. Adj., Add or Deduct	Per 1 Ft.	2.40	2.00	1.60	1.50	1.35	1.40	1.30	1.10	1.15

For Basement, add $39.30 per square foot of basement area

The above costs were calculated using the basic specifications shown on the facing page. These costs should be adjusted where necessary for design alternatives and owner's requirements.

Common additives

Description	Unit	$ Cost
Bar, Front bar	L.F.	450
Back bar	L.F.	360
Booth, Upholstered, custom, straight	L.F.	258 - 475
"L" or "U" shaped	L.F.	267 - 450
Closed Circuit Surveillance, One station		
Camera and monitor	Ea.	1475
For additional camera stations, add	Ea.	665
Directory Boards, Plastic, glass covered		
30" x 20"	Ea.	640
36" x 48"	Ea.	1625
Aluminum, 24" x 18"	Ea.	630
48" x 32"	Ea.	1025
48" x 60"	Ea.	2200
Elevators, Electric passenger, 10 stops		
3500# capacity	Ea.	423,500
5000# capacity	Ea.	432,500
Additional stop, add	Ea.	10,900
Emergency Lighting, 25 watt, battery operated		
Lead battery	Ea.	330
Nickel cadmium	Ea.	550

Description	Unit	$ Cost
Laundry Equipment		
Folders, blankets & sheets, king size	Ea.	77,000
Ironers, 110" single roll	Ea.	40,900
Combination washer & extractor 50#	Ea.	15,600
125#	Ea.	39,000
Sauna, Prefabricated, complete		
6' x 4'	Ea.	6600
6' x 6'	Ea.	8850
6' x 9'	Ea.	10,000
8' x 8'	Ea.	10,600
10' x 12'	Ea.	15,500
Smoke Detectors		
Ceiling type	Ea.	244
Duct type	Ea.	585
Sound System		
Amplifier, 250 watts	Ea.	2025
Speaker, ceiling or wall	Ea.	240
Trumpet	Ea.	460
TV Antenna, Master system, 12 outlet	Outlet	242
30 outlet	Outlet	335
100 outlet	Outlet	370

Important: See the Reference Section for Location Factors.

Model costs calculated for a 15 story building with 10' story height and 450,000 square feet of floor area

				Unit	Unit Cost	Cost Per S.F.	% Of Sub-Total
A. SUBSTRUCTURE							
1010	Standard Foundations	CIP concrete pile caps; 4' foundation wall		S.F. Ground	15.90	1.06	
1020	Special Foundations	Steel H piles, concrete grade beams		S.F. Ground	204	13.60	
1030	Slab on Grade	4" reinforced slab on grade		S.F. Slab	5.80	.38	9.8%
2010	Basement Excavation	Site preparation for slab, piles and grade beams		S.F. Ground	.18	.01	
2020	Basement Walls	N/A		—	—	—	
B. SHELL							
	B10 Superstructure						
1010	Floor Construction	Cast in place concrete columns, beams, and slab		S.F. Floor	20.63	19.25	13.2%
1020	Roof Construction	Precast double T with 2" topping		S.F. Roof	16.50	1.10	
	B20 Exterior Enclosure						
2010	Exterior Walls	Precast concrete panels, insulated	80% of wall	S.F. Wall	60	12.90	
2020	Exterior Windows	Aluminum awning type	20% of wall	Each	755	1.75	9.7%
2030	Exterior Doors	Aluminum & glass doors and entrances		Each	4450	.31	
	B30 Roofing						
3010	Roof Coverings	Single ply membrane, stone ballast, rigid insulation		S.F. Roof	7.05	.47	0.3%
3020	Roof Openings	Roof hatches		S.F. Roof	.15	.01	
C. INTERIORS							
1010	Partitions	Gypsum board on metal studs, sound attenuation insulation	10 S.F. Floor/L.F. Partition	S.F. Partition	6.72	5.97	
1020	Interior Doors	Single leaf hollow metal	90 S.F. Floor/Door	Each	1222	13.58	
1030	Fittings	N/A		—	—	—	
2010	Stair Construction	Concrete filled metal pan with rails		Flight	15,800	2.17	23.4%
3010	Wall Finishes	80% Paint, 20% ceramic tile		S.F. Surface	2.44	4.34	
3020	Floor Finishes	80% carpet , 10% vinyl composition tile, 10% ceramic tile		S.F. Floor	5.69	5.69	
3030	Ceiling Finishes	Gypsum board on resilient channel		S.F. Ceiling	4.19	4.19	
D. SERVICES							
	D10 Conveying						
1010	Elevators & Lifts	Five traction geared passenger elevators		Each	517,500	6.90	4.5%
1020	Escalators & Moving Walks	N/A		—	—	—	
	D20 Plumbing						
2010	Plumbing Fixtures	Kitchen, bath, laundry and service fixtures, supply and drainage	1 Fixture/165 S.F. Floor	Each	3213	19.47	
2020	Domestic Water Distribution	Electric water heaters, gas fired water heaters		S.F. Floor	.21	.21	13.0%
2040	Rain Water Drainage	Roof drains		S.F. Roof	3.60	.24	
	D30 HVAC						
3010	Energy Supply	Gas fired hot water, wall fin radiation		S.F. Floor	2.84	2.84	
3020	Heat Generating Systems	N/A		—	—	—	
3030	Cooling Generating Systems	Chilled water, fan coil units		S.F. Floor	14	14	11.0%
3050	Terminal & Package Units	N/A		—	—	—	
3090	Other HVAC Sys. & Equipment	N/A		—	—	—	
	D40 Fire Protection						
4010	Sprinklers	Wet pipe spinkler system, light hazard and accessory package		S.F. Floor	4.15	4.15	5.5%
4020	Standpipes	Standpipes and hose systems, with pumps		S.F. Floor	4.26	4.26	
	D50 Electrical						
5010	Electrical Service/Distribution	2000 Ampere service, panel boards and feeders		S.F. Floor	.96	.96	
5020	Lighting & Branch Wiring	Fluorescent fixtures, receptacles, switches, A.C. and misc. power		S.F. Floor	9	9	9.6%
5030	Communications & Security	Addressable alarm systems, emergency lighting, internet and phone wiring		S.F. Floor	4.47	4.47	
5090	Other Electrical Systems	Emergency generator muffler and transfer switch, 250 kW		S.F. Floor	.33	.33	
E. EQUIPMENT & FURNISHINGS							
1010	Commercial Equipment	N/A		—	—	—	
1020	Institutional Equipment	N/A		—	—	—	
1030	Vehicular Equipment	N/A		—	—	—	0.0 %
1090	Other Equipment	N/A		—	—	—	
F. SPECIAL CONSTRUCTION							
1020	Integrated Construction	N/A		—	—	—	0.0 %
1040	Special Facilities	N/A		—	—	—	
G. BUILDING SITEWORK	**N/A**						

		Sub-Total	153.61	100%
	CONTRACTOR FEES (General Requirements: 10%, Overhead: 5%, Profit: 10%)	25%	38.37	
	ARCHITECT FEES	6%	11.52	
	Total Building Cost		**203.50**	

For customer support on your Square Foot Costs with RSMeans data, call 800.448.8182.

155

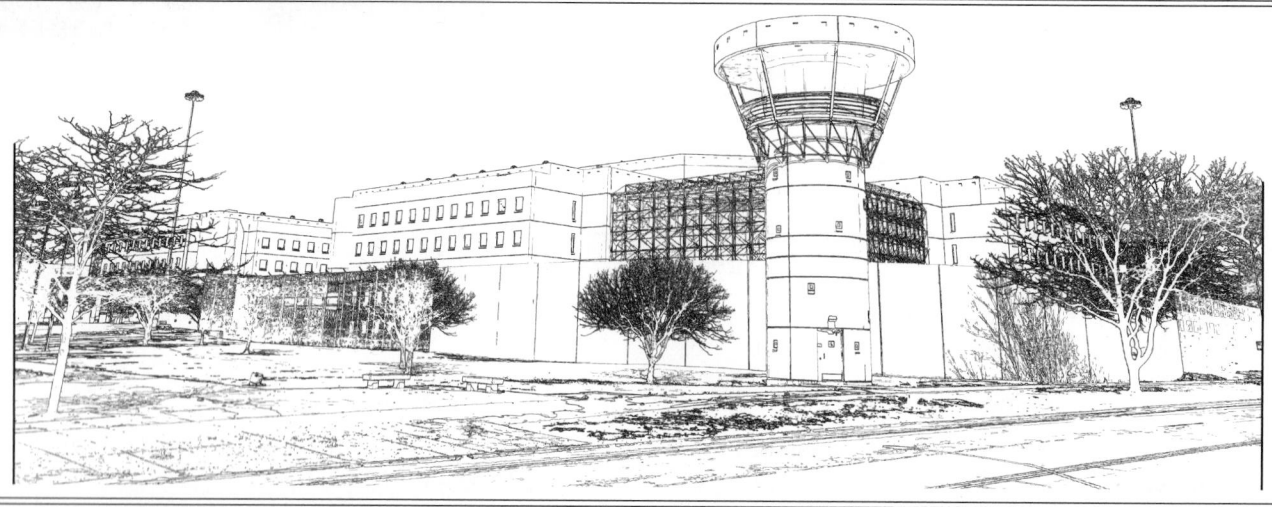

Costs per square foot of floor area

Exterior Wall	S.F. Area	5500	12000	20000	30000	40000	60000	80000	100000	145000
	L.F. Perimeter	310	330	400	490	530	590	680	760	920
Face Brick and Concrete Block	Rigid Steel	569.15	415.70	370.55	348.30	333.40	317.45	310.60	306.20	300.35
	Reinforced Concrete	560.35	406.90	361.80	339.50	324.65	308.65	301.85	297.40	291.60
Stucco and Concrete Block	Rigid Steel	549.60	406.15	363.65	342.65	328.85	314.05	307.65	303.50	298.10
	Reinforced Concrete	540.85	397.35	354.90	333.85	320.05	305.30	298.90	294.75	289.35
Cast in Place Concrete	Rigid Steel	543.55	403.15	361.50	340.85	327.40	313.00	306.75	302.70	297.40
	Reinforced Concrete	534.80	394.40	352.70	332.10	318.55	304.20	297.95	293.95	288.65
Perimeter Adj., Add or Deduct	Per 100 L.F.	55.00	25.20	15.15	10.05	7.55	5.00	3.80	3.00	2.05
Story Hgt. Adj., Add or Deduct	Per 1 Ft.	13.15	6.40	4.70	3.85	3.10	2.30	2.00	1.80	1.45
For Basement, add $54.40 per square foot of basement area										

The above costs were calculated using the basic specifications shown on the facing page. These costs should be adjusted where necessary for design alternatives and owner's requirements.

Common additives

Description	Unit	$ Cost	Description	Unit	$ Cost
Clock System			Emergency Lighting, 25 watt, battery operated		
20 room	Ea.	20,200	Nickel cadmium	Ea.	550
50 room	Ea.	48,200	Flagpoles, Complete		
Closed Circuit Surveillance, One station			Aluminum, 20' high	Ea.	1900
Camera and monitor	Ea.	1475	40' High	Ea.	4500
For additional camera stations, add	Ea.	665	70' High	Ea.	11,600
Elevators, Hydraulic passenger, 2 stops			Fiberglass, 23' High	Ea.	1450
1500# capacity	Ea.	73,100	39'-5" High	Ea.	3325
2500# capacity	Ea.	76,600	59' High	Ea.	7300
3500# capacity	Ea.	81,600	Laundry Equipment		
Additional stop, add	Ea.	8375	Folders, blankets, & sheets	Ea.	77,000
Emergency Generators, Complete system, gas			Ironers, 110" single roll	Ea.	40,900
15 KW	Ea.	18,700	Combination washer & extractor, 50#	Ea.	15,600
70 KW	Ea.	40,200	125#	Ea.	39,000
85 KW	Ea.	45,100	Safe, Office type, 1 hour rating		
115 KW	Ea.	82,500	30" x 18" x 18"	Ea.	2475
170 KW	Ea.	106,000	60" x 36" x 18", double door	Ea.	10,200
Diesel, 50 KW	Ea.	26,500	Sound System		
100 KW	Ea.	36,800	Amplifier, 250 watts	Ea.	2025
150 KW	Ea.	48,600	Speaker, ceiling or wall	Ea.	240
350 KW	Ea.	74,000	Trumpet	Ea.	460

Important: See the Reference Section for Location Factors.

Model costs calculated for a 3 story building with 12' story height and 40,000 square feet of floor area

			Unit	Unit Cost	Cost Per S.F.	% Of Sub-Total
A.	**SUBSTRUCTURE**					
1010	Standard Foundations	Poured concrete; strip and spread footings; 4' foundation wall	S.F. Ground	6.48	2.16	
1020	Special Foundations	N/A	—	—	—	
1030	Slab on Grade	4" reinforced concrete with vapor barrier and granular base	S.F. Slab	5.80	1.93	1.7%
2010	Basement Excavation	Site preparation for slab and trench for foundation wall and footing	S.F. Ground	.33	.11	
2020	Basement Walls	N/A	—	—	—	
B.	**SHELL**					
	B10 Superstructure					
1010	Floor Construction	Concrete slab with metal deck and beams	S.F. Floor	31.31	20.87	11.7%
1020	Roof Construction	Concrete slab with metal deck and beams	S.F. Roof	24.63	8.21	
	B20 Exterior Enclosure					
2010	Exterior Walls	Face brick with reinforced concrete block backup 85% of wall	S.F. Wall	35.24	14.29	
2020	Exterior Windows	Bullet resisting and metal horizontal pivoted 15% of wall	Each	167	11.95	10.7%
2030	Exterior Doors	Metal	Each	7343	.36	
	B30 Roofing					
3010	Roof Coverings	Built-up tar and gravel with flashing; perlite/EPS composite insulation	S.F. Roof	6.93	2.31	0.9%
3020	Roof Openings	Roof hatches	S.F. Roof	.09	.03	
C.	**INTERIORS**					
1010	Partitions	Concrete block 45 S.F. of Floor/L.F. Partition	S.F. Partition	11.59	3.09	
1020	Interior Doors	Hollow metal fire doors 930 S.F. Floor/Door	Each	1222	1.31	
1030	Fittings	N/A	—	—	—	
2010	Stair Construction	Concrete filled metal pan	Flight	18,900	2.37	7.3%
3010	Wall Finishes	Paint	S.F. Surface	4.58	2.44	
3020	Floor Finishes	70% vinyl composition tile, 20% carpet, 10% ceramic tile	S.F. Floor	3.83	3.83	
3030	Ceiling Finishes	Mineral tile on zee runners 30% of area	S.F. Ceiling	11.79	5.23	
D.	**SERVICES**					
	D10 Conveying					
1010	Elevators & Lifts	One hydraulic passenger elevator	Each	180,400	4.51	1.8%
1020	Escalators & Moving Walks	N/A	—	—	—	
	D20 Plumbing					
2010	Plumbing Fixtures	Toilet and service fixtures, supply and drainage 1 Fixture/78 S.F. Floor	Each	4546	58.28	
2020	Domestic Water Distribution	Electric water heater	S.F. Floor	9.82	9.82	27.7%
2040	Rain Water Drainage	Roof drains	S.F. Roof	2.79	.93	
	D30 HVAC					
3010	Energy Supply	N/A	—	—	—	
3020	Heat Generating Systems	Included in D3050	—	—	—	
3030	Cooling Generating Systems	N/A	—	—	—	6.8 %
3050	Terminal & Package Units	Rooftop multizone unit systems	S.F. Floor	16.94	16.94	
3090	Other HVAC Sys. & Equipment	N/A	—	—	—	
	D40 Fire Protection					
4010	Sprinklers	Sprinkler system, light hazard	S.F. Floor	3.12	3.12	1.5%
4020	Standpipes	Standpipes and hose systems	S.F. Floor	.70	.70	
	D50 Electrical					
5010	Electrical Service/Distribution	600 ampere service, panel board and feeders	S.F. Floor	.99	.99	
5020	Lighting & Branch Wiring	Fluorescent fixtures, receptacles, switches, A.C. and misc. power	S.F. Floor	10.30	10.30	6.7%
5030	Communications & Security	Addressable alarm systems, intercom, and emergency lighting	S.F. Floor	4.64	4.64	
5090	Other Electrical Systems	Emergency generator, 80 kW	S.F. Floor	.88	.88	
E.	**EQUIPMENT & FURNISHINGS**					
1010	Commercial Equipment	N/A	—	—	—	
1020	Institutional Equipment	Prefabricated cells, visitor cubicles	S.F. Floor	58	57.67	23.1 %
1030	Vehicular Equipment	N/A	—	—	—	
1090	Other Equipment	N/A	—	—	—	
F.	**SPECIAL CONSTRUCTION**					
1020	Integrated Construction	N/A	—	—	—	0.0 %
1040	Special Facilities	N/A	—	—	—	
G.	**BUILDING SITEWORK** N/A					

		Sub-Total	249.27	100%
CONTRACTOR FEES (General Requirements: 10%, Overhead: 5%, Profit: 10%)		25%	62.32	
ARCHITECT FEES		7%	21.81	
	Total Building Cost		**333.40**	

For customer support on your Square Foot Costs with RSMeans data, call 800.448.8182.

157

Costs per square foot of floor area

Exterior Wall	S.F. Area	1000	2000	3000	4000	5000	10000	15000	20000	25000
	L.F. Perimeter	126	179	219	253	283	400	490	568	632
Decorative Concrete Block	Rigid Steel	296.35	251.60	234.80	225.80	220.00	207.05	201.90	199.05	197.20
	Bearing Walls	318.95	266.15	245.75	234.55	227.35	210.70	203.95	200.20	197.60
Face Brick and Concrete Block	Rigid Steel	353.80	292.15	267.75	254.25	245.40	224.75	216.25	211.45	208.10
	Bearing Walls	351.10	289.00	264.40	250.75	241.75	220.90	212.30	207.45	204.05
Metal Sandwich Panel	Rigid Steel	320.85	268.95	248.95	238.05	230.90	214.70	208.15	204.50	201.90
Precast Concrete	Bearing Walls	313.10	262.05	242.40	231.70	224.70	208.85	202.45	198.90	196.45
Perimeter Adj., Add or Deduct	Per 100 L.F.	43.55	21.80	14.60	10.90	8.70	4.40	2.90	2.20	1.70
Story Hgt. Adj., Add or Deduct	Per 1 Ft.	2.10	1.50	1.25	1.05	0.95	0.65	0.55	0.50	0.35
For Basement, add $36.00 per square foot of basement area										

The above costs were calculated using the basic specifications shown on the facing page. These costs should be adjusted where necessary for design alternatives and owner's requirements.

Common additives

Description	Unit	$ Cost
Closed Circuit Surveillance, One station		
Camera and monitor	Ea.	1475
For additional camera stations, add	Ea.	665
Emergency Lighting, 25 watt, battery operated		
Lead battery	Ea.	330
Nickel cadmium	Ea.	550
Laundry Equipment		
Dryers, coin operated 30 lb.	Ea.	4100
Double stacked	Ea.	9400
50 lb.	Ea.	4650
Dry cleaner 20 lb.	Ea.	44,300
30 lb.	Ea.	69,000
Washers, coin operated	Ea.	1700
Washer/extractor 20 lb.	Ea.	7950
30 lb.	Ea.	12,400
50 lb.	Ea.	15,600
75 lb.	Ea.	24,300
Smoke Detectors		
Ceiling type	Ea.	244
Duct type	Ea.	585

Important: See the Reference Section for Location Factors.

Model costs calculated for a 1 story building with 12′ story height and 3,000 square feet of floor area

Laundromat

				Unit	Unit Cost	Cost Per S.F.	% Of Sub-Total
A. SUBSTRUCTURE							
1010	Standard Foundations	Poured concrete; strip and spread footings; 4′ foundation wall		S.F. Ground	9.60	9.60	
1020	Special Foundations	N/A		—	—	—	
1030	Slab on Grade	5″ reinforced concrete with vapor barrier and granular base		S.F. Slab	6.30	6.30	9.4%
2010	Basement Excavation	Site preparation for slab and trench for foundation wall and footing		S.F. Ground	.57	.57	
2020	Basement Walls	N/A		—	—	—	
B. SHELL							
	B10 Superstructure						
1010	Floor Construction	Steel column fireproofing		S.F. Floor	1	1	6.8%
1020	Roof Construction	Metal deck, open web steel joists, beams, columns		S.F. Roof	10.90	10.90	
	B20 Exterior Enclosure						
2010	Exterior Walls	Decorative concrete block	90% of wall	S.F. Wall			
2020	Exterior Windows	Store front	10% of wall	Each	77	6.69	5.8%
2030	Exterior Doors	Double aluminum and glass		Each	5238	3.49	
	B30 Roofing						
3010	Roof Coverings	Built-up tar and gravel with flashing; perlite/EPS composite insulation		S.F. Roof	9.38	9.38	5.3%
3020	Roof Openings	N/A		—	—	—	
C. INTERIORS							
1010	Partitions	Gypsum board on metal studs	60 S.F. Floor/L.F. Partition	S.F. Partition	29.10	4.85	
1020	Interior Doors	Single leaf wood	750 S.F. Floor/Door	Each	720	.96	
1030	Fittings	N/A		—	—	—	
2010	Stair Construction	N/A		—	—	—	7.3%
3010	Wall Finishes	Paint		S.F. Surface	.87	.29	
3020	Floor Finishes	Vinyl composition tile		S.F. Floor	2.45	2.45	
3030	Ceiling Finishes	Fiberglass board on exposed grid system		S.F. Ceiling	4.35	4.35	
D. SERVICES							
	D10 Conveying						
1010	Elevators & Lifts	N/A		—	—	—	0.0 %
1020	Escalators & Moving Walks	N/A		—	—	—	
	D20 Plumbing						
2010	Plumbing Fixtures	Toilet and service fixtures, supply and drainage	1 Fixture/600 S.F. Floor	Each	7374	12.29	
2020	Domestic Water Distribution	Gas fired hot water heater		S.F. Floor	54	54.23	39.1%
2040	Rain Water Drainage	Roof drains		S.F. Roof	2.18	2.18	
	D30 HVAC						
3010	Energy Supply	N/A		—	—	—	
3020	Heat Generating Systems	Included in D3050		—	—	—	
3030	Cooling Generating Systems	N/A		—	—	—	5.6 %
3050	Terminal & Package Units	Rooftop single zone unit systems		S.F. Floor	9.76	9.76	
3090	Other HVAC Sys. & Equipment	N/A		—	—	—	
	D40 Fire Protection						
4010	Sprinklers	Sprinkler, ordinary hazard		S.F. Floor	5.67	5.67	3.2%
4020	Standpipes	N/A		—	—	—	
	D50 Electrical						
5010	Electrical Service/Distribution	200 ampere service, panel board and feeders		S.F. Floor	4.17	4.17	
5020	Lighting & Branch Wiring	High efficiency fluorescent fixtures, receptacles, switches, A.C. and misc. power		S.F. Floor	22.71	22.71	17.4%
5030	Communications & Security	Addressable alarm systems and emergency lighting		S.F. Floor	3.21	3.21	
5090	Other Electrical Systems	Emergency generator, 7.5 kW		S.F. Floor	.52	.52	
E. EQUIPMENT & FURNISHINGS							
1010	Commercial Equipment	N/A		—	—	—	
1020	Institutional Equipment	N/A		—	—	—	0.0 %
1030	Vehicular Equipment	N/A		—	—	—	
1090	Other Equipment	N/A		—	—	—	
F. SPECIAL CONSTRUCTION							
1020	Integrated Construction	N/A		—	—	—	0.0 %
1040	Special Facilities	N/A		—	—	—	
G. BUILDING SITEWORK	**N/A**						

			Sub-Total	175.57	100%
CONTRACTOR FEES (General Requirements: 10%, Overhead: 5%, Profit: 10%)			25%	43.87	
ARCHITECT FEES			7%	15.36	
		Total Building Cost		**234.80**	

For customer support on your Square Foot Costs with RSMeans data, call 800.448.8182.

159

Costs per square foot of floor area

Exterior Wall	S.F. Area	7000	10000	13000	16000	19000	22000	25000	28000	31000
	L.F. Perimeter	240	300	336	386	411	435	472	510	524
Face Brick and Concrete Block	Reinforced Concrete	210.10	199.95	191.20	187.30	182.20	178.50	176.60	175.15	172.55
	Rigid Steel	207.10	196.90	188.15	184.25	179.20	175.40	173.60	172.10	169.55
Limestone and Concrete Block	Reinforced Concrete	242.75	226.50	213.05	206.85	199.20	193.55	190.65	188.35	184.60
	Rigid Steel	233.30	219.80	207.90	202.65	195.75	190.55	188.00	186.00	182.45
Precast Concrete	Reinforced Concrete	246.60	229.85	215.95	209.50	201.60	195.75	192.80	190.40	186.50
	Rigid Steel	243.55	226.85	212.85	206.50	198.60	192.75	189.70	187.40	183.50
Perimeter Adj., Add or Deduct	Per 100 L.F.	25.60	17.90	13.80	11.15	9.50	8.10	7.15	6.35	5.75
Story Hgt. Adj., Add or Deduct	Per 1 Ft.	3.85	3.35	2.90	2.70	2.45	2.20	2.15	2.05	1.90

For Basement, add $51.50 per square foot of basement area

The above costs were calculated using the basic specifications shown on the facing page. These costs should be adjusted where necessary for design alternatives and owner's requirements.

Common additives

Description	Unit	$ Cost
Carrels Hardwood	Ea.	740 - 1925
Closed Circuit Surveillance, One station		
Camera and monitor	Ea.	1475
For additional camera stations, add	Ea.	665
Elevators, Hydraulic passenger, 2 stops		
1500# capacity	Ea.	73,100
2500# capacity	Ea.	76,600
3500# capacity	Ea.	81,600
Emergency Lighting, 25 watt, battery operated		
Lead battery	Ea.	330
Nickel cadmium	Ea.	550
Flagpoles, Complete		
Aluminum, 20' high	Ea.	1900
40' high	Ea.	4500
70' high	Ea.	11,600
Fiberglass, 23' high	Ea.	1450
39'-5" high	Ea.	3325
59' high	Ea.	7300

Description	Unit	$ Cost
Library Furnishings		
Bookshelf, 90" high, 10" shelf double face	L.F.	260
single face	L.F.	181
Charging desk, built-in with counter		
Plastic laminated top	L.F.	600
Reading table, laminated		
top 60" x 36"	Ea.	505

Important: See the Reference Section for Location Factors.

Model costs calculated for a 2 story building with 14' story height and 22,000 square feet of floor area

				Unit	Unit Cost	Cost Per S.F.	% Of Sub-Total
A.	**SUBSTRUCTURE**						
1010	Standard Foundations	Poured concrete; strip and spread footings; 4' foundation wall		S.F. Ground	7.80	3.90	
1020	Special Foundations	N/A		—	—	—	
1030	Slab on Grade	4" reinforced concrete with vapor barrier and granular base		S.F. Slab	5.80	2.90	5.3%
2010	Basement Excavation	Site preparation for slab and trench for foundation wall and footing		S.F. Ground	.33	.17	
2020	Basement Walls	N/A		—	—	—	
B.	**SHELL**						
	B10 Superstructure						
1010	Floor Construction	Concrete waffle slab		S.F. Floor	29.24	14.62	19.7%
1020	Roof Construction	Concrete waffle slab		S.F. Roof	22.80	11.40	
	B20 Exterior Enclosure						
2010	Exterior Walls	Face brick with concrete block backup	90% of wall	S.F. Wall	34.76	17.32	
2020	Exterior Windows	Window wall	10% of wall	Each	58	3.19	16.0%
2030	Exterior Doors	Double aluminum and glass, single leaf hollow metal		Each	7200	.66	
	B30 Roofing						
3010	Roof Coverings	Single ply membrane, EPDM, fully adhered; perlite/EPS composite insulation		S.F. Roof	5.82	2.91	2.2%
3020	Roof Openings	Roof hatches		S.F. Roof	.10	.05	
C.	**INTERIORS**						
1010	Partitions	Gypsum board on metal studs	30 S.F. Floor/L.F. Partition	S.F. Partition	13.18	5.27	
1020	Interior Doors	Single leaf wood	300 S.F. Floor/Door	Each	1222	4.08	
1030	Fittings	N/A		—	—	—	
2010	Stair Construction	Concrete filled metal pan		Flight	10,475	.96	16.2%
3010	Wall Finishes	Paint		S.F. Surface	.90	.72	
3020	Floor Finishes	50% carpet, 50% vinyl tile		S.F. Floor	4.15	4.15	
3030	Ceiling Finishes	Mineral fiber on concealed zee bars		S.F. Ceiling	6.30	6.30	
D.	**SERVICES**						
	D10 Conveying						
1010	Elevators & Lifts	One hydraulic passenger elevator		Each	94,820	4.31	3.3%
1020	Escalators & Moving Walks	N/A		—	—	—	
	D20 Plumbing						
2010	Plumbing Fixtures	Toilet and service fixtures, supply and drainage	1 Fixture/1835 S.F. Floor	Each	8294	4.52	
2020	Domestic Water Distribution	Gas fired water heater		S.F. Floor	1.44	1.44	5.1%
2040	Rain Water Drainage	Roof drains		S.F. Roof	1.62	.81	
	D30 HVAC						
3010	Energy Supply	N/A		—	—	—	
3020	Heat Generating Systems	Included in D3050		—	—	—	
3030	Cooling Generating Systems	N/A		—	—	—	17.2 %
3050	Terminal & Package Units	Multizone unit, gas heating, electric cooling		S.F. Floor	22.75	22.75	
3090	Other HVAC Sys. & Equipment	N/A		—	—	—	
	D40 Fire Protection						
4010	Sprinklers	Wet pipe sprinkler system		S.F. Floor	3.30	3.30	3.3%
4020	Standpipes	Standpipe		S.F. Floor	1.09	1.09	
	D50 Electrical						
5010	Electrical Service/Distribution	400 ampere service, panel board and feeders		S.F. Floor	1.19	1.19	
5020	Lighting & Branch Wiring	High efficiency fluorescent fixtures, receptacles, switches, A.C. and misc. power		S.F. Floor	11.94	11.94	11.6%
5030	Communications & Security	Addressable alarm systems, internet wiring, and emergency lighting		S.F. Floor	2.15	2.15	
5090	Other Electrical Systems	Emergency generator, 7.5 kW, Uninterruptible power supply		S.F. Floor	.10	.10	
E.	**EQUIPMENT & FURNISHINGS**						
1010	Commercial Equipment	N/A		—	—	—	
1020	Institutional Equipment	N/A		—	—	—	
1030	Vehicular Equipment	N/A		—	—	—	0.0 %
1090	Other Equipment	N/A		—	—	—	
F.	**SPECIAL CONSTRUCTION**						
1020	Integrated Construction	N/A		—	—	—	0.0 %
1040	Special Facilities	N/A		—	—	—	
G.	**BUILDING SITEWORK**	**N/A**					

			Sub-Total	132.20	**100%**
CONTRACTOR FEES (General Requirements: 10%, Overhead: 5%, Profit: 10%)		25%	33.08	
ARCHITECT FEES		8%	13.22	

	Total Building Cost	**178.50**

For customer support on your Square Foot Costs with RSMeans data, call 800.448.8182.

161

Costs per square foot of floor area

Exterior Wall	S.F. Area	4000	5500	7000	8500	10000	11500	13000	14500	16000
	L.F. Perimeter	280	320	380	440	453	503	510	522	560
Vinyl Clapboard	Wood Frame	197.65	188.45	185.15	183.00	178.30	177.00	173.70	171.40	170.50
Stone Veneer	Wood Frame	203.65	191.60	187.35	184.65	178.40	176.75	172.40	169.30	168.25
Fiber Cement	Rigid Steel	201.85	192.20	188.85	186.60	181.70	180.30	176.85	174.45	173.55
E.I.F.S.	Rigid Steel	203.65	193.70	190.20	187.90	182.85	181.45	177.90	175.35	174.45
Precast Concrete	Reinforced Concrete	242.20	227.05	221.80	218.40	210.55	208.40	202.90	198.95	197.60
Brick Veneer	Reinforced Concrete	253.25	236.35	230.55	226.80	217.90	215.55	209.35	204.95	203.45
Perimeter Adj., Add or Deduct	Per 100 L.F.	17.25	12.55	9.90	8.15	6.90	6.00	5.35	4.75	4.35
Story Hgt. Adj., Add or Deduct	Per 1 Ft.	2.30	1.90	1.75	1.75	1.50	1.40	1.30	1.10	1.20
For Basement, add $33.50 per square foot of basement area										

The above costs were calculated using the basic specifications shown on the facing page. These costs should be adjusted where necessary for design alternatives and owner's requirements.

Common additives

Description	Unit	$ Cost
Cabinets, Hospital, base		
Laminated plastic	L.F.	530
Stainless steel	L.F.	870
Counter top, laminated plastic	L.F.	87.50
Stainless steel	L.F.	213
For drop-in sink, add	Ea.	1500
Nurses station, door type		
Laminated plastic	L.F.	595
Enameled steel	L.F.	575
Stainless steel	L.F.	1025
Wall cabinets, laminated plastic	L.F.	390
Enameled steel	L.F.	460
Stainless steel	L.F.	825

Description	Unit	$ Cost
Directory Boards, Plastic, glass covered		
30" x 20"	Ea.	640
36" x 48"	Ea.	1625
Aluminum, 24" x 18"	Ea.	630
36" x 24"	Ea.	790
48" x 32"	Ea.	1025
48" x 60"	Ea.	2200
Heat Therapy Unit		
Humidified, 26" x 78" x 28"	Ea.	3675
Smoke Detectors		
Ceiling type	Ea.	244
Duct type	Ea.	585
Tables, Examining, vinyl top with base cabinets	Ea.	1575 - 6525
Utensil Washer, Sanitizer	Ea.	10,300
X-Ray, Mobile	Ea.	20,800 - 105,000

Important: See the Reference Section for Location Factors.

Model costs calculated for a 1 story building with 10' story height and 7,000 square feet of floor area

	Unit	Unit Cost	Cost Per S.F.	% Of Sub-Total

A. SUBSTRUCTURE

			Unit	Unit Cost	Cost Per S.F.	% Of Sub-Total
1010	Standard Foundations	Poured concrete; strip and spread footings; 4' foundation wall	S.F. Ground	8.21	8.21	
1020	Special Foundations	N/A	—	—	—	
1030	Slab on Grade	4" reinforced slab on grade	S.F. Slab	5.80	5.80	10.4%
2010	Basement Excavation	Site preparation for slab and trench for foundation wall and footing	S.F. Ground	.33	.33	
2020	Basement Walls	N/A	—	—	—	

B. SHELL

B10 Superstructure

				Unit	Unit Cost	Cost Per S.F.	% Of Sub-Total
1010	Floor Construction	Wood columns, fireproofed		S.F. Floor	3.66	3.66	8.6%
1020	Roof Construction	Wood roof trusses, 4:12 slope		S.F. Roof	8.10	8.10	

B20 Exterior Enclosure

				Unit	Unit Cost	Cost Per S.F.	% Of Sub-Total
2010	Exterior Walls	Ashlar stone veneer on wood stud back up	70% of wall	S.F. Wall	43.95	16.70	
2020	Exterior Windows	Double hung, insulated wood windows	30% of wall	Each	665	6.37	18.3%
2030	Exterior Doors	Single leaf solid core wood doors		Each	2385	2.05	

B30 Roofing

			Unit	Unit Cost	Cost Per S.F.	% Of Sub-Total
3010	Roof Coverings	Asphalt shingles with flashing, gutters and downspouts	S.F. Roof	2.82	2.82	2.1%
3020	Roof Openings	N/A	—	—	—	

C. INTERIORS

				Unit	Unit Cost	Cost Per S.F.	% Of Sub-Total
1010	Partitions	Fire rated gypsum board on wood studs	6 S.F. Floor/L.F. Partition	S.F. Partition	8.15	10.87	
1020	Interior Doors	Single leaf solid core wood doors	240 S.F. Floor/Door	Each	1222	5.09	
1030	Fittings	N/A		—	—	—	
2010	Stair Construction	N/A		—	—	—	21.6%
3010	Wall Finishes	Paint on drywall		S.F. Surface	1.02	2.71	
3020	Floor Finishes	50% carpet , 50% vinyl composition tile		S.F. Floor	4.15	4.15	
3030	Ceiling Finishes	Acoustic ceiling tiles on suspended support system		S.F. Ceiling	6.86	6.86	

D. SERVICES

D10 Conveying

			Unit	Unit Cost	Cost Per S.F.	% Of Sub-Total
1010	Elevators & Lifts	N/A	—	—	—	0.0 %
1020	Escalators & Moving Walks	N/A	—	—	—	

D20 Plumbing

				Unit	Unit Cost	Cost Per S.F.	% Of Sub-Total
2010	Plumbing Fixtures	Restroom, exam room and service fixtures, supply and drainage	1 Fixture/350 S.F. Floor	Each	2104	6.01	
2020	Domestic Water Distribution	Gas fired water heater		S.F. Floor	2.25	2.25	6.0%
2040	Rain Water Drainage	Roof drains		S.F. Roof			

D30 HVAC

			Unit	Unit Cost	Cost Per S.F.	% Of Sub-Total
3010	Energy Supply	N/A	—	—	—	
3020	Heat Generating Systems	Included in D3050	—	—	—	
3030	Cooling Generating Systems	N/A	—	—	—	11.9 %
3050	Terminal & Package Units	Multizone unit, gas heating , electric cooling	S.F. Floor	16.30	16.30	
3090	Other HVAC Sys. & Equipment	N/A	—	—	—	

D40 Fire Protection

			Unit	Unit Cost	Cost Per S.F.	% Of Sub-Total
4010	Sprinklers	Wet pipe spinkler system	S.F. Floor	5.13	5.13	4.7%
4020	Standpipes	Standpipes and hose systems, with pumps	S.F. Floor	1.36	1.36	

D50 Electrical

			Unit	Unit Cost	Cost Per S.F.	% Of Sub-Total
5010	Electrical Service/Distribution	200 Ampere service, panel boards and feeders	S.F. Floor	1.65	1.65	
5020	Lighting & Branch Wiring	Fluorescent fixtures, receptacles, switches, A.C. and misc. power	S.F. Floor	8.17	8.17	13.1%
5030	Communications & Security	Alarm systems, emergency lighting, internet and phone wiring	S.F. Floor	8.24	8.24	
5090	Other Electrical Systems	N/A	—	—	—	

E. EQUIPMENT & FURNISHINGS

			Unit	Unit Cost	Cost Per S.F.	% Of Sub-Total
1010	Commercial Equipment	N/A	—	—	—	
1020	Institutional Equipment	Exam room equipment, cabinets and countertops	S.F. Floor	4.70	4.70	3.4 %
1030	Vehicular Equipment	N/A	—	—	—	
1090	Other Equipment	N/A	—	—	—	

F. SPECIAL CONSTRUCTION

			Unit	Unit Cost	Cost Per S.F.	% Of Sub-Total
1020	Integrated Construction	N/A	—	—	—	0.0 %
1040	Special Facilities	N/A	—	—	—	

G. BUILDING SITEWORK   N/A

		Sub-Total	137.53	100%
CONTRACTOR FEES (General Requirements: 10%, Overhead: 5%, Profit: 10%)		25%	34.35	
ARCHITECT FEES		9%	15.47	

	Total Building Cost	**187.35**

For customer support on your Square Foot Costs with RSMeans data, call 800.448.8182.

163

Costs per square foot of floor area

Exterior Wall	S.F. Area	4000	5500	7000	8500	10000	11500	13000	14500	16000
	L.F. Perimeter	180	210	240	270	286	311	336	361	386
Vinyl Clapboard	Wood Frame	228.50	218.70	212.95	209.40	205.45	203.20	201.55	200.30	199.25
Stone Veneer	Wood Frame	249.05	236.00	228.40	223.55	218.15	215.15	212.95	211.25	209.80
Fiber Cement	Rigid Steel	233.40	222.95	217.00	213.15	208.90	206.60	204.85	203.45	202.40
E.I.F.S.	Rigid Steel	238.05	227.35	221.25	217.35	212.95	210.60	208.80	207.35	206.25
Precast Concrete	Reinforced Concrete	278.70	262.40	253.10	247.15	240.15	236.55	233.75	231.60	229.80
Brick Veneer	Reinforced Concrete	247.80	234.30	226.55	221.60	216.00	213.00	210.65	208.85	207.45
Perimeter Adj., Add or Deduct	Per 100 L.F.	25.10	18.15	14.35	11.80	10.00	8.75	7.75	6.95	6.20
Story Hgt. Adj., Add or Deduct	Per 1 Ft.	2.85	2.35	2.20	1.95	1.80	1.75	1.65	1.55	1.45

For Basement, add $37.20 per square foot of basement area

The above costs were calculated using the basic specifications shown on the facing page. These costs should be adjusted where necessary for design alternatives and owner's requirements.

Common additives

Description	Unit	$ Cost
Cabinets, Hospital, base		
Laminated plastic	L.F.	530
Stainless steel	L.F.	870
Counter top, laminated plastic	L.F.	87.50
Stainless steel	L.F.	213
For drop-in sink, add	Ea.	1500
Nurses station, door type		
Laminated plastic	L.F.	595
Enameled steel	L.F.	575
Stainless steel	L.F.	1025
Wall cabinets, laminated plastic	L.F.	390
Enameled steel	L.F.	460
Stainless steel	L.F.	825
Elevators, Hydraulic passenger, 2 stops		
1500# capacity	Ea.	73,100
2500# capacity	Ea.	76,600
3500# capacity	Ea.	81,600

Description	Unit	$ Cost
Directory Boards, Plastic, glass covered		
30" x 20"	Ea.	640
36" x 48"	Ea.	1625
Aluminum, 24" x 18"	Ea.	630
36" x 24"	Ea.	790
48" x 32"	Ea.	1025
48" x 60"	Ea.	2200
Emergency Lighting, 25 watt, battery operated		
Lead battery	Ea.	330
Nickel cadmium	Ea.	550
Heat Therapy Unit		
Humidified, 26" x 78" x 28"	Ea.	3675
Smoke Detectors		
Ceiling type	Ea.	244
Duct type	Ea.	585
Tables, Examining, vinyl top		
with base cabinets	Ea.	1575 - 6525
Utensil Washer, Sanitizer	Ea.	10,300
X-Ray, Mobile	Ea.	20,800 - 105,000

Important: See the Reference Section for Location Factors.

Model costs calculated for a 2 story building with 10' story height and 7,000 square feet of floor area

				Unit	Unit Cost	Cost Per S.F.	% Of Sub-Total
A.	**SUBSTRUCTURE**						
1010	Standard Foundations	Poured concrete; strip and spread footings; 4' foundation wall		S.F. Ground	11.82	5.91	
1020	Special Foundations	N/A		—	—	—	
1030	Slab on Grade	4" reinforced slab on grade		S.F. Slab	5.80	2.90	5.7%
2010	Basement Excavation	Site preparation for slab and trench for foundation wall and footing		S.F. Ground	.57	.29	
2020	Basement Walls	N/A		—	—	—	
B.	**SHELL**						
	B10 Superstructure						
1010	Floor Construction	Open web bar joist, slab form, fireproofed columns		S.F. Floor	18.52	9.26	9.2%
1020	Roof Construction	Metal deck, open web steel joists, beams, columns		S.F. Roof	10.92	5.46	
	B20 Exterior Enclosure						
2010	Exterior Walls	Fiber cement siding on metal studs	70% of wall	S.F. Wall	16.85	8.09	
2020	Exterior Windows	Aluminum sliding windows	30% of wall	Each	529	7.25	12.2%
2030	Exterior Doors	Aluminum & glass entrance doors		Each	7200	4.12	
	B30 Roofing						
3010	Roof Coverings	Single ply membrane, stone ballast, rigid insulation		S.F. Roof	9.14	4.57	3.6%
3020	Roof Openings	Roof and smoke hatches		S.F. Roof	2.22	1.11	
C.	**INTERIORS**						
1010	Partitions	Fire rated gypsum board on metal studs	6 S.F. Floor/L.F. Partition	S.F. Partition	6.06	8.08	
1020	Interior Doors	Single leaf solid core wood doors	240 S.F. Floor/Door	Each	1222	5.76	
1030	Fittings	N/A		—	—	—	
2010	Stair Construction	Concrete filled metal pan, with rails		Flight	15,800	11.28	24.5%
3010	Wall Finishes	Paint on drywall		S.F. Surface	1.05	2.81	
3020	Floor Finishes	50% carpet , 50% vinyl composition tile		S.F. Floor	4.15	4.15	
3030	Ceiling Finishes	Acoustic ceiling tiles on suspended support system		S.F. Ceiling	6.86	6.86	
D.	**SERVICES**						
	D10 Conveying						
1010	Elevators & Lifts	One hydraulic elevator		Each	116,690	16.67	10.5%
1020	Escalators & Moving Walks	N/A		—	—	—	
	D20 Plumbing						
2010	Plumbing Fixtures	Restroom, exam room and service fixtures, supply and drainage	1 Fixture/350 S.F. Floor	Each	2216	6.33	
2020	Domestic Water Distribution	Gas fired water heater		S.F. Floor	2.25	2.25	6.2%
2040	Rain Water Drainage	Roof drains		S.F. Roof	2.72	1.36	
	D30 HVAC						
3010	Energy Supply	N/A		—	—	—	
3020	Heat Generating Systems	Included in D3050		—	—	—	
3030	Cooling Generating Systems	N/A		—	—	—	10.2 %
3050	Terminal & Package Units	Multizone unit, gas heating , electric cooling		S.F. Floor	16.30	16.30	
3090	Other HVAC Sys. & Equipment	N/A		—	—	—	
	D40 Fire Protection						
4010	Sprinklers	Wet pipe spinkler system		S.F. Floor	4.08	4.08	3.9%
4020	Standpipes	Standpipes and hose systems, with pumps		S.F. Floor	2.06	2.06	
	D50 Electrical						
5010	Electrical Service/Distribution	400 Ampere service, panel boards and feeders		S.F. Floor	2.43	2.43	
5020	Lighting & Branch Wiring	Fluorescent fixtures, receptacles, switches, A.C. and misc. power		S.F. Floor	8.17	8.17	11.8%
5030	Communications & Security	Alarm systems, emergency lighting, internet and phone wiring		S.F. Floor	8.24	8.24	
5090	Other Electrical Systems	N/A		—	—	—	
E.	**EQUIPMENT & FURNISHINGS**						
1010	Commercial Equipment	N/A		—	—	—	
1020	Institutional Equipment	Exam room equipment, cabinets and countertops		S.F. Floor	3.46	3.46	2.2 %
1030	Vehicular Equipment	N/A		—	—	—	
1090	Other Equipment	N/A		—	—	—	
F.	**SPECIAL CONSTRUCTION**						
1020	Integrated Construction	N/A		—	—	—	0.0 %
1040	Special Facilities	N/A		—	—	—	
G.	**BUILDING SITEWORK**	**N/A**					

		Sub-Total	159.25	100%
CONTRACTOR FEES (General Requirements: 10%, Overhead: 5%, Profit: 10%)		25%	39.83	
ARCHITECT FEES		9%	17.92	

Total Building Cost	**217**

For customer support on your Square Foot Costs with RSMeans data, call 800.448.8182.

165

Costs per square foot of floor area

Exterior Wall	S.F. Area	2000	3000	4000	6000	8000	10000	12000	14000	16000
	L.F. Perimeter	240	260	280	380	480	560	580	660	740
Brick Veneer	Wood Frame	210.25	184.65	171.90	164.95	161.45	158.20	153.10	151.95	151.10
Aluminum Clapboard	Wood Frame	191.15	170.85	160.75	154.85	151.80	149.25	145.40	144.45	143.80
Wood Clapboard	Wood Frame	190.80	170.60	160.50	154.65	151.65	149.10	145.25	144.30	143.60
Wood Shingles	Wood Frame	194.65	173.35	162.80	156.65	153.60	150.85	146.80	145.80	145.10
Concrete Block	Wood Truss	190.45	170.35	160.35	154.45	151.50	148.95	145.10	144.15	143.50
Face Brick and Concrete Block	Wood Truss	220.10	191.80	177.65	170.15	166.30	162.75	157.05	155.80	154.95
Perimeter Adj., Add or Deduct	Per 100 L.F.	29.20	19.45	14.60	9.70	7.30	5.80	4.85	4.15	3.70
Story Hgt. Adj., Add or Deduct	Per 1 Ft.	5.35	3.85	3.10	2.80	2.65	2.45	2.15	2.10	2.10
For Basement, add $25.60 per square foot of basement area										

The above costs were calculated using the basic specifications shown on the facing page. These costs should be adjusted where necessary for design alternatives and owner's requirements.

Common additives

Description	Unit	$ Cost
Closed Circuit Surveillance, One station		
Camera and monitor	Ea.	1475
For additional camera stations, add	Ea.	665
Emergency Lighting, 25 watt, battery operated		
Lead battery	Ea.	330
Nickel cadmium	Ea.	550
Laundry Equipment		
Dryer, gas, 16 lb. capacity	Ea.	1025
30 lb. capacity	Ea.	4100
Washer, 4 cycle	Ea.	1325
Commercial	Ea.	1700
Sauna, Prefabricated, complete		
6' x 4'	Ea.	6600
6' x 6'	Ea.	8850
6' x 9'	Ea.	10,000
8' x 8'	Ea.	10,600
8' x 10'	Ea.	11,700
10' x 12'	Ea.	15,500
Smoke Detectors		
Ceiling type	Ea.	244
Duct type	Ea.	585

Description	Unit	$ Cost
Swimming Pools, Complete, gunite	S.F.	106 - 132
TV Antenna, Master system, 12 outlet	Outlet	242
30 outlet	Outlet	335
100 outlet	Outlet	370

Important: See the Reference Section for Location Factors.

				Unit	Unit Cost	Cost Per S.F.	% Of Sub-Total
A. SUBSTRUCTURE							
1010	Standard Foundations	Poured concrete; strip and spread footings; 4′ foundation wall		S.F. Ground	7.63	7.63	
1020	Special Foundations	N/A		—	—	—	
1030	Slab on Grade	4″ reinforced concrete with vapor barrier and granular base		S.F. Slab	5.80	5.80	11.4%
2010	Basement Excavation	Site preparation for slab and trench for foundation wall and footing		S.F. Ground	.33	.33	
2020	Basement Walls	N/A		—	—	—	
B. SHELL							
B10 Superstructure							
1010	Floor Construction	N/A		—	—	—	6.4 %
1020	Roof Construction	Plywood on wood trusses		S.F. Roof	7.76	7.76	
B20 Exterior Enclosure							
2010	Exterior Walls	Face brick on wood studs with sheathing, insulation and paper	80% of wall	S.F. Wall	26.76	11.56	
2020	Exterior Windows	Wood double hung	20% of wall	Each	533	4.11	18.4%
2030	Exterior Doors	Wood solid core		Each	2385	6.56	
B30 Roofing							
3010	Roof Coverings	Asphalt shingles with flashing (pitched); rigid fiberglass insulation		S.F. Roof	4.67	4.67	3.9%
3020	Roof Openings	N/A		—	—	—	
C. INTERIORS							
1010	Partitions	Gypsum bd. and sound deadening bd. on wood studs	9 S.F. Floor/L.F. Partition	S.F. Partition	10.96	9.74	
1020	Interior Doors	Single leaf hollow core wood	300 S.F. Floor/Door	Each	640	2.13	
1030	Fittings	N/A		—	—	—	
2010	Stair Construction	N/A		—	—	—	22.2%
3010	Wall Finishes	90% paint, 10% ceramic tile		S.F. Surface	1.61	2.86	
3020	Floor Finishes	85% carpet, 15% ceramic tile		S.F. Floor	7.23	7.23	
3030	Ceiling Finishes	Painted gypsum board on furring		S.F. Ceiling	4.85	4.85	
D. SERVICES							
D10 Conveying							
1010	Elevators & Lifts	N/A		—	—	—	0.0 %
1020	Escalators & Moving Walks	N/A		—	—	—	
D20 Plumbing							
2010	Plumbing Fixtures	Toilet and service fixtures, supply and drainage	1 Fixture/90 S.F. Floor	Each	1931	21.45	
2020	Domestic Water Distribution	Gas fired water heater		S.F. Floor	2.13	2.13	19.5%
2040	Rain Water Drainage	N/A		—	—	—	
D30 HVAC							
3010	Energy Supply	N/A		—	—	—	
3020	Heat Generating Systems	Included in D3050		—	—	—	
3030	Cooling Generating Systems	N/A		—	—	—	2.6 %
3050	Terminal & Package Units	Through the wall electric heating and cooling units		S.F. Floor	3.08	3.08	
3090	Other HVAC Sys. & Equipment	N/A		—	—	—	
D40 Fire Protection							
4010	Sprinklers	Wet pipe sprinkler system		S.F. Floor	5.13	5.13	5.2%
4020	Standpipes	Standpipe, wet, Class III		S.F. Floor	1.19	1.19	
D50 Electrical							
5010	Electrical Service/Distribution	200 ampere service, panel board and feeders		S.F. Floor	1.33	1.33	
5020	Lighting & Branch Wiring	Fluorescent fixtures, receptacles, switches and misc. power		S.F. Floor	7.33	7.33	9.5%
5030	Communications & Security	Addressable alarm systems		S.F. Floor	2.80	2.80	
5090	Other Electrical Systems	N/A		—	—	—	
E. EQUIPMENT & FURNISHINGS							
1010	Commercial Equipment	Laundry equipment		S.F. Floor	1.02	1.02	
1020	Institutional Equipment	N/A		—	—	—	
1030	Vehicular Equipment	N/A		—	—	—	0.8%
1090	Other Equipment	N/A		—	—	—	
F. SPECIAL CONSTRUCTION							
1020	Integrated Construction	N/A		—	—	—	0.0 %
1040	Special Facilities	N/A		—	—	—	
G. BUILDING SITEWORK	N/A						

		Sub-Total	120.69	100%
CONTRACTOR FEES (General Requirements: 10%, Overhead: 5%, Profit: 10%)		25%	30.20	
ARCHITECT FEES		7%	10.56	
		Total Building Cost	**161.45**	

For customer support on your Square Foot Costs with RSMeans data, call 800.448.8182.

167

Costs per square foot of floor area

Exterior Wall	S.F. Area	25000	37000	49000	61000	73000	81000	88000	96000	104000
	L.F. Perimeter	433	593	606	720	835	911	978	1054	1074
Decorative Concrete Block	Wood Joists	174.95	172.05	167.10	166.00	165.30	164.90	164.65	164.40	163.50
	Precast Concrete	196.25	192.05	186.20	184.75	183.75	183.20	182.85	182.50	181.55
Stucco and Concrete Block	Wood Joists	180.55	176.20	169.85	168.30	167.25	166.70	166.30	165.95	164.90
	Precast Concrete	200.90	195.90	188.95	187.20	186.05	185.40	184.95	184.55	183.35
Wood Clapboard	Wood Frame	173.95	171.15	166.40	165.35	164.65	164.25	164.00	163.75	162.95
Brick Veneer	Wood Frame	180.55	176.20	169.80	168.25	167.25	166.65	166.30	165.95	164.85
Perimeter Adj., Add or Deduct	Per 100 L.F.	4.65	3.15	2.35	1.90	1.60	1.40	1.30	1.20	1.20
Story Hgt. Adj., Add or Deduct	Per 1 Ft.	1.85	1.70	1.30	1.25	1.25	1.20	1.15	1.15	1.15

For Basement, add *$33.60 per square foot of basement area*

The above costs were calculated using the basic specifications shown on the facing page. These costs should be adjusted where necessary for design alternatives and owner's requirements.

Common additives

Description	Unit	$ Cost		Description	Unit	$ Cost
Closed Circuit Surveillance, One station				Sauna, Prefabricated, complete		
Camera and monitor	Ea.	1475		6' x 4'	Ea.	6600
For additional camera station, add	Ea.	665		6' x 6'	Ea.	8850
Elevators, Hydraulic passenger, 2 stops				6' x 9'	Ea.	10,000
1500# capacity	Ea.	73,100		8' x 8'	Ea.	10,600
2500# capacity	Ea.	76,600		8' x 10'	Ea.	11,700
3500# capacity	Ea.	81,600		10' x 12'	Ea.	15,500
Additional stop, add	Ea.	8375		Smoke Detectors		
Emergency Lighting, 25 watt, battery operated				Ceiling type	Ea.	244
Lead battery	Ea.	330		Duct type	Ea.	585
Nickel cadmium	Ea.	550		Swimming Pools, Complete, gunite	S.F.	106 - 132
Laundry Equipment				TV Antenna, Master system, 12 outlet	Outlet	242
Dryer, gas, 16 lb. capacity	Ea.	1025		30 outlet	Outlet	335
30 lb. capacity	Ea.	4100		100 outlet	Outlet	370
Washer, 4 cycle	Ea.	1325				
Commercial	Ea.	1700				

Important: See the Reference Section for Location Factors.

Model costs calculated for a 3 story building with 9' story height and 49,000 square feet of floor area

				Unit	Unit Cost	Cost Per S.F.	% Of Sub-Total
A. SUBSTRUCTURE							
1010	Standard Foundations	Poured concrete; strip and spread footings; 4' foundation wall		S.F. Ground	6.09	2.03	
1020	Special Foundations	N/A		—	—	—	
1030	Slab on Grade	4" reinforced concrete with vapor barrier and granular base		S.F. Slab	5.80	1.93	2.9%
2010	Basement Excavation	Site preparation for slab and trench for foundation wall and footing		S.F. Ground	.33	.11	
2020	Basement Walls	N/A		—	—	—	
B. SHELL							
B10 Superstructure							
1010	Floor Construction	Precast concrete plank		S.F. Floor	13.65	9.10	9.5%
1020	Roof Construction	Precast concrete plank		S.F. Roof	12.87	4.29	
B20 Exterior Enclosure							
2010	Exterior Walls	Decorative concrete block	85% of wall	S.F. Wall	19.45	5.52	
2020	Exterior Windows	Aluminum sliding	15% of wall	Each	589	1.97	12.2%
2030	Exterior Doors	Aluminum and glass doors and entrance with transom		Each	2919	9.66	
B30 Roofing							
3010	Roof Coverings	Built-up tar and gravel with flashing; perlite/EPS composite insulation		S.F. Roof	6.78	2.26	1.6%
3020	Roof Openings	Roof hatches		S.F. Roof	.15	.05	
C. INTERIORS							
1010	Partitions	Concrete block	7 S.F. Floor/L.F. Partition	S.F. Partition	22.35	25.54	
1020	Interior Doors	Wood hollow core	70 S.F. Floor/Door	Each	640	9.14	
1030	Fittings	N/A		—	—	—	
2010	Stair Construction	Concrete filled metal pan		Flight	15,800	3.87	38.7%
3010	Wall Finishes	90% paint, 10% ceramic tile		S.F. Surface	1.61	3.69	
3020	Floor Finishes	85% carpet, 5% vinyl composition tile, 10% ceramic tile		S.F. Floor	7.23	7.23	
3030	Ceiling Finishes	Textured finish		S.F. Ceiling	4.85	4.85	
D. SERVICES							
D10 Conveying							
1010	Elevators & Lifts	Two hydraulic passenger elevators		Each	117,845	4.81	3.4%
1020	Escalators & Moving Walks	N/A		—	—	—	
D20 Plumbing							
2010	Plumbing Fixtures	Toilet and service fixtures, supply and drainage	1 Fixture/180 S.F. Floor	Each	4352	24.18	
2020	Domestic Water Distribution	Gas fired water heater		S.F. Floor	1.38	1.38	18.6%
2040	Rain Water Drainage	Roof drains		S.F. Roof	1.86	.62	
D30 HVAC							
3010	Energy Supply	N/A		—	—	—	
3020	Heat Generating Systems	Included in D3050		—	—	—	
3030	Cooling Generating Systems	N/A		—	—	—	2.0 %
3050	Terminal & Package Units	Through the wall electric heating and cooling units		S.F. Floor	2.79	2.79	
3090	Other HVAC Sys. & Equipment	N/A		—	—	—	
D40 Fire Protection							
4010	Sprinklers	Sprinklers, wet, light hazard		S.F. Floor	3.12	3.12	2.4%
4020	Standpipes	Standpipe, wet, Class III		S.F. Floor	.29	.29	
D50 Electrical							
5010	Electrical Service/Distribution	800 ampere service, panel board and feeders		S.F. Floor	1.56	1.56	
5020	Lighting & Branch Wiring	Fluorescent fixtures, receptacles, switches and misc. power		S.F. Floor	8.12	8.12	8.4%
5030	Communications & Security	Addressable alarm systems and emergency lighting		S.F. Floor	1.93	1.93	
5090	Other Electrical Systems	Emergency generator, 7.5 kW		S.F. Floor	.14	.14	
E. EQUIPMENT & FURNISHINGS							
1010	Commercial Equipment	Commercial laundry equipment		S.F. Floor	.34	.34	
1020	Institutional Equipment	N/A		—	—	—	
1030	Vehicular Equipment	N/A		—	—	—	0.2%
1090	Other Equipment	N/A		—	—	—	
F. SPECIAL CONSTRUCTION							
1020	Integrated Construction	N/A		—	—	—	0.0 %
1040	Special Facilities	N/A		—	—	—	
G. BUILDING SITEWORK	**N/A**						

		Sub-Total	140.52	100%
CONTRACTOR FEES (General Requirements: 10%, Overhead: 5%, Profit: 10%)		25%	35.14	
ARCHITECT FEES		6%	10.54	
	Total Building Cost		**186.20**	

For customer support on your Square Foot Costs with RSMeans data, call 800.448.8182.

169

Costs per square foot of floor area

Exterior Wall	S.F. Area	9000	10000	12000	13000	14000	15000	16000	18000	20000
	L.F. Perimeter	385	410	440	460	480	500	510	547	583
Decorative Concrete Block	Steel Joists	200.90	195.75	186.20	182.95	180.20	177.75	175.00	171.30	168.40
Concrete Block	Steel Joists	194.20	189.30	180.50	177.40	174.85	172.60	170.05	166.60	163.85
Face Brick and Concrete Block	Steel Joists	213.30	207.60	196.85	193.15	190.15	187.45	184.25	180.15	176.85
Precast Concrete	Steel Joists	236.05	229.40	216.25	211.90	208.30	205.10	201.10	196.15	192.25
Tilt-up Concrete Panels	Steel Joists	197.85	192.80	183.60	180.40	177.75	175.40	172.70	169.15	166.35
Metal Sandwich Panel	Steel Joists	196.30	191.35	182.30	179.20	176.55	174.20	171.55	168.10	165.25
Perimeter Adj., Add or Deduct	Per 100 L.F.	12.05	10.85	9.05	8.35	7.75	7.30	6.70	6.15	5.45
Story Hgt. Adj., Add or Deduct	Per 1 Ft.	1.80	1.75	1.60	1.50	1.50	1.50	1.30	1.35	1.25
Basement—Not Applicable										

The above costs were calculated using the basic specifications shown on the facing page. These costs should be adjusted where necessary for design alternatives and owner's requirements.

Common additives

Description	Unit	$ Cost
Emergency Lighting, 25 watt, battery operated		
Lead battery	Ea.	330
Nickel cadmium	Ea.	550
Seating		
Auditorium chair, all veneer	Ea.	355
Veneer back, padded seat	Ea.	380
Upholstered, spring seat	Ea.	335
Classroom, movable chair & desk	Set	81 - 171
Lecture hall, pedestal type	Ea.	350 - 650
Smoke Detectors		
Ceiling type	Ea.	244
Duct type	Ea.	585
Sound System		
Amplifier, 250 watts	Ea.	2025
Speaker, ceiling or wall	Ea.	240
Trumpet	Ea.	460

Important: See the Reference Section for Location Factors.

Model costs calculated for a 1 story building with 20' story height and 12,000 square feet of floor area

Movie Theater

				Unit	Unit Cost	Cost Per S.F.	% Of Sub-Total
A. SUBSTRUCTURE							
1010	Standard Foundations	Poured concrete; strip and spread footings; 4' foundation wall		S.F. Ground	4.92	4.92	
1020	Special Foundations	N/A		—	—	—	
1030	Slab on Grade	4" reinforced concrete with vapor barrier and granular base		S.F. Slab	5.80	5.80	7.9%
2010	Basement Excavation	Site preparation for slab and trench for foundation wall and footing		S.F. Ground	.33	.33	
2020	Basement Walls	N/A		—	—	—	
B. SHELL							
B10 Superstructure							
1010	Floor Construction	Open web steel joists, slab form, concrete	mezzanine 2250 S.F.	S.F. Floor	12.78	2.39	10.0%
1020	Roof Construction	Metal deck on open web steel joists		S.F. Roof	11.59	11.59	
B20 Exterior Enclosure							
2010	Exterior Walls	Decorative concrete block	80% of wall	S.F. Wall	17.36	12.73	
2020	Exterior Windows	Window wall	20% of wall	Each	59	8.70	16.7%
2030	Exterior Doors	Sliding mallfront aluminum and glass and hollow metal		Each	3550	1.77	
B30 Roofing							
3010	Roof Coverings	Built-up tar and gravel with flashing; perlite/EPS composite insulation		S.F. Roof	6.82	6.82	4.9%
3020	Roof Openings	N/A		—	—	—	
C. INTERIORS							
1010	Partitions	Concrete block	40 S.F. Floor/L.F. Partition	S.F. Partition	10.40	4.16	
1020	Interior Doors	Single leaf hollow metal	705 S.F. Floor/Door	Each	1222	1.73	
1030	Fittings	Toilet partitions		S.F. Floor	.66	.66	
2010	Stair Construction	Concrete filled metal pan		Flight	18,900	3.15	19.8%
3010	Wall Finishes	Paint		S.F. Surface	4.49	3.59	
3020	Floor Finishes	Carpet 50%, ceramic tile 5%	50% of area	S.F. Floor	13.32	6.66	
3030	Ceiling Finishes	Mineral fiber tile on concealed zee runners suspended		S.F. Ceiling	7.60	7.60	
D. SERVICES							
D10 Conveying							
1010	Elevators & Lifts	N/A		—	—	—	0.0 %
1020	Escalators & Moving Walks	N/A		—	—	—	
D20 Plumbing							
2010	Plumbing Fixtures	Toilet and service fixtures, supply and drainage	1 Fixture/500 S.F. Floor	Each	5490	10.98	
2020	Domestic Water Distribution	Gas fired water heater		S.F. Floor	.60	.60	9.5%
2040	Rain Water Drainage	Roof drains		S.F. Roof	1.62	1.62	
D30 HVAC							
3010	Energy Supply	N/A		—	—	—	
3020	Heat Generating Systems	Included in D3050		—	—	—	
3030	Cooling Generating Systems	N/A		—	—	—	8.3 %
3050	Terminal & Package Units	Single zone rooftop unit, gas heating, electric cooling		S.F. Floor	11.50	11.50	
3090	Other HVAC Sys. & Equipment	N/A		—	—	—	
D40 Fire Protection							
4010	Sprinklers	Wet pipe sprinkler system		S.F. Floor	4.52	4.52	4.0%
4020	Standpipes	Standpipe		S.F. Floor	.98	.98	
D50 Electrical							
5010	Electrical Service/Distribution	400 ampere service, panel board and feeders		S.F. Floor	1.78	1.78	
5020	Lighting & Branch Wiring	High efficiency fluorescent fixtures, receptacles, switches, A.C. and misc. power		S.F. Floor	7.07	7.07	8.2%
5030	Communications & Security	Addressable alarm systems, sound system and emergency lighting		S.F. Floor	2.39	2.39	
5090	Other Electrical Systems	Emergency generator, 7.5 kW		S.F. Floor	.19	.19	
E. EQUIPMENT & FURNISHINGS							
1010	Commercial Equipment	N/A		—	—	—	
1020	Institutional Equipment	Projection equipment, screen		S.F. Floor	9.93	9.93	10.8 %
1030	Vehicular Equipment	N/A		—	—	—	
2010	Fixed Furnishings	Movie theater seating		S.F. Floor	5.07	5.07	
F. SPECIAL CONSTRUCTION							
1020	Integrated Construction	N/A		—	—	—	0.0 %
1040	Special Facilities	N/A		—	—	—	
G. BUILDING SITEWORK	**N/A**						

		Sub-Total	139.23	100%
CONTRACTOR FEES (General Requirements: 10%, Overhead: 5%, Profit: 10%)		25%	34.79	
ARCHITECT FEES		7%	12.18	
		Total Building Cost	**186.20**	

For customer support on your Square Foot Costs with RSMeans data, call 800.448.8182.

171

Costs per square foot of floor area

Exterior Wall	S.F. Area	10000	15000	20000	25000	30000	35000	40000	45000	50000
	L.F. Perimeter	286	370	453	457	513	568	624	680	735
Precast Concrete	Bearing Walls	246.90	236.65	231.40	222.05	219.30	217.30	215.75	214.55	213.55
	Rigid Steel	250.70	240.50	235.20	226.00	223.25	221.25	219.70	218.50	217.55
Face Brick and Concrete Block	Bearing Walls	230.00	222.05	217.95	211.25	209.15	207.65	206.50	205.60	204.85
	Rigid Steel	234.45	226.60	222.55	215.90	213.80	212.30	211.20	210.30	209.60
Stucco and Concrete Block	Bearing Walls	224.60	217.45	213.70	207.80	205.95	204.60	203.55	202.75	202.10
	Steel Joists	229.10	221.95	218.30	212.45	210.60	209.30	208.25	207.45	206.80
Perimeter Adj., Add or Deduct	Per 100 L.F.	19.45	12.95	9.70	7.80	6.45	5.50	4.85	4.30	3.85
Story Hgt. Adj., Add or Deduct	Per 1 Ft.	4.90	4.20	3.80	3.15	2.85	2.70	2.65	2.55	2.50
For Basement, add $34.90 per square foot of basement area										

The above costs were calculated using the basic specifications shown on the facing page. These costs should be adjusted where necessary for design alternatives and owner's requirements.

Common additives

Description	Unit	$ Cost	Description	Unit	$ Cost
Beds, Manual	Ea.	940 - 3100	Kitchen Equipment, cont.		
Elevators, Hydraulic passenger, 2 stops			Ice cube maker, 50 lb. per day	Ea.	2100
1500# capacity	Ea.	73,100	Range with 1 oven	Ea.	3575
2500# capacity	Ea.	76,600	Laundry Equipment		
3500# capacity	Ea.	81,600	Dryer, gas, 16 lb. capacity	Ea.	1025
Emergency Lighting, 25 watt, battery operated			30 lb. capacity	Ea.	4100
Lead battery	Ea.	330	Washer, 4 cycle	Ea.	1325
Nickel cadmium	Ea.	550	Commercial	Ea.	1700
Intercom System, 25 station capacity			Nurses Call System		
Master station	Ea.	3225	Single bedside call station	Ea.	266
Intercom outlets	Ea.	212	Pillow speaker	Ea.	325
Handset	Ea.	590	Refrigerator, Prefabricated, walk-in		
Kitchen Equipment			7'-6" high, 6' x 6'	S.F.	145
Broiler	Ea.	4425	10' x 10'	S.F.	114
Coffee urn, twin 6 gallon	Ea.	2900	12' x 14'	S.F.	99.50
Cooler, 6 ft. long	Ea.	3750	12' x 20'	S.F.	137
Dishwasher, 10-12 racks per hr.	Ea.	4150	TV Antenna, Master system, 12 outlet	Outlet	242
Food warmer	Ea.	790	30 outlet	Outlet	335
Freezer, 44 C.F., reach-in	Ea.	5525	100 outlet	Outlet	370
			Whirlpool Bath, Mobile, 18" x 24" x 60"	Ea.	5975
			X-Ray, Mobile	Ea.	20,800 - 105,000

Important: See the Reference Section for Location Factors.

Model costs calculated for a 2 story building with 10' story height and 25,000 square feet of floor area

Nursing Home

			Unit	Unit Cost	Cost Per S.F.	% Of Sub-Total
A. SUBSTRUCTURE						
1010	Standard Foundations	Poured concrete; strip and spread footings; 4' foundation wall	S.F. Ground	4.86	2.43	
1020	Special Foundations	N/A	—	—	—	
1030	Slab on Grade	4" reinforced concrete with vapor barrier and granular base	S.F. Slab	5.80	2.90	3.4%
2010	Basement Excavation	Site preparation for slab and trench for foundation wall and footing	S.F. Ground	.33	.17	
2020	Basement Walls	N/A	—	—	—	
B. SHELL						
B10 Superstructure						
1010	Floor Construction	Pre-cast double tees with concrete topping	S.F. Floor	16.24	8.12	9.5%
1020	Roof Construction	Pre-cast double tees	S.F. Roof	14.14	7.07	
B20 Exterior Enclosure						
2010	Exterior Walls	Precast concrete panels	S.F. Wall	60	18.60	
2020	Exterior Windows	Wood double hung	15% of wall	Each	589	2.15	13.4%
2030	Exterior Doors	Double aluminum & glass doors, single leaf hollow metal	Each	3348	.67	
B30 Roofing						
3010	Roof Coverings	Built-up tar and gravel with flashing; perlite/EPS composite insulation	S.F. Roof	6.80	3.40	2.2%
3020	Roof Openings	Roof hatches	S.F. Roof	.10	.05	
C. INTERIORS						
1010	Partitions	Gypsum board on metal studs	8 S.F. Floor/L.F. Partition	S.F. Partition	9.84	9.84	
1020	Interior Doors	Single leaf wood	80 S.F. Floor/Door	Each	720	9.01	
1030	Fittings	N/A	—	—	—	
2010	Stair Construction	Concrete stair	Flight	15,800	2.53	20.7%
3010	Wall Finishes	50% vinyl wall coverings, 45% paint, 5% ceramic tile	S.F. Surface	1.85	3.69	
3020	Floor Finishes	95% vinyl tile, 5% ceramic tile	S.F. Floor	3.50	3.50	
3030	Ceiling Finishes	Painted gypsum board	S.F. Ceiling	4.54	4.54	
D. SERVICES						
D10 Conveying						
1010	Elevators & Lifts	One hydraulic hospital elevator	Each	116,750	4.67	2.9%
1020	Escalators & Moving Walks	N/A	—	—	—	
D20 Plumbing						
2010	Plumbing Fixtures	Kitchen, toilet and service fixtures, supply and drainage	1 Fixture/230 S.F. Floor	Each	9085	39.50	
2020	Domestic Water Distribution	Oil fired water heater	S.F. Floor	1.46	1.46	26.1%
2040	Rain Water Drainage	Roof drains	S.F. Roof	1.54	.77	
D30 HVAC						
3010	Energy Supply	Oil fired hot water, wall fin radiation	S.F. Floor	8.90	8.90	
3020	Heat Generating Systems	N/A	—	—	—	
3030	Cooling Generating Systems	N/A	—	—	—	10.6%
3050	Terminal & Package Units	Split systems with air cooled condensing units	S.F. Floor	8.03	8.03	
3090	Other HVAC Sys. & Equipment	N/A	—	—	—	
D40 Fire Protection						
4010	Sprinklers	Sprinkler, light hazard	S.F. Floor	3.30	3.30	2.4%
4020	Standpipes	Standpipe	S.F. Floor	.50	.50	
D50 Electrical						
5010	Electrical Service/Distribution	800 ampere service, panel board and feeders	S.F. Floor	1.62	1.62	
5020	Lighting & Branch Wiring	High efficiency fluorescent fixtures, receptacles, switches, A.C. and misc. power	S.F. Floor	10.62	10.62	8.9%
5030	Communications & Security	Addressable alarm systems and emergency lighting	S.F. Floor	1.45	1.45	
5090	Other Electrical Systems	Emergency generator, 15 kW	S.F. Floor	.56	.56	
E. EQUIPMENT & FURNISHINGS						
1010	Commercial Equipment	N/A	—	—	—	
1020	Institutional Equipment	N/A	—	—	—	
1030	Vehicular Equipment	N/A	—	—	—	0.0 %
1090	Other Equipment	N/A	—	—	—	
F. SPECIAL CONSTRUCTION						
1020	Integrated Construction	N/A	—	—	—	0.0 %
1040	Special Facilities	N/A	—	—	—	
G. BUILDING SITEWORK	**N/A**					

		Sub-Total	160.05	100%
CONTRACTOR FEES (General Requirements: 10%, Overhead: 5%, Profit: 10%)		25%	39.99	
ARCHITECT FEES		11%	22.01	

Total Building Cost | **222.05**

For customer support on your Square Foot Costs with RSMeans data, call 800.448.8182.

173

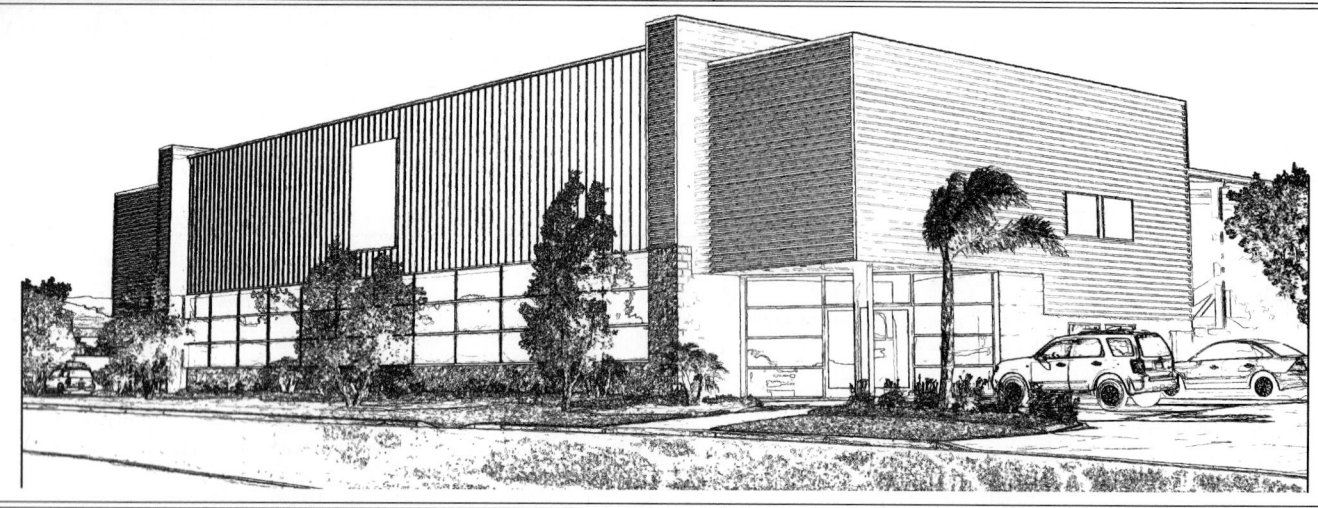

Costs per square foot of floor area

Exterior Wall	S.F. Area	2000	3000	5000	7000	9000	12000	15000	20000	25000
	L.F. Perimeter	220	260	320	360	420	480	520	640	700
Vinyl Clapboard	Wood Frame	220.05	197.40	177.00	166.50	162.05	156.65	152.65	150.15	147.30
Stone Veneer	Wood Frame	255.50	225.40	197.80	183.30	177.35	169.85	164.15	160.80	156.65
Fiber Cement	Rigid Steel	236.90	212.30	190.00	178.50	173.60	167.65	163.20	160.55	157.35
E.I.F.S.	Rigid Steel	239.50	214.40	191.55	179.75	174.75	168.60	164.05	161.30	158.00
Precast Concrete	Reinforced Concrete	306.65	268.85	233.85	215.20	207.60	197.85	190.45	186.25	180.75
Brick Veneer	Reinforced Concrete	313.65	274.40	237.90	218.55	210.70	200.55	192.80	188.50	182.80
Perimeter Adj., Add or Deduct	Per 100 L.F.	29.35	19.55	11.70	8.35	6.50	4.85	3.90	3.00	2.35
Story Hgt. Adj., Add or Deduct	Per 1 Ft.	3.05	2.40	1.75	1.45	1.25	1.10	0.90	0.90	0.80

For Basement, add $35.40 per square foot of basement area

The above costs were calculated using the basic specifications shown on the facing page. These costs should be adjusted where necessary for design alternatives and owner's requirements.

Common additives

Description	Unit	$ Cost
Closed circuit surveillance, one station		
Camera and monitor	Ea.	1475
For additional camera stations, add	Ea.	665
Directory boards, plastic, glass covered		
30" x 20"	Ea.	640
36" x 48"	Ea.	1625
Aluminum, 24" x 18"	Ea.	630
36" x 24"	Ea.	790
48" x 32"	Ea.	1025
48" x 60"	Ea.	2200
Electronic, wall mounted	S.F.	4625
Free standing	S.F.	4075
Pedestal access floor system w/plastic laminate cover		
Computer room, less than 6000 SF	S.F.	22
Greater than 6000 SF	S.F.	21.50
Office, greater than 6000 S.F.	S.F.	17.95

Description	Unit	$ Cost
Security access systems		
Metal detectors, wand type	Ea.	138
Walk-through portal type, single-zone	Ea.	3825
Multi-zone	Ea.	5425
X-ray equipment		
Desk top, for mail, small packages	Ea.	4000
Conveyer type, including monitor, minimum	Ea.	18,000
Maximum	Ea.	31,200
Explosive detection equipment		
Hand held, battery operated	Ea.	28,100
Walk-through portal type	Ea.	49,000
Uninterruptible power supply, 15 kVA/12.75 kW	kW	1.28

Model costs calculated for a 1 story building with 12' story height and 7,000 square feet of floor area

Office, 1 Story

					Unit	Unit Cost	Cost Per S.F.	% Of Sub-Total
A.	**SUBSTRUCTURE**							
1010	Standard Foundations	Poured concrete; strip and spread footings; 4' foundation wall			S.F. Ground	8.32	8.32	
1020	Special Foundations	N/A			—	—	—	
1030	Slab on Grade	4" reinforced concrete			S.F. Slab	5.80	5.80	9.0%
2010	Basement Excavation	Site preparation for slab and trench for foundation wall and footing			S.F. Ground	.33	.33	
2020	Basement Walls	N/A			—	—	—	
B.	**SHELL**							
	B10 Superstructure							
1010	Floor Construction	Cast in place concrete columns			S.F. Floor	2.86	2.86	12.1%
1020	Roof Construction	Concrete beam and slab			S.F. Roof	16.60	16.60	
	B20 Exterior Enclosure							
2010	Exterior Walls	Precast concrete panels, insulated	80% of wall		S.F. Wall	60	29.86	
2020	Exterior Windows	Aluminum awning windows	20% of wall		Each	755	4.05	23.5%
2030	Exterior Doors	Aluminum and glass single and double doors, hollow metal doors			Each	4582	3.93	
	B30 Roofing							
3010	Roof Coverings	Single ply membrane, stone ballast, rigid insulation			S.F. Roof	8.45	8.45	5.7%
3020	Roof Openings	Smoke hatches			S.F. Roof	.73	.73	
C.	**INTERIORS**							
1010	Partitions	Gypsum board on metal studs	20 S.F. Floor/L.F. Partition		S.F. Partition	6.60	3.30	
1020	Interior Doors	70% Single leaf solid wood doors, 30% hollow metal single leaf	320 S.F. Floor/Door		Each	1222	3.84	
1030	Fittings	Plastic laminate toilet partitions			S.F. Floor	.39	.39	
2010	Stair Construction	N/A			—	—	—	12.7%
3010	Wall Finishes	90% Paint, 10% ceramic wall tile			S.F. Surface	1.33	1.33	
3020	Floor Finishes	65% Carpet, 10% porcelain tile, 10% quarry tile			S.F. Floor	5.25	5.25	
3030	Ceiling Finishes	Acoustic ceiling tiles on suspended channel grid			S.F. Ceiling	6.30	6.30	
D.	**SERVICES**							
	D10 Conveying							
1010	Elevators & Lifts	N/A			—	—	—	0.0 %
1020	Escalators & Moving Walks	N/A			—	—	—	
	D20 Plumbing							
2010	Plumbing Fixtures	Restroom and service fixtures, supply and drainage	1 Fixture/700 S.F. Floor		Each	2212	3.16	
2020	Domestic Water Distribution	Gas fired water heater			S.F. Floor	1.70	1.70	3.6%
2040	Rain Water Drainage	Roof drains			S.F. Roof	.87	.87	
	D30 HVAC							
3010	Energy Supply	N/A			—	—	—	
3020	Heat Generating Systems	Included in D3050			—	—	—	
3030	Cooling Generating Systems	N/A			—	—	—	14.2 %
3050	Terminal & Package Units	Multizone gas heating, electric cooling unit			S.F. Floor	22.80	22.80	
3090	Other HVAC Sys. & Equipment	N/A			—	—	—	
	D40 Fire Protection							
4010	Sprinklers	Wet pipe light hazard sprinkler system			S.F. Floor	3.79	3.79	3.4%
4020	Standpipes	Standpipes			S.F. Floor	1.63	1.63	
	D50 Electrical							
5010	Electrical Service/Distribution	400 ampere service, panel boards and feeders			S.F. Floor	4.98	4.98	
5020	Lighting & Branch Wiring	Fluorescent fixtures, receptacles, switches, A.C. and misc. power			S.F. Floor	13.12	13.12	15.9%
5030	Communications & Security	Telephone systems, internet wiring, and addressable alarm systems			S.F. Floor	7.49	7.49	
5090	Other Electrical Systems	Emergency generator, 7.5 kW			S.F. Floor			
E.	**EQUIPMENT & FURNISHINGS**							
1010	Commercial Equipment	N/A			—	—	—	
1020	Institutional Equipment	N/A			—	—	—	0.0 %
1030	Vehicular Equipment	N/A			—	—	—	
1090	Other Equipment	N/A			—	—	—	
F.	**SPECIAL CONSTRUCTION**							
1020	Integrated Construction	N/A			—	—	—	0.0 %
1040	Special Facilities	N/A			—	—	—	
G.	**BUILDING SITEWORK**	**N/A**						

			Sub-Total	160.88	100%
CONTRACTOR FEES (General Requirements: 10%, Overhead: 5%, Profit: 10%)			25%	40.24	
ARCHITECT FEES			7%	14.08	
		Total Building Cost		**215.20**	

For customer support on your Square Foot Costs with RSMeans data, call 800.448.8182.

175

Costs per square foot of floor area

Exterior Wall	S.F. Area	5000	8000	12000	16000	20000	35000	50000	65000	80000
	L.F. Perimeter	220	260	310	330	360	440	550	600	675
Vinyl Clapboard	Wood Frame	213.00	189.60	176.25	167.00	162.15	152.60	149.55	146.65	145.30
Stone Veneer	Wood Frame	255.15	220.75	201.05	186.85	179.45	164.70	160.20	155.65	153.50
Fiber Cement	Rigid Steel	221.70	196.60	182.25	172.15	166.90	156.50	153.25	150.10	148.60
E.I.F.S.	Rigid Steel	227.85	201.90	187.05	176.60	171.20	160.35	157.05	153.75	152.20
Precast Concrete	Reinforced Concrete	299.50	255.70	230.50	212.05	202.50	183.30	177.50	171.45	168.70
Brick Veneer	Reinforced Concrete	292.30	250.55	226.50	208.90	199.85	181.60	176.10	170.40	167.70
Perimeter Adj., Add or Deduct	Per 100 L.F.	27.10	16.95	11.25	8.45	6.80	3.90	2.70	2.10	1.65
Story Hgt. Adj., Add or Deduct	Per 1 Ft.	3.65	2.70	2.15	1.65	1.50	1.00	0.90	0.75	0.70
For Basement, add $39.60 per square foot of basement area										

The above costs were calculated using the basic specifications shown on the facing page. These costs should be adjusted where necessary for design alternatives and owner's requirements.

Common additives

Description	Unit	$ Cost
Closed circuit surveillance, one station		
Camera and monitor	Ea.	1475
For additional camera stations, add	Ea.	665
Directory boards, plastic, glass covered		
30" x 20"	Ea.	640
36" x 48"	Ea.	1625
Aluminum, 24" x 18"	Ea.	630
36" x 24"	Ea.	790
48" x 32"	Ea.	1025
48" x 60"	Ea.	2200
Electronic, wall mounted	S.F.	4625
Free standing	S.F.	4075
Escalators, 10' rise, 32" wide, glass balustrade	Ea.	153,500
Metal balustrade	Ea.	162,000
48" wide, glass balustrade	Ea.	161,000
Metal balustrade	Ea.	171,000
Pedestal access floor system w/plastic laminate cover		
Computer room, less than 6000 SF	S.F.	22
Greater than 6000 SF	S.F.	21.50
Office, greater than 6000 S.F.	S.F.	17.95

Description	Unit	$ Cost
Security access systems		
Metal detectors, wand type	Ea.	138
Walk-through portal type, single-zone	Ea.	3825
Multi-zone	Ea.	5425
X-ray equipment		
Desk top, for mail, small packages	Ea.	4000
Conveyer type, including monitor, minimum	Ea.	18,000
Maximum	Ea.	31,200
Explosive detection equipment		
Hand held, battery operated	Ea.	28,100
Walk-through portal type	Ea.	49,000
Uninterruptible power supply, 15 kVA/12.75 kW	kW	1.28

Important: See the Reference Section for Location Factors.

Model costs calculated for a 3 story building with 12′ story height and 20,000 square feet of floor area

				Unit	Unit Cost	Cost Per S.F.	% Of Sub-Total
A. SUBSTRUCTURE							
1010	Standard Foundations	Poured concrete; strip and spread footings; 4′ foundation wall		S.F. Ground	10.20	3.40	
1020	Special Foundations	N/A					
1030	Slab on Grade	4″ reinforced slab on grade		S.F. Slab	5.80	1.93	4.1%
2010	Basement Excavation	Site preparation for slab and trench for foundation wall and footing		S.F. Ground	.33	.11	
2020	Basement Walls	N/A		—			
B. SHELL							
	B10 Superstructure						
1010	Floor Construction	Wood Beam and joist on wood columns, fireproofed		S.F. Floor	18.48	12.32	11.2%
1020	Roof Construction	Wood roof, truss, 4/12 slope		S.F. Roof	8.10	2.70	
	B20 Exterior Enclosure						
2010	Exterior Walls	Ashlar stone veneer on wood stud back up, insulated	80% of wall	S.F. Wall	40.76	21.13	
2020	Exterior Windows	Aluminum awning type windows	20% of wall	Each	755	4.25	20.0%
2030	Exterior Doors	Aluminum and glass single and double doors, hollow metal doors		Each	4582	1.40	
	B30 Roofing						
3010	Roof Coverings	Asphalt shingles with flashing, gutters and downspouts		S.F. Roof	2.52	.84	0.6%
3020	Roof Openings	N/A		—	—	—	
C. INTERIORS							
1010	Partitions	Gypsum board on wood studs	20 S.F. Floor/L.F. Partition	S.F. Partition	8.43	3.37	
1020	Interior Doors	Single leaf solid core wood doors, hollow metal egress doors	315 S.F. Floor/Door	Each	1222	3.92	
1030	Fittings	Plastic laminate toilet partitions		S.F. Floor	.23	.23	
2010	Stair Construction	Concrete filled metal pan with rails		Flight	12,600	4.41	18.4%
3010	Wall Finishes	Paint on drywall		S.F. Surface	1.48	1.18	
3020	Floor Finishes	60% Carpet tile, 30% vinyl composition tile, 10% ceramic tile		S.F. Floor	5.25	5.25	
3030	Ceiling Finishes	Acoustic ceiling tiles on suspended support system		S.F. Ceiling	6.30	6.30	
D. SERVICES							
	D10 Conveying						
1010	Elevators & Lifts	Two hydraulic passenger elevators		Each	125,900	12.59	9.4%
1020	Escalators & Moving Walks	N/A		—	—	—	
	D20 Plumbing						
2010	Plumbing Fixtures	Restroom and service fixtures, supply and drainage	1 Fixture/1050 S.F. Floor	Each	2296	2.18	
2020	Domestic Water Distribution	Gas fired water heater		S.F. Floor	.60	.60	2.1%
2040	Rain Water Drainage	Roof drains		S.F. Roof			
	D30 HVAC						
3010	Energy Supply	N/A		—	—	—	
3020	Heat Generating Systems	Included in D3050		—	—	—	
3030	Cooling Generating Systems	N/A		—	—	—	13.8 %
3050	Terminal & Package Units	Multizone unit, gas heating , electric cooling		S.F. Floor	18.45	18.45	
3090	Other HVAC Sys. & Equipment	N/A		—	—	—	
	D40 Fire Protection						
4010	Sprinklers	Wet pipe spinkler system		S.F. Floor	4.02	4.02	3.8%
4020	Standpipes	Standpipes and hose systems		S.F. Floor	1.07	1.07	
	D50 Electrical						
5010	Electrical Service/Distribution	1000 Ampere service, panel boards and feeders		S.F. Floor	3.62	3.62	
5020	Lighting & Branch Wiring	Fluorescent fixtures, receptacles, switches, A.C. and misc. power		S.F. Floor	12.64	12.64	16.8%
5030	Communications & Security	Addressable alarm systems, emergency lighting, internet and phone wiring		S.F. Floor	6.26	6.26	
5090	Other Electrical Systems	Uninterruptible power supply with battery pack, 12.75kW		S.F. Floor			
E. EQUIPMENT & FURNISHINGS							
1010	Commercial Equipment	N/A		—	—	—	
1020	Institutional Equipment	N/A		—	—	—	
1030	Vehicular Equipment	N/A		—	—	—	0.0 %
1090	Other Equipment	N/A		—	—	—	
F. SPECIAL CONSTRUCTION							
1020	Integrated Construction	N/A		—	—	—	
1040	Special Facilities	N/A		—	—	—	0.0 %
G. BUILDING SITEWORK	**N/A**						

				Sub-Total		134.17	**100%**
	CONTRACTOR FEES (General Requirements: 10%, Overhead: 5%, Profit: 10%)				25%	33.54	
	ARCHITECT FEES				7%	11.74	

Total Building Cost — **179.45**

For customer support on your Square Foot Costs with RSMeans data, call 800.448.8182.

177

Costs per square foot of floor area

Exterior Wall	S.F. Area	20000	40000	60000	80000	100000	150000	200000	250000	300000
	L.F. Perimeter	260	360	400	420	460	520	650	720	800
Stone Veneer	Rigid Steel	231.60	197.95	179.50	168.45	163.20	154.40	152.55	149.65	148.00
E.I.F.S.	Rigid Steel	184.30	165.20	155.20	149.35	146.50	141.80	140.70	139.15	138.30
Curtain Wall	Rigid Steel	231.50	198.85	180.90	170.20	165.15	156.55	154.75	151.95	150.35
Brick Veneer	Reinforced Concrete	315.80	257.30	224.40	204.55	195.40	179.50	176.30	171.15	168.15
Metal Panel	Reinforced Concrete	223.55	194.20	178.20	168.70	164.30	156.60	154.95	152.50	151.05
E.I.F.S.	Reinforced Concrete	295.75	243.45	214.15	196.50	188.30	174.20	171.25	166.70	164.00
Perimeter Adj., Add or Deduct	Per 100 L.F.	36.30	18.15	12.10	9.05	7.30	4.85	3.60	2.95	2.35
Story Hgt. Adj., Add or Deduct	Per 1 Ft.	7.35	5.10	3.75	2.95	2.65	1.95	1.85	1.65	1.50

For Basement, add $42.60 per square foot of basement area

The above costs were calculated using the basic specifications shown on the facing page. These costs should be adjusted where necessary for design alternatives and owner's requirements.

Common additives

Description	Unit	$ Cost
Closed circuit surveillance, one station		
Camera and monitor	Ea.	1475
For additional camera stations, add	Ea.	665
Directory boards, plastic, glass covered		
30" x 20"	Ea.	640
36" x 48"	Ea.	1625
Aluminum, 24" x 18"	Ea.	630
36" x 24"	Ea.	790
48" x 32"	Ea.	1025
48" x 60"	Ea.	2200
Electronic, wall mounted	S.F.	4625
Free standing	S.F.	4075
Escalators, 10' rise, 32" wide, glass balustrade	Ea.	153,500
Metal balustrade	Ea.	162,000
48" wide, glass balustrade	Ea.	161,000
Metal balustrade	Ea.	171,000
Pedestal access floor system w/plastic laminate cover		
Computer room, less than 6000 SF	S.F.	22
Greater than 6000 SF	S.F.	21.50
Office, greater than 6000 S.F.	S.F.	17.95

Description	Unit	$ Cost
Security access systems		
Metal detectors, wand type	Ea.	138
Walk-through portal type, single-zone	Ea.	3825
Multi-zone	Ea.	5425
X-ray equipment		
Desk top, for mail, small packages	Ea.	4000
Conveyer type, including monitor, minimum	Ea.	18,000
Maximum	Ea.	31,200
Explosive detection equipment		
Hand held, battery operated	Ea.	28,100
Walk-through portal type	Ea.	49,000
Uninterruptible power supply, 15 kVA/12.75 kW	kW	1.28

Important: See the Reference Section for Location Factors.

Model costs calculated for a 8 story building with 12' story height and 80,000 square feet of floor area

				Unit	Unit Cost	Cost Per S.F.	% Of Sub-Total
A. SUBSTRUCTURE							
1010	Standard Foundations	Poured concrete; strip and spread footings; 4' foundation wall		S.F. Ground	7.68	.96	
1020	Special Foundations	N/A		—	—	—	
1030	Slab on Grade	4" reinforced slab on grade		S.F. Slab	5.80	.73	1.4%
2010	Basement Excavation	Site preparation for slab and trench for foundation wall and footing		S.F. Ground	.33	.04	
2020	Basement Walls	N/A		—	—	—	
B. SHELL							
B10 Superstructure							
1010	Floor Construction	Open web steel joists, slab form, concrete, fireproofed steel columns		S.F. Floor	14.90	13.04	11.3%
1020	Roof Construction	Metal deck, open web steel joists, beams, columns		S.F. Roof	10.88	1.36	
B20 Exterior Enclosure							
2010	Exterior Walls	Limestone stone veneer on metal studs	80% of wall	S.F. Wall	54	21.96	
2020	Exterior Windows	Aluminum sliding windows	20% of wall	Each	589	3.96	20.7%
2030	Exterior Doors	Aluminum and glass double doors, hollow metal egress doors		Each	5010	.37	
B30 Roofing							
3010	Roof Coverings	Single ply membrane, stone ballast, rigid insulation		S.F. Roof	7.92	.99	0.9%
3020	Roof Openings	Roof and smoke hatches		S.F. Roof	2587	.19	
C. INTERIORS							
1010	Partitions	Gypsum board on metal studs	30 S.F. Floor/L.F. Partition	S.F. Partition	10.35	3.45	
1020	Interior Doors	Single leaf solid core wood doors, hollow metal egress doors	350 S.F. Floor/Door	Each	1222	3.47	
1030	Fittings	Plastic laminate toilet partitions		S.F. Floor	.20	.20	
2010	Stair Construction	Concrete filled metal pan, with rails		Flight	12,600	4.10	18.7%
3010	Wall Finishes	Paint on drywall		S.F. Surface	1.44	.96	
3020	Floor Finishes	60% Carpet tile, 30% vinyl composition tile, 10% ceramic tile		S.F. Floor	5.25	5.25	
3030	Ceiling Finishes	Acoustic ceiling tiles on suspended support system		S.F. Ceiling	6.30	6.30	
D. SERVICES							
D10 Conveying							
1010	Elevators & Lifts	Traction geared passenger elevators		Each	324,600	16.23	12.8%
1020	Escalators & Moving Walks	N/A		—	—	—	
D20 Plumbing							
2010	Plumbing Fixtures	Restroom and service fixtures, supply and drainage	1 Fixture/1600 S.F. Floor	Each	2576	1.61	
2020	Domestic Water Distribution	Gas fired water heater		S.F. Floor	.63	.63	2.0%
2040	Rain Water Drainage	Roof drains		S.F. Roof	2.48	.31	
D30 HVAC							
3010	Energy Supply	N/A		—	—	—	
3020	Heat Generating Systems	Included in D3050		—	—	—	
3030	Cooling Generating Systems	N/A		—	—	—	14.5 %
3050	Terminal & Package Units	Multizone unit, gas heating , electric cooling		S.F. Floor	18.45	18.45	
3090	Other HVAC Sys. & Equipment	N/A		—	—	—	
D40 Fire Protection							
4010	Sprinklers	Wet pipe spinkler system		S.F. Floor	3.06	3.06	3.3%
4020	Standpipes	Standpipes and hose systems, with pumps		S.F. Floor	1.17	1.17	
D50 Electrical							
5010	Electrical Service/Distribution	800 Ampere service, panel boards and feeders		S.F. Floor	1.10	1.10	
5020	Lighting & Branch Wiring	Fluorescent fixtures, receptacles, switches, A.C. and misc. power		S.F. Floor	12.55	12.55	14.4%
5030	Communications & Security	Alarm systems, emergency lighting, internet and phone wiring		S.F. Floor	4.68	4.68	
5090	Other Electrical Systems	Uninterruptible power supply with battery pack, 12.75kW		S.F. Floor			
E. EQUIPMENT & FURNISHINGS							
1010	Commercial Equipment	N/A		—	—	—	
1020	Institutional Equipment	N/A		—	—	—	
1030	Vehicular Equipment	N/A		—	—	—	0.0 %
1090	Other Equipment	N/A		—	—	—	
F. SPECIAL CONSTRUCTION							
1020	Integrated Construction	N/A		—	—	—	0.0 %
1040	Special Facilities	N/A		—	—	—	
G. BUILDING SITEWORK	**N/A**						

		Sub-Total	127.12	100%
CONTRACTOR FEES (General Requirements: 10%, Overhead: 5%, Profit: 10%)		25%	31.80	
ARCHITECT FEES		6%	9.53	
	Total Building Cost		**168.45**	

For customer support on your Square Foot Costs with RSMeans data, call 800.448.8182.

179

Costs per square foot of floor area

Exterior Wall	S.F. Area	120000	145000	170000	200000	230000	260000	400000	600000	800000
	L.F. Perimeter	420	450	470	490	510	530	650	800	900
Stone Veneer	Rigid Steel	188.65	180.85	174.50	168.80	164.60	161.35	153.50	148.15	144.60
E.I.F.S.	Rigid Steel	163.15	158.25	154.40	151.00	148.45	146.55	141.70	138.40	136.40
Curtain Wall	Rigid Steel	194.05	185.95	179.45	173.50	169.10	165.75	157.60	152.10	148.40
Brick Veneer	Reinforced Concrete	207.20	197.65	189.90	182.85	177.60	173.60	163.90	157.30	152.85
Metal Panel	Reinforced Concrete	182.95	176.45	171.25	166.60	163.10	160.50	153.95	149.60	146.75
E.I.F.S.	Reinforced Concrete	196.40	188.10	181.35	175.30	170.75	167.30	158.90	153.20	149.40
Perimeter Adj., Add or Deduct	Per 100 L.F.	11.35	9.35	8.10	6.80	5.90	5.25	3.40	2.25	1.70
Story Hgt. Adj., Add or Deduct	Per 1 Ft.	3.70	3.30	3.00	2.60	2.35	2.15	1.75	1.40	1.20
For Basement, add $42.60 per square foot of basement area										

The above costs were calculated using the basic specifications shown on the facing page. These costs should be adjusted where necessary for design alternatives and owner's requirements.

Common additives

Description	Unit	$ Cost		Description	Unit	$ Cost
Closed circuit surveillance, one station				Security access systems		
Camera and monitor	Ea.	1475		Metal detectors, wand type	Ea.	138
For additional camera stations, add	Ea.	665		Walk-through portal type, single-zone	Ea.	3825
Directory boards, plastic, glass covered				Multi-zone	Ea.	5425
30" x 20"	Ea.	640		X-ray equipment		
36" x 48"	Ea.	1625		Desk top, for mail, small packages	Ea.	4000
Aluminum, 24" x 18"	Ea.	630		Conveyer type, including monitor, minimum	Ea.	18,000
36" x 24"	Ea.	790		Maximum	Ea.	31,200
48" x 32"	Ea.	1025		Explosive detection equipment		
48" x 60"	Ea.	2200		Hand held, battery operated	Ea.	28,100
Electronic, wall mounted	S.F.	4625		Walk-through portal type	Ea.	49,000
Free standing	S.F.	4075		Uninterruptible power supply, 15 kVA/12.75 kW	kW	1.28
Escalators, 10' rise, 32" wide, glass balustrade	Ea.	153,500				
Metal balustrade	Ea.	162,000				
48" wide, glass balustrade	Ea.	161,000				
Metal balustrade	Ea.	171,000				
Pedestal access floor system w/plastic laminate cover						
Computer room, less than 6000 SF	S.F.	22				
Greater than 6000 SF	S.F.	21.50				
Office, greater than 6000 S.F.	S.F.	17.95				

Model costs calculated for a 16 story building with 12' story height and 260,000 square feet of floor area

				Unit	Unit Cost	Cost Per S.F.	% Of Sub-Total
A.	**SUBSTRUCTURE**						
1010	Standard Foundations	CIP concrete pile caps; 4' foundation wall		S.F. Ground	18.24	1.14	
1020	Special Foundations	Steel H-piles, concrete grade beams		S.F. Ground	178	11.13	
1030	Slab on Grade	4" reinforced concrete with vapor barrier and granular base		S.F. Slab	5.80	.36	10.4%
2010	Basement Excavation	Site preparation for slab and trench for foundation wall and footing		S.F. Ground	.33	.02	
2020	Basement Walls	N/A		—	—	—	
B.	**SHELL**						
	B10 Superstructure						
1010	Floor Construction	Open web steel joists, slab form, concrete, fireproofed steel columns		S.F. Floor	14.37	13.47	11.6%
1020	Roof Construction	Metal deck, open web steel joists, beams, columns		S.F. Roof	10.88	.68	
	B20 Exterior Enclosure						
2010	Exterior Walls	Limestone stone veneer on metal studs	80% of wall	S.F. Wall	43.56	17.05	
2020	Exterior Windows	Curtain wall glazing system, thermo-break frame	20% of wall	Each	51	1.82	16.2%
2030	Exterior Doors	Double aluminum & glass doors		Each	8111	.88	
	B30 Roofing						
3010	Roof Coverings	Single ply membrane, stone ballast, rigid insulation		S.F. Roof	7.36	.46	0.4%
3020	Roof Openings	Roof and smoke hatches.		S.F. Roof	2587	.07	
C.	**INTERIORS**						
1010	Partitions	Concrete block partitions, gypsum board on metal studs	30 S.F. Floor/L.F. Partition	S.F. Partition	9.90	3.30	
1020	Interior Doors	Single leaf solid core wood doors, hollow metal egress doors	350 S.F. Floor/Door	Each	1222	3.43	
1030	Fittings	Plastic laminate toilet partitions		S.F. Floor	.18	.18	
2010	Stair Construction	Concrete filled metal pan, with rails		Flight	15,800	4.01	19.2%
3010	Wall Finishes	Paint on drywall		S.F. Surface	1.32	.88	
3020	Floor Finishes	60% Carpet tile, 30% vinyl composition tile, 10% ceramic tile		S.F. Floor	5.25	5.25	
3030	Ceiling Finishes	Acoustic ceiling tiles on suspended support system		S.F. Ceiling	6.30	6.30	
D.	**SERVICES**						
	D10 Conveying						
1010	Elevators & Lifts	Traction geared passenger elevators		Each	505,050	7.77	6.4%
1020	Escalators & Moving Walks	N/A		—	—	—	
	D20 Plumbing						
2010	Plumbing Fixtures	Restroom and service fixtures, supply and drainage	1 Fixture/1600 S.F. Floor	Each	2249	1.41	
2020	Domestic Water Distribution	Gas fired water heater		S.F. Floor	.36	.36	1.5%
2040	Rain Water Drainage	Roof drains		S.F. Roof	1.76	.11	
	D30 HVAC						
3010	Energy Supply	N/A		—	—	—	
3020	Heat Generating Systems	Boiler, heat exchanger and fans		Each	518,125	2.70	
3030	Cooling Generating Systems	Chiled water, fan coil units		S.F. Floor	16.84	16.84	16.0 %
3050	Terminal & Package Units	N/A		—	—	—	
3090	Other HVAC Sys. & Equipment	N/A		—	—	—	
	D40 Fire Protection						
4010	Sprinklers	Sprinkler system and accessory package		S.F. Floor	2.95	2.95	2.9%
4020	Standpipes	Standpipes, pumps and hose systems		S.F. Floor	.60	.60	
	D50 Electrical						
5010	Electrical Service/Distribution	1200 Ampere service, panel boards and feeders		S.F. Floor	.56	.56	
5020	Lighting & Branch Wiring	Fluorescent fixtures, receptacles, switches, A.C. and misc. power		S.F. Floor	12.44	12.44	15.3%
5030	Communications & Security	Alarm systems, emergency lighting, internet and phone wiring		S.F. Floor	5.60	5.60	
5090	Other Electrical Systems	Uninterruptible power supply with battery pack, 12.75kW		S.F. Floor			
E.	**EQUIPMENT & FURNISHINGS**						
1010	Commercial Equipment	N/A		—	—	—	
1020	Institutional Equipment	N/A		—	—	—	
1030	Vehicular Equipment	N/A		—	—	—	0.0 %
1090	Other Equipment	N/A		—	—	—	
F.	**SPECIAL CONSTRUCTION**						
1020	Integrated Construction	N/A		—	—	—	
1040	Special Facilities	N/A		—	—	—	0.0 %
G.	**BUILDING SITEWORK**	**N/A**					

		Sub-Total	121.77	100%
CONTRACTOR FEES (General Requirements: 10%, Overhead: 5%, Profit: 10%)		25%	30.45	
ARCHITECT FEES		6%	9.13	
		Total Building Cost	**161.35**	

For customer support on your Square Foot Costs with RSMeans data, call 800.448.8182.

181

Costs per square foot of floor area

Exterior Wall	S.F. Area	5000	7500	10000	12500	15000	17500	20000	22500	25000
	L.F. Perimeter	284	350	410	450	490	530	570	600	632
Brick Veneer	Wood Truss	480.15	466.50	459.00	452.80	448.70	445.75	443.50	441.30	439.65
Brick Veneer	Rigid Steel	480.80	467.65	460.50	454.60	450.65	447.90	445.75	443.65	442.05
Decorative Concrete Block	Rigid Steel	483.75	470.05	462.60	456.45	452.35	449.40	447.25	445.00	443.35
Decorative Concrete Block	Bearing Walls	473.85	461.95	455.45	450.20	446.65	444.15	442.25	440.40	438.95
Tilt-up Concrete Panels	Rigid Steel	474.55	462.50	456.00	450.60	447.05	444.50	442.60	440.75	439.25
E.I.F.S. and Concrete Block	Rigid Steel	474.95	462.85	456.25	450.85	447.25	444.70	442.80	440.90	439.45
Perimeter Adj., Add or Deduct	Per 100 L.F.	21.30	14.15	10.65	8.50	7.10	6.05	5.35	4.75	4.35
Story Hgt. Adj., Add or Deduct	Per 1 Ft.	3.20	2.60	2.30	2.05	1.80	1.70	1.65	1.55	1.40
For Basement, add $33.50 per square foot of basement area										

The above costs were calculated using the basic specifications shown on the facing page. These costs should be adjusted where necessary for design alternatives and owner's requirements.

Common additives

Description	Unit	$ Cost
Cabinets, base, door units, metal	L.F.	350
Drawer units	L.F.	690
Tall storage cabinets, open, 7' high	L.F.	640
With glazed doors	L.F.	1025
Wall cabinets, metal, 12-1/2" deep, open	L.F.	294
With doors	L.F.	520
Counter top, laminated plastic, no backsplash	L.F.	87.50
Stainless steel counter top	L.F.	213
For drop-in stainless 43" x 21" sink, add	Ea.	1500
Nurse Call Station		
Single bedside call station	Ea.	266
Ceiling speaker station	Ea.	178
Emergency call station	Ea.	185
Pillow speaker	Ea.	325
Standard call button	Ea.	205
Master control station for 20 stations	Total	6025

Description	Unit	$ Cost
Directory Boards, Plastic, glass covered		
Plastic, glass covered, 30" x 20"	Ea.	640
36" x 48"	Ea.	1625
Building directory, alum., black felt panels, 1 door, 24" x 18"	Ea.	630
36" x 24"	Ea.	790
48" x 32"	Ea.	1025
48" x 60"	Ea.	2200
Tables, Examining, vinyl top		
with base cabinets	Ea.	1575
Utensil washer-sanitizer	Ea.	10,300
X-ray, mobile, minimum	Ea.	20,800

Important: See the Reference Section for Location Factors.

Model costs calculated for a 1 story building with 15'-4" story height and 12,500 square feet of floor area

Outpatient Surgery Center

				Unit	Unit Cost	Cost Per S.F.	% Of Sub-Total
A. SUBSTRUCTURE							
1010	Standard Foundations	Poured concrete; strip and spread footings; 4' foundation wall		S.F. Ground	5.23	5.23	
1020	Special Foundations	N/A		—	—	—	
1030	Slab on Grade	4" reinforced concrete with vapor barrier and granular base		S.F. Slab	6.97	6.97	3.8%
2010	Basement Excavation	Site preparation for slab and trench for foundation wall and footing		S.F. Ground	.33	.33	
2020	Basement Walls	N/A		—	—	—	
B. SHELL							
	B10 Superstructure						
1010	Floor Construction	N/A		—	—	—	
1020	Roof Construction	Steel Joists and girders on columns		S.F. Roof	8.31	8.31	2.5 %
	B20 Exterior Enclosure						
2010	Exterior Walls	Brick veneer	85% of wall	S.F. Wall	20.06	9.41	
2020	Exterior Windows	Aluminum frame, fixed	15% of wall	Each	875	4.83	6.4%
2030	Exterior Doors	Aluminum and glass entry; steel flush doors		Each	8786	7.03	
	B30 Roofing						
3010	Roof Coverings	Single-ply membrane, loose laid and ballasted; polyisocyanurate insulation		S.F. Roof	6.71	6.71	2.1%
3020	Roof Openings	Roof hatches		S.F. Roof	.11	.11	
C. INTERIORS							
1010	Partitions	Gypsum board on metal studs	7 S.F. Floor/L.F. Partition	S.F. Partition	8.32	11.89	
1020	Interior Doors	Single and double leaf fire doors	200 S.F. Floor/Door	Each	1377	6.89	
1030	Fittings	Stainless steel toilet partitions, steel lockers		S.F. Floor	1.70	1.70	
2010	Stair Construction	N/A		—	—	—	12.7%
3010	Wall Finishes	80% Paint, 10% vinyl wall covering, 10% ceramic tile		S.F. Surface	2.18	6.23	
3020	Floor Finishes	10% Carpet, 80% vinyl sheet, 10% ceramic tile		S.F. Floor	10.71	10.71	
3030	Ceiling Finishes	Mineral fiber tile and plastic coated tile on suspension system		S.F. Ceiling	4.64	4.63	
D. SERVICES							
	D10 Conveying						
1010	Elevators & Lifts	N/A		—	—	—	0.0 %
1020	Escalators & Moving Walks	N/A		—	—	—	
	D20 Plumbing						
2010	Plumbing Fixtures	Medical, patient & specialty fixtures, supply and drainage	1 fixture/350 S.F. Floor	Each	3609	10.31	
2020	Domestic Water Distribution	Electric water heater		S.F. Floor	2.45	2.45	4.6%
2040	Rain Water Drainage	Roof drains		S.F. Roof	2.49	2.49	
	D30 HVAC						
3010	Energy Supply	Hot water reheat for surgery		S.F. Floor	12.05	12.05	
3020	Heat Generating Systems	Boiler		Each	64,875	7.25	
3030	Cooling Generating Systems	N/A		—	—	—	38.1%
3050	Terminal & Package Units	N/A		—	—	—	
3090	Other HVAC Sys. & Equipment	Rooftop AC, CAV, roof vents, heat exchanger, surgical air curtains		S.F. Floor	107	106.59	
	D40 Fire Protection						
4010	Sprinklers	Wet pipe sprinkler system		S.F. Floor	3.79	3.79	1.7%
4020	Standpipes	Standpipe		S.F. Floor	1.80	1.80	
	D50 Electrical						
5010	Electrical Service/Distribution	1200 ampere service, panel board and feeders		S.F. Floor	6.64	6.64	
5020	Lighting & Branch Wiring	T-8 fluorescent light fixtures, receptacles, switches, A.C. and misc. power		S.F. Floor	17.91	17.91	12.5%
5030	Communications & Security	Addressable alarm system, internet wiring, communications system, emergency lighting		S.F. Floor	10.44	10.44	
5090	Other Electrical Systems	Emergency generator, 300 kW with fuel tank		S.F. Floor	6.37	6.37	
E. EQUIPMENT & FURNISHINGS							
1010	Commercial Equipment	N/A		—	—	—	
1020	Institutional Equipment	Medical gas system, cabinetry, scrub sink, sterilizer		S.F. Floor	51	50.62	15.6 %
1030	Vehicular Equipment	N/A		—	—	—	
1090	Other Equipment	Ice maker, washing machine		S.F. Floor	1.04	1.04	
F. SPECIAL CONSTRUCTION & DEMOLITION							
1020	Integrated Construction	N/A		—	—	—	0.0 %
1040	Special Facilities	N/A		—	—	—	
G. BUILDING SITEWORK	**N/A**						

		Sub-Total	330.73	100%
CONTRACTOR FEES (General Requirements: 10%, Overhead: 5%, Profit: 5%)		20%	82.66	
ARCHITECT FEES		9%	37.21	

Total Building Cost 450.60

For customer support on your Square Foot Costs with RSMeans data, call 800.448.8182.

183

Costs per square foot of floor area

Exterior Wall	S.F. Area	7000	9000	11000	13000	15000	17000	19000	21000	23000
	L.F. Perimeter	240	280	303	325	354	372	397	422	447
Limestone and Concrete Block	Bearing Walls	299.25	280.35	264.30	253.00	245.95	238.90	234.20	230.45	227.35
	Reinforced Concrete	315.90	297.10	281.15	269.90	262.90	255.85	251.15	247.40	244.40
Face Brick and Concrete Block	Bearing Walls	270.75	254.50	241.45	232.25	226.35	220.65	216.85	213.75	211.20
	Reinforced Concrete	287.45	271.25	258.30	249.15	243.30	237.65	233.80	230.75	228.25
Decorative Concrete Block	Bearing Walls	258.55	243.45	231.65	223.40	218.05	212.95	209.40	206.65	204.30
	Reinforced Concrete	275.25	260.25	248.55	240.25	234.90	229.90	226.40	223.65	221.30
Perimeter Adj., Add or Deduct	Per 100 L.F.	37.25	29.00	23.80	20.10	17.40	15.30	13.70	12.45	11.35
Story Hgt. Adj., Add or Deduct	Per 1 Ft.	6.70	6.05	5.40	4.90	4.60	4.25	4.10	3.90	3.85

For Basement, add $ 33.50 per square foot of basement area

The above costs were calculated using the basic specifications shown on the facing page. These costs should be adjusted where necessary for design alternatives and owner's requirements.

Common additives

Description	Unit	$ Cost
Cells Prefabricated, 5'-6' wide, 7'-8' high, 7'-8' deep	Ea.	12,700
Elevators, Hydraulic passenger, 2 stops		
1500# capacity	Ea.	73,100
2500# capacity	Ea.	76,600
3500# capacity	Ea.	81,600
Emergency Lighting, 25 watt, battery operated		
Lead battery	Ea.	330
Nickel cadmium	Ea.	550
Flagpoles, Complete		
Aluminum, 20' high	Ea.	1900
40' high	Ea.	4500
70' high	Ea.	11,600
Fiberglass, 23' high	Ea.	1450
39'-5" high	Ea.	3325
59' high	Ea.	7300

Description	Unit	$ Cost
Lockers, Steel, Single tier, 60" to 72"	Opng.	232 - 395
2 tier, 60" or 72" total	Opng.	139 - 172
5 tier, box lockers	Opng.	73 - 87
Locker bench, lam. maple top only	L.F.	36.50
Pedestals, steel pipe	Ea.	79
Safe, Office type, 1 hour rating		
30" x 18" x 18"	Ea.	2475
60" x 36" x 18", double door	Ea.	10,200
Shooting Range, Incl. bullet traps, target provisions, and contols, not incl. structural shell	Ea.	57,500
Smoke Detectors		
Ceiling type	Ea.	244
Duct type	Ea.	585
Sound System		
Amplifier, 250 watts	Ea.	2025
Speaker, ceiling or wall	Ea.	240
Trumpet	Ea.	460

Important: See the Reference Section for Location Factors.

Model costs calculated for a 2 story building with 12' story height and 11,000 square feet of floor area

Police Station

				Unit	Unit Cost	Cost Per S.F.	% Of Sub-Total
A. SUBSTRUCTURE							
1010	Standard Foundations	Poured concrete; strip and spread footings; 4' foundation wall		S.F. Ground	8.08	4.04	
1020	Special Foundations	N/A		—	—	—	
1030	Slab on Grade	4" reinforced concrete with vapor barrier and granular base		S.F. Slab	5.80	2.90	3.7%
2010	Basement Excavation	Site preparation for slab and trench for foundation wall and footing		S.F. Ground	.33	.17	
2020	Basement Walls	N/A		—	—	—	
B. SHELL							
B10 Superstructure							
1010	Floor Construction	Open web steel joists, slab form, concrete		S.F. Floor	13.46	6.73	4.9%
1020	Roof Construction	Metal deck on open web steel joists		S.F. Roof	5.40	2.70	
B20 Exterior Enclosure							
2010	Exterior Walls	Limestone with concrete block backup	80% of wall	S.F. Wall	67	35.43	
2020	Exterior Windows	Metal horizontal sliding	20% of wall	Each	1162	10.25	24.9%
2030	Exterior Doors	Hollow metal		Each	3531	2.56	
B30 Roofing							
3010	Roof Coverings	Built-up tar and gravel with flashing; perlite/EPS composite insulation		S.F. Roof	7.88	3.94	2.0%
3020	Roof Openings	N/A		—	—	—	
C. INTERIORS							
1010	Partitions	Concrete block	20 S.F. Floor/L.F. Partition	S.F. Partition	15.94	7.97	
1020	Interior Doors	Single leaf kalamein fire door	200 S.F. Floor/Door	Each	1222	6.11	
1030	Fittings	Toilet partitions		S.F. Floor	.98	.98	
2010	Stair Construction	Concrete filled metal pan		Flight	18,900	5.16	18.0%
3010	Wall Finishes	90% paint, 10% ceramic tile		S.F. Surface	4.07	4.07	
3020	Floor Finishes	70% vinyl composition tile, 20% carpet tile, 10% ceramic tile		S.F. Floor	4.40	4.40	
3030	Ceiling Finishes	Mineral fiber tile on concealed zee bars		S.F. Ceiling	6.30	6.30	
D. SERVICES							
D10 Conveying							
1010	Elevators & Lifts	One hydraulic passenger elevator		Each	91,190	8.29	4.3%
1020	Escalators & Moving Walks	N/A		—	—	—	
D20 Plumbing							
2010	Plumbing Fixtures	Toilet and service fixtures, supply and drainage	1 Fixture/580 S.F. Floor	Each	4849	8.36	
2020	Domestic Water Distribution	Oil fired water heater		S.F. Floor	4.79	4.79	7.7%
2040	Rain Water Drainage	Roof drains		S.F. Roof	3.44	1.72	
D30 HVAC							
3010	Energy Supply	N/A		—	—	—	
3020	Heat Generating Systems	N/A		—	—	—	
3030	Cooling Generating Systems	N/A		—	—	—	11.8 %
3050	Terminal & Package Units	Multizone HVAC air cooled system		S.F. Floor	22.80	22.80	
3090	Other HVAC Sys. & Equipment	N/A		—	—	—	
D40 Fire Protection							
4010	Sprinklers	Wet pipe sprinkler system		S.F. Floor	4.08	4.08	2.7%
4020	Standpipes	Standpipe		S.F. Floor	1.15	1.15	
D50 Electrical							
5010	Electrical Service/Distribution	400 ampere service, panel board and feeders		S.F. Floor	2.04	2.04	
5020	Lighting & Branch Wiring	T-8 fluorescent fixtures, receptacles, switches, A.C. and misc. power		S.F. Floor	12.73	12.73	11.6%
5030	Communications & Security	Addressable alarm systems, internet wiring, intercom and emergency lighting		S.F. Floor	7.52	7.52	
5090	Other Electrical Systems	Emergency generator, 15 kW		S.F. Floor	.24	.24	
E. EQUIPMENT & FURNISHINGS							
1010	Commercial Equipment	N/A		—	—	—	
1020	Institutional Equipment	Lockers, detention rooms, cells, gasoline dispensers		S.F. Floor	13.65	13.65	8.3 %
1030	Vehicular Equipment	Gasoline dispenser system		S.F. Floor	2.54	2.54	
1090	Other Equipment	N/A		—	—	—	
F. SPECIAL CONSTRUCTION							
1020	Integrated Construction	N/A		—	—	—	0.0 %
1040	Special Facilities	N/A		—	—	—	
G. BUILDING SITEWORK	**N/A**						

		Sub-Total	194	100%
CONTRACTOR FEES (General Requirements: 10%, Overhead: 5%, Profit: 10%)		25%	48.47	
ARCHITECT FEES		9%	21.83	

Total Building Cost 264.30

For customer support on your Square Foot Costs with RSMeans data, call 800.448.8182.

185

Costs per square foot of floor area

Exterior Wall	S.F. Area	600	750	925	1150	1450	1800	2100	2600	3200
	L.F. Perimeter	100	112	128	150	164	182	200	216	240
Metal Panel and Wood Studs	Wood Truss	85.95	80.45	76.85	74.15	68.90	65.30	63.50	59.95	57.55
Aluminum Clapboard	Wood Truss	79.15	74.10	70.85	68.40	63.80	60.50	58.90	55.70	53.55
Wood Clapboard	Wood Truss	84.10	78.80	75.30	72.65	67.65	64.20	62.45	59.00	56.70
Plywood Siding	Wood Truss	78.70	73.70	70.45	68.10	63.40	60.20	58.60	55.45	53.35
Perimeter Adj., Add or Deduct	Per 100 L.F.	46.30	37.00	30.00	24.10	19.20	15.45	13.25	10.65	8.65
Story Hgt. Adj., Add or Deduct	Per 1 Ft.	3.35	2.95	2.70	2.60	2.25	2.00	1.90	1.60	1.45
Basement—Not Applicable										

The above costs were calculated using the basic specifications shown on the facing page. These costs should be adjusted where necessary for design alternatives and owner's requirements.

Common additives

Description	Unit	$ Cost
Slab on Grade		
4" thick, non-industrial, non-reinforced	S.F.	5.25
Reinforced	S.F.	5.80
Light industrial, non-reinforced	S.F.	6.42
Reinforced	S.F.	6.97
Industrial, non-reinforced	S.F.	10.99
Reinforced	S.F.	11.56
6" thick, non-industrial, non-reinforced	S.F.	6.23
Reinforced	S.F.	7.01
Light industrial, non-reinforced	S.F.	7.42
Reinforced	S.F.	8.15
Heavy industrial, non-reinforced	S.F.	14.10
Reinforced	S.F.	14.85
Insulation		
Fiberboard, low density, 1" thick, R2.78	S.F.	1.52
Foil on reinforced scrim, single bubble air space, R8.8	C.S.F.	70
Vinyl faced fiberglass, 1-1/2" thick, R5	S.F.	0.96
Vinyl/scrim/foil (VSF), 1-1/2" thick, R5	S.F.	1.18

Description	Unit	$ Cost
Roof ventilation, ridge vent, aluminum	L.F.	8.30
Mushroom vent, aluminum	Ea.	91
Gutter, aluminum, 5" K type, enameled	L.F.	9.05
Steel, half round or box, 5", enameled	L.F.	8.35
Wood, clear treated cedar, fir or hemlock, 4" x 5"	L.F.	30
Downspout, aluminum, 3" x 4", enameled	L.F.	7.35
Steel, round, 4" diameter	L.F.	7.35

Important: See the Reference Section for Location Factors.

Post Frame Barn

Model costs calculated for a 1 story building with 14' story height and 1,800 square feet of floor area

			Unit	Unit Cost	Cost Per S.F.	% Of Sub-Total
A. SUBSTRUCTURE						
1010	Standard Foundations	Included in B2010 Exterior Walls	—	—	—	
1020	Special Foundations	N/A	—	—	—	
1030	Slab on Grade	N/A	—	—	—	0.0%
2010	Basement Excavation	N/A	—	—	—	
2020	Basement Walls	N/A	—	—	—	
B. SHELL						
B10 Superstructure						
1010	Floor Construction	N/A	—	—	—	16.0 %
1020	Roof Construction	Wood truss, 4/12 pitch	S.F. Roof	7.76	7.76	
B20 Exterior Enclosure						
2010	Exterior Walls	2 x 6 Wood framing, 6 x 6 wood posts in concrete; steel siding panels	S.F. Wall	12.37	16.63	
2020	Exterior Windows	Steel windows, fixed	Each	528	4.15	54.3%
2030	Exterior Doors	Steel overhead, hollow metal doors	Each	2465	5.47	
B30 Roofing						
3010	Roof Coverings	Metal panel roof	S.F. Roof	4.55	4.55	9.4%
3020	Roof Openings	N/A	—	—	—	
C. INTERIORS						
1010	Partitions	N/A	—	—	—	
1020	Interior Doors	N/A	—	—	—	
1030	Fittings	N/A	—	—	—	
2010	Stair Construction	N/A	—	—	—	0.0 %
3010	Wall Finishes	N/A	—	—	—	
3020	Floor Finishes	N/A	—	—	—	
3030	Ceiling Finishes	N/A	—	—	—	
D. SERVICES						
D10 Conveying						
1010	Elevators & Lifts	N/A	—	—	—	0.0 %
1020	Escalators & Moving Walks	N/A	—	—	—	
D20 Plumbing						
2010	Plumbing Fixtures	N/A	—	—	—	
2020	Domestic Water Distribution	N/A	—	—	—	0.0 %
2040	Rain Water Drainage	N/A	—	—	—	
D30 HVAC						
3010	Energy Supply	N/A	—	—	—	
3020	Heat Generating Systems	N/A	—	—	—	
3030	Cooling Generating Systems	N/A	—	—	—	0.0 %
3050	Terminal & Package Units	N/A	—	—	—	
3090	Other HVAC Sys. & Equipment	N/A	—	—	—	
D40 Fire Protection						
4010	Sprinklers	N/A	—	—	—	0.0 %
4020	Standpipes	N/A	—	—	—	
D50 Electrical						
5010	Electrical Service/Distribution	60 Ampere service, panel board & feeders	S.F. Floor	2.02	2.02	
5020	Lighting & Branch Wiring	High bay fixtures, receptacles & switches	S.F. Floor	7.79	7.79	20.3%
5030	Communications & Security	N/A	—	—	—	
5090	Other Electrical Systems	N/A	—	—	—	
E. EQUIPMENT & FURNISHINGS						
1010	Commercial Equipment	N/A	—	—	—	
1020	Institutional Equipment	N/A	—	—	—	0.0 %
1030	Vehicular Equipment	N/A	—	—	—	
1090	Other Equipment	N/A	—	—	—	
F. SPECIAL CONSTRUCTION & DEMOLITION						
1020	Integrated Construction	N/A	—	—	—	0.0 %
1040	Special Facilities	N/A	—	—	—	
G. BUILDING SITEWORK	N/A					

		Sub-Total	48.37	100%
CONTRACTOR FEES (General Requirements: 10%, Overhead: 5%, Profit: 5%)		20%	12.09	
ARCHITECT FEES		8%	4.84	
	Total Building Cost		**65.30**	

For customer support on your Square Foot Costs with RSMeans data, call 800.448.8182.

Costs per square foot of floor area

Exterior Wall	S.F. Area	5000	7000	9000	11000	13000	15000	17000	19000	21000
	L.F. Perimeter	300	380	420	468	486	513	540	580	620
Face Brick and Concrete Block	Rigid Steel	185.15	174.80	164.75	159.00	152.90	148.95	145.95	144.25	142.80
	Bearing Walls	185.65	175.15	165.00	159.20	153.05	149.10	146.05	144.25	142.85
Limestone and Concrete Block	Rigid Steel	214.25	201.05	187.30	179.65	171.05	165.55	161.35	159.05	157.15
	Bearing Walls	214.70	201.45	187.60	179.85	171.15	165.60	161.40	159.10	157.15
Decorative Concrete Block	Rigid Steel	172.75	163.55	155.05	150.25	145.20	141.90	139.40	137.95	136.70
	Bearing Walls	173.25	163.90	155.35	150.45	145.30	142.00	139.45	137.95	136.75
Perimeter Adj., Add or Deduct	Per 100 L.F.	19.20	13.70	10.65	8.70	7.40	6.40	5.65	5.05	4.60
Story Hgt. Adj., Add or Deduct	Per 1 Ft.	3.10	2.70	2.35	2.20	1.90	1.75	1.65	1.55	1.55

For Basement, add $32.70 per square foot of basement area

The above costs were calculated using the basic specifications shown on the facing page. These costs should be adjusted where necessary for design alternatives and owner's requirements.

Common additives

Description	Unit	$ Cost
Closed Circuit Surveillance, One station		
Camera and monitor	Ea.	1475
For additional camera stations, add	Ea.	665
Emergency Lighting, 25 watt, battery operated		
Lead battery	Ea.	330
Nickel cadmium	Ea.	550
Flagpoles, Complete		
Aluminum, 20' high	Ea.	1900
40' high	Ea.	4500
70' high	Ea.	11,600
Fiberglass, 23' high	Ea.	1450
39'-5" high	Ea.	3325
59' high	Ea.	7300

Description	Unit	$ Cost
Mail Boxes, Horizontal, key lock, 15" x 6" x 5"	Ea.	65
Double 15" x 12" x 5"	Ea.	97
Quadruple 15" x 12" x 10"	Ea.	158
Vertical, 6" x 5" x 15", aluminum	Ea.	67.50
Bronze	Ea.	71.50
Steel, enameled	Ea.	67.50
Scales, Dial type, 5 ton cap.		
8' x 6' platform	Ea.	13,800
9' x 7' platform	Ea.	15,800
Smoke Detectors		
Ceiling type	Ea.	244
Duct type	Ea.	585

Model costs calculated for a 1 story building with 14' story height and 13,000 square feet of floor area

					Unit	Unit Cost	Cost Per S.F.	% Of Sub-Total
A. SUBSTRUCTURE								
1010	Standard Foundations	Poured concrete; strip and spread footings; 4' foundation wall			S.F. Ground	5.14	5.14	
1020	Special Foundations	N/A			—	—	—	
1030	Slab on Grade	4" reinforced concrete with vapor barrier and granular base			S.F. Slab	5.80	5.80	10.0%
2010	Basement Excavation	Site preparation for slab and trench for foundation wall and footing			S.F. Ground	.33	.33	
2020	Basement Walls	N/A			—	—	—	
B. SHELL								
B10 Superstructure								
1010	Floor Construction	Steel column fireproofing			L.F. Column	38.10	.29	9.2%
1020	Roof Construction	Metal deck, open web steel joists, columns			S.F. Roof	10.06	10.06	
B20 Exterior Enclosure								
2010	Exterior Walls	Face brick with concrete block backup	80% of wall		S.F. Wall	35.25	14.76	
2020	Exterior Windows	Double strength window glass	20% of wall		Each	755	3.44	17.7%
2030	Exterior Doors	Double aluminum & glass, single aluminum, hollow metal, steel overhead			Each	3701	1.71	
B30 Roofing								
3010	Roof Coverings	Built-up tar and gravel with flashing; perlite/EPS composite insulation			S.F. Roof	7.06	7.06	6.3%
3020	Roof Openings	N/A			—	—	—	
C. INTERIORS								
1010	Partitions	Concrete block	15 S.F. Floor/L.F. Partition		S.F. Partition	10.40	8.32	
1020	Interior Doors	Single leaf hollow metal	150 S.F. Floor/Door		Each	1222	8.14	
1030	Fittings	Toilet partitions, cabinets, shelving, lockers			S.F. Floor	1.66	1.66	
2010	Stair Construction	N/A			—	—	—	23.6%
3010	Wall Finishes	Paint			S.F. Surface	2.49	3.98	
3020	Floor Finishes	50% vinyl tile, 50% paint			S.F. Floor	2.52	2.52	
3030	Ceiling Finishes	Mineral fiber tile on concealed zee bars	25% of area		S.F. Ceiling	7.60	1.90	
D. SERVICES								
D10 Conveying								
1010	Elevators & Lifts	N/A			—	—	—	0.0%
1020	Escalators & Moving Walks	N/A			—	—	—	
D20 Plumbing								
2010	Plumbing Fixtures	Toilet and service fixtures, supply and drainage	1 Fixture/1180 S.F. Floor		Each	2443	2.07	
2020	Domestic Water Distribution	Gas fired water heater			S.F. Floor	.55	.55	4.0%
2040	Rain Water Drainage	Roof drains			S.F. Roof	1.87	1.87	
D30 HVAC								
3010	Energy Supply	N/A			—	—	—	
3020	Heat Generating Systems	Included in D3050			—	—	—	
3030	Cooling Generating Systems	N/A			—	—	—	8.8 %
3050	Terminal & Package Units	Single zone, gas heating, electric cooling			S.F. Floor	9.86	9.86	
3090	Other HVAC Sys. & Equipment	N/A			—	—	—	
D40 Fire Protection								
4010	Sprinklers	Wet pipe sprinkler system			S.F. Floor	4.96	4.96	5.3%
4020	Standpipes	Standpipe			S.F. Floor	1.03	1.03	
D50 Electrical								
5010	Electrical Service/Distribution	600 ampere service, panel board and feeders			S.F. Floor	2.46	2.46	
5020	Lighting & Branch Wiring	T-8 fluorescent fixtures, receptacles, switches, A.C. and misc. power			S.F. Floor	9.44	9.44	15.0%
5030	Communications & Security	Addressable alarm systems, internet wiring and emergency lighting			S.F. Floor	4.75	4.75	
5090	Other Electrical Systems	Emergency generator, 15 kW			S.F. Floor	.14	.14	
E. EQUIPMENT & FURNISHINGS								
1010	Commercial Equipment	N/A			—	—	—	
1020	Institutional Equipment	N/A			—	—	—	0.0 %
1030	Vehicular Equipment	N/A			—	—	—	
1090	Other Equipment	N/A			—	—	—	
F. SPECIAL CONSTRUCTION								
1020	Integrated Construction	N/A			—	—	—	0.0 %
1040	Special Facilities	N/A			—	—	—	
G. BUILDING SITEWORK	**N/A**							

				Sub-Total	112.24	100%
CONTRACTOR FEES (General Requirements: 10%, Overhead: 5%, Profit: 10%)				25%	28.03	
ARCHITECT FEES				9%	12.63	

Total Building Cost 152.90

For customer support on your Square Foot Costs with RSMeans data, call 800.448.8182.

189

Costs per square foot of floor area

Exterior Wall	S.F. Area	5000	10000	15000	21000	25000	30000	40000	50000	60000
	L.F. Perimeter	287	400	500	600	660	700	834	900	1000
Face Brick and Concrete Block	Rigid Steel	261.15	223.35	209.40	199.80	195.60	190.00	185.20	180.00	177.55
	Bearing Walls	259.00	220.75	206.55	196.90	192.55	186.85	182.00	176.75	174.20
Concrete Block	Rigid Steel	225.15	198.30	188.50	181.95	179.05	175.40	172.05	168.75	167.05
Brick Veneer	Rigid Steel	246.10	212.90	200.70	192.35	188.70	183.90	179.70	175.30	173.15
Metal Panel	Rigid Steel	213.00	189.80	181.45	175.85	173.45	170.40	167.65	164.95	163.55
Metal Sandwich Panel	Rigid Steel	267.25	227.65	212.95	202.90	198.40	192.50	187.40	181.95	179.30
Perimeter Adj., Add or Deduct	Per 100 L.F.	32.60	16.35	10.85	7.80	6.50	5.40	4.00	3.25	2.70
Story Hgt. Adj., Add or Deduct	Per 1 Ft.	6.65	4.70	3.85	3.35	3.05	2.65	2.35	2.10	1.95
For Basement, add $31.40 per square foot of basement area										

The above costs were calculated using the basic specifications shown on the facing page. These costs should be adjusted where necessary for design alternatives and owner's requirements.

Common additives

Description	Unit	$ Cost	Description	Unit	$ Cost
Bar, Front Bar	L.F.	450	Lockers, Steel, single tier, 60" or 72"	Opng.	232 - 395
Back Bar	L.F.	360	2 tier, 60" or 72" total	Opng.	139 - 172
Booth, Upholstered, custom straight	L.F.	258 - 475	5 tier, box lockers	Opng.	73 - 87
"L" or "U" shaped	L.F.	267 - 450	Locker bench, lam. maple top only	L.F.	36.50
Bleachers, Telescoping, manual			Pedestals, steel pipe	Ea.	79
To 15 tier	Seat	164 - 222	Sauna, Prefabricated, complete		
21-30 tier	Seat	340 - 460	6' x 4'	Ea.	6600
Courts			6' x 9'	Ea.	10,000
Ceiling	Court	10,700	8' x 8'	Ea.	10,600
Floor	Court	17,900	8' x 10'	Ea.	11,700
Walls	Court	30,600	10' x 12'	Ea.	15,500
Emergency Lighting, 25 watt, battery operated			Sound System		
Lead battery	Ea.	330	Amplifier, 250 watts	Ea.	2025
Nickel cadmium	Ea.	550	Speaker, ceiling or wall	Ea.	240
Kitchen Equipment			Trumpet	Ea.	460
Broiler	Ea.	4425	Steam Bath, Complete, to 140 C.F.	Ea.	3175
Cooler, 6 ft. long, reach-in	Ea.	3750	To 300 C.F.	Ea.	3525
Dishwasher, 10-12 racks per hr.	Ea.	4150	To 800 C.F.	Ea.	7125
Food warmer, counter 1.2 KW	Ea.	790	To 2500 C.F.	Ea.	9725
Freezer, reach-in, 44 C.F.	Ea.	5525			
Ice cube maker, 50 lb. per day	Ea.	2100			

Important: See the Reference Section for Location Factors.

Model costs calculated for a 2 story building with 12' story height and 30,000 square feet of floor area

				Unit	Unit Cost	Cost Per S.F.	% Of Sub-Total
A. SUBSTRUCTURE							
1010	Standard Foundations	Poured concrete; strip and spread footings; 4' foundation wall		S.F. Ground	6.62	3.31	
1020	Special Foundations	N/A		—	—	—	
1030	Slab on Grade	5" reinforced concrete with vapor barrier and granular base		S.F. Slab	7.47	3.74	5.1%
2010	Basement Excavation	Site preparation for slab and trench for foundation wall and footing		S.F. Ground	.33	.17	
2020	Basement Walls	N/A		—	—	—	
B. SHELL							
	B10 Superstructure						
1010	Floor Construction	Open web steel joists, slab form, concrete, columns	50% of area	S.F. Floor	19.76	9.88	10.0%
1020	Roof Construction	Metal deck on open web steel joists, columns		S.F. Roof	8.32	4.16	
	B20 Exterior Enclosure						
2010	Exterior Walls	Face brick with concrete block backup	95% of wall	S.F. Wall	34.76	18.49	
2020	Exterior Windows	Storefront	5% of wall	S.F. Window	105	2.94	15.6%
2030	Exterior Doors	Aluminum and glass and hollow metal		Each	4109	.56	
	B30 Roofing						
3010	Roof Coverings	Built-up tar and gravel with flashing; perlite/EPS composite insulation		S.F. Roof	7.24	3.62	2.6%
3020	Roof Openings	Roof hatches		S.F. Roof	.16	.08	
C. INTERIORS							
1010	Partitions	Concrete block, gypsum board on metal studs	25 S.F. Floor/L.F. Partition	S.F. Partition	15.80	6.32	
1020	Interior Doors	Single leaf hollow metal	810 S.F. Floor/Door	Each	1222	1.51	
1030	Fittings	Toilet partitions		S.F. Floor	.41	.41	
2010	Stair Construction	Concrete filled metal pan		Flight	18,900	1.89	13.6%
3010	Wall Finishes	Paint		S.F. Surface	1.51	1.21	
3020	Floor Finishes	80% carpet tile, 20% ceramic tile	50% of floor area	S.F. Floor	6.60	3.30	
3030	Ceiling Finishes	Mineral fiber tile on concealed zee bars	60% of area	S.F. Ceiling	7.60	4.56	
D. SERVICES							
	D10 Conveying						
1010	Elevators & Lifts	N/A		—	—	—	0.0 %
1020	Escalators & Moving Walks	N/A		—	—	—	
	D20 Plumbing						
2010	Plumbing Fixtures	Kitchen, bathroom and service fixtures, supply and drainage	1 Fixture/1000 S.F. Floor	Each	3930	3.93	
2020	Domestic Water Distribution	Gas fired water heater		S.F. Floor	3.62	3.62	5.7%
2040	Rain Water Drainage	Roof drains		S.F. Roof	.82	.41	
	D30 HVAC						
3010	Energy Supply	N/A		—	—	—	
3020	Heat Generating Systems	Included in D3050		—	—	—	
3030	Cooling Generating Systems	N/A		—	—	—	22.0 %
3050	Terminal & Package Units	Multizone unit, gas heating, electric cooling		S.F. Floor	30.95	30.95	
3090	Other HVAC Sys. & Equipment	N/A		—	—	—	
	D40 Fire Protection						
4010	Sprinklers	Sprinklers, light hazard		S.F. Floor	3.55	3.55	2.8%
4020	Standpipes	Standpipe		S.F. Floor	.41	.41	
	D50 Electrical						
5010	Electrical Service/Distribution	400 ampere service, panel board and feeders		S.F. Floor	.92	.92	
5020	Lighting & Branch Wiring	Fluorescent and high intensity discharge fixtures, receptacles, switches, A.C. and misc. power		S.F. Floor	8.46	8.46	7.5%
5030	Communications & Security	Addressable alarm systems and emergency lighting		S.F. Floor	1.12	1.12	
5090	Other Electrical Systems	Emergency generator, 15 kW		S.F. Floor	.06	.06	
E. EQUIPMENT & FURNISHINGS							
1010	Commercial Equipment	N/A		—	—	—	
1020	Institutional Equipment	N/A		—	—	—	15.0 %
1030	Vehicular Equipment	N/A		—	—	—	
1090	Other Equipment	Courts, sauna baths		S.F. Floor	21.14	21.14	
F. SPECIAL CONSTRUCTION							
1020	Integrated Construction	N/A		—	—	—	0.0 %
1040	Special Facilities	N/A		—	—	—	
G. BUILDING SITEWORK	**N/A**						

		Sub-Total	140.72	100%
CONTRACTOR FEES (General Requirements: 10%, Overhead: 5%, Profit: 10%)		25%	35.21	
ARCHITECT FEES		8%	14.07	
	Total Building Cost		**190**	

For customer support on your Square Foot Costs with RSMeans data, call 800.448.8182.

Costs per square foot of floor area

Exterior Wall	S.F. Area	5000	6000	7000	8000	9000	10000	11000	12000	13000
	L.F. Perimeter	286	320	353	386	397	425	454	460	486
Face Brick and Concrete Block	Steel Joists	212.40	207.60	203.95	201.30	197.00	195.10	193.65	190.60	189.55
	Wood Joists	207.00	202.20	198.50	195.85	191.55	189.60	188.10	185.10	184.00
Stucco and Concrete Block	Steel Joists	206.85	202.45	199.05	196.65	192.75	191.00	189.60	186.90	185.95
	Wood Joists	201.50	197.05	193.60	191.20	187.30	185.50	184.10	181.40	180.40
Limestone and Concrete Block	Steel Joists	233.15	226.95	222.25	218.85	213.05	210.55	208.60	204.55	203.15
	Wood Joists	227.80	221.60	216.85	213.40	207.55	205.05	203.10	199.05	197.55
Perimeter Adj., Add or Deduct	Per 100 L.F.	18.05	15.00	12.85	11.30	10.05	9.05	8.15	7.55	6.95
Story Hgt. Adj., Add or Deduct	Per 1 Ft.	3.20	2.95	2.80	2.65	2.45	2.40	2.25	2.15	2.10

For Basement, add $33.10 per square foot of basement area

The above costs were calculated using the basic specifications shown on the facing page. These costs should be adjusted where necessary for design alternatives and owner's requirements.

Common additives

Description	Unit	$ Cost
Carrels Hardwood	Ea.	740 - 1925
Emergency Lighting, 25 watt, battery operated		
Lead battery	Ea.	330
Nickel cadmium	Ea.	550
Flagpoles, Complete		
Aluminum, 20' high	Ea.	1900
40' high	Ea.	4500
70' high	Ea.	11,600
Fiberglass, 23' high	Ea.	1450
39'-5" high	Ea.	3325
59' high	Ea.	7300
Gym Floor, Incl. sleepers and finish, maple	S.F.	16.70
Intercom System, 25 Station capacity		
Master station	Ea.	3225
Intercom outlets	Ea.	212
Handset	Ea.	590

Description	Unit	$ Cost
Lockers, Steel, single tier, 60" to 72"	Opng.	232 - 395
2 tier, 60" to 72" total	Opng.	139 - 172
5 tier, box lockers	Opng.	73 - 87
Locker bench, lam. maple top only	L.F.	36.50
Pedestals, steel pipe	Ea.	79
Seating		
Auditorium chair, all veneer	Ea.	355
Veneer back, padded seat	Ea.	380
Upholstered, spring seat	Ea.	335
Classroom, movable chair & desk	Set	81 - 171
Lecture hall, pedestal type	Ea.	350 - 650
Smoke Detectors		
Ceiling type	Ea.	244
Duct type	Ea.	585
Sound System		
Amplifier, 250 watts	Ea.	2025
Speaker, ceiling or wall	Ea.	240
Trumpet	Ea.	460
Swimming Pools, Complete, gunite	S.F.	106 - 132

Important: See the Reference Section for Location Factors.

Model costs calculated for a 1 story building with 12' story height and 10,000 square feet of floor area

				Unit	Unit Cost	Cost Per S.F.	% Of Sub-Total
A. SUBSTRUCTURE							
1010	Standard Foundations	Poured concrete; strip and spread footings; 4' foundation wall		S.F. Ground	6.23	6.23	
1020	Special Foundations	N/A		—	—	—	
1030	Slab on Grade	4" reinforced concrete with vapor barrier and granular base		S.F. Slab	5.80	5.80	8.6%
2010	Basement Excavation	Site preparation for slab and trench for foundation wall and footing		S.F. Ground	.33	.33	
2020	Basement Walls	N/A		—	—	—	
B. SHELL							
	B10 Superstructure						
1010	Floor Construction	N/A		—	—	—	6.9 %
1020	Roof Construction	Metal deck, open web steel joists, beams, interior columns		S.F. Roof	9.90	9.90	
	B20 Exterior Enclosure						
2010	Exterior Walls	Face brick with concrete block backup	85% of wall	S.F. Wall	29.04	12.59	
2020	Exterior Windows	Window wall	15% of wall	Each	585	6.97	15.9%
2030	Exterior Doors	Double aluminum and glass		Each	8163	3.27	
	B30 Roofing						
3010	Roof Coverings	Built-up tar and gravel with flashing; perlite/EPS composite insulation		S.F. Roof	7.05	7.05	4.9%
3020	Roof Openings	N/A		—	—	—	
C. INTERIORS							
1010	Partitions	Concrete block	8 S.F. Floor/S.F. Partition	S.F. Partition	10.41	13.01	
1020	Interior Doors	Single leaf hollow metal	700 S.F. Floor/Door	Each	1222	1.75	
1030	Fittings	Toilet partitions		S.F. Floor	1.25	1.25	
2010	Stair Construction	N/A		—	—	—	24.1%
3010	Wall Finishes	Paint		S.F. Surface	2.44	6.11	
3020	Floor Finishes	50% vinyl tile, 50% carpet		S.F. Floor	4.74	4.74	
3030	Ceiling Finishes	Mineral fiber tile on concealed zee bars		S.F. Ceiling	7.60	7.60	
D. SERVICES							
	D10 Conveying						
1010	Elevators & Lifts	N/A		—	—	—	0.0 %
1020	Escalators & Moving Walks	N/A		—	—	—	
	D20 Plumbing						
2010	Plumbing Fixtures	Toilet and service fixtures, supply and drainage	1 Fixture/455 S.F. Floor	Each	4195	9.22	
2020	Domestic Water Distribution	Gas fired water heater		S.F. Floor	1.80	1.80	8.5%
2040	Rain Water Drainage	Roof drains		S.F. Roof	1.16	1.16	
	D30 HVAC						
3010	Energy Supply	N/A		—	—	—	
3020	Heat Generating Systems	N/A		—	—	—	
3030	Cooling Generating Systems	N/A		—	—	—	18.5 %
3050	Terminal & Package Units	Multizone HVAC air cooled system		S.F. Floor	26.55	26.55	
3090	Other HVAC Sys. & Equipment	N/A		—	—	—	
	D40 Fire Protection						
4010	Sprinklers	Sprinkler, light hazard		S.F. Floor	3.79	3.79	3.1%
4020	Standpipes	Standpipe		S.F. Floor	.65	.65	
	D50 Electrical						
5010	Electrical Service/Distribution	200 ampere service, panel board and feeders		S.F. Floor	1.14	1.14	
5020	Lighting & Branch Wiring	T-8 fluorescent fixtures, receptacles, switches, A.C. and misc. power		S.F. Floor	8.56	8.56	9.4%
5030	Communications & Security	Addressable alarm systems, partial internet wiring and emergency lighting		S.F. Floor	3.64	3.64	
5090	Other Electrical Systems	Emergency generator, 15 kW		S.F. Floor	.09	.09	
E. EQUIPMENT & FURNISHINGS							
1010	Commercial Equipment	N/A		—	—	—	
1020	Institutional Equipment	N/A		—	—	—	
1030	Vehicular Equipment	N/A		—	—	—	0.0 %
1090	Other Equipment	N/A		—	—	—	
F. SPECIAL CONSTRUCTION							
1020	Integrated Construction	N/A		—	—	—	0.0 %
1040	Special Facilities	N/A		—	—	—	
G. BUILDING SITEWORK	**N/A**						

	Sub-Total	143.20	100%
CONTRACTOR FEES (General Requirements: 10%, Overhead: 5%, Profit: 10%)	25%	35.79	
ARCHITECT FEES	9%	16.11	
Total Building Cost		**195.10**	

For customer support on your Square Foot Costs with RSMeans data, call 800.448.8182.

193

Costs per square foot of floor area

Exterior Wall	S.F. Area	2000	2800	3500	4200	5000	5800	6500	7200	8000
	L.F. Perimeter	180	212	240	268	300	314	336	344	368
Wood Clapboard	Wood Frame	234.80	222.20	215.95	211.70	208.45	204.25	202.30	199.60	198.20
Brick Veneer	Wood Frame	244.30	230.55	223.70	219.10	215.50	210.85	208.70	205.70	204.15
Stone Veneer	Rigid Steel	279.75	261.55	252.50	246.40	241.65	235.10	232.20	228.00	225.90
Fiber Cement	Rigid Steel	241.70	229.55	223.50	219.45	216.30	212.25	210.40	207.80	206.50
E.I.F.S.	Reinforced Concrete	271.80	257.25	250.05	245.20	241.40	236.40	234.10	230.90	229.25
Stucco	Reinforced Concrete	272.55	257.90	250.60	245.75	241.85	236.85	234.50	231.30	229.60
Perimeter Adj., Add or Deduct	Per 100 L.F.	28.20	20.15	16.15	13.55	11.25	9.75	8.65	7.85	7.00
Story Hgt. Adj., Add or Deduct	Per 1 Ft.	2.80	2.40	2.15	2.00	1.90	1.70	1.60	1.50	1.45

For Basement, add $38.30 per square foot of basement area

The above costs were calculated using the basic specifications shown on the facing page. These costs should be adjusted where necessary for design alternatives and owner's requirements.

Common additives

Description	Unit	$ Cost
Bar, Front Bar	L.F.	450
Back bar	L.F.	360
Booth, Upholstered, custom straight	L.F.	258 - 475
"L" or "U" shaped	L.F.	267 - 450
Cupola, Stock unit, redwood		
30" square, 37" high, aluminum roof	Ea.	825
Copper roof	Ea.	485
Fiberglass, 5'-0" base, 63" high	Ea.	4450 - 4825
6'-0" base, 63" high	Ea.	5225 - 7725
Decorative Wood Beams, Non load bearing		
Rough sawn, 4" x 6"	L.F.	9.20
4" x 8"	L.F.	10.25
4" x 10"	L.F.	12.70
4" x 12"	L.F.	13.95
8" x 8"	L.F.	14.75
Emergency Lighting, 25 watt, battery operated		
Lead battery	Ea.	330
Nickel cadmium	Ea.	550

Description	Unit	$ Cost
Fireplace, Brick, not incl. chimney or foundation		
30" x 29" opening	Ea.	3400
Chimney, standard brick		
Single flue, 16" x 20"	V.L.F.	113
20" x 20"	V.L.F.	133
2 Flue, 20" x 24"	V.L.F.	165
20" x 32"	V.L.F.	203
Kitchen Equipment		
Broiler	Ea.	4425
Coffee urn, twin 6 gallon	Ea.	2900
Cooler, 6 ft. long	Ea.	3750
Dishwasher, 10-12 racks per hr.	Ea.	4150
Food warmer, counter, 1.2 KW	Ea.	790
Freezer, 44 C.F., reach-in	Ea.	5525
Ice cube maker, 50 lb. per day	Ea.	2100
Range with 1 oven	Ea.	3575
Refrigerators, Prefabricated, walk-in		
7'-6" high, 6' x 6'	S.F.	145
10' x 10'	S.F.	114
12' x 14'	S.F.	99.50
12' x 20'	S.F.	137

Important: See the Reference Section for Location Factors.

Model costs calculated for a 1 story building with 12' story height and 5,000 square feet of floor area

Restaurant

				Unit	Unit Cost	Cost Per S.F.	% Of Sub-Total
A. SUBSTRUCTURE							
1010	Standard Foundations	Poured concrete; strip and spread footings; 4' foundation wall		S.F. Ground	10.16	10.16	
1020	Special Foundations	N/A		—	—	—	
1030	Slab on Grade	4" reinforced concrete		S.F. Slab	5.80	5.80	10.6%
2010	Basement Excavation	Site preparation for slab and trench for foundation wall and footing		S.F. Ground	.57	.57	
2020	Basement Walls	N/A		—	—	—	
B. SHELL							
	B10 Superstructure						
1010	Floor Construction	Wood columns		S.F. Floor	.18	.46	5.3%
1020	Roof Construction	Wood roof truss, 4:12 slope		S.F. Roof	6.93	7.76	
	B20 Exterior Enclosure						
2010	Exterior Walls	Wood siding on wood studs, insulated	70% of wall	S.F. Wall	26.17	13.19	
2020	Exterior Windows	Curtain glazing wall system	30% of wall	Each	44.85	4.10	15.6%
2030	Exterior Doors	Aluminum & glass double doors, hollow metal doors		Each	6949	6.95	
	B30 Roofing						
3010	Roof Coverings	Asphalt strip shingles, gutters and downspouts		S.F. Roof	16.31	2.48	1.6%
3020	Roof Openings	N/A		—	—	—	
C. INTERIORS							
1010	Partitions	Fire rated gypsum board on wood studs	25 S.F. Floor/L.F. Partition	S.F. Partition	7.25	2.90	
1020	Interior Doors	Single leaf hollow core wood doors	1000 S.F. Floor/Door	Each	640	.64	
1030	Fittings	Plastic laminate toilet partitions		S.F. Floor	.92	.92	
2010	Stair Construction	N/A		—	—	—	13.3%
3010	Wall Finishes	80% Paint, 20% ceramic tile		S.F. Surface	2.90	2.32	
3020	Floor Finishes	70% Carpet tile, 30% quarry tile		S.F. Floor	9.03	9.03	
3030	Ceiling Finishes	Gypsum board ceiling on wood furring		S.F. Ceiling	4.84	4.84	
D. SERVICES							
	D10 Conveying						
1010	Elevators & Lifts	N/A		—	—	—	0.0 %
1020	Escalators & Moving Walks	N/A		—	—	—	
	D20 Plumbing						
2010	Plumbing Fixtures	Restroom, kitchen and service fixtures, supply and drainage	1 Fixture/350 S.F. Floor	Each	3294	9.28	
2020	Domestic Water Distribution	Gas fired water heater		S.F. Floor	5.58	5.58	9.5%
2040	Rain Water Drainage	Roof drains		S.F. Roof			
	D30 HVAC						
3010	Energy Supply	N/A		—	—	—	
3020	Heat Generating Systems	Included in D3050		—	—	—	
3030	Cooling Generating Systems	N/A		—	—	—	25.0 %
3050	Terminal & Package Units	Multizone unit gas heating, electric cooling, kitchen exhaust system		S.F. Floor	38.94	38.94	
3090	Other HVAC Sys. & Equipment	N/A		—	—	—	
	D40 Fire Protection						
4010	Sprinklers	Wet pipe sprinkler system		S.F. Floor	9.18	9.18	7.4%
4020	Standpipes	Standpipe		S.F. Floor	2.29	2.29	
	D50 Electrical						
5010	Electrical Service/Distribution	400 Ampere service, panelboards and feeders		S.F. Floor	4.71	4.71	
5020	Lighting & Branch Wiring	Fluorescent fixtures, receptacles, switches, a.c. and misc. power		S.F. Floor	10.04	10.04	11.8%
5030	Communications & Security	Addressable alarm system		S.F. Floor	3.70	3.70	
5090	Other Electrical Systems	N/A		—	—	—	
E. EQUIPMENT & FURNISHINGS							
1010	Commercial Equipment	N/A		—	—	—	
1020	Institutional Equipment	N/A		—	—	—	
1030	Vehicular Equipment	N/A		—	—	—	0.0 %
1090	Other Equipment	N/A		—	—	—	
F. SPECIAL CONSTRUCTION							
1020	Integrated Construction	N/A		—	—	—	0.0 %
1040	Special Facilities	N/A		—	—	—	
G. BUILDING SITEWORK	**N/A**						

		Sub-Total	155.84	**100%**
CONTRACTOR FEES (General Requirements: 10%, Overhead: 5%, Profit: 10%)		25%	38.97	
ARCHITECT FEES		7%	13.64	

Total Building Cost	**208.45**

For customer support on your Square Foot Costs with RSMeans data, call 800.448.8182.

195

Costs per square foot of floor area

Exterior Wall	S.F. Area	2000	2800	3500	4000	5000	5800	6500	7200	8000
	L.F. Perimeter	180	212	240	260	300	314	336	344	368
Wood Clapboard	Wood Frame	232.15	224.10	220.00	218.00	215.10	211.80	210.45	208.20	207.25
Brick Veneer	Wood Frame	235.75	227.15	222.80	220.65	217.65	214.15	212.65	210.30	209.25
Stone Veneer	Rigid Steel	266.35	254.05	247.85	244.80	240.45	235.35	233.25	229.85	228.35
Fiber Cement	Rigid Steel	234.40	227.10	223.50	221.65	219.10	216.15	214.85	212.90	212.05
E.I.F.S.	Reinforced Concrete	261.55	252.15	247.40	245.10	241.80	237.95	236.35	233.70	232.55
Stucco	Reinforced Concrete	262.20	252.70	247.90	245.55	242.15	238.30	236.70	234.05	232.90
Perimeter Adj., Add or Deduct	Per 100 L.F.	28.35	20.15	16.20	14.15	11.40	9.80	8.65	7.90	7.05
Story Hgt. Adj., Add or Deduct	Per 1 Ft.	3.05	2.50	2.35	2.25	2.10	1.90	1.80	1.70	1.55
Basement—Not Applicable										

The above costs were calculated using the basic specifications shown on the facing page. These costs should be adjusted where necessary for design alternatives and owner's requirements.

Common additives

Description	Unit	$ Cost
Bar, Front Bar	L.F.	450
Back bar	L.F.	360
Booth, Upholstered, custom straight	L.F.	258 - 475
"L" or "U" shaped	L.F.	267 - 450
Drive-up Window	Ea.	9525 - 13,300
Emergency Lighting, 25 watt, battery operated		
Lead battery	Ea.	330
Nickel cadmium	Ea.	550
Kitchen Equipment		
Broiler	Ea.	4425
Coffee urn, twin 6 gallon	Ea.	2900
Cooler, 6 ft. long	Ea.	3750
Dishwasher, 10-12 racks per hr.	Ea.	4150
Food warmer, counter, 1.2 KW	Ea.	790
Freezer, 44 C.F., reach-in	Ea.	5525
Ice cube maker, 50 lb. per day	Ea.	2100
Range with 1 oven	Ea.	3575

Description	Unit	$ Cost
Refrigerators, Prefabricated, walk-in		
7'-6" High, 6' x 6'	S.F.	145
10' x 10'	S.F.	114
12' x 14'	S.F.	99.50
12' x 20'	S.F.	137
Serving		
Counter top (Stainless steel)	L.F.	213
Base cabinets	L.F.	530 - 870
Sound System		
Amplifier, 250 watts	Ea.	2025
Speaker, ceiling or wall	Ea.	240
Trumpet	Ea.	460
Storage		
Shelving	S.F.	9.80
Washing		
Stainless steel counter	L.F.	213

Important: See the Reference Section for Location Factors.

Model costs calculated for a 1 story building with 10' story height and 4,000 square feet of floor area

Restaurant, Fast Food

			Unit	Unit Cost	Cost Per S.F.	% Of Sub-Total
A. SUBSTRUCTURE						
1010	Standard Foundations	Poured concrete; strip and spread footings; 4' foundation wall	S.F. Ground	9.61	9.61	
1020	Special Foundations	N/A	—	—	—	
1030	Slab on Grade	4" reinforced concrete	S.F. Slab	5.80	5.80	9.9%
2010	Basement Excavation	Site preparation for slab and trench for foundation wall and footing	S.F. Ground	.57	.57	
2020	Basement Walls	N/A	—	—	—	
B. SHELL						
B10 Superstructure						
1010	Floor Construction	Wood columns	S.F. Floor	.46	.46	3.2%
1020	Roof Construction	Flat wood rafter roof	S.F. Roof	4.67	4.67	
B20 Exterior Enclosure						
2010	Exterior Walls	Wood siding on wood studs, insulated 70% of wall	S.F. Wall	22.99	10.46	
2020	Exterior Windows	Curtain glazing wall system 30% of wall	Each	53	4.82	16.2%
2030	Exterior Doors	Aluminum & glass double doors, hollow metal doors	Each	5460	10.93	
B30 Roofing						
3010	Roof Coverings	Single ply membrane, stone ballast, rigid insulation	S.F. Roof	8.10	8.10	5.4%
3020	Roof Openings	Roof and smoke hatches	S.F. Roof	.64	.64	
C. INTERIORS						
1010	Partitions	Water resistant gypsum board on wood studs 25 S.F. Floor/L.F. Partition	S.F. Partition	7.18	2.87	
1020	Interior Doors	Single leaf hollow core wood doors 1000 S.F. Floor/Door	Each	640	.64	
1030	Fittings	Stainless steel toilet partitions	S.F. Floor	1.56	1.56	
2010	Stair Construction	N/A	—	—	—	19.2%
3010	Wall Finishes	90% Paint, 10% ceramic tile	S.F. Surface	2.11	1.69	
3020	Floor Finishes	Quarry tile	S.F. Floor	17.88	17.88	
3030	Ceiling Finishes	Acoustic ceiling tiles on suspended support grid	S.F. Ceiling	6.30	6.30	
D. SERVICES						
D10 Conveying						
1010	Elevators & Lifts	N/A	—	—	—	0.0 %
1020	Escalators & Moving Walks	N/A	—	—	—	
D20 Plumbing						
2010	Plumbing Fixtures	Restroom, kitchen and service fixtures, supply and drainage 1 Fixture/330 S.F. Floor	Each	2900	8.71	
2020	Domestic Water Distribution	Gas fired water heater	S.F. Floor	3.93	3.93	9.0%
2040	Rain Water Drainage	Roof drains	S.F. Roof	1.91	1.91	
D30 HVAC						
3010	Energy Supply	N/A	—	—	—	
3020	Heat Generating Systems	Included in D3050	—	—	—	
3030	Cooling Generating Systems	N/A	—	—	—	12.2 %
3050	Terminal & Package Units	Single zone unit gas heating, electric cooling	S.F. Floor	19.68	19.68	
3090	Other HVAC Sys. & Equipment	N/A	—	—	—	
D40 Fire Protection						
4010	Sprinklers	Wet pipe sprinkler system, ordinary hazard	S.F. Floor	9.18	9.18	7.2%
4020	Standpipes	Standpipe	S.F. Floor	2.39	2.39	
D50 Electrical						
5010	Electrical Service/Distribution	400 Ampere service, panelboards and feeders	S.F. Floor	4.93	4.93	
5020	Lighting & Branch Wiring	Fluorescent fixtures, receptacles, switches, a.c. and misc. power	S.F. Floor	10.87	10.87	12.6%
5030	Communications & Security	Addressable alarm, emergency lighting	S.F. Floor	4.62	4.62	
5090	Other Electrical Systems	N/A	—	—	—	
E. EQUIPMENT & FURNISHINGS						
1010	Commercial Equipment	N/A	—	—	—	
1020	Institutional Equipment	N/A	—	—	—	
1030	Vehicular Equipment	N/A	—	—	—	5.1 %
1090	Other Equipment	Prefabricated walk in refrigerators	S.F. Floor	8.24	8.24	
F. SPECIAL CONSTRUCTION						
1020	Integrated Construction	N/A	—	—	—	0.0 %
1040	Special Facilities	N/A	—	—	—	
G. BUILDING SITEWORK	**N/A**					

		Sub-Total	161.46	**100%**
CONTRACTOR FEES (General Requirements: 10%, Overhead: 5%, Profit: 10%)		25%	40.39	
ARCHITECT FEES		8%	16.15	
	Total Building Cost		**218**	

For customer support on your Square Foot Costs with RSMeans data, call 800.448.8182.

197

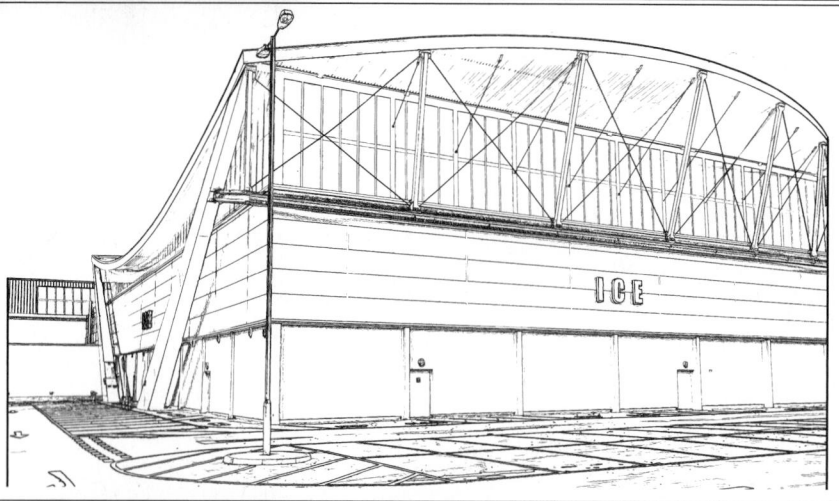

Costs per square foot of floor area

Exterior Wall	S.F. Area	10000	15000	20000	25000	30000	35000	40000	45000	50000
	L.F. Perimeter	450	500	600	700	740	822	890	920	966
Face Brick and Concrete Block	Steel Joists	213.40	194.10	188.15	184.55	179.20	177.20	175.15	172.30	170.45
	Wood Truss	225.30	207.30	202.05	198.90	193.85	192.05	190.10	187.35	185.65
Concrete Block	Rigid Steel	206.10	193.70	189.70	187.25	183.85	182.60	181.25	179.40	178.30
	Wood Truss	198.85	187.80	184.50	182.50	179.35	178.25	177.05	175.35	174.30
Metal Panel	Rigid Steel	205.05	192.85	188.95	186.60	183.30	182.00	180.70	178.90	177.80
Metal Sandwich Panel	Rigid Steel	209.60	196.20	191.95	189.40	185.75	184.35	182.90	181.00	179.75
Perimeter Adj., Add or Deduct	Per 100 L.F.	14.85	9.85	7.45	5.95	5.00	4.25	3.70	3.30	3.00
Story Hgt. Adj., Add or Deduct	Per 1 Ft.	2.35	1.75	1.60	1.45	1.30	1.25	1.15	1.05	1.05
Basement—Not Applicable										

The above costs were calculated using the basic specifications shown on the facing page. These costs should be adjusted where necessary for design alternatives and owner's requirements.

Common additives

Description	Unit	$ Cost
Bar, Front Bar	L.F.	450
Back bar	L.F.	360
Booth, Upholstered, custom straight	L.F.	258 - 475
"L" or "U" shaped	L.F.	267 - 450
Bleachers, Telescoping, manual		
To 15 tier	Seat	164 - 222
16-20 tier	Seat	340 - 405
21-30 tier	Seat	340 - 460
For power operation, add	Seat	63 - 101
Emergency Lighting, 25 watt, battery operated		
Lead battery	Ea.	330
Nickel cadmium	Ea.	550
Lockers, Steel, single tier, 60" or 72"	Opng.	232 - 395
2 tier, 60" or 72" total	Opng.	139 - 172
5 tier, box lockers	Opng.	73 - 87
Locker bench, lam. maple top only	L.F.	36.50
Pedestals, steel pipe	Ea.	79

Description	Unit	$ Cost
Rink		
Dasher boards & top guard	Ea.	197,500
Mats, rubber	S.F.	25
Score Board	Ea.	17,700 - 30,400

Important: See the Reference Section for Location Factors.

Rink, Hockey/Indoor Soccer

Model costs calculated for a 1 story building with 24' story height and 30,000 square feet of floor area

				Unit	Unit Cost	Cost Per S.F.	% Of Sub-Total
A. SUBSTRUCTURE							
1010	Standard Foundations	Poured concrete; strip and spread footings; 4' foundation wall		S.F. Ground	3.71	3.71	
1020	Special Foundations	N/A		—	—	—	
1030	Slab on Grade	6" reinforced concrete with vapor barrier and granular base		S.F. Slab	7.01	7.01	7.9%
2010	Basement Excavation	Site preparation for slab and trench for foundation wall and footing		S.F. Ground	.18	.18	
2020	Basement Walls	N/A		—	—	—	
B. SHELL							
B10 Superstructure							
1010	Floor Construction	Wide flange beams and columns		S.F. Floor	45.43	31.80	23.1%
1020	Roof Construction	(incl. in B1010)					
B20 Exterior Enclosure							
2010	Exterior Walls	Concrete block	95% of wall	S.F. Wall	15.99	8.99	
2020	Exterior Windows	Store front	5% of wall	Each	53	1.54	8.5%
2030	Exterior Doors	Aluminum and glass, hollow metal, overhead		Each	4346	1.17	
B30 Roofing							
3010	Roof Coverings	Elastomeric neoprene membrane with flashing; perlite/EPS composite insulation		S.F. Roof	5.33	5.33	4.1%
3020	Roof Openings	Roof hatches		S.F. Roof	.30	.30	
C. INTERIORS							
1010	Partitions	Concrete block	140 S.F. Floor/L.F. Partition	S.F. Partition	10.38	.89	
1020	Interior Doors	Hollow metal	2500 S.F. Floor/Door	Each	1222	.49	
1030	Fittings	N/A		—	—	—	
2010	Stair Construction	N/A		—	—	—	6.6%
3010	Wall Finishes	Paint		S.F. Surface	13.30	2.28	
3020	Floor Finishes	80% rubber mat, 20% paint	50% of floor area	S.F. Floor	9.20	4.60	
3030	Ceiling Finishes	Mineral fiber tile on concealed zee bar	10% of area	S.F. Ceiling	7.60	.77	
D. SERVICES							
D10 Conveying							
1010	Elevators & Lifts	N/A		—	—	—	0.0 %
1020	Escalators & Moving Walks	N/A		—	—	—	
D20 Plumbing							
2010	Plumbing Fixtures	Toilet and service fixtures, supply and drainage	1 Fixture/1070 S.F. Floor	Each	6944	6.49	
2020	Domestic Water Distribution	Oil fired water heater		S.F. Floor	3.62	3.62	8.1%
2040	Rain Water Drainage	Roof drains		S.F. Roof	1.08	1.08	
D30 HVAC							
3010	Energy Supply	Oil fired hot water, unit heaters	10% of area	S.F. Floor	.93	.93	
3020	Heat Generating Systems	N/A		—	—	—	
3030	Cooling Generating Systems	N/A		—	—	—	11.4%
3050	Terminal & Package Units	Single zone, electric cooling	90% of area	S.F. Floor	14.81	14.81	
3090	Other HVAC Sys. & Equipment	N/A		—	—	—	
D40 Fire Protection							
4010	Sprinklers	N/A		—	—	—	0.5 %
4020	Standpipes	Standpipe		S.F. Floor	.75	.75	
D50 Electrical							
5010	Electrical Service/Distribution	600 ampere service, panel board and feeders		S.F. Floor	1.49	1.49	
5020	Lighting & Branch Wiring	High intensity discharge and fluorescent fixtures, receptacles, switches, A.C. and misc. power		S.F. Floor	8.16	8.16	8.7%
5030	Communications & Security	Addressable alarm systems, emergency lighting and public address		S.F. Floor	2.15	2.15	
5090	Other Electrical Systems	Emergency generator		S.F. Floor	.22	.22	
E. EQUIPMENT & FURNISHINGS							
1010	Commercial Equipment	N/A		—	—	—	
1020	Institutional Equipment	N/A		—	—	—	
1030	Vehicular Equipment	N/A		—	—	—	0.0 %
1090	Other Equipment	N/A		—	—	—	
F. SPECIAL CONSTRUCTION							
1020	Integrated Construction	N/A		—	—	—	20.9 %
1040	Special Facilities	Dasher boards and rink (including ice making system)		S.F. Floor	28.72	28.72	
G. BUILDING SITEWORK	**N/A**						

		Sub-Total	137.48	**100%**
CONTRACTOR FEES (General Requirements: 10%, Overhead: 5%, Profit: 10%)		25%	34.34	
ARCHITECT FEES		7%	12.03	
	Total Building Cost		**183.85**	

For customer support on your Square Foot Costs with RSMeans data, call 800.448.8182.

199

Costs per square foot of floor area

Exterior Wall	S.F. Area	25000	30000	35000	40000	45000	50000	55000	60000	65000
	L.F. Perimeter	900	1050	1200	1350	1510	1650	1800	1970	2100
Fiber Cement	Rigid Steel	165.75	164.50	163.55	162.95	162.55	161.95	161.60	161.55	161.10
Metal Panel	Rigid Steel	165.60	164.30	163.40	162.70	162.35	161.70	161.40	161.35	160.90
E.I.F.S.	Rigid Steel	169.60	168.30	167.40	166.70	166.35	165.70	165.35	165.35	164.85
Tilt-up Concrete Panels	Reinforced Concrete	212.60	210.45	208.85	207.75	207.20	206.10	205.50	205.55	204.60
Brick Veneer	Reinforced Concrete	192.95	191.30	190.10	189.30	188.90	188.05	187.65	187.60	186.95
Stone Veneer	Reinforced Concrete	238.05	235.15	233.10	231.60	230.90	229.45	228.70	228.75	227.45
Perimeter Adj., Add or Deduct	Per 100 L.F.	3.35	2.80	2.45	2.05	1.85	1.70	1.50	1.40	1.20
Story Hgt. Adj., Add or Deduct	Per 1 Ft.	1.30	1.15	1.25	1.10	1.15	1.15	1.10	1.15	1.10
For Basement, add *$28.40 per square foot of basement area*										

The above costs were calculated using the basic specifications shown on the facing page. These costs should be adjusted where necessary for design alternatives and owner's requirements.

Common additives

Description	Unit	$ Cost
Bleachers, Telescoping, manual		
To 15 tier	Seat	164 - 222
16-20 tier	Seat	340 - 405
21-30 tier	Seat	340 - 460
For power operation, add	Seat	63 - 101
Carrels Hardwood	Ea.	740 - 1925
Clock System		
20 room	Ea.	20,200
50 room	Ea.	48,200
Emergency Lighting, 25 watt, battery operated		
Lead battery	Ea.	330
Nickel cadmium	Ea.	550
Flagpoles, Complete		
Aluminum, 20' high	Ea.	1900
40' high	Ea.	4500
Fiberglass, 23' high	Ea.	1450
39'-5" high	Ea.	3325
Kitchen Equipment		
Broiler	Ea.	4425
Cooler, 6 ft. long, reach-in	Ea.	3750

Description	Unit	$ Cost
Kitchen Equipment, cont.		
Dishwasher, 10-12 racks per hr.	Ea.	4150
Food warmer, counter, 1.2 KW	Ea.	790
Freezer, 44 C.F., reach-in	Ea.	5525
Ice cube maker, 50 lb. per day	Ea.	2100
Range with 1 oven	Ea.	3575
Lockers, Steel, single tier, 60" to 72"	Opng.	232 - 395
2 tier, 60" to 72" total	Opng.	139 - 172
5 tier, box lockers	Opng.	73 - 87
Locker bench, lam. maple top only	L.F.	36.50
Pedestals, steel pipe	Ea.	79
Seating		
Auditorium chair, all veneer	Ea.	355
Veneer back, padded seat	Ea.	380
Upholstered, spring seat	Ea.	335
Classroom, movable chair & desk	Set	81 - 171
Lecture hall, pedestal type	Ea.	350 - 650
Sound System		
Amplifier, 250 watts	Ea.	2025
Speaker, ceiling or wall	Ea.	240
Trumpet	Ea.	460

Important: See the Reference Section for Location Factors.

Model costs calculated for a 1 story building with 15′ story height and 45,000 square feet of floor area

				Unit	Unit Cost	Cost Per S.F.	% Of Sub-Total
A. SUBSTRUCTURE							
1010	Standard Foundations	Poured concrete; strip and spread footings; 4′ foundation wall		S.F. Ground	5.37	5.37	
1020	Special Foundations	N/A		—	—	—	
1030	Slab on Grade	4″ reinforced concrete		S.F. Slab	5.80	5.80	9.5%
2010	Basement Excavation	Site preparation for slab and trench for foundation wall and footing		S.F. Ground	.18	.32	
2020	Basement Walls	N/A		—	—	—	
B. SHELL							
B10 Superstructure							
1010	Floor Construction	Fireproofing for structural columns		S.F. Floor	1.11	1.11	9.9%
1020	Roof Construction	Metal deck, open web steel joists, beams, columns		S.F. Roof	10.90	10.90	
B20 Exterior Enclosure							
2010	Exterior Walls	Textured metal panel on metal stud, insulated	70% of wall	S.F. Wall	17.99	6.34	
2020	Exterior Windows	Curtain wall system, thermo-break frame, aluminum awning windows	30% of wall	Each	814	6.28	11.1%
2030	Exterior Doors	Aluminum & glass double doors, hollow metal doors		Each	5010	.89	
B30 Roofing							
3010	Roof Coverings	Single ply membrane, stone ballast, rigid insulation		S.F. Roof	8.08	8.08	6.8%
3020	Roof Openings	Roof and smoke hatches		S.F. Roof	2587	.23	
C. INTERIORS							
1010	Partitions	Gypsum board on metal studs, sound attentuation insulation	20 S.F. Floor/L.F. Partition	S.F. Partition	5.94	2.97	
1020	Interior Doors	Hollow metal doors	700 S.F. Floor/Door	Each	1222	1.75	
1030	Fittings	Chalkboards, lockers, toilet partitions		S.F. Floor	1.28	1.28	
2010	Stair Construction	N/A		—	—	—	19.1%
3010	Wall Finishes	90% Paint, 10% ceramic tile		S.F. Surface	1.94	1.94	
3020	Floor Finishes	10% Carpet, 10% terrazzo, 60% VCT, 20% wood		S.F. Floor	7.70	7.70	
3030	Ceiling Finishes	Acoustic ceiling tiles on suspended channel grid		S.F. Ceiling	7.60	7.60	
D. SERVICES							
D10 Conveying							
1010	Elevators & Lifts	N/A		—	—	—	0.0 %
1020	Escalators & Moving Walks	N/A		—	—	—	
D20 Plumbing							
2010	Plumbing Fixtures	Restroom, kitchen, and service fixtures, supply and drainage	1 Fixture/325 S.F. Floor	Each	2311	7.11	
2020	Domestic Water Distribution	Gas fired water heater		S.F. Floor	.71	.71	7.4%
2040	Rain Water Drainage	Roof drains		S.F. Roof	1.12	1.12	
D30 HVAC							
3010	Energy Supply	Forced hot water heating system, fin tube radiation		S.F. Floor	10.89	10.89	
3020	Heat Generating Systems	N/A		—	—	—	
3030	Cooling Generating Systems	N/A		—	—	—	19.7%
3050	Terminal & Package Units	Split system with air cooled condensing unit		S.F. Floor	13.02	13.02	
3090	Other HVAC Sys. & Equipment	N/A		—	—	—	
D40 Fire Protection							
4010	Sprinklers	Wet pipe spinkler system		S.F. Floor	2.93	2.93	2.8%
4020	Standpipes	Standpipe		S.F. Floor	.46	.46	
D50 Electrical							
5010	Electrical Service/Distribution	800 Ampere service, panelboards and feeders		S.F. Floor	1.10	1.10	
5020	Lighting & Branch Wiring	Fluorescent fixtures, receptacles, switches, a.c. and misc. power		S.F. Floor	11.30	11.30	13.5%
5030	Communications & Security	Addressable alarm system, emergency lighting, internet & phone wiring		S.F. Floor	3.87	3.87	
5090	Other Electrical Systems	Emergency generator, 15kW		S.F. Floor	.10	.10	
E. EQUIPMENT & FURNISHINGS							
1010	Commercial Equipment	N/A		—	—	—	
1020	Institutional Equipment	Laboratory equipment		S.F. Floor	.22	.22	0.2 %
1030	Vehicular Equipment	N/A		—	—	—	
1090	Other Equipment	N/A		—	—	—	
F. SPECIAL CONSTRUCTION							
1020	Integrated Construction	N/A		—	—	—	0.0 %
1040	Special Facilities	N/A		—	—	—	
G. BUILDING SITEWORK	**N/A**						

		Sub-Total	121.39	100%
CONTRACTOR FEES (General Requirements: 10%, Overhead: 5%, Profit: 10%)		25%	30.34	
ARCHITECT FEES		7%	10.62	
		Total Building Cost	**162.35**	

For customer support on your Square Foot Costs with RSMeans data, call 800.448.8182.

201

Costs per square foot of floor area

Exterior Wall	S.F. Area	50000	70000	90000	110000	130000	150000	170000	190000	210000
	L.F. Perimeter	850	1140	1420	1700	1980	2280	2560	2840	3120
Fiber Cement	Rigid Steel	187.00	183.20	180.85	179.35	178.35	177.85	177.25	176.70	176.35
Metal Panel	Rigid Steel	191.95	188.10	185.75	184.30	183.25	182.75	182.15	181.60	181.20
E.I.F.S.	Rigid Steel	194.20	190.45	188.05	186.60	185.55	185.10	184.45	183.95	183.55
Tilt-up Concrete Panels	Reinforced Concrete	231.10	225.95	222.65	220.60	219.15	218.60	217.75	217.00	216.45
Brick Veneer	Reinforced Concrete	214.20	209.80	207.00	205.30	204.05	203.55	202.85	202.20	201.80
Stone Veneer	Reinforced Concrete	253.80	247.70	243.75	241.20	239.45	238.85	237.80	236.95	236.25
Perimeter Adj., Add or Deduct	Per 100 L.F.	4.00	2.80	2.15	1.80	1.45	1.30	1.15	1.10	0.90
Story Hgt. Adj., Add or Deduct	Per 1 Ft.	1.55	1.45	1.40	1.40	1.35	1.40	1.35	1.40	1.35

For Basement, add $ 37.00 per square foot of basement area

The above costs were calculated using the basic specifications shown on the facing page. These costs should be adjusted where necessary for design alternatives and owner's requirements.

Common additives

Description	Unit	$ Cost
Bleachers, Telescoping, manual		
To 15 tier	Seat	164 - 222
16-20 tier	Seat	340 - 405
21-30 tier	Seat	340 - 460
For power operation, add	Seat	63 - 101
Carrels Hardwood	Ea.	740 - 1925
Clock System		
20 room	Ea.	20,200
50 room	Ea.	48,200
Elevators, Hydraulic passenger, 2 stops		
1500# capacity	Ea.	73,100
2500# capacity	Ea.	76,600
Emergency Lighting, 25 watt, battery operated		
Lead battery	Ea.	330
Nickel cadmium	Ea.	550
Flagpoles, Complete		
Aluminum, 20' high	Ea.	1900
40' high	Ea.	4500
Fiberglass, 23' high	Ea.	1450
39'-5" high	Ea.	3325

Description	Unit	$ Cost
Kitchen Equipment		
Broiler	Ea.	4425
Cooler, 6 ft. long, reach-in	Ea.	3750
Dishwasher, 10-12 racks per hr.	Ea.	4150
Food warmer, counter, 1.2 KW	Ea.	790
Freezer, 44 C.F., reach-in	Ea.	5525
Lockers, Steel, single tier, 60" or 72"	Opng.	232 - 395
2 tier, 60" or 72" total	Opng.	139 - 172
5 tier, box lockers	Opng.	73 - 87
Locker bench, lam. maple top only	L.F.	36.50
Pedestals, steel pipe	Ea.	79
Seating		
Auditorium chair, all veneer	Ea.	355
Veneer back, padded seat	Ea.	380
Upholstered, spring seat	Ea.	335
Classroom, movable chair & desk	Set	81 - 171
Lecture hall, pedestal type	Ea.	350 - 650
Sound System		
Amplifier, 250 watts	Ea.	2025
Speaker, ceiling or wall	Ea.	240
Trumpet	Ea.	460

Important: See the Reference Section for Location Factors.

Model costs calculated for a 2 story building with 15' story height and 130,000 square feet of floor area

				Unit	Unit Cost	Cost Per S.F.	% Of Sub-Total
A. SUBSTRUCTURE							
1010	Standard Foundations	Poured concrete; strip and spread footings; 4' foundation wall		S.F. Ground	5.88	2.94	
1020	Special Foundations	N/A		—	—	—	
1030	Slab on Grade	4" reinforced concrete		S.F. Slab	5.80	2.90	4.4%
2010	Basement Excavation	Site preparation for slab and trench for foundation wall and footing		S.F. Ground	.18	.14	
2020	Basement Walls	N/A		—	—	—	
B. SHELL							
B10 Superstructure							
1010	Floor Construction	Open web steel joists, slab form, concrete, fireproofed steel columns		S.F. Floor	16.46	8.23	10.0%
1020	Roof Construction	Metal deck, open web steel joists, beams, columns		S.F. Roof	10.92	5.46	
B20 Exterior Enclosure							
2010	Exterior Walls	Textured metal panel on metal stud, insulated	75% of wall	S.F. Wall	18	6.17	
2020	Exterior Windows	Curtain wall system, thermo-break frame	25% of wall	Each	77	8.73	11.4%
2030	Exterior Doors	Aluminum & glass double doors, hollow metal doors, overhead doors		Each	3197	.72	
B30 Roofing							
3010	Roof Coverings	Single ply membrane, stone ballast, rigid insulation		S.F. Roof	7.68	3.84	2.9%
3020	Roof Openings	Roof and smoke hatches		S.F. Roof	.40	.20	
C. INTERIORS							
1010	Partitions	Gypsum board on metal studs, sound attentuation insulation	25 S.F. Floor/L.F. Partition	S.F. Partition	7.02	3.37	
1020	Interior Doors	Hollow metal doors	700 S.F. Floor/Door	Each	1222	1.75	
1030	Fittings	Chalkboards, lockers, toilet partitions		S.F. Floor	9.89	9.89	
2010	Stair Construction	Cement filled metal pan, picket rail		Flight	12,600	.97	23.2%
3010	Wall Finishes	90% Paint, 10% ceramic tile		S.F. Surface	1.94	1.86	
3020	Floor Finishes	10% Carpet, 10% terrazzo, 60% VCT, 20% wood		S.F. Floor	7.70	7.70	
3030	Ceiling Finishes	Acoustic ceiling tiles on suspended channel grid		S.F. Ceiling	6.30	6.30	
D. SERVICES							
D10 Conveying							
1010	Elevators & Lifts	Hydraulic passenger elevator		Each	91,000	.70	0.5%
1020	Escalators & Moving Walks	N/A		—	—	—	
D20 Plumbing							
2010	Plumbing Fixtures	Restroom, kitchen, and service fixtures, supply and drainage	1 Fixture/340 S.F. Floor	Each	2152	6.33	
2020	Domestic Water Distribution	Gas fired water heater		S.F. Floor	1.23	1.23	5.9%
2040	Rain Water Drainage	Roof drains		S.F. Roof	1.04	.52	
D30 HVAC							
3010	Energy Supply	Forced hot water heating system, fin tube radiation		S.F. Floor	5.69	5.69	
3020	Heat Generating Systems	N/A		—	—	—	
3030	Cooling Generating Systems	Packaged chiller		S.F. Floor	18.17	18.17	17.4%
3050	Terminal & Package Units	N/A		—	—	—	
3090	Other HVAC Sys. & Equipment	N/A		—	—	—	
D40 Fire Protection							
4010	Sprinklers	Wet pipe spinkler system		S.F. Floor	2.57	2.57	2.2%
4020	Standpipes	Standpipe		S.F. Floor	.39	.39	
D50 Electrical							
5010	Electrical Service/Distribution	2000 Ampere service, panelboards and feeders		S.F. Floor	1.07	1.07	
5020	Lighting & Branch Wiring	Fluorescent fixtures, receptacles, switches, a.c. and misc. power		S.F. Floor	10.50	10.50	11.9%
5030	Communications & Security	Addressable alarm system, emergency lighting, internet & phone wiring		S.F. Floor	4.32	4.32	
5090	Other Electrical Systems	Emergency generator, 250kW		S.F. Floor	.47	.47	
E. EQUIPMENT & FURNISHINGS							
1010	Commercial Equipment	N/A		—	—	—	
1020	Institutional Equipment	Laboratory equipment		S.F. Floor	9.71	9.71	
1030	Vehicular Equipment	N/A		—	—	—	10.1 %
1090	Other Equipment	Gym equipment, bleachers, scoreboards		S.F. Floor	4.15	4.15	
F. SPECIAL CONSTRUCTION							
1020	Integrated Construction	N/A		—	—	—	0.0 %
1040	Special Facilities	N/A		—	—	—	
G. BUILDING SITEWORK	**N/A**						

		Sub-Total	136.99	**100%**
	CONTRACTOR FEES (General Requirements: 10%, Overhead: 5%, Profit: 10%)	25%	34.27	
	ARCHITECT FEES	7%	11.99	
	Total Building Cost		**183.25**	

For customer support on your Square Foot Costs with RSMeans data, call 800.448.8182.

203

Costs per square foot of floor area

Exterior Wall	S.F. Area	50000	65000	80000	95000	110000	125000	140000	155000	170000
	L.F. Perimeter	850	1060	1280	1490	1700	1920	2140	2340	2560
Fiber Cement	Rigid Steel	168.65	166.75	165.70	164.85	164.25	163.85	163.60	163.15	163.00
Metal Panel	Rigid Steel	169.30	167.40	166.30	165.45	164.80	164.45	164.15	163.70	163.55
E.I.F.S.	Rigid Steel	172.60	170.65	169.55	168.70	168.05	167.70	167.45	166.95	166.80
Tilt-up Concrete Panels	Reinforced Concrete	202.40	199.65	198.35	197.10	196.20	195.70	195.35	194.70	194.55
Brick Veneer	Reinforced Concrete	197.40	194.90	193.60	192.55	191.70	191.25	191.00	190.35	190.15
Stone Veneer	Reinforced Concrete	223.05	219.50	217.75	216.15	215.00	214.40	213.95	213.05	212.80
Perimeter Adj., Add or Deduct	Per 100 L.F.	3.30	2.55	2.10	1.70	1.50	1.35	1.15	1.05	1.00
Story Hgt. Adj., Add or Deduct	Per 1 Ft.	1.30	1.25	1.25	1.20	1.20	1.20	1.15	1.20	1.20
For Basement, add $38.60 per square foot of basement area										

The above costs were calculated using the basic specifications shown on the facing page. These costs should be adjusted where necessary for design alternatives and owner's requirements.

Common additives

Description	Unit	$ Cost
Bleachers, Telescoping, manual		
To 15 tier	Seat	164 - 222
16-20 tier	Seat	340 - 405
21-30 tier	Seat	340 - 460
For power operation, add	Seat	63 - 101
Carrels Hardwood	Ea.	740 - 1925
Clock System		
20 room	Ea.	20,200
50 room	Ea.	48,200
Elevators, Hydraulic passenger, 2 stops		
1500# capacity	Ea.	73,100
2500# capacity	Ea.	76,600
Emergency Lighting, 25 watt, battery operated		
Lead battery	Ea.	330
Nickel cadmium	Ea.	550
Flagpoles, Complete		
Aluminum, 20' high	Ea.	1900
40' high	Ea.	4500
Fiberglass, 23' high	Ea.	1450
39'-5" high	Ea.	3325

Description	Unit	$ Cost
Kitchen Equipment		
Broiler	Ea.	4425
Cooler, 6 ft. long, reach-in	Ea.	3750
Dishwasher, 10-12 racks per hr.	Ea.	4150
Food warmer, counter, 1.2 KW	Ea.	790
Freezer, 44 C.F., reach-in	Ea.	5525
Lockers, Steel, single tier, 60" to 72"	Opng.	232 - 395
2 tier, 60" to 72" total	Opng.	139 - 172
5 tier, box lockers	Opng.	73 - 87
Locker bench, lam. maple top only	L.F.	36.50
Pedestals, steel pipe	Ea.	79
Seating		
Auditorium chair, all veneer	Ea.	355
Veneer back, padded seat	Ea.	380
Upholstered, spring seat	Ea.	335
Classroom, movable chair & desk	Set	81 - 171
Lecture hall, pedestal type	Ea.	350 - 650
Sound System		
Amplifier, 250 watts	Ea.	2025
Speaker, ceiling or wall	Ea.	240
Trumpet	Ea.	460

Important: See the Reference Section for Location Factors.

Model costs calculated for a 2 story building with 15' story height and 110,000 square feet of floor area

				Unit	Unit Cost	Cost Per S.F.	% Of Sub-Total
A. SUBSTRUCTURE							
1010	Standard Foundations	Poured concrete; strip and spread footings; 4' foundation wall		S.F. Ground	5.34	2.67	
1020	Special Foundations	N/A		—	—	—	
1030	Slab on Grade	4" reinforced concrete		S.F. Slab	5.80	2.90	4.7%
2010	Basement Excavation	Site preparation for slab and trench for foundation wall and footing		S.F. Ground	.18	.16	
2020	Basement Walls	N/A		—	—	—	
B. SHELL							
B10 Superstructure							
1010	Floor Construction	Open web steel joists, slab form, concrete, fireproofed steel columns		S.F. Floor	16.70	8.35	11.2%
1020	Roof Construction	Metal deck, open web steel joists, beams, columns		S.F. Roof	10.92	5.46	
B20 Exterior Enclosure							
2010	Exterior Walls	Fiber cement siding on wood studs, insulated	75% of wall	S.F. Wall	16.85	5.86	
2020	Exterior Windows	Curtain wall system, thermo-break frame	25% of wall	Each	58	6.74	11.0%
2030	Exterior Doors	Aluminum & glass double doors, hollow metal doors, overhead doors		Each	3396	.89	
B30 Roofing							
3010	Roof Coverings	Single ply membrane, stone ballast, rigid insulation		S.F. Roof	7.86	3.93	3.4%
3020	Roof Openings	Roof and smoke hatches		S.F. Roof	.38	.19	
C. INTERIORS							
1010	Partitions	Gypsum board on metal studs, sound attentuation insulation	20 S.F. Floor/L.F. Partition	S.F. Partition	6.75	4.05	
1020	Interior Doors	Hollow metal doors	750 S.F. Floor/Door	Each	1222	1.63	
1030	Fittings	Chalkboards, lockers, toilet partitions		S.F. Floor	7.61	7.61	
2010	Stair Construction	Cement filled metal pan, picket rail		Flight	12,600	.92	24.0%
3010	Wall Finishes	90% Paint, 10% ceramic tile		S.F. Surface	1.87	2.24	
3020	Floor Finishes	10% Carpet, 10% terrazzo, 70% VCT, 10% wood		S.F. Floor	6.77	6.77	
3030	Ceiling Finishes	Acoustic ceiling tiles on suspended channel grid		S.F. Ceiling	6.30	6.30	
D. SERVICES							
D10 Conveying							
1010	Elevators & Lifts	Hydraulic passenger elevator		Each	91,300	.83	0.7%
1020	Escalators & Moving Walks	N/A		—	—	—	
D20 Plumbing							
2010	Plumbing Fixtures	Restroom, kitchen, and service fixtures, supply and drainage	1 Fixture/350 S.F. Floor	Each	2258	6.45	
2020	Domestic Water Distribution	Gas fired water heater		S.F. Floor	.72	.72	6.2%
2040	Rain Water Drainage	Roof drains		S.F. Roof	.92	.46	
D30 HVAC							
3010	Energy Supply	N/A		—	—	—	
3020	Heat Generating Systems	Included in D3050		—	—	—	
3030	Cooling Generating Systems	N/A		—	—	—	17.5 %
3050	Terminal & Package Units	Single zone unit gas heating, electric cooling		S.F. Floor	21.45	21.45	
3090	Other HVAC Sys. & Equipment	N/A		—	—	—	
D40 Fire Protection							
4010	Sprinklers	Wet pipe spinkler system		S.F. Floor	2.57	2.57	2.5%
4020	Standpipes	Standpipe		S.F. Floor	.45	.45	
D50 Electrical							
5010	Electrical Service/Distribution	1600 Ampere service, panelboards and feeders		S.F. Floor	1.03	1.03	
5020	Lighting & Branch Wiring	Fluorescent fixtures, receptacles, switches, a.c. and misc. power		S.F. Floor	10.50	10.50	13.3%
5030	Communications & Security	Addressable alarm system, emergency lighting, internet & phone wiring		S.F. Floor	4.51	4.51	
5090	Other Electrical Systems	Emergency generator, 100kW		S.F. Floor	.34	.34	
E. EQUIPMENT & FURNISHINGS							
1010	Commercial Equipment	N/A		—	—	—	
1020	Institutional Equipment	Laboratory equipment		S.F. Floor	1.89	1.89	5.6 %
1030	Vehicular Equipment	N/A		—	—	—	
1090	Other Equipment	Gym equipment, bleachers, scoreboards		S.F. Floor	4.93	4.93	
F. SPECIAL CONSTRUCTION							
1020	Integrated Construction	N/A		—	—	—	0.0 %
1040	Special Facilities	N/A		—	—	—	
G. BUILDING SITEWORK	**N/A**						

	Sub-Total	122.80	**100%**
CONTRACTOR FEES (General Requirements: 10%, Overhead: 5%, Profit: 10%)		25%	30.70
ARCHITECT FEES		7%	10.75
	Total Building Cost	164.25	

For customer support on your Square Foot Costs with RSMeans data, call 800.448.8182.

Costs per square foot of floor area

Exterior Wall	S.F. Area	20000	30000	40000	50000	60000	70000	80000	90000	100000
	L.F. Perimeter	440	590	740	900	1050	1200	1360	1510	1660
Face Brick and Concrete Block	Rigid Steel	197.80	190.90	187.50	185.75	184.30	183.30	182.80	182.15	181.65
	Bearing Walls	193.35	186.55	183.10	181.35	179.90	178.90	178.35	177.75	177.20
Decorative Concrete Block	Rigid Steel	183.70	178.30	175.60	174.20	173.05	172.25	171.90	171.35	171.00
	Bearing Walls	179.25	173.85	171.20	169.75	168.70	167.85	167.45	166.95	166.55
Metal Panel and Metal Studs	Rigid Steel	178.15	173.30	170.90	169.65	168.70	168.00	167.60	167.15	166.80
Metal Sandwich Panel	Rigid Steel	177.70	172.95	170.60	169.30	168.35	167.60	167.30	166.85	166.45
Perimeter Adj., Add or Deduct	Per 100 L.F.	10.05	6.70	4.95	4.05	3.40	2.90	2.50	2.20	2.05
Story Hgt. Adj., Add or Deduct	Per 1 Ft.	2.45	2.20	2.05	2.00	2.00	1.95	1.85	1.85	1.85

For Basement, add $37.40 per square foot of basement area

The above costs were calculated using the basic specifications shown on the facing page. These costs should be adjusted where necessary for design alternatives and owner's requirements.

Common additives

Description	Unit	$ Cost	Description	Unit	$ Cost
Carrels Hardwood	Ea.	740 - 1925	Flagpoles, Complete		
Clock System			Aluminum, 20' high	Ea.	1900
20 room	Ea.	20,200	40' high	Ea.	4500
50 room	Ea.	48,200	Fiberglass, 23' high	Ea.	1450
Directory Boards, Plastic, glass covered			39'-5" high	Ea.	3325
30" x 20"	Ea.	640	Seating		
36" x 48"	Ea.	1625	Auditorium chair, all veneer	Ea.	355
Aluminum, 24" x 18"	Ea.	630	Veneer back, padded seat	Ea.	380
36" x 24"	Ea.	790	Upholstered, spring seat	Ea.	335
48" x 32"	Ea.	1025	Classroom, movable chair & desk	Set	81 - 171
48" x 60"	Ea.	2200	Lecture hall, pedestal type	Ea.	350 - 650
Elevators, Hydraulic passenger, 2 stops			Shops & Workroom:		
1500# capacity	Ea.	73,100	Benches, metal	Ea.	740
2500# capacity	Ea.	76,600	Parts bins 6'-3" high, 3' wide, 12" deep, 72 bins	Ea.	710
3500# capacity	Ea.	81,600	Shelving, metal 1' x 3'	S.F.	13
Emergency Lighting, 25 watt, battery operated			Wide span 6' wide x 24" deep	S.F.	9.80
Lead battery	Ea.	330	Sound System		
Nickel cadmium	Ea.	550	Amplifier, 250 watts	Ea.	2025
			Speaker, ceiling or wall	Ea.	240
			Trumpet	Ea.	460

Important: See the Reference Section for Location Factors.

Model costs calculated for a 2 story building with 16' story height and 40,000 square feet of floor area

School, Vocational

				Unit	Unit Cost	Cost Per S.F.	% Of Sub-Total
A. SUBSTRUCTURE							
1010	Standard Foundations	Poured concrete; strip and spread footings; 4' foundation wall		S.F. Ground	6.56	3.28	
1020	Special Foundations	N/A		—	—	—	
1030	Slab on Grade	5" reinforced concrete with vapor barrier and granular base		S.F. Slab	13.95	6.98	7.6%
2010	Basement Excavation	Site preparation for slab and trench for foundation wall and footing		S.F. Ground	.18	.09	
2020	Basement Walls	N/A		—	—	—	
B. SHELL							
	B10 Superstructure						
1010	Floor Construction	Open web steel joists, slab form, concrete, beams, columns		S.F. Floor	22.66	11.33	10.9%
1020	Roof Construction	Metal deck, open web steel joists, beams, columns		S.F. Roof	7.08	3.54	
	B20 Exterior Enclosure						
2010	Exterior Walls	Face brick with concrete block backup	85% of wall	S.F. Wall	35.25	17.74	
2020	Exterior Windows	Tubular aluminum framing with insulated glass	15% of wall	Each	58	5.16	17.3%
2030	Exterior Doors	Metal and glass doors		Each	3701	.74	
	B30 Roofing						
3010	Roof Coverings	Single-ply membrane with polyisocyanurate insulation		S.F. Roof	6.86	3.43	2.7%
3020	Roof Openings	Roof hatches		S.F. Roof	.40	.20	
C. INTERIORS							
1010	Partitions	Concrete block	20 S.F. Floor/L.F. Partition	S.F. Partition	10.40	6.24	
1020	Interior Doors	Single leaf kalamein fire doors	600 S.F. Floor/Door	Each	1222	2.03	
1030	Fittings	Toilet partitions, chalkboards		S.F. Floor	.83	.83	
2010	Stair Construction	Concrete filled metal pan		Flight	15,800	2.37	20.6%
3010	Wall Finishes	50% paint, 40% glazed coating, 10% ceramic tile		S.F. Surface	4.98	5.98	
3020	Floor Finishes	70% vinyl composition tile, 20% carpet, 10% terrazzo		S.F. Floor	6.17	6.17	
3030	Ceiling Finishes	Mineral fiber tile on concealed zee bars	60% of area	S.F. Ceiling	7.60	4.56	
D. SERVICES							
	D10 Conveying						
1010	Elevators & Lifts	One hydraulic passenger elevator		Each	91,200	2.28	1.7%
1020	Escalators & Moving Walks	N/A		—	—	—	
	D20 Plumbing						
2010	Plumbing Fixtures	Toilet and service fixtures, supply and drainage	1 Fixture/700 S.F. Floor	Each	3465	4.95	
2020	Domestic Water Distribution	Gas fired water heater		S.F. Floor	1.18	1.18	5.4%
2040	Rain Water Drainage	Roof drains		S.F. Roof	2.62	1.31	
	D30 HVAC						
3010	Energy Supply	Oil fired hot water, wall fin radiation		S.F. Floor	10.24	10.24	
3020	Heat Generating Systems	N/A		—	—	—	
3030	Cooling Generating Systems	Chilled water, cooling tower systems		S.F. Floor	15.45	15.45	18.8%
3050	Terminal & Package Units	N/A		—	—	—	
3090	Other HVAC Sys. & Equipment	N/A		—	—	—	
	D40 Fire Protection						
4010	Sprinklers	Sprinklers, light hazard		S.F. Floor	3.30	3.30	3.3%
4020	Standpipes	Standpipe, wet, Class III		S.F. Floor	1.24	1.24	
	D50 Electrical						
5010	Electrical Service/Distribution	800 ampere service, panel board and feeders		S.F. Floor	1.24	1.24	
5020	Lighting & Branch Wiring	High efficiency fluorescent fixtures, receptacles, switches, A.C. and misc. power		S.F. Floor	11.34	11.34	11.7%
5030	Communications & Security	Addressable alarm systems, internet wiring, communications systems and emergency lighting		S.F. Floor	3.35	3.35	
5090	Other Electrical Systems	Emergency generator, 11.5 kW		S.F. Floor	.12	.12	
E. EQUIPMENT & FURNISHINGS							
1010	Commercial Equipment	N/A		—	—	—	
1020	Institutional Equipment	Stainless steel countertops		S.F. Floor	.20	.20	0.1%
1030	Vehicular Equipment	N/A		—	—	—	
1090	Other Equipment	N/A		—	—	—	
F. SPECIAL CONSTRUCTION							
1020	Integrated Construction	N/A		—	—	—	0.0%
1040	Special Facilities	N/A		—	—	—	
G. BUILDING SITEWORK	**N/A**						

		Sub-Total	136.87	100%
CONTRACTOR FEES (General Requirements: 10%, Overhead: 5%, Profit: 10%)		25%	34.25	
ARCHITECT FEES		7%	11.98	
		Total Building Cost	**183.10**	

For customer support on your Square Foot Costs with RSMeans data, call 800.448.8182.

207

Costs per square foot of floor area

Exterior Wall	S.F. Area	1000	2000	3000	4000	6000	8000	10000	12000	15000
	L.F. Perimeter	126	179	219	253	310	358	400	438	490
Wood Clapboard	Wood Frame	170.45	142.65	131.35	125.00	117.80	113.60	110.90	108.85	106.75
Face Brick	Wood Frame	190.60	155.40	140.80	132.45	122.85	117.30	113.60	110.90	108.00
Stucco and Concrete Block	Rigid Steel	193.00	161.00	147.80	140.30	131.80	126.75	123.50	121.10	118.50
	Bearing Walls	187.65	155.60	142.40	134.95	126.40	121.40	118.10	115.70	113.10
Metal Sandwich Panel	Rigid Steel	205.50	169.85	155.00	146.60	136.85	131.25	127.45	124.70	121.70
Precast Concrete	Rigid Steel	264.65	211.85	189.30	176.25	161.10	152.20	146.20	141.85	137.05
Perimeter Adj., Add or Deduct	Per 100 L.F.	47.20	23.60	15.75	11.85	7.85	5.95	4.75	3.95	3.05
Story Hgt. Adj., Add or Deduct	Per 1 Ft.	3.50	2.50	2.05	1.80	1.40	1.30	1.10	1.00	0.90
For Basement, add $ 28.20 per square foot of basement area										

The above costs were calculated using the basic specifications shown on the facing page. These costs should be adjusted where necessary for design alternatives and owner's requirements.

Common additives

Description	Unit	$ Cost
Check Out Counter		
Single belt	Ea.	3850
Double belt	Ea.	5475
Emergency Lighting, 25 watt, battery operated		
Lead battery	Ea.	330
Nickel cadmium	Ea.	550
Refrigerators, Prefabricated, walk-in		
7'-6" high, 6' x 6'	S.F.	145
10' x 10'	S.F.	114
12' x 14'	S.F.	99.50
12' x 20'	S.F.	137
Refrigerated Food Cases		
Dairy, multi deck, 12' long	Ea.	12,300
Delicatessen case, single deck, 12' long	Ea.	9125
Multi deck, 18 S.F. shelf display	Ea.	9200
Freezer, self-contained chest type, 30 C.F.	Ea.	5150
Glass door upright, 78 C.F.	Ea.	10,600

Description	Unit	$ Cost
Refrigerated Food Cases, cont.		
Frozen food, chest type, 12' long	Ea.	9025
Glass door reach-in, 5 door	Ea.	11,300
Island case 12' long, single deck	Ea.	8825
Multi deck	Ea.	10,300
Meat cases, 12' long, single deck	Ea.	8825
Multi deck	Ea.	11,300
Produce, 12' long single deck	Ea.	7625
Multi deck	Ea.	9575
Safe, Office type, 1 hour rating		
30" x 18" x 18"	Ea.	2475
60" x 36" x 18", double door	Ea.	10,200
Smoke Detectors		
Ceiling type	Ea.	244
Duct type	Ea.	585
Sound System		
Amplifier, 250 watts	Ea.	2025
Speaker, ceiling or wall	Ea.	240
Trumpet	Ea.	460

Important: See the Reference Section for Location Factors.

Model costs calculated for a 1 story building with 12' story height and 4,000 square feet of floor area

Store, Convenience

			Unit	Unit Cost	Cost Per S.F.	% Of Sub-Total
A. SUBSTRUCTURE						
1010	Standard Foundations	Poured concrete; strip and spread footings; 4' foundation wall	S.F. Ground	7.07	7.07	
1020	Special Foundations	N/A	—	—	—	
1030	Slab on Grade	4" reinforced concrete with vapor barrier and granular base	S.F. Slab	5.80	5.80	14.4%
2010	Basement Excavation	Site preparation for slab and trench for foundation wall and footing	S.F. Ground	.57	.57	
2020	Basement Walls	N/A	—	—	—	
B. SHELL						
B10 Superstructure						
1010	Floor Construction	N/A	—	—	—	
1020	Roof Construction	Wood truss with plywood sheathing	S.F. Roof	7.76	7.76	8.6 %
B20 Exterior Enclosure						
2010	Exterior Walls	Wood siding on wood studs, insulated — 80% of wall	S.F. Wall	10.51	6.38	
2020	Exterior Windows	Storefront — 20% of wall	Each	51	7.81	18.6%
2030	Exterior Doors	Double aluminum and glass, solid core wood	Each	4290	3.22	
B30 Roofing						
3010	Roof Coverings	Asphalt shingles; rigid fiberglass insulation	S.F. Roof	3.96	3.96	4.2%
3020	Roof Openings	N/A	—	—	—	
C. INTERIORS						
1010	Partitions	Gypsum board on wood studs — 60 S.F. Floor/L.F. Partition	S.F. Partition	14.46	2.41	
1020	Interior Doors	Single leaf wood, hollow metal — 1300 S.F. Floor/Door	Each	1682	1.29	
1030	Fittings	N/A	—	—	—	
2010	Stair Construction	N/A	—	—	—	12.3%
3010	Wall Finishes	Paint	S.F. Surface	1.26	.42	
3020	Floor Finishes	Vinyl composition tile	S.F. Floor	3.06	3.06	
3030	Ceiling Finishes	Mineral fiber tile on wood furring	S.F. Ceiling	4.35	4.35	
D. SERVICES						
D10 Conveying						
1010	Elevators & Lifts	N/A	—	—	—	0.0 %
1020	Escalators & Moving Walks	N/A	—	—	—	
D20 Plumbing						
2010	Plumbing Fixtures	Toilet and service fixtures, supply and drainage — 1 Fixture/1000 S.F. Floor	Each	4210	4.21	
2020	Domestic Water Distribution	Gas fired water heater	S.F. Floor	1.79	1.79	6.4%
2040	Rain Water Drainage	N/A	—	—	—	
D30 HVAC						
3010	Energy Supply	N/A	—	—	—	
3020	Heat Generating Systems	Included in D3050	—	—	—	
3030	Cooling Generating Systems	N/A	—	—	—	8.5 %
3050	Terminal & Package Units	Single zone rooftop unit, gas heating, electric cooling	S.F. Floor	7.95	7.95	
3090	Other HVAC Sys. & Equipment	N/A	—	—	—	
D40 Fire Protection						
4010	Sprinklers	Sprinkler, ordinary hazard	S.F. Floor	5.13	5.13	8.0%
4020	Standpipes	Standpipe	S.F. floor	2.39	2.39	
D50 Electrical						
5010	Electrical Service/Distribution	200 ampere service, panel board and feeders	S.F. Floor	3.11	3.11	
5020	Lighting & Branch Wiring	High efficiency fluorescent fixtures, receptacles, switches, A.C. and misc. power	S.F. Floor	11.20	11.20	18.8%
5030	Communications & Security	Addressable alarm systems and emergency lighting	S.F. Floor	2.89	2.89	
5090	Other Electrical Systems	Emergency generator, 7.5 kW	S.F. Floor	.39	.39	
E. EQUIPMENT & FURNISHINGS						
1010	Commercial Equipment	N/A	—	—	—	
1020	Institutional Equipment	N/A	—	—	—	0.0 %
1030	Vehicular Equipment	N/A	—	—	—	
1090	Other Equipment	N/A	—	—	—	
F. SPECIAL CONSTRUCTION						
1020	Integrated Construction	N/A	—	—	—	0.0 %
1040	Special Facilities	N/A	—	—	—	
G. BUILDING SITEWORK	**N/A**					

		Sub-Total	93.47	100%
CONTRACTOR FEES (General Requirements: 10%, Overhead: 5%, Profit: 10%)		25%	23.35	
ARCHITECT FEES		7%	8.18	
	Total Building Cost		**125**	

For customer support on your Square Foot Costs with RSMeans data, call 800.448.8182.

Costs per square foot of floor area

Exterior Wall	S.F. Area	50000	65000	80000	95000	110000	125000	140000	155000	170000
	L.F. Perimeter	920	1065	1167	1303	1333	1433	1533	1633	1733
Face Brick and Concrete Block	Reinforced Concrete	135.65	133.10	130.90	129.75	127.95	127.15	126.55	125.95	125.55
	Rigid Steel	111.10	108.55	106.30	105.15	103.35	102.55	101.95	101.40	101.00
Decorative Concrete Block	Reinforced Concrete	130.95	128.90	127.15	126.25	124.85	124.20	123.75	123.30	122.95
	Steel Joists	106.40	104.35	102.60	101.65	100.25	99.60	99.15	98.70	98.40
Precast Concrete	Reinforced Concrete	135.00	132.50	130.35	129.30	127.50	126.75	126.10	125.60	125.20
	Steel Joists	110.40	107.90	105.75	104.65	102.90	102.15	101.55	101.00	100.60
Perimeter Adj., Add or Deduct	Per 100 L.F.	2.15	1.60	1.30	1.10	0.95	0.85	0.70	0.70	0.65
Story Hgt. Adj., Add or Deduct	Per 1 Ft.	1.10	0.95	0.85	0.75	0.70	0.60	0.60	0.65	0.55
For Basement, add $ 27.80 per square foot of basement area										

The above costs were calculated using the basic specifications shown on the facing page. These costs should be adjusted where necessary for design alternatives and owner's requirements.

Common additives

Description	Unit	$ Cost
Closed Circuit Surveillance, One station		
Camera and monitor	Ea.	1475
For additional camera stations, add	Ea.	665
Directory Boards, Plastic, glass covered		
30" x 20"	Ea.	640
36" x 48"	Ea.	1625
Aluminum, 24" x 18"	Ea.	630
36" x 24"	Ea.	790
48" x 32"	Ea.	1025
48" x 60"	Ea.	2200
Emergency Lighting, 25 watt, battery operated		
Lead battery	Ea.	330
Nickel cadmium	Ea.	550
Safe, Office type, 1 hour rating		
30" x 18" x 18"	Ea.	2475
60" x 36" x 18", double door	Ea.	10,200
Sound System		
Amplifier, 250 watts	Ea.	2025
Speaker, ceiling or wall	Ea.	240
Trumpet	Ea.	460

Important: See the Reference Section for Location Factors.

Model costs calculated for a 1 story building with 14' story height and 110,000 square feet of floor area

Store, Department, 1 Story

				Unit	Unit Cost	Cost Per S.F.	% Of Sub-Total
A. SUBSTRUCTURE							
1010	Standard Foundations	Poured concrete; strip and spread footings; 4' foundation wall		S.F. Ground	2.10	2.10	
1020	Special Foundations	N/A		—	—	—	
1030	Slab on Grade	4" reinforced concrete with vapor barrier and granular base		S.F. Slab	5.80	5.80	8.5%
2010	Basement Excavation	Site preparation for slab and trench for foundation wall and footing		S.F. Ground	.31	.31	
2020	Basement Walls	N/A		—	—	—	
B. SHELL							
	B10 Superstructure						
1010	Floor Construction	Cast-in-place concrete columns		L.F. Column	78	.99	31.4%
1020	Roof Construction	Precast concrete beam and plank, concrete columns		S.F. Roof	29.37	29.37	
	B20 Exterior Enclosure						
2010	Exterior Walls	Face brick with concrete block backup	90% of wall	S.F. Wall	34.71	5.30	
2020	Exterior Windows	Storefront	10% of wall	Each	72	1.22	7.2%
2030	Exterior Doors	Sliding electric operated entrance, hollow metal		Each	8713	.47	
	B30 Roofing						
3010	Roof Coverings	Built-up tar and gravel with flashing; perlite/EPS composite insulation		S.F. Roof	5.82	5.82	6.1%
3020	Roof Openings	Roof hatches		S.F. Roof	.07	.07	
C. INTERIORS							
1010	Partitions	Gypsum board on metal studs	60 S.F. Floor/L.F. Partition	S.F. Partition	11.52	1.92	
1020	Interior Doors	Single leaf hollow metal	600 S.F. Floor/Door	Each	1222	2.03	
1030	Fittings	N/A		—	—	—	
2010	Stair Construction	N/A		—	—	—	18.8%
3010	Wall Finishes	Paint		S.F. Surface	.87	.29	
3020	Floor Finishes	50% ceramic tile, 50% carpet tile		S.F. Floor	8.87	8.87	
3030	Ceiling Finishes	Mineral fiber board on exposed grid system, suspended		S.F. Ceiling	5.07	5.07	
D. SERVICES							
	D10 Conveying						
1010	Elevators & Lifts	N/A		—	—	—	0.0 %
1020	Escalators & Moving Walks	N/A		—	—	—	
	D20 Plumbing						
2010	Plumbing Fixtures	Toilet and service fixtures, supply and drainage	1 Fixture/4075 S.F. Floor	Each	7580	1.86	
2020	Domestic Water Distribution	Gas fired water heater		S.F. Floor	.43	.43	3.4%
2040	Rain Water Drainage	Roof drains		S.F. Roof	1	1	
	D30 HVAC						
3010	Energy Supply	N/A		—	—	—	
3020	Heat Generating Systems	Included in D3050		—	—	—	
3030	Cooling Generating Systems	N/A		—	—	—	9.4 %
3050	Terminal & Package Units	Single zone unit gas heating, electric cooling		S.F. Floor	9.08	9.08	
3090	Other HVAC Sys. & Equipment	N/A		—	—	—	
	D40 Fire Protection						
4010	Sprinklers	Sprinklers, light hazard		S.F. Floor	2.93	2.93	3.4%
4020	Standpipes	Standpipe		S.F. Floor	.31	.31	
	D50 Electrical						
5010	Electrical Service/Distribution	1200 ampere service, panel board and feeders		S.F. Floor	.75	.75	
5020	Lighting & Branch Wiring	High efficiency fluorescent fixtures, receptacles, switches, A.C. and misc. power		S.F. Floor	9.24	9.24	11.7%
5030	Communications & Security	Addressable alarm systems, internet wiring and emergency lighting		S.F. Floor	1.28	1.28	
5090	Other Electrical Systems	Emergency generator, 7.5 kW		S.F. Floor	.04	.04	
E. EQUIPMENT & FURNISHINGS							
1010	Commercial Equipment	N/A		—	—	—	
1020	Institutional Equipment	N/A		—	—	—	
1030	Vehicular Equipment	N/A		—	—	—	0.0 %
1090	Other Equipment	N/A		—	—	—	
F. SPECIAL CONSTRUCTION							
1020	Integrated Construction	N/A		—	—	—	0.0 %
1040	Special Facilities	N/A		—	—	—	
G. BUILDING SITEWORK	**N/A**						

		Sub-Total	96.55	100%
CONTRACTOR FEES (General Requirements: 10%, Overhead: 5%, Profit: 10%)		25%	24.16	
ARCHITECT FEES		6%	7.24	
	Total Building Cost		**127.95**	

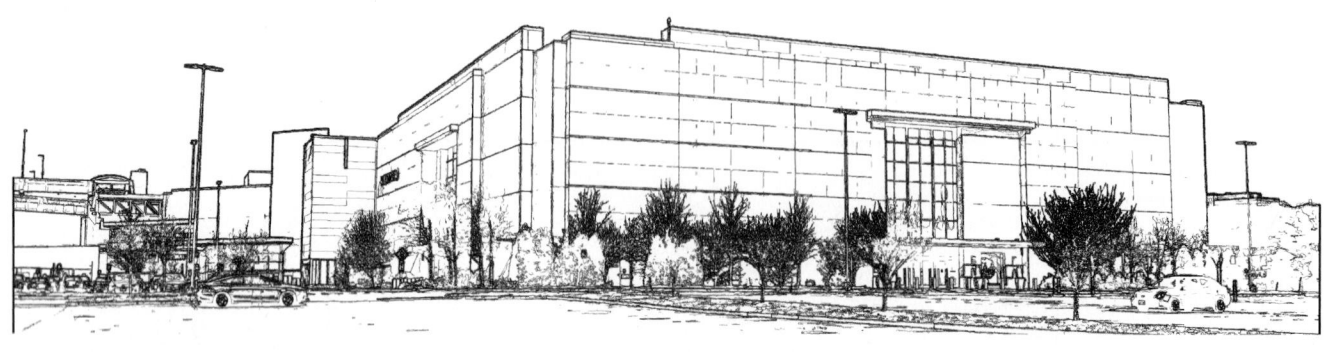

Costs per square foot of floor area

Exterior Wall	S.F. Area	50000	65000	80000	95000	110000	125000	140000	155000	170000
	L.F. Perimeter	533	593	670	715	778	840	871	923	976
Face Brick and Concrete Block	Rigid Steel	161.05	154.70	151.30	148.15	146.20	144.80	142.95	141.95	141.05
	Reinforced Concrete	162.00	155.70	152.25	149.10	147.15	145.70	143.90	142.90	142.00
Face Brick and Metal Studs	Rigid Steel	155.85	150.30	147.25	144.45	142.80	141.45	139.95	139.00	138.25
	Reinforced Concrete	158.85	153.30	150.25	147.45	145.80	144.45	142.90	142.00	141.25
Precast Concrete	Rigid Steel	173.70	165.90	161.75	157.65	155.35	153.50	151.20	149.90	148.80
	Reinforced Concrete	174.20	166.35	162.20	158.20	155.80	154.00	151.70	150.45	149.30
Perimeter Adj., Add or Deduct	Per 100 L.F.	5.45	4.25	3.45	2.85	2.50	2.15	1.95	1.80	1.65
Story Hgt. Adj., Add or Deduct	Per 1 Ft.	1.65	1.45	1.35	1.10	1.15	1.05	0.95	0.95	0.90
For Basement, add $47.10 per square foot of basement area										

The above costs were calculated using the basic specifications shown on the facing page. These costs should be adjusted where necessary for design alternatives and owner's requirements.

Common additives

Description	Unit	$ Cost	Description	Unit	$ Cost
Closed Circuit Surveillance, One station			Escalators, Metal		
Camera and monitor	Ea.	1475	32" wide, 10' story height	Ea.	162,000
For additional camera stations, add	Ea.	665	20' story height	Ea.	199,000
Directory Boards, Plastic, glass covered			48" wide, 10' story height	Ea.	171,000
30" x 20"	Ea.	640	20' story height	Ea.	209,000
36" x 48"	Ea.	1625	Glass		
Aluminum, 24" x 18"	Ea.	630	32" wide, 10' story height	Ea.	153,500
36" x 24"	Ea.	790	20' story height	Ea.	188,500
48" x 32"	Ea.	1025	48" wide, 10' story height	Ea.	161,000
48" x 60"	Ea.	2200	20' story height	Ea.	198,000
Elevators, Hydraulic passenger, 2 stops			Safe, Office type, 1 hour rating		
1500# capacity	Ea.	73,100	30" x 18" x 18"	Ea.	2475
2500# capacity	Ea.	76,600	60" x 36" x 18", double door	Ea.	10,200
3500# capacity	Ea.	81,600	Sound System		
Additional stop, add	Ea.	8375	Amplifier, 250 watts	Ea.	2025
Emergency Lighting, 25 watt, battery operated			Speaker, ceiling or wall	Ea.	240
Lead battery	Ea.	330	Trumpet	Ea.	460
Nickel cadmium	Ea.	550			

Important: See the Reference Section for Location Factors.

Store, Department, 3 Story

Model costs calculated for a 3 story building with 16' story height and 95,000 square feet of floor area

				Unit	Unit Cost	Cost Per S.F.	% Of Sub-Total
A. SUBSTRUCTURE							
1010	Standard Foundations	Poured concrete; strip and spread footings; 4' foundation wall		S.F. Ground	6.27	2.09	
1020	Special Foundations	N/A		—	—	—	
1030	Slab on Grade	4" reinforced concrete with vapor barrier and granular base		S.F. Slab	5.80	1.93	3.6%
2010	Basement Excavation	Site preparation for slab and trench for foundation wall and footing		S.F. Ground	.18	.06	
2020	Basement Walls	N/A		—	—	—	
B. SHELL							
B10 Superstructure							
1010	Floor Construction	Concrete slab with metal deck and beams, steel columns		S.F. Floor	30.90	20.60	21.7%
1020	Roof Construction	Metal deck, open web steel joists, beams, columns		S.F. Roof	11.13	3.71	
B20 Exterior Enclosure							
2010	Exterior Walls	Face brick with concrete block backup	90% of wall	S.F. Wall	34.75	11.30	
2020	Exterior Windows	Storefront	10% of wall	Each	51	1.86	13.5%
2030	Exterior Doors	Revolving and sliding panel, mall-front		Each	13,041	1.93	
B30 Roofing							
3010	Roof Coverings	Built-up tar and gravel with flashing; perlite/EPS composite insulation		S.F. Roof	6.33	2.11	2.0%
3020	Roof Openings	Roof hatches		S.F. Roof	.27	.09	
C. INTERIORS							
1010	Partitions	Gypsum board on metal studs	60 S.F. Floor/L.F. Partition	S.F. Partition	6.78	1.13	
1020	Interior Doors	Single leaf hollow metal	600 S.F. Floor/Door	Each	1222	2.03	
1030	Fittings	N/A		—	—	—	
2010	Stair Construction	Concrete filled metal pan		Flight	22,075	2.32	22.7%
3010	Wall Finishes	70% paint, 20% vinyl wall covering, 10% ceramic tile		S.F. Surface	5.25	1.75	
3020	Floor Finishes	50% carpet tile, 40% marble tile, 10% terrazzo		S.F. Floor	10.59	10.59	
3030	Ceiling Finishes	Mineral fiber tile on concealed zee bars		S.F. Ceiling	7.60	7.60	
D. SERVICES							
D10 Conveying							
1010	Elevators & Lifts	One hydraulic passenger, one hydraulic freight		Each	320,150	3.37	9.3%
1020	Escalators & Moving Walks	Four escalators		Each	167,913	7.07	
D20 Plumbing							
2010	Plumbing Fixtures	Toilet and service fixtures, supply and drainage	1 Fixture/2570 S.F. Floor	Each	5217	2.03	
2020	Domestic Water Distribution	Gas fired water heater		S.F. Floor	.50	.50	2.9%
2040	Rain Water Drainage	Roof drains		S.F. Roof	2.19	.73	
D30 HVAC							
3010	Energy Supply	N/A		—	—	—	
3020	Heat Generating Systems	Included in D3050		—	—	—	
3030	Cooling Generating Systems	N/A		—	—	—	8.1%
3050	Terminal & Package Units	Multizone rooftop unit, gas heating, electric cooling		S.F. Floor	9.08	9.08	
3090	Other HVAC Sys. & Equipment	N/A		—	—	—	
D40 Fire Protection							
4010	Sprinklers	Sprinklers, light hazard		S.F. Floor	2.47	2.47	3.9%
4020	Standpipes	Standpipe		S.F. Floor	1.92	1.92	
D50 Electrical							
5010	Electrical Service/Distribution	1200 ampere service, panel board and feeders		S.F. Floor	1.22	1.22	
5020	Lighting & Branch Wiring	High efficiency fluorescent fixtures, receptacles, switches, A.C. and misc. power		S.F. Floor	9.46	9.46	12.1%
5030	Communications & Security	Addressable alarm systems, internet wiring and emergency lighting		S.F. Floor	2.56	2.56	
5090	Other Electrical Systems	Emergency generator, 50 kW		S.F. Floor	.28	.28	
E. EQUIPMENT & FURNISHINGS							
1010	Commercial Equipment	N/A		—	—	—	
1020	Institutional Equipment	N/A		—	—	—	0.0%
1030	Vehicular Equipment	N/A		—	—	—	
1090	Other Equipment	N/A		—	—	—	
F. SPECIAL CONSTRUCTION							
1020	Integrated Construction	N/A		—	—	—	0.0%
1040	Special Facilities	N/A		—	—	—	
G. BUILDING SITEWORK	N/A						

		Sub-Total	111.79	100%
CONTRACTOR FEES (General Requirements: 10%, Overhead: 5%, Profit: 10%)		25%	27.98	
ARCHITECT FEES		6%	8.38	
	Total Building Cost		**148.15**	

For customer support on your Square Foot Costs with RSMeans data, call 800.448.8182.

Costs per square foot of floor area

Exterior Wall	S.F. Area	4000	6000	8000	10000	12000	15000	18000	20000	22000
	L.F. Perimeter	260	340	360	410	440	490	540	565	594
Vinyl Clapboard	Wood Frame	142.20	131.35	121.50	117.40	113.65	110.20	107.80	106.35	105.30
Fiber Cement	Wood Frame	144.60	133.70	123.70	119.55	115.75	112.25	109.85	108.40	107.35
E.I.F.S. and Metal Studs	Steel Joists	154.15	142.80	132.30	127.95	123.95	120.20	117.70	116.20	115.00
Stone Veneer	Rigid Steel	198.85	182.70	165.50	158.85	152.35	146.30	142.30	139.70	137.90
Brick Veneer	Reinforced Concrete	191.25	176.15	160.30	154.20	148.25	142.70	138.95	136.65	135.00
Stucco	Reinforced Concrete	208.60	191.35	172.50	165.35	158.30	151.65	147.30	144.50	142.50
Perimeter Adj., Add or Deduct	Per 100 L.F.	14.55	9.70	7.25	5.85	4.90	3.85	3.20	2.95	2.65
Story Hgt. Adj., Add or Deduct	Per 1 Ft.	1.60	1.40	1.15	1.05	1.00	0.80	0.75	0.70	0.70

For Basement, add $39.10 per square foot of basement area

The above costs were calculated using the basic specifications shown on the facing page. These costs should be adjusted where necessary for design alternatives and owner's requirements.

Common additives

Description	Unit	$ Cost
Emergency Lighting, 25 watt, battery operated		
Lead battery	Ea.	330
Nickel cadmium	Ea.	550
Safe, Office type, 1 hour rating		
30" x 18" x 18"	Ea.	2475
60" x 36" x 18", double door	Ea.	10,200
Smoke Detectors		
Ceiling type	Ea.	244
Duct type	Ea.	585
Sound System		
Amplifier, 250 watts	Ea.	2025
Speaker, ceiling or wall	Ea.	240
Trumpet	Ea.	460

Important: See the Reference Section for Location Factors.

Model costs calculated for a 1 story building with 14' story height and 8,000 square feet of floor area

				Unit	Unit Cost	Cost Per S.F.	% Of Sub-Total
A.	**SUBSTRUCTURE**						
1010	Standard Foundations	Poured concrete; strip and spread footings; 4' foundation wall		S.F. Ground	6.29	6.29	
1020	Special Foundations	N/A		—	—	—	
1030	Slab on Grade	4" reinforced concrete		S.F. Slab	5.80	5.80	13.8%
2010	Basement Excavation	Site preparation for slab and trench for foundation wall and footing		S.F. Ground	.33	.33	
2020	Basement Walls	N/A		—	—	—	
B.	**SHELL**						
	B10 Superstructure						
1010	Floor Construction	Wood columns,		S.F. Floor	.46	.46	9.5%
1020	Roof Construction	Wood roof truss 4:12 pitch		S.F. Roof	8.10	8.10	
	B20 Exterior Enclosure						
2010	Exterior Walls	Vinyl siding on wood studs, insulated	85% of wall	S.F. Wall	15	8.03	
2020	Exterior Windows	Storefront glazing	15% of wall	Each	51	2.28	13.3%
2030	Exterior Doors	Aluminum and glass single doors, hollow metal doors		Each	3368	1.69	
	B30 Roofing						
3010	Roof Coverings	Asphalt roofing, strip shingles, gutters and downspouts		S.F. Roof	3.95	3.95	4.6%
3020	Roof Openings	Roof hatches		S.F. Roof	.15	.15	
C.	**INTERIORS**						
1010	Partitions	Sound deadening gypsum board on wood studs	60 S.F. Floor/L.F. Partition	S.F. Partition	10.50	1.75	
1020	Interior Doors	Hollow metal doors	1150 S.F. Floor/Door	Each	1222	1.07	
1030	Fittings	Toilet partitions		S.F. Floor	.20	.20	
2010	Stair Construction	N/A		—	—	—	16.3%
3010	Wall Finishes	90% Paint, 10% ceramic tile		S.F. Surface	3.09	1.03	
3020	Floor Finishes	Vinyl composition tile		S.F. Floor	3.06	3.06	
3030	Ceiling Finishes	Acoustic ceiling tiles on suspended channel grid		S.F. Ceiling	7.60	7.60	
D.	**SERVICES**						
	D10 Conveying						
1010	Elevators & Lifts	N/A		—	—	—	0.0 %
1020	Escalators & Moving Walks	N/A		—	—	—	
	D20 Plumbing						
2010	Plumbing Fixtures	Restroom and service fixtures, supply and drainage	1 Fixture/1350 S.F. Floor	Each	2946	2.21	
2020	Domestic Water Distribution	Gas fired water heater		S.F. Floor	3.49	3.49	8.2%
2040	Rain Water Drainage	Roof Drains		S.F. Roof	1.66	1.66	
	D30 HVAC						
3010	Energy Supply	N/A		—	—	—	
3020	Heat Generating Systems	Included in D3050		—	—	—	
3030	Cooling Generating Systems	N/A		—	—	—	10.1 %
3050	Terminal & Package Units	Single zone unit gas heating, electric cooling		S.F. Floor	9.08	9.08	
3090	Other HVAC Sys. & Equipment	N/A		—	—	—	
	D40 Fire Protection						
4010	Sprinklers	Wet pipe spinkler system		S.F. Floor	4.96	4.96	6.8%
4020	Standpipes	Standpipe		S.F. Floor	1.19	1.19	
	D50 Electrical						
5010	Electrical Service/Distribution	400 Ampere service, panelboards and feeders		S.F. Floor	2.81	2.81	
5020	Lighting & Branch Wiring	Flourescent fixtures, receptacles, a.c. and misc. power		S.F. Floor	11.09	11.09	17.4%
5030	Communications & Security	Addressable alarm system, emergency lighting		S.F. Floor	1.73	1.73	
5090	Other Electrical Systems	N/A		—	—	—	
E.	**EQUIPMENT & FURNISHINGS**						
1010	Commercial Equipment	N/A		—	—	—	
1020	Institutional Equipment	N/A		—	—	—	0.0 %
1030	Vehicular Equipment	N/A		—	—	—	
1090	Other Equipment	N/A		—	—	—	
F.	**SPECIAL CONSTRUCTION**						
1020	Integrated Construction	N/A		—	—	—	0.0 %
1040	Special Facilities	N/A		—	—	—	
G.	**BUILDING SITEWORK**	**N/A**					

		Sub-Total	90.01	**100%**
CONTRACTOR FEES (General Requirements: 10%, Overhead: 5%, Profit: 10%)		25%	22.49	
ARCHITECT FEES		8%	9	
	Total Building Cost		**121.50**	

Costs per square foot of floor area

Exterior Wall	S.F. Area	12000	16000	20000	26000	32000	38000	44000	52000	60000
	L.F. Perimeter	450	510	570	650	716	780	840	920	1020
Stucco	Reinforced Concrete	190.50	174.85	165.40	155.90	148.95	144.10	140.40	136.80	134.85
Curtain Wall	Rigid Steel	156.70	144.80	137.75	130.60	125.45	121.85	119.10	116.50	115.00
E.I.F.S.	Rigid Steel	134.95	126.80	121.95	117.10	113.65	111.30	109.50	107.80	106.80
Metal Panel	Rigid Steel	131.50	123.45	118.60	113.85	110.45	108.10	106.30	104.60	103.60
Tilt-up Concrete Panels	Reinforced Concrete	147.85	138.45	132.80	127.20	123.25	120.50	118.40	116.30	115.15
Brick Veneer	Reinforced Concrete	201.60	184.25	173.80	163.30	155.55	150.15	146.05	141.95	139.85
Perimeter Adj., Add or Deduct	Per 100 L.F.	18.25	13.65	11.05	8.45	6.90	5.80	4.95	4.20	3.65
Story Hgt. Adj., Add or Deduct	Per 1 Ft.	1.70	1.35	1.30	1.10	0.95	0.85	0.85	0.75	0.70

For Basement, add $28.40 per square foot of basement area

The above costs were calculated using the basic specifications shown on the facing page. These costs should be adjusted where necessary for design alternatives and owner's requirements.

Common additives

Description	Unit	$ Cost
Check Out Counter		
Single belt	Ea.	3850
Double belt	Ea.	5475
Scanner, registers, guns & memory 2 lanes	Ea.	19,600
10 lanes	Ea.	186,000
Power take away	Ea.	6475
Emergency Lighting, 25 watt, battery operated		
Lead battery	Ea.	330
Nickel cadmium	Ea.	550
Refrigerators, Prefabricated, walk-in		
7'-6" High, 6' x 6'	S.F.	145
10' x 10'	S.F.	114
12' x 14'	S.F.	99.50
12' x 20'	S.F.	137
Refrigerated Food Cases		
Dairy, multi deck, 12' long	Ea.	12,300
Delicatessen case, single deck, 12' long	Ea.	9125
Multi deck, 18 S.F. shelf display	Ea.	9200
Freezer, self-contained chest type, 30 C.F.	Ea.	5150
Glass door upright, 78 C.F.	Ea.	10,600

Description	Unit	$ Cost
Refrigerated Food Cases, cont.		
Frozen food, chest type, 12' long	Ea.	9025
Glass door reach in, 5 door	Ea.	11,300
Island case 12' long, single deck	Ea.	8825
Multi deck	Ea.	10,300
Meat case 12' long, single deck	Ea.	8825
Multi deck	Ea.	11,300
Produce, 12' long single deck	Ea.	7625
Multi deck	Ea.	9575
Safe, Office type, 1 hour rating		
30" x 18" x 18"	Ea.	2475
60" x 36" x 18", double door	Ea.	10,200
Smoke Detectors		
Ceiling type	Ea.	244
Duct type	Ea.	585
Sound System		
Amplifier, 250 watts	Ea.	2025
Speaker, ceiling or wall	Ea.	240
Trumpet	Ea.	460

Important: See the Reference Section for Location Factors.

Model costs calculated for a 1 story building with 18′ story height and 44,000 square feet of floor area

			Unit	Unit Cost	Cost Per S.F.	% Of Sub-Total
A. SUBSTRUCTURE						
1010	Standard Foundations	Poured concrete; strip and spread footings; 4′ foundation wall	S.F. Ground	2.84	2.84	
1020	Special Foundations	N/A	—	—	—	
1030	Slab on Grade	4″ reinforced slab on grade	S.F. Slab	5.80	5.80	9.9%
2010	Basement Excavation	Site preparation for slab and trench for foundation wall and footing	S.F. Ground	.18	.18	
2020	Basement Walls	N/A	—	—	—	
B. SHELL						
B10 Superstructure						
1010	Floor Construction	N/A	—	—	—	12.2 %
1020	Roof Construction	Metal deck on open web steel joints, columns	S.F. Roof	10.90	10.90	
B20 Exterior Enclosure						
2010	Exterior Walls	N/A	—	—	—	
2020	Exterior Windows	Curtain wall glazing system, thermo-break frame 100% of wall	Each	51	17.50	21.3 %
2030	Exterior Doors	Glass sliding entrance doors, overhead doors, hollow metal doors	Each	6004	1.49	
B30 Roofing						
3010	Roof Coverings	Single ply membrane, rigid insulation, gravel stop	S.F. Roof	6.67	6.67	7.7%
3020	Roof Openings	Roof and smoke hatches	S.F. Roof	.16	.16	
C. INTERIORS						
1010	Partitions	Concrete partitions, gypsum board on metal stud 50 S.F. Floor/L.F. Partition	S.F. Partition	9.75	2.34	
1020	Interior Doors	Hollow metal doors 2000 S.F. Floor/Door	Each	1222	.61	
1030	Fittings	Stainless steel toilet partitions	S.F. Floor	.11	.11	
2010	Stair Construction	N/A	—	—	—	15.1%
3010	Wall Finishes	Paint	S.F. Surface	2.06	.99	
3020	Floor Finishes	Vinyl composition tile	S.F. Floor	3.06	3.06	
3030	Ceiling Finishes	Acoustic ceiling tiles on suspended channel grid	S.F. Ceiling	6.30	6.30	
D. SERVICES						
D10 Conveying						
1010	Elevators & Lifts	N/A	—	—	—	0.0 %
1020	Escalators & Moving Walks	N/A	—	—	—	
D20 Plumbing						
2010	Plumbing Fixtures	Restroom, food prep and service fixtures, supply and drainage 1 Fixture/1750 S.F. Floor	Each	2253	1.28	
2020	Domestic Water Distribution	Gas fired water heater	S.F. Floor	.27	.27	3.1%
2040	Rain Water Drainage	Roof drains	S.F. Roof	1.23	1.23	
D30 HVAC						
3010	Energy Supply	N/A	—	—	—	
3020	Heat Generating Systems	Included in D3050	—	—	—	
3030	Cooling Generating Systems	N/A	—	—	—	5.8 %
3050	Terminal & Package Units	Single zone unit gas heating, electric cooling	S.F. Floor	5.17	5.17	
3090	Other HVAC Sys. & Equipment	N/A	—	—	—	
D40 Fire Protection						
4010	Sprinklers	Wet pipe sprinkler system	S.F. Floor	3.79	3.79	4.9%
4020	Standpipes	Standpipe	S.F. Floor	.60	.60	
D50 Electrical						
5010	Electrical Service/Distribution	1600 ampere service, panel board and feeders	S.F. Floor	2.56	2.56	
5020	Lighting & Branch Wiring	Fluorescent fixtures, receptacles, switches, A.C. and misc. power	S.F. Floor	12.88	12.88	20.0%
5030	Communications & Security	Addressable alarm system, emergency lighting, internet & phone wiring	S.F. Floor	2.30	2.30	
5090	Other Electrical Systems	Emergency generator, 15kW	S.F. Floor	.05	.05	
E. EQUIPMENT & FURNISHINGS						
1010	Commercial Equipment	N/A	—	—	—	
1020	Institutional Equipment	N/A	—	—	—	0.0 %
1030	Vehicular Equipment	N/A	—	—	—	
1090	Other Equipment	N/A	—	—	—	
F. SPECIAL CONSTRUCTION						
1020	Integrated Construction	N/A	—	—	—	0.0 %
1040	Special Facilities	N/A	—	—	—	
G. BUILDING SITEWORK	**N/A**					

		Sub-Total	89.08	100%
CONTRACTOR FEES (General Requirements: 10%, Overhead: 5%, Profit: 10%)		25%	22.23	
ARCHITECT FEES		7%	7.79	
	Total Building Cost		**119.10**	

For customer support on your Square Foot Costs with RSMeans data, call 800.448.8182.

217

Costs per square foot of floor area

Exterior Wall	S.F. Area	10000	16000	20000	22000	24000	26000	28000	30000	32000
	L.F. Perimeter	420	510	600	640	660	684	706	740	737
Face Brick and Concrete Block	Wood Truss	310.40	291.00	287.15	285.30	282.40	280.05	278.05	277.05	274.05
	Precast Concrete	329.10	305.20	300.55	298.25	294.65	291.80	289.30	288.05	284.35
Metal Sandwich Panel	Wood Truss	297.05	280.90	277.60	276.05	273.65	271.75	270.05	269.15	266.75
Precast Concrete	Precast Concrete	341.60	315.35	310.25	307.80	303.85	300.75	298.00	296.60	292.50
Concrete Block	Wood Frame	288.20	274.15	271.30	269.90	267.80	266.15	264.75	263.95	261.85
	Precast Concrete	309.85	291.30	287.60	285.85	283.05	280.85	278.95	277.95	275.10
Perimeter Adj., Add or Deduct	Per 100 L.F.	17.40	10.85	8.70	7.90	7.20	6.75	6.25	5.75	5.40
Story Hgt. Adj., Add or Deduct	Per 1 Ft.	2.30	1.75	1.60	1.60	1.45	1.50	1.45	1.30	1.25
For Basement, add $42.10 per square foot of basement area										

The above costs were calculated using the basic specifications shown on the facing page. These costs should be adjusted where necessary for design alternatives and owner's requirements.

Common additives

Description	Unit	$ Cost	Description	Unit	$ Cost
Bleachers, Telescoping, manual			Sauna, Prefabricated, complete		
To 15 tier	Seat	164 - 222	6' x 4'	Ea.	6600
16-20 tier	Seat	340 - 405	6' x 6'	Ea.	8850
21-30 tier	Seat	340 - 460	6' x 9'	Ea.	10,000
For power operation, add	Seat	63 - 101	8' x 8'	Ea.	10,600
Emergency Lighting, 25 watt, battery operated			8' x 10'	Ea.	11,700
Lead battery	Ea.	330	10' x 12'	Ea.	15,500
Nickel cadmium	Ea.	550	Sound System		
Lockers, Steel, single tier, 60" or 72"	Opng.	232 - 395	Amplifier, 250 watts	Ea.	2025
2 tier, 60" or 72" total	Opng.	139 - 172	Speaker, ceiling or wall	Ea.	240
5 tier, box lockers	Opng.	73 - 87	Trumpet	Ea.	460
Locker bench, lam. maple top only	L.F.	36.50	Steam Bath, Complete, to 140 C.F.	Ea.	3175
Pedestal, steel pipe	Ea.	79	To 300 C.F.	Ea.	3525
Pool Equipment			To 800 C.F.	Ea.	7125
Diving stand, 3 meter	Ea.	21,500	To 2500 C.F.	Ea.	9725
1 meter	Ea.	11,600			
Diving board, 16' aluminum	Ea.	5200			
Fiberglass	Ea.	4150			
Lifeguard chair, fixed	Ea.	4200			
Portable	Ea.	3175			
Lights, underwater, 12 volt, 300 watt	Ea.	1100			

Important: See the Reference Section for Location Factors.

Swimming Pool, Enclosed

Model costs calculated for a 1 story building with 24' story height and 20,000 square feet of floor area

				Unit	Unit Cost	Cost Per S.F.	% Of Sub-Total
A. SUBSTRUCTURE							
1010	Standard Foundations	Poured concrete; strip and spread footings; 4' foundation wall		S.F. Ground	7.20	7.20	
1020	Special Foundations	N/A		—			
1030	Slab on Grade	4" reinforced concrete with vapor barrier and granular base		S.F. Slab	5.80	5.80	7.6%
2010	Basement Excavation	Site preparation for slab and trench for foundation wall and footing		S.F. Ground	.33	.66	
2020	Basement Walls	N/A		—			
B. SHELL							
	B10 Superstructure						
1010	Floor Construction	N/A		—			
1020	Roof Construction	Wood deck on laminated wood truss		S.F. Roof	20.75	20.75	9.8 %
	B20 Exterior Enclosure						
2010	Exterior Walls	Face brick with concrete block backup	80% of wall	S.F. Wall	34.74	20.01	
2020	Exterior Windows	Outward projecting steel	20% of wall	Each	51	7.39	13.3%
2030	Exterior Doors	Double aluminum and glass, hollow metal		Each	3696	.92	
	B30 Roofing						
3010	Roof Coverings	Asphalt shingles with flashing; perlite/EPS composite insulation		S.F. Roof	4.58	4.58	2.2%
3020	Roof Openings	N/A		—	—	—	
C. INTERIORS							
1010	Partitions	Concrete block	100 S.F. Floor/L.F. Partition	S.F. Partition	10.40	1.04	
1020	Interior Doors	Single leaf hollow metal	1000 S.F. Floor/Door	Each	1222	1.22	
1030	Fittings	Toilet partitions		S.F. Floor	.18	.18	
2010	Stair Construction	N/A		—	—	—	14.3%
3010	Wall Finishes	Acrylic glazed coating		S.F. Surface	12.50	2.50	
3020	Floor Finishes	70% terrazzo, 30% ceramic tile		S.F. Floor	24.44	24.44	
3030	Ceiling Finishes	Mineral fiber tile on concealed zee bars	20% of area	S.F. Ceiling	5.07	1.01	
D. SERVICES							
	D10 Conveying						
1010	Elevators & Lifts	N/A		—	—	—	0.0 %
1020	Escalators & Moving Walks	N/A		—	—	—	
	D20 Plumbing						
2010	Plumbing Fixtures	Toilet and service fixtures, supply and drainage	1 Fixture/210 S.F. Floor	Each	1346	6.41	
2020	Domestic Water Distribution	Gas fired water heater		S.F. Floor	4.74	4.74	5.2%
2040	Rain Water Drainage	N/A		—	—	—	
	D30 HVAC						
3010	Energy Supply	Terminal unit heaters	10% of area	S.F. Floor	.93	.93	
3020	Heat Generating Systems	N/A		—	—	—	
3030	Cooling Generating Systems	N/A		—	—	—	15.1%
3050	Terminal & Package Units	Single zone unit gas heating, electric cooling		S.F. Floor	31.26	31.26	
3090	Other HVAC Sys. & Equipment	N/A		—	—	—	
	D40 Fire Protection						
4010	Sprinklers	Sprinklers, light hazard		S.F. Floor	1.90	1.90	1.2%
4020	Standpipes	Standpipe		S.F. Floor	.64	.64	
	D50 Electrical						
5010	Electrical Service/Distribution	400 ampere service, panel board and feeders		S.F. Floor	1.19	1.19	
5020	Lighting & Branch Wiring	High efficiency fluorescent fixtures, receptacles, switches, A.C. and misc. power		S.F. Floor	21.89	21.89	11.4%
5030	Communications & Security	Addressable alarm systems and emergency lighting		S.F. Floor	1.08	1.08	
5090	Other Electrical Systems	Emergency generator, 15 kW		S.F. Floor	.09	.09	
E. EQUIPMENT & FURNISHINGS							
1010	Commercial Equipment	N/A		—	—	—	
1020	Institutional Equipment	N/A		—	—	—	
1030	Vehicular Equipment	N/A		—	—	—	20.0 %
1090	Other Equipment	Swimming pool		S.F. Floor	42.46	42.46	
F. SPECIAL CONSTRUCTION							
1020	Integrated Construction	N/A		—	—	—	0.0 %
1040	Special Facilities	N/A		—	—	—	
G. BUILDING SITEWORK	**N/A**						

			Sub-Total	212.70	**100%**
CONTRACTOR FEES (General Requirements: 10%, Overhead: 5%, Profit: 10%)			25%	53.18	
ARCHITECT FEES			8%	21.27	
		Total Building Cost		**287.15**	

For customer support on your Square Foot Costs with RSMeans data, call 800.448.8182.

219

Costs per square foot of floor area

Exterior Wall	S.F. Area	2000	3000	4000	5000	6000	7000	8000	9000	10000
	L.F. Perimeter	180	220	260	286	320	353	368	397	425
Face Brick and Concrete Block	Rigid Steel	237.40	210.50	197.10	186.10	180.15	175.70	170.05	167.30	165.00
	Bearing Walls	226.75	202.85	190.85	180.75	175.35	171.30	166.00	163.50	161.35
Limestone and Concrete Block	Rigid Steel	270.15	239.15	223.70	210.25	203.30	198.05	190.80	187.50	184.70
	Bearing Walls	270.10	236.15	219.20	204.90	197.35	191.65	184.15	180.60	177.60
Decorative Concrete Block	Rigid Steel	221.15	197.25	185.35	175.80	170.55	166.60	161.80	159.35	157.35
	Bearing Walls	215.80	191.90	180.00	170.45	165.20	161.20	156.45	154.00	152.00
Perimeter Adj., Add or Deduct	Per 100 L.F.	52.70	35.20	26.35	21.10	17.55	15.10	13.20	11.75	10.55
Story Hgt. Adj., Add or Deduct	Per 1 Ft.	6.10	5.00	4.40	3.90	3.60	3.45	3.10	3.00	2.90

For Basement, add $40.80 per square foot of basement area

The above costs were calculated using the basic specifications shown on the facing page. These costs should be adjusted where necessary for design alternatives and owner's requirements.

Common additives

Description	Unit	$ Cost
Emergency Lighting, 25 watt, battery operated		
Lead battery	Ea.	330
Nickel cadmium	Ea.	550
Smoke Detectors		
Ceiling type	Ea.	244
Duct type	Ea.	585
Emergency Generators, complete system, gas		
15 kw	Ea.	18,700
85 kw	Ea.	45,100
170 kw	Ea.	106,000
Diesel, 50 kw	Ea.	26,500
150 kw	Ea.	48,600
350 kw	Ea.	74,000

Important: See the Reference Section for Location Factors.

Model costs calculated for a 1 story building with 12' story height and 5,000 square feet of floor area

Telephone Exchange

				Unit	Unit Cost	Cost Per S.F.	% Of Sub-Total
A. SUBSTRUCTURE							
1010	Standard Foundations	Poured concrete; strip and spread footings; 4' foundation wall		S.F. Ground	7.90	7.90	
1020	Special Foundations	N/A		—	—	—	
1030	Slab on Grade	4" reinforced concrete with vapor barrier and granular base		S.F. Slab	5.80	5.80	10.7%
2010	Basement Excavation	Site preparation for slab and trench for foundation wall and footing		S.F. Ground	.57	.71	
2020	Basement Walls	N/A		—	—	—	
B. SHELL							
	B10 Superstructure						
1010	Floor Construction	Steel column fireproofing		S.F. Floor	.88	.88	8.8%
1020	Roof Construction	Metal deck, open web steel joists, beams, columns		S.F. Roof	10.90	10.90	
	B20 Exterior Enclosure						
2010	Exterior Walls	Face brick with concrete block backup	80% of wall	S.F. Wall	35.24	19.35	
2020	Exterior Windows	Outward projecting steel	20% of wall	Each	1795	12.32	25.2%
2030	Exterior Doors	Single aluminum glass with transom		Each	5310	2.12	
	B30 Roofing						
3010	Roof Coverings	Built-up tar and gravel with flashing; perlite/EPS composite insulation		S.F. Roof	7.65	7.65	5.7%
3020	Roof Openings	N/A		—	—	—	
C. INTERIORS							
1010	Partitions	Double layer gypsum board on metal studs	15 S.F. Floor/L.F. Partition	S.F. Partition	6.77	4.51	
1020	Interior Doors	Single leaf hollow metal	150 S.F. Floor/Door	Each	1222	8.14	
1030	Fittings	Toilet partitions		S.F. Floor	1.25	1.25	
2010	Stair Construction	N/A		—	—	—	22.0%
3010	Wall Finishes	Paint		S.F. Surface	2.31	3.08	
3020	Floor Finishes	90% carpet, 10% terrazzo		S.F. Floor	6.28	6.28	
3030	Ceiling Finishes	Fiberglass board on exposed grid system, suspended		S.F. Ceiling	6.30	6.30	
D. SERVICES							
	D10 Conveying						
1010	Elevators & Lifts	N/A		—	—	—	0.0%
1020	Escalators & Moving Walks	N/A		—	—	—	
	D20 Plumbing						
2010	Plumbing Fixtures	Kitchen, toilet and service fixtures, supply and drainage	1 Fixture/715 S.F. Floor	Each	4033	5.64	
2020	Domestic Water Distribution	Gas fired water heater		S.F. Floor	2.53	2.53	7.6%
2040	Rain Water Drainage	Roof drains		S.F. Roof	1.96	1.96	
	D30 HVAC						
3010	Energy Supply	N/A		—	—	—	
3020	Heat Generating Systems	Included in D3050		—	—	—	
3030	Cooling Generating Systems	N/A		—	—	—	6.6%
3050	Terminal & Package Units	Single zone unit, gas heating, electric cooling		S.F. Floor	8.90	8.90	
3090	Other HVAC Sys. & Equipment	N/A		—	—	—	
	D40 Fire Protection						
4010	Sprinklers	Wet pipe sprinkler system		S.F. Floor	5.67	5.67	5.9%
4020	Standpipes	Standpipe		S.F. Floor	2.29	2.29	
	D50 Electrical						
5010	Electrical Service/Distribution	200 ampere service, panel board and feeders		S.F. Floor	1.99	1.99	
5020	Lighting & Branch Wiring	High efficiency fluorescent fixtures, receptacles, switches, A.C. and misc. power		S.F. Floor	4.72	4.72	7.4%
5030	Communications & Security	Addressable alarm systems and emergency lighting		S.F. Floor	3.04	3.04	
5090	Other Electrical Systems	Emergency generator, 15 kW		S.F. Floor	.19	.19	
E. EQUIPMENT & FURNISHINGS							
1010	Commercial Equipment	N/A		—	—	—	
1020	Institutional Equipment	N/A		—	—	—	0.0%
1030	Vehicular Equipment	N/A		—	—	—	
1090	Other Equipment	N/A		—	—	—	
F. SPECIAL CONSTRUCTION							
1020	Integrated Construction	N/A		—	—	—	0.0%
1040	Special Facilities	N/A		—	—	—	
G. BUILDING SITEWORK	**N/A**						

	Sub-Total	134.12	**100%**
CONTRACTOR FEES (General Requirements: 10%, Overhead: 5%, Profit: 10%)	25%	33.54	
ARCHITECT FEES	11%	18.44	

Total Building Cost	**186.10**

For customer support on your Square Foot Costs with RSMeans data, call 800.448.8182.

221

Costs per square foot of floor area

Exterior Wall	S.F. Area	5000	6500	8000	9500	11000	14000	17500	21000	24000
	L.F. Perimeter	300	360	386	396	435	510	550	620	680
Face Brick and Concrete Block	Steel Joists	178.50	171.80	163.85	157.00	154.30	150.45	145.30	143.10	141.65
	Wood Joists	170.15	163.60	156.20	149.75	147.15	143.50	138.65	136.55	135.20
Stone and Concrete Block	Steel Joists	187.05	179.65	170.70	162.95	159.90	155.60	149.75	147.25	145.70
	Wood Joists	178.65	171.50	163.05	155.60	152.75	148.70	143.10	140.75	139.25
Brick Veneer	Wood Frame	164.30	158.25	151.50	145.70	143.30	139.95	135.60	133.70	132.50
E.I.F.S.	Wood Frame	161.75	155.85	149.40	143.90	141.60	138.40	134.20	132.40	131.25
Perimeter Adj., Add or Deduct	Per 100 L.F.	17.35	13.30	10.85	9.10	7.90	6.20	4.95	4.10	3.60
Story Hgt. Adj., Add or Deduct	Per 1 Ft.	3.15	2.85	2.55	2.20	2.05	1.90	1.65	1.50	1.50

For Basement, add $33.60 per square foot of basement area

The above costs were calculated using the basic specifications shown on the facing page. These costs should be adjusted where necessary for design alternatives and owner's requirements.

Common additives

Description	Unit	$ Cost
Directory Boards, Plastic, glass covered		
30" x 20"	Ea.	640
36" x 48"	Ea.	1625
Aluminum, 24" x 18"	Ea.	630
36" x 24"	Ea.	790
48" x 32"	Ea.	1025
48" x 60"	Ea.	2200
Emergency Lighting, 25 watt, battery operated		
Lead battery	Ea.	330
Nickel cadmium	Ea.	550
Flagpoles, Complete		
Aluminum, 20' high	Ea.	1900
40' high	Ea.	4500
70' high	Ea.	11,600
Fiberglass, 23' high	Ea.	1450
39'-5" high	Ea.	3325
59' high	Ea.	7300
Safe, Office type, 1 hour rating		
30" x 18" x 18"	Ea.	2475
60" x 36" x 18", double door	Ea.	10,200

Description	Unit	$ Cost
Smoke Detectors		
Ceiling type	Ea.	244
Duct type	Ea.	585
Vault Front, Door & frame		
1 Hour test, 32" x 78"	Opng.	8000
2 Hour test, 32" door	Opng.	9800
40" door	Opng.	11,300
4 Hour test, 32" door	Opng.	12,100
40" door	Opng.	13,800
Time lock movement; two movement	Ea.	2400

Important: See the Reference Section for Location Factors.

Model costs calculated for a 1 story building with 12' story height and 11,000 square feet of floor area

				Unit	Unit Cost	Cost Per S.F.	% Of Sub-Total
A. SUBSTRUCTURE							
1010	Standard Foundations	Poured concrete; strip and spread footings; 4' foundation wall		S.F. Ground	5.42	5.42	
1020	Special Foundations	N/A		—	—	—	
1030	Slab on Grade	4" reinforced concrete with vapor barrier and granular base		S.F. Slab	5.80	5.80	10.2%
2010	Basement Excavation	Site preparation for slab and trench for foundation wall and footing		S.F. Ground	.33	.36	
2020	Basement Walls	N/A		—	—	—	
B. SHELL							
B10 Superstructure							
1010	Floor Construction	Steel column fireproofing		S.F. Floor	.31	.31	7.6%
1020	Roof Construction	Metal deck, open web steel joists, beams, interior columns		S.F. Roof	8.32	8.32	
B20 Exterior Enclosure							
2010	Exterior Walls	Face brick with concrete block backup	70% of wall	S.F. Wall	35.25	11.71	
2020	Exterior Windows	Metal outward projecting	30% of wall	Each	755	4.67	15.7%
2030	Exterior Doors	Metal and glass with transom		Each	3915	1.43	
B30 Roofing							
3010	Roof Coverings	Built-up tar and gravel with flashing; perlite/EPS composite insulation		S.F. Roof	7.14	7.14	6.3%
3020	Roof Openings	N/A		—	—	—	
C. INTERIORS							
1010	Partitions	Gypsum board on metal studs	20 S.F. Floor/L.F. Partition	S.F. Partition	10.22	5.11	
1020	Interior Doors	Wood solid core	200 S.F. Floor/Door	Each	720	3.60	
1030	Fittings	Toilet partitions		S.F. Floor	.50	.50	
2010	Stair Construction	N/A		—	—	—	23.9%
3010	Wall Finishes	90% paint, 10% ceramic tile		S.F. Surface	1.62	1.62	
3020	Floor Finishes	70% carpet tile, 15% terrazzo, 15% vinyl composition tile		S.F. Floor	8.58	8.58	
3030	Ceiling Finishes	Mineral fiber tile on concealed zee bars		S.F. Ceiling	7.60	7.60	
D. SERVICES							
D10 Conveying							
1010	Elevators & Lifts	N/A		—	—	—	0.0 %
1020	Escalators & Moving Walks	N/A		—	—	—	
D20 Plumbing							
2010	Plumbing Fixtures	Kitchen, toilet and service fixtures, supply and drainage	1 Fixture/500 S.F. Floor	Each	2675	5.35	
2020	Domestic Water Distribution	Gas fired water heater		S.F. Floor	1.90	1.90	9.0%
2040	Rain Water Drainage	Roof drains		S.F. Roof	2.90	2.90	
D30 HVAC							
3010	Energy Supply	N/A		—	—	—	
3020	Heat Generating Systems	Included in D3050		—	—	—	
3030	Cooling Generating Systems	N/A		—	—	—	8.7 %
3050	Terminal & Package Units	Multizone unit, gas heating, electric cooling		S.F. Floor	9.86	9.86	
3090	Other HVAC Sys. & Equipment	N/A		—	—	—	
D40 Fire Protection							
4010	Sprinklers	Wet pipe sprinkler system		S.F. Floor	3.79	3.79	4.1%
4020	Standpipes	Standpipe, wet, Class III		S.F. Floor	.87	.87	
D50 Electrical							
5010	Electrical Service/Distribution	400 ampere service, panel board and feeders		S.F. Floor	2.04	2.04	
5020	Lighting & Branch Wiring	High efficiency fluorescent fixtures, receptacles, switches, A.C. and misc. power		S.F. Floor	10.85	10.85	14.5%
5030	Communications & Security	Addressable alarm systems, internet wiring, and emergency lighting		S.F. Floor	3.34	3.34	
5090	Other Electrical Systems	Emergency generator, 15 kW		S.F. Floor	.17	.17	
E. EQUIPMENT & FURNISHINGS							
1010	Commercial Equipment	N/A		—	—	—	
1020	Institutional Equipment	N/A		—	—	—	0.0 %
1030	Vehicular Equipment	N/A		—	—	—	
1090	Other Equipment	N/A		—	—	—	
F. SPECIAL CONSTRUCTION							
1020	Integrated Construction	N/A		—	—	—	0.0 %
1040	Special Facilities	N/A		—	—	—	
G. BUILDING SITEWORK	**N/A**						

		Sub-Total	113.24	100%
	CONTRACTOR FEES (General Requirements: 10%, Overhead: 5%, Profit: 10%)	25%	28.32	
	ARCHITECT FEES	9%	12.74	
	Total Building Cost		**154.30**	

For customer support on your Square Foot Costs with RSMeans data, call 800.448.8182.

223

Costs per square foot of floor area

Exterior Wall	S.F. Area	8000	10000	12000	15000	18000	24000	28000	35000	40000
	L.F. Perimeter	206	233	260	300	320	360	393	451	493
Face Brick and Concrete Block	Rigid Steel	233.95	221.45	213.10	204.60	196.65	186.70	182.95	178.40	176.15
	Reinforced Concrete	239.30	226.75	218.45	209.95	202.00	192.05	188.25	183.70	181.50
Stone and Concrete Block	Rigid Steel	244.95	231.35	222.35	213.15	204.25	193.10	188.90	183.85	181.35
	Reinforced Concrete	250.25	236.70	227.65	218.50	209.60	198.40	194.20	189.20	186.70
Limestone and Concrete Block	Rigid Steel	259.75	244.55	234.50	224.20	213.80	200.80	196.00	190.15	187.35
	Reinforced Concrete	267.35	252.15	242.05	231.75	221.40	208.40	203.55	197.75	194.90
Perimeter Adj., Add or Deduct	Per 100 L.F.	26.55	21.20	17.70	14.20	11.85	8.90	7.55	6.00	5.35
Story Hgt. Adj., Add or Deduct	Per 1 Ft.	4.00	3.60	3.40	3.15	2.80	2.35	2.20	2.00	1.95

For Basement, add $32.70 per square foot of basement area

The above costs were calculated using the basic specifications shown on the facing page. These costs should be adjusted where necessary for design alternatives and owner's requirements.

Common additives

Description	Unit	$ Cost	Description	Unit	$ Cost
Directory Boards, Plastic, glass covered			Flagpoles, Complete		
30" x 20"	Ea.	640	Aluminum, 20' high	Ea.	1900
36" x 48"	Ea.	1625	40' high	Ea.	4500
Aluminum, 24" x 18"	Ea.	630	70' high	Ea.	11,600
36" x 24"	Ea.	790	Fiberglass, 23' high	Ea.	1450
48" x 32"	Ea.	1025	39'-5" high	Ea.	3325
48" x 60"	Ea.	2200	59' high	Ea.	7300
Elevators, Hydraulic passenger, 2 stops			Safe, Office type, 1 hour rating		
1500# capacity	Ea.	73,100	30" x 18" x 18"	Ea.	2475
2500# capacity	Ea.	76,600	60" x 36" x 18", double door	Ea.	10,200
3500# capacity	Ea.	81,600	Smoke Detectors		
Additional stop, add	Ea.	8375	Ceiling type	Ea.	244
Emergency Lighting, 25 watt, battery operated			Duct type	Ea.	585
Lead battery	Ea.	330	Vault Front, Door & frame		
Nickel cadmium	Ea.	550	1 Hour test, 32" x 78"	Opng.	8000
			2 Hour test, 32" door	Opng.	9800
			40" door	Opng.	11,300
			4 Hour test, 32" door	Opng.	12,100
			40" door	Opng.	13,800
			Time lock movement; two movement	Ea.	2400

Important: See the Reference Section for Location Factors.

Model costs calculated for a 3 story building with 12' story height and 18,000 square feet of floor area

Town Hall, 2-3 Story

				Unit	Unit Cost	Cost Per S.F.	% Of Sub-Total
A. SUBSTRUCTURE							
1010	Standard Foundations	Poured concrete; strip and spread footings; 4' foundation wall		S.F. Ground	9.63	3.21	
1020	Special Foundations	N/A		—	—	—	
1030	Slab on Grade	4" reinforced concrete with vapor barrier and granular base		S.F. Slab	5.80	1.93	3.6%
2010	Basement Excavation	Site preparation for slab and trench for foundation wall and footing		S.F. Ground	.57	.29	
2020	Basement Walls	N/A		—	—	—	
B. SHELL							
	B10 Superstructure						
1010	Floor Construction	Open web steel joists, slab form, concrete, wide flange steel columns		S.F. Floor	28.61	19.07	14.8%
1020	Roof Construction	Metal deck, open web steel joists, beams, interior columns		S.F. Roof	9.15	3.05	
	B20 Exterior Enclosure						
2010	Exterior Walls	Stone with concrete block backup	70% of wall	S.F. Wall	47.66	21.35	
2020	Exterior Windows	Metal outward projecting	10% of wall	Each	755	6.30	19.1%
2030	Exterior Doors	Metal and glass with transoms		Each	3696	1.02	
	B30 Roofing						
3010	Roof Coverings	Built-up tar and gravel with flashing; perlite/EPS composite insulation		S.F. Roof	7.77	2.59	1.7%
3020	Roof Openings	N/A		—	—	—	
C. INTERIORS							
1010	Partitions	Gypsum board on metal studs	20 S.F. Floor/L.F. Partition	S.F. Partition	11.42	5.71	
1020	Interior Doors	Wood solid core	200 S.F. Floor/Door	Each	720	3.60	
1030	Fittings	Toilet partitions		S.F. Floor	.30	.30	
2010	Stair Construction	Concrete filled metal pan		Flight	18,900	8.40	23.9%
3010	Wall Finishes	90% paint, 10% ceramic tile		S.F. Surface	1.62	1.62	
3020	Floor Finishes	70% carpet tile, 15% terrazzo, 15% vinyl composition tile		S.F. Floor	8.58	8.58	
3030	Ceiling Finishes	Mineral fiber tile on concealed zee bars		S.F. Ceiling	7.60	7.60	
D. SERVICES							
	D10 Conveying						
1010	Elevators & Lifts	Two hydraulic elevators		Each	125,910	13.99	9.3%
1020	Escalators & Moving Walks	N/A		—	—	—	
	D20 Plumbing						
2010	Plumbing Fixtures	Toilet and service fixtures, supply and drainage	1 Fixture/1385 S.F. Floor	Each	7050	5.09	
2020	Domestic Water Distribution	Gas fired water heater		S.F. Floor	.68	.68	4.6%
2040	Rain Water Drainage	Roof drains		S.F. Roof	3.24	1.08	
	D30 HVAC						
3010	Energy Supply	N/A		—	—	—	
3020	Heat Generating Systems	Included in D3050		—	—	—	
3030	Cooling Generating Systems	N/A		—	—	—	6.6 %
3050	Terminal & Package Units	Multizone unit, gas heating, electric cooling		S.F. Floor	9.86	9.86	
3090	Other HVAC Sys. & Equipment	N/A		—	—	—	
	D40 Fire Protection						
4010	Sprinklers	Sprinklers, light hazard		S.F. Floor	3.49	3.49	4.8%
4020	Standpipes	Standpipe, wet, Class III		S.F. Floor	3.69	3.69	
	D50 Electrical						
5010	Electrical Service/Distribution	400 ampere service, panel board and feeders		S.F. Floor	1.73	1.73	
5020	Lighting & Branch Wiring	High efficiency fluorescent fixtures, receptacles, switches, A.C. and misc. power		S.F. Floor	12.01	12.01	11.6%
5030	Communications & Security	Addressable alarm systems, internet wiring, and emergency lighting		S.F. Floor	3.47	3.47	
5090	Other Electrical Systems	Emergency generator, 15 kW		S.F. Floor	.20	.20	
E. EQUIPMENT & FURNISHINGS							
1010	Commercial Equipment	N/A		—	—	—	
1020	Institutional Equipment	N/A		—	—	—	0.0 %
1030	Vehicular Equipment	N/A		—	—	—	
1090	Other Equipment	N/A		—	—	—	
F. SPECIAL CONSTRUCTION							
1020	Integrated Construction	N/A		—	—	—	0.0 %
1040	Special Facilities	N/A		—	—	—	
G. BUILDING SITEWORK	**N/A**						

		Sub-Total	149.91	100%
CONTRACTOR FEES (General Requirements: 10%, Overhead: 5%, Profit: 10%)		25%	37.47	
ARCHITECT FEES		9%	16.87	
	Total Building Cost		**204.25**	

For customer support on your Square Foot Costs with RSMeans data, call 800.448.8182.

225

Costs per square foot of floor area

Exterior Wall	S.F. Area	3500	5000	7500	9000	11000	12500	15000	20000	25000
	L.F. Perimeter	240	290	350	390	430	460	510	580	650
E.I.F.S.	Wood Truss	181.65	172.10	163.15	160.35	157.30	155.60	153.55	150.30	148.35
	Steel Joists	192.80	182.80	173.45	170.60	167.30	165.55	163.40	159.90	157.85
Wood Clapboard	Wood Truss	184.50	175.10	166.30	163.65	160.60	158.95	157.00	153.75	151.85
Brick Veneer	Wood Truss	196.40	185.15	174.45	171.15	167.40	165.35	162.90	158.80	156.40
Face Brick and Concrete Block	Wood Truss	200.70	188.25	176.30	172.65	168.45	166.10	163.35	158.70	155.90
Tilt-up Concrete Panels	Steel Joists	184.95	175.95	167.65	165.10	162.25	160.65	158.80	155.80	154.00
Perimeter Adj., Add or Deduct	Per 100 L.F.	13.75	9.60	6.40	5.35	4.30	3.85	3.25	2.40	1.90
Story Hgt. Adj., Add or Deduct	Per 1 Ft.	1.65	1.40	1.10	1.10	0.95	0.90	0.85	0.70	0.60
For Basement, add $35.40 per square foot of basement area										

The above costs were calculated using the basic specifications shown on the facing page. These costs should be adjusted where necessary for design alternatives and owner's requirements.

Common additives

Description	Unit	$ Cost
Closed circuit surveillance, one station		
Camera and monitor	Ea.	1475
For additional camera stations, add	Ea.	665
Directory boards, plastic, glass covered		
30" x 20"	Ea.	640
36" x 48"	Ea.	1625
Aluminum, 24" x 18"	Ea.	630
36" x 24"	Ea.	790
48" x 32"	Ea.	1025
48" x 60"	Ea.	2200
Electronic, wall mounted	S.F.	4625
Free standing	S.F.	4075
Kennel fencing		
Kennel fencing, 1-1/2" mesh, 6' long, 3'-6" wide, 6'-2" high	Ea.	790
12' long	Ea.	1000
Top covers, 1-1/2" mesh, 6' long	Ea.	210
12' long	Ea.	285

Description	Unit	$ Cost
Kennel doors		
2 way, swinging type, 13" x 19" opening	Opng.	211
17" x 29" opening	Opng.	255
9" x 9" opening, electronic with accessories	Opng.	276

Important: See the Reference Section for Location Factors.

Model costs calculated for a 1 story building with 12' story height and 7,500 square feet of floor area

Veterinary Hospital

				Unit	Unit Cost	Cost Per S.F.	% Of Sub-Total
A. SUBSTRUCTURE							
1010	Standard Foundations	Poured concrete; strip and spread footings; 4' foundation wall		S.F. Ground	6.39	6.39	
1020	Special Foundations	N/A		—	—	—	
1030	Slab on Grade	4" and 5" reinforced concrete with vapor barrier and granular base		S.F. Slab	5.80	5.80	10.9%
2010	Basement Excavation	Site preparation for slab and trench for foundation wall and footing		S.F. Ground	.57	1.07	
2020	Basement Walls	N/A		—	—	—	
B. SHELL							
B10 Superstructure							
1010	Floor Construction	6 x 6 wood columns		S.F. Floor	.25	.25	6.8%
1020	Roof Construction	Wood roof truss with plywood sheathing		S.F. Roof	8.10	8.10	
B20 Exterior Enclosure							
2010	Exterior Walls	Cedar bevel siding on wood studs, insulated	85% of wall	S.F. Wall	14.26	6.79	
2020	Exterior Windows	Wood double hung and awning	15% of wall	Each	1211	4.12	11.3%
2030	Exterior Doors	Aluminum and glass		Each	3590	2.87	
B30 Roofing							
3010	Roof Coverings	Ashphalt shingles with flashing (pitched); al. gutters and downspouts		S.F. Roof	5.50	5.50	4.5%
3020	Roof Openings	N/A		—	—	—	
C. INTERIORS							
1010	Partitions	Gypsum board on wood studs	8 S.F. Floor/L.F. Partition	S.F. Partition	5.50	6.88	
1020	Interior Doors	Solid core wood	180 S.F. Floor/Door	Each	733	2.03	
1030	Fittings	Lockers		S.F. Floor	.46	.46	
2010	Stair Construction	N/A		—	—	—	18.9%
3010	Wall Finishes	Paint on drywall		S.F. Surface	1.06	2.65	
3020	Floor Finishes	90% vinyl composition tile, 5% ceramic tile, 5% sealed concrete		S.F. Floor	3.73	3.73	
3030	Ceiling Finishes	90% acoustic tile, 10% gypsum board		S.F. Ceiling	7.37	7.38	
D. SERVICES							
D10 Conveying							
1010	Elevators & Lifts	N/A		—	—	—	0.0 %
1020	Escalators & Moving Walks	N/A		—	—	—	
D20 Plumbing							
2010	Plumbing Fixtures	Toilet and service fixtures, supply and drainage	1 Fixture/1075 S.F. Floor	Each	1882	3.36	
2020	Domestic Water Distribution	Electric water heater		S.F. Floor	1.68	1.68	4.1%
2040	Rain Water Drainage	N/A		—	—	—	
D30 HVAC							
3010	Energy Supply	N/A		—	—	—	
3020	Heat Generating Systems	N/A		—	—	—	
3030	Cooling Generating Systems	N/A		—	—	—	5.4 %
3050	Terminal & Package Units	Split system with air cooled condensing units		S.F. Floor	6.57	6.57	
3090	Other HVAC Sys. & Equipment	N/A		—	—	—	
D40 Fire Protection							
4010	Sprinklers	Wet pipe sprinkler system		S.F. Floor	3.79	3.79	3.1%
4020	Standpipes	N/A		—	—	—	
D50 Electrical							
5010	Electrical Service/Distribution	800 ampere service, panel board and feeders		S.F. Floor	6.69	6.69	
5020	Lighting & Branch Wiring	High efficiency fluorescent fixtures, receptacles, switches, A.C. and misc. power		S.F. Floor	12.94	12.94	20.2%
5030	Communications & Security	Addressable alarm systems, internet wiring, and emergency lighting		S.F. Floor	4.95	4.95	
5090	Other Electrical Systems	Emergency generator, 100 kW		S.F. Floor	.06	.06	
E. EQUIPMENT & FURNISHINGS							
1010	Commercial Equipment	N/A		—	—	—	
1020	Institutional Equipment	Eye wash station		S.F. Floor	15.92	15.92	13.0 %
1030	Vehicular Equipment	N/A		—	—	—	
2020	Moveable Furnishings	Countertop, plastic laminate		S.F. Floor	2.10	2.10	
F. SPECIAL CONSTRUCTION & DEMOLITION							
1020	Integrated Construction	N/A		—	—	—	0.0 %
1040	Special Facilities	N/A		—	—	—	
G. BUILDING SITEWORK	**N/A**						

		Sub-Total	122.08	100%
CONTRACTOR FEES (General Requirements: 10%, Overhead: 5%, Profit: 10%)		25%	30.49	
ARCHITECT FEES		9%	13.73	
	Total Building Cost		**166.30**	

For customer support on your Square Foot Costs with RSMeans data, call 800.448.8182.

227

Costs per square foot of floor area

Exterior Wall	S.F. Area	10000	15000	20000	25000	30000	35000	40000	50000	60000
	L.F. Perimeter	410	500	600	640	700	766	833	966	1000
Metal Panel	Rigid Steel	127.75	117.85	113.40	108.40	105.75	104.00	102.70	100.85	98.10
Pre-Engineered Metal Building	Rigid Steel	125.90	114.75	109.75	104.00	101.00	99.00	97.55	95.45	92.20
E.I.F.S.	Rigid Steel	128.65	118.60	114.05	108.95	106.25	104.50	103.10	101.30	98.45
Brick Veneer	Reinforced Concrete	175.45	158.05	150.30	140.90	136.10	132.85	130.55	127.20	121.85
Precast Concrete	Reinforced Concrete	198.50	176.45	166.65	154.65	148.45	144.35	141.35	137.15	130.15
Tilt-up Concrete Panels	Reinforced Concrete	151.55	138.55	132.70	125.95	122.45	120.10	118.35	115.95	112.05
Perimeter Adj., Add or Deduct	Per 100 L.F.	9.50	6.35	4.70	3.85	3.20	2.70	2.40	1.90	1.50
Story Hgt. Adj., Add or Deduct	Per 1 Ft.	1.10	0.95	0.80	0.70	0.60	0.65	0.55	0.55	0.45
For Basement, add $31.40 per square foot of basement area										

The above costs were calculated using the basic specifications shown on the facing page. These costs should be adjusted where necessary for design alternatives and owner's requirements.

Common additives

Description	Unit	$ Cost
Dock Leveler, 10 ton cap.		
6' x 8'	Ea.	6250
7' x 8'	Ea.	8625
Emergency Lighting, 25 watt, battery operated		
Lead battery	Ea.	330
Nickel cadmium	Ea.	550
Fence, Chain link, 6' high		
9 ga. wire	L.F.	29
6 ga. wire	L.F.	34.50
Gate	Ea.	400
Flagpoles, Complete		
Aluminum, 20' high	Ea.	1900
40' high	Ea.	4500
70' high	Ea.	11,600
Fiberglass, 23' high	Ea.	1450
39'-5" high	Ea.	3325
59' high	Ea.	7300
Paving, Bituminous		
Wearing course plus base course	S.Y.	7.40
Sidewalks, Concrete 4" thick	S.F.	4.99

Description	Unit	$ Cost
Sound System		
Amplifier, 250 watts	Ea.	2025
Speaker, ceiling or wall	Ea.	240
Trumpet	Ea.	460
Yard Lighting, 20' aluminum pole with 400 watt high pressure sodium fixture	Ea.	3600

Important: See the Reference Section for Location Factors.

Model costs calculated for a 1 story building with 24' story height and 30,000 square feet of floor area

Warehouse

				Unit	Unit Cost	Cost Per S.F.	% Of Sub-Total
A. SUBSTRUCTURE							
1010	Standard Foundations	Poured concrete; strip and spread footings; 4' foundation wall		S.F. Ground	3.60	3.60	
1020	Special Foundations	N/A		—	—	—	
1030	Slab on Grade	5" reinforced concrete		S.F. Slab	6.30	6.30	9.9%
2010	Basement Excavation	Site preparation for slab and trench for foundation wall and footing		S.F. Ground	.18	.18	
2020	Basement Walls	N/A		—	—	—	
B. SHELL							
	B10 Superstructure						
1010	Floor Construction	Cast in place concrete columns, beams, and slab, fireproofed		S.F. Floor	70	7.02	25.0%
1020	Roof Construction	Precast double T with 2" topping		S.F. Roof	18.37	18.37	
	B20 Exterior Enclosure						
2010	Exterior Walls	Face brick with concrete block back up	98% of wall	S.F. Wall	41.20	22.61	
2020	Exterior Windows	Aluminum horizontal sliding	2% of wall	Each	529	.40	24.0%
2030	Exterior Doors	Aluminum and glass entry doors, hollow steel doors, overhead doors		Each	3875	1.43	
	B30 Roofing						
3010	Roof Coverings	Single ply membrane, stone ballast, rigid insulation		S.F. Roof	6.84	6.84	7.1%
3020	Roof Openings	Roof and smoke hatches		S.F. Roof	.42	.42	
C. INTERIORS							
1010	Partitions	Concrete block partitions, gypsum board on metal stud	100 S.F. Floor/L.F. Partition	S.F. Partition	18.25	1.46	
1020	Interior Doors	Single leaf wood hollow core doors, metal doors	2500 S.F. Floor/Door	Each	1222	.49	
1030	Fittings	N/A		—	—	—	
2010	Stair Construction	Steel grate type with rails		Flight	13,950	.93	6.2%
3010	Wall Finishes	Paint		S.F. Surface	4.88	.78	
3020	Floor Finishes	90% hardener, 10% vinyl composition tile		S.F. Floor	1.90	1.90	
3030	Ceiling Finishes	Acoustic ceiling tiles on suspended channel grid		S.F. Ceiling	7.60	.77	
D. SERVICES							
	D10 Conveying						
1010	Elevators & Lifts	N/A		—	—	—	0.0 %
1020	Escalators & Moving Walks	N/A		—	—	—	
	D20 Plumbing						
2010	Plumbing Fixtures	Restroom and service fixtures, supply and drainage	1 Fixture/2500 S.F. Floor	Each	1475	.59	
2020	Domestic Water Distribution	Gas fired water heater		S.F. Floor	.25	.25	1.5%
2040	Rain Water Drainage	Roof drains		S.F. Roof	.66	.66	
	D30 HVAC						
3010	Energy Supply	N/A		—	—	—	
3020	Heat Generating Systems	Ventilation with heat system		Each	151,400	5.45	
3030	Cooling Generating Systems	N/A		—	—	—	6.3 %
3050	Terminal & Package Units	Single zone unit gas heating, electric cooling		S.F. Floor	.93	.93	
3090	Other HVAC Sys. & Equipment	N/A		—	—	—	
	D40 Fire Protection						
4010	Sprinklers	Wet pipe sprinkler system, ordinary hazard		S.F. Floor	4.36	4.36	4.8%
4020	Standpipes	Standpipe		S.F. Floor	.54	.54	
	D50 Electrical						
5010	Electrical Service/Distribution	200 ampere service, panel board and feeders		S.F. Floor	.58	.58	
5020	Lighting & Branch Wiring	Fluorescent fixtures, receptacles, switches, A.C. and misc. power		S.F. Floor	4.61	4.61	7.9%
5030	Communications & Security	Addressable alarm system		S.F. Floor	2.83	2.83	
5090	Other Electrical Systems	N/A		—	—	—	
E. EQUIPMENT & FURNISHINGS							
1010	Commercial Equipment	N/A		—	—	—	
1020	Institutional Equipment	N/A		—	—	—	1.8 %
1030	Vehicular Equipment	Dock boards and levelers		S.F. Floor	1.87	1.87	
1090	Other Equipment	N/A		—	—	—	
F. SPECIAL CONSTRUCTION							
1020	Integrated Construction	Shipping and receiving air curtain		S.F. Floor	5.56	5.56	5.5%
1040	Special Facilities	N/A		—	—	—	
G. BUILDING SITEWORK	**N/A**						

	Sub-Total	101.73	**100%**
CONTRACTOR FEES (General Requirements: 10%, Overhead: 5%, Profit: 10%)	25%	25.47	
ARCHITECT FEES	7%	8.90	
Total Building Cost		**136.10**	

For customer support on your Square Foot Costs with RSMeans data, call 800.448.8182.

Costs per square foot of floor area

Exterior Wall	S.F. Area	2000	3000	5000	8000	12000	20000	30000	50000	100000
	L.F. Perimeter	180	220	300	420	580	900	1300	2100	4100
Concrete Block	Rigid Steel	112.75	99.25	88.50	82.45	79.05	76.40	75.05	73.95	73.20
	Reinforced Concrete	124.10	111.30	101.20	95.55	92.30	89.75	88.50	87.45	86.75
Metal Sandwich Panel	Rigid Steel	112.60	99.10	88.40	82.40	79.00	76.30	74.95	73.85	73.10
Tilt-up Concrete Panels	Reinforced Concrete	114.40	101.25	90.80	84.90	81.60	79.00	77.70	76.65	75.90
Precast Concrete	Rigid Steel	136.90	118.85	104.55	96.50	92.00	88.45	86.65	85.20	84.15
	Reinforced Concrete	170.25	151.20	136.05	127.50	122.75	118.95	117.10	115.50	114.40
Perimeter Adj., Add or Deduct	Per 100 L.F.	22.30	14.85	9.00	5.60	3.75	2.20	1.50	0.90	0.40
Story Hgt. Adj., Add or Deduct	Per 1 Ft.	1.50	1.20	1.00	0.90	0.80	0.70	0.70	0.70	0.65
Basement—Not Applicable										

The above costs were calculated using the basic specifications shown on the facing page. These costs should be adjusted where necessary for design alternatives and owner's requirements.

Common additives

Description	Unit	$ Cost
Dock Leveler, 10 ton cap.		
6' x 8'	Ea.	6250
7' x 8'	Ea.	8625
Emergency Lighting, 25 watt, battery operated		
Lead battery	Ea.	330
Nickel cadmium	Ea.	550
Fence, Chain link, 6' high		
9 ga. wire	L.F.	29
6 ga. wire	L.F.	34.50
Gate	Ea.	400
Flagpoles, Complete		
Aluminum, 20' high	Ea.	1900
40' high	Ea.	4500
70' high	Ea.	11,600
Fiberglass, 23' high	Ea.	1450
39'-5" high	Ea.	3325
59' high	Ea.	7300
Paving, Bituminous		
Wearing course plus base course	S.Y.	7.40
Sidewalks, Concrete 4" thick	S.F.	4.99

Description	Unit	$ Cost
Sound System		
Amplifier, 250 watts	Ea.	2025
Speaker, ceiling or wall	Ea.	240
Trumpet	Ea.	460
Yard Lighting, 20' aluminum pole with 400 watt high pressure sodium fixture	Ea.	3600

Important: See the Reference Section for Location Factors.

Model costs calculated for a 1 story building with 12' story height and 20,000 square feet of floor area

Warehouse Self Storage

					Unit	Unit Cost	Cost Per S.F.	% Of Sub-Total
A. SUBSTRUCTURE								
1010	Standard Foundations	Poured concrete; strip and spread footings; 4' foundation wall			S.F. Ground	6.53	6.53	
1020	Special Foundations	N/A			—	—	—	
1030	Slab on Grade	4" reinforced concrete with vapor barrier and granular base			S.F. Slab	6.97	6.97	24.8%
2010	Basement Excavation	Site preparation for slab and trench for foundation wall and footing			S.F. Ground	.33	.66	
2020	Basement Walls	N/A			—	—	—	
B. SHELL								
B10 Superstructure								
1010	Floor Construction	Steel column fireproofing			S.F. Floor	.78	.78	13.9%
1020	Roof Construction	Metal deck, open web steel joists, beams, columns			S.F. Roof	7.17	7.17	
B20 Exterior Enclosure								
2010	Exterior Walls	Concrete block	53% of wall		S.F. Wall	23.41	6.70	
2020	Exterior Windows	Aluminum projecting	1% of wall		Each	930	.09	24.2%
2030	Exterior Doors	Steel overhead, hollow metal	46% of wall		Each	1734	7.02	
B30 Roofing								
3010	Roof Coverings	Metal panel roof; perlite/EPS composite insulation			S.F. Roof	7.39	7.39	12.9%
3020	Roof Openings	N/A			—	—	—	
C. INTERIORS								
1010	Partitions	Metal panels on metal studs	11 S.F. Floor/L.F. Partition		S.F. Partition	3.01	3.28	
1020	Interior Doors	Single leaf hollow metal	20,000 S.F. Floor/Door		Each	600	.06	
1030	Fittings	N/A			—	—	—	
2010	Stair Construction	N/A			—	—	—	5.8%
3010	Wall Finishes	N/A			—	—	—	
3020	Floor Finishes	N/A			—	—	—	
3030	Ceiling Finishes	N/A			—	—	—	
D. SERVICES								
D10 Conveying								
1010	Elevators & Lifts	N/A			—	—	—	0.0 %
1020	Escalators & Moving Walks	N/A			—	—	—	
D20 Plumbing								
2010	Plumbing Fixtures	Toilet and service fixtures, supply and drainage	1 Fixture/10,000 S.F. Floor		Each	2800	.28	
2020	Domestic Water Distribution	Gas fired water heater			S.F. Floor	.36	.36	2.4%
2040	Rain Water Drainage	Roof drain			S.F. Roof	.73	.73	
D30 HVAC								
3010	Energy Supply	N/A			—	—	—	
3020	Heat Generating Systems	N/A			—	—	—	
3030	Cooling Generating Systems	N/A			—	—	—	2.3 %
3050	Terminal & Package Units	Single zone rooftop unit			S.F. Floor	1.31	1.31	
3090	Other HVAC Sys. & Equipment	N/A			—	—	—	
D40 Fire Protection								
4010	Sprinklers	N/A			—	—	—	0.0 %
4020	Standpipes	N/A			—	—	—	
D50 Electrical								
5010	Electrical Service/Distribution	60 ampere service, panel board and feeders			S.F. Floor	.39	.39	
5020	Lighting & Branch Wiring	High bay fixtures, receptacles, switches and misc. power			S.F. Floor	6	6	13.6%
5030	Communications & Security	Addressable alarm systems			S.F. Floor	1.38	1.38	
5090	Other Electrical Systems	N/A			—	—	—	
E. EQUIPMENT & FURNISHINGS								
1010	Commercial Equipment	N/A			—	—	—	
1020	Institutional Equipment	N/A			—	—	—	
1030	Vehicular Equipment	N/A			—	—	—	0.0 %
1090	Other Equipment	N/A			—	—	—	
F. SPECIAL CONSTRUCTION								
1020	Integrated Construction	N/A			—	—	—	0.0 %
1040	Special Facilities	N/A			—	—	—	
G. BUILDING SITEWORK	**N/A**							

		Sub-Total	57.10	100%
CONTRACTOR FEES (General Requirements: 10%, Overhead: 5%, Profit: 10%)			25%	14.30	
ARCHITECT FEES			7%	5	
		Total Building Cost	**76.40**	

For customer support on your Square Foot Costs with RSMeans data, call 800.448.8182.

231

Deterioration occurs within the structure itself and is determined by the observation of both materials and equipment.

Curable deterioration can be remedied either by maintenance, repair or replacement, within prudent economic limits.

Incurable deterioration that has progressed to the point of actually affecting the structural integrity of the structure, making repair or replacement not economically feasible.

Actual Versus Observed Age

The observed age of a structure refers to the age that the structure appears to be. Periodic maintenance, remodeling, and renovation all tend to reduce the amount of deterioration that has taken place, thereby decreasing the observed age. Actual age, on the other hand, relates solely to the year that the structure was built.

The Depreciation Table shown here relates to the observed age.

Obsolescence arises from conditions either occurring within the structure (functional) or caused by factors outside the limits of the structure (economic).

Functional obsolescence is any inadequacy caused by outmoded design, dated construction materials, or oversized or undersized areas, all of which cause excessive operation costs.

Incurable is so costly as not to economically justify the capital expenditure required to correct the deficiency.

Economic obsolescence is caused by factors outside the limits of the structure. The prime causes of economic obsolescence are:

- zoning and environmental laws
- government legislation
- negative neighborhood influences
- business climate
- proximity to transportation facilities

Depreciation Table
Commercial/Industrial/Institutional

Building Material			
Observed Age (Years)	Frame	Masonry On Wood	Masonry On Masonry Or Steel
1	1%	0%	0%
2	2	1	0
3	3	2	1
4	4	3	2
5	6	5	3
10	20	15	8
15	25	20	15
20	30	25	20
25	35	30	25
30	40	35	30
35	45	40	35
40	50	45	40
45	55	50	45
50	60	55	50
55	65	60	55
60	70	65	60

Green Commercial/Industrial/ Institutional Section

Table of Contents

Costs per square foot of floor area

Exterior Wall	S.F. Area	2000	2700	3400	4100	4800	5500	6200	6900	7600
	L.F. Perimeter	180	208	236	256	280	303	317	337	357
Face Brick and Concrete Block	Rigid Steel	307.90	287.65	275.80	265.65	259.55	254.65	249.20	245.90	243.20
	Reinforced Concrete	325.55	305.30	293.50	283.35	277.20	272.30	266.85	263.55	260.85
Precast Concrete	Rigid Steel	301.40	279.65	266.90	255.90	249.35	244.00	238.10	234.55	231.60
	Reinforced Concrete	336.25	314.45	301.75	290.75	284.15	278.85	272.95	269.35	266.45
Limestone and Concrete Block	Rigid Steel	334.35	310.45	296.50	284.45	277.20	271.40	264.85	260.95	257.70
	Reinforced Concrete	350.50	326.60	312.65	300.60	293.35	287.55	281.00	277.05	273.85
Perimeter Adj., Add or Deduct	Per 100 L.F.	58.70	43.45	34.50	28.65	24.40	21.30	18.90	17.00	15.40
Story Hgt. Adj., Add or Deduct	Per 1 Ft.	5.65	4.80	4.35	3.90	3.65	3.40	3.20	3.05	2.90
For Basement, add $39.57 per square foot of basement area										

The above costs were calculated using the basic specifications shown on the facing page. These costs should be adjusted where necessary for design alternatives and owner's requirements.

Common additives

Description	Unit	$ Cost
Bulletproof Teller Window, 44" x 60"	Ea.	5850
60" x 48"	Ea.	7700
Closed Circuit Surveillance, one station, camera & monitor	Ea.	1475
For additional camera stations, add	Ea.	665
Counters, complete, dr & frame, 3' x 6'-8", bullet resist stl		
with vision panel	Ea.	7375 - 9525
Drive-up Window, drawer & micr., not incl. glass	Ea.	9525 - 13,300
Night Depository	Ea.	10,000 - 15,200
Package Receiver, painted	Ea.	1975
Stainless steel	Ea.	3050
Partitions, bullet resistant to 8' high	L.F.	345 - 545
Pneumatic Tube Systems, 2 station	Ea.	34,200
With TV viewer	Ea.	62,500
Service Windows, pass thru, steel, 24" x 36"	Ea.	4100
48" x 48"	Ea.	4175
Twenty-four Hour Teller, automatic deposit cash & memo	Ea.	54,500
Vault Front, door & frame, 2 hour test, 32" door	Opng.	9800
4 hour test, 40" door	Opng.	13,800
Time lock, two movement, add	Ea.	2400

Description	Unit	$ Cost
Commissioning Fees, sustainable commercial construction	S.F.	0.24 - 3.04
Energy Modelling Fees, banks to 10,000 SF	Ea.	11,000
Green Bldg Cert Fees for comm construction project reg	Project	900
Photovoltaic Pwr Sys, grid connected, 20 kW (~2400 SF), roof	Ea.	249,200
Green Roofs, 6" soil depth, w/treated wd edging & sedum mats	S.F.	12.41
10" Soil depth, with treated wood edging & sedum mats	S.F.	14.10
Greywater Recovery Systems, prepackaged comm, 1530 gal	Ea.	33,855
Rainwater Harvest Sys, prepckged comm, 10,000 gal, sys contrller	Ea.	37,550
20,000 gal. w/system controller	Ea.	61,000
30,000 gal. w/system controller	Ea.	97,350
Solar Domestic HW, closed loop, add-on sys, ext heat exchanger	Ea.	11,300
Draindown, hot water system, 120 gal tank	Ea.	14,425

Important: See the Reference Section for Location Factors.

Model costs calculated for a 1 story building with 14' story height and 4,100 square feet of floor area

G Bank

				Unit	Unit Cost	Cost Per S.F.	% Of Sub-Total
A. SUBSTRUCTURE							
1010	Standard Foundations	Poured concrete; strip and spread footings		S.F. Ground	4.62	4.62	
1020	Special Foundations	N/A		—	—	—	
1030	Slab on Grade	4" reinforced concrete with recycled vapor barrier and granular base		S.F. Slab	5.84	5.84	8.9%
2010	Basement Excavation	Site preparation for slab and trench for foundation wall and footing		S.F. Ground	.33	.33	
2020	Basement Walls	4' Foundation wall		L.F. Wall	98	7.34	
B. SHELL							
B10 Superstructure							
1010	Floor Construction	Cast-in-place columns		L.F. Column	6.39	6.39	11.6%
1020	Roof Construction	Cast-in-place concrete flat plate		S.F. Roof	17.25	17.25	
B20 Exterior Enclosure							
2010	Exterior Walls	Face brick with concrete block backup	80% of wall	S.F. Wall	41.20	28.81	
2020	Exterior Windows	Horizontal aluminum sliding	20% of wall	Each	589	6.86	18.6%
2030	Exterior Doors	Double aluminum and glass and hollow metal, low VOC paint		Each	4560	2.22	
B30 Roofing							
3010	Roof Coverings	Single-ply TPO membrane, 60 mils, heat welded seams w/4" thk. R20 insul.		S.F. Roof	8.77	8.77	4.3%
3020	Roof Openings	N/A		—	—	—	
C. INTERIORS							
1010	Partitions	Gypsum board on metal studs w/sound attenuation	20 SF of Flr./LF Part.	S.F. Partition	13.32	6.66	
1020	Interior Doors	Single leaf hollow metal, low VOC paint	200 S.F. Floor/Door	Each	1222	6.11	
1030	Fittings	N/A		—	—	—	
2010	Stair Construction	N/A		—	—	—	13.7%
3010	Wall Finishes	50% vinyl wall covering, 50% paint, low VOC		S.F. Surface	1.57	1.57	
3020	Floor Finishes	50% carpet tile, 40% vinyl composition tile, recycled, 10% quarry tile		S.F. Floor	6.10	6.10	
3030	Ceiling Finishes	Mineral fiber tile on concealed zee bars		S.F. Ceiling	7.60	7.60	
D. SERVICES							
D10 Conveying							
1010	Elevators & Lifts	N/A		—	—	—	0.0 %
1020	Escalators & Moving Walks	N/A		—	—	—	
D20 Plumbing							
2010	Plumbing Fixtures	Toilet low flow, auto sensor and service fixt., supply and drainage	1 Fixt./580 SF Flr.	Each	6247	10.77	
2020	Domestic Water Distribution	Tankless, on demand water heaters, natural gas/propane		S.F. Floor	1.06	1.06	6.6%
2040	Rain Water Drainage	Roof drains		S.F. Roof	1.67	1.67	
D30 HVAC							
3010	Energy Supply	N/A		—	—	—	
3020	Heat Generating Systems	Included in D3050		—	—	—	
3040	Distribution Systems	Enthalpy heat recovery packages		Each	15,550	3.79	6.1 %
3050	Terminal & Package Units	Single zone rooftop air conditioner		S.F. Floor	8.67	8.67	
3090	Other HVAC Sys. & Equipment	N/A		—	—	—	
D40 Fire Protection							
4010	Sprinklers	Wet pipe sprinkler system		S.F. Floor	5.13	5.13	4.1%
4020	Standpipes	Standpipe		S.F. Floor	3.25	3.25	
D50 Electrical							
5010	Electrical Service/Distribution	200 ampere service, panel board and feeders		S.F. Floor	1.51	1.51	
5020	Lighting & Branch Wiring	LED fixtures, daylt. dimming & ltg. on/off control, receptacles, switches, A.C.		S.F. Floor	11.50	11.50	13.8%
5030	Communications & Security	Alarm systems, internet/phone wiring, and security television		S.F. Floor	11.15	11.15	
5090	Other Electrical Systems	Emergency generator, 15 kW, UPS, energy monitoring systems		S.F. Floor	4	4	
E. EQUIPMENT & FURNISHINGS							
1010	Commercial Equipment	Automatic teller, drive up window, night depository		S.F. Floor	8.37	8.37	
1020	Institutional Equipment	Closed circuit TV monitoring system		S.F. Floor	2.95	2.95	6.3%
1090	Other Equipment	Waste handling recycling tilt truck		S.F. Floor	1.52	1.52	
2020	Moveable Furnishings	No smoking signage		S.F. Floor	.04	.04	
F. SPECIAL CONSTRUCTION							
1020	Integrated Construction	N/A		—	—	—	6.1 %
1040	Special Facilities	Security vault door		S.F. Floor	12.36	12.36	
G. BUILDING SITEWORK	**N/A**						

		Sub-Total	204.21	100%
CONTRACTOR FEES (General Requirements: 10%, Overhead: 5%, Profit: 10%)		25%	51.06	
ARCHITECT FEES		11%	28.08	
	Total Building Cost		**283.35**	

For customer support on your Square Foot Costs with RSMeans data, call 800.448.8182.

Costs per square foot of floor area

Exterior Wall	S.F. Area	15000	20000	28000	38000	50000	65000	85000	100000	150000
	L.F. Perimeter	350	400	480	550	630	660	750	825	1035
Face Brick and Concrete Block	Rigid Steel	266.10	253.35	242.45	233.35	227.05	219.80	215.50	213.70	209.35
	Bearing Walls	257.95	244.95	233.85	224.50	218.10	210.65	206.30	204.35	199.95
Decorative Concrete Block	Rigid Steel	256.75	245.35	235.60	227.55	221.95	215.70	211.95	210.35	206.60
	Bearing Walls	246.75	235.15	225.15	216.90	211.25	204.80	200.95	199.30	195.40
Stucco and Concrete Block	Rigid Steel	269.80	256.50	245.15	235.60	229.00	221.40	216.90	215.00	210.45
	Bearing Walls	261.65	248.10	236.55	226.80	220.10	212.25	207.65	205.70	201.05
Perimeter Adj., Add or Deduct	Per 100 L.F.	16.35	12.30	8.80	6.45	4.90	3.80	2.95	2.40	1.70
Story Hgt. Adj., Add or Deduct	Per 1 Ft.	3.00	2.55	2.25	1.85	1.60	1.30	1.15	1.00	0.90

For Basement, add $42.46 per square foot of basement area

The above costs were calculated using the basic specifications shown on the facing page. These costs should be adjusted where necessary for design alternatives and owner's requirements.

Common additives

Description	Unit	$ Cost
Clock System, 20 room	Ea.	20,200
50 room	Ea.	48,200
Elevators, hydraulic passenger, 2 stops, 2000# capacity	Ea.	74,100
3500# capacity	Ea.	81,600
Additional stop, add	Ea.	8375
Seating, auditorium chair, all veneer	Ea.	355
Veneer back, padded seat	Ea.	380
Upholstered, spring seat	Ea.	335
Classroom, movable chair & desk	Set	81 - 171
Lecture hall, pedestal type	Ea.	350 - 650
Sound System, amplifier, 250 watts	Ea.	2025
Speaker, ceiling or wall	Ea.	240
Trumpet	Ea.	460
TV Antenna, Master system, 30 outlet	Outlet	335
100 outlet	Outlet	370

Description	Unit	$ Cost
Commissioning Fees, sustainable institutional construction	S.F.	0.58 - 2.47
Energy Modelling Fees, academic buildings to 10,000 SF	Ea.	9000
Greater than 10,000 SF add	S.F.	0.20
Green Bldg Cert Fees for school construction project reg	Project	900
Photovoltaic Pwr Sys, grid connected, 20 kW (~2400 SF), roof	Ea.	249,200
Green Roofs, 6" soil depth, w/treated wd edging & sedum mats	S.F.	12.41
10" Soil depth, with treated wood edging & sedum mats	S.F.	14.10
Greywater Recovery Systems, prepackaged comm, 3060 gal.	Ea.	45,900
4590 gal.	Ea.	57,625
Rainwater Harvest Sys, prepckged comm, 10,000 gal, sys contrller	Ea.	37,550
20,000 gal. w/system controller	Ea.	61,000
30,000 gal. w/system controller	Ea.	97,350
Solar Domestic HW, closed loop, add-on sys, ext heat exchanger	Ea.	11,300
Drainback, hot water system, 120 gal tank	Ea.	14,175
Draindown, hot water system, 120 gal tank	Ea.	14,425

Important: See the Reference Section for Location Factors.

Model costs calculated for a 2 story building with 12' story height and 50,000 square feet of floor area

				Unit	Unit Cost	Cost Per S.F.	% Of Sub-Total
A.	**SUBSTRUCTURE**						
1010	Standard Foundations	Poured concrete; strip and spread footings		S.F. Ground	1.38	.69	
1020	Special Foundations	N/A		—	—	—	
1030	Slab on Grade	4" reinforced concrete with vapor barrier and granular base		S.F. Slab	5.84	2.93	3.6%
2010	Basement Excavation	Site preparation for slab and trench for foundation wall and footing		S.F. Ground	.31	.16	
2020	Basement Walls	4' Foundation wall		L.F. Wall	186	1.91	
B.	**SHELL**						
	B10 Superstructure						
1010	Floor Construction	Open web steel joists, slab form, concrete		S.F. Floor	18.66	9.33	9.3%
1020	Roof Construction	Metal deck on open web steel joists, columns		S.F. Roof	10.84	5.42	
	B20 Exterior Enclosure						
2010	Exterior Walls	Decorative concrete block	65% of wall	S.F. Wall	21.83	4.29	
2020	Exterior Windows	Window wall	35% of wall	Each	58	5.77	6.9%
2030	Exterior Doors	Double glass and aluminum with transom		Each	7200	.87	
	B30 Roofing						
3010	Roof Coverings	Built-up tar and gravel with flashing; perlite/EPS composite insulation		S.F. Roof	7.06	3.53	2.2%
3020	Roof Openings	N/A		—	—	—	
C.	**INTERIORS**						
1010	Partitions	Concrete block	20 SF Flr.LF Part.	S.F. Partition	24.16	12.08	
1020	Interior Doors	Single leaf hollow metal	200 S.F. Floor/Door	Each	1222	6.11	
1030	Fittings	Chalkboards, counters, cabinets		S.F. Floor	6.31	6.31	
2010	Stair Construction	Concrete filled metal pan		Flight	18,900	3.78	28.4%
3010	Wall Finishes	95% paint, 5% ceramic tile		S.F. Surface	4.53	4.53	
3020	Floor Finishes	70% vinyl composition tile, 25% carpet, 5% ceramic tile		S.F. Floor	4.42	4.42	
3030	Ceiling Finishes	Mineral fiber tile on concealed zee bars		S.F. Ceiling	7.60	7.60	
D.	**SERVICES**						
	D10 Conveying						
1010	Elevators & Lifts	Two hydraulic passenger elevators		Each	91,250	3.65	2.3%
1020	Escalators & Moving Walks	N/A		—	—	—	
	D20 Plumbing						
2010	Plumbing Fixtures	Toilet, low flow, auto sensor & service fixt., supply & drain.	1 Fixt./455 SF Flr.	Each	8959	19.69	
2020	Domestic Water Distribution	Oil fired hot water heater		S.F. Floor	.84	.84	13.5%
2040	Rain Water Drainage	Roof drains		S.F. Roof	1.66	.83	
	D30 HVAC						
3010	Energy Supply	N/A		—	—	—	
3020	Heat Generating Systems	Included in D3050		—	—	—	
3040	Distribution Systems	Enthalpy heat recovery packages		Each	38,725	2.33	14.3 %
3050	Terminal & Package Units	Multizone unit, gas heating, electric cooling, SEER 14		S.F. Floor	20.20	20.20	
3090	Other HVAC Sys. & Equipment	N/A		—	—	—	
	D40 Fire Protection						
4010	Sprinklers	Sprinklers, light hazard		S.F. Floor	3.30	3.30	2.3%
4020	Standpipes	Standpipe, dry, Class III		S.F. Floor	.39	.39	
	D50 Electrical						
5010	Electrical Service/Distribution	2000 ampere service, panel board and feeders		S.F. Floor	3.02	3.02	
5020	Lighting & Branch Wiring	LED fixtures, receptacles, switches, A.C. & misc. power		S.F. Floor	14.62	14.62	17.0%
5030	Communications & Security	Addressable alarm sys., internet wiring, comm. sys. & emerg. ltg.		S.F. Floor	7.43	7.43	
5090	Other Electrical Systems	Emergency generator, 100 kW		S.F. Floor	1.77	1.77	
E.	**EQUIPMENT & FURNISHINGS**						
1010	Commercial Equipment	N/A		—	—	—	
1020	Institutional Equipment	N/A		—	—	—	
1090	Other Equipment	Waste handling recycling tilt truck		S.F. Floor	.13	.13	0.1 %
2020	Moveable Furnishings	No smoking signage		S.F. Floor			
F.	**SPECIAL CONSTRUCTION**						
1020	Integrated Construction	N/A		—	—	—	
1040	Special Facilities	N/A		—	—	—	0.0 %
G.	**BUILDING SITEWORK**	**N/A**					

			Sub-Total	157.93	**100%**
	CONTRACTOR FEES (General Requirements: 10%, Overhead: 5%, Profit: 10%)		25%	39.50	
	ARCHITECT FEES		7%	13.82	
		Total Building Cost		**211.25**	

For customer support on your Square Foot Costs with RSMeans data, call 800.448.8182.

237

Costs per square foot of floor area

Exterior Wall	S.F. Area	10000	15000	25000	40000	55000	70000	80000	90000	100000
	L.F. Perimeter	260	320	400	476	575	628	684	721	772
Face Brick and Concrete Block	Reinforced Concrete	270.95	255.50	239.45	227.95	223.70	219.75	218.45	217.05	216.20
	Rigid Steel	283.10	267.65	251.65	240.15	235.85	231.90	230.65	229.20	228.35
Decorative Concrete Block	Reinforced Concrete	251.55	239.60	227.55	219.10	215.85	213.05	212.10	211.10	210.45
	Rigid Steel	263.70	251.75	239.70	231.25	228.00	225.25	224.30	223.25	222.60
Precast Concrete	Reinforced Concrete	290.25	271.35	251.35	236.80	231.40	226.40	224.85	223.00	221.90
	Rigid Steel	302.45	283.50	263.50	249.00	243.60	238.60	237.05	235.15	234.10
Perimeter Adj., Add or Deduct	Per 100 L.F.	22.85	15.25	9.15	5.75	4.10	3.25	2.95	2.55	2.30
Story Hgt. Adj., Add or Deduct	Per 1 Ft.	4.40	3.65	2.75	2.05	1.75	1.50	1.50	1.35	1.30

For Basement, add $41.33 per square foot of basement area

The above costs were calculated using the basic specifications shown on the facing page. These costs should be adjusted where necessary for design alternatives and owner's requirements.

Common additives

Description	Unit	$ Cost
Closed Circuit Surveillance, one station, camera & monitor	Ea.	1475
For additional camera stations, add	Ea.	665
Elevators, hydraulic passenger, 2 stops, 3500# capacity	Ea.	81,600
Additional stop, add	Ea.	8375
Furniture	Student	2800 - 5350
Intercom System, 25 station capacity		
Master station	Ea.	3225
Intercom outlets	Ea.	212
Kitchen Equipment		
Broiler	Ea.	4425
Coffee urn, twin 6 gallon	Ea.	2900
Cooler, 6 ft. long	Ea.	3750
Dishwasher, 10-12 racks per hr.	Ea.	4150
Food warmer	Ea.	790
Freezer, 44 C.F., reach-in	Ea.	5525
Ice cube maker, 50 lb. per day	Ea.	2100
Range with 1 oven	Ea.	3575
Laundry Equipment, dryer, 30 lb. capacity	Ea.	4100
Washer, commercial	Ea.	1700
TV Antenna, Master system, 30 outlet	Outlet	335
100 outlet	Outlet	370

Description	Unit	$ Cost
Commissioning Fees, sustainable institutional construction	S.F.	0.58 - 2.47
Energy Modelling Fees, academic buildings to 10,000 SF	Ea.	9000
Greater than 10,000 SF add	S.F.	0.20
Green Bldg Cert Fees for school construction project reg	Project	900
Photovoltaic Pwr Sys, grid connected, 20 kW (~2400 SF), roof	Ea.	249,200
Green Roofs, 6" soil depth, w/treated wd edging & sedum mats	S.F.	12.41
10" Soil depth, with treated wood edging & sedum mats	S.F.	14.10
Greywater Recovery Systems, prepackaged comm, 3060 gal.	Ea.	45,900
4590 gal.	Ea.	57,625
Rainwater Harvest Sys, prepckged comm, 30,000 gal, sys contrller	Ea.	97,350
Solar Domestic HW, closed loop, add-on sys, ext heat exchanger	Ea.	11,300
Drainback, hot water system, 120 gal tank	Ea.	14,175
Draindown, hot water system, 120 gal tank	Ea.	14,425

Important: See the Reference Section for Location Factors.

Model costs calculated for a 3 story building with 12′ story height and 25,000 square feet of floor area

G College, Dormitory, 2-3 Story

				Unit	Unit Cost	Cost Per S.F.	% Of Sub-Total
A. SUBSTRUCTURE							
1010	Standard Foundations	Poured concrete; strip and spread footings		S.F. Ground	6.81	2.27	
1020	Special Foundations	N/A		—	—	—	
1030	Slab on Grade	4″ reinforced concrete with recycled vapor barrier and granular base		S.F. Slab	5.84	1.95	3.3%
2010	Basement Excavation	Site preparation for slab and trench for foundation wall and footing		S.F. Ground	.18	.06	
2020	Basement Walls	4′ Foundation wall		L.F. Wall	98	1.57	
B. SHELL							
	B10 Superstructure						
1010	Floor Construction	Concrete flat plate		S.F. Floor	27.21	18.14	12.9%
1020	Roof Construction	Concrete flat plate		S.F. Roof	14.88	4.96	
	B20 Exterior Enclosure						
2010	Exterior Walls	Face brick with concrete block backup	80% of wall	S.F. Wall	41.19	18.98	
2020	Exterior Windows	Aluminum horizontal sliding	20% of wall	Each	755	3.78	13.8%
2030	Exterior Doors	Double glass & aluminum doors		Each	7800	1.88	
	B30 Roofing						
3010	Roof Coverings	Single-ply TPO membrane, heat welded w/R-20 insul.		S.F. Roof	8.10	2.70	1.5%
3020	Roof Openings	N/A		—	—	—	
C. INTERIORS							
1010	Partitions	Gypsum board on metal studs, CMU w/foamed-in insul.	9 SF Flr./LF Part.	S.F. Partition	8.20	9.11	
1020	Interior Doors	Single leaf wood, low VOC paint	90 S.F. Floor/Door	Each	720	8	
1030	Fittings	Closet shelving, mirrors, bathroom accessories		S.F. Floor	1.78	1.78	
2010	Stair Construction	Cast in place concrete		Flight	5850	3.05	18.6%
3010	Wall Finishes	95% paint, low VOC paint, 5% ceramic tile		S.F. Surface	2.20	4.89	
3020	Floor Finishes	80% carpet tile, 10% vinyl comp. tile, recycled content, 10% ceramic tile		S.F. Floor	5.23	5.23	
3030	Ceiling Finishes	90% paint, low VOC, 10% suspended fiberglass board		S.F. Ceiling	1.31	1.32	
D. SERVICES							
	D10 Conveying						
1010	Elevators & Lifts	One hydraulic passenger elevator		Each	124,000	4.96	2.8%
1020	Escalators & Moving Walks	N/A		—	—	—	
	D20 Plumbing						
2010	Plumbing Fixtures	Toilet, low flow, auto sensor, & service fixt., supply & drain.	1 Fixt./455 SF Flr.	Each	12,221	26.86	
2020	Domestic Water Distribution	Electric, point-of-use water heater		S.F. Floor	1.31	1.31	16.1%
2040	Rain Water Drainage	Roof drains		S.F. Roof	1.89	.63	
	D30 HVAC						
3010	Energy Supply	N/A		—	—	—	
3020	Heat Generating Systems	Included in D3050		—	—	—	
3040	Distribution Systems	Enthalpy heat recovery packages		Each	31,850	1.27	8.1 %
3050	Terminal & Package Units	Rooftop multizone unit system, SEER 14		S.F. Floor	13.31	13.31	
3090	Other HVAC Sys. & Equipment	N/A		—	—	—	
	D40 Fire Protection						
4010	Sprinklers	Wet pipe sprinkler system		S.F. Floor	3.12	3.12	2.3%
4020	Standpipes	Standpipe, dry, Class III		S.F. Floor	1	1	
	D50 Electrical						
5010	Electrical Service/Distribution	800 ampere service, panel board and feeders		S.F. Floor	1.58	1.58	
5020	Lighting & Branch Wiring	LED fixtures, daylt. dim., ltg. on/off, recept., switches, and A.C. power		S.F. Floor	19.83	19.83	18.3%
5030	Communications & Security	Alarm sys., internet & phone wiring, comm. sys. and emergency ltg.		S.F. Floor	10.28	10.28	
5090	Other Electrical Systems	Emergency generator, 7.5 kW, energy monitoring systems		S.F. Floor	1.11	1.11	
E. EQUIPMENT & FURNISHINGS							
1010	Commercial Equipment	N/A		—	—	—	
1020	Institutional Equipment	N/A		—	—	—	2.3 %
1090	Other Equipment	Waste handling recycling tilt truck		S.F. Floor	.25	.25	
2020	Moveable Furnishings	Dormitory furniture, no smoking signage		S.F. Floor	3.86	3.86	
F. SPECIAL CONSTRUCTION							
1020	Integrated Construction	N/A		—	—	—	0.0 %
1040	Special Facilities	N/A		—	—	—	
G. BUILDING SITEWORK	**N/A**						

		Sub-Total	179.04	100%
CONTRACTOR FEES (General Requirements: 10%, Overhead: 5%, Profit: 10%)		25%	44.74	
ARCHITECT FEES		7%	15.67	

Total Building Cost	**239.45**

For customer support on your Square Foot Costs with RSMeans data, call 800.448.8182.

Costs per square foot of floor area

Exterior Wall	S.F. Area	20000	35000	45000	65000	85000	110000	135000	160000	200000
	L.F. Perimeter	260	340	400	440	500	540	560	590	640
Face Brick and Concrete Block	Reinforced Concrete	269.60	250.20	245.05	234.10	229.35	224.40	220.65	218.35	215.90
	Rigid Steel	282.85	263.50	258.35	247.35	242.60	237.70	233.90	231.60	229.15
Decorative Concrete Block	Reinforced Concrete	252.05	237.05	233.05	224.90	221.40	217.80	215.05	213.35	211.60
	Rigid Steel	265.30	250.35	246.30	238.20	234.65	231.05	228.30	226.60	224.85
Precast Concrete	Reinforced Concrete	274.95	254.15	248.65	236.85	231.75	226.45	222.30	219.85	217.25
	Rigid Steel	288.20	267.45	261.90	250.10	245.00	239.70	235.55	233.10	230.50
Perimeter Adj., Add or Deduct	Per 100 L.F.	22.25	12.75	9.90	6.85	5.20	4.10	3.30	2.80	2.25
Story Hgt. Adj., Add or Deduct	Per 1 Ft.	4.45	3.30	3.05	2.35	2.00	1.70	1.45	1.30	1.10

For Basement, add $41.33 per square foot of basement area

The above costs were calculated using the basic specifications shown on the facing page. These costs should be adjusted where necessary for design alternatives and owner's requirements.

Common additives

Description	Unit	$ Cost
Closed Circuit Surveillance, one station, camera & monitor	Ea.	1475
For additional camera stations, add	Ea.	665
Elevators, electric passenger, 5 stops, 3500# capacity	Ea.	193,000
Additional stop, add	Ea.	10,900
Furniture	Student	2800 - 5350
Intercom System, 25 station capacity		
Master station	Ea.	3225
Intercom outlets	Ea.	212
Kitchen Equipment		
Broiler	Ea.	4425
Coffee urn, twin, 6 gallon	Ea.	2900
Cooler, 6 ft. long	Ea.	3750
Dishwasher, 10-12 racks per hr.	Ea.	4150
Food warmer	Ea.	790
Freezer, 44 C.F., reach-in	Ea.	5525
Ice cube maker, 50 lb. per day	Ea.	2100
Range with 1 oven	Ea.	3575
Laundry Equipment, dryer, 30 lb. capacity	Ea.	4100
Washer, commercial	Ea.	1700
TV Antenna, Master system, 30 outlet	Outlet	335
100 outlet	Outlet	370

Description	Unit	$ Cost
Commissioning Fees, sustainable institutional construction	S.F.	0.58 - 2.47
Energy Modelling Fees, academic buildings to 10,000 SF	Ea.	9000
Greater than 10,000 SF add	S.F.	0.20
Green Bldg Cert Fees for school construction project reg	Project	900
Photovoltaic Pwr Sys, grid connected, 20 kW (~2400 SF), roof	Ea.	249,200
Green Roofs, 6" soil depth, w/treated wd edging & sedum mats	S.F.	12.41
10" Soil depth, with treated wood edging & sedum mats	S.F.	14.10
Greywater Recovery Systems, prepackaged comm, 3060 gal.	Ea.	45,900
4590 gal.	Ea.	57,625
Rainwater Harvest Sys, prepckged comm, 30,000 gal, sys contrller	Ea.	97,350
Solar Domestic HW, closed loop, add-on sys, ext heat exchanger	Ea.	11,300
Drainback, hot water system, 120 gal tank	Ea.	14,175
Draindown, hot water system, 120 gal tank	Ea.	14,425

Important: See the Reference Section for Location Factors.

Model costs calculated for a 6 story building with 12' story height and 85,000 square feet of floor area

G College, Dormitory, 4-8 Story

				Unit	Unit Cost	Cost Per S.F.	% Of Sub-Total
A. SUBSTRUCTURE							
1010	Standard Foundations	Poured concrete; strip and spread footings		S.F. Ground	10.14	1.69	
1020	Special Foundations	N/A		—	—	—	
1030	Slab on Grade	4" reinforced concrete with recycled vapor barrier and granular base		S.F. Slab	5.84	.98	2.0%
2010	Basement Excavation	Site preparation for slab and trench for foundation wall and footing		S.F. Ground	.31	.05	
2020	Basement Walls	4' foundation wall		L.F. Wall	186	.81	
B. SHELL							
	B10 Superstructure						
1010	Floor Construction	Concrete slab with metal deck and beams		S.F. Floor	30.46	25.38	16.6%
1020	Roof Construction	Concrete slab with metal deck and beams		S.F. Roof	22.86	3.81	
	B20 Exterior Enclosure						
2010	Exterior Walls	Decorative concrete block	80% of wall	S.F. Wall	23.67	8.02	
2020	Exterior Windows	Aluminum horizontal sliding	20% of wall	Each	529	2.99	6.5%
2030	Exterior Doors	Double glass & aluminum doors		Each	4663	.44	
	B30 Roofing						
3010	Roof Coverings	Single-ply TPO membrane, 60 mils, heat welded seams w/R-20 insul.		S.F. Roof	7.56	1.26	0.7%
3020	Roof Openings	N/A		—	—	—	
C. INTERIORS							
1010	Partitions	Concrete block w/foamed-in insulation	9 SF Floor/LF Part.	S.F. Partition	13.20	14.67	
1020	Interior Doors	Single leaf wood, low VOC paint	90 S.F. Floor/Door	Each	725	8.06	
1030	Fittings	Closet shelving, mirrors, bathroom accessories		S.F. Floor	1.46	1.46	
2010	Stair Construction	Concrete filled metal pan		Flight	18,900	4	23.5%
3010	Wall Finishes	95% paint, low VOC, 5% ceramic tile		S.F. Surface	2.95	6.55	
3020	Floor Finishes	80% carpet tile, 10% vinyl comp. tile, recycled, 10% ceramic tile		S.F. Floor	5.23	5.23	
3030	Ceiling Finishes	Mineral fiber tile on concealed zee bars, paint		S.F. Ceiling	1.31	1.32	
D. SERVICES							
	D10 Conveying						
1010	Elevators & Lifts	Four geared passenger elevators		Each	267,113	12.57	7.2%
1020	Escalators & Moving Walks	N/A		—	—	—	
	D20 Plumbing						
2010	Plumbing Fixtures	Toilet, low flow, auto sensor, & service fixt., supply & drain.	1 Fixt./390 SF Flr.	Each	10,417	26.71	
2020	Domestic Water Distribution	Electric water heater, point-of-use, energy saver		S.F. Floor	.74	.74	15.9%
2040	Rain Water Drainage	Roof drains		S.F. Roof	2.28	.38	
	D30 HVAC						
3010	Energy Supply	N/A		—	—	—	
3020	Heat Generating Systems	N/A		—	—	—	
3040	Distribution Systems	Enthalpy heat recovery packages		Each	31,850	.37	7.8 %
3050	Terminal & Package Units	Multizone rooftop air conditioner, SEER 14		S.F. Floor	13.31	13.31	
3090	Other HVAC Sys. & Equipment	N/A		—	—	—	
	D40 Fire Protection						
4010	Sprinklers	Sprinklers, light hazard		S.F. Floor	3.06	3.06	2.1%
4020	Standpipes	Standpipe, dry, Class III		S.F. Floor	.71	.71	
	D50 Electrical						
5010	Electrical Service/Distribution	1200 ampere service, panel board and feeders		S.F. Floor	.83	.83	
5020	Lighting & Branch Wiring	LED fixtures, daylt. dim., ltg. on/off, recept. switches, and A.C. power		S.F. Floor	18.50	18.50	15.8%
5030	Communications & Security	Alarm systems, internet & phone wiring, comm. systems and emerg. ltg.		S.F. Floor	7.44	7.44	
5090	Other Electrical Systems	Emergency generator, 30 kW, energy monitoring systems		S.F. Floor	.88	.88	
E. EQUIPMENT & FURNISHINGS							
1010	Commercial Equipment	N/A		—	—	—	
1020	Institutional Equipment	N/A		—	—	—	
1090	Other Equipment	Waste handling recycling tilt truck		S.F. Floor	.08	.08	1.8 %
2020	Moveable Furnishings	Built-in dormitory furnishings, no smoking signage		S.F. Floor	3.14	3.14	
F. SPECIAL CONSTRUCTION							
1020	Integrated Construction	N/A		—	—	—	0.0 %
1040	Special Facilities	N/A		—	—	—	
G. BUILDING SITEWORK	**N/A**						

			Sub-Total	175.44	100%
CONTRACTOR FEES (General Requirements: 10%, Overhead: 5%, Profit: 10%)			25%	43.86	
ARCHITECT FEES			7%	15.35	
			Total Building Cost	**234.65**	

For customer support on your Square Foot Costs with RSMeans data, call 800.448.8182.

Costs per square foot of floor area

Exterior Wall	S.F. Area	12000	20000	28000	37000	45000	57000	68000	80000	92000
	L.F. Perimeter	470	600	698	793	900	1000	1075	1180	1275
Face Brick and Concrete Block	Rigid Steel	353.50	297.50	271.55	255.20	246.95	237.20	230.85	226.50	223.05
	Bearing Walls	348.20	292.25	266.25	249.90	241.65	231.90	225.55	221.25	217.75
Decorative Concrete Block	Rigid Steel	344.95	290.95	266.15	250.50	242.60	233.35	227.40	223.25	220.00
	Bearing Walls	339.95	285.95	261.10	245.50	237.60	228.35	222.40	218.25	215.00
Stucco and Concrete Block	Rigid Steel	357.20	300.35	273.95	257.20	248.80	238.90	232.40	227.90	224.35
	Bearing Walls	352.15	295.30	268.85	252.15	243.80	233.85	227.30	222.85	219.35
Perimeter Adj., Add or Deduct	Per 100 L.F.	13.40	8.00	5.75	4.35	3.55	2.85	2.40	2.00	1.70
Story Hgt. Adj., Add or Deduct	Per 1 Ft.	2.60	1.95	1.65	1.35	1.30	1.20	1.05	1.00	0.90

For Basement, add $ 26.69 per square foot of basement area

The above costs were calculated using the basic specifications shown on the facing page. These costs should be adjusted where necessary for design alternatives and owner's requirements.

Common additives

Description	Unit	$ Cost
Cabinets, Base, door units, metal	L.F.	350
Drawer units	L.F.	690
Tall storage cabinets, open	L.F.	640
With doors	L.F.	1025
Wall, metal 12-1/2" deep, open	L.F.	294
With doors	L.F.	520
Countertops, not incl. base cabinets, acid proof	S.F.	67 - 70
Stainless steel	S.F.	196
Fume Hood, Not incl. ductwork	L.F.	845 - 2000
Ductwork	Hood	6450 - 10,200
Glassware Washer, Distilled water rinse	Ea.	8150 - 17,500
Seating, auditorium chair, veneer back, padded seat	Ea.	380
Upholstered, spring seat	Ea.	335
Classroom, movable chair & desk	Set	81 - 171
Lecture hall, pedestal type	Ea.	350 - 650
Safety Equipment, Eye wash, hand held	Ea.	455
Deluge shower	Ea.	900
Tables, acid resist. top, drawers	L.F.	229
Titration Unit, Four 2000 ml reservoirs	Ea.	6650

Description	Unit	$ Cost
Commissioning Fees, sustainable institutional construction	S.F.	0.58 - 2.47
Energy Modelling Fees, academic buildings to 10,000 SF	Ea.	9000
Greater than 10,000 SF add	S.F.	0.20
Green Bldg Cert Fees for school construction project reg	Project	900
Photovoltaic Pwr Sys, grid connected, 20 kW (~2400 SF), roof	Ea.	249,200
Green Roofs, 6" soil depth, w/treated wd edging & sedum mats	S.F.	12.41
10" Soil depth, with treated wood edging & sedum mats	S.F.	14.10
Greywater Recovery Systems, prepackaged comm, 3060 gal.	Ea.	45,900
4590 gal.	Ea.	57,625
Rainwater Harvest Sys, prepckged comm, 30,000 gal, sys contrller	Ea.	97,350
Solar Domestic HW, closed loop, add-on sys, ext heat exchanger	Ea.	11,300
Drainback, hot water system, 120 gal tank	Ea.	14,175
Draindown, hot water system, 120 gal tank	Ea.	14,425

Important: See the Reference Section for Location Factors.

G College, Laboratory

				Unit	Unit Cost	Cost Per S.F.	% Of Sub-Total
A. SUBSTRUCTURE							
1010	Standard Foundations	Poured concrete; strip and spread footings		S.F. Ground	3.41	3.41	
1020	Special Foundations	N/A		—	—	—	
1030	Slab on Grade	4" reinforced concrete with recycled vapor barrier and granular base		S.F. Slab	5.84	5.84	9.7%
2010	Basement Excavation	Site preparation for slab and trench for foundation wall and footing		S.F. Ground	.18	.18	
2020	Basement Walls	4' foundation wall		L.F. Wall	186	7.64	
B. SHELL							
B10 Superstructure							
1010	Floor Construction	Metal deck on open web steel joists	(5680 S.F.)	S.F. Floor	2.42	2.42	4.6%
1020	Roof Construction	Metal deck on open web steel joists		S.F. Roof	5.59	5.59	
B20 Exterior Enclosure							
2010	Exterior Walls	Face brick with concrete block backup	75% of wall	S.F. Wall	41.22	7.42	
2020	Exterior Windows	Window wall	25% of wall	Each	58	3.27	7.6%
2030	Exterior Doors	Glass and metal doors and entrances with transom		Each	6168	2.74	
B30 Roofing							
3010	Roof Coverings	Single-ply TPO membrane, 60 mils, heat welded w/R-20 insul.		S.F. Roof	6.80	6.80	4.1%
3020	Roof Openings	Skylight		S.F. Roof	.43	.43	
C. INTERIORS							
1010	Partitions	Concrete block, foamed-in insulation	10 SF Flr./LF Part.	S.F. Partition	15.15	15.15	
1020	Interior Doors	Single leaf-kalamein fire doors, low VOC paint	820 S.F. Floor/Door	Each	1452	1.77	
1030	Fittings	Lockers		S.F. Floor	.05	.05	
2010	Stair Construction	N/A		—	—	—	21.0%
3010	Wall Finishes	60% paint, low VOC, 40% epoxy coating		S.F. Surface	2.92	5.83	
3020	Floor Finishes	60% epoxy, 20% carpet, 20% vinyl composition tile, recycled content		S.F. Floor	6.57	6.57	
3030	Ceiling Finishes	Mineral fiber tile on concealed zee runners		S.F. Ceiling	7.60	7.60	
D. SERVICES							
D10 Conveying							
1010	Elevators & Lifts	N/A		—	—	—	0.0 %
1020	Escalators & Moving Walks	N/A		—	—	—	
D20 Plumbing							
2010	Plumbing Fixtures	Toilet, low flow, auto sensor & service fixt., supply & drain.	1 Fixt./260 SF Flr.	Each	8445	32.48	
2020	Domestic Water Distribution	Electric, point-of-use water heater		S.F. Floor	1.67	1.67	19.9%
2040	Rain Water Drainage	Roof drains		S.F. Roof	.90	.90	
D30 HVAC							
3010	Energy Supply	N/A		—	—	—	
3020	Heat Generating Systems	Included in D3050		—	—	—	
3040	Distribution Systems	Enthalpy heat recovery packages		Each	38,725	.87	12.0 %
3050	Terminal & Package Units	Multizone rooftop air conditioner, SEER 14		S.F. Floor	20.20	20.20	
3090	Other HVAC Sys. & Equipment	N/A		—	—	—	
D40 Fire Protection							
4010	Sprinklers	Sprinklers, light hazard		S.F. Floor	2.93	2.93	1.9%
4020	Standpipes	Standpipe, wet, Class III		S.F. Floor	.35	.35	
D50 Electrical							
5010	Electrical Service/Distribution	1000 ampere service, panel board and feeders		S.F. Floor	1.38	1.38	
5020	Lighting & Branch Wiring	LED fixtures, daylt. dim., ltg. on/off, recept., switches, & A.C. power		S.F. Floor	26.37	26.37	18.2%
5030	Communications & Security	Addressable alarm systems, internet wiring, & emergency lighting		S.F. Floor	3.39	3.39	
5090	Other Electrical Systems	Emergency generator, 11.5 kW, UPS, and energy monitoring systems		S.F. Floor	.79	.79	
E. EQUIPMENT & FURNISHINGS							
1010	Commercial Equipment	N/A		—	—	—	
1020	Institutional Equipment	Cabinets, fume hoods, lockers, glassware washer		S.F. Floor	1.54	1.54	1.0 %
1090	Other Equipment	Waste handling recycling tilt truck		S.F. Floor	.14	.14	
2020	Moveable Furnishings	No smoking signage		S.F. Floor	.02	.02	
F. SPECIAL CONSTRUCTION							
1020	Integrated Construction	N/A		—	—	—	0.0 %
1040	Special Facilities	N/A		—	—	—	
G. BUILDING SITEWORK	**N/A**						

		Sub-Total	175.74	100%
CONTRACTOR FEES (General Requirements: 10%, Overhead: 5%, Profit: 10%)		25%	43.94	
ARCHITECT FEES		10%	21.97	
	Total Building Cost		**241.65**	

For customer support on your Square Foot Costs with RSMeans data, call 800.448.8182.

243

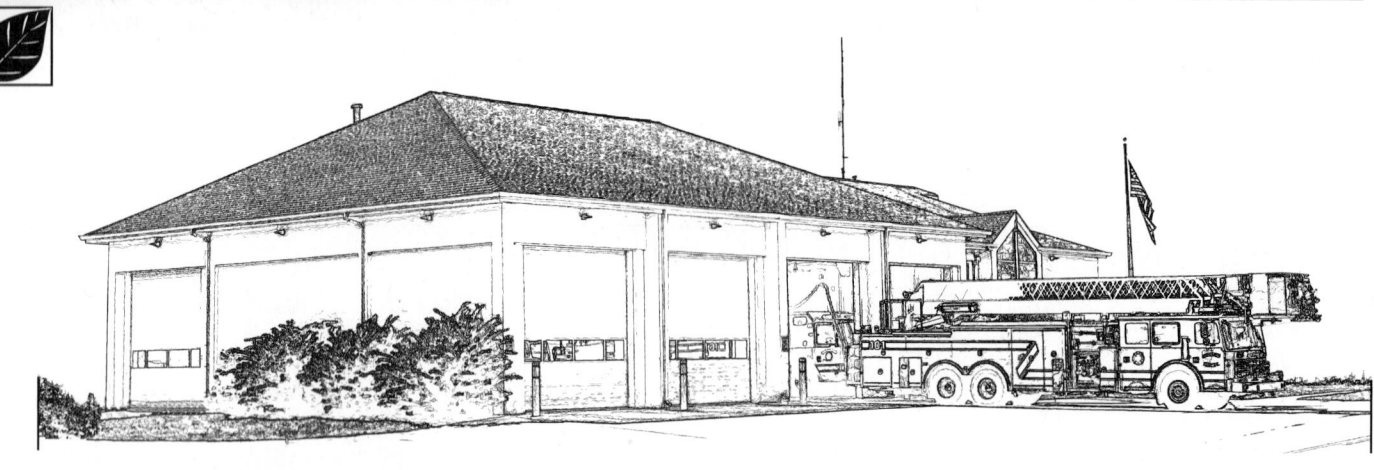

Costs per square foot of floor area

Exterior Wall	S.F. Area	4000	4500	5000	5500	6000	6500	7000	7500	8000
	L.F. Perimeter	260	280	300	310	320	336	353	370	386
Face Brick and Concrete Block	Steel Joists	232.05	227.55	224.00	219.15	215.15	212.75	210.85	209.15	207.50
	Bearing Walls	224.35	219.80	216.30	211.45	207.45	205.05	203.15	201.45	199.80
Decorative Concrete Block	Steel Joists	214.25	210.45	207.55	203.70	200.55	198.60	197.00	195.60	194.30
	Bearing Walls	208.15	204.35	201.45	197.60	194.40	192.50	190.90	189.50	188.20
Limestone and Concrete Block	Steel Joists	248.45	243.25	239.20	233.35	228.60	225.85	223.55	221.60	219.70
	Bearing Walls	242.35	237.15	233.10	227.25	222.50	219.70	217.45	215.50	213.60
Perimeter Adj., Add or Deduct	Per 100 L.F.	25.35	22.60	20.35	18.45	16.95	15.60	14.50	13.50	12.75
Story Hgt. Adj., Add or Deduct	Per 1 Ft.	3.40	3.20	3.10	2.90	2.75	2.70	2.60	2.55	2.55

For Basement, add $45.29 per square foot of basement area

The above costs were calculated using the basic specifications shown on the facing page. These costs should be adjusted where necessary for design alternatives and owner's requirements.

Common additives

Description	Unit	$ Cost
Appliances, cooking range, 30" free standing		
1 oven	Ea.	610 - 2775
2 oven	Ea.	1200 - 3725
Microwave oven	Ea.	281 - 855
Compactor, residential, 4-1 compaction	Ea.	885 - 1475
Dishwasher, built-in, 2 cycles	Ea.	685 - 1225
4 cycles	Ea.	800 - 2025
Garbage disposer, sink type	Ea.	260 - 370
Hood for range, 2 speed, vented, 30" wide	Ea.	370 - 1450
Refrigerator, no frost 10-12 C.F.	Ea.	580 - 750
14-16 C.F.	Ea.	720 - 1025
18-20 C.F.	Ea.	895 - 2075
Lockers, Steel, single tier, 60" or 72"	Opng.	232 - 395
Locker bench, lam. maple top only	L.F.	36.50
Pedestals, steel pipe	Ea.	79
Sound System, amplifier, 250 watts	Ea.	2025
Speaker, ceiling or wall	Ea.	240
Trumpet	Ea.	460

Description	Unit	$ Cost
Commissioning Fees, sustainable commercial construction	S.F.	0.24 - 3.04
Energy Modelling Fees, Fire Stations to 10,000 SF	Ea.	11,000
Green Bldg Cert Fees for comm construction project reg	Project	900
Photovoltaic Pwr Sys, grid connected, 20 kW (~2400 SF), roof	Ea.	249,200
Green Roofs, 6" soil depth, w/treated wd edging & sedum mats	S.F.	12.41
10" Soil depth, with treated wood edging & sedum mats	S.F.	14.10
Greywater Recovery Systems, prepackaged comm, 3060 gal.	Ea.	45,900
4590 gal.	Ea.	57,625
Rainwater Harvest Sys, prepckged comm, 10,000 gal, sys contrller	Ea.	37,550
20,000 gal. w/system controller	Ea.	61,000
30,000 gal. w/system controller	Ea.	97,350
Solar Domestic HW, closed loop, add-on sys, ext heat exchanger	Ea.	11,300
Drainback, hot water system, 120 gal tank	Ea.	14,175

Important: See the Reference Section for Location Factors.

Model costs calculated for a 1 story building with 14′ story height and 6,000 square feet of floor area

				Unit	Unit Cost	Cost Per S.F.	% Of Sub-Total
A. SUBSTRUCTURE							
1010	Standard Foundations	Poured concrete; strip and spread footings		S.F. Ground	3.63	3.63	
1020	Special Foundations	N/A		—	—	—	
1030	Slab on Grade	6″ reinforced concrete with recycled vapor barrier and granular base		S.F. Slab	7.51	7.51	10.6%
2010	Basement Excavation	Site preparation for slab and trench for foundation wall and footing		S.F. Ground	.57	.57	
2020	Basement Walls	4′ foundation wall		L.F. Wall	98	5.22	
B. SHELL							
B10 Superstructure							
1010	Floor Construction	N/A		—	—	—	
1020	Roof Construction	Metal deck, open web steel joists, beams on columns		S.F. Roof	10.90	10.90	6.8 %
B20 Exterior Enclosure							
2010	Exterior Walls	Face brick with concrete block backup	75% of wall	S.F. Wall	41.20	23.07	
2020	Exterior Windows	Aluminum insulated glass	10% of wall	Each	873	2.04	19.3%
2030	Exterior Doors	Single aluminum and glass, overhead, hollow metal, low VOC paint	15% of wall	S.F. Door	3678	5.58	
B30 Roofing							
3010	Roof Coverings	Single-ply TPO membrane, 60 mil, with flashing, R-20 insulation and roof edges		S.F. Roof	8.34	8.34	5.4%
3020	Roof Openings	Skylights, roof hatches		S.F. Roof	.19	.19	
C. INTERIORS							
1010	Partitions	Concrete block w/foamed-in insulation	17 S.F. Floor/LF Part.	S.F. Partition	12.04	7.08	
1020	Interior Doors	Single leaf hollow metal, low VOC paint	500 S.F. Floor/Door	Each	1222	2.44	
1030	Fittings	Toilet partitions		S.F. Floor	.52	.52	
2010	Stair Construction	N/A		—	—	—	12.4%
3010	Wall Finishes	Paint, low VOC		S.F. Surface	2.55	3	
3020	Floor Finishes	50% vinyl tile with recycled content, 50% paint, low VOC		S.F. Floor	2.99	2.99	
3030	Ceiling Finishes	Mineral board acoustic ceiling tiles, concealed grid, suspended	50% of area	S.F. Ceiling	3.80	3.81	
D. SERVICES							
D10 Conveying							
1010	Elevators & Lifts	N/A		—	—	—	0.0 %
1020	Escalators & Moving Walks	N/A		—	—	—	
D20 Plumbing							
2010	Plumbing Fixtures	Kitchen, toilet, low flow, auto sensor & service fixt., supply & drain.	1 Fixt./375 SF Flr.	Each	4586	12.23	
2020	Domestic Water Distribution	Water heater, tankless, on-demand, natural gas/propane		S.F. Floor	2.23	2.23	10.2%
2040	Rain Water Drainage	Roof drains		S.F. Roof	1.75	1.75	
D30 HVAC							
3010	Energy Supply	N/A		—	—	—	
3020	Heat Generating Systems	Included in D3050		—	—	—	
3040	Distribution Systems	Enthalpy heat recovery packages		Each	15,550	2.59	15.9 %
3050	Terminal & Package Units	Multizone rooftop air conditioner, SEER 14		S.F. Floor	22.78	22.78	
3090	Other HVAC Sys. & Equipment	N/A		—	—	—	
D40 Fire Protection							
4010	Sprinklers	Wet pipe sprinkler system		S.F. Floor	5.13	5.13	4.2%
4020	Standpipes	Standpipe, wet, Class III		S.F. Floor	1.59	1.59	
D50 Electrical							
5010	Electrical Service/Distribution	200 ampere service, panel board and feeders		S.F. Floor	1.41	1.41	
5020	Lighting & Branch Wiring	LED fixt., daylt. dimming & ltg. on/off control, receptacles, switches, A.C. & misc. pwr.		S.F. Floor	15.38	15.38	14.5%
5030	Communications & Security	Addressable alarm systems		S.F. Floor	2.03	2.03	
5090	Other Electrical Systems	Energy monitoring systems		S.F. Floor	4.28	4.28	
E. EQUIPMENT & FURNISHINGS							
1010	Commercial Equipment	N/A		—	—	—	
1020	Institutional Equipment	N/A		—	—	—	
1090	Other Equipment	Waste handling recycling tilt truck		S.F. Floor	1.04	1.04	0.7 %
2020	Moveable Furnishings	No smoking signage		S.F. Floor	.04	.04	
F. SPECIAL CONSTRUCTION							
1020	Integrated Construction	N/A		—	—	—	0.0 %
1040	Special Facilities	N/A		—	—	—	
G. BUILDING SITEWORK	N/A						

	Sub-Total	159.37	**100%**
CONTRACTOR FEES (General Requirements: 10%, Overhead: 5%, Profit: 10%)	25%	39.84	
ARCHITECT FEES	8%	15.94	
	Total Building Cost	215.15	

For customer support on your Square Foot Costs with RSMeans data, call 800.448.8182.

245

Costs per square foot of floor area

Exterior Wall	S.F. Area	6000	7000	8000	9000	10000	11000	12000	13000	14000
	L.F. Perimeter	220	240	260	280	286	303	320	336	353
Face Brick and Concrete Block	Steel Joists	240.10	233.15	227.85	223.80	218.10	215.15	212.75	210.65	208.85
	Precast Concrete	253.60	246.65	241.45	237.40	231.75	228.85	226.45	224.35	222.65
Decorative Concrete Block	Steel Joists	214.55	208.65	204.20	200.80	196.20	193.75	191.70	190.00	188.45
	Precast Concrete	236.95	231.10	226.70	223.30	218.85	216.35	214.35	212.65	211.15
Limestone and Concrete Block	Steel Joists	258.60	250.45	244.25	239.55	232.50	229.10	226.20	223.70	221.60
	Precast Concrete	272.10	263.95	257.85	253.10	246.20	242.80	239.90	237.40	235.30
Perimeter Adj., Add or Deduct	Per 100 L.F.	29.40	25.20	22.05	19.65	17.70	16.10	14.70	13.50	12.60
Story Hgt. Adj., Add or Deduct	Per 1 Ft.	3.85	3.65	3.45	3.30	3.05	2.90	2.80	2.65	2.70

For Basement, add $44.01 per square foot of basement area

The above costs were calculated using the basic specifications shown on the facing page. These costs should be adjusted where necessary for design alternatives and owner's requirements.

Common additives

Description	Unit	$ Cost
Appliances		
Cooking range, 30" free standing		
1 oven	Ea.	610 - 2775
2 oven	Ea.	1200 - 3725
Microwave oven	Ea.	281 - 855
Compactor, residential, 4-1 compaction	Ea.	885 - 1475
Dishwasher, built-in, 2 cycles	Ea.	685 - 1225
4 cycles	Ea.	800 - 2025
Garbage diposer, sink type	Ea.	260 - 370
Hood for range, 2 speed, vented, 30" wide	Ea.	370 - 1450
Refrigerator, no frost 10-12 C.F.	Ea.	580 - 750
14-16 C.F.	Ea.	720 - 1025
18-20 C.F.	Ea.	895 - 2075
Elevators, hydraulic passenger, 2 stops, 2500# capacity	Ea.	76,600
3500# capacity	Ea.	81,600
Lockers, Steel, single tier, 60" or 72"	Opng.	232 - 395
Locker bench, lam. maple top only	L.F.	36.50
Pedestals, steel pipe	Ea.	79
Sound System, amplifier, 250 watts	Ea.	2025
Speaker, ceiling or wall	Ea.	240
Trumpet	Ea.	460

Description	Unit	$ Cost
Commissioning Fees, sustainable commercial construction	S.F.	0.24 - 3.04
Energy Modelling Fees, Fire Stations to 10,000 SF	Ea.	11,000
Greater than 10,000 SF add	S.F.	0.04
Green Bldg Cert Fees for comm construction project reg	Project	900
Photovoltaic Pwr Sys, grid connected, 20 kW (~2400 SF), roof	Ea.	249,200
Green Roofs, 6" soil depth, w/treated wd edging & sedum mats	S.F.	12.41
10" Soil depth, with treated wood edging & sedum mats	S.F.	14.10
Greywater Recovery Systems, prepackaged comm, 3060 gal.	Ea.	45,900
4590 gal.	Ea.	57,625
Rainwater Harvest Sys, prepckged comm, 10,000 gal, sys contrller	Ea.	37,550
20,000 gal. w/system controller	Ea.	61,000
30,000 gal. w/system controller	Ea.	97,350
Solar Domestic HW, closed loop, add-on sys, ext heat exchanger	Ea.	11,300
Drainback, hot water system, 120 gal tank	Ea.	14,175

Important: See the Reference Section for Location Factors.

Model costs calculated for a 2 story building with 14' story height and 10,000 square feet of floor area

				Unit	Unit Cost	Cost Per S.F.	% Of Sub-Total
A. SUBSTRUCTURE							
1010	Standard Foundations	Poured concrete; strip and spread footings		S.F. Ground	3.40	1.70	
1020	Special Foundations	N/A		—	—	—	
1030	Slab on Grade	6" reinforced concrete with recycled vapor barrier and granular base		S.F. Slab	7.51	3.76	6.1%
2010	Basement Excavation	Site preparation for slab and trench for foundation wall and footing		S.F. Ground	.57	.29	
2020	Basement Walls	4' foundation wall with insulation		L.F. Wall	108	3.07	
B. SHELL							
B10 Superstructure							
1010	Floor Construction	Open web steel joists, slab form, concrete		S.F. Floor	15.96	7.98	7.4%
1020	Roof Construction	Metal deck on open web steel joists		S.F. Roof	5.60	2.80	
B20 Exterior Enclosure							
2010	Exterior Walls	Decorative concrete block, insul.	75% of wall	S.F. Wall	25.21	15.14	
2020	Exterior Windows	Aluminum insulated glass	10% of wall	Each	755	2.63	15.3%
2030	Exterior Doors	Single aluminum and glass, steel overhead, hollow metal, low VOC paint	15% of wall	S.F. Door	2815	4.47	
B30 Roofing							
3010	Roof Coverings	Single-ply TPO, 60 mils, heat welded seams w/R-20 insul.		S.F. Roof	8.52	4.26	2.9%
3020	Roof Openings	N/A		—	—	—	
C. INTERIORS							
1010	Partitions	Concrete block, foamed-in insul.	10 SF Floor/LF Part.	S.F. Partition	12.04	7.08	
1020	Interior Doors	Single leaf hollow metal, low VOC paint	500 S.F. Floor/Door	Each	1452	2.90	
1030	Fittings	Toilet partitions		S.F. Floor	.47	.47	
2010	Stair Construction	Concrete filled metal pan		Flight	22,075	4.42	17.8%
3010	Wall Finishes	Paint, low VOC		S.F. Surface	4.10	4.82	
3020	Floor Finishes	50% vinyl tile, (recycled content) 50% paint (low VOC)		S.F. Floor	2.99	2.99	
3030	Ceiling Finishes	Fiberglass board on exposed grid, suspended	50% of area	S.F. Ceiling	3.15	3.16	
D. SERVICES							
D10 Conveying							
1010	Elevators & Lifts	One hydraulic passenger elevator		Each	91,200	9.12	6.3%
1020	Escalators & Moving Walks	N/A		—	—	—	
D20 Plumbing							
2010	Plumbing Fixtures	Kitchen, toilet low flow, auto sensor & service fixt., supply & drain.	1 Fixt./400 SF Flr.	Each	4528	11.32	
2020	Domestic Water Distribution	Gas fired tankless water heater		S.F. Floor	.88	.88	9.1%
2040	Rain Water Drainage	Roof drains		S.F. Roof	2.18	1.09	
D30 HVAC							
3010	Energy Supply	N/A		—	—	—	
3020	Heat Generating Systems	Included in D3050		—	—	—	
3040	Distribution Systems	Enthalpy heat recovery packages		Each	15,550	1.56	16.7 %
3050	Terminal & Package Units	Multizone rooftop air conditioner, SEER 14		S.F. Floor	22.78	22.78	
3090	Other HVAC Sys. & Equipment	N/A		—	—	—	
D40 Fire Protection							
4010	Sprinklers	Wet pipe sprinkler system		S.F. Floor	4.08	4.08	3.9%
4020	Standpipes	Standpipe, wet, Class III		S.F. Floor	1.64	1.64	
D50 Electrical							
5010	Electrical Service/Distribution	400 ampere service, panel board and feeders		S.F. Floor	1.14	1.14	
5020	Lighting & Branch Wiring	LED fixt., daylt. dimming & ltg. on/off control., receptacles, switches, A.C. and misc. pwr.		S.F. Floor	15.61	15.61	14.1%
5030	Communications & Security	Addressable alarm systems and emergency lighting		S.F. Floor	1.58	1.58	
5090	Other Electrical Systems	Emergency generator, 15 kW, and energy monitoring systems		S.F. Floor	2.21	2.21	
E. EQUIPMENT & FURNISHINGS							
1010	Commercial Equipment	N/A		—	—	—	
1020	Institutional Equipment	N/A		—	—	—	
1090	Other Equipment	Waste handling recycling tilt truck		S.F. Floor	.37	.37	0.3 %
2020	Moveable Furnishings	No smoking signage		S.F. Floor	.02	.02	
F. SPECIAL CONSTRUCTION							
1020	Integrated Construction	N/A		—	—	—	0.0 %
1040	Special Facilities	N/A		—	—	—	
G. BUILDING SITEWORK	**N/A**						

		Sub-Total	145.34	100%
CONTRACTOR FEES (General Requirements: 10%, Overhead: 5%, Profit: 10%)		25%	36.33	
ARCHITECT FEES		8%	14.53	
		Total Building Cost	**196.20**	

For customer support on your Square Foot Costs with RSMeans data, call 800.448.8182.

247

Costs per square foot of floor area

Exterior Wall	S.F. Area	4000	6000	8000	10000	12000	14000	16000	18000	20000
	L.F. Perimeter	260	320	384	424	460	484	510	540	576
Wood Board and Batten	Wood Frame	175.80	163.50	157.70	152.70	149.20	146.20	143.90	142.45	141.30
Brick Veneer	Wood Frame	186.05	172.20	165.65	159.90	155.85	152.35	149.70	147.95	146.70
Aluminum Clapboard	Wood Frame	170.70	159.60	154.30	149.90	146.80	144.15	142.15	140.85	139.90
Face Brick and Concrete Block	Wood Truss	201.15	185.25	177.80	171.05	166.30	162.10	159.05	157.00	155.50
Limestone and Concrete Block	Wood Truss	221.00	201.55	192.45	184.00	178.05	172.65	168.75	166.10	164.25
Stucco and Concrete Block	Wood Truss	209.25	191.65	183.35	175.80	170.50	165.70	162.20	159.90	158.25
Perimeter Adj., Add or Deduct	Per 100 L.F.	15.30	10.20	7.65	6.15	5.10	4.40	3.90	3.35	3.05
Story Hgt. Adj., Add or Deduct	Per 1 Ft.	2.00	1.65	1.45	1.30	1.20	1.05	1.00	0.90	0.90
For Basement, add $ 37.31 per square foot of basement area										

The above costs were calculated using the basic specifications shown on the facing page. These costs should be adjusted where necessary for design alternatives and owner's requirements.

Common additives

Description	Unit	$ Cost
Autopsy Table, standard	Ea.	11,600
Deluxe	Ea.	19,400
Directory Boards, plastic, glass covered, 30" x 20"	Ea.	640
36" x 48"	Ea.	1625
Aluminum, 24" x 18"	Ea.	630
36" x 24"	Ea.	790
48" x 32"	Ea.	1025
Mortuary Refrigerator, end operated		
Two capacity	Ea.	9800
Six capacity	Ea.	18,400
Planters, precast concrete, 48" diam., 24" high	Ea.	790
7" diam., 36" high	Ea.	1850
Fiberglass, 36" diam., 24" high	Ea.	825
60" diam., 24" high	Ea.	1350

Description	Unit	$ Cost
Commissioning Fees, sustainable commercial construction	S.F.	0.24 - 3.04
Energy Modelling Fees, commercial buildings to 10,000 SF	Ea.	11,000
Greater than 10,000 SF add	S.F.	0.04
Green Bldg Cert Fees for comm construction project reg	Project	900
Photovoltaic Pwr Sys, grid connected, 20 kW (~2400 SF), roof	Ea.	249,200
Green Roofs, 6" soil depth, w/treated wd edging & sedum mats	S.F.	12.41
10" Soil depth, with treated wood edging & sedum mats	S.F.	14.10
Greywater Recovery Systems, prepackaged comm, 3060 gal.	Ea.	45,900
4590 gal.	Ea.	57,625
Rainwater Harvest Sys, prepckged comm, 10,000 gal, sys contrller	Ea.	37,550
20,000 gal. w/system controller	Ea.	61,000
30,000 gal. w/system controller	Ea.	97,350
Solar Domestic HW, closed loop, add-on sys, ext heat exchanger	Ea.	11,300
Drainback, hot water system, 120 gal tank	Ea.	14,175
Draindown, hot water system, 120 gal tank	Ea.	14,425

Important: See the Reference Section for Location Factors.

Model costs calculated for a 1 story building with 12' story height and 10,000 square feet of floor area

G Funeral Home

				Unit	Unit Cost	Cost Per S.F.	% Of Sub-Total
A. SUBSTRUCTURE							
1010	Standard Foundations	Poured concrete; strip and spread footings		S.F. Ground	1.98	1.98	
1020	Special Foundations	N/A		—	—	—	
1030	Slab on Grade	4" reinforced concrete with recycled plastic vapor barrier and granular base		S.F. Slab	5.84	5.84	10.9%
2010	Basement Excavation	Site preparation for slab and trench for foundation wall and footing		S.F. Ground	.33	.33	
2020	Basement Walls	4' foundation wall		L.F. Wall	88	4.10	
B. SHELL							
B10 Superstructure							
1010	Floor Construction	N/A		—	—	—	
1020	Roof Construction	Plywood on wood truss		S.F. Roof	3.91	3.91	3.5 %
B20 Exterior Enclosure							
2010	Exterior Walls	1" x 4" vertical T & G redwood siding on 2x6 wood studs with insul.	90% of wall	S.F. Wall	17.71	8.11	
2020	Exterior Windows	Double hung wood	10% of wall	Each	533	2.26	11.1%
2030	Exterior Doors	Wood swinging double doors, single leaf hollow metal, low VOC paint		Each	3360	2.02	
B30 Roofing							
3010	Roof Coverings	Single ply membrane, TPO, 45 mil, fully adhered; polyisocyanurate sheets		S.F. Roof	6.27	6.27	5.6%
3020	Roof Openings	N/A		—	—	—	
C. INTERIORS							
1010	Partitions	Gypsum board on wood studs with sound deadening board	15 SF Flr./LF Part.	S.F. Partition	10.58	5.64	
1020	Interior Doors	Single leaf wood, low VOC paint	150 S.F. Floor/Door	Each	725	4.83	
1030	Fittings	N/A		—	—	—	
2010	Stair Construction	N/A		—	—	—	25.0%
3010	Wall Finishes	50% wallpaper, 25% wood paneling, 25% paint (low VOC)		S.F. Surface	3.46	3.69	
3020	Floor Finishes	70% carpet tile, 30% ceramic tile		S.F. Floor	7.61	7.61	
3030	Ceiling Finishes	Fiberglass board on exposed grid, suspended		S.F. Ceiling	6.30	6.30	
D. SERVICES							
D10 Conveying							
1010	Elevators & Lifts	N/A		—	—	—	0.0 %
1020	Escalators & Moving Walks	N/A		—	—	—	
D20 Plumbing							
2010	Plumbing Fixtures	Toilets, low flow, auto sensor, urinals & service fixt., supply & drain.	1 Fixt./770 SF Flr.	Each	3573	4.64	
2020	Domestic Water Distribution	Electric water heater, point-of-use		S.F. Floor	.49	.49	5.5%
2040	Rain Water Drainage	Roof drain		S.F. Roof	1.08	1.08	
D30 HVAC							
3010	Energy Supply	N/A		—	—	—	
3020	Heat Generating Systems	Included in D3050		—	—	—	
3040	Distribution Systems	Enthalpy heat recovery packages		Each	38,725	3.87	19.5 %
3050	Terminal & Package Units	Multizone rooftop air conditioner, SEER 14		S.F. Floor	18.02	18.02	
3090	Other HVAC Sys. & Equipment	N/A		—	—	—	
D40 Fire Protection							
4010	Sprinklers	Wet pipe sprinkler system		S.F. Floor	3.79	3.79	4.2%
4020	Standpipes	Standpipe, wet, Class III		S.F. Floor	.95	.95	
D50 Electrical							
5010	Electrical Service/Distribution	400 ampere service, panel board and feeders		S.F. Floor	1.60	1.60	
5020	Lighting & Branch Wiring	LED light fixtures, daylt. dimming control & ltg. on/off sys., recept., switches, & A.C. power		S.F. Floor	10.53	10.53	14.0%
5030	Communications & Security	Addressable alarm systems		S.F. Floor	1.58	1.58	
5090	Other Electrical Systems	Emergency generator, 15 kW and energy monitoring systems		S.F. Floor	1.96	1.96	
E. EQUIPMENT & FURNISHINGS							
1010	Commercial Equipment	N/A		—	—	—	
1020	Institutional Equipment	N/A		—	—	—	0.6 %
1090	Other Equipment	Waste handling recycling tilt truck		S.F. Floor	.63	.63	
2020	Moveable Furnishings	No smoking signage		S.F. Floor	.04	.04	
F. SPECIAL CONSTRUCTION							
1020	Integrated Construction	N/A		—	—	—	0.0 %
1040	Special Facilities	N/A		—	—	—	
G. BUILDING SITEWORK	**N/A**						

			Sub-Total	112.07	100%
CONTRACTOR FEES (General Requirements: 10%, Overhead: 5%, Profit: 10%)		25%		28.02	
ARCHITECT FEES		9%		12.61	
			Total Building Cost	**152.70**	

For customer support on your Square Foot Costs with RSMeans data, call 800.448.8182.

249

Costs per square foot of floor area

Exterior Wall	S.F. Area	7000	10000	13000	16000	19000	22000	25000	28000	31000
	L.F. Perimeter	240	300	336	386	411	435	472	510	524
Face Brick and Concrete Block	Reinforced Concrete	225.00	213.65	203.95	199.55	193.95	189.85	187.75	186.10	183.30
	Rigid Steel	221.95	210.55	200.85	196.45	190.85	186.75	184.65	183.00	180.20
Limestone and Concrete Block	Reinforced Concrete	251.60	236.10	222.95	217.00	209.45	203.80	200.95	198.75	194.95
	Rigid Steel	246.15	231.70	219.05	213.45	206.10	200.70	198.00	195.85	192.15
Precast Concrete	Reinforced Concrete	252.70	237.10	223.80	217.80	210.10	204.50	201.60	199.30	195.50
	Rigid Steel	249.60	234.00	220.65	214.70	207.00	201.40	198.50	196.20	192.40
Perimeter Adj., Add or Deduct	Per 100 L.F.	27.80	19.45	14.95	12.15	10.30	8.85	7.80	6.90	6.25
Story Hgt. Adj., Add or Deduct	Per 1 Ft.	4.20	3.65	3.15	3.00	2.70	2.45	2.30	2.25	2.10
For Basement, add $55.65 per square foot of basement area										

The above costs were calculated using the basic specifications shown on the facing page. These costs should be adjusted where necessary for design alternatives and owner's requirements.

Common additives

Description	Unit	$ Cost
Carrels Hardwood	Ea.	740 - 1925
Closed Circuit Surveillance, one station, camera and monitor	Ea.	1475
For additional camera stations, add	Ea.	665
Elevators, hydraulic passenger, 2 stops, 1500# capacity	Ea.	73,100
2500# capacity	Ea.	76,600
3500# capacity	Ea.	81,600
Library Furnishings, bookshelf, 90" high, 10" shelf double face	L.F.	260
Single face	L.F.	181
Charging desk, built-in with counter		
Plastic laminated top	L.F.	600
Reading table, laminated		
Top 60" x 36"	Ea.	505

Description	Unit	$ Cost
Commissioning Fees, sustainable institutional construction	S.F.	0.58 - 2.47
Energy Modelling Fees, commercial buildings to 10,000 SF	Ea.	11,000
Greater than 10,000 SF add	S.F.	0.04
Green Bldg Cert Fees for comm construction project reg	Project	900
Photovoltaic Pwr Sys, grid connected, 20 kW (~2400 SF), roof	Ea.	249,200
Green Roofs, 6" soil depth, w/treated wd edging & sedum mats	S.F.	12.41
10" Soil depth, with treated wood edging & sedum mats	S.F.	14.10
Greywater Recovery Systems, prepackaged comm, 3060 gal.	Ea.	45,900
4590 gal.	Ea.	57,625
Rainwater Harvest Sys, prepckged comm, 10,000 gal, sys contrller	Ea.	37,550
20,000 gal. w/system controller	Ea.	61,000
30,000 gal. w/system controller	Ea.	97,350
Solar Domestic HW, closed loop, add-on sys, ext heat exchanger	Ea.	11,300
Drainback, hot water system, 120 gal tank	Ea.	14,175
Draindown, hot water system, 120 gal tank	Ea.	14,425

Important: See the Reference Section for Location Factors.

Model costs calculated for a 2 story building with 14' story height and 22,000 square feet of floor area

				Unit	Unit Cost	Cost Per S.F.	% Of Sub-Total
A.	**SUBSTRUCTURE**						
1010	Standard Foundations	Poured concrete; strip and spread footings		S.F. Ground	4.24	2.12	
1020	Special Foundations	N/A		—	—	—	
1030	Slab on Grade	4" reinforced concrete with recycled vapor barrier and granular base		S.F. Slab	5.84	2.93	5.1%
2010	Basement Excavation	Site preparation for slab and trench for foundation wall and footing		S.F. Ground	.33	.17	
2020	Basement Walls	4' foundation wall		L.F. Wall	98	1.94	
B.	**SHELL**						
	B10 Superstructure						
1010	Floor Construction	Concrete waffle slab		S.F. Floor	30.96	15.48	19.1%
1020	Roof Construction	Concrete waffle slab		S.F. Roof	22.80	11.40	
	B20 Exterior Enclosure						
2010	Exterior Walls	Face brick with concrete block backup	90% of wall	S.F. Wall	38.73	19.30	
2020	Exterior Windows	Window wall	10% of wall	Each	58	3.19	16.5%
2030	Exterior Doors	Double aluminum and glass, single leaf hollow metal, low VOC paint		Each	7200	.66	
	B30 Roofing						
3010	Roof Coverings	Single-ply TPO membrane, 60 mil, heat welded w/R-20 insulation		S.F. Roof	7.50	3.75	2.7%
3020	Roof Openings	Roof hatches		S.F. Roof	.10	.05	
C.	**INTERIORS**						
1010	Partitions	Gypsum board on metal studs w/sound attenuation insul.	30 SF Flr./LF Part.	S.F. Partition	14.70	5.88	
1020	Interior Doors	Single leaf wood, low VOC paint	300 S.F. Floor/Door	Each	725	2.42	
1030	Fittings	N/A		—	—	—	
2010	Stair Construction	Concrete filled metal pan		Flight	10,475	.96	16.4%
3010	Wall Finishes	Paint, low VOC		S.F. Surface	2.10	1.68	
3020	Floor Finishes	50% carpet tile, 50% vinyl composition tile, recycled content		S.F. Floor	4.54	4.54	
3030	Ceiling Finishes	Mineral fiber on concealed zee bars		S.F. Ceiling	7.60	7.60	
D.	**SERVICES**						
	D10 Conveying						
1010	Elevators & Lifts	One hydraulic passenger elevator		Each	94,820	4.31	3.1%
1020	Escalators & Moving Walks	N/A		—	—	—	
	D20 Plumbing						
2010	Plumbing Fixtures	Toilet, low flow, auto sensor & service fixt., supply & drain.	1 Fixt./1835 SF.Flr.	Each	8680	4.73	
2020	Domestic Water Distribution	Tankless, on demand water heater, gas/propane		S.F. Floor	.60	.60	4.4%
2040	Rain Water Drainage	Roof drains		S.F. Roof	1.62	.81	
	D30 HVAC						
3010	Energy Supply	N/A		—	—	—	
3020	Heat Generating Systems	Included in D3050		—	—	—	
3040	Distribution Systems	Enthalpy heat recovery packages		Each	38,725	1.76	16.0 %
3050	Terminal & Package Units	Multizone rooftop air conditioner, SEER 14		S.F. Floor	20.80	20.80	
3090	Other HVAC Sys. & Equipment	N/A		—	—	—	
	D40 Fire Protection						
4010	Sprinklers	Wet pipe sprinkler system		S.F. Floor	3.30	3.30	3.1%
4020	Standpipes	Standpipe		S.F. Floor	1.09	1.09	
	D50 Electrical						
5010	Electrical Service/Distribution	400 ampere service, panel board and feeders		S.F. Floor	.55	.55	
5020	Lighting & Branch Wiring	LED fixtures, daylt. dim., ltg. on/off, recept., switches, and A.C. power		S.F. Floor	14.95	14.95	13.4%
5030	Communications & Security	Addressable alarm systems, internet wiring, and emergency lighting		S.F. Floor	2.15	2.15	
5090	Other Electrical Systems	Emergency generator, 7.5 kW, UPS, energy monitoring systems		S.F. Floor	1.23	1.23	
E.	**EQUIPMENT & FURNISHINGS**						
1010	Commercial Equipment	N/A		—	—	—	
1020	Institutional Equipment	N/A		—	—	—	0.2 %
1090	Other Equipment	Waste handling recycling tilt truck		S.F. Floor	.28	.28	
2020	Moveable Furnishings	No smoking signage		S.F. Floor			
F.	**SPECIAL CONSTRUCTION**						
1020	Integrated Construction	N/A		—	—	—	0.0 %
1040	Special Facilities	N/A		—	—	—	
G.	**BUILDING SITEWORK**	**N/A**					

		Sub-Total	140.63	100%
CONTRACTOR FEES (General Requirements: 10%, Overhead: 5%, Profit: 10%)		25%	35.16	
ARCHITECT FEES		8%	14.06	
	Total Building Cost		**189.85**	

For customer support on your Square Foot Costs with RSMeans data, call 800.448.8182.

251

Costs per square foot of floor area

Exterior Wall	S.F. Area	4000	5500	7000	8500	10000	11500	13000	14500	16000
	L.F. Perimeter	280	320	380	440	453	503	510	522	560
Face Brick and Concrete Block	Steel Joists	269.60	257.80	253.45	250.55	244.75	243.05	239.15	236.30	235.25
	Wood Truss	266.60	254.80	250.45	247.55	241.75	240.10	236.15	233.30	232.25
Stucco and Concrete Block	Steel Joists	274.10	261.15	256.45	253.35	246.95	245.05	240.70	237.55	236.40
	Wood Truss	273.30	260.40	255.60	252.55	246.05	244.30	239.90	236.75	235.60
Brick Veneer	Wood Truss	257.45	247.05	243.10	240.50	235.50	233.95	230.50	228.05	227.10
Wood Clapboard	Wood Frame	247.75	239.15	235.80	233.65	229.60	228.35	225.60	223.60	222.80
Perimeter Adj., Add or Deduct	Per 100 L.F.	19.95	14.55	11.40	9.40	8.05	7.00	6.10	5.50	5.00
Story Hgt. Adj., Add or Deduct	Per 1 Ft.	4.00	3.40	3.10	3.00	2.65	2.55	2.25	2.10	2.05
For Basement, add $36.22 per square foot of basement area										

The above costs were calculated using the basic specifications shown on the facing page. These costs should be adjusted where necessary for design alternatives and owner's requirements.

Common additives

Description	Unit	$ Cost
Cabinets, hospital, base, laminated plastic	L.F.	530
Counter top, laminated plastic	L.F.	87.50
Nurses station, door type , laminated plastic	L.F.	595
Wall cabinets, laminated plastic	L.F.	390
Directory Boards, plastic, glass covered, 30" x 20"	Ea.	640
36" x 48"	Ea.	1625
Aluminum, 36" x 24"	Ea.	790
48" x 32"	Ea.	1025
Heat Therapy Unit, humidified, 26" x 78" x 28"	Ea.	3675
Tables, examining, vinyl top, with base cabinets	Ea.	1575 - 6525
Utensil Washer, Sanitizer	Ea.	10,300
X-Ray, Mobile	Ea.	20,800 - 105,000

Description	Unit	$ Cost
Commissioning Fees, sustainable commercial construction	S.F.	0.24 - 3.04
Energy Modelling Fees, commercial buildings to 10,000 SF	Ea.	11,000
Greater than 10,000 SF add	S.F.	0.04
Green Bldg Cert Fees for comm construction project reg	Project	900
Photovoltaic Pwr Sys, grid connected, 20 kW (~2400 SF), roof	Ea.	249,200
Green Roofs, 6" soil depth, w/treated wd edging & sedum mats	S.F.	12.41
10" Soil depth, with treated wood edging & sedum mats	S.F.	14.10
Greywater Recovery Systems, prepackaged comm, 3060 gal.	Ea.	45,900
4590 gal.	Ea.	57,625
Rainwater Harvest Sys, prepckged comm, 10,000 gal, sys contrller	Ea.	37,550
20,000 gal. w/system controller	Ea.	61,000
30,000 gal. w/system controller	Ea.	97,350
Solar Domestic HW, closed loop, add-on sys, ext heat exchanger	Ea.	11,300
Drainback, hot water system, 120 gal tank	Ea.	14,175
Draindown, hot water system, 120 gal tank	Ea.	14,425

Important: See the Reference Section for Location Factors.

				Unit	Unit Cost	Cost Per S.F.	% Of Sub-Total
A. SUBSTRUCTURE							
1010	Standard Foundations	Poured concrete; strip and spread footings		S.F. Ground	2.31	2.31	
1020	Special Foundations	N/A		—	—	—	
1030	Slab on Grade	4" reinforced concrete with recycled vapor barrier and granular base		S.F. Slab	5.84	5.84	7.5%
2010	Basement Excavation	Site preparation for slab and trench for foundation wall and footing		S.F. Ground	.33	.33	
2020	Basement Walls	4' foundation wall		L.F. Wall	98	5.32	
B. SHELL							
B10 Superstructure							
1010	Floor Construction	N/A		—	—	—	4.2 %
1020	Roof Construction	Plywood on wood trusses		S.F. Roof	7.76	7.76	
B20 Exterior Enclosure							
2010	Exterior Walls	Face brick with concrete block backup	70% of wall	S.F. Wall	42.95	16.32	
2020	Exterior Windows	Wood double hung	30% of wall	Each	665	6.37	14.6%
2030	Exterior Doors	Aluminum and glass doors and entrance with transoms		Each	2385	4.08	
B30 Roofing							
3010	Roof Coverings	Asphalt shingles with flashing (Pitched); rigid fiber glass insulation, gutters		S.F. Roof	3.33	3.33	1.8%
3020	Roof Openings	N/A		—	—	—	
C. INTERIORS							
1010	Partitions	Gypsum bd. & sound deadening bd. on wood studs w/insul.	6 SF Flr./LF Part.	S.F. Partition	10.29	13.72	
1020	Interior Doors	Single leaf wood, low VOC paint	60 S.F. Floor/Door	Each	725	12.08	
1030	Fittings	N/A		—	—	—	
2010	Stair Construction	N/A		—	—	—	22.2%
3010	Wall Finishes	50% paint, low VOC, 50% vinyl wall covering		S.F. Surface	1.48	3.94	
3020	Floor Finishes	50% carpet tile, 50% vinyl composition tile, recycled content		S.F. Floor	4.21	4.21	
3030	Ceiling Finishes	Mineral fiber tile on concealed zee bars		S.F. Ceiling	6.86	6.86	
D. SERVICES							
D10 Conveying							
1010	Elevators & Lifts	N/A		—	—	—	0.0 %
1020	Escalators & Moving Walks	N/A		—	—	—	
D20 Plumbing							
2010	Plumbing Fixtures	Toilet, low flow, auto sensor, exam room & service fixt., supply & drain.	1 Fixt./195 SF Flr.	Each	5117	26.24	
2020	Domestic Water Distribution	Gas fired tankless water heater		S.F. Floor	3.02	3.02	16.7%
2040	Rain Water Drainage	Roof drains		S.F. Roof	1.37	1.37	
D30 HVAC							
3010	Energy Supply	N/A		—	—	—	
3020	Heat Generating Systems	Included in D3050		—	—	—	
3040	Distribution Systems	Enthalpy heat recovery packages		Each	14,075	2.01	8.9 %
3050	Terminal & Package Units	Multizone rooftop air conditioner, SEER 14		S.F. Floor	14.43	14.43	
3090	Other HVAC Sys. & Equipment	N/A		—	—	—	
D40 Fire Protection							
4010	Sprinklers	Wet pipe sprinkler system		S.F. Floor	5.13	5.13	3.5%
4020	Standpipes	Standpipe		S.F. Floor	1.36	1.36	
D50 Electrical							
5010	Electrical Service/Distribution	200 ampere service, panel board and feeders		S.F. Floor	1.33	1.33	
5020	Lighting & Branch Wiring	LED fixtures, daylt. dim., ltg. on/off, recept., switches, and A.C. power		S.F. Floor	13.46	13.46	14.3%
5030	Communications & Security	Alarm systems, internet & phone wiring, intercom system, & emerg. ltg.		S.F. Floor	8.24	8.24	
5090	Other Electrical Systems	Emergency generator, 7.5 kW, and energy monitoring systems		S.F. Floor	3.17	3.17	
E. EQUIPMENT & FURNISHINGS							
1010	Commercial Equipment	N/A		—	—	—	
1020	Institutional Equipment	Exam room casework and countertops		S.F. Floor	10.58	10.58	6.3 %
1090	Other Equipment	Waste handling recycling tilt truck		S.F. Floor	.89	.89	
2020	Moveable Furnishings	No smoking signage		S.F. Floor	.11	.11	
F. SPECIAL CONSTRUCTION							
1020	Integrated Construction	N/A		—	—	—	0.0 %
1040	Special Facilities	N/A		—	—	—	
G. BUILDING SITEWORK	**N/A**						

			Sub-Total	183.81	100%
CONTRACTOR FEES (General Requirements: 10%, Overhead: 5%, Profit: 10%)			25%	45.96	
ARCHITECT FEES			9%	20.68	
		Total Building Cost		**250.45**	

For customer support on your Square Foot Costs with RSMeans data, call 800.448.8182.

253

Costs per square foot of floor area

Exterior Wall	S.F. Area	4000	5500	7000	8500	10000	11500	13000	14500	16000
	L.F. Perimeter	180	210	240	270	286	311	336	361	386
Face Brick and Concrete Block	Steel Joists	302.35	289.95	282.85	278.25	273.00	270.25	268.15	266.45	265.10
	Wood Joists	301.15	288.75	281.70	277.05	271.80	269.05	266.95	265.25	263.90
Stucco and Concrete Block	Steel Joists	284.80	275.05	269.50	265.85	261.85	259.65	258.05	256.70	255.70
	Wood Joists	283.65	273.85	268.30	264.65	260.70	258.55	256.85	255.55	254.50
Brick Veneer	Wood Frame	288.90	278.30	272.25	268.35	264.00	261.70	259.90	258.45	257.30
Wood Clapboard	Wood Frame	283.75	273.90	268.35	264.75	260.75	258.60	256.90	255.55	254.55
Perimeter Adj., Add or Deduct	Per 100 L.F.	35.95	26.05	20.50	16.85	14.35	12.45	11.00	9.85	8.95
Story Hgt. Adj., Add or Deduct	Per 1 Ft.	5.30	4.45	4.00	3.70	3.35	3.20	3.00	2.90	2.85

For Basement, add $40.19 per square foot of basement area

The above costs were calculated using the basic specifications shown on the facing page. These costs should be adjusted where necessary for design alternatives and owner's requirements.

Common additives

Description	Unit	$ Cost
Cabinets, hospital, base, laminated plastic	L.F.	530
Counter top, laminated plastic	L.F.	87.50
Nurses station, door type , laminated plastic	L.F.	595
Wall cabinets, laminated plastic	L.F.	390
Elevators, hydraulic passenger, 2 stops, 2500# capacity	Ea.	76,600
3500# capacity	Ea.	81,600
Directory Boards, plastic, glass covered, 30" x 20"	Ea.	640
36" x 48"	Ea.	1625
Aluminum, 36" x 24"	Ea.	790
48" x 32"	Ea.	1025
Heat Therapy Unit, humidified, 26" x 78" x 28"	Ea.	3675
Tables, examining, vinyl top, with base cabinets	Ea.	1575 - 6525
Utensil Washer, Sanitizer	Ea.	10,300
X-Ray, Mobile	Ea.	20,800 - 105,000

Description	Unit	$ Cost
Commissioning Fees, sustainable commercial construction	S.F.	0.24 - 3.04
Energy Modelling Fees, commercial buildings to 10,000 SF	Ea.	11,000
Greater than 10,000 SF add	S.F.	0.04
Green Bldg Cert Fees for comm construction project reg	Project	900
Photovoltaic Pwr Sys, grid connected, 20 kW (~2400 SF), roof	Ea.	249,200
Green Roofs, 6" soil depth, w/treated wd edging & sedum mats	S.F.	12.41
10" Soil depth, with treated wood edging & sedum mats	S.F.	14.10
Greywater Recovery Systems, prepackaged comm, 3060 gal.	Ea.	45,900
4590 gal.	Ea.	57,625
Rainwater Harvest Sys, prepckged comm, 10,000 gal, sys contrller	Ea.	37,550
20,000 gal. w/system controller	Ea.	61,000
30,000 gal. w/system controller	Ea.	97,350
Solar Domestic HW, closed loop, add-on sys, ext heat exchanger	Ea.	11,300
Drainback, hot water system, 120 gal tank	Ea.	14,175
Draindown, hot water system, 120 gal tank	Ea.	14,425

Important: See the Reference Section for Location Factors.

Model costs calculated for a 2 story building with 10' story height and 7,000 square feet of floor area

				Unit	Unit Cost	Cost Per S.F.	% Of Sub-Total
A. SUBSTRUCTURE							
1010	Standard Foundations	Poured concrete; strip and spread footings		S.F. Ground	3.72	1.86	
1020	Special Foundations	N/A		—	—	—	
1030	Slab on Grade	4" reinforced concrete with vapor barrier and granular base		S.F. Slab	5.84	2.93	4.3%
2010	Basement Excavation	Site preparation for slab and trench for foundation wall and footing		S.F. Ground	.57	.29	
2020	Basement Walls	4' foundation wall		L.F. Wall	98	3.36	
B. SHELL							
	B10 Superstructure						
1010	Floor Construction	Open web steel joists, slab form, concrete, columns		S.F. Floor	14.62	7.31	5.4%
1020	Roof Construction	Metal deck, open web steel joists, beams, columns		S.F. Roof	6.64	3.32	
	B20 Exterior Enclosure						
2010	Exterior Walls	Stucco on concrete block		S.F. Wall	20.75	9.96	
2020	Exterior Windows	Outward projecting metal	30% of wall	Each	529	7.25	9.7%
2030	Exterior Doors	Aluminum and glass doors with transoms		Each	7200	2.05	
	B30 Roofing						
3010	Roof Coverings	Built-up tar and gravel with flashing; perlite/EPS composite insulation		S.F. Roof	8.68	4.34	2.3%
3020	Roof Openings	Roof hatches		S.F. Roof	.32	.16	
C. INTERIORS							
1010	Partitions	Gypsum bd. & acous. insul. on metal studs	6 SF Floor/LF Part.	S.F. Partition	8.44	11.25	
1020	Interior Doors	Single leaf wood	60 S.F. Floor/Door	Each	725	12.08	
1030	Fittings	N/A		—	—	—	
2010	Stair Construction	Concrete filled metal pan		Flight	15,800	4.52	21.7%
3010	Wall Finishes	45% paint, 50% vinyl wall coating, 5% ceramic tile		S.F. Surface	1.48	3.94	
3020	Floor Finishes	50% carpet, 50% vinyl composition tile		S.F. Floor	4.21	4.21	
3030	Ceiling Finishes	Mineral fiber tile on concealed zee bars		S.F. Ceiling	6.86	6.86	
D. SERVICES							
	D10 Conveying						
1010	Elevators & Lifts	One hydraulic hospital elevator		Each	116,690	16.67	8.4%
1020	Escalators & Moving Walks	N/A		—	—	—	
	D20 Plumbing						
2010	Plumbing Fixtures	Toilet, exam room & service fixt., supply & drain.	1 Fixt./160 SF Flr.	Each	4752	29.70	
2020	Domestic Water Distribution	Gas fired water heater		S.F. Floor	3.02	3.02	17.3%
2040	Rain Water Drainage	Roof drains		S.F. Roof	2.96	1.48	
	D30 HVAC						
3010	Energy Supply	N/A		—	—	—	
3020	Heat Generating Systems	Included in D3050		—	—	—	
3040	Distribution Systems	Enthalpy heat recovery packages		Each	14,075	2.01	8.3 %
3050	Terminal & Package Units	Multizone rooftop air conditioner, SEER 14		S.F. Floor	14.43	14.43	
3090	Other HVAC Sys. & Equipment	N/A		—	—	—	
	D40 Fire Protection						
4010	Sprinklers	Wet pipe sprinkler system		S.F. Floor	4.08	4.08	3.1%
4020	Standpipes	Standpipe		S.F. Floor	2.06	2.06	
	D50 Electrical						
5010	Electrical Service/Distribution	400 ampere service, panel board and feeders		S.F. Floor	1.65	1.65	
5020	Lighting & Branch Wiring	LED fixtures, receptacles, switches, A.C. and misc. power		S.F. Floor	14.36	14.36	13.9%
5030	Communications & Security	Alarm system, internet & phone wiring, intercom system, & emergency ltg.		S.F. Floor	8.24	8.24	
5090	Other Electrical Systems	Emergency generator, 7.5 kW		S.F. Floor	3.26	3.26	
E. EQUIPMENT & FURNISHINGS							
1010	Commercial Equipment	N/A		—	—	—	
1020	Institutional Equipment	Exam room casework and contertops		S.F. Floor	10.58	10.58	5.6 %
1090	Other Equipment	Waste handling recycling tilt truck		S.F. Floor	.52	.52	
2020	Moveable Furnishings	No smoking signage		S.F. Floor	.02	.02	
F. SPECIAL CONSTRUCTION							
1020	Integrated Construction	N/A		—	—	—	0.0 %
1040	Special Facilities	N/A		—	—	—	
G. BUILDING SITEWORK	**N/A**						

			Sub-Total	197.77	100%
CONTRACTOR FEES (General Requirements: 10%, Overhead: 5%, Profit: 10%)			25%	49.48	
ARCHITECT FEES			9%	22.25	
		Total Building Cost		**269.50**	

Costs per square foot of floor area

Exterior Wall		S.F. Area	2000	3000	4000	6000	8000	10000	12000	14000	16000
		L.F. Perimeter	240	260	280	380	480	560	580	660	740
Brick Veneer	Wood Frame		239.70	210.75	196.25	188.50	184.60	181.00	175.10	173.95	172.90
Aluminum Clapboard	Wood Frame		221.15	197.40	185.45	178.70	175.35	172.35	167.70	166.60	165.80
Wood Clapboard	Wood Frame		220.80	197.15	185.25	178.50	175.20	172.20	167.50	166.45	165.60
Wood Shingles	Wood Frame		224.65	199.90	187.50	180.60	177.10	173.95	169.10	168.05	167.10
Precast Concrete	Wood Truss		224.65	199.85	187.45	180.55	177.10	173.95	169.05	168.00	167.10
Face Brick and Concrete Block	Wood Truss		256.25	222.70	205.95	197.25	192.90	188.75	181.80	180.40	179.30
Perimeter Adj., Add or Deduct	Per 100 L.F.		33.55	22.40	16.75	11.20	8.40	6.70	5.65	4.80	4.20
Story Hgt. Adj., Add or Deduct	Per 1 Ft.		5.30	3.80	3.10	2.80	2.70	2.40	2.20	2.00	2.05

For Basement, add $27.67 per square foot of basement area

The above costs were calculated using the basic specifications shown on the facing page. These costs should be adjusted where necessary for design alternatives and owner's requirements.

Common additives

Description	Unit	$ Cost
Closed Circuit Surveillance, one station, camera & monitor	Ea.	1475
For additional camera stations, add	Ea.	665
Laundry Equipment, dryer, gas, 30 lb. capacity	Ea.	4100
Washer, commercial	Ea.	1700
Sauna, prefabricated, complete, 6' x 9'	Ea.	10,000
8' x 8'	Ea.	10,600
Swimming Pools, Complete, gunite	S.F.	106 - 132
TV Antenna, Master system, 12 outlet	Outlet	242
30 outlet	Outlet	335
100 outlet	Outlet	370

Description	Unit	$ Cost
Commissioning Fees, sustainable commercial construction	S.F.	0.24 - 3.04
Energy Modelling Fees, commercial buildings to 10,000 SF	Ea.	11,000
Greater than 10,000 SF add	S.F.	0.04
Green Bldg Cert Fees for comm construction project reg	Project	900
Photovoltaic Pwr Sys, grid connected, 20 kW (~2400 SF), roof	Ea.	249,200
Green Roofs, 6" soil depth, w/treated wd edging & sedum mats	S.F.	12.41
10" Soil depth, with treated wood edging & sedum mats	S.F.	14.10
Greywater Recovery Systems, prepackaged comm, 3060 gal.	Ea.	45,900
4590 gal.	Ea.	57,625
Rainwater Harvest Sys, prepckged comm, 10,000 gal, sys contrller	Ea.	37,550
20,000 gal. w/system controller	Ea.	61,000
30,000 gal. w/system controller	Ea.	97,350
Solar Domestic HW, closed loop, add-on sys, ext heat exchanger	Ea.	11,300
Drainback, hot water system, 120 gal tank	Ea.	14,175
Draindown, hot water system, 120 gal tank	Ea.	14,425

Important: See the Reference Section for Location Factors.

Model costs calculated for a 1 story building with 9' story height and 8,000 square feet of floor area

G Motel, 1 Story

				Unit	Unit Cost	Cost Per S.F.	% Of Sub-Total
A.	**SUBSTRUCTURE**						
1010	Standard Foundations	Poured concrete; strip and spread footings		S.F. Ground	3.50	3.50	
1020	Special Foundations	N/A		—	—	—	
1030	Slab on Grade	4" reinforced concrete with recycled vapor barrier and granular base		S.F. Slab	5.84	5.84	12.8%
2010	Basement Excavation	Site preparation for slab and trench for foundation wall and footing		S.F. Ground	.33	.33	
2020	Basement Walls	4' foundation wall		L.F. Wall	98	8.06	
B.	**SHELL**						
	B10 Superstructure						
1010	Floor Construction	N/A		—	—	—	5.6 %
1020	Roof Construction	Plywood on wood trusses		S.F. Roof	7.76	7.76	
	B20 Exterior Enclosure						
2010	Exterior Walls	Face brick on wood studs with sheathing, insulation and paper	80% of wall	S.F. Wall	26.85	11.60	
2020	Exterior Windows	Wood double hung	20% of wall	Each	533	4.11	16.1%
2030	Exterior Doors	Wood solid core		Each	2385	6.56	
	B30 Roofing						
3010	Roof Coverings	Asphalt strip shingles, 210-235 lbs/Sq., R-10 insul. & weather barrier-recycled		S.F. Roof	8.65	8.65	6.3%
3020	Roof Openings	N/A		—	—	—	
C.	**INTERIORS**						
1010	Partitions	Gypsum bd. and sound deadening bd. on wood studs	9 SF Flr./LF Part.	S.F. Partition	11.04	9.81	
1020	Interior Doors	Single leaf hollow core wood, low VOC paint	300 S.F. Floor/Door	Each	645	2.15	
1030	Fittings	N/A		—	—	—	
2010	Stair Construction	N/A		—	—	—	18.6%
3010	Wall Finishes	90% paint, low VOC, 10% ceramic tile		S.F. Surface	1.76	3.12	
3020	Floor Finishes	85% carpet tile, 15% ceramic tile		S.F. Floor	5.65	5.65	
3030	Ceiling Finishes	Painted gypsum board on furring, low VOC paint		S.F. Ceiling	4.94	4.94	
D.	**SERVICES**						
	D10 Conveying						
1010	Elevators & Lifts	N/A		—	—	—	0.0 %
1020	Escalators & Moving Walks	N/A		—	—	—	
	D20 Plumbing						
2010	Plumbing Fixtures	Toilet, low flow, auto sensor, & service fixt., supply & drainage	1 Fixt./90 SF Flr.	Each	2276	25.29	
2020	Domestic Water Distribution	Gas fired tankless water heater		S.F. Floor	.43	.43	20.1%
2040	Rain Water Drainage	Roof Drains		S.F. Roof	2.05	2.05	
	D30 HVAC						
3010	Energy Supply	N/A		—	—	—	
3020	Heat Generating Systems	Included in D3050		—	—	—	
3030	Cooling Generating Systems	N/A		—	—	—	2.2 %
3050	Terminal & Package Units	Through the wall electric heating and cooling units		S.F. Floor	3.08	3.08	
3090	Other HVAC Sys. & Equipment	N/A		—	—	—	
	D40 Fire Protection						
4010	Sprinklers	Wet pipe sprinkler system		S.F. Floor	5.13	5.13	4.6%
4020	Standpipes	Standpipe, wet, Class III		S.F. Floor	1.19	1.19	
	D50 Electrical						
5010	Electrical Service/Distribution	200 ampere service, panel board and feeders		S.F. Floor	1.31	1.31	
5020	Lighting & Branch Wiring	LED fixt., daylit. dim., ltg. on/off control, recept., switches and misc. pwr.		S.F. Floor	10.56	10.56	12.2%
5030	Communications & Security	Addressable alarm systems		S.F. Floor	2.80	2.80	
5090	Other Electrical Systems	Emergency generator, 7.5 kW, and energy monitoring systems		S.F. Floor	2.14	2.14	
E.	**EQUIPMENT & FURNISHINGS**						
1010	Commercial Equipment	Laundry equipment		S.F. Floor	1.02	1.02	
1020	Institutional Equipment	N/A		—	—	—	1.4%
1090	Other Equipment	Waste handling recycling tilt truck		S.F. Floor	.78	.78	
2020	Moveable Furnishings	No smoking signage		S.F. Floor	.18	.18	
F.	**SPECIAL CONSTRUCTION**						
1020	Integrated Construction	N/A		—	—	—	0.0 %
1040	Special Facilities	N/A		—	—	—	
G.	**BUILDING SITEWORK**	**N/A**					

				Sub-Total		138.04	100%
	CONTRACTOR FEES (General Requirements:10%, Overhead: 5%, Profit:10%)				25%	34.48	
	ARCHITECT FEES				7%	12.08	

Total Building Cost **184.60**

For customer support on your Square Foot Costs with RSMeans data, call 800.448.8182.

257

Costs per square foot of floor area

Exterior Wall	S.F. Area	25000	37000	49000	61000	73000	81000	88000	96000	104000
	L.F. Perimeter	433	593	606	720	835	911	978	1054	1074
Decorative Concrete Block	Wood Joists	197.45	192.85	185.95	184.35	183.20	182.65	182.25	181.80	180.65
	Precast Concrete	213.45	208.80	202.00	200.30	199.20	198.65	198.20	197.85	196.65
Stucco and Concrete Block	Wood Joists	198.20	193.45	186.30	184.60	183.40	182.85	182.40	182.00	180.80
	Precast Concrete	215.10	210.35	203.20	201.50	200.35	199.75	199.30	198.90	197.65
Wood Clapboard	Wood Frame	193.70	189.30	183.10	181.60	180.50	179.95	179.50	179.15	178.10
Brick Veneer	Wood Frame	198.25	193.50	186.35	184.65	183.50	182.90	182.50	182.10	180.85
Perimeter Adj., Add or Deduct	Per 100 L.F.	5.95	3.90	3.05	2.35	2.00	1.80	1.65	1.55	1.45
Story Hgt. Adj., Add or Deduct	Per 1 Ft.	2.05	1.90	1.50	1.40	1.40	1.35	1.30	1.35	1.25

For Basement, add $ 36.28 per square foot of basement area

The above costs were calculated using the basic specifications shown on the facing page. These costs should be adjusted where necessary for design alternatives and owner's requirements.

Common additives

Description	Unit	$ Cost
Closed Circuit Surveillance, one station, camera & monitor	Ea.	1475
For additional camera station, add	Ea.	665
Elevators, hydraulic passenger, 2 stops, 2500# capacity	Ea.	76,600
3500# capacity	Ea.	81,600
Additional stop, add	Ea.	8375
Laundry Equipment, dryer, gas, 30 lb. capacity	Ea.	4100
Washer, commercial	Ea.	1700
Sauna, prefabricated, complete, 6' x 9'	Ea.	10,000
8' x 8'	Ea.	10,600
Swimming Pools, Complete, gunite	S.F.	106 - 132
TV Antenna, Master system, 12 outlet	Outlet	242
30 outlet	Outlet	335
100 outlet	Outlet	370

Description	Unit	$ Cost
Commissioning Fees, sustainable commercial construction	S.F.	0.24 - 3.04
Energy Modelling Fees, commercial buildings to 10,000 SF	Ea.	11,000
Greater than 10,000 SF add	S.F.	0.04
Green Bldg Cert Fees for comm construction project reg	Project	900
Photovoltaic Pwr Sys, grid connected, 20 kW (~2400 SF), roof	Ea.	249,200
Green Roofs, 6" soil depth, w/treated wd edging & sedum mats	S.F.	12.41
10" Soil depth, with treated wood edging & sedum mats	S.F.	14.10
Greywater Recovery Systems, prepackaged comm, 3060 gal.	Ea.	45,900
4590 gal.	Ea.	57,625
Rainwater Harvest Sys, prepckged comm, 10,000 gal, sys contrller	Ea.	37,550
20,000 gal. w/system controller	Ea.	61,000
30,000 gal. w/system controller	Ea.	97,350
Solar Domestic HW, closed loop, add-on sys, ext heat exchanger	Ea.	11,300
Drainback, hot water system, 120 gal tank	Ea.	14,175
Draindown, hot water system, 120 gal tank	Ea.	14,425

Important: See the Reference Section for Location Factors.

Model costs calculated for a 3 story building with 9' story height and 49,000 square feet of floor area

G Motel, 2-3 Story

			Unit	Unit Cost	Cost Per S.F.	% Of Sub-Total
A.	**SUBSTRUCTURE**					
1010	Standard Foundations	Poured concrete; strip and spread footings	S.F. Ground	1.62	.54	
1020	Special Foundations	N/A	—	—	—	
1030	Slab on Grade	4" reinforced concrete with recycled vapor barrier and granular base	S.F. Slab	5.84	1.95	3.3%
2010	Basement Excavation	Site preparation for slab and trench for foundation wall and footing	S.F. Ground	.18	.06	
2020	Basement Walls	4' foundation wall	L.F. Wall	98	2.43	
B.	**SHELL**					
	B10 Superstructure					
1010	Floor Construction	Precast concrete plank	S.F. Floor	13.65	9.10	8.8%
1020	Roof Construction	Precast concrete plank	S.F. Roof	12.87	4.29	
	B20 Exterior Enclosure					
2010	Exterior Walls	Decorative concrete block 85% of wall	S.F. Wall	23.64	6.71	
2020	Exterior Windows	Aluminum sliding 15% of wall	Each	589	1.97	11.9%
2030	Exterior Doors	Aluminum and glass doors and entrance with transom	Each	2872	9.50	
	B30 Roofing					
3010	Roof Coverings	Single-ply TPO membrane, 60 mils w/R-20 insul.	S.F. Roof	7.86	2.62	1.8%
3020	Roof Openings	Roof hatches	S.F. Roof	.15	.05	
C.	**INTERIORS**					
1010	Partitions	Concrete block w/foamed-in insul. 7 SF Flr./LF Part.	S.F. Partition	24.16	27.61	
1020	Interior Doors	Wood hollow core, low VOC paint 70 S.F. Floor/Door	Each	645	9.21	
1030	Fittings	N/A	—	—	—	
2010	Stair Construction	Concrete filled metal pan	Flight	15,800	3.87	36.1%
3010	Wall Finishes	90% paint, low VOC, 10% ceramic tile	S.F. Surface	1.76	4.02	
3020	Floor Finishes	85% carpet tile, 5% vinyl composition tile, recycled content, 10% ceramic tile	S.F. Floor	5.65	5.65	
3030	Ceiling Finishes	Textured finish	S.F. Ceiling	4.66	4.66	
D.	**SERVICES**					
	D10 Conveying					
1010	Elevators & Lifts	Two hydraulic passenger elevators	Each	117,845	4.81	3.2%
1020	Escalators & Moving Walks	N/A	—	—	—	
	D20 Plumbing					
2010	Plumbing Fixtures	Toilet, low flow, auto sensor & service fixtures, supply & drain. 1 Fixt./180 SF Flr.	Each	5443	30.24	
2020	Domestic Water Distribution	Gas fired, tankless water heater	S.F. Floor	.32	.32	20.5%
2040	Rain Water Drainage	Roof drains	S.F. Roof	1.86	.62	
	D30 HVAC					
3010	Energy Supply	N/A	—	—	—	
3020	Heat Generating Systems	Included in D3050	—	—	—	
3030	Cooling Generating Systems	N/A	—	—	—	1.8 %
3050	Terminal & Package Units	Through the wall electric heating and cooling units	S.F. Floor	2.79	2.79	
3090	Other HVAC Sys. & Equipment	N/A	—	—	—	
	D40 Fire Protection					
4010	Sprinklers	Sprinklers, wet, light hazard	S.F. Floor	3.12	3.12	2.2%
4020	Standpipes	Standpipe, wet, Class III	S.F. Floor	.29	.29	
	D50 Electrical					
5010	Electrical Service/Distribution	800 ampere service, panel board and feeders	S.F. Floor	1.23	1.23	
5020	Lighting & Branch Wiring	LED fixtures, daylt. dim., ltg. on/off, recept., switches and misc. power	S.F. Floor	11.35	11.35	10.2%
5030	Communications & Security	Addressable alarm systems and emergency lighting	S.F. Floor	1.93	1.93	
5090	Other Electrical Systems	Emergency generator, 7.5 kW, energy monitoring systems	S.F. Floor	1.07	1.07	
E.	**EQUIPMENT & FURNISHINGS**					
1010	Commercial Equipment	Commercial laundry equipment	S.F. Floor	.34	.34	
1020	Institutional Equipment	N/A	—	—	—	
1090	Other Equipment	Waste handling recycling tilt truck	S.F. Floor	.07	.07	0.3%
2020	Moveable Furnishings	No smoking signage	S.F. Floor	.02	.02	
F.	**SPECIAL CONSTRUCTION**					
1020	Integrated Construction	N/A	—	—	—	0.0 %
1040	Special Facilities	N/A	—	—	—	
G.	**BUILDING SITEWORK**	**N/A**				

				Sub-Total	152.44	100%
	CONTRACTOR FEES (General Requirements: 10%, Overhead: 5%, Profit: 10%)			25%	38.13	
	ARCHITECT FEES			6%	11.43	

		Total Building Cost	**202**

For customer support on your Square Foot Costs with RSMeans data, call 800.448.8182.

259

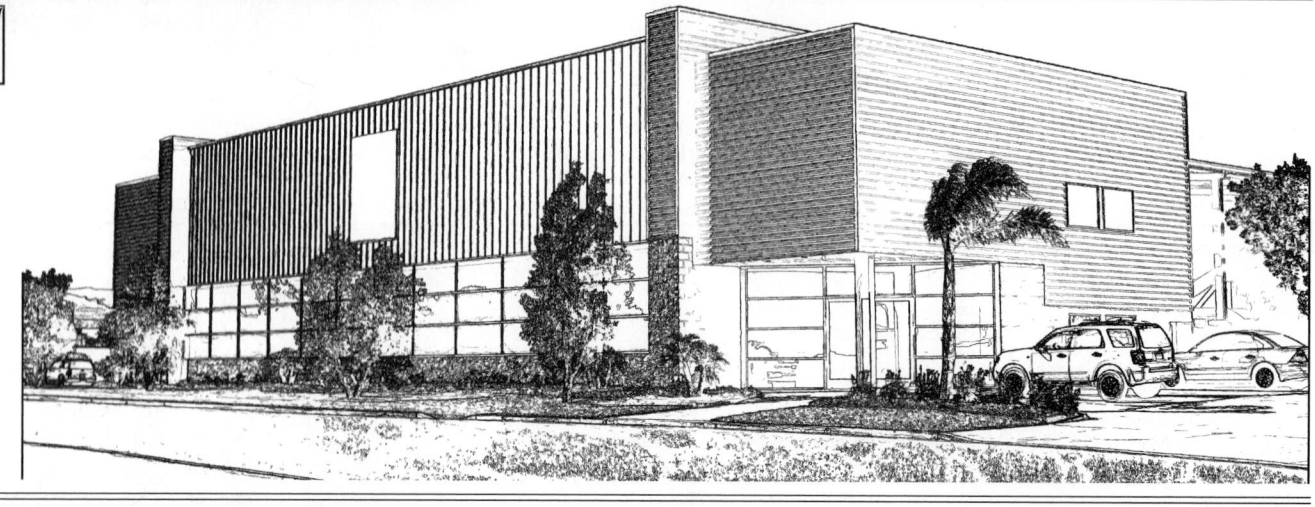

Costs per square foot of floor area

Exterior Wall	S.F. Area	2000	3000	5000	7000	9000	12000	15000	20000	25000
	L.F. Perimeter	220	260	320	360	420	480	520	640	700
Wood Clapboard	Wood Truss	236.75	213.95	193.30	182.55	178.00	172.40	168.25	165.75	162.75
Brick Veneer	Wood Truss	243.70	219.30	197.05	185.45	180.60	174.50	170.00	167.30	164.05
Face Brick and Concrete Block	Wood Truss	262.00	233.85	208.00	194.35	188.75	181.60	176.25	173.10	169.15
	Steel Roof Deck	279.05	249.25	221.80	207.35	201.40	193.75	188.10	184.75	180.60
E.I.F.S. and Metal Studs	Steel Roof Deck	255.55	231.70	210.10	198.90	194.15	188.25	183.95	181.30	178.20
Tilt-up Concrete Panels	Steel Roof Deck	262.45	237.45	214.65	202.85	197.80	191.60	187.05	184.25	180.95
Perimeter Adj., Add or Deduct	Per 100 L.F.	31.85	21.30	12.70	9.10	7.10	5.30	4.35	3.20	2.55
Story Hgt. Adj., Add or Deduct	Per 1 Ft.	4.00	3.20	2.35	1.85	1.70	1.45	1.35	1.20	1.05

For Basement, add $38.18 per square foot of basement area

The above costs were calculated using the basic specifications shown on the facing page. These costs should be adjusted where necessary for design alternatives and owner's requirements.

Common additives

Description	Unit	$ Cost
Closed Circuit Surveillance, one station, camera & monitor	Ea.	1475
For additional camera stations, add	Ea.	665
Directory boards, plastic, glass covered, 30" x 20"	Ea.	640
36" x 48"	Ea.	1625
Aluminum, 36" x 24"	Ea.	790
48" x 32"	Ea.	1025
Electronic, wall mounted	S.F.	4625
Pedestal access flr sys w/PLAM cover, comp rm,		
Less than 6000 SF	S.F.	22
Greater than 6000 SF	S.F.	21.50
Office, greater than 6000 S.F.	S.F.	17.95
Uninterruptible power supply, 15 kVA/12.75 kW	kW	1.28

Description	Unit	$ Cost
Commissioning Fees, sustainable commercial construction	S.F.	0.24 - 3.04
Energy Modelling Fees, commercial buildings to 10,000 SF	Ea.	11,000
Greater than 10,000 SF add	S.F.	0.04
Green Bldg Cert Fees for comm construction project reg	Project	900
Photovoltaic Pwr Sys, grid connected, 20 kW (~2400 SF), roof	Ea.	249,200
Green Roofs, 6" soil depth, w/treated wd edging & sedum mats	S.F.	12.41
10" Soil depth, with treated wood edging & sedum mats	S.F.	14.10
Greywater Recovery Systems, prepackaged comm, 3060 gal.	Ea.	45,900
4590 gal.	Ea.	57,625
Rainwater Harvest Sys, prepckged comm, 10,000 gal, sys contrller	Ea.	37,550
20,000 gal. w/system controller	Ea.	61,000
30,000 gal. w/system controller	Ea.	97,350
Solar Domestic HW, closed loop, add-on sys, ext heat exchanger	Ea.	11,300
Drainback, hot water system, 120 gal tank	Ea.	14,175
Draindown, hot water system, 120 gal tank	Ea.	14,425

Important: See the Reference Section for Location Factors.

Model costs calculated for a 1 story building with 12' story height and 7,000 square feet of floor area

				Unit	Unit Cost	Cost Per S.F.	% Of Sub-Total
A. SUBSTRUCTURE							
1010	Standard Foundations	Poured concrete; strip and spread footings		S.F. Ground	3.04	3.04	
1020	Special Foundations	N/A		—	—	—	
1030	Slab on Grade	4" reinforced concrete with recycled vapor barrier and granular base		S.F. Slab	5.84	5.84	9.8%
2010	Basement Excavation	Site preparation for slab and trench for foundation wall and footing		S.F. Ground	.33	.33	
2020	Basement Walls	4' foundation wall		L.F. Wall	186	5.29	
B. SHELL							
	B10 Superstructure						
1010	Floor Construction	N/A		—	—	—	7.4 %
1020	Roof Construction	Steel joists, girders & deck on columns		S.F. Roof	10.96	10.96	
	B20 Exterior Enclosure						
2010	Exterior Walls	E.I.F.S. on metal studs		S.F. Wall	18.70	9.23	
2020	Exterior Windows	Aluminum outward projecting	20% of wall	Each	755	4.05	11.6%
2030	Exterior Doors	Aluminum and glass, hollow metal		Each	4565	3.91	
	B30 Roofing						
3010	Roof Coverings	Single-ply TPO membrane, 45 mils, heat welded seams w/R-20 insul.		S.F. Roof	13	13	8.9%
3020	Roof Openings	Roof hatch		S.F. Roof	.21	.21	
C. INTERIORS							
1010	Partitions	Gyp. bd. on mtl. studs w/sound attenuation	20 SF Flr/LF Part.	S.F. Partition	11.08	5.54	
1020	Interior Doors	Single leaf hollow metal, low VOC paint	200 S.F. Floor/Door	Each	1222	6.11	
1030	Fittings	Toilet partitions		S.F. Floor	.89	.89	
2010	Stair Construction	N/A		—	—	—	18.1%
3010	Wall Finishes	60% vinyl wall covering, 40% paint, low VOC		S.F. Surface	1.66	1.66	
3020	Floor Finishes	60% carpet tile, 30% vinyl composition tile, recycled content, 10% ceramic tile		S.F. Floor	5.08	5.08	
3030	Ceiling Finishes	Mineral fiber tile on concealed zee bars		S.F. Ceiling	7.60	7.60	
D. SERVICES							
	D10 Conveying						
1010	Elevators & Lifts	N/A		—	—	—	0.0 %
1020	Escalators & Moving Walks	N/A		—	—	—	
	D20 Plumbing						
2010	Plumbing Fixtures	Toilet, low flow, auto sensor, & service fixt., supply & drain.	1 Fixt./1320 SF Flr.	Each	10,019	7.59	
2020	Domestic Water Distribution	Tankless, on-demand water heater, gas/propane		S.F. Floor	.96	.96	6.4%
2040	Rain Water Drainage	Roof drains		S.F. Roof	.90	.90	
	D30 HVAC						
3010	Energy Supply	N/A		—	—	—	
3020	Heat Generating Systems	Included in D3050		—	—	—	
3040	Distribution Systems	Enthalpy heat recovery packages		Each	10,375	1.48	14.9 %
3050	Terminal & Package Units	Multizone rooftop air conditioner, SEER 14		S.F. Floor	20.70	20.70	
3090	Other HVAC Sys. & Equipment	N/A		—	—	—	
	D40 Fire Protection						
4010	Sprinklers	Sprinkler system, light hazard		S.F. Floor	3.79	3.79	3.6%
4020	Standpipes	Standpipes and hose systems		S.F. Floor	1.63	1.63	
	D50 Electrical						
5010	Electrical Service/Distribution	400 ampere service, panel board and feeders		S.F. Floor	2.06	2.06	
5020	Lighting & Branch Wiring	LED fixtures, daylt. dim., ltg. on/off, recept., switches, and A.C. power		S.F. Floor	15.96	15.96	18.8%
5030	Communications & Security	Addressable alarm systems, internet and phone wiring, and emergency lighting		S.F. Floor	7.49	7.49	
5090	Other Electrical Systems	Emergency generator, 7.5 kW, energy monitoring systems		S.F. Floor	2.45	2.45	
E. EQUIPMENT & FURNISHINGS							
1010	Commercial Equipment	N/A		—	—	—	
1020	Institutional Equipment	N/A		—	—	—	0.6 %
1090	Other Equipment	Waste handling recycling tilt truck		S.F. Floor	.89	.89	
2020	Moveable Furnishings	No smoking signage		S.F. Floor	.06	.06	
F. SPECIAL CONSTRUCTION							
1020	Integrated Construction	N/A		—	—	—	0.0 %
1040	Special Facilities	N/A		—	—	—	
G. BUILDING SITEWORK	**N/A**						

		Sub-Total	148.70	100%
CONTRACTOR FEES (General Requirements: 10%, Overhead: 5%, Profit: 10%)		25%	37.19	
ARCHITECT FEES		7%	13.01	
	Total Building Cost		**198.90**	

For customer support on your Square Foot Costs with RSMeans data, call 800.448.8182.

261

Costs per square foot of floor area

Exterior Wall	S.F. Area	5000	8000	12000	16000	20000	35000	50000	65000	80000
	L.F. Perimeter	220	260	310	330	360	440	550	600	675
Face Brick and Concrete Block	Wood Joists	279.60	243.65	223.05	208.35	200.70	185.35	180.65	175.90	173.70
	Steel Joists	291.30	255.35	234.75	220.05	212.40	197.05	192.30	187.60	185.40
Curtain Wall	Rigid Steel	323.95	280.95	256.30	238.40	229.10	210.45	204.85	199.00	196.35
	Reinforced Concrete	326.65	283.65	259.00	241.10	231.80	213.15	207.50	201.75	199.05
Wood Clapboard	Wood Frame	236.55	209.60	194.25	183.65	178.05	166.95	163.45	160.15	158.60
Brick Veneer	Wood Frame	246.75	217.15	200.20	188.40	182.20	169.85	166.00	162.30	160.55
Perimeter Adj., Add or Deduct	Per 100 L.F.	47.20	29.45	19.65	14.75	11.75	6.75	4.70	3.60	2.95
Story Hgt. Adj., Add or Deduct	Per 1 Ft.	7.75	5.70	4.50	3.55	3.10	2.20	1.90	1.65	1.50

For Basement, add $42.72 per square foot of basement area

The above costs were calculated using the basic specifications shown on the facing page. These costs should be adjusted where necessary for design alternatives and owner's requirements.

Common additives

Description	Unit	$ Cost
Closed Circuit Surveillance, one station, camera and monitor	Ea.	1475
For additional camera stations, add	Ea.	665
Directory Boards, plastic, glass covered, 30" x 20"	Ea.	640
36" x 48"	Ea.	1625
Aluminum, 36" x 24"	Ea.	790
48" x 32"	Ea.	1025
Electronic, wall mounted	S.F.	4625
Escalators, 10' rise, 32" wide, glass balustrade	Ea.	153,500
48" wide, glass balustrade	Ea.	161,000
Pedestal access flr sys w/PLAM cover, comp rm,		
Less than 6000 SF	S.F.	22
Greater than 6000 SF	S.F.	21.50
Office, greater than 6000 S.F.	S.F.	17.95
Uninterruptible power supply, 15 kVA/12.75 kW	kW	1.28

Description	Unit	$ Cost
Commissioning Fees, sustainable commercial construction	S.F.	0.24 - 3.04
Energy Modelling Fees, commercial buildings to 10,000 SF	Ea.	11,000
Greater than 10,000 SF add	S.F.	0.04
Green Bldg Cert Fees for comm construction project reg	Project	900
Photovoltaic Pwr Sys, grid connected, 20 kW (~2400 SF), roof	Ea.	249,200
Green Roofs, 6" soil depth, w/treated wd edging & sedum mats	S.F.	12.41
10" Soil depth, with treated wood edging & sedum mats	S.F.	14.10
Greywater Recovery Systems, prepackaged comm, 3060 gal.	Ea.	45,900
4590 gal.	Ea.	57,625
Rainwater Harvest Sys, prepckged comm, 10,000 gal, sys contrller	Ea.	37,550
20,000 gal. w/system controller	Ea.	61,000
30,000 gal. w/system controller	Ea.	97,350
Solar Domestic HW, closed loop, add-on sys, ext heat exchanger	Ea.	11,300
Drainback, hot water system, 120 gal tank	Ea.	14,175
Draindown, hot water system, 120 gal tank	Ea.	14,425

Important: See the Reference Section for Location Factors.

Model costs calculated for a 3 story building with 12' story height and 20,000 square feet of floor area

				Unit	Unit Cost	Cost Per S.F.	% Of Sub-Total
A.	**SUBSTRUCTURE**						
1010	Standard Foundations	Poured concrete; strip and spread footings		S.F. Ground	8.31	2.77	
1020	Special Foundations	N/A		—	—	—	
1030	Slab on Grade	4" reinforced concrete with recycled vapor barrier and granular base		S.F. Slab	5.84	1.95	4.4%
2010	Basement Excavation	Site preparation for slab and trench for foundation wall and footing		S.F. Ground	.33	.11	
2020	Basement Walls	4' foundation wall		L.F. Wall	186	2.08	
B.	**SHELL**						
	B10 Superstructure						
1010	Floor Construction	Open web steel joists, slab form, concrete, columns		S.F. Floor	20.63	13.75	10.3%
1020	Roof Construction	Metal deck, open web steel joists, columns		S.F. Roof	7.92	2.64	
	B20 Exterior Enclosure						
2010	Exterior Walls	Face brick with concrete block backup	80% of wall	S.F. Wall	41.20	21.36	
2020	Exterior Windows	Aluminum outward projecting	20% of wall	Each	755	4.25	17.0%
2030	Exterior Doors	Aluminum and glass, hollow metal		Each	4565	1.38	
	B30 Roofing						
3010	Roof Coverings	Single-ply, TPO membrane, 60 mils, heat welded seams w/R-20 insul.		S.F. Roof	8.88	2.96	1.9%
3020	Roof Openings	N/A		—	—	—	
C.	**INTERIORS**						
1010	Partitions	Gypsum board on metal studs	20 SF Flr/LF Part.	S.F. Partition	11.43	4.57	
1020	Interior Doors	Single leaf hollow metal, low VOC paint	200 S.F. Floor/Door	Each	1222	6.11	
1030	Fittings	Toilet partitions		S.F. Floor	1.87	1.87	
2010	Stair Construction	Concrete filled metal pan		Flight	15,800	5.53	20.2%
3010	Wall Finishes	60% vinyl wall covering, 40% paint, low VOC		S.F. Surface	1.65	1.32	
3020	Floor Finishes	60% carpet tile, 30% vinyl composition tile, recycled content, 10% ceramic tile		S.F. Floor	5.08	5.08	
3030	Ceiling Finishes	Mineral fiber tile on concealed zee bars		S.F. Ceiling	7.60	7.60	
D.	**SERVICES**						
	D10 Conveying						
1010	Elevators & Lifts	Two hydraulic passenger elevators		Each	125,900	12.59	7.9%
1020	Escalators & Moving Walks	N/A		—	—	—	
	D20 Plumbing						
2010	Plumbing Fixtures	Toilet, low flow, auto sensor, & service fixt., supply & drain.	1 Fixt./1320 SF Flr.	Each	5900	4.47	
2020	Domestic Water Distribution	Gas fired, tankless water heater		S.F. Floor	.22	.22	3.5%
2040	Rain Water Drainage	Roof drains		S.F. Roof	2.82	.94	
	D30 HVAC						
3010	Energy Supply	N/A		—	—	—	
3020	Heat Generating Systems	Included in D3050		—	—	—	
3040	Distribution Systems	Enthalpy heat recovery packages		Each	18,475	2.77	12.2 %
3050	Terminal & Package Units	Multizone rooftop air conditioner. SEER 14		S.F. Floor	16.65	16.65	
3090	Other HVAC Sys. & Equipment	N/A		—	—	—	
	D40 Fire Protection						
4010	Sprinklers	Wet pipe sprinkler system		S.F. Floor	4.02	4.02	3.2%
4020	Standpipes	Standpipes and hose systems		S.F. Floor	1.07	1.07	
	D50 Electrical						
5010	Electrical Service/Distribution	1000 ampere service, panel board and feeders		S.F. Floor	2.86	2.86	
5020	Lighting & Branch Wiring	LED fixtures, daylt. dim., ltg. on/off, recept., switches, and A.C. power		S.F. Floor	16.63	16.63	19.1%
5030	Communications & Security	Addressable alarm systems, internet and phone wiring, & emergency ltg.		S.F. Floor	9.57	9.57	
5090	Other Electrical Systems	Emergency generator, 7.5 kW, UPS, energy monitoring systems		S.F. Floor	1.33	1.33	
E.	**EQUIPMENT & FURNISHINGS**						
1010	Commercial Equipment	N/A		—	—	—	
1020	Institutional Equipment	N/A		—	—	—	
1090	Other Equipment	Waste handling recycling tilt truck		S.F. Floor	.31	.31	0.2 %
2020	Moveable Furnishings	No smoking signage		S.F. Floor	.02	.02	
F.	**SPECIAL CONSTRUCTION**						
1020	Integrated Construction	N/A		—	—	—	0.0 %
1040	Special Facilities	N/A		—	—	—	
G.	**BUILDING SITEWORK**	**N/A**					

				Sub-Total	158.78	**100%**
CONTRACTOR FEES (General Requirements: 10%, Overhead: 5%, Profit: 10%)				25%	39.73	
ARCHITECT FEES				7%	13.89	

Total Building Cost		**212.40**

For customer support on your Square Foot Costs with RSMeans data, call 800.448.8182.

Costs per square foot of floor area

Exterior Wall	S.F. Area	20000	40000	60000	80000	100000	150000	200000	250000	300000
	L.F. Perimeter	260	360	400	420	460	520	650	720	800
Precast Concrete	Rigid Steel	269.15	230.10	209.00	196.40	190.45	180.40	178.20	174.90	172.95
	Reinforced Concrete	268.35	229.35	208.15	195.60	189.65	179.55	177.40	174.05	172.10
Face Brick and Concrete Block	Rigid Steel	241.50	210.95	194.80	185.25	180.65	173.00	171.25	168.75	167.25
	Reinforced Concrete	240.75	210.15	194.00	184.40	179.85	172.15	170.45	167.95	166.45
Limestone and Concrete Block	Rigid Steel	271.00	231.35	209.90	197.15	191.15	180.85	178.60	175.30	173.30
	Reinforced Concrete	270.20	230.60	209.15	196.30	190.30	180.05	177.85	174.50	172.50
Perimeter Adj., Add or Deduct	Per 100 L.F.	40.75	20.40	13.55	10.20	8.10	5.40	4.05	3.25	2.70
Story Hgt. Adj., Add or Deduct	Per 1 Ft.	8.50	5.90	4.35	3.45	3.00	2.25	2.10	1.85	1.75

For Basement, add $46.07 per square foot of basement area

The above costs were calculated using the basic specifications shown on the facing page. These costs should be adjusted where necessary for design alternatives and owner's requirements.

Common additives

Description	Unit	$ Cost
Closed Circuit Surveillance, one station, camera and monitor	Ea.	1475
For additional camera stations, add	Ea.	665
Directory Boards, plastic, glass covered, 30" x 20"	Ea.	640
36" x 48"	Ea.	1625
Aluminum, 36" x 24"	Ea.	790
48" x 32"	Ea.	1025
Electronic, wall mounted	S.F.	4625
Escalators, 10' rise, 32" wide, glass balustrade	Ea.	153,500
48" wide, glass balustrade	Ea.	161,000
Pedestal access flr sys w/PLAM cover, comp rm,		
Less than 6000 SF	S.F.	22
Greater than 6000 SF	S.F.	21.50
Office, greater than 6000 S.F.	S.F.	17.95
Uninterruptible power supply, 15 kVA/12.75 kW	kW	1.28

Description	Unit	$ Cost
Commissioning Fees, sustainable commercial construction	S.F.	0.24 - 3.04
Energy Modelling Fees, commercial buildings to 10,000 SF	Ea.	11,000
Greater than 10,000 SF add	S.F.	0.04
Green Bldg Cert Fees for comm construction project reg	Project	900
Photovoltaic Pwr Sys, grid connected, 20 kW (~2400 SF), roof	Ea.	249,200
Green Roofs, 6" soil depth, w/treated wd edging & sedum mats	S.F.	12.41
10" Soil depth, with treated wood edging & sedum mats	S.F.	14.10
Greywater Recovery Systems, prepackaged comm, 3060 gal.	Ea.	45,900
4590 gal.	Ea.	57,625
Rainwater Harvest Sys, prepckged comm, 10,000 gal, sys contrller	Ea.	37,550
20,000 gal. w/system controller	Ea.	61,000
30,000 gal. w/system controller	Ea.	97,350
Solar Domestic HW, closed loop, add-on sys, ext heat exchanger	Ea.	11,300
Drainback, hot water system, 120 gal tank	Ea.	14,175
Draindown, hot water system, 120 gal tank	Ea.	14,425

Important: See the Reference Section for Location Factors.

Model costs calculated for a 8 story building with 12' story height and 80,000 square feet of floor area

			Unit	Unit Cost	Cost Per S.F.	% Of Sub-Total

A. SUBSTRUCTURE

1010	Standard Foundations	Poured concrete; strip and spread footings	S.F. Ground	13.36	1.67	
1020	Special Foundations	N/A	—	—	—	
1030	Slab on Grade	4" reinforced concrete with recycled vapor barrier and granular base	S.F. Slab	5.84	.73	2.1%
2010	Basement Excavation	Site preparation for slab and trench for foundation wall and footing	S.F. Ground	.33	.04	
2020	Basement Walls	4' foundation wall	L.F. Wall	98	.66	

B. SHELL

B10 Superstructure

1010	Floor Construction	Concrete slab with metal deck and beams	S.F. Floor	25.26	22.10	15.6%
1020	Roof Construction	Metal deck, open web steel joists, interior columns	S.F. Roof	7.60	.95	

B20 Exterior Enclosure

2010	Exterior Walls	Precast concrete panels 80% of wall	S.F. Wall	62	25.04	
2020	Exterior Windows	Vertical pivoted steel 20% of wall	Each	589	3.96	19.8%
2030	Exterior Doors	Double aluminum and glass doors and entrance with transoms	Each	5428	.34	

B30 Roofing

3010	Roof Coverings	Single-ply TPO membrane, 60 mils, welded seams w/R-20 insul.	S.F. Roof	7.36	.92	0.6%
3020	Roof Openings	N/A	—	—	—	

C. INTERIORS

1010	Partitions	Gypsum board on metal studs w/sound attenuation 30 SF Flr./LF Part.	S.F. Partition	12.27	4.09	
1020	Interior Doors	Single leaf hollow metal, low VOC paint 400 S.F. Floor/Door	Each	1222	3.06	
1030	Fittings	Toilet Partitions	S.F. Floor	1.25	1.25	
2010	Stair Construction	Concrete filled metal pan	Flight	15,800	3.36	17.2%
3010	Wall Finishes	60% vinyl wall covering, 40% paint, low VOC paint	S.F. Surface	1.67	1.11	
3020	Floor Finishes	60% carpet tile, 30% vinyl composition tile, recycled content, 10% ceramic tile	S.F. Floor	5.08	5.08	
3030	Ceiling Finishes	Mineral fiber tile on concealed zee bars	S.F. Ceiling	7.60	7.60	

D. SERVICES

D10 Conveying

1010	Elevators & Lifts	Four geared passenger elevators	Each	324,600	16.23	10.9%
1020	Escalators & Moving Walks	N/A	—	—	—	

D20 Plumbing

2010	Plumbing Fixtures	Toilet, low flow, auto sensor, & service fixt., supply & drain. 1 Fixt./1370 SF Flr.	Each	4302	3.14	
2020	Domestic Water Distribution	Gas fired tankless water heater	S.F. Floor	.42	.42	2.7%
2040	Rain Water Drainage	Roof drains	S.F. Roof	3.36	.42	

D30 HVAC

3010	Energy Supply	N/A	—	—	—	
3020	Heat Generating Systems	Included in D3050	—	—	—	
3040	Distribution Systems	Enthalpy heat recovery packages	Each	38,725	1.94	12.5 %
3050	Terminal & Package Units	Multizone rooftop air conditioner, SEER 14	S.F. Floor	16.65	16.65	
3090	Other HVAC Sys. & Equipment	N/A	—	—	—	

D40 Fire Protection

4010	Sprinklers	Wet pipe sprinkler system	S.F. Floor	3.06	3.06	2.9%
4020	Standpipes	Standpipes and hose systems	S.F. Floor	1.17	1.17	

D50 Electrical

5010	Electrical Service/Distribution	1600 ampere service, panel board and feeders	S.F. Floor	1.10	1.10	
5020	Lighting & Branch Wiring	LED fixtures, daylt. dim., ltg. on/off, recept., switches, and A.C. power	S.F. Floor	15.39	15.39	15.6%
5030	Communications & Security	Addressable alarm systems, internet and phone wiring, emergency lighting	S.F. Floor	5.51	5.51	
5090	Other Electrical Systems	Emergency generator, 100 kW, UPS, energy monitoring systems	S.F. Floor	1.19	1.19	

E. EQUIPMENT & FURNISHINGS

1010	Commercial Equipment	N/A	—	—	—	
1020	Institutional Equipment	N/A	—	—	—	
1090	Other Equipment	Waste handling recycling tilt truck	S.F. Floor	.04	.04	0.0 %
2020	Moveable Furnishings	No smoking signage	S.F. Floor			

F. SPECIAL CONSTRUCTION

1020	Integrated Construction	N/A	—	—	—	0.0 %
1040	Special Facilities	N/A	—	—	—	

G. BUILDING SITEWORK   N/A

	Unit	Unit Cost	Cost Per S.F.	% Of Sub-Total
Sub-Total			148.22	**100%**
CONTRACTOR FEES (General Requirements: 10%, Overhead: 5%, Profit: 10%)		25%	37.06	
ARCHITECT FEES		6%	11.12	
Total Building Cost			**196.40**	

Costs per square foot of floor area

Exterior Wall	S.F. Area	120000	145000	170000	200000	230000	260000	400000	600000	800000
	L.F. Perimeter	420	450	470	490	510	530	650	800	900
Curtain Wall	Rigid Steel	224.35	215.00	207.40	200.40	195.20	191.20	181.75	175.20	170.75
	Reinforced Concrete	215.00	205.90	198.40	191.50	186.40	182.50	173.20	166.85	162.45
Face Brick and Concrete Block	Rigid Steel	196.95	190.75	185.80	181.25	177.90	175.25	169.00	164.80	162.00
	Reinforced Concrete	187.60	181.60	176.80	172.35	169.10	166.55	160.50	156.40	153.70
Precast Concrete	Rigid Steel	207.10	199.75	193.80	188.30	184.25	181.15	173.75	168.65	165.25
	Reinforced Concrete	197.75	190.60	184.80	179.50	175.50	172.50	165.00	160.30	156.95
Perimeter Adj., Add or Deduct	Per 100 L.F.	15.40	12.80	10.95	9.20	8.05	7.15	4.60	3.15	2.40
Story Hgt. Adj., Add or Deduct	Per 1 Ft.	6.20	5.55	4.95	4.35	3.95	3.65	2.90	2.40	2.05

For Basement, add $46.07 per square foot of basement area

The above costs were calculated using the basic specifications shown on the facing page. These costs should be adjusted where necessary for design alternatives and owner's requirements.

Common additives

Description	Unit	$ Cost
Closed Circuit Surveillance, one station, camera and monitor	Ea.	1475
For additional camera stations, add	Ea.	665
Directory Boards, plastic, glass covered, 30" x 20"	Ea.	640
36" x 48"	Ea.	1625
Aluminum, 36" x 24"	Ea.	790
48" x 32"	Ea.	1025
Electronic, wall mounted	S.F.	4625
Escalators, 10' rise, 32" wide, glass balustrade	Ea.	153,500
48" wide, glass balustrade	Ea.	161,000
Pedestal access flr sys w/PLAM cover, comp rm,		
Less than 6000 SF	S.F.	22
Greater than 6000 SF	S.F.	21.50
Office, greater than 6000 S.F.	S.F.	17.95
Uninterruptible power supply, 15 kVA/12.75 kW	kW	1.28

Description	Unit	$ Cost
Commissioning Fees, sustainable commercial construction	S.F.	0.24 - 3.04
Energy Modelling Fees, commercial buildings to 10,000 SF	Ea.	11,000
Greater than 10,000 SF add	S.F.	0.04
Green Bldg Cert Fees for comm construction project reg	Project	900
Photovoltaic Pwr Sys, grid connected, 20 kW (~2400 SF), roof	Ea.	249,200
Green Roofs, 6" soil depth, w/treated wd edging & sedum mats	S.F.	12.41
10" Soil depth, with treated wood edging & sedum mats	S.F.	14.10
Greywater Recovery Systems, prepackaged comm, 3060 gal.	Ea.	45,900
4590 gal.	Ea.	57,625
Rainwater Harvest Sys, prepckged comm, 10,000 gal, sys contrller	Ea.	37,550
20,000 gal. w/system controller	Ea.	61,000
30,000 gal. w/system controller	Ea.	97,350
Solar Domestic HW, closed loop, add-on sys, ext heat exchanger	Ea.	11,300
Drainback, hot water system, 120 gal tank	Ea.	14,175
Draindown, hot water system, 120 gal tank	Ea.	14,425

Important: See the Reference Section for Location Factors.

Model costs calculated for a 16 story building with 10' story height and 260,000 square feet of floor area

			Unit	Unit Cost	Cost Per S.F.	% Of Sub-Total
A. SUBSTRUCTURE						
1010	Standard Foundations	CIP concrete pile caps	S.F. Ground	11.20	.70	
1020	Special Foundations	Steel H-piles, concrete grade beams	S.F. Ground	65	4.06	
1030	Slab on Grade	4" reinforced concrete with recycled vapor barrier and granular base	S.F. Slab	5.84	.36	3.9%
2010	Basement Excavation	Site preparation for slab, piles and grade beams	S.F. Ground	.33	.02	
2020	Basement Walls	4' foundation wall	L.F. Wall	98	.48	
B. SHELL						
B10 Superstructure						
1010	Floor Construction	Concrete slab, metal deck, beams	S.F. Floor	29.69	27.83	19.7%
1020	Roof Construction	Metal deck, open web steel joists, beams, columns	S.F. Roof	9.12	.57	
B20 Exterior Enclosure						
2010	Exterior Walls	N/A	—	—	—	
2020	Exterior Windows	Double glazed heat absorbing, tinted plate glass wall panels 100% of wall	Each	86	26.17	18.7 %
2030	Exterior Doors	Double aluminum & glass doors	Each	8111	.88	
B30 Roofing						
3010	Roof Coverings	Single ply membrane, fully adhered; perlite/EPS R-20 insulation	S.F. Roof	8	.50	0.3%
3020	Roof Openings	N/A	—	—	—	
C. INTERIORS						
1010	Partitions	Gypsum board on metal studs, sound attenuation 30 SF Flr./LF Part.	S.F. Partition	13.88	3.70	
1020	Interior Doors	Single leaf hollow metal, low VOC paint 400 S.F. Floor/Door	Each	1222	3.06	
1030	Fittings	Toilet partitions	S.F. Floor	.72	.72	
2010	Stair Construction	Concrete filled metal pan	Flight	18,900	2.54	16.6%
3010	Wall Finishes	60% vinyl wall covering, 40% paint, low VOC	S.F. Surface	1.65	.88	
3020	Floor Finishes	60% carpet tile, 30% vinyl composition tile, recycled content, 10% ceramic tile	S.F. Floor	5.48	5.48	
3030	Ceiling Finishes	Mineral fiber tile on concealed zee bars	S.F. Ceiling	7.60	7.60	
D. SERVICES						
D10 Conveying						
1010	Elevators & Lifts	Four geared passenger elevators	Each	505,050	7.77	5.4%
1020	Escalators & Moving Walks	N/A	—	—	—	
D20 Plumbing						
2010	Plumbing Fixtures	Toilet, low flow, auto sensor, & service fixt., supply & drain. 1 Fixt./1345 SF Flr.	Each	7021	5.22	
2020	Domestic Water Distribution	Gas fired, tankless water heater	S.F. Floor	.19	.19	3.9%
2040	Rain Water Drainage	Roof drains	S.F. Roof	4.32	.27	
D30 HVAC						
3010	Energy Supply	N/A	—	—	—	
3020	Heat Generating Systems	N/A	—	—	—	
3040	Distribution Systems	Enthalpy heat recovery packages	Each	38,725	1.63	12.7 %
3050	Terminal & Package Units	Multizone rooftop air conditioner	S.F. Floor	16.65	16.65	
3090	Other HVAC Sys. & Equipment	N/A	—	—	—	
D40 Fire Protection						
4010	Sprinklers	Sprinkler system, light hazard	S.F. Floor	2.95	2.95	2.5%
4020	Standpipes	Standpipes and hose systems	S.F. Floor	.60	.60	
D50 Electrical						
5010	Electrical Service/Distribution	2400 ampere service, panel board and feeders	S.F. Floor	.69	.69	
5020	Lighting & Branch Wiring	LED fixtures, daylt. dim., ltg. on/off, recept., switches, and A.C. power	S.F. Floor	16.27	16.27	16.3%
5030	Communications & Security	Addressable alarm systems, internet & phone wiring, emergency ltg.	S.F. Floor	5.85	5.85	
5090	Other Electrical Systems	Emergency generator, 200 kW, UPS, energy monitoring systems	S.F. Floor	.65	.65	
E. EQUIPMENT & FURNISHINGS						
1010	Commercial Equipment	N/A	—	—	—	
1020	Institutional Equipment	N/A	—	—	—	
1090	Other Equipment	Waste handling recycling tilt truck	S.F. Floor	.01	.01	0.0 %
2020	Moveable Furnishings	No smoking signage	S.F. Floor			
F. SPECIAL CONSTRUCTION						
1020	Integrated Construction	N/A	—	—	—	0.0 %
1040	Special Facilities	N/A	—	—	—	
G. BUILDING SITEWORK	**N/A**					

		Sub-Total	144.30	100%
CONTRACTOR FEES (General Requirements: 10%, Overhead: 5%, Profit: 10%)		25%	36.08	
ARCHITECT FEES		6%	10.82	
		Total Building Cost	**191.20**	

For customer support on your Square Foot Costs with RSMeans data, call 800.448.8182.

Costs per square foot of floor area

Exterior Wall	S.F. Area	7000	9000	11000	13000	15000	17000	19000	21000	23000
	L.F. Perimeter	240	280	303	325	354	372	397	422	447
Limestone and Concrete Block	Bearing Walls	280.65	264.85	251.35	241.80	235.85	229.85	225.95	222.75	220.20
	Reinforced Concrete	303.75	288.00	274.45	264.90	259.00	253.00	249.05	245.90	243.30
Face Brick and Concrete Block	Bearing Walls	260.60	246.75	235.30	227.20	222.10	217.10	213.75	211.05	208.85
	Reinforced Concrete	283.75	269.80	258.40	250.30	245.20	240.25	236.90	234.15	232.00
Decorative Concrete Block	Bearing Walls	246.25	233.65	223.75	216.75	212.25	207.95	205.00	202.65	200.75
	Reinforced Concrete	269.40	256.80	246.85	239.85	235.30	231.05	228.15	225.80	223.85
Perimeter Adj., Add or Deduct	Per 100 L.F.	31.90	24.90	20.35	17.20	14.95	13.15	11.75	10.70	9.70
Story Hgt. Adj., Add or Deduct	Per 1 Ft.	5.45	5.00	4.40	4.05	3.80	3.50	3.35	3.25	3.05
For Basement, add $36.22 per square foot of basement area										

The above costs were calculated using the basic specifications shown on the facing page. These costs should be adjusted where necessary for design alternatives and owner's requirements.

Common additives

Description	Unit	$ Cost
Cells Prefabricated, 5'-6' x 7'-8", 7'-8' high	Ea.	12,700
Elevators, hydraulic passenger, 2 stops, 2500# capacity	Ea.	76,600
3500# capacity	Ea.	81,600
Lockers, Steel, Single tier, 60" to 72"	Opng.	232 - 395
Locker bench, lam. maple top only	L.F.	36.50
Pedestals, steel pipe	Ea.	79
Safe, Office type, 1 hr rating, 60"x36"x18", double door	Ea.	10,200
Shooting Range, incl. bullet traps,		
Target provisions & controls, excl. struct shell	Ea.	57,500
Sound System, amplifier, 250 watts	Ea.	2025
Speaker, ceiling or wall	Ea.	240
Trumpet	Ea.	460

Description	Unit	$ Cost
Commissioning Fees, sustainable commercial construction	S.F.	0.24 - 3.04
Energy Modelling Fees, commercial buildings to 10,000 SF	Ea.	11,000
Greater than 10,000 SF add	S.F.	0.04
Green Bldg Cert Fees for comm construction project reg	Project	900
Photovoltaic Pwr Sys, grid connected, 20 kW (~2400 SF), roof	Ea.	249,200
Green Roofs, 6" soil depth, w/treated wd edging & sedum mats	S.F.	12.41
10" Soil depth, with treated wood edging & sedum mats	S.F.	14.10
Greywater Recovery Systems, prepackaged comm, 3060 gal.	Ea.	45,900
4590 gal.	Ea.	57,625
Rainwater Harvest Sys, prepckged comm, 10,000 gal, sys contrller	Ea.	37,550
20,000 gal. w/system controller	Ea.	61,000
30,000 gal. w/system controller	Ea.	97,350
Solar Domestic HW, closed loop, add-on sys, ext heat exchanger	Ea.	11,300
Drainback, hot water system, 120 gal tank	Ea.	14,175
Draindown, hot water system, 120 gal tank	Ea.	14,425

Model costs calculated for a 2 story building with 12' story height and 11,000 square feet of floor area

				Unit	Unit Cost	Cost Per S.F.	% Of Sub-Total
A. SUBSTRUCTURE							
1010	Standard Foundations	Poured concrete; strip and spread footings		S.F. Ground	3.12	1.56	
1020	Special Foundations	N/A		—	—	—	
1030	Slab on Grade	4" reinforced concrete with recycled vapor barrier and granular base		S.F. Slab	5.84	2.93	4.5%
2010	Basement Excavation	Site preparation for slab and trench for foundation wall and footing		S.F. Ground	.33	.17	
2020	Basement Walls	4' foundation wall		L.F. Wall	98	3.59	
B. SHELL							
	B10 Superstructure						
1010	Floor Construction	Open web steel joists, slab form, concrete		S.F. Floor	13.46	6.73	5.1%
1020	Roof Construction	Metal deck on open web steel joists		S.F. Roof	5.40	2.70	
	B20 Exterior Enclosure						
2010	Exterior Walls	Limestone with concrete block backup	80% of wall	S.F. Wall	63	33.58	
2020	Exterior Windows	Metal horizontal sliding	20% of wall	Each	529	4.66	22.1%
2030	Exterior Doors	Hollow metal		Each	3506	2.55	
	B30 Roofing						
3010	Roof Coverings	Single-ply TPO membrane, 60 mils, heat welded w/R-20 insul.		S.F. Roof	8.44	4.22	2.3%
3020	Roof Openings	N/A		—	—	—	
C. INTERIORS							
1010	Partitions	Concrete block w/foamed-in insul.	20 SF Flr./LF Part.	S.F. Partition	12.06	6.03	
1020	Interior Doors	Single leaf kalamein fire door, low VOC paint	200 S.F. Floor/Door	Each	1222	6.11	
1030	Fittings	Toilet partitions		S.F. Floor	.98	.98	
2010	Stair Construction	Concrete filled metal pan		Flight	18,900	3.43	17.4%
3010	Wall Finishes	90% paint, low VOC, 10% ceramic tile		S.F. Surface	2.97	2.97	
3020	Floor Finishes	70% vinyl composition tile, recycled content, 20% carpet tile, 10% ceramic tile		S.F. Floor	4.94	4.94	
3030	Ceiling Finishes	Mineral fiber tile on concealed zee bars		S.F. Ceiling	7.60	7.60	
D. SERVICES							
	D10 Conveying						
1010	Elevators & Lifts	One hydraulic passenger elevator		Each	91,190	8.29	4.5%
1020	Escalators & Moving Walks	N/A		—	—	—	
	D20 Plumbing						
2010	Plumbing Fixtures	Toilet, low flow, auto sensor, & service fixt., supply & drainage	1 Fixt./580 SF Flr.	Each	5823	10.04	
2020	Domestic Water Distribution	Electric water heater, point-of-use, energy saver		S.F. Floor	1.28	1.28	7.1%
2040	Rain Water Drainage	Roof drains		S.F. Roof	3.44	1.72	
	D30 HVAC						
3010	Energy Supply	N/A		—	—	—	
3020	Heat Generating Systems	N/A		—	—	—	
3040	Distribution Systems	Enthalpy heat recovery packages		Each	15,550	1.42	12.0 %
3050	Terminal & Package Units	Multizone rooftop air conditioner, SEER 14		S.F. Floor	20.70	20.70	
3090	Other HVAC Sys. & Equipment	N/A		—	—	—.	
	D40 Fire Protection						
4010	Sprinklers	Wet pipe sprinkler system		S.F. Floor	4.08	4.08	2.8%
4020	Standpipes	Standpipe		S.F. Floor	1.15	1.15	
	D50 Electrical						
5010	Electrical Service/Distribution	400 ampere service, panel board and feeders		S.F. Floor	1.03	1.03	
5020	Lighting & Branch Wiring	LED fixtures, daylt. dim., ltg. on/off cntrl., recept., switches, & A.C. pwr.		S.F. Floor	13.60	13.60	12.9%
5030	Communications & Security	Addressable alarm systems, internet wiring, intercom &emergency ltg.		S.F. Floor	7.52	7.52	
5090	Other Electrical Systems	Emergency generator, 15 kW, energy monitoring systems		S.F. Floor	1.71	1.71	
E. EQUIPMENT & FURNISHINGS							
1020	Institutional Equipment	Lockers, detention rooms, cells, gasoline dispensers		S.F. Floor	13.65	13.65	
1030	Vehicular Equipment	Gasoline dispenser system		S.F. Floor	2.54	2.54	9.1%
1090	Other Equipment	Waste handling recycling tilt truck		S.F. Floor	.57	.57	
2020	Moveable Furnishings	No smoking signage		S.F. Floor	.04	.04	
F. SPECIAL CONSTRUCTION							
1020	Integrated Construction	N/A		—	—	—	0.0 %
1040	Special Facilities	N/A		—	—	—	
G. BUILDING SITEWORK	**N/A**						

				Sub-Total	184.47	**100%**
CONTRACTOR FEES (General Requirements: 10%, Overhead: 5%, Profit: 10%)				25%	46.13	
ARCHITECT FEES				9%	20.75	

Total Building Cost   **251.35**

For customer support on your Square Foot Costs with RSMeans data, call 800.448.8182.

269

Costs per square foot of floor area

Exterior Wall	S.F. Area	2000	2800	3500	4200	5000	5800	6500	7200	8000
	L.F. Perimeter	180	212	240	268	300	314	336	344	368
Wood Clapboard	Wood Frame	284.55	269.70	262.25	257.30	253.40	248.40	246.15	242.95	241.20
Brick Veneer	Wood Frame	297.75	280.75	272.30	266.60	262.15	256.30	253.65	249.95	247.95
Face Brick and Concrete Block	Wood Joists	312.25	293.00	283.35	276.90	271.85	265.00	262.00	257.65	255.35
	Steel Joists	305.40	286.10	276.50	270.05	264.95	258.15	255.15	250.75	248.45
Stucco and Concrete Block	Wood Joists	300.40	283.50	275.00	269.35	264.95	259.05	256.40	252.70	250.70
	Steel Joists	290.70	273.80	265.30	259.65	255.20	249.35	246.75	242.95	241.00
Perimeter Adj., Add or Deduct	Per 100 L.F.	33.40	23.85	19.15	15.95	13.35	11.50	10.25	9.25	8.40
Story Hgt. Adj., Add or Deduct	Per 1 Ft.	3.45	2.85	2.65	2.45	2.30	2.10	1.95	1.80	1.85

For Basement, add $41.33 per square foot of basement area

The above costs were calculated using the basic specifications shown on the facing page. These costs should be adjusted where necessary for design alternatives and owner's requirements.

Common additives

Description	Unit	$ Cost
Bar, Front Bar	L.F.	450
Back bar	L.F.	360
Booth, Upholstered, custom straight	L.F.	258 - 475
"L" or "U" shaped	L.F.	267 - 450
Fireplace, brick, excl. chimney or foundation, 30"x29" opening	Ea.	3400
Chimney, standard brick, single flue, 16" x 20"	V.L.F.	113
2 Flue, 20" x 24"	V.L.F.	165
Kitchen Equipment		
Broiler	Ea.	4425
Coffee urn, twin 6 gallon	Ea.	2900
Cooler, 6 ft. long	Ea.	3750
Dishwasher, 10-12 racks per hr.	Ea.	4150
Food warmer, counter, 1.2 KW	Ea.	790
Freezer, 44 C.F., reach-in	Ea.	5525
Ice cube maker, 50 lb. per day	Ea.	2100
Range with 1 oven	Ea.	3575
Refrigerators, prefabricated, walk-in, 7'-6" high, 6' x 6'	S.F.	145
10' x 10'	S.F.	114
12' x 14'	S.F.	99.50
12' x 20'	S.F.	137

Description	Unit	$ Cost
Commissioning Fees, sustainable commercial construction	S.F.	0.24 - 3.04
Energy Modelling Fees, commercial buildings to 10,000 SF	Ea.	11,000
Greater than 10,000 SF add	S.F.	0.04
Green Bldg Cert Fees for comm construction project reg	Project	900
Photovoltaic Pwr Sys, grid connected, 20 kW (~2400 SF), roof	Ea.	249,200
Green Roofs, 6" soil depth, w/treated wd edging & sedum mats	S.F.	12.41
10" Soil depth, with treated wood edging & sedum mats	S.F.	14.10
Greywater Recovery Systems, prepackaged comm, 3060 gal.	Ea.	45,900
4590 gal.	Ea.	57,625
Rainwater Harvest Sys, prepckged comm, 10,000 gal, sys contrller	Ea.	37,550
20,000 gal. w/system controller	Ea.	61,000
30,000 gal. w/system controller	Ea.	97,350
Solar Domestic HW, closed loop, add-on sys, ext heat exchanger	Ea.	11,300
Drainback, hot water system, 120 gal tank	Ea.	14,175
Draindown, hot water system, 120 gal tank	Ea.	14,425

Important: See the Reference Section for Location Factors.

G Restaurant

Model costs calculated for a 1 story building with 12' story height and 5,000 square feet of floor area

				Unit	Unit Cost	Cost Per S.F.	% Of Sub-Total
A. SUBSTRUCTURE							
1010	Standard Foundations	Poured concrete; strip and spread footings		S.F. Ground	2.87	2.87	
1020	Special Foundations	N/A		—	—	—	
1030	Slab on Grade	4" reinforced concrete with recycled vapor barrier and granular base		S.F. Slab	5.84	5.84	8.0%
2010	Basement Excavation	Site preparation for slab and trench for foundation wall and footing		S.F. Ground	.57	.57	
2020	Basement Walls	4' foundation wall		L.F. Wall	98	5.88	
B. SHELL							
B10 Superstructure							
1010	Floor Construction	Wood columns		S.F. Floor	.93	.93	5.1%
1020	Roof Construction	Plywood on wood truss (pitched)		S.F. Roof	8.68	8.68	
B20 Exterior Enclosure							
2010	Exterior Walls	Cedar siding on wood studs with insulation	70% of wall	S.F. Wall	17	8.57	
2020	Exterior Windows	Storefront windows	30% of wall	Each	54	11.06	14.0%
2030	Exterior Doors	Aluminum and glass doors and entrance with transom		Each	6939	6.95	
B30 Roofing							
3010	Roof Coverings	Cedar shingles with flashing (pitched); rigid fiberglass insulation		S.F. Roof	7.59	7.59	4.1%
3020	Roof Openings	Skylights		S.F. Roof	.10	.10	
C. INTERIORS							
1010	Partitions	Gypsum board on wood studs	25 SF Floor/LF Part.	S.F. Partition	13.15	5.26	
1020	Interior Doors	Hollow core wood, low VOC paint	250 S.F. Floor/Door	Each	645	2.58	
1030	Fittings	Toilet partitions		S.F. Floor	1.25	1.25	
2010	Stair Construction	N/A		—	—	—	15.1%
3010	Wall Finishes	75% paint, low VOC, 25% ceramic tile		S.F. Surface	2.81	2.25	
3020	Floor Finishes	65% carpet, 35% quarry tile		S.F. Floor	9.64	9.64	
3030	Ceiling Finishes	Mineral fiber tile on concealed zee bars		S.F. Ceiling	7.60	7.60	
D. SERVICES							
D10 Conveying							
1010	Elevators & Lifts	N/A		—	—	—	0.0 %
1020	Escalators & Moving Walks	N/A		—	—	—	
D20 Plumbing							
2010	Plumbing Fixtures	Kitchen, toilet, low flow, auto sensor, & service fixt., supply & drain.	1 Fixt./335 SF Flr.	Each	3969	11.18	
2020	Domestic Water Distribution	Tankless, on-demand water heater gas/propane		S.F. Floor	2.81	2.81	8.6%
2040	Rain Water Drainage	Roof drains		S.F. Roof	2.34	2.34	
D30 HVAC							
3010	Energy Supply	N/A		—	—	—	
3020	Heat Generating Systems	Included in D3050		—	—	—	
3040	Distribution Systems	Enthalpy heat recovery packages and kitchen exhaust/make-up air system		Each	46,050	9.21	26.2 %
3050	Terminal & Package Units	Multizone rooftop air conditioner, SEER 14		S.F. Floor	40.50	40.50	
3090	Other HVAC Sys. & Equipment	N/A		—	—	—	
D40 Fire Protection							
4010	Sprinklers	Sprinklers, light hazard and ordinary hazard kitchen		S.F. Floor	9.18	9.18	6.1%
4020	Standpipes	Standpipe		S.F. Floor	2.29	2.29	
D50 Electrical							
5010	Electrical Service/Distribution	400 ampere service, panel board and feeders		S.F. Floor	2.37	2.37	
5020	Lighting & Branch Wiring	LED fixtures, daylt. dim., ltg. on/off, receptacles, switches, and A.C. power		S.F. Floor	13.48	13.48	12.2%
5030	Communications & Security	Addressable alarm systems and emergency lighting		S.F. Floor	3.70	3.70	
5090	Other Electrical Systems	Emergency generator, 15 kW and energy monitoring systems		S.F. Floor	3.48	3.48	
E. EQUIPMENT & FURNISHINGS							
1010	Commercial Equipment	N/A		—	—	—	
1020	Institutional Equipment	N/A		—	—	—	0.7 %
1090	Other Equipment	Waste handling recycling tilt truck		S.F. Floor	1.24	1.24	
2020	Moveable Furnishings	No smoking signage		S.F. Floor	.06	.06	
F. SPECIAL CONSTRUCTION							
1020	Integrated Construction	N/A		—	—	—	0.0 %
1040	Special Facilities	N/A		—	—	—	
G. BUILDING SITEWORK	**N/A**						

	Sub-Total	189.46	100%
CONTRACTOR FEES (General Requirements: 10%, Overhead: 5%, Profit: 10%)	25%	47.36	
ARCHITECT FEES	7%	16.58	
Total Building Cost		**253.40**	

For customer support on your Square Foot Costs with RSMeans data, call 800.448.8182.

271

Costs per square foot of floor area

Exterior Wall	S.F. Area	25000	30000	35000	40000	45000	50000	55000	60000	65000
	L.F. Perimeter	900	1050	1200	1350	1510	1650	1800	1970	2100
Face Brick and Concrete Block	Rigid Steel	208.80	206.50	204.85	203.55	202.90	201.85	201.15	201.10	200.25
	Bearing Walls	200.65	198.50	197.10	196.00	195.45	194.45	193.90	193.85	193.05
Stucco and Concrete Block	Rigid Steel	201.50	199.40	197.80	196.70	196.05	195.10	194.50	194.40	193.65
	Bearing Walls	194.85	192.75	191.20	190.05	189.45	188.50	187.85	187.75	187.00
Decorative Concrete Block	Rigid Steel	200.75	198.60	197.10	196.00	195.35	194.40	193.85	193.70	192.95
	Bearing Walls	194.10	191.95	190.50	189.35	188.70	187.75	187.20	187.05	186.35
Perimeter Adj., Add or Deduct	Per 100 L.F.	5.30	4.40	3.75	3.30	2.95	2.60	2.45	2.20	2.00
Story Hgt. Adj., Add or Deduct	Per 1 Ft.	2.05	1.95	1.85	1.85	1.85	1.80	1.85	1.75	1.75

For Basement, add $30.61 per square foot of basement area

The above costs were calculated using the basic specifications shown on the facing page. These costs should be adjusted where necessary for design alternatives and owner's requirements.

Common additives

Description	Unit	$ Cost
Bleachers, Telescoping, manual, 16-20 tier	Seat	340 - 405
21-30 tier	Seat	340 - 460
For power operation, add	Seat	63 - 101
Carrels Hardwood	Ea.	740 - 1925
Clock System, 20 room	Ea.	20,200
50 room	Ea.	48,200
Kitchen Equipment		
Broiler	Ea.	4425
Cooler, 6 ft. long, reach-in	Ea.	3750
Dishwasher, 10-12 racks per hr.	Ea.	4150
Food warmer, counter, 1.2 KW	Ea.	790
Freezer, 44 C.F., reach-in	Ea.	5525
Ice cube maker, 50 lb. per day	Ea.	2100
Range with 1 oven	Ea.	3575
Lockers, Steel, single tier, 60" to 72"	Opng.	232 - 395
2 tier, 60" to 72" total	Opng.	139 - 172
Locker bench, lam. maple top only	L.F.	36.50
Pedestals, steel pipe	Ea.	79
Seating, auditorium chair, veneer back, padded seat	Ea.	380
Classroom, movable chair & desk	Set	81 - 171
Lecture hall, pedestal type	Ea.	350 - 650

Description	Unit	$ Cost
Sound System, amplifier, 250 watts	Ea.	2025
Speaker, ceiling or wall	Ea.	240
Commissioning Fees, sustainable institutional construction	S.F.	0.58 - 2.47
Energy Modelling Fees, academic buildings to 10,000 SF	Ea.	9000
Greater than 10,000 SF add	S.F.	0.20
Green Bldg Cert Fees for school construction project reg	Project	900
Photovoltaic Pwr Sys, grid connected, 20 kW (~2400 SF), roof	Ea.	249,200
Green Roofs, 6" soil depth, w/treated wd edging & sedum mats	S.F.	12.41
10" Soil depth, with treated wood edging & sedum mats	S.F.	14.10
Greywater Recovery Systems, prepackaged comm, 3060 gal.	Ea.	45,900
4590 gal.	Ea.	57,625
Rainwater Harvest Sys, prepckged comm, 10,000 gal, sys contrller	Ea.	37,550
20,000 gal. w/system controller	Ea.	61,000
30,000 gal. w/system controller	Ea.	97,350
Solar Domestic HW, closed loop, add-on sys, ext heat exchanger	Ea.	11,300
Drainback, hot water system, 120 gal tank	Ea.	14,175
Draindown, hot water system, 120 gal tank	Ea.	14,425

Important: See the Reference Section for Location Factors.

Model costs calculated for a 1 story building with 15' story height and 45,000 square feet of floor area

G School, Elementary

				Unit	Unit Cost	Cost Per S.F.	% Of Sub-Total
A. SUBSTRUCTURE							
1010	Standard Foundations	Poured concrete; strip and spread footings		S.F. Ground	5.43	5.43	
1020	Special Foundations	N/A		—	—	—	
1030	Slab on Grade	4" reinforced concrete with recycled vapor barrier and granular base		S.F. Slab	5.84	5.84	11.7%
2010	Basement Excavation	Site preparation for slab and trench for foundation wall and footing		S.F. Ground	.18	.18	
2020	Basement Walls	4' foundation wall		L.F. Wall	98	5.60	
B. SHELL							
B10 Superstructure							
1010	Floor Construction	N/A		—	—	—	
1020	Roof Construction	Metal deck on open web steel joists		S.F. Roof	4.92	4.92	3.4 %
B20 Exterior Enclosure							
2010	Exterior Walls	Face brick with concrete block backup	70% of wall	S.F. Wall	41.21	14.52	
2020	Exterior Windows	Steel outward projecting	25% of wall	Each	755	4.96	13.9%
2030	Exterior Doors	Metal and glass	5% of wall	Each	4985	.89	
B30 Roofing							
3010	Roof Coverings	Single-ply TPO membrane, 60 mils, w/flashing; polyiso. insulation		S.F. Roof	10.08	10.08	6.9%
3020	Roof Openings	N/A		—	—	—	
C. INTERIORS							
1010	Partitions	Concrete block w/foamed-in insul.	20 SF Flr./LF Part.	S.F. Partition	12.56	6.28	
1020	Interior Doors	Single leaf kalamein fire doors, low VOC paint	700 S.F. Floor/Door	Each	1222	1.75	
1030	Fittings	Toilet partitions		S.F. Floor	2.04	2.04	
2010	Stair Construction	N/A		—	—	—	20.1%
3010	Wall Finishes	75% paint, low VOC, 15% glazed coating, 10% ceramic tile		S.F. Surface	5.17	5.17	
3020	Floor Finishes	65% vinyl composition tile, recycled content, 25% carpet tile, 10% terrazzo		S.F. Floor	6.60	6.60	
3030	Ceiling Finishes	Mineral fiber tile on concealed zee bars		S.F. Ceiling	7.60	7.60	
D. SERVICES							
D10 Conveying							
1010	Elevators & Lifts	N/A		—	—	—	0.0 %
1020	Escalators & Moving Walks	N/A		—	—	—	
D20 Plumbing							
2010	Plumbing Fixtures	Kitchen, toilet, low flow, auto sensor, & service fixt., supply & drain.	1 Fixt./625 SF Flr.	Each	9994	15.99	
2020	Domestic Water Distribution	Gas fired, tankless water heater		S.F. Floor	.17	.17	12.2%
2040	Rain Water Drainage	Roof drains		S.F. Roof	1.72	1.72	
D30 HVAC							
3010	Energy Supply	N/A		—	—	—	
3020	Heat Generating Systems	N/A		—	—	—	
3040	Distribution Systems	Enthalpy heat recovery packages		Each	31,850	.71	14.3 %
3050	Terminal & Package Units	Multizone rooftop air conditioner		S.F. Floor	20.20	20.20	
3090	Other HVAC Sys. & Equipment	N/A		—	—	—	
D40 Fire Protection							
4010	Sprinklers	Sprinklers, light hazard		S.F. Floor	2.93	2.93	2.3%
4020	Standpipes	Standpipe		S.F. Floor	.46	.46	
D50 Electrical							
5010	Electrical Service/Distribution	800 ampere service, panel board and feeders		S.F. Floor	.86	.86	
5020	Lighting & Branch Wiring	LED fixtures, daylt. dim., ltg. on/off, recept., switches, and A.C. power		S.F. Floor	14.51	14.51	14.9%
5030	Communications & Security	Addressable alarm systems, internet wiring, comm. systems & emerg. ltg.		S.F. Floor	5.34	5.34	
5090	Other Electrical Systems	Emergency generator, 15 kW, energy monitoring systems		S.F. Floor	1.06	1.06	
E. EQUIPMENT & FURNISHINGS							
1010	Commercial Equipment	N/A		—	—	—	
1020	Institutional Equipment	Chalkboards		S.F. Floor	.22	.22	0.2 %
1090	Other Equipment	Waste handling recycling tilt truck		S.F. Floor	.08	.08	
2020	Moveable Furnishings	No smoking signage		S.F. Floor	.02	.02	
F. SPECIAL CONSTRUCTION							
1020	Integrated Construction	N/A		—	—	—	0.0 %
1040	Special Facilities	N/A		—	—	—	
G. BUILDING SITEWORK	**N/A**						

		Sub-Total	146.13	100%
CONTRACTOR FEES (General Requirements: 10%, Overhead: 5%, Profit: 10%)		25%	36.53	
ARCHITECT FEES		7%	12.79	
		Total Building Cost	**195.45**	

For customer support on your Square Foot Costs with RSMeans data, call 800.448.8182.

273

Costs per square foot of floor area

Exterior Wall	S.F. Area	50000	70000	90000	110000	130000	150000	170000	190000	210000
	L.F. Perimeter	850	1140	1420	1700	1980	2280	2560	2840	3120
Face Brick and Concrete Block	Rigid Steel	195.45	192.45	190.45	189.25	188.30	188.05	187.45	187.05	186.75
	Reinforced Concrete	199.35	196.30	194.35	193.10	192.20	191.95	191.35	190.95	190.65
Decorative Concrete Block	Rigid Steel	187.30	184.60	182.85	181.80	181.00	180.70	180.20	179.85	179.60
	Reinforced Concrete	191.20	188.45	186.70	185.65	184.80	184.60	184.10	183.75	183.45
Limestone and Concrete Block	Rigid Steel	206.85	203.35	201.00	199.60	198.50	198.25	197.60	197.10	196.70
	Reinforced Concrete	215.15	211.60	209.30	207.85	206.80	206.50	205.90	205.35	205.00
Perimeter Adj., Add or Deduct	Per 100 L.F.	4.85	3.40	2.70	2.15	1.85	1.55	1.45	1.35	1.10
Story Hgt. Adj., Add or Deduct	Per 1 Ft.	2.45	2.30	2.20	2.20	2.20	2.15	2.20	2.15	2.15

For Basement, add $39.99 per square foot of basement area

The above costs were calculated using the basic specifications shown on the facing page. These costs should be adjusted where necessary for design alternatives and owner's requirements.

Common additives

Description	Unit	$ Cost
Bleachers, Telescoping, manual, 16-20 tier	Seat	340 - 405
21-30 tier	Seat	340 - 460
For power operation, add	Seat	63 - 101
Carrels Hardwood	Ea.	740 - 1925
Clock System, 20 room	Ea.	20,200
50 room	Ea.	48,200
Elevators, hydraulic passenger, 2 stops, 2500# capacity	Ea.	76,600
3500# capacity	Ea.	81,600
Kitchen Equipment		
Broiler	Ea.	4425
Cooler, 6 ft. long, reach-in	Ea.	3750
Dishwasher, 10-12 racks per hr.	Ea.	4150
Food warmer, counter, 1.2 KW	Ea.	790
Freezer, 44 C.F., reach-in	Ea.	5525
Lockers, Steel, single tier, 60" or 72"	Opng.	232 - 395
2 tier, 60" or 72" total	Opng.	139 - 172
Locker bench, lam. maple top only	L.F.	36.50
Pedestals, steel pipe	Ea.	79
Seat, auditorium chair, veneer back, padded seat	Ea.	380
Classroom, movable chair & desk	Set	81 - 171
Lecture hall, pedestal type	Ea.	350 - 650

Description	Unit	$ Cost
Sound System, amplifier, 250 watts	Ea.	2025
Speaker, ceiling or wall	Ea.	240
Commissioning Fees, sustainable institutional construction	S.F.	0.58 - 2.47
Energy Modelling Fees, academic buildings to 10,000 SF	Ea.	9000
Greater than 10,000 SF add	S.F.	0.20
Green Bldg Cert Fees for school construction project reg	Project	900
Photovoltaic Pwr Sys, grid connected, 20 kW (~2400 SF), roof	Ea.	249,200
Green Roofs, 6" soil depth, w/treated wd edging & sedum mats	S.F.	12.41
10" Soil depth, with treated wood edging & sedum mats	S.F.	14.10
Greywater Recovery Systems, prepackaged comm, 3060 gal.	Ea.	45,900
4590 gal.	Ea.	57,625
Rainwater Harvest Sys, prepckged comm, 10,000 gal, sys contrller	Ea.	37,550
20,000 gal. w/system controller	Ea.	61,000
30,000 gal. w/system controller	Ea.	97,350
Solar Domestic HW, closed loop, add-on sys, ext heat exchanger	Ea.	11,300
Drainback, hot water system, 120 gal tank	Ea.	14,175
Draindown, hot water system, 120 gal tank	Ea.	14,425

Important: See the Reference Section for Location Factors.

Model costs calculated for a 2 story building with 15' story height and 130,000 square feet of floor area

				Unit	Unit Cost	Cost Per S.F.	% Of Sub-Total
A. SUBSTRUCTURE							
1010	Standard Foundations	Poured concrete; strip and spread footings		S.F. Ground	2.94	1.47	
1020	Special Foundations	N/A		—	—	—	
1030	Slab on Grade	4" reinforced concrete with recycled vapor barrier and granular base		S.F. Slab	5.84	2.93	4.2%
2010	Basement Excavation	Site preparation for slab and trench for foundation wall and footing		S.F. Ground	.18	.09	
2020	Basement Walls	4' foundation wall		L.F. Wall	98	1.49	
B. SHELL							
	B10 Superstructure						
1010	Floor Construction	Concrete slab without drop panel, concrete columns		S.F. Floor	23.92	11.96	14.1%
1020	Roof Construction	Concrete slab without drop panel		S.F. Roof	16.62	8.31	
	B20 Exterior Enclosure						
2010	Exterior Walls	Face brick with concrete block backup	75% of wall	S.F. Wall	41.20	14.12	
2020	Exterior Windows	Window wall	25% of wall	Each	77	8.73	16.4%
2030	Exterior Doors	Metal and glass		Each	3156	.71	
	B30 Roofing						
3010	Roof Coverings	Single-ply TPO membrane and standing seam metal; polyiso. insulation		S.F. Roof	13.82	6.91	4.9%
3020	Roof Openings	Roof hatches		S.F. Roof	.14	.07	
C. INTERIORS							
1010	Partitions	Concrete block w/foamed-in insul.	25 SF Flr./LF Part.	S.F. Partition	15.85	7.61	
1020	Interior Doors	Single leaf kalamein fire doors, low VOC paint	700 S.F. Floor/Door	Each	1222	1.75	
1030	Fittings	Toilet partitions, chalkboards		S.F. Floor	1.60	1.60	
2010	Stair Construction	Concrete filled metal pan		Flight	15,800	.73	20.4%
3010	Wall Finishes	75% paint, low VOC. 15% glazed coating, 10% ceramic tile		S.F. Surface	3.66	3.51	
3020	Floor Finishes	70% vinyl comp. tile, recycled content, 20% carpet tile, 10% terrazzo		S.F. Floor	6.57	6.57	
3030	Ceiling Finishes	Mineral fiber tile on concealed zee bars		S.F. Ceiling	7.60	7.60	
D. SERVICES							
	D10 Conveying						
1010	Elevators & Lifts	One hydraulic passenger elevator		Each	91,000	.70	0.5%
1020	Escalators & Moving Walks	N/A		—	—	—	
	D20 Plumbing						
2010	Plumbing Fixtures	Kitchen, toilet, low flow, auto sensor & service fixt., supply & drain.	1 Fixt./860 SF Flr.	Each	5315	6.18	
2020	Domestic Water Distribution	Gas fired, tankless water heater		S.F. Floor	.32	.32	5.2%
2040	Rain Water Drainage	Roof drains		S.F. Roof	2	1	
	D30 HVAC						
3010	Energy Supply	N/A		—	—	—	
3020	Heat Generating Systems	N/A		—	—	—	
3040	Distribution Systems	Enthalpy heat recovery packages		Each	31,850	.74	14.6 %
3050	Terminal & Package Units	Multizone rooftop air conditioner, SEER 14		S.F. Floor	20.20	20.20	
3090	Other HVAC Sys. & Equipment	N/A		—	—	—	
	D40 Fire Protection						
4010	Sprinklers	Sprinklers, light hazard		S.F. Floor	2.57	2.57	2.1%
4020	Standpipes	Standpipe, wet, Class III		S.F. Floor	.39	.39	
	D50 Electrical						
5010	Electrical Service/Distribution	2000 ampere service, panel board and feeders		S.F. Floor	.86	.86	
5020	Lighting & Branch Wiring	LED fixtures, daylt. dim., ltg. on/off, recept., switches, and A.C. power		S.F. Floor	15.42	15.42	15.3%
5030	Communications & Security	Addressable alarm systems, internet wiring, comm. systems and emerg. ltg.		S.F. Floor	4.83	4.83	
5090	Other Electrical Systems	Emergency generator, 250 kW, and energy monitoring systems		S.F. Floor	.85	.85	
E. EQUIPMENT & FURNISHINGS							
1010	Commercial Equipment	N/A		—	—	—	
1020	Institutional Equipment	Laboratory casework and counters		S.F. Floor	2.25	2.25	2.4 %
1090	Other Equipment	Waste handlg. recyc. tilt truck, built-in athletic equip., bleachers & backstp.		S.F. Floor	1.15	1.15	
2020	Moveable Furnishings	No smoking signage		S.F. Floor	.02	.02	
F. SPECIAL CONSTRUCTION							
1020	Integrated Construction	N/A		—	—	—	0.0 %
1040	Special Facilities	N/A		—	—	—	
G. BUILDING SITEWORK	**N/A**						
				Sub-Total		143.69	**100%**
	CONTRACTOR FEES (General Requirements: 10%, Overhead: 5%, Profit: 10%)				25%	35.94	
	ARCHITECT FEES				7%	12.57	
				Total Building Cost		**192.20**	

For customer support on your Square Foot Costs with RSMeans data, call 800.448.8182.

Costs per square foot of floor area

Exterior Wall	S.F. Area	50000	65000	80000	95000	110000	125000	140000	155000	170000
	L.F. Perimeter	850	1060	1280	1490	1700	1920	2140	2340	2560
Face Brick and Concrete Block	Rigid Steel	202.75	200.20	198.90	197.70	197.00	196.50	196.15	195.55	195.35
	Bearing Walls	194.90	192.30	191.00	189.85	189.10	188.65	188.25	187.70	187.50
Stone and Concrete Block	Rigid Steel	208.85	206.05	204.65	203.35	202.55	202.00	201.60	200.95	200.75
	Bearing Walls	201.00	198.20	196.75	195.50	194.65	194.15	193.75	193.10	192.90
Decorative Concrete Block	Rigid Steel	194.55	192.35	191.20	190.20	189.55	189.10	188.80	188.30	188.10
	Bearing Walls	187.00	184.80	183.60	182.55	181.95	181.55	181.15	180.65	180.50
Perimeter Adj., Add or Deduct	Per 100 L.F.	4.45	3.45	2.80	2.40	2.05	1.75	1.60	1.45	1.35
Story Hgt. Adj., Add or Deduct	Per 1 Ft.	2.25	2.20	2.10	2.05	2.05	2.05	1.95	2.05	2.05

For Basement, add $41.63 per square foot of basement area

The above costs were calculated using the basic specifications shown on the facing page. These costs should be adjusted where necessary for design alternatives and owner's requirements.

Common additives

Description	Unit	$ Cost
Bleachers, Telescoping, manual, 16-20 tier	Seat	340 - 405
21-30 tier	Seat	340 - 460
For power operation, add	Seat	63 - 101
Carrels Hardwood	Ea.	740 - 1925
Clock System, 20 room	Ea.	20,200
50 room	Ea.	48,200
Elevators, hydraulic passenger, 2 stops, 2500# capacity	Ea.	76,600
3500# capacity	Ea.	81,600
Kitchen Equipment		
Broiler	Ea.	4425
Cooler, 6 ft. long, reach-in	Ea.	3750
Dishwasher, 10-12 racks per hr.	Ea.	4150
Food warmer, counter, 1.2 KW	Ea.	790
Freezer, 44 C.F., reach-in	Ea.	5525
Lockers, Steel, single tier, 60" to 72"	Opng.	232 - 395
2 tier, 60" to 72" total	Opng.	139 - 172
Locker bench, lam. maple top only	L.F.	36.50
Pedestals, steel pipe	Ea.	79

Description	Unit	$ Cost
Seating, auditorium chair, veneer back, padded seat	Ea.	380
Upholstered, spring seat	Ea.	335
Classroom, movable chair & desk	Set	81 - 171
Lecture hall, pedestal type	Ea.	350 - 650
Sound System, amplifier, 250 watts	Ea.	2025
Speaker, ceiling or wall	Ea.	240
Commissioning Fees, sustainable institutional construction	S.F.	0.58 - 2.47
Energy Modelling Fees, academic buildings to 10,000 SF	Ea.	9000
Greater than 10,000 SF add	S.F.	0.20
Green Bldg Cert Fees for school construction project reg	Project	900
Photovoltaic Pwr Sys, grid connected, 20 kW (~2400 SF), roof	Ea.	249,200
Green Roofs, 6" soil depth, w/treated wd edging & sedum mats	S.F.	12.41
10" Soil depth, with treated wood edging & sedum mats	S.F.	14.10
Greywater Recovery Systems, prepackaged comm, 3060 gal.	Ea.	45,900
4590 gal.	Ea.	57,625
Rainwater Harvest Sys, prepckged comm, 10,000 gal, sys contrller	Ea.	37,550
20,000 gal. w/system controller	Ea.	61,000
30,000 gal. w/system controller	Ea.	97,350
Solar Domestic HW, closed loop, add-on sys, ext heat exchanger	Ea.	11,300
Drainback, hot water system, 120 gal tank	Ea.	14,175
Draindown, hot water system, 120 gal tank	Ea.	14,425

Important: See the Reference Section for Location Factors.

Model costs calculated for a 2 story building with 15' story height and 110,000 square feet of floor area

G School, Jr High, 2-3 Story

				Unit	Unit Cost	Cost Per S.F.	% Of Sub-Total
A. SUBSTRUCTURE							
1010	Standard Foundations	Poured concrete; strip and spread footings		S.F. Ground	2.72	1.36	
1020	Special Foundations	N/A		—	—	—	
1030	Slab on Grade	4" reinforced concrete with recycled vapor barrier and granular base		S.F. Slab	5.84	2.93	4.0%
2010	Basement Excavation	Site preparation for slab and trench for foundation wall and footing		S.F. Ground	.18	.09	
2020	Basement Walls	4' foundation wall		L.F. Wall	98	1.52	
B. SHELL							
B10 Superstructure							
1010	Floor Construction	Open web steel joists, slab form, concrete, columns		S.F. Floor	30.22	15.11	14.5%
1020	Roof Construction	Metal deck, open web steel joists, columns		S.F. Roof	12.58	6.29	
B20 Exterior Enclosure							
2010	Exterior Walls	Face brick with concrete block backup	75% of wall	S.F. Wall	41.21	14.33	
2020	Exterior Windows	Window wall	25% of wall	Each	65	7.21	15.2%
2030	Exterior Doors	Double aluminum & glass		Each	3355	.89	
B30 Roofing							
3010	Roof Coverings	Single-ply TPO membrane & standing seam metal; polyiso. insulation		S.F. Roof	14	7	4.8%
3020	Roof Openings	Roof hatches		S.F. Roof	.06	.03	
C. INTERIORS							
1010	Partitions	Concrete block w/foamed-in insul.	20 SF Flr./LF Part.	S.F. Partition	13.22	7.93	
1020	Interior Doors	Single leaf kalamein fire doors, low VOC paint	750 S.F. Floor/Door	Each	1222	1.63	
1030	Fittings	Toilet partitions, chalkboards		S.F. Floor	1.75	1.75	
2010	Stair Construction	Concrete filled metal pan		Flight	15,800	.86	23.0%
3010	Wall Finishes	50% paint, low VOC, 40% glazed coatings, 10% ceramic tile		S.F. Surface	4.04	4.85	
3020	Floor Finishes	50% vinyl comp. tile, recycled content, 30% carpet tile, 20% terrrazzo		S.F. Floor	9.24	9.24	
3030	Ceiling Finishes	Mineral fiberboard on concealed zee bars		S.F. Ceiling	7.60	7.60	
D. SERVICES							
D10 Conveying							
1010	Elevators & Lifts	One hydraulic passenger elevator		Each	91,300	.83	0.6%
1020	Escalators & Moving Walks	N/A		—	—	—	
D20 Plumbing							
2010	Plumbing Fixtures	Kitchen, toilet, low flow, auto sensor, & service fixt., supply & drain.	1 Fixt./1170 SF Flr.	Each	7675	6.56	
2020	Domestic Water Distribution	Gas fired, tankless water heater		S.F. Floor	.38	.38	5.4%
2040	Rain Water Drainage	Roof drains		S.F. Roof	2.02	1.01	
D30 HVAC							
3010	Energy Supply	N/A		—	—	—	
3020	Heat Generating Systems	Included in D3050		—	—	—	
3040	Distribution Systems	Enthalpy heat recovery packages		Each	38,725	.70	14.2 %
3050	Terminal & Package Units	Multizone rooftop air conditioner, SEER 14		S.F. Floor	20.20	20.20	
3090	Other HVAC Sys. & Equipment	N/A		—	—	—	
D40 Fire Protection							
4010	Sprinklers	Sprinklers, light hazard	10% of area	S.F. Floor	2.57	2.57	2.1%
4020	Standpipes	Standpipe, wet, Class III		S.F. Floor	.45	.45	
D50 Electrical							
5010	Electrical Service/Distribution	1600 ampere service, panel board and feeders		S.F. Floor	.75	.75	
5020	Lighting & Branch Wiring	LED fixtures, daylt. dim., ltg. on/off, recept., switches, and A.C. power		S.F. Floor	13.28	13.28	13.5%
5030	Communications & Security	Addressable alarm systems, internet wiring, comm. systems and emerg. ltg.		S.F. Floor	5.11	5.11	
5090	Other Electrical Systems	Emergency generator, 100 kW, and energy monitoring systems		S.F. Floor	.80	.80	
E. EQUIPMENT & FURNISHINGS							
1010	Commercial Equipment	N/A		—	—	—	
1020	Institutional Equipment	Laboratory casework and counters		S.F. Floor	2.65	2.65	2.7 %
1090	Other Equipment	Waste handlg. recyc. tilt truck, built-in athletic equip., bleachers & backstps.		S.F. Floor	1.35	1.35	
2020	Moveable Furnishings	No smoking signage		S.F. Floor	.02	.02	
F. SPECIAL CONSTRUCTION							
1020	Integrated Construction	N/A		—	—	—	0.0 %
1040	Special Facilities	N/A		—	—	—	
G. BUILDING SITEWORK	**N/A**						

	Sub-Total	147.28	100%
CONTRACTOR FEES (General Requirements: 10%, Overhead: 5%, Profit: 10%)	25%	36.83	
ARCHITECT FEES	7%	12.89	
Total Building Cost		**197**	

For customer support on your Square Foot Costs with RSMeans data, call 800.448.8182.

277

Costs per square foot of floor area

Exterior Wall	S.F. Area	20000	30000	40000	50000	60000	70000	80000	90000	100000
	L.F. Perimeter	440	590	740	900	1050	1200	1360	1510	1660
Face Brick and Concrete Block	Rigid Steel	206.90	199.40	195.70	193.85	192.30	191.15	190.60	189.85	189.35
	Bearing Walls	200.55	193.10	189.35	187.55	186.00	184.85	184.35	183.60	183.05
Decorative Concrete Block	Rigid Steel	191.40	185.55	182.60	181.20	179.95	179.10	178.65	178.10	177.65
	Bearing Walls	184.40	178.55	175.60	174.20	172.95	172.10	171.60	171.10	170.65
Metal Panel and Metal Studs	Rigid Steel	181.85	177.00	174.65	173.40	172.40	171.65	171.30	170.85	170.50
Metal Sandwich Panel	Rigid Steel	209.00	201.30	197.45	195.60	193.95	192.80	192.25	191.50	191.00
Perimeter Adj., Add or Deduct	Per 100 L.F.	11.45	7.60	5.60	4.50	3.80	3.25	2.85	2.55	2.30
Story Hgt. Adj., Add or Deduct	Per 1 Ft.	2.80	2.50	2.35	2.25	2.20	2.15	2.15	2.20	2.10
For Basement, add $40.40 per square foot of basement area										

The above costs were calculated using the basic specifications shown on the facing page. These costs should be adjusted where necessary for design alternatives and owner's requirements.

Common additives

Description	Unit	$ Cost
Carrels Hardwood	Ea.	740 - 1925
Clock System, 20 room	Ea.	20,200
50 room	Ea.	48,200
Directory Boards, plastic, glass covered, 36" x 48"	Ea.	1625
Aluminum, 36" x 24"	Ea.	790
48" x 60"	Ea.	2200
Elevators, hydraulic passenger, 2 stops, 2500# capacity	Ea.	76,600
3500# capacity	Ea.	81,600
Seating, auditorium chair, veneer back, padded seat	Ea.	380
Classroom, movable chair & desk	Set	81 - 171
Lecture hall, pedestal type	Ea.	350 - 650
Shops & Workroom, benches, metal	Ea.	740
Parts bins 6'-3" high, 3' wide, 12" deep, 72 bins	Ea.	710
Shelving, metal 1' x 3'	S.F.	13
Wide span 6' wide x 24" deep	S.F.	9.80

Description	Unit	$ Cost
Sound System, amplifier, 250 watts	Ea.	2025
Speaker, ceiling or wall	Ea.	240
Commissioning Fees, sustainable institutional construction	S.F.	0.58 - 2.47
Energy Modelling Fees, academic buildings to 10,000 SF	Ea.	9000
Greater than 10,000 SF add	S.F.	0.20
Green Bldg Cert Fees for school construction project reg	Project	900
Photovoltaic Pwr Sys, grid connected, 20 kW (~2400 SF), roof	Ea.	249,200
Green Roofs, 6" soil depth, w/treated wd edging & sedum mats	S.F.	12.41
10" Soil depth, with treated wood edging & sedum mats	S.F.	14.10
Greywater Recovery Systems, prepackaged comm, 3060 gal.	Ea.	45,900
4590 gal.	Ea.	57,625
Rainwater Harvest Sys, prepckged comm, 10,000 gal, sys contrller	Ea.	37,550
20,000 gal. w/system controller	Ea.	61,000
30,000 gal. w/system controller	Ea.	97,350
Solar Domestic HW, closed loop, add-on sys, ext heat exchanger	Ea.	11,300
Drainback, hot water system, 120 gal tank	Ea.	14,175
Draindown, hot water system, 120 gal tank	Ea.	14,425

Model costs calculated for a 2 story building with 16' story height and 40,000 square feet of floor area

				Unit	Unit Cost	Cost Per S.F.	% Of Sub-Total
A. SUBSTRUCTURE							
1010	Standard Foundations	Poured concrete; strip and spread footings		S.F. Ground	3.04	1.52	
1020	Special Foundations	N/A		—	—	—	
1030	Slab on Grade	5" reinforced concrete with recycled vapor barrier and granular base		S.F. Slab	13.99	7	7.4%
2010	Basement Excavation	Site preparation for slab and trench for foundation wall and footing		S.F. Ground	.18	.09	
2020	Basement Walls	4' foundation wall		L.F. Wall	98	1.82	
B. SHELL							
B10 Superstructure							
1010	Floor Construction	Open web steel joists, slab form, concrete, beams, columns		S.F. Floor	20.10	10.05	9.6%
1020	Roof Construction	Metal deck, open web steel joists, beams, columns		S.F. Roof	7.08	3.54	
B20 Exterior Enclosure							
2010	Exterior Walls	Face brick with concrete block backup	85% of wall	S.F. Wall	41.20	20.73	
2020	Exterior Windows	Tubular aluminum framing with insulated glass	15% of wall	Each	65	5.53	19.1%
2030	Exterior Doors	Metal and glass doors, low VOC paint		Each	3682	.74	
B30 Roofing							
3010	Roof Coverings	Single-ply TPO membrane, 60 mils, heat welded seams w/polyiso. insul.		S.F. Roof	9.16	4.58	3.4%
3020	Roof Openings	Roof hatches		S.F. Roof	.40	.20	
C. INTERIORS							
1010	Partitions	Concrete block w/foamed-in insul.	20 SF Flr./LF Part.	S.F. Partition	12.03	7.22	
1020	Interior Doors	Single leaf kalamein fire doors, low VOC paint	600 S.F. Floor/Door	Each	1222	2.03	
1030	Fittings	Toilet partitions, chalkboards		S.F. Floor	1.70	1.70	
2010	Stair Construction	Concrete filled metal pan		Flight	15,800	1.58	20.9%
3010	Wall Finishes	50% paint, low VOC, 40% glazed coating, 10% ceramic tile		S.F. Surface	4.90	5.88	
3020	Floor Finishes	70% vinyl comp. tile, recycled content,20% carpet tile, 10% terrazzo		S.F. Floor	6.57	6.57	
3030	Ceiling Finishes	Mineral fiber tile on concealed zee bars	60% of area	S.F. Ceiling	4.56	4.56	
D. SERVICES							
D10 Conveying							
1010	Elevators & Lifts	One hydraulic passenger elevator		Each	91,200	2.28	1.6%
1020	Escalators & Moving Walks	N/A		—	—	—	
D20 Plumbing							
2010	Plumbing Fixtures	Toilet, low flow, auto sensor, & service fixt., supply & drain.	1 Fixt./700 SF.Flr.	Each	3626	5.18	
2020	Domestic Water Distribution	Gas fired, tankless water heater		S.F. Floor	.33	.33	4.8%
2040	Rain Water Drainage	Roof drains		S.F. Roof	2.62	1.31	
D30 HVAC							
3010	Energy Supply	N/A		—	—	—	
3020	Heat Generating Systems	N/A		—	—	—	
3040	Distribution Systems	Enthalpy heat recovery packages		Each	31,850	.79	14.8 %
3050	Terminal & Package Units	Multizone rooftop air conditioner, SEER 14		S.F. Floor	20.20	20.20	
3090	Other HVAC Sys. & Equipment	N/A		—	—	—	
D40 Fire Protection							
4010	Sprinklers	Sprinklers, light hazard		S.F. Floor	3.30	3.30	3.1%
4020	Standpipes	Standpipe, wet, Class III		S.F. Floor	1.05	1.05	
D50 Electrical							
5010	Electrical Service/Distribution	800 ampere service, panel board and feeders		S.F. Floor	.96	.96	
5020	Lighting & Branch Wiring	LED fixtures, daylt. dim., ltg. on/off, recept., switches, and A.C. power		S.F. Floor	14.24	14.24	15.1%
5030	Communications & Security	Addressable alarm systems, internet wiring, comm. systems & emerg. ltg.		S.F. Floor	5	5	
5090	Other Electrical Systems	Emergency generator, 11.5 kW, and energy monitoring systems		S.F. Floor	1.23	1.23	
E. EQUIPMENT & FURNISHINGS							
1010	Commercial Equipment	N/A		—	—	—	
1020	Institutional Equipment	Stainless steel countertops		S.F. Floor	.20	.20	0.3 %
1090	Other Equipment	Waste handling recycling tilt truck		S.F. Floor	.15	.15	
2020	Moveable Furnishings	No smoking signage		S.F. Floor	.02	.02	
F. SPECIAL CONSTRUCTION							
1020	Integrated Construction	N/A		—	—	—	0.0 %
1040	Special Facilities	N/A		—	—	—	
G. BUILDING SITEWORK	**N/A**						

			Sub-Total	141.58	100%
CONTRACTOR FEES (General Requirements: 10%, Overhead: 5%, Profit: 10%)			25%	35.38	
ARCHITECT FEES			7%	12.39	
		Total Building Cost		**189.35**	

For customer support on your Square Foot Costs with RSMeans data, call 800.448.8182.

279

Costs per square foot of floor area

Exterior Wall	S.F. Area	12000	16000	20000	26000	32000	38000	44000	52000	60000
	L.F. Perimeter	450	510	570	650	716	780	840	920	1020
Face Brick and Concrete Block	Rigid Steel	190.30	178.80	171.85	164.85	159.90	156.40	153.80	151.20	149.75
	Bearing Walls	184.45	172.95	166.00	159.00	154.05	150.50	147.95	145.35	143.90
Stucco and Concrete Block	Rigid Steel	179.15	169.35	163.40	157.45	153.25	150.30	148.15	145.95	144.75
	Bearing Walls	176.60	166.75	160.80	154.90	150.70	147.70	145.55	143.40	142.10
Precast Concrete	Rigid Steel	202.75	189.35	181.30	173.20	167.30	163.20	160.15	157.10	155.40
Metal Sandwich Panel	Rigid Steel	188.75	177.50	170.65	163.85	158.95	155.55	153.05	150.50	149.05
Perimeter Adj., Add or Deduct	Per 100 L.F.	11.65	8.60	6.95	5.40	4.30	3.70	3.15	2.70	2.35
Story Hgt. Adj., Add or Deduct	Per 1 Ft.	2.45	2.00	1.80	1.65	1.40	1.30	1.25	1.15	1.10

For Basement, add $ 30.66 per square foot of basement area

The above costs were calculated using the basic specifications shown on the facing page. These costs should be adjusted where necessary for design alternatives and owner's requirements.

Common additives

Description	Unit	$ Cost
Check Out Counter, single belt	Ea.	3850
Scanner, registers, guns & memory 10 lanes	Ea.	186,000
Power take away	Ea.	6475
Refrigerators, prefabricated, walk-in, 7'-6" High, 10' x 10'	S.F.	114
12' x 14'	S.F.	99.50
12' x 20'	S.F.	137
Refrigerated Food Cases, dairy, multi deck, 12' long	Ea.	12,300
Delicatessen case, multi deck, 18 S.F. shelf display	Ea.	9200
Freezer, glass door upright, 78 C.F.	Ea.	10,600
Frozen food, chest type, 12' long	Ea.	9025
Glass door reach in, 5 door	Ea.	11,300
Island case 12' long, multi deck	Ea.	10,300
Meat case, 12' long, multi deck	Ea.	11,300
Produce, 12' long multi deck	Ea.	9575
Safe, Office type, 1 hr rating, 60"x36"x18", double door	Ea.	10,200

Description	Unit	$ Cost
Sound System, amplifier, 250 watts	Ea.	2025
Speaker, ceiling or wall	Ea.	240
Commissioning Fees, sustainable commercial construction	S.F.	0.24 - 3.04
Energy Modelling Fees, commercial buildings to 10,000 SF	Ea.	11,000
Greater than 10,000 SF add	S.F.	0.04
Green Bldg Cert Fees for comm construction project reg	Project	900
Photovoltaic Pwr Sys, grid connected, 20 kW (~2400 SF), roof	Ea.	249,200
Green Roofs, 6" soil depth, w/treated wd edging & sedum mats	S.F.	12.41
10" Soil depth, with treated wood edging & sedum mats	S.F.	14.10
Greywater Recovery Systems, prepackaged comm, 3060 gal.	Ea.	45,900
4590 gal.	Ea.	57,625
Rainwater Harvest Sys, prepckged comm, 10,000 gal, sys contrller	Ea.	37,550
20,000 gal. w/system controller	Ea.	61,000
30,000 gal. w/system controller	Ea.	97,350
Solar Domestic HW, closed loop, add-on sys, ext heat exchanger	Ea.	11,300
Drainback, hot water system, 120 gal tank	Ea.	14,175
Draindown, hot water system, 120 gal tank	Ea.	14,425

Important: See the Reference Section for Location Factors.

Model costs calculated for a 1 story building with 18' story height and 44,000 square feet of floor area

G Supermarket

				Unit	Unit Cost	Cost Per S.F.	% Of Sub-Total
A. SUBSTRUCTURE							
1010	Standard Foundations	Poured concrete; strip and spread footings		S.F. Ground	1.25	1.25	
1020	Special Foundations	N/A		—	—	—	
1030	Slab on Grade	4" reinforced concrete with recycled vapor barrier and granular base		S.F. Slab	5.84	5.84	8.3%
2010	Basement Excavation	Site preparation for slab and trench for foundation wall and footing		S.F. Ground	.18	.18	
2020	Basement Walls	4' foundation wall		L.F. Wall	98	1.87	
B. SHELL							
B10 Superstructure							
1010	Floor Construction	N/A		—	—	—	7.9 %
1020	Roof Construction	Metal deck, open web steel joists, beams, interior columns		S.F. Roof	8.70	8.70	
B20 Exterior Enclosure							
2010	Exterior Walls	Face brick with concrete block backup	85% of wall	S.F. Wall	41.22	12.04	
2020	Exterior Windows	Storefront windows	15% of wall	Each	65	3.36	15.3%
2030	Exterior Doors	Sliding entrance doors with electrical operator, hollow metal		Each	5986	1.49	
B30 Roofing							
3010	Roof Coverings	Single-ply TPO membrane, 60 mils, heat welded seams w/R-20 insul.		S.F. Roof	6.67	6.67	6.2%
3020	Roof Openings	Roof hatches		S.F. Roof	.16	.16	
C. INTERIORS							
1010	Partitions	50% CMU w/foamed-in insul., 50% gyp. bd. on mtl. studs	50 SF Flr./LF Part.	S.F. Partition	10.04	2.41	
1020	Interior Doors	Single leaf hollow metal, low VOC paint	2000 S.F. Floor/Door	Each	1222	.61	
1030	Fittings	N/A		—	—	—	
2010	Stair Construction	N/A		—	—	—	15.6%
3010	Wall Finishes	Paint, low VOC		S.F. Surface	5.79	2.78	
3020	Floor Finishes	Vinyl composition tile, recycled content		S.F. Floor	3.84	3.84	
3030	Ceiling Finishes	Mineral fiber tile on concealed zee bars		S.F. Ceiling	7.60	7.60	
D. SERVICES							
D10 Conveying							
1010	Elevators & Lifts	N/A		—	—	—	0.0 %
1020	Escalators & Moving Walks	N/A		—	—	—	
D20 Plumbing							
2010	Plumbing Fixtures	Toilet, low flow, auto sensor, & service fixt., supply & drain.	1 Fixt./1820 SF Flr.	Each	2639	1.45	
2020	Domestic Water Distribution	Point of use water heater, electric, energy saver		S.F. Floor	.22	.22	2.6%
2040	Rain Water Drainage	Roof drains		S.F. Roof	1.23	1.23	
D30 HVAC							
3010	Energy Supply	N/A		—	—	—	
3020	Heat Generating Systems	Included in D3050		—	—	—	
3040	Distribution Systems	Enthalpy heat recovery packages		Each	31,850	.79	4.7 %
3050	Terminal & Package Units	Single zone rooftop air conditioner		S.F. Floor	4.39	4.39	
3090	Other HVAC Sys. & Equipment	N/A		—	—	—	
D40 Fire Protection							
4010	Sprinklers	Sprinklers, light hazard		S.F. Floor	3.79	3.79	4.0%
4020	Standpipes	Standpipe		S.F. Floor	.60	.60	
D50 Electrical							
5010	Electrical Service/Distribution	1600 ampere service, panel board and feeders		S.F. Floor	1.89	1.89	
5020	Lighting & Branch Wiring	LED fixtures, daylt. dim., ltg. on/off, receptacles, switches, and A.C.power		S.F. Floor	32.43	32.43	35.4%
5030	Communications & Security	Addressable alarm systems, partial internet wiring and emergency lighting		S.F. Floor	3.80	3.80	
5090	Other Electrical Systems	Emergency generator, 15 kW and energy monitoring systems		S.F. Floor	1.06	1.06	
E. EQUIPMENT & FURNISHINGS							
1010	Commercial Equipment	N/A		—	—	—	
1020	Institutional Equipment	N/A		—	—	—	0.1 %
1090	Other Equipment	Waste handling recycling tilt truck		S.F. Floor	.14	.14	
2020	Moveable Furnishings	No smoking signage		S.F. Floor	.02	.02	
F. SPECIAL CONSTRUCTION							
1020	Integrated Construction	N/A		—	—	—	0.0 %
1040	Special Facilities	N/A		—	—	—	
G. BUILDING SITEWORK	**N/A**						

	Sub-Total	110.61	100%
CONTRACTOR FEES (General Requirements: 10%, Overhead: 5%, Profit: 10%)		25%	27.66
ARCHITECT FEES		7%	9.68
	Total Building Cost	**147.95**	

For customer support on your Square Foot Costs with RSMeans data, call 800.448.8182.

Costs per square foot of floor area

Exterior Wall	S.F. Area	10000	15000	20000	25000	30000	35000	40000	50000	60000
	L.F. Perimeter	410	500	600	640	700	766	833	966	1000
Tilt-up Concrete Panels	Rigid Steel	161.50	150.35	145.15	139.80	136.85	134.95	133.50	131.50	128.55
Face Brick and Concrete Block	Bearing Walls	186.00	169.40	161.90	153.45	148.90	145.95	143.75	140.65	135.85
Concrete Block	Rigid Steel	161.50	150.15	145.00	139.55	136.55	134.60	133.15	131.10	128.15
	Bearing Walls	157.20	145.90	140.70	135.30	132.30	130.35	128.90	126.85	123.90
Metal Panel	Rigid Steel	162.35	150.90	145.60	140.10	137.05	135.10	133.60	131.55	128.50
Metal Sandwich Panel	Rigid Steel	169.30	156.50	150.75	144.45	141.00	138.80	137.10	134.80	131.30
Perimeter Adj., Add or Deduct	Per 100 L.F.	9.60	6.35	4.80	3.85	3.20	2.75	2.45	1.85	1.55
Story Hgt. Adj., Add or Deduct	Per 1 Ft.	1.15	0.90	0.85	0.70	0.60	0.55	0.55	0.50	0.40

For Basement, add $33.96 per square foot of basement area

The above costs were calculated using the basic specifications shown on the facing page. These costs should be adjusted where necessary for design alternatives and owner's requirements.

Common additives

Description	Unit	$ Cost
Dock Leveler, 10 ton cap., 6' x 8'	Ea.	6250
7' x 8'	Ea.	8625
Fence, Chain link, 6' high, 9 ga. wire	L.F.	29
6 ga. wire	L.F.	34.50
Gate	Ea.	400
Paving, Bituminous, wearing course plus base course	S.Y.	7.40
Sidewalks, Concrete 4" thick	S.F.	4.99
Sound System, amplifier, 250 watts	Ea.	2025
Speaker, ceiling or wall	Ea.	240
Yard Lighting, 20' aluminum pole, 400W, HP sodium fixture	Ea.	3600

Description	Unit	$ Cost
Commissioning Fees, sustainable industrial construction	S.F.	0.20 - 1.06
Energy Modelling Fees, commercial buildings to 10,000 SF	Ea.	11,000
Greater than 10,000 SF add	S.F.	0.04
Green Bldg Cert Fees for comm construction project reg	Project	900
Photovoltaic Pwr Sys, grid connected, 20 kW (~2400 SF), roof	Ea.	249,200
Green Roofs, 6" soil depth, w/treated wd edging & sedum mats	S.F.	12.41
10" Soil depth, with treated wood edging & sedum mats	S.F.	14.10
Greywater Recovery Systems, prepackaged comm, 3060 gal.	Ea.	45,900
4590 gal.	Ea.	57,625
Rainwater Harvest Sys, prepckged comm, 10,000 gal, sys contrller	Ea.	37,550
20,000 gal. w/system controller	Ea.	61,000
30,000 gal. w/system controller	Ea.	97,350
Solar Domestic HW, closed loop, add-on sys, ext heat exchanger	Ea.	11,300
Drainback, hot water system, 120 gal tank	Ea.	14,175
Draindown, hot water system, 120 gal tank	Ea.	14,425

Important: See the Reference Section for Location Factors.

Model costs calculated for a 1 story building with 24' story height and 30,000 square feet of floor area

				Unit	Unit Cost	Cost Per S.F.	% Of Sub-Total
A. SUBSTRUCTURE							
1010	Standard Foundations	Poured concrete; strip and spread footings		S.F. Ground	1.57	1.57	
1020	Special Foundations	N/A		—	—	—	
1030	Slab on Grade	5" reinforced concrete with recycled vapor barrier and granular base		S.F. Slab	13.99	13.99	18.9%
2010	Basement Excavation	Site preparation for slab and trench for foundation wall and footing		S.F. Ground	.18	.18	
2020	Basement Walls	4' foundation wall		L.F. Wall	98	2.98	
B. SHELL							
B10 Superstructure							
1010	Floor Construction	Mezzanine: open web steel joists, slab form, concrete beams, columns	10% of area	S.F. Floor	2.33	2.33	10.4%
1020	Roof Construction	Metal deck, open web steel joists, beams, columns		S.F. Roof	7.92	7.92	
B20 Exterior Enclosure							
2010	Exterior Walls	Concrete block	95% of wall	S.F. Wall	18.59	9.89	
2020	Exterior Windows	N/A		—	—	—	11.4%
2030	Exterior Doors	Steel overhead, hollow metal	5% of wall	Each	3857	1.42	
B30 Roofing							
3010	Roof Coverings	Single-ply TPO membrane, 60 mils, heat welded w/R-20 insul.		S.F. Roof	6.84	6.84	7.3%
3020	Roof Openings	Roof hatches and skylight		S.F. Roof	.42	.42	
C. INTERIORS							
1010	Partitions	Concrete block w/foamed-in insul.. (office and washrooms)	100 SF Flr./LF Part.	S.F. Partition	12.13	.97	
1020	Interior Doors	Single leaf hollow metal, low VOC paint	5000 S.F. Floor/Door	Each	1222	.25	
1030	Fittings	N/A		—	—	—	
2010	Stair Construction	Steel gate with rails		Flight	13,950	.93	7.4%
3010	Wall Finishes	Paint, low VOC		S.F. Surface	15.06	2.41	
3020	Floor Finishes	90% hardener, 10% vinyl composition tile, recycled content		S.F. Floor	1.97	1.97	
3030	Ceiling Finishes	Suspended mineral tile on zee channels in office area	10% of area	S.F. Ceiling	.76	.77	
D. SERVICES							
D10 Conveying							
1010	Elevators & Lifts	N/A		—	—	—	0.0 %
1020	Escalators & Moving Walks	N/A		—	—	—	
D20 Plumbing							
2010	Plumbing Fixtures	Toilet, low flow, auto sensor & service fixt., supply & drainage	1 Fixt./2500 SF Flr.	Each	4625	1.85	
2020	Domestic Water Distribution	Tankless, on-demand water heater, gas/propane		S.F. Floor	.45	.45	3.8%
2040	Rain Water Drainage	Roof drains		S.F. Roof	1.46	1.46	
D30 HVAC							
3010	Energy Supply	N/A		—	—	—	
3020	Heat Generating Systems	Ventilation with heat		Each	151,400	5.45	
3030	Cooling Generating Systems	N/A		—	—	—	6.3 %
3050	Terminal & Package Units	Single zone rooftop air conditioner, SEER 14		S.F. Floor	.75	.75	
3090	Other HVAC Sys. & Equipment	N/A		—	—	—	
D40 Fire Protection							
4010	Sprinklers	Sprinklers, ordinary hazard		S.F. Floor	4.36	4.36	5.0%
4020	Standpipes	Standpipe		S.F. Floor	.54	.54	
D50 Electrical							
5010	Electrical Service/Distribution	200 ampere service, panel board and feeders		S.F. Floor	.30	.30	
5020	Lighting & Branch Wiring	LED fixtures, daylt. dim., ltg. on/off, receptacles, switches, and A.C. power		S.F. Floor	17.18	17.18	21.8%
5030	Communications & Security	Addressable alarm systems		S.F. Floor	2.83	2.83	
5090	Other Electrical Systems	Energy monitoring systems		S.F. Floor	1.24	1.24	
E. EQUIPMENT & FURNISHINGS							
1010	Commercial Equipment	N/A		—	—	—	
1030	Vehicular Equipment	Dock boards, dock levelers		S.F. Floor	1.87	1.87	2.1 %
1090	Other Equipment	Waste handling recycling tilt truck		S.F. Floor	.21	.21	
2020	Moveable Furnishings	No smoking signage		S.F. Floor	.02	.02	
F. SPECIAL CONSTRUCTION							
1020	Integrated Construction	Shipping & receiving air curtain		S.F. Floor	5.56	5.56	5.6%
1040	Special Facilities	N/A		—	—	—	
G. BUILDING SITEWORK	**N/A**						
				Sub-Total		98.91	100%
	CONTRACTOR FEES (General Requirements: 10%, Overhead: 5%, Profit: 10%)				25%	24.74	
	ARCHITECT FEES				7%	8.65	
				Total Building Cost		132.30	

Assemblies Section

Table of Contents

Table of Contents

Table of Contents

Table of Contents

288

Introduction to the Assemblies Section

This section contains the technical description and installed cost of all components used in the commercial/industrial/institutional section. In most cases, there is a graphic of the component to aid in identification.

All Costs Include:

Materials purchased in lot sizes typical of normal construction projects with discounts appropriate to established contractors.

Installation work performed by a union labor working under standard conditions at a normal pace.

Installing Contractor's Overhead & Profit

Standard installing contractor's overhead & profit are included in the assemblies costs.

These assemblies costs contain no provisions for the general contractor's overhead and profit or the architect's fees, nor do they include premiums for labor or materials, or savings which may be realized under certain economic situations.

The assemblies tables are arranged by physical size (dimensions) of the component wherever possible.

RSMeans data: Assemblies—How They Work

Assemblies estimating provides a fast and reasonably accurate way to develop construction costs. An assembly is the grouping of individual work items—with appropriate quantities—to provide a cost for a major construction component in a convenient unit of measure.

An assemblies estimate is often used during early stages of design development to compare the cost impact of various design alternatives on total building cost.

Assemblies estimates are also used as an efficient tool to verify construction estimates.

Assemblies estimates do not require a completed design or detailed drawings. Instead, they are based on the general size of the structure and other known parameters of the project. The degree of accuracy of an assemblies estimate is generally within +/- 15%.

Most assemblies consist of three major elements: a graphic, the system components, and the cost data itself. The **Graphic** is a visual representation showing the typical appearance of the assembly

① Unique 12-character Identifier

Our assemblies are identified by a **unique 12-character identifier**. The assemblies are numbered using UNIFORMAT II, ASTM Standard E1557. The first 5 characters represent this system to Level 3. The last 7 characters represent further breakdown in order to arrange items in understandable groups of similar tasks. Line numbers are consistent across all of our publications, so a line number in any assemblies data set will always refer to the same item.

② Narrative Descriptions

Our assemblies descriptions appear in two formats: narrative and table. **Narrative descriptions** are shown in a hierarchical structure to make them readable. In order to read a complete description, read up through the indents to the top of the section. Include everything that is above and to the left that is not contradicted by information below.

Narrative Format

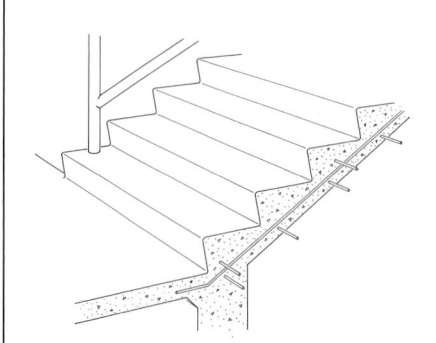

C20 Stairs

C2010 Stair Construction

The table below lists the cost per flight for 4'-0" wide stairs. Side walls are not included. Railings are included in the prices.

C2010 110	Stairs	COST PER FLIGHT		
		MAT.	INST.	TOTAL
0470	Stairs, C.I.P. concrete, w/o landing, 12 risers, w/o nosing	1,475	2,500	3,975
0480	With nosing	2,400	2,725	5,125
0550	W/landing, 12 risers, w/o nosing	1,650	3,075	4,725
0560	With nosing	2,550	3,300	5,850
0570	16 risers, w/o nosing	2,000	3,900	5,900
0580	With nosing	3,225	4,175	7,400
0590	20 risers, w/o nosing	2,375	4,675	7,050
0600	With nosing	3,900	5,050	8,950
0610	24 risers, w/o nosing	2,750	5,475	8,225
0620	With nosing	4,575	5,900	10,475
0630	Steel, grate type w/nosing & rails, 12 risers, w/o landing	5,575	1,200	6,775
0640	With landing	7,825	1,625	9,450
0660	16 risers, with landing	9,675	2,050	11,725
0680	20 risers, with landing	11,500	2,450	13,950
0700	24 risers, with landing	13,400	2,825	16,225
0701				
0710	Metal pan stairs for concrete in-fill, picket rail, 12 risers, w/o landing	8,225	1,200	9,425
0720	With landing	10,800	1,800	12,600
0740	16 risers, with landing	13,600	2,200	15,800
0760	20 risers, with landing	16,300	2,600	18,900
0780	24 risers, with landing	19,100	2,975	22,075
0790	Cast iron tread & pipe rail, 12 risers, w/o landing	8,275	1,200	9,475
0800	With landing	10,900	1,800	12,700
1120	Wood, prefab box type, oak treads, wood rails 3'-6" wide, 14 risers	2,400	515	2,915
1150	Prefab basement type, oak treads, wood rails 3'-0" wide, 14 risers	1,150	127	1,277

For supplemental customizable square foot estimating forms, visit: **www.RSMeans.com/2018books**

in question. It is frequently accompanied by additional explanatory technical information describing the class of items. The **Assemblies data** below lists prices for other similar systems with dimensional and/or size variations.

All of our assemblies costs represent the cost for the installing contractor. An allowance for profit has been added to all material, labor, and equipment rental costs. A markup for labor burdens, including workers' compensation, fixed overhead, and business overhead, is included with installation costs.

The information in RSMeans cost data represents a "national average" cost. This data should be modified to the project location using the **Location Factors** tables found in the Reference Section.

Table Format

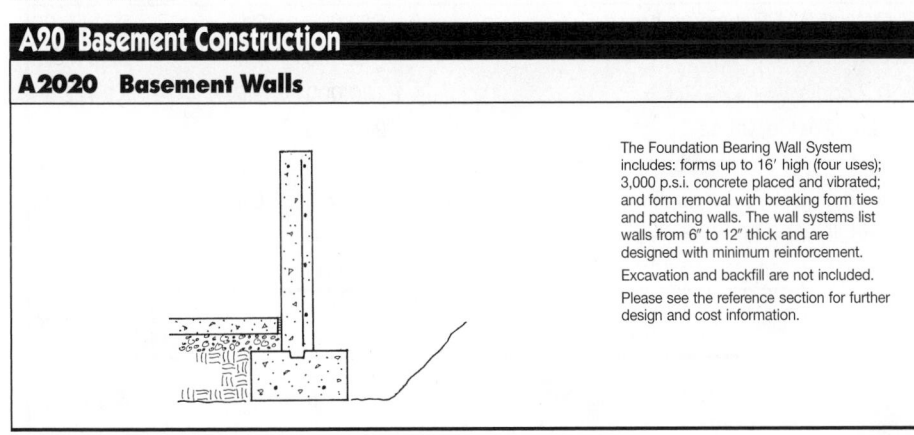

A20 Basement Construction

A2020 Basement Walls

The Foundation Bearing Wall System includes: forms up to 16' high (four uses); 3,000 p.s.i. concrete placed and vibrated; and form removal with breaking form ties and patching walls. The wall systems list walls from 6" to 12" thick and are designed with minimum reinforcement.

Excavation and backfill are not included.

Please see the reference section for further design and cost information.

3 Unit of Measure

All RSMeans data: Assemblies include a typical **Unit of Measure** used for estimating that item. For instance, for continuous footings or foundation walls the unit is linear feet (L.F.). For spread footings the unit is each (Ea.). The estimator needs to take special care that the unit in the data matches the unit in the takeoff. Abbreviations and unit conversions can be found in the Reference Section.

4 Table Descriptions

Table descriptions work similar to Narrative Descriptions, except that if there is a blank in the column at a particular line number, read up to the description above in the same column.

A2020 110		Walls, Cast in Place						
	WALL HEIGHT (FT.)	PLACING METHOD	CONCRETE (C.Y. per L.F.)	REINFORCING (LBS. per L.F.)	WALL THICKNESS (IN.)	COST PER L.F.		
						MAT.	INST.	TOTAL
5000	8'	direct chute	.148	6.6	6	36.50	112	148.50
5020			.199	9.6	8	44.50	119	163.50
5040			.250	12	10	53	117	170
5061			.296	14.39	12	60	120	180
6020	10'	direct chute	.248	12	8	56	144	200
6040			.307	14.99	10	65	147	212
6061			.370	17.99	12	75	150	225
7220	12'	pumped	.298	14.39	8	67	179	246
7240			.369	17.99	10	78.50	183	261.50
7260			.444	21.59	12	90	187	277
9220	16'	pumped	.397	19.19	8	89.50	238	327.50
9240			.492	23.99	10	104	243	347
9260			.593	28.79	12	120	250	370

Sample Estimate

This sample demonstrates the elements of an estimate, including a tally of the RSMeans data lines. Published assemblies costs include all markups for labor burden and profit for the installing contractor. This estimate adds a summary of the markups applied by a general contractor on the installing contractor's work. These figures represent the total cost to the owner. The location factor with RSMeans data is applied at the bottom of the estimate to adjust the cost of the work to a specific location.

Project Name:	Interior Fit-out, ABC Office			
Location:	Anywhere, USA	Date: 1/1/2018		STD
Assembly Number	**Description**	**Qty.**	**Unit**	**Subtotal**
C1010 34 1200 ①	Wood partition, 2 x 4 @ 16" OC w/5/8" FR gypsum board	560.000	S.F.	$2,856.00
C1020 114 1800	Metal door & frame, flush hollow core, 3'-0" x 7'-0"	2.000	Ea.	$2,510.00
C3010 230 0080	Painting, brushwork, primer & 2 coats	1,120.000	S.F.	$1,422.40
C3020 410 0140	Carpet, tufted, nylon, roll goods, 12' wide, 26 oz	240.000	S.F.	$842.40
C3030 210 6000	Acoustic ceilings, 24" x 48" tile, tee grid suspension	200.000	S.F.	$1,260.00
D5020 125 0560	Receptacles incl plate, box, conduit, wire, 20 A duplex	8.000	Ea.	$2,320.00
D5020 125 0720	Light switch incl plate, box, conduit, wire, 20 A single pole	2.000	Ea.	$566.00
D5020 210 0560	Fluorescent fixtures, recess mounted, 20 per 1000 SF	200.000	S.F.	$2,226.00
	Assembly Subtotal			**$14,002.80**
	Sales Tax @ ②	5 %		$ 350.07
	General Requirements @ ③	7 %		$ 980.20
	Subtotal A			**$15,333.07**
	GC Overhead @ ④	5 %		$ 766.65
	Subtotal B			**$16,099.72**
	GC Profit @ ⑤	5 %		$ 804.99
	Subtotal C			**$16,904.71**
	Adjusted by Location Factor ⑥	115.2		$ 19,474.22
	Architects Fee @ ⑦	8 %		$ 1,557.94
	Contingency @ ⑧	15 %		$ 2,921.13
	Project Total Cost			**$ 23,953.29**

This estimate is based on an interactive spreadsheet. You are free to download it and adjust it to your methodology. A copy of this spreadsheet is available at **www.RSMeans.com/2018books.**

1 Work Performed

The body of the estimate shows the RSMeans data selected, including line numbers, a brief description of each item, its takeoff quantity and unit, and the total installed cost, including the installing contractor's overhead and profit.

2 Sales Tax

If the work is subject to state or local sales taxes, the amount must be added to the estimate. In a conceptual estimate it can be assumed that one half of the total represents material costs. Therefore, apply the sales tax rate to 50% of the assembly subtotal.

3 General Requirements

This item covers project-wide needs provided by the general contractor. These items vary by project but may include temporary facilities and utilities, security, testing, project cleanup, etc. In assemblies estimates a percentage is used—typically between 5% and 15% of the project cost.

4 General Contractor Overhead

This entry represents the general contractor's markup on all work to cover project administration costs.

5 General Contractor Profit

This entry represents the GC's profit on all work performed. The value included here can vary widely by project and is influenced by the GC's perception of the project's financial risk and market conditions.

6 Location Factor

RSMeans published data are based on national average costs. If necessary, adjust the total cost of the project using a location factor from the "Location Factor" table or the "City Cost Indexes" table found in the Reference Section. Use location factors if the work is general, covering the work of multiple trades. If the work is by a single trade (e.g., masonry) use the more specific data found in the City Cost Indexes.

To adjust costs by location factors, multiply the base cost by the factor and divide by 100.

7 Architect's Fee

If appropriate, add the design cost to the project estimate. These fees vary based on project complexity and size. Typical design and engineering fees can be found in the Reference Section.

8 Contingency

A factor for contingency may be added to any estimate to represent the cost of unknowns that may occur between the time that the estimate is performed and the time the project is constructed. The amount of the allowance will depend on the stage of design at which the estimate is done, and the contractor's assessment of the risk involved.

Example

Assemblies costs can be used to calculate the cost for each component of a building and, accumulated with appropriate markups, produce a cost for the complete structure. Components from a model building with similar characteristics in conjunction with Assemblies components may be used to calculate costs for a complete structure.

Example:

Outline Specifications

General
Building size—60' x 100', 8 suspended floors, 12' floor to floor, 4' high parapet above roof, full basement 11'-8" floor to floor, bay size 25' x 30', ceiling heights 9' in office area and 8' in core area.

A Substructure
Concrete, spread and strip footings, concrete walls waterproofed, 4" slab on grade.

B Shell

B10 Superstructure
Columns, wide flange; 3 hr. fire rated; Floors, composite steel frame & deck with concrete slab; Roof, steel beams, open web joists and deck.

B20 Exterior Enclosure
Walls; North, East & West, brick & lightweight concrete block with 2" cavity insulation, 25% window; South, 8" lightweight concrete block insulated, 10% window. Doors, aluminum & glass. Windows, aluminum, 3'-0" x 5'-4", insulating glass.

B30 Roofing
Tar & gravel, 2" rigid insulation, R 12.5.

C Interiors
Core—6" lightweight concrete block, full height with plaster on exposed faces to ceiling height.

Corridors—1st & 2nd floor; 3-5/8" steel studs with F.R. gypsum board, full height.

Exterior Wall—plaster, ceiling height.

Fittings—Toilet accessories, directory boards.

Doors—hollow metal.

Wall Finishes—lobby, mahogany paneling on furring, remainder plaster & gypsum board, paint.

Floor Finishes—1st floor lobby, corridors & toilet rooms, terrazzo; remainder concrete, tenant developed. 2nd through 8th, toilet rooms, ceramic tile; office & corridor, carpet.

Ceiling Finishes—24" x 48" fiberglass board on Tee grid.

D Services

D10 Conveying
2-2500 lb. capacity, 200 F.P.M. geared elevators, 9 stops.

D20 Plumbing
Fixtures, see sketch.

Roof Drains—2-4" C.I. pipe.

D30 HVAC
Heating—fin tube radiation, forced hot water.

Air Conditioning—chilled water with water cooled condenser.

D40 Fire Protection
Fire Protection—4" standpipe, 9 hose cabinets.

D50 Electrical
Lighting, 1st through 8th, 15 fluorescent fixtures/1000 S.F., 3 Watts/S.F.

Basement, 10 fluorescent fixtures/1000 S.F., 2 Watts/S.F.

Receptacles, 1st through 8th, 16.5/1000 S.F., 2 Watts/S.F.

Basement, 10 receptacles/1000 S.F., 1.2 Watts/S.F.

Air Conditioning, 4 Watts/S.F.

Miscellaneous Connections, 1.2 Watts/S.F.

Elevator Power, 2-10 H.P., 230 volt motors.

Wall Switches, 2/1000 S.F.

Service, panel board & feeder, 2000 Amp.

Fire Detection System, pull stations, signals, smoke and heat detectors.

Emergency Lighting Generator, 30 KW.

E Equipment & Furnishings
NA.

F Special Construction
NA.

G Building Sitework
NA.

Front Elevation

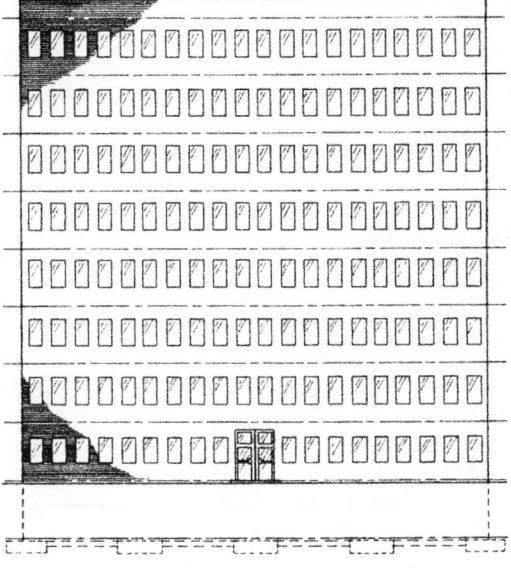

Basement Plan

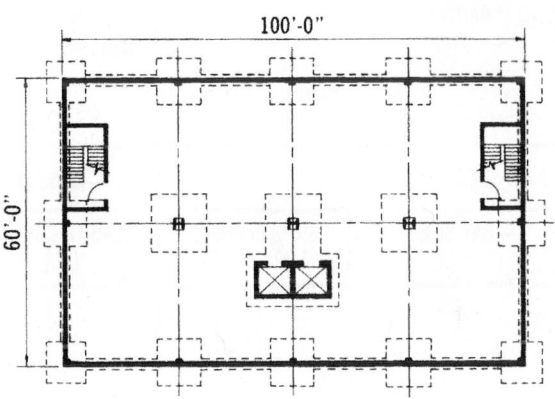

Ground Floor Plan

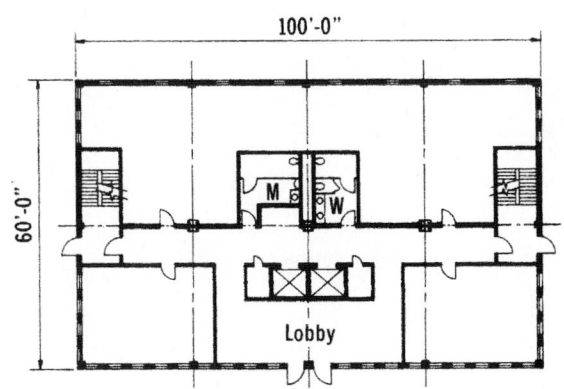

Typical Floor Plan

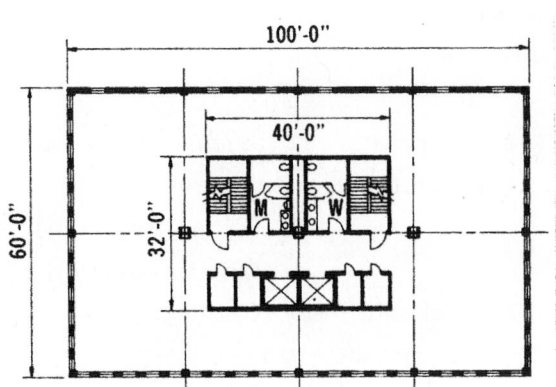

Example

If spread footing & column sizes are unknown, develop approximate loads as follows. Enter tables with these loads to determine costs.

Approximate loads/S.F. for roof & floors.

Roof. Assume 40 psf superimposed load.

Steel joists, beams & deck.

Table B1020 112—Line 3900

Superimposed Load Ranges

Apartments & Residential Structures	65	to	75 psf
Assembly Areas & Retail Stores	110	to	125 psf
Commercial & Manufacturing	150	to	250 psf
Offices	75	to	100 psf

B1010 256		**Composite Beams, Deck & Slab**						
	BAY SIZE (FT.)	SUPERIMPOSED LOAD (P.S.F.)	SLAB THICKNESS (IN.)	TOTAL DEPTH (FT.-IN.)	TOTAL LOAD (P.S.F.)	COST PER S.F.		
						MAT.	INST.	TOTAL
3400	25x30	40	5-1/2	1 - 11-1/2	83	12.80	6.05	18.85
3600		75	5-1/2	1 - 11-1/2	119	13.75	6.10	19.85
3900		125	5-1/2	1 - 11-1/2	170	16	6.85	22.85
4000		200	6-1/4	2 - 6-1/4	252	19.85	7.80	27.65

B1020 112		**Steel Joists, Beams, & Deck on Columns**						
	BAY SIZE (FT.)	SUPERIMPOSED LOAD (P.S.F.)	DEPTH (IN.)	TOTAL LOAD (P.S.F.)	COLUMN ADD	COST PER S.F.		
						MAT.	INST.	TOTAL
3500	25x30	20	22	40		6	1.64	7.64
3600					columns	1.22	.29	1.51
3900		40	25	60		7.20	1.94	9.14
4000					columns	1.47	.35	1.82

Floors—Total load, 119 psf.

Interior foundation load.

Roof

[(25' x 30' x 60 psf) + 8 floors x (25' x 30' x 119 psf)] x 1/1000 lb./Kip	=	759 Kips
Approximate Footing Loads, Interior footing	=	759 Kips
Exterior footing (1/2 bay) 759 k x .6	=	455 Kips
Corner footing (1/4 bay) 759 k x .45	=	342 Kips

[Factors to convert Interior load to Exterior & Corner loads]

Approximate average Column load 759 k/2	=	379 Kips

PRELIMINARY ESTIMATE

PROJECT	**Office Building**	TOTAL SITE AREA		
BUILDING TYPE		OWNER		
LOCATION		ARCHITECT		
DATE OF CONSTRUCTION		ESTIMATED CONSTRUCTION PERIOD		

BRIEF DESCRIPTION **Building size: 60' x 100' - 8 structural floors - 12' floor to floor**

4' high parapet above roof, full basement 11' - 8" floor to floor, bay size

25' x 30', ceiling heights - 9' in office area + 8' in core area

TYPE OF PLAN		TYPE OF CONSTRUCTION	
QUALITY		BUILDING CAPACITY	

Floor				**Wall Area**						
Below Grade Levels				Foundation Walls	L.F.		Ht.		S.F.	
Area			S.F.	Frost Walls	L.F.		Ht.		S.F.	
Area			S.F.	Exterior Closure		Total			S.F.	
Total Area			S.F.	Comment						
Ground Floor				Fenestration		%			S.F.	
Area			S.F.			%			S.F.	
Area			S.F.	Exterior Wall		%			S.F.	
Total Area			S.F.			%			S.F.	
Supported Levels				**Site Work**						
Area				Parking		S.F. (For			Cars)	
Area			S.F.	Access Roads		L.F. (X			Ft. Wide)	
Area			S.F.	Sidewalk		L.F. (X			Ft. Wide)	
Area			S.F.	Landscaping		S.F. (			% Unbuilt Site)	
Area			S.F.	**Building Codes**						
Total Area			S.F.	City		Country				
Miscellaneous				National		Other				
Area			S.F.	**Loading**						
Area			S.F.	Roof	psf	Ground Floor			psf	
Area			S.F.	Supported Floors	psf	Corridor			psf	
Area			S.F.	Balcony	psf	Partition, allow			psf	
Total Area			S.F.	Miscellaneous					psf	
Net Finished Area			S.F.	Live Load Reduction						
Net Floor Area			S.F.	Wind						
Gross Floor Area		**54,000**	S.F.	Earthquake			Zone			
Roof				Comment						
Total Area			S.F.	Soil Type						
Comments				Bearing Capacity					K.S.F.	
				Frost Depth					Ft.	
Volume				**Frame**						
Depth of Floor System				Type		Bay Spacing				
Minimum			In.	Foundation						
Maximum			In.	Special						
Foundation Wall Height			Ft.	Substructure						
Floor to Floor Height			Ft.	Comment						
Floor to Ceiling Height			Ft.	Superstructure, Vertical						
Subgrade Volume			C.F.	Fireproofing			☐ Columns			Hrs.
Above Grade Volume			C.F.	☐ Girders	Hrs.		☐ Beams			Hrs.
Total Building Volume		**648,000**	C.F.	☐ Floor	Hrs.		☐ None			

NUMBER		QTY.	UNIT	TOTAL COST UNIT	TOTAL	COST PER S.F.
A	**SUBSTRUCTURE**					
A1010 210 7900	Corner Footings 8'-6" SQ. x 27"	4	Ea.	2,100.00	8,400	
8010	Exterior 9' -6" SQ. x 30"	8		2,825.00	22,600	
8300	Interior 12' SQ.	3		5,175.00	15,525	
A1010 110 2700	Strip 2' Wide x 1' Thick					
	320 L.F.- [(4 x 8.5) + 8 x 9.5)] =	210	L.F.	42.60	8,946	
A2020 110 7260	Foundation Wall 12' High, 1' Thick	320		277.00	88,640	
A1010 320 2800	Foundation Waterproofing	320		24.90	7,968	
A1030 120 2240	4", non-industrial, reinf. slab on grade	6000	S.F.	6.97	41,820	
A2010 110 3440	Building Excavation + Backfill	6000	S.F.	4.40	26,400	
3500	(Interpolated ; 12' Between					
4620	8' and 16' ; 6,000 Between					
4680	4,000 and 10,000 S.F.					
	Total				220,299	4.08
B	**SHELL**					
B10	**Superstructure**					
B1010 208 5800	Columns: Load 379K Use 400K					
	Exterior 12 x 96' + Interior 3 x 108'	1476	V.L.F.	152.25	224,721	
B1010 720 3650	Column Fireproofing - Interior 4 sides					
3700	Exterior 1/2 (Interpolated for 12")	900	V.L.F	38.10	34,290	
B1010 256 3600	Floors: Composite steel + lt. wt. conc.	48000	S.F.	19.85	952,800	
B1020 112 3900	Roof: Open web joists, beams + decks	6000	S.F.	9.14	54,840	
	Total				1,266,651	23.46
B20	**Exterior Enclosure**					
B2010 134 1200	4" Brick + 6" Block - Insulated					
	75% NE + W Walls 220' x 100' x .75	16500	S.F.	32.85	542,025	
B2010 110 3410	8" Block - Insulated 90%					
	100' x 100' x .9	9000	S.F.	13.10	117,900	
B2030 110 6950	Double Aluminum + Glass Door	1	Pr.	7,200.00	7,200	
6300	Single	2	Ea.	3,800.00	7,600	
B2020 106 8800	Aluminum windows - Insul. glass					
	(5,500 S.F. + 1,000 S.F.)/(3 x 5.33)	406	Ea.	923.00	374,738	
B2010 105 8480	Precast concrete coping	320	L.F.	51.30	16,416	
	Total				1,065,879	19.74
B30	**Roofing**					
B3010 320 4020	Insulation: Perlite/Polyisocyanurate	6000	S.F.	1.91	11,460	
B3010 105 1400	Roof: 4 Ply T & G composite	6000		3.29	19,740	
B3010 430 0300	Flashing: Aluminum - fabric backed	640		3.55	2,272	
B3020 210 0500	Roof hatch	1	Ea.	1,470.00	1,470	
	Total				34,942	0.65

ASSEMBLY NUMBER		QTY.	UNIT	TOTAL COST		COST PER S.F.
				UNIT	TOTAL	
C	**INTERIORS**					
C1010 104 5500	Core partitions: 6" Lt. Wt. concrete					
	block 288 L.F. x 11.5' x 8 Floors	26,449	S.F.	10.40	275,070	
C1010 144 0920	Plaster: [(196 L.F. x 8') + (144 L.F. x 9')] 8					
	+ Ext. Wall [(320 x 9 x 8 Floors) - 6500]	39,452	S.F.	3.39	133,742	
C1010 126 5400	Corridor partitions 1st + 2nd floors					
	steel studs + F.R. gypsum board	2,530	S.F.	4.52	11,436	
C1020 102 2600	Interior doors	66	Ea.	1,222.00	80,652	
	Wall finishes - Lobby: Mahogany w/furring					
C1010 128 0652	Furring: 150 L.F. x 9' + Ht.	1,350	S.F.	1.76	2,376	
C1030 710 0120	Towel Dispenser	16	Ea.	85.50	1,368	
C1030 710 0140	Grab Bar	32	Ea.	58.50	1,872	
C1030 710 0160	Mirror	31	Ea.	40.35	1,251	
C1030 710 0180	Toilet Tissue Dispenser	31	Ea.	42.00	1,302	
C1030 510 0140	Directory Board	8	Ea.	367.00	2,936	
C2010 110 0780	Stairs: Steel w/conc. fill	18	Flt.	22,075.00	397,350	
C3010 230 1662	Mahogany paneling	1,350	S.F.	5.79	7,817	
C3010 230 0080	Paint, primer + 2 coats finish, on plaster or gyp. board					
	39,452 + [(220' x 9' x 2) - 1350] + 72	42,134	S.F.	1.27	53,510	
	Floor Finishes					
C3020 410 1100	1st floor lobby + terrazzo	2,175	S.F.	22.08	48,024	
	Remainder: Concrete - Tenant finished					
C3020 410 0060	2nd - 8th: Carpet - 4725 S.F. x 7	33,075	S.F.	4.54	150,161	
C3020 410 1720	Ceramic Tile - 300 S.F. x 7	2,100	S.F.	12.00	25,200	
C3030 240 2780	Suspended Ceiling Tiles- 5200 S.F. x 8	41,600	S.F.	2.32	96,512	
C3030 240 3260	Suspended Ceiling Grid- 5200 S.F. x 8	41,600	S.F.	1.64	68,224	
C1030 110 0700	Toilet partitions	31	Ea.	781.00	24,211	
0760	Handicapped addition	16	Ea.	360.00	5,760	
	Total				1,388,773	25.72
D	**SERVICES**					
D10	**Conveying**					
	2 Elevators - 2500# 200'/min. Geared					
D1010 140 1600	5 Floors $190,500					
1800	15 Floors $429,000					
	Diff. $238,500/10 = 23,850/Flr.					
	$190,500 + (4 x23,850) = $285,900/Ea.	2	Ea.	285,900.00	571,800	
	Total				571,800	10.59
D20	**Plumbing**					
D2010 310 1600	Plumbing - Lavatories	31	Ea.	1,190.00	36,890	
D2010 440 4340	-Service Sink	8	Ea.	3,875.00	31,000	
D2010 210 2000	-Urinals	8	Ea.	1,480.00	11,840	

ASSEMBLY NUMBER		QTY.	UNIT	TOTAL COST UNIT	TOTAL	COST PER S.F.
D2010 110 2080	-Water Closets	31	Ea.	3,400.00	105,400	
	Water Control, Waste Vent Piping	45%			83,309	
D2040 210 4200	- Roof Drains, 4" C.I.	2	Ea.	2,395.00	4,790	
4240	- Pipe, 9 Flr. x 12' x 12' Ea.	216	L.F.	63.00	13,608	
	Total				286,837	5.31
D30	HVAC					
D3010 520 2000	10,000 S.F. @ $10.89/S.F. Interpolate					
2040	100,000 S.F. @ $4.94/S.F.	48,000	S.F.	7.92	380,160	
	Cooling - Chilled Water, Air Cool, Cond.					
D3030 115 3840	4,000 S.F. @20.15/S.F. Interpolate	48,000	S.F.	17.73	851,040	
4040	60,000 S.F. @15.30/S.F.					
	Total				1,231,200	22.80
D40	Fire Protection					
D4020 310 0560	Wet Stand Pipe: 4" x 10' 1st floor	12/10	Ea.	9,400.00	11,280	
0580	Additional Floors: (12'/10 x 8)	9.6	Ea.	2,550.00	24,480	
D4020 410 8400	Cabinet Assembly	9	Ea.	1,335.00	12,015	
	Total				47,775	0.88
D50	Electrical					
D5020 210 0280	Office Lighting, 15/1000 S.F. - 3 Watts/S.F.	48,000	S.F.	8.17	392,160	
0240	Basement Lighting, 10/1000 S.F. - 2 Watts/S.F.	6,000	S.F.	5.43	32,580	
D5020 110 0640	Office Receptacles 16.5/1000 S.F. - 2 Watts/S.F.	48,000	S.F.	5.01	240,480	
0560	Basement Receptacles 10/1000 S.F. - 1.2 Watts/S.F.	6,000	S.F.	3.81	22,860	
D5020 140 0280	Central A.C. - 4 Watts/S.F.	48,000	S.F.	0.63	30,240	
D5020 135 0320	Misc. Connections - 1.2 Watts/S.F.	48,000	S.F.	0.35	16,800	
D5020 145 0680	Elevator Motor Power - 10 H.P.	2	Ea.	2,875.00	5,750	
D5020 130 0280	Wall Switches - 2/1000 S.F.	54,000	S.F.	0.48	25,920	
D5010 120 0560	2000 Amp Service	1	Ea.	29,900.00	29,900	
D5010 240 0400	Switchgear	1	Ea.	44,225.00	44,225	
D5010 230 0560	Feeder	50	L.F.	491.00	24,550	
D5030 910 0400	Fire Detection System - 50 Detectors	1	Ea.	36,200.00	36,200	
D5090 210 0320	Emergency Generator - 30 kW	30	kW	632.50	18,975	
	Total				920,640	17.05
E	EQUIPMENT & FURNISHINGS					
G	BUILDING SITEWORK					
	Miscellaneous					

Preliminary Estimate Cost Summary

PROJECT: Office Building	TOTAL AREA	54,000 S.F.	SHEET NO.
LOCATION	TOTAL VOLUME	648,000 C.F.	ESTIMATE NO.
ARCHITECT	COST PER S.F.		DATE
OWNER	COST PER C.F.		NO OF STORIES
QUANTITIES BY	EXTENSIONS BY		CHECKED BY

DIV	DESCRIPTION	SUBTOTAL COST	COST/S.F.	PERCENTAGE
A	SUBSTRUCTURE	220,299	$ 4.08	
B10	SHELL: SUPERSTRUCTURE	1,266,651	$ 23.46	
B20	SHELL: EXTERIOR ENCLOSURE	1,065,879	$ 19.74	
B30	SHELL: ROOFING	34,942	$ 0.65	
C	INTERIORS	1,388,773	$ 25.72	
D10	SERVICES: CONVEYING	571,800	$ 10.59	
D20	SERVICES: PLUMBING	286,837	$ 5.31	
D30	SERVICES: HVAC	1,231,200	$ 22.80	
D40	SERVICES: FIRE PROTECTION	47,775	$ 0.88	
D50	SERVICES: ELECTRICAL	920,640	$ 17.05	
E	EQUIPMENT & FURNISHINGS			
F	SPECIAL CONSTRUCTION & DEMO.			
G	BUILDING SITEWORK			

BUILDING SUBTOTAL	$ 7,034,795		$	7,034,795
Sales Tax % x Subtotal /2	N/A		$	-
General Conditions 10 % x Subtotal			$	703,480
		Subtotal "A"	$	7,738,275
Overhead 5 % x Subtotal "A"			$	386,914
		Subtotal "B"	$	8,125,188
Profit 10% x Subtotal "B"			$	812,519
		Subtotal "C"	$	8,937,707
Location Factor % x Subtotal "C"	N/A	Localized Cost		
Architects Fee 7% x Localized Cost =			$	625,639
Contingency x Localized Cost =	N/A		$	-
		Project Total Cost	$	9,563,347
Square Foot Cost $9,081,534/ 54,000 S.F. =		S.F. Cost	$	177.10
Cubic Foot Cost $9,081,534/ 648,000 C.F. =		C.F. Cost	$	14.76

A1010 Standard Foundations

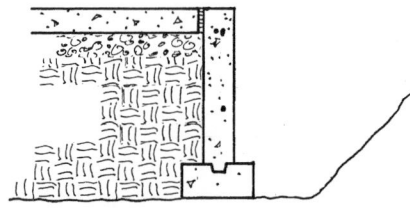

The Foundation Bearing Wall System includes: forms up to 6' high (four uses); 3,000 p.s.i. concrete placed and vibrated; and form removal with breaking form ties and patching walls. The wall systems list walls from 6" to 16" thick and are designed with minimum reinforcement.

Excavation and backfill are not included.

Please see the reference section for further design and cost information.

A1010 105 — Wall Foundations

	WALL HEIGHT (FT.)	PLACING METHOD	CONCRETE (C.Y. per L.F.)	REINFORCING (LBS. per L.F.)	WALL THICKNESS (IN.)	COST PER L.F.		
						MAT.	INST.	TOTAL
1500	4'	direct chute	.074	3.3	6	18.15	56	74.15
1520			.099	4.8	8	22.50	57.50	80
1540			.123	6.0	10	26	58.50	84.50
1560			.148	7.2	12	30	60	90
1580			.173	8.1	14	34	61	95
1600			.197	9.44	16	38	62	100
1700		pumped	.074	3.3	6	18.15	58	76.15
1720			.099	4.8	8	22.50	59.50	82
1740			.123	6.0	10	26	60.50	86.50
1760			.148	7.2	12	30	62.50	92.50
1780			.173	8.1	14	34	63.50	97.50
1800			.197	9.44	16	38	65	103
3000	6'	direct chute	.111	4.95	6	27.50	84.50	112
3020			.149	7.20	8	33.50	86.50	120
3040			.184	9.00	10	39	88	127
3060			.222	10.8	12	45	90	135
3080			.260	12.15	14	51	91.50	142.50
3100			.300	14.39	16	57.50	93.50	151
3200		pumped	.111	4.95	6	27.50	86	113.50
3220			.149	7.20	8	33.50	89.50	123
3240			.184	9.00	10	39	91.50	130.50
3260			.222	10.8	12	45	93.50	138.50
3280			.260	12.15	14	51	95.50	146.50
3300			.300	14.39	16	57.50	98	155.50

A1010 Standard Foundations

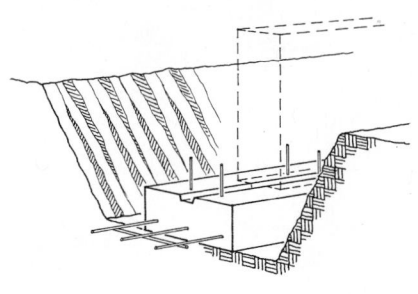

The Strip Footing System includes: excavation; hand trim; all forms needed for footing placement; forms for 2″ x 6″ keyway (four uses); dowels; and 3,000 p.s.i. concrete.

The footing size required varies for different soils. Soil bearing capacities are listed for 3 KSF and 6 KSF. Depths of the system range from 8″ and deeper. Widths range from 16″ and wider. Smaller strip footings may not require reinforcement.

Please see the reference section for further design and cost information.

A1010 110	Strip Footings	COST PER L.F.		
		MAT.	INST.	TOTAL
2100	Strip footing, load 2.6 KLF, soil capacity 3 KSF, 16″ wide x 8″ deep, plain	7.90	11.55	19.45
2300	Load 3.9 KLF, soil capacity 3 KSF, 24″ wide x 8″ deep, plain	10.15	13	23.15
2500	Load 5.1 KLF, soil capacity 3 KSF, 24″ wide x 12″ deep, reinf.	18.10	24.50	42.60
2700	Load 11.1 KLF, soil capacity 6 KSF, 24″ wide x 12″ deep, reinf.	18.10	24.50	42.60
2900	Load 6.8 KLF, soil capacity 3 KSF, 32″ wide x 12″ deep, reinf.	22	27	49
3100	Load 14.8 KLF, soil capacity 6 KSF, 32″ wide x 12″ deep, reinf.	22	27	49
3300	Load 9.3 KLF, soil capacity 3 KSF, 40″ wide x 12″ deep, reinf.	26	29	55
3500	Load 18.4 KLF, soil capacity 6 KSF, 40″ wide x 12″ deep, reinf.	26	29.50	55.50
4500	Load 10 KLF, soil capacity 3 KSF, 48″ wide x 16″ deep, reinf.	37	36.50	73.50
4700	Load 22 KLF, soil capacity 6 KSF, 48″ wide, 16″ deep, reinf.	38	37.50	75.50
5700	Load 15 KLF, soil capacity 3 KSF, 72″ wide x 20″ deep, reinf.	64.50	52.50	117
5900	Load 33 KLF, soil capacity 6 KSF, 72″ wide x 20″ deep, reinf.	67.50	56	123.50

A1010 Standard Foundations

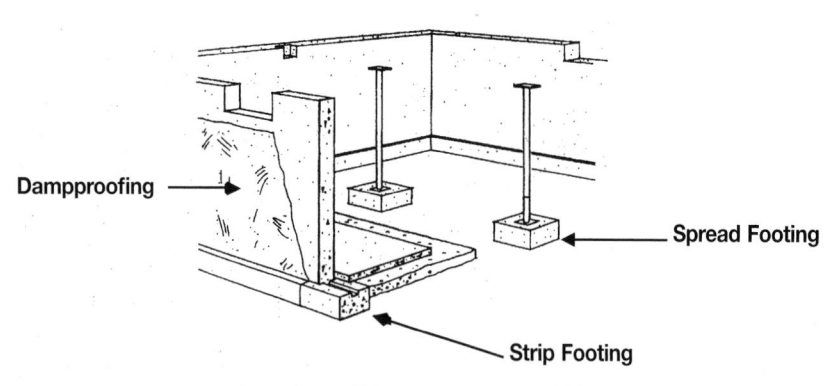

Dampproofing

Spread Footing

Strip Footing

A1010 210	Spread Footings	COST EACH		
		MAT.	INST.	TOTAL
7090	Spread footings, 3000 psi concrete, chute delivered			
7100	Load 25K, soil capacity 3 KSF, 3'-0" sq. x 12" deep	65.50	130	195.50
7150	Load 50K, soil capacity 3 KSF, 4'-6" sq. x 12" deep	140	224	364
7200	Load 50K, soil capacity 6 KSF, 3'-0" sq. x 12" deep	65.50	130	195.50
7250	Load 75K, soil capacity 3 KSF, 5'-6" sq. x 13" deep	223	315	538
7300	Load 75K, soil capacity 6 KSF, 4'-0" sq. x 12" deep	113	193	306
7350	Load 100K, soil capacity 3 KSF, 6'-0" sq. x 14" deep	283	375	658
7410	Load 100K, soil capacity 6 KSF, 4'-6" sq. x 15" deep	173	265	438
7450	Load 125K, soil capacity 3 KSF, 7'-0" sq. x 17" deep	450	545	995
7500	Load 125K, soil capacity 6 KSF, 5'-0" sq. x 16" deep	223	320	543
7550	Load 150K, soil capacity 3 KSF 7'-6" sq. x 18" deep	545	635	1,180
7610	Load 150K, soil capacity 6 KSF, 5'-6" sq. x 18" deep	298	400	698
7650	Load 200K, soil capacity 3 KSF, 8'-6" sq. x 20" deep	775	835	1,610
7700	Load 200K, soil capacity 6 KSF, 6'-0" sq. x 20" deep	390	495	885
7750	Load 300K, soil capacity 3 KSF, 10'-6" sq. x 25" deep	1,425	1,375	2,800
7810	Load 300K, soil capacity 6 KSF, 7'-6" sq. x 25" deep	745	825	1,570
7850	Load 400K, soil capacity 3 KSF, 12'-6" sq. x 28" deep	2,275	2,025	4,300
7900	Load 400K, soil capacity 6 KSF, 8'-6" sq. x 27" deep	1,025	1,075	2,100
8010	Load 500K, soil capacity 6 KSF, 9'-6" sq. x 30" deep	1,425	1,400	2,825
8100	Load 600K, soil capacity 6 KSF, 10'-6" sq. x 33" deep	1,925	1,775	3,700
8200	Load 700K, soil capacity 6 KSF, 11'-6" sq. x 36" deep	2,475	2,175	4,650
8300	Load 800K, soil capacity 6 KSF, 12'-0" sq. x 37" deep	2,775	2,400	5,175
8400	Load 900K, soil capacity 6 KSF, 13'-0" sq. x 39" deep	3,425	2,850	6,275
8500	Load 1000K, soil capacity 6 KSF, 13'-6" sq. x 41" deep	3,875	3,150	7,025

A1010 320	Foundation Dampproofing	COST PER L.F.		
		MAT.	INST.	TOTAL
1000	Foundation dampproofing, bituminous, 1 coat, 4' high	.96	5	5.96
1400	8' high	1.92	10.05	11.97
1800	12' high	2.88	15.55	18.43
2000	2 coats, 4' high	1.96	6.20	8.16
2400	8' high	3.92	12.35	16.27
2800	12' high	5.90	19	24.90
3000	Asphalt with fibers, 1/16" thick, 4' high	1.60	6.20	7.80
3400	8' high	3.20	12.35	15.55
3800	12' high	4.80	19	23.80
4000	1/8" thick, 4' high	2.84	7.40	10.24
4400	8' high	5.70	14.75	20.45
4800	12' high	8.50	22.50	31
5000	Asphalt coated board and mastic, 1/4" thick, 4' high	5.15	6.70	11.85
5400	8' high	10.30	13.40	23.70

A10 Foundations

A1010 Standard Foundations

A1010 320	Foundation Dampproofing	COST PER L.F.		
		MAT.	INST.	TOTAL
5800	12' high	15.50	20.50	36
6000	1/2" thick, 4' high	7.75	9.30	17.05
6400	8' high	15.50	18.65	34.15
6800	12' high	23	28.50	51.50
7000	Cementitious coating, on walls, 1/8" thick coating, 4' high	2.96	8.80	11.76
7400	8' high	5.90	17.60	23.50
7800	12' high	8.90	26.50	35.40
8000	Cementitious/metallic slurry, 4 coat, 1/2"thick, 2' high	.86	9.40	10.26
8400	4' high	1.72	18.80	20.52
8800	6' high	2.58	28	30.58

A1030 Slab on Grade

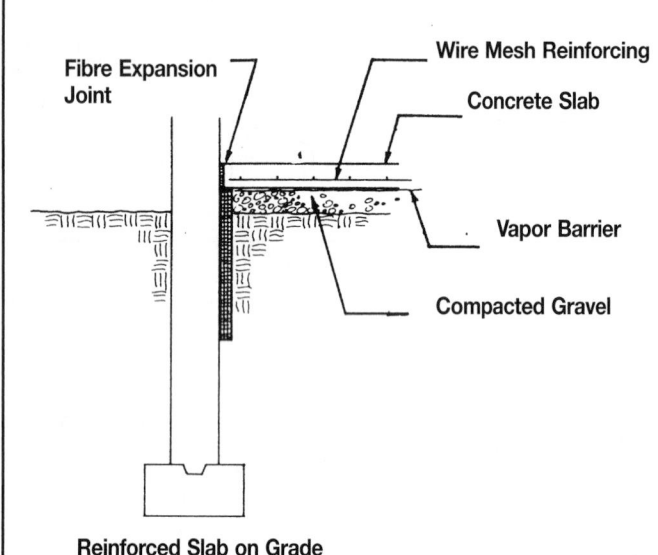

Fibre Expansion Joint

Wire Mesh Reinforcing

Concrete Slab

Vapor Barrier

Compacted Gravel

Reinforced Slab on Grade

A Slab on Grade system includes fine grading; 6″ of compacted gravel; vapor barrier; 3500 p.s.i. concrete; bituminous fiber expansion joint; all necessary edge forms 4 uses; steel trowel finish; and sprayed on membrane curing compound. Wire mesh reinforcing used in all reinforced slabs.

Non-industrial slabs are for foot traffic only with negligible abrasion. Light industrial slabs are for pneumatic wheels and light abrasion. Industrial slabs are for solid rubber wheels and moderate abrasion. Heavy industrial slabs are for steel wheels and severe abrasion.

A1030 120	Plain & Reinforced	COST PER S.F.		
		MAT.	INST.	TOTAL
2220	Slab on grade, 4″ thick, non industrial, non reinforced	2.36	2.89	5.25
2240	Reinforced	2.46	3.34	5.80
2260	Light industrial, non reinforced	2.87	3.55	6.42
2280	Reinforced	3.04	3.93	6.97
2300	Industrial, non reinforced	3.64	7.35	10.99
2320	Reinforced	3.81	7.75	11.56
3340	5″ thick, non industrial, non reinforced	2.78	2.97	5.75
3360	Reinforced	2.95	3.35	6.30
3380	Light industrial, non reinforced	3.29	3.63	6.92
3400	Reinforced	3.46	4.01	7.47
3420	Heavy industrial, non reinforced	4.63	8.80	13.43
3440	Reinforced	4.75	9.20	13.95
4460	6″ thick, non industrial, non reinforced	3.32	2.91	6.23
4480	Reinforced	3.58	3.43	7.01
4500	Light industrial, non reinforced	3.85	3.57	7.42
4520	Reinforced	4.18	3.97	8.15
4540	Heavy industrial, non reinforced	5.20	8.90	14.10
4560	Reinforced	5.45	9.40	14.85
5580	7″ thick, non industrial, non reinforced	3.74	3	6.74
5600	Reinforced	4.08	3.55	7.63
5620	Light industrial, non reinforced	4.28	3.66	7.94
5640	Reinforced	4.62	4.21	8.83
5660	Heavy industrial, non reinforced	5.65	8.75	14.40
5680	Reinforced	5.90	9.30	15.20
6700	8″ thick, non industrial, non reinforced	4.16	3.06	7.22
6720	Reinforced	4.44	3.52	7.96
6740	Light industrial, non reinforced	4.70	3.72	8.42
6760	Reinforced	4.98	4.18	9.16
6780	Heavy industrial, non reinforced	6.10	8.85	14.95
6800	Reinforced	6.45	9.35	15.80

A20 Basement Construction

A2010 Basement Excavation

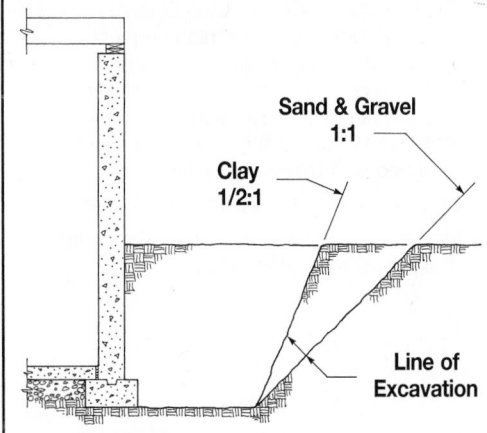

Sand & Gravel 1:1

Clay 1/2:1

Line of Excavation

In general, the following items are accounted for in the table below.

Costs:
1) Excavation for building or other structure to depth and extent indicated.
2) Backfill compacted in place.
3) Haul of excavated waste.
4) Replacement of unsuitable backfill material with bank run gravel.

A2010 110	Building Excavation & Backfill	COST PER S.F.		
		MAT.	INST.	TOTAL
2220	Excav & fill, 1000 S.F., 4' sand, gravel, or common earth, on site storage		1.06	1.06
2240	Off site storage		1.56	1.56
2260	Clay excavation, bank run gravel borrow for backfill	2.05	2.35	4.40
2280	8' deep, sand, gravel, or common earth, on site storage		6.30	6.30
2300	Off site storage		13.45	13.45
2320	Clay excavation, bank run gravel borrow for backfill	8.30	11.10	19.40
2340	16' deep, sand, gravel, or common earth, on site storage		17.15	17.15
2350	Off site storage		33.50	33.50
2360	Clay excavation, bank run gravel borrow for backfill	23	28.50	51.50
3380	4000 S.F., 4' deep, sand, gravel, or common earth, on site storage		.57	.57
3400	Off site storage		1.09	1.09
3420	Clay excavation, bank run gravel borrow for backfill	1.05	1.20	2.25
3440	8' deep, sand, gravel, or common earth, on site storage		4.40	4.40
3460	Off site storage		7.60	7.60
3480	Clay excavation, bank run gravel borrow for backfill	3.79	6.65	10.44
3500	16' deep, sand, gravel, or common earth, on site storage		10.50	10.50
3520	Off site storage		20.50	20.50
3540	Clay, excavation, bank run gravel borrow for backfill	10.15	15.75	25.90
4560	10,000 S.F., 4' deep, sand, gravel, or common earth, on site storage		.33	.33
4580	Off site storage		.65	.65
4600	Clay excavation, bank run gravel borrow for backfill	.64	.72	1.36
4620	8' deep, sand, gravel, or common earth, on site storage		3.80	3.80
4640	Off site storage		5.70	5.70
4660	Clay excavation, bank run gravel borrow for backfill	2.30	5.15	7.45
4680	16' deep, sand, gravel, or common earth, on site storage		8.55	8.55
4700	Off site storage		14.40	14.40
4720	Clay excavation, bank run gravel borrow for backfill	6.10	11.80	17.90
5740	30,000 S.F., 4' deep, sand, gravel, or common earth, on site storage		.18	.18
5760	Off site storage		.37	.37
5780	Clay excavation, bank run gravel borrow for backfill	.37	.44	.81
5860	16' deep, sand, gravel, or common earth, on site storage		7.30	7.30
5880	Off site storage		10.40	10.40
5900	Clay excavation, bank run gravel borrow for backfill	3.38	9.15	12.53
6910	100,000 S.F., 4' deep, sand, gravel, or common earth, on site storage		.11	.11
6940	8' deep, sand, gravel, or common earth, on site storage		3.18	3.18
6970	16' deep, sand, gravel, or common earth, on site storage		6.60	6.60
6980	Off site storage		8.25	8.25
6990	Clay excavation, bank run gravel borrow for backfill	1.82	7.60	9.42

309

A2020 Basement Walls

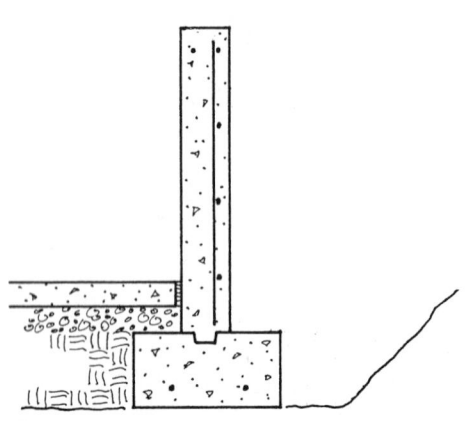

The Foundation Bearing Wall System includes: forms up to 16' high (four uses); 3,000 p.s.i. concrete placed and vibrated; and form removal with breaking form ties and patching walls. The wall systems list walls from 6" to 12" thick and are designed with minimum reinforcement.

Excavation and backfill are not included.

Please see the reference section for further design and cost information.

A2020 110		Walls, Cast in Place						
	WALL HEIGHT (FT.)	PLACING METHOD	CONCRETE (C.Y. per L.F.)	REINFORCING (LBS. per L.F.)	WALL THICKNESS (IN.)	COST PER L.F.		
						MAT.	INST.	TOTAL
5000	8'	direct chute	.148	6.6	6	36.50	112	148.50
5020			.199	9.6	8	44.50	119	163.50
5040			.250	12	10	53	117	170
5061			.296	14.39	12	60	120	180
6020	10'	direct chute	.248	12	8	56	144	200
6040			.307	14.99	10	65	147	212
6061			.370	17.99	12	75	150	225
7220	12'	pumped	.298	14.39	8	67	179	246
7240			.369	17.99	10	78.50	183	261.50
7260			.444	21.59	12	90	187	277
9220	16'	pumped	.397	19.19	8	89.50	238	327.50
9240			.492	23.99	10	104	243	347
9260			.593	28.79	12	120	250	370

B1010 Floor Construction

General: It is desirable for purposes of consistency and simplicity to maintain constant column sizes throughout building height. To do this, concrete strength may be varied (higher strength concrete at lower stories and lower strength concrete at upper stories), as well as varying the amount of reinforcing.

The table provides probably minimum column sizes with related costs and weight per lineal foot of story height.

B1010 203		C.I.P. Column, Square Tied						
	LOAD (KIPS)	STORY HEIGHT (FT.)	COLUMN SIZE (IN.)	COLUMN WEIGHT (P.L.F.)	CONCRETE STRENGTH (PSI)	COST PER V.L.F.		
						MAT.	INST.	TOTAL
0640	100	10	10	96	4000	11.35	52	63.35
0680		12	10	97	4000	11.55	52	63.55
0700		14	12	142	4000	15.15	63	78.15
0840	200	10	12	140	4000	16.50	65.50	82
0860		12	12	142	4000	16.80	65.50	82.30
0900		14	14	196	4000	19.75	73.50	93.25
0920	300	10	14	192	4000	21	75.50	96.50
0960		12	14	194	4000	21	76	97
0980		14	16	253	4000	25	84	109
1020	400	10	16	248	4000	26	86	112
1060		12	16	251	4000	26.50	87	113.50
1080		14	16	253	4000	27	87.50	114.50
1200	500	10	18	315	4000	31	96	127
1250		12	20	394	4000	37.50	109	146.50
1300		14	20	397	4000	38	110	148
1350	600	10	20	388	4000	39	111	150
1400		12	20	394	4000	39.50	112	151.50
1600		14	20	397	4000	40.50	114	154.50
3400	900	10	24	560	4000	53	139	192
3800		12	24	567	4000	54	141	195
4000		14	24	571	4000	55	142	197
7300	300	10	14	192	6000	21	74.50	95.50
7500		12	14	194	6000	21.50	75.50	97
7600		14	14	196	6000	22	76	98
8000	500	10	16	248	6000	27	86	113
8050		12	16	251	6000	27.50	87	114.50
8100		14	16	253	6000	28	87.50	115.50
8200	600	10	18	315	6000	31.50	96.50	128
8300		12	18	319	6000	32	97.50	129.50
8400		14	18	321	6000	32.50	98.50	131
8800	800	10	20	388	6000	38.50	109	147.50
8900		12	20	394	6000	39.50	110	149.50
9000		14	20	397	6000	40	111	151

B1010 Floor Construction

B1010 203	C.I.P. Column, Square Tied

	LOAD (KIPS)	STORY HEIGHT (FT.)	COLUMN SIZE (IN.)	COLUMN WEIGHT (P.L.F.)	CONCRETE STRENGTH (PSI)	COST PER V.L.F.		
						MAT.	INST.	TOTAL
9100	900	10	20	388	6000	41.50	114	155.50
9300		12	20	394	6000	42.50	115	157.50
9600		14	20	397	6000	43	116	159

B1010 204	C.I.P. Column, Square Tied-Minimum Reinforcing

	LOAD (KIPS)	STORY HEIGHT (FT.)	COLUMN SIZE (IN.)	COLUMN WEIGHT (P.L.F.)	CONCRETE STRENGTH (PSI)	COST PER V.L.F.		
						MAT.	INST.	TOTAL
9913	150	10-14	12	135	4000	13.60	56	69.60
9918	300	10-14	16	240	4000	22	73.50	95.50
9924	500	10-14	20	375	4000	33.50	103	136.50
9930	700	10-14	24	540	4000	46.50	128	174.50
9936	1000	10-14	28	740	4000	61	155	216
9942	1400	10-14	32	965	4000	75	174	249
9948	1800	10-14	36	1220	4000	93.50	203	296.50

B1010 Floor Construction

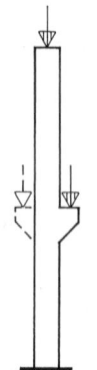

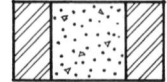

Concentric Load

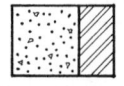

Eccentric Load

General: Data presented here is for plant produced members transported 50 miles to 100 miles to the site and erected.

Design and pricing assumptions:
Normal wt. concrete, f'c = 5 KSI

Main reinforcement, fy = 60 KSI
Ties, fy = 40 KSI

Minimum design eccentricity, 0.1t.

Concrete encased structural steel haunches are assumed where practical; otherwise galvanized rebar haunches are assumed.

Base plates are integral with columns.

Foundation anchor bolts, nuts and washers are included in price.

B1010 206		Tied, Concentric Loaded Precast Concrete Columns						
	LOAD (KIPS)	STORY HEIGHT (FT.)	COLUMN SIZE (IN.)	COLUMN WEIGHT (P.L.F.)	LOAD LEVELS	COST PER V.L.F.		
						MAT.	INST.	TOTAL
0560	100	10	12x12	164	2	223	14.55	237.55
0570		12	12x12	162	2	218	12.15	230.15
0580		14	12x12	161	2	218	11.90	229.90
0590	150	10	12x12	166	3	217	11.80	228.80
0600		12	12x12	169	3	217	10.60	227.60
0610		14	12x12	162	3	217	10.45	227.45
0620	200	10	12x12	168	4	217	12.60	229.60
0630		12	12x12	170	4	218	11.35	229.35
0640		14	12x12	220	4	217	11.25	228.25

B1010 207		Tied, Eccentric Loaded Precast Concrete Columns						
	LOAD (KIPS)	STORY HEIGHT (FT.)	COLUMN SIZE (IN.)	COLUMN WEIGHT (P.L.F.)	LOAD LEVELS	COST PER V.L.F.		
						MAT.	INST.	TOTAL
1130	100	10	12x12	161	2	213	14.55	227.55
1140		12	12x12	159	2	213	12.15	225.15
1150		14	12x12	159	2	213	11.90	224.90
1390	600	10	18x18	385	4	315	12.60	327.60
1400		12	18x18	380	4	315	11.35	326.35
1410		14	18x18	375	4	315	11.25	326.25
1480	800	10	20x20	490	4	286	12.60	298.60
1490		12	20x20	480	4	284	11.35	295.35
1500		14	20x20	475	4	285	11.25	296.25

B10 Superstructure

B1010 Floor Construction

 (A) Wide Flange

 (B) Pipe

 (C) Pipe, Concrete Filled

 (G) Rectangular Tube, Concrete Filled

B1010 208					Steel Columns			
	LOAD (KIPS)	UNSUPPORTED HEIGHT (FT.)	WEIGHT (P.L.F.)	SIZE (IN.)	TYPE	COST PER V.L.F.		
						MAT.	INST.	TOTAL
1000	25	10	13	4	A	23.50	11.25	34.75
1020			7.58	3	B	13.55	11.25	24.80
1040			15	3-1/2	C	16.60	11.25	27.85
1120			20	4x3	G	19.15	11.25	30.40
1200		16	16	5	A	26.50	8.40	34.90
1220			10.79	4	B	17.90	8.40	26.30
1240			36	5-1/2	C	25	8.40	33.40
1320			64	8x6	G	38.50	8.40	46.90
1600	50	10	16	5	A	28.50	11.25	39.75
1620			14.62	5	B	26	11.25	37.25
1640			24	4-1/2	C	19.80	11.25	31.05
1720			28	6x3	G	25.50	11.25	36.75
1800		16	24	8	A	40	8.40	48.40
1840			36	5-1/2	C	25	8.40	33.40
1920			64	8x6	G	38.50	8.40	46.90
2000		20	28	8	A	44	8.40	52.40
2040			49	6-5/8	C	31	8.40	39.40
2120			64	8x6	G	36.50	8.40	44.90
2200	75	10	20	6	A	36	11.25	47.25
2240			36	4-1/2	C	49.50	11.25	60.75
2320			35	6x4	G	28.50	11.25	39.75
2400		16	31	8	A	51.50	8.40	59.90
2440			49	6-5/8	C	32.50	8.40	40.90
2520			64	8x6	G	38.50	8.40	46.90
2600		20	31	8	A	48.50	8.40	56.90
2640			81	8-5/8	C	46.50	8.40	54.90
2720			64	8x6	G	36.50	8.40	44.90
2800	100	10	24	8	A	43	11.25	54.25
2840			35	4-1/2	C	49.50	11.25	60.75
2920			46	8x4	G	35	11.25	46.25

315

B1010 Floor Construction

B1010 208	Steel Columns

	LOAD (KIPS)	UNSUPPORTED HEIGHT (FT.)	WEIGHT (P.L.F.)	SIZE (IN.)	TYPE	COST PER V.L.F.		
						MAT.	INST.	TOTAL
3000	100	16	31	8	A	51.50	8.40	59.90
3040			56	6-5/8	C	48.50	8.40	56.90
3120			64	8x6	G	38.50	8.40	46.90
3200		20	40	8	A	63	8.40	71.40
3240			81	8-5/8	C	46.50	8.40	54.90
3320			70	8x6	G	52.50	8.40	60.90
3400	125	10	31	8	A	55.50	11.25	66.75
3440			81	8	C	53	11.25	64.25
3520			64	8x6	G	41.50	11.25	52.75
3600		16	40	8	A	66.50	8.40	74.90
3640			81	8	C	49	8.40	57.40
3720			64	8x6	G	38.50	8.40	46.90
3800		20	48	8	A	75.50	8.40	83.90
3840			81	8	C	46.50	8.40	54.90
3920			60	8x6	G	52.50	8.40	60.90
4000	150	10	35	8	A	62.50	11.25	73.75
4040			81	8-5/8	C	53	11.25	64.25
4120			64	8x6	G	41.50	11.25	52.75
4200		16	45	10	A	74.50	8.40	82.90
4240			81	8-5/8	C	49	8.40	57.40
4320			70	8x6	G	55.50	8.40	63.90
4400		20	49	10	A	77	8.40	85.40
4440			123	10-3/4	C	66.50	8.40	74.90
4520			86	10x6	G	52	8.40	60.40
4600	200	10	45	10	A	80.50	11.25	91.75
4640			81	8-5/8	C	53	11.25	64.25
4720			70	8x6	G	60	11.25	71.25
4800		16	49	10	A	81	8.40	89.40
4840			123	10-3/4	C	70	8.40	78.40
4920			85	10x6	G	64.50	8.40	72.90
5200	300	10	61	14	A	109	11.25	120.25
5240			169	12-3/4	C	93	11.25	104.25
5320			86	10x6	G	89	11.25	100.25
5400		16	72	12	A	119	8.40	127.40
5440			169	12-3/4	C	86.50	8.40	94.90
5600		20	79	12	A	124	8.40	132.40
5640			169	12-3/4	C	82	8.40	90.40
5800	400	10	79	12	A	141	11.25	152.25
5840			178	12-3/4	C	121	11.25	132.25
6000		16	87	12	A	144	8.40	152.40
6040			178	12-3/4	C	113	8.40	121.40
6400	500	10	99	14	A	177	11.25	188.25
6600		16	109	14	A	181	8.40	189.40
6800		20	120	12	A	188	8.40	196.40
7000	600	10	120	12	A	215	11.25	226.25
7200		16	132	14	A	219	8.40	227.40
7400		20	132	14	A	207	8.40	215.40
7600	700	10	136	12	A	244	11.25	255.25
7800		16	145	14	A	240	8.40	248.40
8000		20	145	14	A	227	8.40	235.40
8200	800	10	145	14	A	260	11.25	271.25
8300		16	159	14	A	264	8.40	272.40
8400		20	176	14	A	276	8.40	284.40

B1010 Floor Construction

B1010 208					Steel Columns				
	LOAD (KIPS)	UNSUPPORTED HEIGHT (FT.)	WEIGHT (P.L.F.)	SIZE (IN.)	TYPE	COST PER V.L.F.			
						MAT.	INST.	TOTAL	
8800	900	10	159	14	A	285	11.25	296.25	
8900		16	176	14	A	292	8.40	300.40	
9000		20	193	14	A	305	8.40	313.40	
9100	1000	10	176	14	A	315	11.25	326.25	
9200		16	193	14	A	320	8.40	328.40	
9300		20	211	14	A	330	8.40	338.40	

B1010 Floor Construction

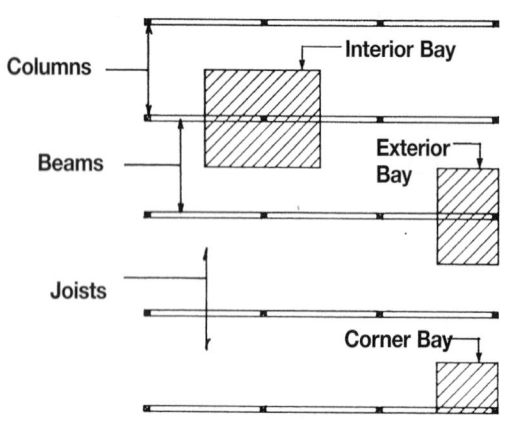

Description: Table below lists costs per S.F. of bay size for wood columns of various sizes and unsupported heights and the maximum allowable total load per S.F. per bay size.

B1010 210	Wood Columns							
	NOMINAL COLUMN SIZE (IN.)	BAY SIZE (FT.)	UNSUPPORTED HEIGHT (FT.)	MATERIAL (BF per M.S.F.)	TOTAL LOAD (P.S.F.)	COST PER S.F.		
						MAT.	INST.	TOTAL
1000	4 x 4	10 x 8	8	133	100	.20	.27	.47
1050			10	167	60	.25	.34	.59
1200		10 x 10	8	106	80	.16	.22	.38
1250			10	133	50	.20	.27	.47
1400		10 x 15	8	71	50	.10	.15	.25
1450			10	88	30	.13	.18	.31
1600		15 x 15	8	47	30	.07	.10	.17
1650			10	59	15	.09	.12	.21
2000	6 x 6	10 x 15	8	160	230	.24	.30	.54
2050			10	200	210	.31	.38	.69
2200		15 x 15	8	107	150	.16	.20	.36
2250			10	133	140	.20	.25	.45
2400		15 x 20	8	80	110	.12	.15	.27
2450			10	100	100	.15	.19	.34
2500			12	120	70	.18	.23	.41
2600		20 x 20	8	60	80	.09	.11	.20
2650			10	75	70	.11	.14	.25
2800		20 x 25	8	48	60	.07	.09	.16
2850			10	60	50	.09	.11	.20
3400	8 x 8	20 x 20	8	107	160	.18	.19	.37
3450			10	133	160	.22	.24	.46
3600		20 x 25	8	85	130	.13	.17	.30
3650			10	107	130	.16	.22	.38
3800		25 x 25	8	68	100	.12	.11	.23
3850			10	85	100	.15	.14	.29
4200	10 x 10	20 x 25	8	133	210	.24	.22	.46
4250			10	167	210	.30	.28	.58
4400		25 x 25	8	107	160	.19	.18	.37
4450			10	133	160	.24	.22	.46
4700	12 x 12	20 x 25	8	192	310	.33	.30	.63
4750			10	240	310	.41	.37	.78
4900		25 x 25	8	154	240	.26	.24	.50
4950			10	192	240	.33	.30	.63

B1010 Floor Construction

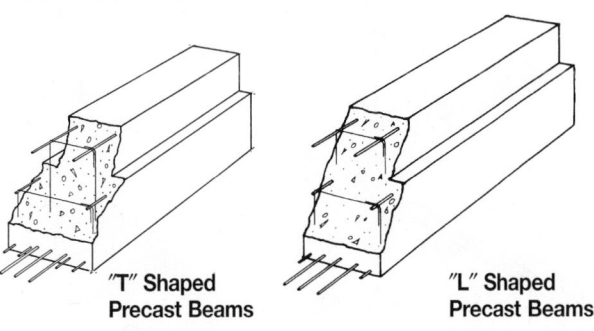

"T" Shaped
Precast Beams

"L" Shaped
Precast Beams

B1010 214		**"T" Shaped Precast Beams**						
	SPAN (FT.)	SUPERIMPOSED LOAD (K.L.F.)	SIZE W X D (IN.)	BEAM WEIGHT (P.L.F.)	TOTAL LOAD (K.L.F.)	COST PER L.F.		
						MAT.	INST.	TOTAL
2300	15	2.8	12x16	260	3.06	240	21	261
2500		8.37	12x28	515	8.89	285	22.50	307.50
8900	45	3.34	12x60	1165	4.51	460	13.50	473.50
9900		6.5	24x60	1915	8.42	495	18.95	513.95

B1010 215		**"L" Shaped Precast Beams**						
	SPAN (FT.)	SUPERIMPOSED LOAD (K.L.F.)	SIZE W X D (IN.)	BEAM WEIGHT (P.L.F.)	TOTAL LOAD (K.L.F.)	COST PER L.F.		
						MAT.	INST.	TOTAL
2250	15	2.58	12x16	230	2.81	226	21	247
2400		5.92	12x24	370	6.29	248	21	269
4000	25	2.64	12x28	435	3.08	249	13.60	262.60
4450		6.44	18x36	790	7.23	310	16.25	326.25
5300	30	2.80	12x36	565	3.37	278	12.30	290.30
6400		8.66	24x44	1245	9.90	370	17.70	387.70

B10 Superstructure

B1010 Floor Construction

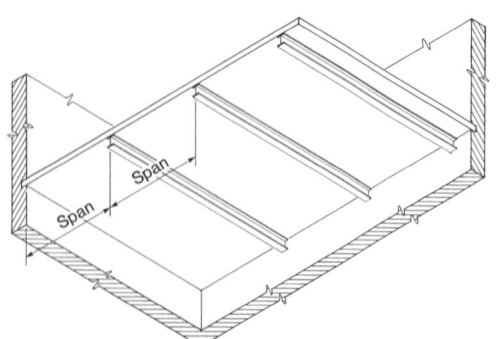

Cast in Place Floor Slab, One Way
General: Solid concrete slabs of uniform depth reinforced for flexure in one direction and for temperature and shrinkage in the other direction.

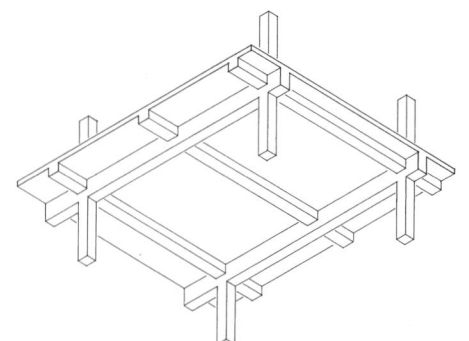

Cast in Place Beam & Slab, One Way
General: Solid concrete one way slab cast monolithically with reinforced concrete beams and girders.

B1010 217	Cast in Place Slabs, One Way							
	SLAB DESIGN & SPAN (FT.)	SUPERIMPOSED LOAD (P.S.F.)	THICKNESS (IN.)	TOTAL LOAD (P.S.F.)		COST PER S.F.		
						MAT.	INST.	TOTAL
2500	Single 8	40	4	90		4.39	8.85	13.24
2600		75	4	125		4.43	9.65	14.08
2700		125	4-1/2	181		4.64	9.70	14.34
2800		200	5	262		5.15	10.05	15.20
3000	Single 10	40	4	90		4.57	9.55	14.12
3100		75	4	125		4.57	9.55	14.12
3200		125	5	188		5.10	9.80	14.90
3300		200	7-1/2	293		6.80	10.95	17.75
3500	Single 15	40	5-1/2	90		5.50	9.70	15.20
3600		75	6-1/2	156		6.10	10.20	16.30
3700		125	7-1/2	219		6.65	10.40	17.05
3800		200	8-1/2	306		7.25	10.70	17.95
4000	Single 20	40	7-1/2	115		6.65	10.15	16.80
4100		75	9	200		7.30	10.20	17.50
4200		125	10	250		8	10.55	18.55
4300		200	10	324		8.55	11	19.55

B1010 219	Cast in Place Beam & Slab, One Way							
	BAY SIZE (FT.)	SUPERIMPOSED LOAD (P.S.F.)	MINIMUM COL. SIZE (IN.)	SLAB THICKNESS (IN.)	TOTAL LOAD (P.S.F.)	COST PER S.F.		
						MAT.	INST.	TOTAL
5000	20x25	40	12	5-1/2	121	5.60	12.15	17.75
5200		125	16	5-1/2	215	6.60	14	20.60
5500	25x25	40	12	6	129	5.95	11.95	17.90
5700		125	18	6	227	7.50	14.70	22.20
7500	30x35	40	16	8	158	7.45	13.50	20.95
7600		75	18	8	196	7.95	13.90	21.85
7700		125	22	8	254	8.85	15.50	24.35
7800		200	26	8	332	9.60	16	25.60
8000	35x35	40	16	9	169	8.35	13.95	22.30
8200		75	20	9	213	9	15.20	24.20
8400		125	24	9	272	9.85	15.70	25.55
8600		200	26	9	355	10.80	16.80	27.60

320

B1010 Floor Construction

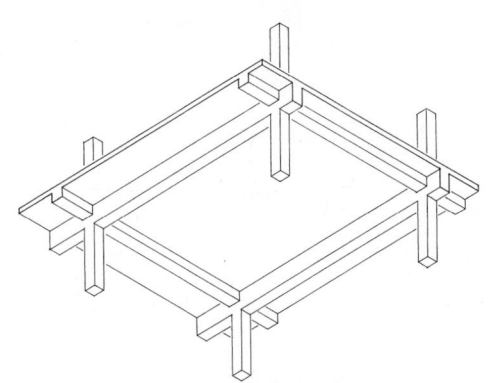

General: Solid concrete two way slab cast monolithically with reinforced concrete support beams and girders.

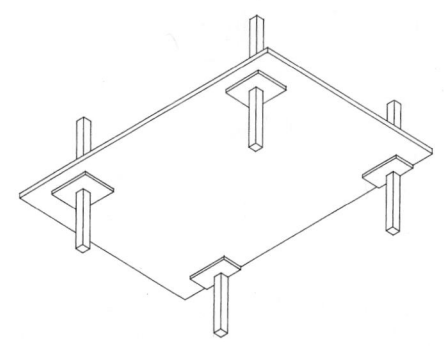

General: Flat Slab: Solid uniform depth concrete two way slabs with drop panels at columns and no column capitals.

B1010 220 — Cast in Place Beam & Slab, Two Way

	BAY SIZE (FT.)	SUPERIMPOSED LOAD (P.S.F.)	MINIMUM COL. SIZE (IN.)	SLAB THICKNESS (IN.)	TOTAL LOAD (P.S.F.)	COST PER S.F.		
						MAT.	INST.	TOTAL
4000	20 x 25	40	12	7	141	6.55	12.25	18.80
4300		75	14	7	181	7.45	13.40	20.85
4500		125	16	7	236	7.60	13.80	21.40
5100	25 x 25	40	12	7-1/2	149	6.85	11.45	18.30
5200		75	16	7-1/2	185	7.40	13.30	20.70
5300		125	18	7-1/2	250	8.10	14.40	22.50
7600	30 x 35	40	16	10	188	9.10	14	23.10
7700		75	18	10	225	9.65	14.50	24.15
8000		125	22	10	282	10.65	15.65	26.30
8500	35 x 35	40	16	10-1/2	193	9.75	14.35	24.10
8600		75	20	10-1/2	233	10.20	14.85	25.05
9000		125	24	10-1/2	287	11.35	15.90	27.25

B1010 222 — Cast in Place Flat Slab with Drop Panels

	BAY SIZE (FT.)	SUPERIMPOSED LOAD (P.S.F.)	MINIMUM COL. SIZE (IN.)	SLAB & DROP (IN.)	TOTAL LOAD (P.S.F.)	COST PER S.F.		
						MAT.	INST.	TOTAL
1960	20 x 20	40	12	7 - 3	132	5.85	9.75	15.60
1980		75	16	7 - 4	168	6.15	10	16.15
2000		125	18	7 - 6	221	6.90	10.40	17.30
3200	25 x 25	40	12	8-1/2 - 5-1/2	154	6.90	10.30	17.20
4000		125	20	8-1/2 - 8-1/2	243	7.75	10.95	18.70
4400		200	24	9 - 8-1/2	329	8.15	11.20	19.35
5000	25 x 30	40	14	9-1/2 - 7	168	7.55	10.65	18.20
5200		75	18	9-1/2 - 7	203	8	11.05	19.05
5600		125	22	9-1/2 - 8	256	8.35	11.30	19.65
6400	30 x 30	40	14	10-1/2 - 7-1/2	182	8.15	10.90	19.05
6600		75	18	10-1/2 - 7-1/2	217	8.65	11.30	19.95
6800		125	22	10-1/2 - 9	269	9.05	11.60	20.65
7400	30 x 35	40	16	11-1/2 - 9	196	8.90	11.35	20.25
7900		75	20	11-1/2 - 9	231	9.45	11.80	21.25
8000		125	24	11-1/2 - 11	284	9.85	12.05	21.90
9000	35 x 35	40	16	12 - 9	202	9.15	11.45	20.60
9400		75	20	12 - 11	240	9.80	11.95	21.75
9600		125	24	12 - 11	290	10.10	12.15	22.25

B10 Superstructure

B1010 Floor Construction

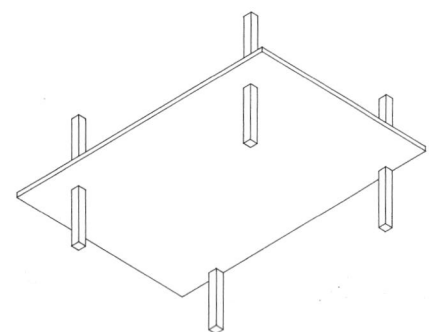

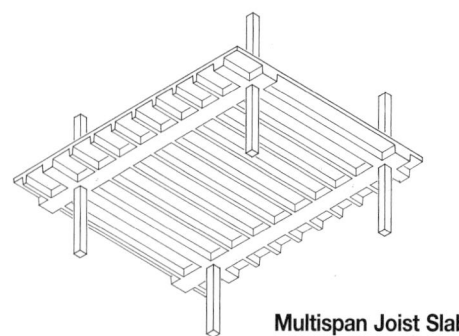

Multispan Joist Slab

General: Flat Plates: Solid uniform depth concrete two way slab without drops or interior beams. Primary design limit is shear at columns.

General: Combination of thin concrete slab and monolithic ribs at uniform spacing to reduce dead weight and increase rigidity.

B1010 223			Cast in Place Flat Plate					
	BAY SIZE (FT.)	SUPERIMPOSED LOAD (P.S.F.)	MINIMUM COL. SIZE (IN.)	SLAB THICKNESS (IN.)	TOTAL LOAD (P.S.F.)	COST PER S.F.		
						MAT.	INST.	TOTAL
3000	15 x 20	40	14	7	127	5.50	9.40	14.90
3400		75	16	7-1/2	169	5.85	9.60	15.45
3600		125	22	8-1/2	231	6.45	9.90	16.35
3800		175	24	8-1/2	281	6.50	9.90	16.40
4200	20 x 20	40	16	7	127	5.50	9.40	14.90
4400		75	20	7-1/2	175	5.90	9.60	15.50
4600		125	24	8-1/2	231	6.50	9.85	16.35
5000		175	24	8-1/2	281	6.55	9.90	16.45
5600	20 x 25	40	18	8-1/2	146	6.45	9.90	16.35
6000		75	20	9	188	6.65	9.95	16.60
6400		125	26	9-1/2	244	7.20	10.20	17.40
6600		175	30	10	300	7.50	10.35	17.85
7000	25 x 25	40	20	9	152	6.65	9.95	16.60
7400		75	24	9-1/2	194	7.10	10.15	17.25
7600		125	30	10	250	7.50	10.35	17.85

B1010 226			Cast in Place Multispan Joist Slab					
	BAY SIZE (FT.)	SUPERIMPOSED LOAD (P.S.F.)	MINIMUM COL. SIZE (IN.)	RIB DEPTH (IN.)	TOTAL LOAD (P.S.F.)	COST PER S.F.		
						MAT.	INST.	TOTAL
2000	15 x 15	40	12	8	115	8.60	11.65	20.25
2100		75	12	8	150	8.65	11.70	20.35
2200		125	12	8	200	8.75	11.80	20.55
2300		200	14	8	275	8.95	12.25	21.20
2600	15 x 20	40	12	8	115	8.75	11.65	20.40
2800		75	12	8	150	8.90	12.25	21.15
3000		125	14	8	200	9.10	12.50	21.60
3300		200	16	8	275	9.45	12.65	22.10
3600	20 x 20	40	12	10	120	8.90	11.40	20.30
3900		75	14	10	155	9.20	12.05	21.25
4000		125	16	10	205	9.25	12.30	21.55
4100		200	18	10	280	9.55	12.80	22.35
6200	30 x 30	40	14	14	131	9.75	11.95	21.70
6400		75	18	14	166	9.95	12.35	22.30
6600		125	20	14	216	10.45	12.95	23.40
6700		200	24	16	297	11.05	13.40	24.45

B10 Superstructure

B1010 Floor Construction

Waffle Slab

B1010 227		Cast in Place Waffle Slab						
	BAY SIZE (FT.)	SUPERIMPOSED LOAD (P.S.F.)	MINIMUM COL. SIZE (IN.)	RIB DEPTH (IN.)	TOTAL LOAD (P.S.F.)	COST PER S.F.		
						MAT.	INST.	TOTAL
3900	20 x 20	40	12	8	144	10.80	11.45	22.25
4000		75	12	8	179	10.90	11.60	22.50
4100		125	16	8	229	11.10	11.70	22.80
4200		200	18	8	304	11.55	12.10	23.65
4400	20 x 25	40	12	8	146	10.95	11.50	22.45
4500		75	14	8	181	11.15	11.65	22.80
4600		125	16	8	231	11.35	11.80	23.15
4700		200	18	8	306	11.75	12.15	23.90
4900	25 x 25	40	12	10	150	11.20	11.60	22.80
5000		75	16	10	185	11.45	11.85	23.30
5300		125	18	10	235	11.70	12.05	23.75
5500		200	20	10	310	12	12.25	24.25
5700	25 x 30	40	14	10	154	11.40	11.65	23.05
5800		75	16	10	189	11.65	11.90	23.55
5900		125	18	10	239	11.90	12.10	24
6000		200	20	12	329	12.90	12.65	25.55
6400	30 x 30	40	14	12	169	12.05	11.95	24
6500		75	18	12	204	12.25	12.15	24.40
6600		125	20	12	254	12.40	12.25	24.65
6700		200	24	12	329	13.30	12.95	26.25
6900	30 x 35	40	16	12	169	12.25	12.05	24.30
7000		75	18	12	204	12.25	12.05	24.30
7100		125	22	12	254	12.75	12.45	25.20
7200		200	26	14	334	13.90	13.20	27.10
7400	35 x 35	40	16	14	174	12.90	12.35	25.25
7500		75	20	14	209	13.10	12.55	25.65
7600		125	24	14	259	13.35	12.70	26.05
7700		200	26	16	346	14.10	13.35	27.45
8000	35 x 40	40	18	14	176	13.15	12.55	25.70
8300		75	22	14	211	13.50	12.85	26.35
8500		125	26	16	271	14.10	13.15	27.25
8750		200	30	20	372	15.25	13.80	29.05
9200	40 x 40	40	18	14	176	13.50	12.85	26.35
9400		75	24	14	211	13.90	13.20	27.10
9500		125	26	16	271	14.30	13.25	27.55
9700	40 x 45	40	20	16	186	14	13	27
9800		75	24	16	221	14.40	13.35	27.75
9900		125	28	16	271	14.60	13.55	28.15

B1010 Floor Construction

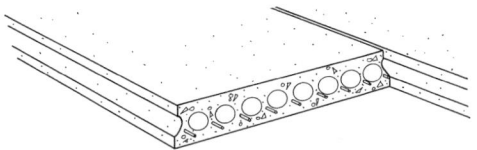

Precast Plank with No Topping

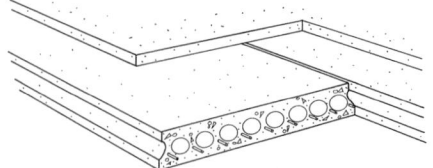

Precast Plank with 2″ Concrete Topping

B1010 229		Precast Plank with No Topping						
	SPAN (FT.)	SUPERIMPOSED LOAD (P.S.F.)	TOTAL DEPTH (IN.)	DEAD LOAD (P.S.F.)	TOTAL LOAD (P.S.F.)	COST PER S.F.		
						MAT.	INST.	TOTAL
0720	10	40	4	50	90	8.80	3.54	12.34
0750		75	6	50	125	9.85	3.04	12.89
0770		100	6	50	150	9.85	3.04	12.89
0800	15	40	6	50	90	9.85	3.04	12.89
0820		75	6	50	125	9.85	3.04	12.89
0850		100	6	50	150	9.85	3.04	12.89
0950	25	40	6	50	90	9.85	3.04	12.89
0970		75	8	55	130	11	2.66	13.66
1000		100	8	55	155	11	2.66	13.66
1200	30	40	8	55	95	11	2.66	13.66
1300		75	8	55	130	11	2.66	13.66
1400		100	10	70	170	11.40	2.37	13.77
1500	40	40	10	70	110	11.40	2.37	13.77
1600		75	12	70	145	11.80	2.12	13.92
1700	45	40	12	70	110	11.80	2.12	13.92

B1010 230		Precast Plank with 2″ Concrete Topping						
	SPAN (FT.)	SUPERIMPOSED LOAD (P.S.F.)	TOTAL DEPTH (IN.)	DEAD LOAD (P.S.F.)	TOTAL LOAD (P.S.F.)	COST PER S.F.		
						MAT.	INST.	TOTAL
2000	10	40	6	75	115	9.95	6.10	16.05
2100		75	8	75	150	11	5.55	16.55
2200		100	8	75	175	11	5.55	16.55
2500	15	40	8	75	115	11	5.55	16.55
2600		75	8	75	150	11	5.55	16.55
2700		100	8	75	175	11	5.55	16.55
3100	25	40	8	75	115	11	5.55	16.55
3200		75	8	75	150	11	5.55	16.55
3300		100	10	80	180	12.15	5.20	17.35
3400	30	40	10	80	120	12.15	5.20	17.35
3500		75	10	80	155	12.15	5.20	17.35
3600		100	10	80	180	12.15	5.20	17.35
4000	40	40	12	95	135	12.55	4.89	17.44
4500		75	14	95	170	12.95	4.64	17.59
5000	45	40	14	95	135	12.95	4.64	17.59

B1010 Floor Construction

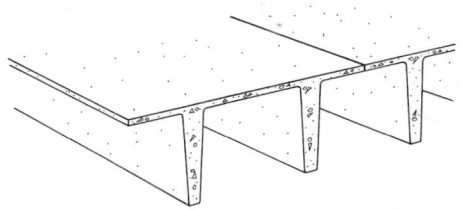

Precast Double "T" Beams with No Topping

Most widely used for moderate span floors and moderate and long span roofs. At shorter spans, they tend to be competitive with hollow core slabs. They are also used as wall panels.

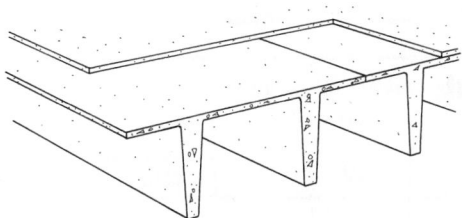

Precast Double "T" Beams with 2″ Topping

B1010 234	Precast Double "T" Beams with No Topping							
	SPAN (FT.)	SUPERIMPOSED LOAD (P.S.F.)	DBL. "T" SIZE D (IN.) W (FT.)	CONCRETE "T" TYPE	TOTAL LOAD (P.S.F.)	COST PER S.F.		
						MAT.	INST.	TOTAL
4300	50	30	20x8	Lt. Wt.	66	12.65	1.62	14.27
4400		40	20x8	Lt. Wt.	76	12.65	1.48	14.13
4500		50	20x8	Lt. Wt.	86	12.75	1.78	14.53
4600		75	20x8	Lt. Wt.	111	12.85	2.04	14.89
5600	70	30	32x10	Lt. Wt.	78	12.65	1.12	13.77
5750		40	32x10	Lt. Wt.	88	12.75	1.36	14.11
5900		50	32x10	Lt. Wt.	98	12.80	1.56	14.36
6000		75	32x10	Lt. Wt.	123	12.90	1.89	14.79
6100		100	32x10	Lt. Wt.	148	13.10	2.57	15.67
6200	80	30	32x10	Lt. Wt.	78	12.80	1.56	14.36
6300		40	32x10	Lt. Wt.	88	12.90	1.89	14.79
6400		50	32x10	Lt. Wt.	98	13	2.23	15.23

B1010 235	Precast Double "T" Beams With 2″ Topping							
	SPAN (FT.)	SUPERIMPOSED LOAD (P.S.F.)	DBL. "T" SIZE D (IN.) W (FT.)	CONCRETE "T" TYPE	TOTAL LOAD (P.S.F.)	COST PER S.F.		
						MAT.	INST.	TOTAL
7100	40	30	18x8	Reg. Wt.	120	10.90	3.36	14.26
7200		40	20x8	Reg. Wt.	130	12.80	3.21	16.01
7300		50	20x8	Reg. Wt.	140	12.85	3.50	16.35
7400		75	20x8	Reg. Wt.	165	12.90	3.64	16.54
7500		100	20x8	Reg. Wt.	190	13.05	4.06	17.11
7550	50	30	24x8	Reg. Wt.	120	12.90	3.37	16.27
7600		40	24x8	Reg. Wt.	130	12.95	3.48	16.43
7750		50	24x8	Reg. Wt.	140	12.95	3.51	16.46
7800		75	24x8	Reg. Wt.	165	13.10	3.93	17.03
7900		100	32x10	Reg. Wt.	189	13	3.23	16.23

325

For customer support on your Square Foot Costs with RSMeans data, call 800.448.8182.

B10 Superstructure

B1010 Floor Construction

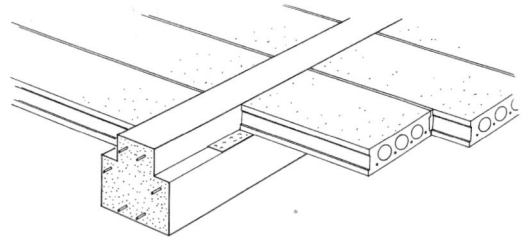

Precast Beam and Plank with No Topping

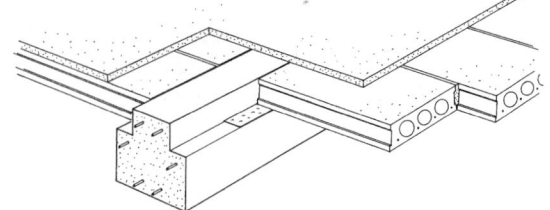

Precast Beam and Plank with 2″ Topping

B1010 236		Precast Beam & Plank with No Topping						
	BAY SIZE (FT.)	SUPERIMPOSED LOAD (P.S.F.)	PLANK THICKNESS (IN.)	TOTAL DEPTH (IN.)	TOTAL LOAD (P.S.F.)	COST PER S.F.		
						MAT.	INST.	TOTAL
6200	25x25	40	8	28	118	28.50	3.54	32.04
6400		75	8	36	158	28.50	3.55	32.05
6500		100	8	36	183	28.50	3.55	32.05
7000	25x30	40	8	36	110	26	3.41	29.41
7200		75	10	36	159	27	3.14	30.14
7400		100	12	36	188	27	2.85	29.85
7600	30x30	40	8	36	121	26.50	3.32	29.82
8000		75	10	44	140	26.50	3.33	29.83
8250		100	12	52	206	29	2.80	31.80
8500	30x35	40	12	44	135	25.50	2.71	28.21
8750		75	12	52	176	26.50	2.70	29.20
9000	35x35	40	12	52	141	26.50	2.64	29.14
9250		75	12	60	181	27.50	2.64	30.14
9500	35x40	40	12	52	137	26.50	2.64	29.14

B1010 238		Precast Beam & Plank with 2″ Topping						
	BAY SIZE (FT.)	SUPERIMPOSED LOAD (P.S.F.)	PLANK THICKNESS (IN.)	TOTAL DEPTH (IN.)	TOTAL LOAD (P.S.F.)	COST PER S.F.		
						MAT.	INST.	TOTAL
4300	20x20	40	6	22	135	30.50	6.35	36.85
4400		75	6	24	173	31.50	6.40	37.90
4500		100	6	28	200	31.50	6.40	37.90
4600	20x25	40	6	26	134	28.50	6.25	34.75
5000		75	8	30	177	29	5.75	34.75
5200		100	8	30	202	29	5.75	34.75
5400	25x25	40	6	38	143	29	5.95	34.95
5600		75	8	38	183	23	5.65	28.65
6000		100	8	46	216	23.50	5.20	28.70
6200	25x30	40	8	38	144	27	5.40	32.40
6400		75	10	46	200	27.50	5.10	32.60
6600		100	10	46	225	28	5.10	33.10
7000	30x30	40	8	46	150	28	5.30	33.30
7200		75	10	54	181	29.50	5.30	34.80
7600		100	10	54	231	29.50	5.30	34.80
7800	30x35	40	10	54	166	27.50	4.91	32.41
8000		75	12	54	200	27.50	4.64	32.14

B1010   Floor Construction

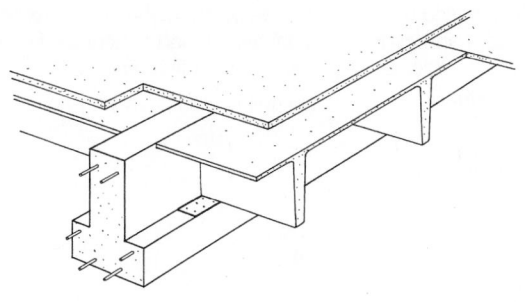

With Topping

General: Beams and double tees priced here are for plant produced prestressed members transported to the site and erected.

The 2″ structural topping is applied after the beams and double tees are in place and is reinforced with W.W.F.

B1010 239		Precast Double "T" & 2″ Topping on Precast Beams					
	BAY SIZE (FT.)	SUPERIMPOSED LOAD (P.S.F.)	DEPTH (IN.)		TOTAL LOAD (P.S.F.)	COST PER S.F.		
						MAT.	INST.	TOTAL
3000	25x30	40	38		130	28	4.50	32.50
3100		75	38		168	28	4.50	32.50
3300		100	46		196	28	4.48	32.48
3600	30x30	40	46		150	28.50	4.39	32.89
3750		75	46		174	28.50	4.39	32.89
4000		100	54		203	30	4.37	34.37
4100	30x40	40	46		136	25	3.96	28.96
4300		75	54		173	26	3.95	29.95
4400		100	62		204	27	3.96	30.96
4600	30x50	40	54		138	23.50	3.67	27.17
4800		75	54		181	23.50	3.71	27.21
5000		100	54		219	23.50	3.38	26.88
5200	30x60	40	62		151	23	3.34	26.34
5400		75	62		192	22.50	3.39	25.89
5600		100	62		215	22.50	3.39	25.89
5800	35x40	40	54		139	26.50	3.89	30.39
6000		75	62		179	26.50	3.72	30.22
6250		100	62		212	26.50	3.87	30.37
6500	35x50	40	62		142	24.50	3.63	28.13
6750		75	62		186	24.50	3.76	28.26
7300		100	62		231	24.50	3.42	27.92
7600	35x60	40	54		154	23.50	3.19	26.69
7750		75	54		179	23.50	3.21	26.71
8000		100	62		224	24.50	3.24	27.74
8250	40x40	40	62		145	28.50	3.89	32.39
8400		75	62		187	27.50	3.91	31.41
8750		100	62		223	28.50	3.99	32.49
9000	40x50	40	62		151	25.50	3.67	29.17
9300		75	62		193	24.50	2.71	27.21
9800	40x60	40	62		164	28	2.59	30.59

B10 Superstructure

B1010 Floor Construction

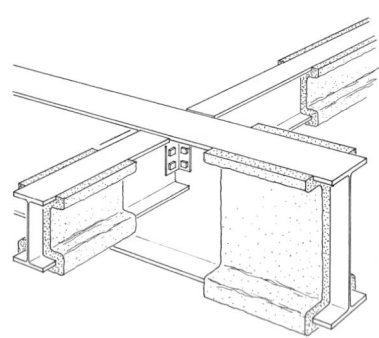

General: The following table is based upon structural wide flange (WF) beam and girder framing. Non-composite action is assumed between WF framing and decking. Deck costs not included.

The deck spans the short direction. The steel beams and girders are fireproofed with sprayed fiber fireproofing.

No columns included in price.

B1010 241	W Shape Beams & Girders							
	BAY SIZE (FT.) BEAM X GIRD	SUPERIMPOSED LOAD (P.S.F.)	STEEL FRAMING DEPTH (IN.)	FIREPROOFING (S.F. PER S.F.)	TOTAL LOAD (P.S.F.)	COST PER S.F.		
						MAT.	INST.	TOTAL
1350	15x20	40	12	.535	50	5.15	2.03	7.18
1400		40	16	.65	90	6.75	2.61	9.36
1450		75	18	.694	125	8.90	3.28	12.18
1500		125	24	.796	175	12.25	4.46	16.71
2000	20x20	40	12	.55	50	5.80	2.23	8.03
2050		40	14	.579	90	7.90	2.88	10.78
2100		75	16	.672	125	9.50	3.44	12.94
2150		125	16	.714	175	11.30	4.09	15.39
2900	20x25	40	16	.53	50	6.35	2.39	8.74
2950		40	18	.621	96	10.05	3.57	13.62
3000		75	18	.651	131	11.55	4.05	15.60
3050		125	24	.77	200	15.30	5.35	20.65
5200	25x25	40	18	.486	50	6.95	2.51	9.46
5300		40	18	.592	96	10.35	3.62	13.97
5400		75	21	.668	131	12.50	4.33	16.83
5450		125	24	.738	191	16.45	5.70	22.15
6300	25x30	40	21	.568	50	7.90	2.87	10.77
6350		40	21	.694	90	10.70	3.82	14.52
6400		75	24	.776	125	13.75	4.81	18.56
6450		125	30	.904	175	16.55	5.85	22.40
7700	30x30	40	21	.52	50	8.45	3.01	11.46
7750		40	24	.629	103	13.05	4.48	17.53
7800		75	30	.715	138	15.55	5.30	20.85
7850		125	36	.822	206	20.50	7	27.50
8250	30x35	40	21	.508	50	9.65	3.35	13
8300		40	24	.651	109	14.30	4.86	19.16
8350		75	33	.732	150	17.35	5.85	23.20
8400		125	36	.802	225	21.50	7.35	28.85
9550	35x35	40	27	.560	50	10	3.50	13.50
9600		40	36	.706	109	18.25	6.10	24.35
9650		75	36	.750	150	18.90	6.35	25.25
9820		125	36	.797	225	25.50	8.40	33.90

328

For customer support on your Square Foot Costs with RSMeans data, call 800.448.8182.

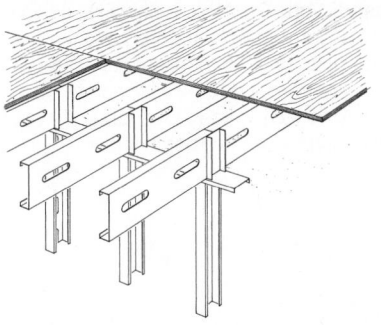

Description: Table below lists costs for light gage CEE or PUNCHED DOUBLE joists to suit the span and loading with the minimum thickness subfloor required by the joist spacing.

B1010 244	Light Gauge Steel Floor Systems							
	SPAN (FT.)	SUPERIMPOSED LOAD (P.S.F.)	FRAMING DEPTH (IN.)	FRAMING SPAC. (IN.)	TOTAL LOAD (P.S.F.)	COST PER S.F.		
						MAT.	INST.	TOTAL
1500	15	40	8	16	54	2.86	2.30	5.16
1550			8	24	54	3.02	2.36	5.38
1600		65	10	16	80	3.73	2.76	6.49
1650			10	24	80	3.71	2.70	6.41
1700		75	10	16	90	3.73	2.76	6.49
1750			10	24	90	3.71	2.70	6.41
1800		100	10	16	116	4.86	3.47	8.33
1850			10	24	116	3.71	2.70	6.41
1900		125	10	16	141	4.86	3.47	8.33
1950			10	24	141	3.71	2.70	6.41
2500	20	40	8	16	55	3.29	2.59	5.88
2550			8	24	55	3.02	2.36	5.38
2600		65	8	16	80	3.81	2.95	6.76
2650			10	24	80	3.72	2.72	6.44
2700		75	10	16	90	3.52	2.62	6.14
2750			12	24	90	4.14	2.41	6.55
2800		100	10	16	115	3.52	2.62	6.14
2850			10	24	116	4.77	3.08	7.85
2900		125	12	16	142	5.50	3.01	8.51
2950			12	24	141	5.15	3.27	8.42
3500	25	40	10	16	55	3.73	2.76	6.49
3550			10	24	55	3.71	2.70	6.41
3600		65	10	16	81	4.86	3.47	8.33
3650			12	24	81	4.59	2.60	7.19
3700		75	12	16	92	5.50	3.01	8.51
3750			12	24	91	5.15	3.27	8.42
3800		100	12	16	117	6.20	3.31	9.51
3850		125	12	16	143	7.05	4.33	11.38
4500	30	40	12	16	57	5.50	3.01	8.51
4550			12	24	56	4.57	2.60	7.17
4600		65	12	16	82	6.15	3.30	9.45
4650		75	12	16	92	6.15	3.30	9.45

329

B10 Superstructure

B1010 Floor Construction

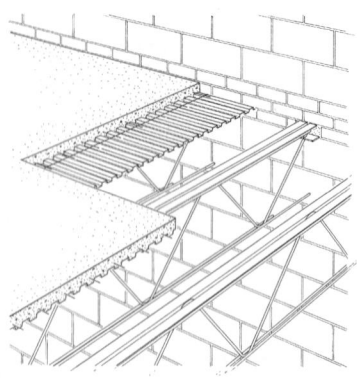

Description: The table below lists costs per S.F. for a floor system on bearing walls using open web steel joists, galvanized steel slab form and 2-1/2″ concrete slab reinforced with welded wire fabric. Costs of the bearing walls are not included.

B1010 246		Deck & Joists on Bearing Walls						
	SPAN (FT.)	SUPERIMPOSED LOAD (P.S.F.)	JOIST SPACING FT. - IN.	DEPTH (IN.)	TOTAL LOAD (P.S.F.)	COST PER S.F.		
						MAT.	INST.	TOTAL
1050	20	40	2-0	14-1/2	83	6.75	4	10.75
1070		65	2-0	16-1/2	109	7.25	4.24	11.49
1100		75	2-0	16-1/2	119	7.25	4.24	11.49
1120		100	2-0	18-1/2	145	7.30	4.25	11.55
1150		125	1-9	18-1/2	170	7.95	4.51	12.46
1170	25	40	2-0	18-1/2	84	7.30	4.25	11.55
1200		65	2-0	20-1/2	109	7.50	4.34	11.84
1220		75	2-0	20-1/2	119	8	4.53	12.53
1250		100	2-0	22-1/2	145	8.15	4.58	12.73
1270		125	1-9	22-1/2	170	9.20	5	14.20
1300	30	40	2-0	22-1/2	84	8.10	4.14	12.24
1320		65	2-0	24-1/2	110	8.75	4.32	13.07
1350		75	2-0	26-1/2	121	9.05	4.40	13.45
1370		100	2-0	26-1/2	146	9.65	4.56	14.21
1400		125	2-0	24-1/2	172	11.30	5	16.30
1420	35	40	2-0	26-1/2	85	10.65	4.84	15.49
1450		65	2-0	28-1/2	111	11	4.95	15.95
1470		75	2-0	28-1/2	121	11	4.95	15.95
1500		100	1-11	28-1/2	147	11.95	5.20	17.15
1520		125	1-8	28-1/2	172	13.10	5.55	18.65
1550	5/8″ gyp. fireproof.							
1560	On metal furring, add					.88	3.40	4.28

B1010 Floor Construction

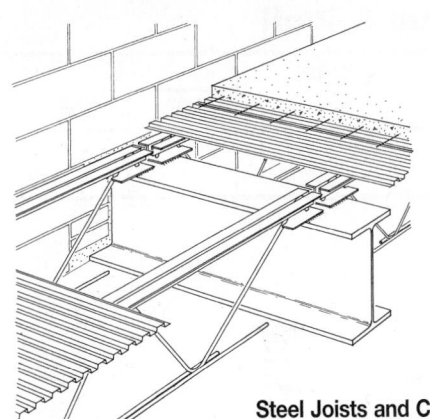

The table below lists costs for a floor system on exterior bearing walls and interior columns and beams using open web steel joists, galvanized steel slab form, 2-1/2″ concrete slab reinforced with welded wire fabric.

Columns are sized to accommodate one floor plus roof loading but costs for columns are from suspended floor to slab on grade only. Costs of the bearing walls are not included.

Steel Joists and Concrete Slab on Walls and Beams

B1010 248					Steel Joists on Beam & Wall			
	BAY SIZE (FT.)	SUPERIMPOSED LOAD (P.S.F.)	DEPTH (IN.)	TOTAL LOAD (P.S.F.)	COLUMN ADD	COST PER S.F.		
						MAT.	INST.	TOTAL
1720	25x25	40	23	84		8.85	4.60	13.45
1730					columns	.42	.13	.55
1750	25x25	65	29	110		9.30	4.75	14.05
1760					columns	.42	.13	.55
1770	25x25	75	26	120		10.30	5.10	15.40
1780					columns	.49	.14	.63
1800	25x25	100	29	145		11.60	5.55	17.15
1810					columns	.49	.14	.63
1820	25x25	125	29	170		12.15	5.75	17.90
1830					columns	.54	.17	.71
2020	30x30	40	29	84		10.10	4.61	14.71
2030					columns	.38	.12	.50
2050	30x30	65	29	110		11.40	4.97	16.37
2060					columns	.38	.12	.50
2070	30x30	75	32	120		11.70	5.05	16.75
2080					columns	.44	.13	.57
2100	30x30	100	35	145		12.95	5.40	18.35
2110					columns	.51	.15	.66
2120	30x30	125	35	172		14.95	6	20.95
2130					columns	.61	.18	.79
2170	30x35	40	29	85		11.40	4.97	16.37
2180					columns	.37	.10	.47
2200	30x35	65	29	111		13.55	5.55	19.10
2210					columns	.42	.13	.55
2220	30x35	75	32	121		13.55	5.55	19.10
2230					columns	.43	.13	.56
2250	30x35	100	35	148		13.75	5.65	19.40
2260					columns	.52	.16	.68
2270	30x35	125	38	173		15.40	6.10	21.50
2280					columns	.53	.16	.69
2320	35x35	40	32	85		11.70	5.05	16.75
2330					columns	.37	.12	.49
2350	35x35	65	35	111		14.25	5.80	20.05
2360					columns	.45	.13	.58
2370	35x35	75	35	121		14.55	5.85	20.40
2380					columns	.45	.13	.58
2400	35x35	100	38	148		15.15	6.05	21.20
2410					columns	.55	.17	.72

B1010 Floor Construction

B1010 248		Steel Joists on Beam & Wall						
	BAY SIZE (FT.)	SUPERIMPOSED LOAD (P.S.F.)	DEPTH (IN.)	TOTAL LOAD (P.S.F.)	COLUMN ADD	COST PER S.F.		
						MAT.	INST.	TOTAL
2460	5/8 gyp. fireproof.							
2475	On metal furring, add					.88	3.40	4.28

For customer support on your Square Foot Costs with RSMeans data, call 800.448.8182.

B1010 Floor Construction

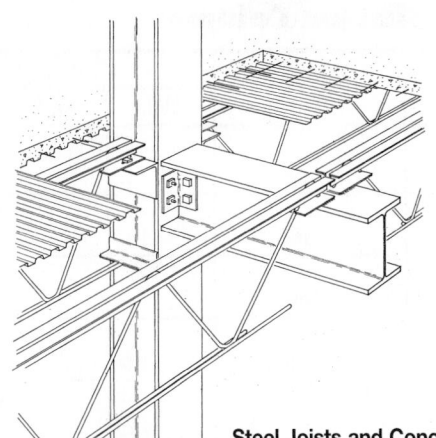

The table below lists costs per S.F. for a floor system on steel columns and beams using open web steel joists, galvanized steel slab form, 2-1/2″ concrete slab reinforced with welded wire fabric.

Columns are sized to accommodate one floor plus roof loading but costs for columns are from floor to grade only.

Steel Joists and Concrete Slab on Steel Columns and Beams

B1010 250					Steel Joists, Beams & Slab on Columns			
	BAY SIZE (FT.)	SUPERIMPOSED LOAD (P.S.F.)	DEPTH (IN.)	TOTAL LOAD (P.S.F.)	COLUMN ADD	COST PER S.F.		
						MAT.	INST.	TOTAL
2350	15x20	40	17	83		9.15	4.63	13.78
2400					column	1.40	.41	1.81
2450	15x20	65	19	108		10.10	4.91	15.01
2500					column	1.40	.41	1.81
2550	15x20	75	19	119		10.50	5.05	15.55
2600					column	1.53	.45	1.98
2650	15x20	100	19	144		11.10	5.30	16.40
2700					column	1.53	.45	1.98
2750	15x20	125	19	170		11.65	5.45	17.10
2800					column	2.04	.61	2.65
2850	20x20	40	19	83		9.90	4.85	14.75
2900					column	1.15	.34	1.49
2950	20x20	65	23	109		10.90	5.20	16.10
3000					column	1.53	.45	1.98
3100	20x20	75	26	119		11.45	5.35	16.80
3200					column	1.53	.45	1.98
3400	20x20	100	23	144		11.90	5.45	17.35
3450					column	1.53	.45	1.98
3500	20x20	125	23	170		13.20	5.90	19.10
3600					column	1.83	.54	2.37
3700	20x25	40	44	83		10.60	5.10	15.70
3800					column	1.22	.36	1.58
3900	20x25	65	26	110		11.55	5.40	16.95
4000					column	1.22	.36	1.58
4100	20x25	75	26	120		12.05	5.60	17.65
4200					column	1.47	.44	1.91
4300	20x25	100	26	145		12.75	5.80	18.55
4400					column	1.47	.44	1.91
4500	20x25	125	29	170		14.15	6.30	20.45
4600					column	1.71	.51	2.22
4700	25x25	40	23	84		11.45	5.35	16.80
4800					column	1.17	.35	1.52
4900	25x25	65	29	110		12.10	5.55	17.65
5000					column	1.17	.35	1.52
5100	25x25	75	26	120		13.30	5.95	19.25
5200					column	1.37	.40	1.77

B1010 Floor Construction

B1010 250	Steel Joists, Beams & Slab on Columns

	BAY SIZE (FT.)	SUPERIMPOSED LOAD (P.S.F.)	DEPTH (IN.)	TOTAL LOAD (P.S.F.)	COLUMN ADD	COST PER S.F.		
						MAT.	INST.	TOTAL
5300	25x25	100	29	145		14.80	6.50	21.30
5400					column	1.37	.40	1.77
5500	25x25	125	32	170		15.60	6.75	22.35
5600					column	1.52	.45	1.97
5700	25x30	40	29	84		13.10	5.45	18.55
5800					column	1.14	.34	1.48
5900	25x30	65	29	110		13.65	5.60	19.25
6000					column	1.14	.34	1.48
6050	25x30	75	29	120		13.95	5.70	19.65
6100					column	1.26	.38	1.64
6150	25x30	100	29	145		15.10	6.05	21.15
6200					column	1.26	.38	1.64
6250	25x30	125	32	170		17	6.55	23.55
6300					column	1.45	.43	1.88
6350	30x30	40	29	84		13.05	5.45	18.50
6400					column	1.05	.31	1.36
6500	30x30	65	29	110		14.80	5.95	20.75
6600					column	1.05	.31	1.36
6700	30x30	75	32	120		15.10	6	21.10
6800					column	1.21	.36	1.57
6900	30x30	100	35	145		16.70	6.50	23.20
7000					column	1.41	.42	1.83
7100	30x30	125	35	172		19.05	7.15	26.20
7200					column	1.57	.47	2.04
7300	30x35	40	29	85		14.65	5.95	20.60
7400					column	.90	.27	1.17
7500	30x35	65	29	111		17.10	6.60	23.70
7600					column	1.16	.35	1.51
7700	30x35	75	32	121		17.10	6.60	23.70
7800					column	1.19	.35	1.54
7900	30x35	100	35	148		17.60	6.70	24.30
8000					column	1.45	.43	1.88
8100	30x35	125	38	173		19.50	7.30	26.80
8200					column	1.48	.44	1.92
8300	35x35	40	32	85		15.05	6	21.05
8400					column	1.04	.31	1.35
8500	35x35	65	35	111		17.95	6.85	24.80
8600					column	1.25	.37	1.62
9300	35x35	75	38	121		18.40	6.95	25.35
9400					column	1.25	.37	1.62
9500	35x35	100	38	148		19.80	7.35	27.15
9600					column	1.54	.45	1.99
9750	35x35	125	41	173		20.50	7.55	28.05
9800					column	1.57	.47	2.04
9810	5/8 gyp. fireproof.							
9815	On metal furring, add					.88	3.40	4.28

B1010 Floor Construction

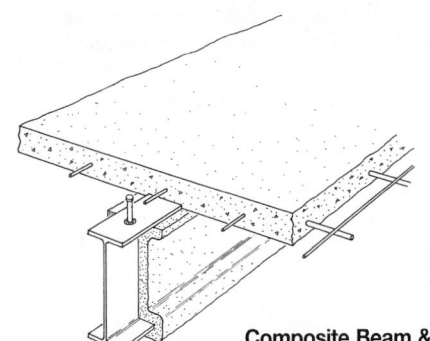

General: Composite construction of wide flange beams and concrete slabs is most efficiently used when loads are heavy and spans are moderately long. It is stiffer with less deflection than non-composite construction of similar depth and spans.

Composite Beam & Cast in Place Slab

B1010 252		Composite Beam & Cast in Place Slab						
	BAY SIZE (FT.)	SUPERIMPOSED LOAD (P.S.F.)	SLAB THICKNESS (IN.)	TOTAL DEPTH (FT.-IN.)	TOTAL LOAD (P.S.F.)	COST PER S.F.		
						MAT.	INST.	TOTAL
3800	20x25	40	4	1 - 8	94	10.65	11.65	22.30
3900		75	4	1 - 8	130	12.35	12.35	24.70
4000		125	4	1 - 10	181	14.40	13.15	27.55
4100		200	5	2 - 2	272	18.75	14.80	33.55
4200	25x25	40	4-1/2	1 - 8-1/2	99	11.10	11.75	22.85
4300		75	4-1/2	1 - 10-1/2	136	13.40	12.70	26.10
4400		125	5-1/2	2 - 0-1/2	200	15.90	13.65	29.55
4500		200	5-1/2	2 - 2-1/2	278	19.70	15.10	34.80
4600	25x30	40	4-1/2	1 - 8-1/2	100	12.40	12.25	24.65
4700		75	4-1/2	1 - 10-1/2	136	14.55	13.05	27.60
4800		125	5-1/2	2 - 2-1/2	202	17.90	14.30	32.20
4900		200	5-1/2	2 - 5-1/2	279	21.50	15.60	37.10
5000	30x30	40	4	1 - 8	95	12.45	12.20	24.65
5200		75	4	2 - 1	131	14.45	12.95	27.40
5400		125	4	2 - 4	183	17.80	14.20	32
5600		200	5	2 - 10	274	23	16	39
5800	30x35	40	4	2 - 1	95	13.20	12.45	25.65
6000		75	4	2 - 4	139	15.90	13.40	29.30
6250		125	4	2 - 4	185	20.50	15.10	35.60
6500		200	5	2 - 11	276	25	16.85	41.85
8000	35x35	40	4	2 - 1	96	13.90	12.70	26.60
8250		75	4	2 - 4	133	16.50	13.70	30.20
8500		125	4	2 - 10	185	19.85	14.90	34.75
8750		200	5	3 - 5	276	26	17	43
9000	35x40	40	4	2 - 4	97	15.45	13.30	28.75
9250		75	4	2 - 4	134	18.10	14.10	32.20
9500		125	4	2 - 7	186	22	15.50	37.50
9750		200	5	3 - 5	278	28.50	17.85	46.35

B1010 Floor Construction

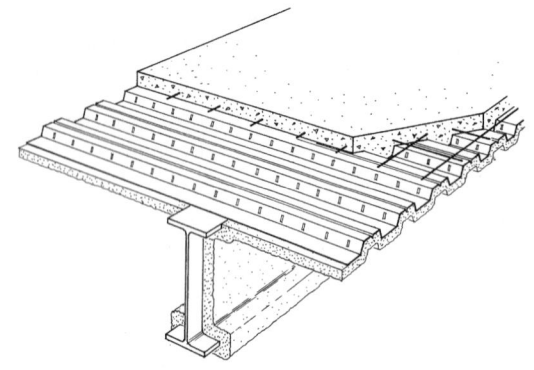

The table below lists costs per S.F. for floors using steel beams and girders, composite steel deck, concrete slab reinforced with W.W.F. and sprayed fiber fireproofing (non- asbestos) on the steel beams and girders and on the steel deck.

Wide Flange, Composite Deck & Slab

B1010 254				W Shape, Composite Deck, & Slab				
	BAY SIZE (FT.) BEAM X GIRD	SUPERIMPOSED LOAD (P.S.F.)	SLAB THICKNESS (IN.)	TOTAL DEPTH (FT.-IN.)	TOTAL LOAD (P.S.F.)	COST PER S.F.		
						MAT.	INST.	TOTAL
0540	15x20	40	5	1-7	89	12.75	6.90	19.65
0560		75	5	1-9	125	13.95	7.35	21.30
0580		125	5	1-11	176	16.25	8.05	24.30
0600		200	5	2-2	254	19.90	9.15	29.05
0700	20x20	40	5	1-9	90	13.80	7.20	21
0720		75	5	1-9	126	15.55	7.80	23.35
0740		125	5	1-11	177	17.60	8.50	26.10
0760		200	5	2-5	255	21	9.45	30.45
0780	20x25	40	5	1-9	90	14.15	7.15	21.30
0800		75	5	1-9	127	17.50	8.20	25.70
0820		125	5	1-11	180	21	9	30
0840		200	5	2-5	256	23	10.20	33.20
0940	25x25	40	5	1-11	91	15.95	7.65	23.60
0960		75	5	2-5	178	18.30	8.40	26.70
0980		125	5	2-5	181	22.50	9.40	31.90
1000		200	5-1/2	2-8-1/2	263	25	10.90	35.90
1400	25x30	40	5	2-5	91	15.75	7.90	23.65
1500		75	5	2-5	128	19.15	8.80	27.95
1600		125	5	2-8	180	21.50	9.75	31.25
1700		200	5	2-11	259	27	11.20	38.20
2200	30x30	40	5	2-2	92	17.40	8.35	25.75
2300		75	5	2-5	129	20.50	9.30	29.80
2400		125	5	2-11	182	24	10.45	34.45
2500		200	5	3-2	263	32.50	13.10	45.60
2600	30x35	40	5	2-5	94	18.95	8.85	27.80
2700		75	5	2-11	131	22	9.80	31.80
2800		125	5	3-2	183	25.50	10.95	36.45
2900		200	5-1/2	3-5-1/2	268	31.50	12.60	44.10
3800	35x35	40	5	2-8	94	19.55	8.85	28.40
3900		75	5	2-11	131	22.50	9.80	32.30
4000		125	5	3-5	184	27	11.25	38.25
4100		200	5-1/2	3-5-1/2	270	35	13.30	48.30

For customer support on your Square Foot Costs with RSMeans data, call 800.448.8182.

B1010 Floor Construction

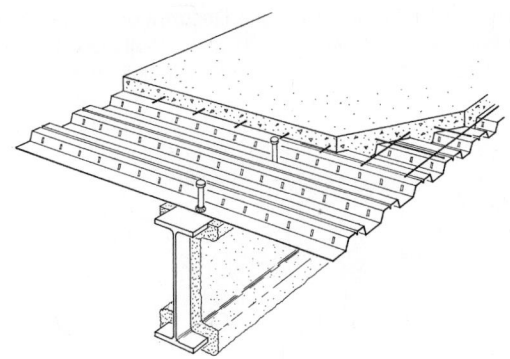

The table below lists costs per S.F. for floors using composite steel beams with welded shear studs, composite steel deck and light weight concrete slab reinforced with W.W.F.

Composite Beam, Deck & Slab

B1010 256		Composite Beams, Deck & Slab						
	BAY SIZE (FT.)	SUPERIMPOSED LOAD (P.S.F.)	SLAB THICKNESS (IN.)	TOTAL DEPTH (FT.-IN.)	TOTAL LOAD (P.S.F.)	COST PER S.F.		
						MAT.	INST.	TOTAL
2400	20x25	40	5-1/2	1 - 5-1/2	80	12.55	6.40	18.95
2500		75	5-1/2	1 - 9-1/2	115	13.05	6.40	19.45
2750		125	5-1/2	1 - 9-1/2	167	16.10	7.45	23.55
2900		200	6-1/4	1 - 11-1/2	251	18.05	8.05	26.10
3000	25x25	40	5-1/2	1 - 9-1/2	82	12.55	6.10	18.65
3100		75	5-1/2	1 - 11-1/2	118	13.95	6.15	20.10
3200		125	5-1/2	2 - 2-1/2	169	14.60	6.65	21.25
3300		200	6-1/4	2 - 6-1/4	252	19.80	7.75	27.55
3400	25x30	40	5-1/2	1 - 11-1/2	83	12.80	6.05	18.85
3600		75	5-1/2	1 - 11-1/2	119	13.75	6.10	19.85
3900		125	5-1/2	1 - 11-1/2	170	16	6.85	22.85
4000		200	6-1/4	2 - 6-1/4	252	19.85	7.80	27.65
4200	30x30	40	5-1/2	1 - 11-1/2	81	12.70	6.25	18.95
4400		75	5-1/2	2 - 2-1/2	116	13.80	6.55	20.35
4500		125	5-1/2	2 - 5-1/2	168	16.75	7.25	24
4700		200	6-1/4	2 - 9-1/4	252	20	8.40	28.40
4900	30x35	40	5-1/2	2 - 2-1/2	82	13.30	6.40	19.70
5100		75	5-1/2	2 - 5-1/2	117	14.55	6.55	21.10
5300		125	5-1/2	2 - 5-1/2	169	17.25	7.45	24.70
5500		200	6-1/4	2 - 9-1/4	254	20	8.45	28.45
5750	35x35	40	5-1/2	2 - 5-1/2	84	14.35	6.45	20.80
6000		75	5-1/2	2 - 5-1/2	121	16.40	6.90	23.30
7000	35x35	125	5-1/2	2 - 8-1/2	170	19.25	7.85	27.10
7200		200	5-1/2	2 - 11-1/2	254	22	8.75	30.75
7400	35x40	40	5-1/2	2 - 5-1/2	85	15.85	7	22.85
7600		75	5-1/2	2 - 5-1/2	121	17.20	7.15	24.35
8000		125	5-1/2	2 - 5-1/2	171	19.75	8	27.75
9000		200	5-1/2	2 - 11-1/2	255	24	9.10	33.10

B1010 Floor Construction

Description: Table B1010 258 lists S.F. costs for steel deck and concrete slabs for various spans.

Description: Table B1010 261 lists the S.F. costs for wood joists and a minimum thickness plywood subfloor.

Description: Table B1010 264 lists the S.F. costs, total load, and member sizes, for various bay sizes and loading conditions.

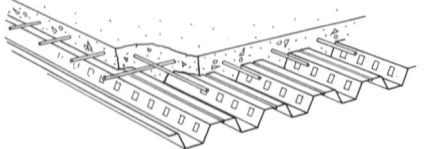

Metal Deck/Concrete Fill

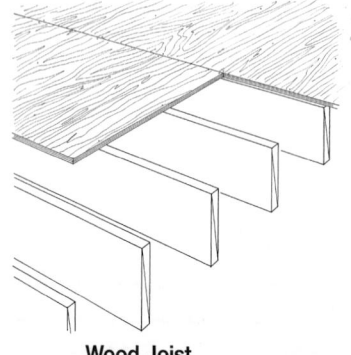

Wood Joist

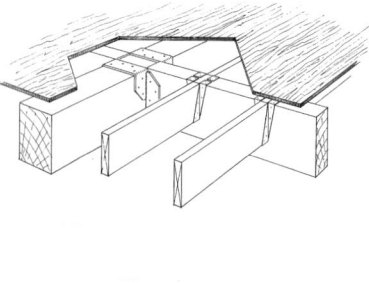

Wood Beam and Joist

B1010 258		**Metal Deck/Concrete Fill**						
	SUPERIMPOSED LOAD (P.S.F.)	DECK SPAN (FT.)	DECK GAGE DEPTH	SLAB THICKNESS (IN.)	TOTAL LOAD (P.S.F.)	COST PER S.F.		
						MAT.	INST.	TOTAL
0900	125	6	22 1-1/2	4	164	3.49	2.64	6.13
0920		7	20 1-1/2	4	164	3.83	2.79	6.62
0950		8	20 1-1/2	4	165	3.83	2.79	6.62
0970		9	18 1-1/2	4	165	4.49	2.79	7.28
1000		10	18 2	4	165	4.39	2.93	7.32
1020		11	18 3	5	169	5.40	3.16	8.56

B1010 261		**Wood Joist**	COST PER S.F.		
			MAT.	INST.	TOTAL
2900	2"x8", 12" O.C.		1.78	2.09	3.87
2950	16" O.C.		1.52	1.79	3.31
3000	24" O.C.		1.61	1.67	3.28
3300	2"x10", 12" O.C.		2.42	2.39	4.81
3350	16" O.C.		1.99	2.01	4
3400	24" O.C.		1.92	1.80	3.72
3700	2"x12", 12" O.C.		2.88	2.41	5.29
3750	16" O.C.		2.34	2.02	4.36
3800	24" O.C.		2.15	1.82	3.97
4100	2"x14", 12" O.C.		3.46	2.63	6.09
4150	16" O.C.		2.78	2.19	4.97
4200	24" O.C.		2.46	1.93	4.39
7101	Note: Subfloor cost is included in these prices.				

B1010 264		**Wood Beam & Joist**						
	BAY SIZE (FT.)	SUPERIMPOSED LOAD (P.S.F.)	GIRDER BEAM (IN.)	JOISTS (IN.)	TOTAL LOAD (P.S.F.)	COST PER S.F.		
						MAT.	INST.	TOTAL
2000	15x15	40	8 x 12 4 x 12	2 x 6 @ 16	53	9.70	4.03	13.73
2050		75	8 x 16 4 x 16	2 x 8 @ 16	90	13.60	4.28	17.88
2100		125	12 x 16 6 x 16	2 x 8 @ 12	144	20.50	5.30	25.80
2150		200	14 x 22 12 x 16	2 x 10 @ 12	227	42	6.75	48.75
3000	20x20	40	10 x 14 10 x 12	2 x 8 @ 16	63	12.65	4.09	16.74
3050		75	12 x 16 8 x 16	2 x 10 @ 16	102	10.35	4.05	14.40
3100		125	14 x 22 12 x 16	2 x 10 @ 12	163	35.50	5.50	41

B1010 Floor Construction

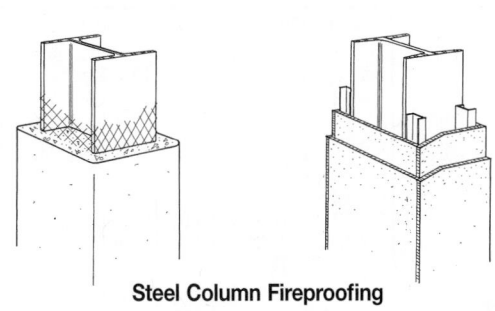

Steel Column Fireproofing

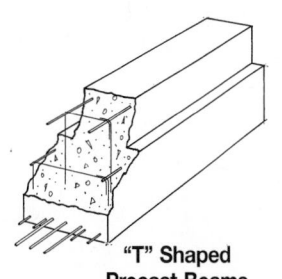

"T" Shaped
Precast Beams

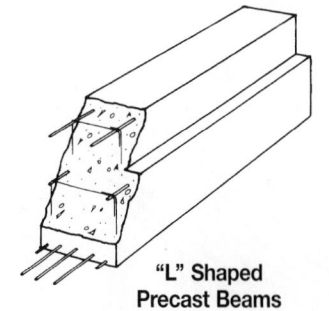

"L" Shaped
Precast Beams

B1010 720						Steel Column Fireproofing		
	ENCASEMENT SYSTEM	COLUMN SIZE (IN.)	THICKNESS (IN.)	FIRE RATING (HRS.)	WEIGHT (P.L.F.)	COST PER V.L.F.		
						MAT.	INST.	TOTAL
3000	Concrete	8	1	1	110	7	39	46
3300		14	1	1	258	11.80	56.50	68.30
3400			2	3	325	14.15	63.50	77.65
3450	Gypsum board	8	1/2	2	8	3.75	24	27.75
3550	1 layer	14	1/2	2	18	4.06	25.50	29.56
3600	Gypsum board	8	1	3	14	5.20	30.50	35.70
3650	1/2" fire rated	10	1	3	17	5.60	32.50	38.10
3700	2 layers	14	1	3	22	5.80	33.50	39.30
3750	Gypsum board	8	1-1/2	3	23	6.95	38.50	45.45
3800	1/2" fire rated	10	1-1/2	3	27	7.80	42.50	50.30
3850	3 layers	14	1-1/2	3	35	8.65	46.50	55.15
3900	Sprayed fiber	8	1-1/2	2	6.3	4.20	7.45	11.65
3950	Direct application		2	3	8.3	5.80	10.30	16.10
4050		10	1-1/2	2	7.9	5.10	9	14.10
4200	Sprayed fiber	14	1-1/2	2	10.8	6.30	11.20	17.50

339

For customer support on your Square Foot Costs with RSMeans data, call 800.448.8182.

The table below lists prices per S.F. for roof rafters and sheathing by nominal size and spacing. Sheathing is 5/16″ CDX for 12″ and 16″ spacing and 3/8″ CDX for 24″ spacing.

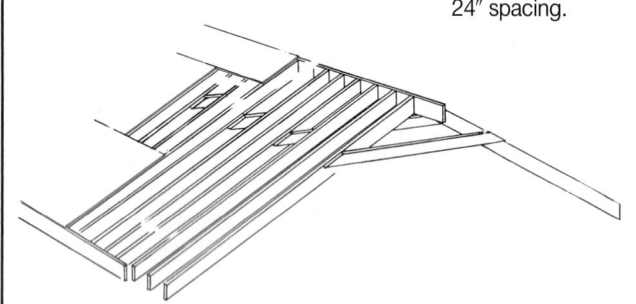

Factors for Converting Inclined to Horizontal

Roof Slope	Approx. Angle	Factor	Roof Slope	Approx. Angle	Factor
Flat	0°	1.000	12 in 12	45.0°	1.414
1 in 12	4.8°	1.003	13 in 12	47.3°	1.474
2 in 12	9.5°	1.014	14 in 12	49.4°	1.537
3 in 12	14.0°	1.031	15 in 12	51.3°	1.601
4 in 12	18.4°	1.054	16 in 12	53.1°	1.667
5 in 12	22.6°	1.083	17 in 12	54.8°	1.734
6 in 12	26.6°	1.118	18 in 12	56.3°	1.803
7 in 12	30.3°	1.158	19 in 12	57.7°	1.873
8 in 12	33.7°	1.202	20 in 12	59.0°	1.943
9 in 12	36.9°	1.250	21 in 12	60.3°	2.015
10 in 12	39.8°	1.302	22 in 12	61.4°	2.088
11 in 12	42.5°	1.357	23 in 12	62.4°	2.162

B1020 102	Wood/Flat or Pitched	COST PER S.F.		
		MAT.	INST.	TOTAL
2500	Flat rafter, 2″x4″, 12″ O.C.	1.17	1.91	3.08
2550	16″ O.C.	1.05	1.64	2.69
2600	24″ O.C.	.95	1.41	2.36
2900	2″x6″, 12″ O.C.	1.45	1.90	3.35
2950	16″ O.C.	1.26	1.63	2.89
3000	24″ O.C.	1.08	1.39	2.47
3300	2″x8″, 12″ O.C.	1.74	2.04	3.78
3350	16″ O.C.	1.48	1.74	3.22
3400	24″ O.C.	1.23	1.48	2.71
3700	2″x10″, 12″ O.C.	2.38	2.34	4.72
3750	16″ O.C.	1.95	1.96	3.91
3800	24″ O.C.	1.54	1.61	3.15
4100	2″x12″, 12″ O.C.	2.84	2.36	5.20
4150	16″ O.C.	2.51	2.16	4.67
4200	24″ O.C.	1.77	1.63	3.40
4500	2″x14″, 12″ O.C.	3.42	3.43	6.85
4550	16″ O.C.	2.74	2.79	5.53
4600	24″ O.C.	2.08	2.17	4.25
4900	3″x6″, 12″ O.C.	3.06	2.01	5.07
4950	16″ O.C.	2.47	1.71	4.18
5000	24″ O.C.	1.89	1.45	3.34
5300	3″x8″, 12″ O.C.	4.14	2.24	6.38
5350	16″ O.C.	3.27	1.88	5.15
5400	24″ O.C.	2.42	1.57	3.99
5700	3″x10″, 12″ O.C.	5	2.56	7.56
5750	16″ O.C.	3.93	2.12	6.05
5800	24″ O.C.	2.86	1.73	4.59
6100	3″x12″, 12″ O.C.	5.85	3.07	8.92
6150	16″ O.C.	4.58	2.51	7.09
6200	24″ O.C.	3.29	1.98	5.27
7001	Wood truss, 4 in 12 slope, 24″ O.C., 24′ to 29′ span	3.77	2.77	6.54
7100	30′ to 43′ span	4.71	3.05	7.76
7200	44′ to 60′ span	4.55	2.77	7.32

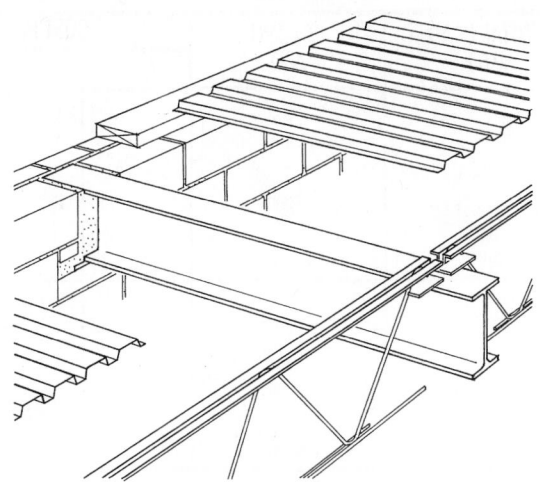

The table below lists the cost per S.F. for a roof system with steel columns, beams and deck using open web steel joists and 1-1/2″ galvanized metal deck. Perimeter of system is supported on bearing walls.

Fireproofing is not included. Costs/S.F. are based on a building 4 bays long and 4 bays wide.

Column costs are additive. Costs for the bearing walls are not included.

Steel Joists, Beams and Deck on Bearing Walls

B1020 108			Steel Joists, Beams & Deck on Columns & Walls					
	BAY SIZE (FT.)	SUPERIMPOSED LOAD (P.S.F.)	DEPTH (IN.)	TOTAL LOAD (P.S.F.)	COLUMN ADD	COST PER S.F.		
						MAT.	INST.	TOTAL
1800	20x20	20	16	40		4.35	1.33	5.68
1900					columns	.62	.14	.76
2000		30	18	50		4.35	1.33	5.68
2100					columns	.62	.14	.76
2200		40	18	60		4.86	1.50	6.36
2300					columns	.62	.14	.76
2400	20x25	20	18	40		4.35	1.38	5.73
2500					columns	.49	.12	.61
2600		30	18	50		4.55	1.42	5.97
2700					columns	.49	.12	.61
2800		40	20	60		4.83	1.52	6.35
2900					columns	.66	.16	.82
3000	25x25	20	18	40		4.69	1.44	6.13
3100					columns	.40	.09	.49
3200		30	22	50		5	1.53	6.53
3300					columns	.53	.13	.66
3400		40	20	60		5.55	1.71	7.26
3500					columns	.53	.13	.66
3600	25x30	20	22	40		4.95	1.40	6.35
3700					columns	.44	.10	.54
3800		30	20	50		5.60	1.56	7.16
3900					columns	.44	.10	.54
4000		40	25	60		5.80	1.61	7.41
4100					columns	.53	.13	.66
4200	30x30	20	25	42		5.40	1.51	6.91
4300					columns	.37	.09	.46
4400		30	22	52		5.90	1.63	7.53
4500					columns	.44	.10	.54
4600		40	28	62		6.10	1.68	7.78
4700					columns	.44	.10	.54

B1020 Roof Construction

B1020 108	Steel Joists, Beams & Deck on Columns & Walls

	BAY SIZE (FT.)	SUPERIMPOSED LOAD (P.S.F.)	DEPTH (IN.)	TOTAL LOAD (P.S.F.)	COLUMN ADD	COST PER S.F.		
						MAT.	INST.	TOTAL
4800	30x35	20	22	42		5.60	1.56	7.16
4900					columns	.38	.09	.47
5000		30	28	52		5.90	1.63	7.53
5100					columns	.38	.09	.47
5200		40	25	62		6.40	1.76	8.16
5300					columns	.44	.10	.54
5400	35x35	20	28	42		5.65	1.57	7.22
5500					columns	.32	.08	.40
5600		30	25	52		6.80	1.86	8.66
5700					columns	.38	.09	.47
5800		40	28	62		6.90	1.88	8.78
5900					columns	.42	.10	.52

B1020 Roof Construction

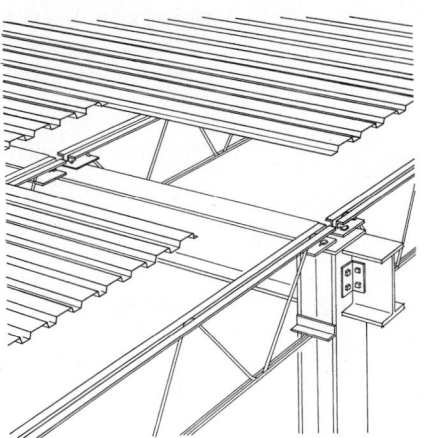

Description: The table below lists the cost per S.F. for a roof system with steel columns, beams, and deck, using open web steel joists and 1-1/2″ galvanized metal deck.

Column costs are additive. Fireproofing is not included.

Steel Joists, Beams and Deck on Columns

B1020 112		Steel Joists, Beams, & Deck on Columns						
	BAY SIZE (FT.)	SUPERIMPOSED LOAD (P.S.F.)	DEPTH (IN.)	TOTAL LOAD (P.S.F.)	COLUMN ADD	COST PER S.F.		
						MAT.	INST.	TOTAL
1100	15x20	20	16	40		4.64	1.44	6.08
1200					columns	2.29	.54	2.83
1500		40	18	60		5.15	1.60	6.75
1600					columns	2.29	.54	2.83
2300	20x25	20	18	40		5.25	1.61	6.86
2400					columns	1.37	.32	1.69
2700		40	20	60		5.85	1.79	7.64
2800					columns	1.83	.43	2.26
2900	25x25	20	18	40		6.10	1.81	7.91
3000					columns	1.10	.26	1.36
3100		30	22	50		6.30	1.85	8.15
3200					columns	1.47	.35	1.82
3300		40	20	60		7	2.08	9.08
3400					columns	1.47	.35	1.82
3500	25x30	20	22	40		6	1.64	7.64
3600					columns	1.22	.29	1.51
3900		40	25	60		7.20	1.94	9.14
4000					columns	1.47	.35	1.82
4100	30x30	20	25	42		6.75	1.82	8.57
4200					columns	1.02	.25	1.27
4300		30	22	52		7.40	1.99	9.39
4400					columns	1.22	.29	1.51
4500		40	28	62		7.75	2.08	9.83
4600					columns	1.22	.29	1.51
5300	35x35	20	28	42		7.35	1.97	9.32
5400					columns	.90	.21	1.11
5500		30	25	52		8.20	2.18	10.38
5600					columns	1.05	.25	1.30
5700		40	28	62		8.80	2.34	11.14
5800					columns	1.16	.27	1.43

343

For customer support on your Square Foot Costs with RSMeans data, call 800.448.8182.

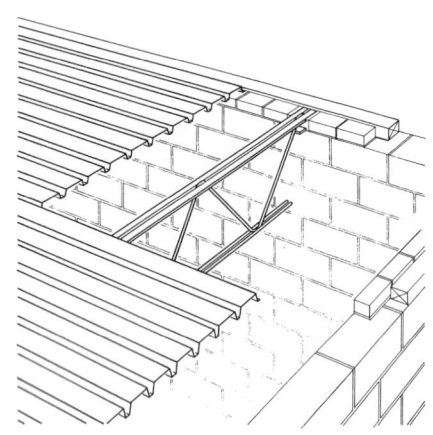

Description: The table below lists the cost per S.F. for a roof system using open web steel joists and 1-1/2″ galvanized metal deck. The system is assumed supported on bearing walls or other suitable support. Costs for the supports are not included.

B1020 116	Steel Joists & Deck on Bearing Walls							
	BAY SIZE (FT.)	SUPERIMPOSED LOAD (P.S.F.)	DEPTH (IN.)	TOTAL LOAD (P.S.F.)		COST PER S.F.		
						MAT.	INST.	TOTAL
1100	20	20	13-1/2	40		3.08	1.04	4.12
1200		30	15-1/2	50		3.15	1.07	4.22
1300		40	15-1/2	60		3.36	1.15	4.51
1400	25	20	17-1/2	40		3.37	1.16	4.53
1500		30	17-1/2	50		3.61	1.25	4.86
1600		40	19-1/2	60		3.66	1.26	4.92
1700	30	20	19-1/2	40		3.65	1.26	4.91
1800		30	21-1/2	50		3.73	1.29	5.02
1900		40	23-1/2	60		4	1.39	5.39
2000	35	20	23-1/2	40		3.96	1.17	5.13
2100		30	25-1/2	50		4.08	1.20	5.28
2200		40	25-1/2	60		4.32	1.27	5.59
2300	40	20	25-1/2	41		4.38	1.29	5.67
2400		30	25-1/2	51		4.64	1.36	6
2500		40	25-1/2	61		4.79	1.40	6.19
2600	45	20	27-1/2	41		5.25	1.53	6.78
2700		30	31-1/2	51		5.55	1.61	7.16
2800		40	31-1/2	61		5.85	1.69	7.54
2900	50	20	29-1/2	42		5.85	1.70	7.55
3000		30	31-1/2	52		6.40	1.84	8.24
3100		40	31-1/2	62		6.75	1.96	8.71
3200	60	20	37-1/2	42		7	2.19	9.19
3300		30	37-1/2	52		7.75	2.42	10.17
3400		40	37-1/2	62		7.75	2.42	10.17
3500	70	20	41-1/2	42		7.75	2.42	10.17
3600		30	41-1/2	52		8.25	2.58	10.83
3700		40	41-1/2	64		10	3.12	13.12
3800	80	20	45-1/2	44		9.70	3.03	12.73
3900		30	45-1/2	54		9.70	3.03	12.73
4000		40	45-1/2	64		10.70	3.35	14.05
4400	100	20	57-1/2	44		9.50	2.88	12.38
4500		30	57-1/2	54		11.35	3.42	14.77
4600		40	57-1/2	65		12.50	3.77	16.27
4700	125	20	69-1/2	44		11.30	3.41	14.71
4800		30	69-1/2	56		13.15	3.96	17.11
4900		40	69-1/2	67		14.95	4.50	19.45

B1020 Roof Construction

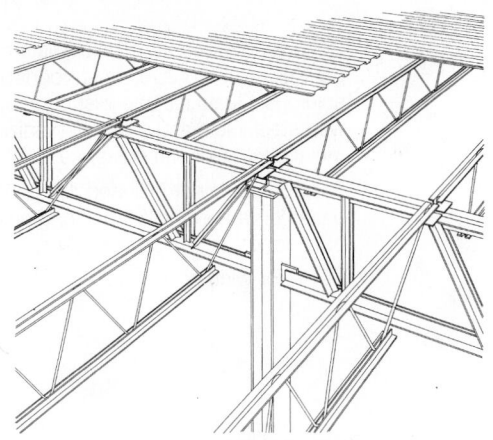

Description: The table below lists costs for a roof system supported on exterior bearing walls and interior columns. Costs include bracing, joist girders, open web steel joists and 1-1/2″ galvanized metal deck.

Column costs are additive. Fireproofing is not included.

Costs/S.F. are based on a building 4 bays long and 4 bays wide. Costs for bearing walls are not included.

B1020 120		Steel Joists, Joist Girders & Deck on Columns & Walls						
	BAY SIZE (FT.) GIRD X JOISTS	SUPERIMPOSED LOAD (P.S.F.)	DEPTH (IN.)	TOTAL LOAD (P.S.F.)	COLUMN ADD	COST PER S.F.		
						MAT.	INST.	TOTAL
2350	30x35	20	32-1/2	40		4.41	1.44	5.85
2400					columns	.38	.09	.47
2550		40	36-1/2	60		4.93	1.60	6.53
2600					columns	.44	.10	.54
2650	35x30	20	36-1/2	40		4.08	1.38	5.46
2700					columns	.38	.09	.47
2750		30	36-1/2	50		4.51	1.49	6
2800					columns	.38	.09	.47
2850		40	36-1/2	60		5.95	1.91	7.86
2900					columns	.44	.10	.54
3000	35x35	20	36-1/2	40		4.77	1.78	6.55
3050					columns	.32	.08	.40
3200		40	36-1/2	60		5.35	1.97	7.32
3250					columns	.42	.10	.52
3300	35x40	20	36-1/2	40		4.88	1.63	6.51
3350					columns	.33	.08	.41
3500		40	36-1/2	60		5.45	1.81	7.26
3550					columns	.36	.09	.45
3900	40x40	20	40-1/2	41		5.65	1.84	7.49
3950					columns	.32	.08	.40
4100		40	40-1/2	61		6.30	2	8.30
4150					columns	.32	.08	.40
5100	45x50	20	52-1/2	41		6.30	2.08	8.38
5150					columns	.23	.05	.28
5300		40	52-1/2	61		7.65	2.46	10.11
5350					columns	.33	.08	.41
5400	50x45	20	56-1/2	41		6.10	2.02	8.12
5450					columns	.23	.05	.28
5600		40	56-1/2	61		7.85	2.52	10.37
5650					columns	.31	.08	.39
5700	50x50	20	56-1/2	42		6.85	2.25	9.10
5750					columns	.23	.05	.28
5900		40	59	64		8	2.75	10.75
5950					columns	.32	.08	.40
6300	60x50	20	62-1/2	43		6.90	2.31	9.21
6350					columns	.37	.09	.46
6500		40	71	65		8.15	2.85	11
6550					columns	.48	.11	.59

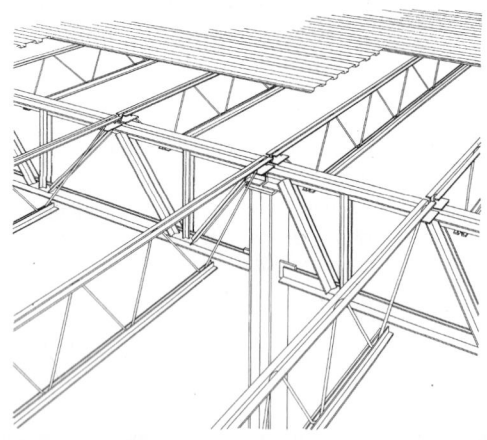

Description: The table below lists the cost per S.F. for a roof system supported on columns. Costs include joist girders, open web steel joists and 1-1/2″ galvanized metal deck. Column costs are additive.

Fireproofing is not included. Costs/S.F. are based on a building 4 bays long and 4 bays wide.

Costs for wind columns are not included.

B1020 124	Steel Joists & Joist Girders on Columns						
	BAY SIZE (FT.) GIRD X JOISTS	SUPERIMPOSED LOAD (P.S.F.)	DEPTH (IN.)	TOTAL LOAD (P.S.F.)	COLUMN ADD	COST PER S.F.	
						MAT.	INST.	TOTAL
2000	30x30	20	17-1/2	40		4.42	1.72	6.14
2050					columns	1.22	.28	1.50
2100		30	17-1/2	50		4.78	1.86	6.64
2150					columns	1.22	.28	1.50
2200		40	21-1/2	60		4.91	1.90	6.81
2250					columns	1.22	.28	1.50
2300	30x35	20	32-1/2	40		4.73	1.61	6.34
2350					columns	1.04	.24	1.28
2500		40	36-1/2	60		5.30	1.78	7.08
2550					columns	1.22	.28	1.50
3200	35x40	20	36-1/2	40		6.10	2.19	8.29
3250					columns	.91	.22	1.13
3400		40	36-1/2	60		6.85	2.41	9.26
3450					columns	1.01	.24	1.25
3800	40x40	20	40-1/2	41		6.15	2.19	8.34
3850					columns	.88	.21	1.09
4000		40	40-1/2	61		6.90	2.41	9.31
4050					columns	.88	.21	1.09
4100	40x45	20	40-1/2	41		6.60	2.48	9.08
4150					columns	.79	.19	.98
4200		30	40-1/2	51		7.20	2.64	9.84
4250					columns	.79	.19	.98
4300		40	40-1/2	61		8	2.88	10.88
4350					columns	.89	.21	1.10
5000	45x50	20	52-1/2	41		6.95	2.52	9.47
5050					columns	.63	.15	.78
5200		40	52-1/2	61		8.50	2.97	11.47
5250					columns	.81	.19	1
5300	50x45	20	56-1/2	41		6.80	2.48	9.28
5350					columns	.63	.15	.78
5500		40	56-1/2	61		8.75	3.05	11.80
5550					columns	.81	.19	1
5600	50x50	20	56-1/2	42		7.55	2.71	10.26
5650					columns	.64	.15	.79
5800		40	59	64		8.75	3.22	11.97
5850					columns	.89	.22	1.11

For customer support on your Square Foot Costs with RSMeans data, call 800.448.8182.

B20 Exterior Enclosure

B2010 Exterior Walls

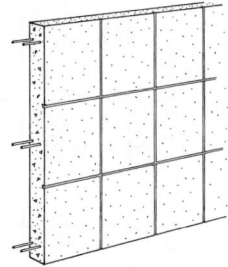

Cast in Place Concrete Wall

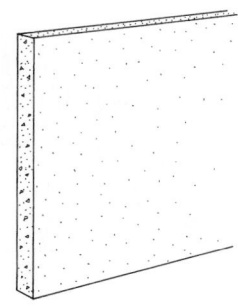

Flat Precast Concrete Wall

Table B2010 101 describes a concrete wall system for exterior closure. There are several types of wall finishes priced from plain finish to a finish with 3/4″ rustication strip.

In Table B2010 103, precast concrete wall panels are either solid or insulated. Prices below are based on delivery within 50 miles of a plant. Small jobs can double the prices below. For large, highly repetitive jobs, deduct up to 15%.

B2010 101	Cast In Place Concrete	COST PER S.F.		
		MAT.	INST.	TOTAL
2100	Conc wall reinforced, 8′ high, 6″ thick, plain finish, 3000 PSI	5.20	17.55	22.75
2700	Aged wood liner, 3000 PSI	6.40	19.95	26.35
3000	Sand blast light 1 side, 3000 PSI	5.80	19.80	25.60
3700	3/4″ bevel rustication strip, 3000 PSI	5.30	18.65	23.95
4000	8″ thick, plain finish, 3000 PSI	6.15	18	24.15
4100	4000 PSI	6.35	18	24.35
4300	Rub concrete 1 side, 3000 PSI	6.15	21.50	27.65
4550	8″ thick, aged wood liner, 3000 PSI	7.40	20.50	27.90
4750	Sand blast light 1 side, 3000 PSI	6.75	20.50	27.25
5300	3/4″ bevel rustication strip, 3000 PSI	6.30	19.05	25.35
5600	10″ thick, plain finish, 3000 PSI	7.10	18.40	25.50
5900	Rub concrete 1 side, 3000 PSI	7.10	22	29.10
6200	Aged wood liner, 3000 PSI	8.35	21	29.35
6500	Sand blast light 1 side, 3000 PSI	7.70	20.50	28.20
7100	3/4″ bevel rustication strip, 3000 PSI	7.20	19.50	26.70
7400	12″ thick, plain finish, 3000 PSI	8.20	18.95	27.15
7700	Rub concrete 1 side, 3000 PSI	8.20	22.50	30.70
8000	Aged wood liner, 3000 PSI	9.45	21.50	30.95
8300	Sand blast light 1 side, 3000 PSI	8.80	21	29.80
8900	3/4″ bevel rustication strip, 3000 PSI	8.35	20	28.35

B2010 102		Flat Precast Concrete							
	THICKNESS (IN.)	PANEL SIZE (FT.)	FINISHES	RIGID INSULATION (IN)	TYPE	COST PER S.F.		
						MAT.	INST.	TOTAL
3000	4	5x18	smooth gray	none	low rise	16.50	8.10	24.60
3050		6x18				13.80	6.80	20.60
3100		8x20				27.50	3.29	30.79
3150		12x20				26	3.11	29.11
3200	6	5x18	smooth gray	2	low rise	17.55	8.75	26.30
3250		6x18				14.85	7.40	22.25
3300		8x20				28.50	4.01	32.51
3350		12x20				26	3.68	29.68
3400	8	5x18	smooth gray	2	low rise	37.50	5.05	42.55
3450		6x18				35.50	4.80	40.30
3500		8x20				32	4.43	36.43
3550		12x20				29	4.07	33.07

347

B2010 Exterior Walls

B2010 102 — Flat Precast Concrete

	THICKNESS (IN.)	PANEL SIZE (FT.)	FINISHES	RIGID INSULATION (IN)	TYPE	COST PER S.F.		
						MAT.	INST.	TOTAL
3600	4	4x8	white face	none	low rise	62.50	4.71	67.21
3650		8x8				47	5.15	52.15
3700		10x10				41	3.10	44.10
3750		20x10				37.50	2.80	40.30
3800	5	4x8	white face	none	low rise	64	4.81	68.81
3850		8x8				48.50	3.63	52.13
3900		10x10				43	3.22	46.22
3950		20x20				39	2.95	41.95
4000	6	4x8	white face	none	low rise	66.50	4.98	71.48
4050		8x8				50.50	3.79	54.29
4100		10x10				44.50	3.33	47.83
4150		20x10				40.50	3.06	43.56
4200	6	4x8	white face	2	low rise	67.50	5.65	73.15
4250		8x8				51.50	4.48	55.98
4300		10x10				45.50	4.02	49.52
4350		20x10				40.50	3.06	43.56
4400	7	4x8	white face	none	low rise	68	5.10	73.10
4450		8x8				52	3.92	55.92
4500		10x10				46.50	3.52	50.02
4550		20x10				43	3.22	46.22
4600	7	4x8	white face	2	low rise	69	5.80	74.80
4650		8x8				53.50	4.61	58.11
4700		10x10				48	4.21	52.21
4750		20x10				44	3.91	47.91
4800	8	4x8	white face	none	low rise	69.50	5.20	74.70
4850		8x8				53.50	4.02	57.52
4900		10x10				48	3.61	51.61
4950		20x10				44.50	3.33	47.83
5000	8	4x8	white face	2	low rise	70.50	5.90	76.40
5050		8x8				54.50	4.71	59.21
5100		10x10				55.50	4.98	60.48
5150		20x10				55.50	4.98	60.48

B2010 103 — Fluted Window or Mullion Precast Concrete

	THICKNESS (IN.)	PANEL SIZE (FT.)	FINISHES	RIGID INSULATION (IN)	TYPE	COST PER S.F.		
						MAT.	INST.	TOTAL
5200	4	4x8	smooth gray	none	high rise	38	18.65	56.65
5250		8x8				27	13.30	40.30
5300		10x10				52	6.30	58.30
5350		20x10				45.50	5.50	51
5400	5	4x8	smooth gray	none	high rise	38.50	18.95	57.45
5450		8x8				28	13.80	41.80
5500		10x10				54.50	6.55	61.05
5550		20x10				48	5.80	53.80
5600	6	4x8	smooth gray	none	high rise	39.50	19.45	58.95
5650		8x8				29	14.25	43.25
5700		10x10				56	6.75	62.75
5750		20x10				49.50	6	55.50
5800	6	4x8	smooth gray	2	high rise	40.50	20	60.50
5850		8x8				30	14.95	44.95
5900		10x10				57	7.40	64.40
5950		20x10				51	6.65	57.65

B2010 Exterior Walls

B2010 103	Fluted Window or Mullion Precast Concrete

	THICKNESS (IN.)	PANEL SIZE (FT.)	FINISHES	RIGID INSULATION (IN)	TYPE	COST PER S.F.		
						MAT.	INST.	TOTAL
6000	7	4x8	smooth gray	none	high rise	40.50	19.85	60.35
6050		8x8				29.50	14.65	44.15
6100		10x10				58.50	7.05	65.55
6150		20x10				51	6.15	57.15
6200	7	4x8	smooth gray	2	high rise	41.50	20.50	62
6250		8x8				31	15.35	46.35
6300		10x10				59.50	7.75	67.25
6350		20x10				52.50	6.85	59.35
6400	8	4x8	smooth gray	none	high rise	41	20	61
6450		8x8				30.50	15	45.50
6500		10x10				60	7.25	67.25
6550		20x10				53.50	6.45	59.95
6600	8	4x8	smooth gray	2	high rise	42	21	63
6650		8x8				31.50	15.70	47.20
6700		10x10				61.50	7.95	69.45
6750		20x10				55	7.15	62.15

B2010 104	Ribbed Precast Concrete

	THICKNESS (IN.)	PANEL SIZE (FT.)	FINISHES	RIGID INSULATION(IN.)	TYPE	COST PER S.F.		
						MAT.	INST.	TOTAL
6800	4	4x8	aggregate	none	high rise	44.50	18.65	63.15
6850		8x8				32.50	13.60	46.10
6900		10x10				55.50	4.92	60.42
6950		20x10				49	4.35	53.35
7000	5	4x8	aggregate	none	high rise	45.50	18.95	64.45
7050		8x8				33.50	14	47.50
7100		10x10				58	5.10	63.10
7150		20x10				51.50	4.56	56.06
7200	6	4x8	aggregate	none	high rise	46.50	19.35	65.85
7250		8x8				34.50	14.40	48.90
7300		10x10				59.50	5.25	64.75
7350		20x10				53.50	4.73	58.23
7400	6	4x8	aggregate	2	high rise	47.50	20	67.50
7450		8x8				35.50	15.10	50.60
7500		10x10				61	5.95	66.95
7550		20x10				54.50	5.40	59.90
7600	7	4x8	aggregate	none	high rise	47.50	19.80	67.30
7650		8x8				35.50	14.75	50.25
7700		10x10				61.50	5.45	66.95
7750		20x10				55	4.86	59.86
7800	7	4x8	aggregate	2	high rise	48.50	20.50	69
7850		8x8				36.50	15.45	51.95
7900		10x10				62.50	6.15	68.65
7950		20x10				56	5.55	61.55
8000	8	4x8	aggregate	none	high rise	48	20	68
8050		8x8				36.50	15.15	51.65
8100		10x10				63.50	5.65	69.15
8150		20x10				57.50	5.10	62.60
8200	8	4x8	aggregate	2	high rise	49.50	21	70.50
8250		8x8				37.50	15.85	53.35
8300		10x10				65	7.95	72.95
8350		20x10				58.50	5.75	64.25

B2010 Exterior Walls

B2010 105	Precast Concrete Specialties

	TYPE	SIZE				COST PER L.F.		
						MAT.	INST.	TOTAL
8400	Coping, precast	6" wide				25	14.65	39.65
8450	Stock units	10" wide				27	15.70	42.70
8460		12" wide				30.50	16.90	47.40
8480		14" wide				33	18.30	51.30
8500	Window sills	6" wide				21	15.70	36.70
8550	Precast	10" wide				34	18.30	52.30
8600		14" wide				34	22	56
8610								

B2010 Exterior Walls

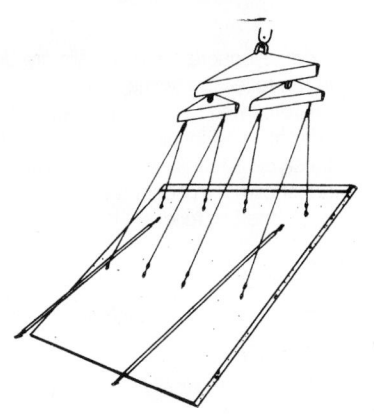

The advantage of tilt up construction is in the low cost of forms and placing of concrete and reinforcing. Tilt up has been used for several types of buildings, including warehouses, stores, offices, and schools. The panels are cast in forms on the ground, or floor slab. Most jobs use 5-1/2" thick solid reinforced concrete panels.

Design Assumptions:
Conc. f'c = 3000 psi
Reinf. fy = 60,000

B2010 106	Tilt-Up Concrete Panel	COST PER S.F.		
		MAT.	INST.	TOTAL
3200	Tilt-up conc panels, broom finish, 5-1/2" thick, 3000 PSI	4.60	6.45	11.05
3250	5000 PSI	4.75	6.30	11.05
3300	6" thick, 3000 PSI	5.05	6.60	11.65
3350	5000 PSI	5.25	6.50	11.75
3400	7-1/2" thick, 3000 PSI	6.45	6.85	13.30
3450	5000 PSI	6.70	6.70	13.40
3500	8" thick, 3000 PSI	6.90	7	13.90
3550	5000 PSI	7.20	6.90	14.10
3700	Steel trowel finish, 5-1/2" thick, 3000 PSI	4.60	6.50	11.10
3750	5000 PSI	4.75	6.40	11.15
3800	6" thick, 3000 PSI	5.05	6.65	11.70
3850	5000 PSI	5.25	6.55	11.80
3900	7-1/2" thick, 3000 PSI	6.45	6.90	13.35
3950	5000 PSI	6.70	6.80	13.50
4000	8" thick, 3000 PSI	6.90	7.05	13.95
4050	5000 PSI	7.20	6.95	14.15
4200	Exp. aggregate finish, 5-1/2" thick, 3000 PSI	4.92	6.60	11.52
4250	5000 PSI	5.10	6.45	11.55
4300	6" thick, 3000 PSI	5.40	6.75	12.15
4350	5000 PSI	5.55	6.60	12.15
4400	7-1/2" thick, 3000 PSI	6.80	7	13.80
4450	5000 PSI	7	6.85	13.85
4500	8" thick, 3000 PSI	7.25	7.15	14.40
4550	5000 PSI	7.50	7	14.50
4600	Exposed aggregate & vert. rustication 5-1/2" thick, 3000 PSI	7.30	8.20	15.50
4650	5000 PSI	7.45	8.10	15.55
4700	6" thick, 3000 PSI	7.75	8.35	16.10
4750	5000 PSI	7.90	8.25	16.15
4800	7-1/2" thick, 3000 PSI	9.15	8.60	17.75
4850	5000 PSI	9.40	8.50	17.90
4900	8" thick, 3000 PSI	9.60	8.75	18.35
4950	5000 PSI	9.85	8.65	18.50
5000	Vertical rib & light sandblast, 5-1/2" thick, 3000 PSI	7.10	10.70	17.80
5100	6" thick, 3000 PSI	7.55	10.85	18.40
5200	7-1/2" thick, 3000 PSI	8.95	11.10	20.05
5300	8" thick, 3000 PSI	9.45	11.25	20.70
6000	Broom finish w/2" polystyrene insulation, 6" thick, 3000 PSI	4.44	8	12.44
6100	Broom finish 2" fiberplank insulation, 6" thick, 3000 PSI	4.99	7.90	12.89
6200	Exposed aggregate w/2" polystyrene insulation, 6" thick, 3000 PSI	4.66	7.95	12.61
6300	Exposed aggregate 2" fiberplank insulation, 6" thick, 3000 PSI	5.20	7.90	13.10

B2010   Exterior Walls

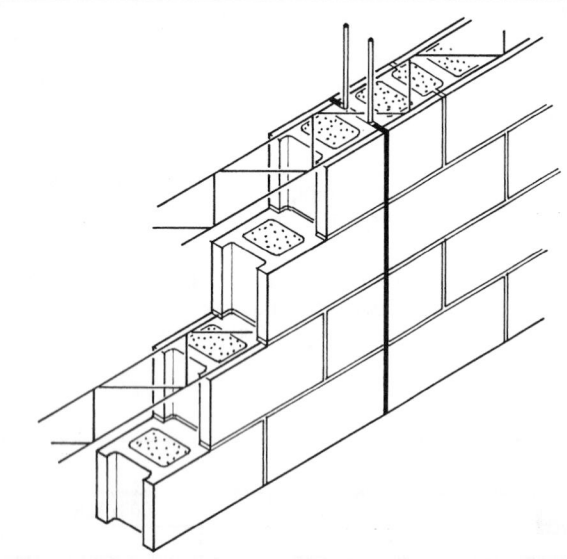

Exterior concrete block walls are defined in the following terms; structural reinforcement, weight, percent solid, size, strength and insulation. Within each of these categories, two to four variations are shown. No costs are included for brick shelf or relieving angles.

B2010 109			Concrete Block Wall - Regular Weight					
	TYPE	SIZE (IN.)	STRENGTH (P.S.I.)	CORE FILL		COST PER S.F.		
						MAT.	INST.	TOTAL
1200	Hollow	4x8x16	2,000	none		2.30	6.80	9.10
1250			4,500	none		2.81	6.80	9.61
1400		8x8x16	2,000	perlite		4.60	8.25	12.85
1410				styrofoam		4.61	7.80	12.41
1440				none		3.10	7.80	10.90
1450			4,500	perlite		5.65	8.25	13.90
1460				styrofoam		5.65	7.80	13.45
1490				none		4.16	7.80	11.96
2000	75% solid	4x8x16	2,000	none		2.77	6.85	9.62
2100		6x8x16	2,000	perlite		4.04	7.55	11.59
2500	Solid	4x8x16	2,000	none		2.58	7	9.58
2700		8x8x16	2,000	none		4.33	8.10	12.43

B2010 110			Concrete Block Wall - Lightweight					
	TYPE	SIZE (IN.)	WEIGHT (P.C.F.)	CORE FILL		COST PER S.F.		
						MAT.	INST.	TOTAL
3100	Hollow	8x4x16	105	perlite		4.54	7.80	12.34
3110				styrofoam		4.55	7.35	11.90
3200		4x8x16	105	none		2.43	6.65	9.08
3250			85	none		3.09	6.50	9.59
3300		6x8x16	105	perlite		4.32	7.45	11.77
3310				styrofoam		4.64	7.10	11.74
3340				none		3.30	7.10	10.40
3400		8x8x16	105	perlite		5.45	8.05	13.50
3410				styrofoam		5.50	7.60	13.10
3440				none		3.97	7.60	11.57
3450			85	perlite		5.65	7.85	13.50
4000	75% solid	4x8x16	105	none		3.51	6.70	10.21
4050			85	none		3.81	6.55	10.36
4100		6x8x16	105	perlite		5.35	7.40	12.75
4500	Solid	4x8x16	105	none		2.18	6.95	9.13
4700		8x8x16	105	none		4.32	8	12.32

B2010 Exterior Walls

B2010 111 — Reinforced Concrete Block Wall - Regular Weight

	TYPE	SIZE (IN.)	STRENGTH (P.S.I.)	VERT. REINF & GROUT SPACING		COST PER S.F.		
						MAT.	INST.	TOTAL
5200	Hollow	4x8x16	2,000	#4 @ 48"		2.46	7.55	10.01
5300		6x8x16	2,000	#4 @ 48"		3.19	8.05	11.24
5330				#5 @ 32"		3.41	8.35	11.76
5340				#5 @ 16"		3.90	9.40	13.30
5350			4,500	#4 @ 28"		3.71	8.05	11.76
5390				#5 @ 16"		4.42	9.40	13.82
5400		8x8x16	2,000	#4 @ 48"		3.53	8.55	12.08
5430				#5 @ 32"		3.69	9.05	12.74
5440				#5 @ 16"		4.27	10.30	14.57
5450		8x8x16	4,500	#4 @ 48"		4.50	8.70	13.20
5490				#5 @ 16"		5.35	10.30	15.65
5500		12x8x16	2,000	#4 @ 48"		5.50	10.95	16.45
5540				#5 @ 16"		6.60	12.65	19.25
6100	75% solid	6x8x16	2,000	#4 @ 48"		3.68	8	11.68
6140				#5 @ 16"		4.12	9.10	13.22
6150			4,500	#4 @ 48"		4.37	8	12.37
6190				#5 @ 16"		4.81	9.10	13.91
6200		8x8x16	2,000	#4 @ 48"		3.89	8.65	12.54
6230				#5 @ 32"		4.05	8.95	13
6240				#5 @ 16"		4.37	9.95	14.32
6250			4,500	#4 @ 48"		5.25	8.65	13.90
6280				#5 @ 32"		5.45	8.95	14.40
6290				#5 @ 16"		5.75	9.95	15.70
6500	Solid-double	2-4x8x16	2,000	#4 @ 48" E.W.		6.05	15.80	21.85
6530	Wythe			#5 @ 16" E.W.		6.75	16.80	23.55
6550			4,500	#4 @ 48" E.W.		8.60	15.70	24.30
6580				#5 @ 16" E.W.		9.35	16.70	26.05

B2010 112 — Reinforced Concrete Block Wall - Lightweight

	TYPE	SIZE (IN.)	WEIGHT (P.C.F.)	VERT REINF. & GROUT SPACING		COST PER S.F.		
						MAT.	INST.	TOTAL
7100	Hollow	8x4x16	105	#4 @ 48"		3.38	8.25	11.63
7140				#5 @ 16"		4.21	9.85	14.06
7150			85	#4 @ 48"		5	8.05	13.05
7190				#5 @ 16"		5.85	9.65	15.50
7400		8x8x16	105	#4 @ 48"		4.31	8.50	12.81
7440				#5 @ 16"		5.15	10.10	15.25
7450		8x8x16	85	#4 @ 48"		4.49	8.30	12.79
7490				#5 @ 16"		5.30	9.90	15.20
7800		8x8x24	105	#4 @ 48"		3.31	9.15	12.46
7840				#5 @ 16"		4.14	10.75	14.89
7850			85	#4 @ 48"		7.45	7.80	15.25
7890				#5 @ 16"		8.30	9.40	17.70
8100	75% solid	6x8x16	105	#4 @ 48"		4.97	7.85	12.82
8130				#5 @ 32"		5.10	8.10	13.20
8150			85	#4 @ 48"		5.05	7.65	12.70
8180				#5 @ 32"		5.20	7.90	13.10
8200		8x8x16	105	#4 @ 48"		6	8.45	14.45
8230				#5 @ 32"		6.15	8.75	14.90
8250			85	#4 @ 48"		5.25	8.25	13.50
8280				#5 @ 32"		5.40	8.55	13.95

353

For customer support on your Square Foot Costs with RSMeans data, call 800.448.8182.

B2010 Exterior Walls

B2010 112	Reinforced Concrete Block Wall - Lightweight

	TYPE	SIZE (IN.)	WEIGHT (P.C.F.)	VERT REINF. & GROUT SPACING		COST PER S.F.		
						MAT.	INST.	TOTAL
8500	Solid-double	2-4x8x16	105	#4 @ 48"		5.25	15.70	20.95
8530	Wythe		105	#5 @ 16"		5.95	16.70	22.65
8600		2-6x8x16	105	#4 @ 48"		8.60	16.75	25.35
8630				#5 @ 16"		9.30	17.75	27.05
8650			85	#4 @ 48"		12.20	16.05	28.25

B2010 Exterior Walls

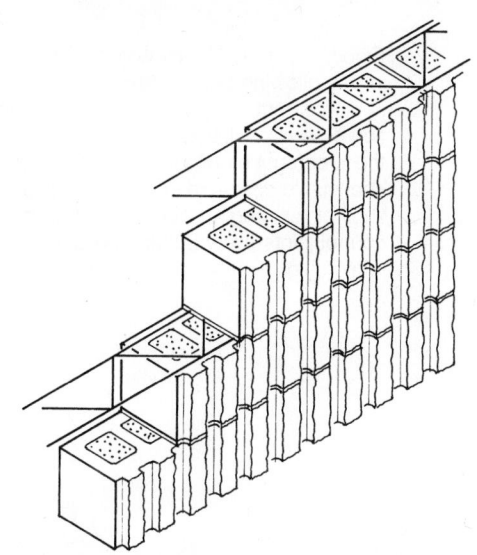

Exterior split ribbed block walls are defined in the following terms; structural reinforcement, weight, percent solid, size, number of ribs and insulation. Within each of these categories two to four variations are shown. No costs are included for brick shelf or relieving angles. Costs include control joints every 20′ and horizontal reinforcing.

B2010 113 — Split Ribbed Block Wall - Regular Weight

	TYPE	SIZE (IN.)	RIBS	CORE FILL		MAT.	INST.	TOTAL
1220	Hollow	4x8x16	4	none		4.52	8.40	12.92
1250			8	none		4.90	8.40	13.30
1280			16	none		5.30	8.50	13.80
1430		8x8x16	8	perlite		7.80	9.95	17.75
1440				styrofoam		7.60	9.50	17.10
1450				none		6.30	9.50	15.80
1530		12x8x16	8	perlite		9.75	13.20	22.95
1540				styrofoam		9.05	12.30	21.35
1550				none		7.30	12.30	19.60
2120	75% solid	4x8x16	4	none		5.65	8.50	14.15
2150			8	none		6.15	8.50	14.65
2180			16	none		6.65	8.65	15.30
2520	Solid	4x8x16	4	none		6.45	8.65	15.10
2550			8	none		7.60	8.65	16.25
2580			16	none		7.60	8.75	16.35

B2010 115 — Reinforced Split Ribbed Block Wall - Regular Weight

	TYPE	SIZE (IN.)	RIBS	VERT. REINF. & GROUT SPACING		MAT.	INST.	TOTAL
5200	Hollow	4x8x16	4	#4 @ 48″		4.68	9.15	13.83
5230			8	#4 @ 48″		5.05	9.15	14.20
5260			16	#4 @ 48″		5.45	9.25	14.70
5430		8x8x16	8	#4 @ 48″		6.60	10.40	17
5440				#5 @ 32″		6.85	10.75	17.60
5450				#5 @ 16″		7.45	12	19.45
5530		12x8x16	8	#4 @ 48″		7.80	13.20	21
5540				#5 @ 32″		8.10	13.60	21.70
5550				#5 @ 16″		8.95	14.90	23.85
6230	75% solid	6x8x16	8	#4 @ 48″		7.15	9.70	16.85
6240				#5 @ 32″		7.30	9.95	17.25
6250				#5 @ 16″		7.60	10.80	18.40
6330		8x8x16	8	#4 @ 48″		8	10.40	18.40
6340				#5 @ 32″		8.15	10.70	18.85
6350				#5 @ 16″		8.45	11.70	20.15

355

For customer support on your Square Foot Costs with RSMeans data, call 800.448.8182.

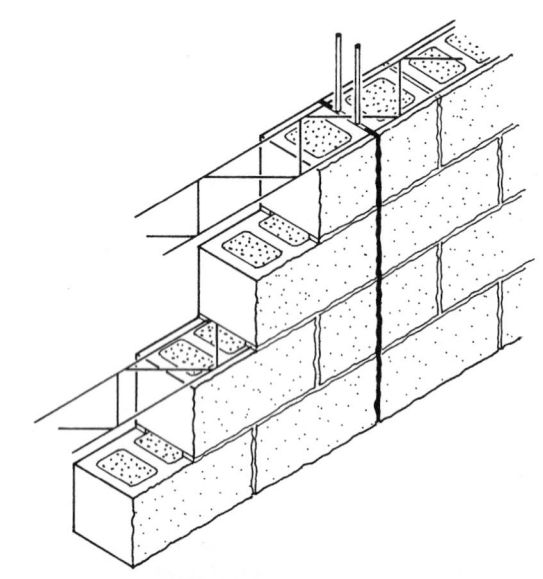

Exterior split face block walls are defined in the following terms; structural reinforcement, weight, percent solid, size, scores and insulation. Within each of these categories two to four variations are shown. No costs are included for brick shelf or relieving angles. Costs include control joints every 20' and horizontal reinforcing.

B2010 117			Split Face Block Wall - Regular Weight					
	TYPE	SIZE (IN.)	SCORES	CORE FILL		COST PER S.F.		
						MAT.	INST.	TOTAL
1200	Hollow	8x4x16	0	perlite		7.60	10.95	18.55
1210				styrofoam		7.60	10.50	18.10
1240				none		6.10	10.50	16.60
1250			1	perlite		8.05	10.95	19
1260				styrofoam		8.05	10.50	18.55
1290				none		6.55	10.50	17.05
1300		12x4x16	0	perlite		9.95	12.50	22.45
1310				styrofoam		9.25	11.60	20.85
1340				none		7.50	11.60	19.10
1350			1	perlite		10.45	12.50	22.95
1360				styrofoam		9.75	11.60	21.35
1390				none		8	11.60	19.60
1400		4x8x16	0	none		4.23	8.30	12.53
1450			1	none		4.59	8.40	12.99
1500		6x8x16	0	perlite		5.80	9.50	15.30
1510				styrofoam		6.15	9.15	15.30
1540				none		4.79	9.15	13.94
1550			1	perlite		6.20	9.65	15.85
1560				styrofoam		6.55	9.30	15.85
1590				none		5.20	9.30	14.50
1600		8x8x16	0	perlite		6.90	10.30	17.20
1610				styrofoam		6.90	9.85	16.75
1640				none		5.40	9.85	15.25
1650		8x8x16	1	perlite		7.35	10.45	17.80
1660				styrofoam		7.35	10	17.35
1690				none		5.85	10	15.85
1700		12x8x16	0	perlite		8.95	13.45	22.40
1710				styrofoam		8.25	12.55	20.80
1740				none		6.50	12.55	19.05
1750			1	perlite		9.45	13.65	23.10
1760				styrofoam		8.75	12.75	21.50
1790				none		7	12.75	19.75

B2010 Exterior Walls

B2010 117 — Split Face Block Wall - Regular Weight

	TYPE	SIZE (IN.)	SCORES	CORE FILL		MAT.	INST.	TOTAL
						COST PER S.F.		
1800	75% solid	8x4x16	0	perlite		8.20	10.90	19.10
1840				none		7.45	10.70	18.15
1850			1	perlite		8.75	10.90	19.65
1890				none		8	10.70	18.70
2000		4x8x16	0	none		5.30	8.40	13.70
2050			1	none		5.75	8.50	14.25
2400	Solid	8x4x16	0	none		8.30	10.90	19.20
2450			1	none		8.95	10.90	19.85
2900		12x8x16	0	none		9.05	13	22.05
2950			1	none		9.85	13.25	23.10

B2010 119 — Reinforced Split Face Block Wall - Regular Weight

	TYPE	SIZE (IN.)	SCORES	VERT. REINF. & GROUT SPACING		MAT.	INST.	TOTAL
						COST PER S.F.		
5200	Hollow	8x4x16	0	#4 @ 48"		6.40	11.40	17.80
5210				#5 @ 32"		6.65	11.75	18.40
5240				#5 @ 16"		7.25	13	20.25
5250			1	#4 @ 48"		6.85	11.40	18.25
5260				#5 @ 32"		7.10	11.75	18.85
5290				#5 @ 16"		7.70	13	20.70
5700		12x8x16	0	#4 @ 48"		7	13.45	20.45
5710				#5 @ 32"		7.30	13.85	21.15
5740				#5 @ 16"		8.15	15.15	23.30
5750			1	#4 @ 48"		7.50	13.65	21.15
5760				#5 @ 32"		7.80	14.05	21.85
5790				#5 @ 16"		8.65	15.35	24
6000	75% solid	8x4x16	0	#4 @ 48"		7.60	11.45	19.05
6010				#5 @ 32"		7.75	11.75	19.50
6040				#5 @ 16"		8.05	12.75	20.80
6050			1	#4 @ 48"		8.15	11.45	19.60
6060				#5 @ 32"		8.30	11.75	20.05
6090				#5 @ 16"		8.60	12.75	21.35
6100		12x4x16	0	#4 @ 48"		9.35	12.55	21.90
6110				#5 @ 32"		9.55	12.85	22.40
6140				#5 @ 16"		9.95	13.70	23.65
6150			1	#4 @ 48"		10	12.55	22.55
6160				#5 @ 32"		10.20	12.85	23.05
6190				#5 @ 16"		10.70	13.90	24.60
6700	Solid-double	2-4x8x16	0	#4 @ 48" E.W.		12.85	18.80	31.65
6710	Wythe			#5 @ 32" E.W.		13.10	19	32.10
6740				#5 @ 16" E.W.		13.55	19.80	33.35
6750			1	#4 @ 48" E.W.		13.95	19.10	33.05
6760				#5 @ 32" E.W.		14.20	19.30	33.50
6790				#5 @ 16" E.W.		14.65	20	34.65
6800		2-6x8x16	0	#4 @ 48" E.W.		14.40	20.50	34.90
6810				#5 @ 32" E.W.		14.70	21	35.70
6840				#5 @ 16" E.W.		15.15	21.50	36.65
6850			1	#4 @ 48" E.W.		15.60	21	36.60
6860				#5 @ 32" E.W.		15.90	21	36.90
6890				#5 @ 16" E.W.		16.35	22	38.35

357

B2010 Exterior Walls

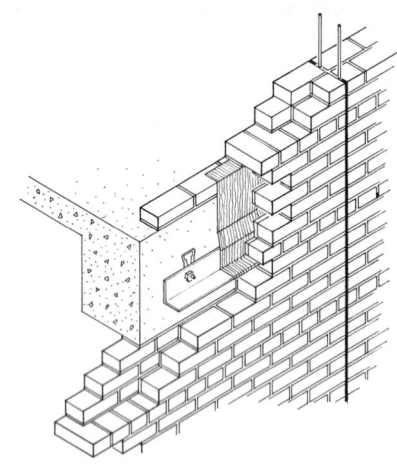

Exterior solid brick walls are defined in the following terms; structural reinforcement, type of brick, thickness, and bond. Nine different types of face bricks are presented, with single wythes shown in four different bonds. Shelf angles are included in system components. These walls do not include ties and as such are not tied to a backup wall.

B2010 125				Solid Brick Walls - Single Wythe				
	TYPE	THICKNESS (IN.)	BOND			COST PER S.F.		
						MAT.	INST.	TOTAL
1000	Common	4	running			6.20	15.25	21.45
1010			common			7.50	17.45	24.95
1050			Flemish			8.25	21	29.25
1100			English			9.25	22.50	31.75
1150	Standard	4	running			5.75	15.80	21.55
1160			common			6.80	18.25	25.05
1200			Flemish			7.50	22	29.50
1250			English			8.40	23	31.40
1300	Glazed	4	running			22.50	16.45	38.95
1310			common			27	19.10	46.10
1350			Flemish			30	23	53
1400			English			33.50	24.50	58
1600	Economy	4	running			8.50	12.10	20.60
1610			common			10	13.85	23.85
1650			Flemish			11	16.45	27.45
1700			English			21	17.45	38.45
1900	Fire	4-1/2	running			14.40	13.85	28.25
1910			common			17.10	15.80	32.90
1950			Flemish			18.95	19.10	38.05
2000			English			21.50	20	41.50
2050	King	3-1/2	running			4.74	12.40	17.14
2060			common			5.50	14.30	19.80
2100			Flemish			6.65	17.10	23.75
2150			English			7.35	17.85	25.20
2200	Roman	4	running			9	14.30	23.30
2210			common			10.70	16.45	27.15
2250			Flemish			11.80	19.60	31.40
2300			English			13.35	21	34.35
2350	Norman	4	running			6.70	11.85	18.55
2360			common			7.85	13.45	21.30
2400			Flemish			17.30	16.10	33.40
2450			English			9.75	17.10	26.85

B20 Exterior Enclosure

B2010 Exterior Walls

B2010 125 — Solid Brick Walls - Single Wythe

	TYPE	THICKNESS (IN.)	BOND			COST PER S.F. MAT.	COST PER S.F. INST.	COST PER S.F. TOTAL
2500	Norwegian	4	running			6.15	10.55	16.70
2510			common			7.25	11.95	19.20
2550			Flemish			8	14.30	22.30
2600			English			8.95	15	23.95

B2010 126 — Solid Brick Walls - Double Wythe

	TYPE	THICKNESS (IN.)	COLLAR JOINT THICKNESS (IN.)			COST PER S.F. MAT.	COST PER S.F. INST.	COST PER S.F. TOTAL
4100	Common	8	3/4			12.10	25.50	37.60
4150	Standard	8	1/2			10.95	26.50	37.45
4200	Glazed	8	1/2			44.50	27.50	72
4250	Engineer	8	1/2			10.65	23.50	34.15
4300	Economy	8	3/4			16.35	20	36.35
4350	Double	8	3/4			15.10	16.20	31.30
4400	Fire	8	3/4			28	23.50	51.50
4450	King	7	3/4			8.90	20.50	29.40
4500	Roman	8	1			17.40	24	41.40
4550	Norman	8	3/4			12.80	19.60	32.40
4600	Norwegian	8	3/4			11.80	17.40	29.20
4650	Utility	8	3/4			12.65	15	27.65
4700	Triple	8	3/4			9.85	13.55	23.40
4750	SCR	12	3/4			15.95	20	35.95
4800	Norwegian	12	3/4			13.90	17.75	31.65

B2010 127 — Reinforced Brick Walls

	TYPE	THICKNESS (IN.)	WYTHE	REINF. & SPACING		COST PER S.F. MAT.	COST PER S.F. INST.	COST PER S.F. TOTAL
7200	Common	4"	1	#4 @ 48"vert		6.40	15.85	22.25
7220				#5 @ 32"vert		6.55	16.05	22.60
7230				#5 @ 16"vert		6.85	16.75	23.60
7700	King	9"	2	#4 @ 48" E.W.		9.30	21.50	30.80
7720				#5 @ 32" E.W.		9.55	21.50	31.05
7730				#5 @ 16" E.W.		10	22.50	32.50
7800	Common	10"	2	#4 @ 48" E.W.		12.50	26.50	39
7820				#5 @ 32" E.W.		12.75	26.50	39.25
7830				#5 @ 16" E.W.		13.20	27	40.20
7900	Standard	10"	2	#4 @ 48" E.W.		11.35	27.50	38.85
7920				#5 @ 32" E.W.		11.60	27.50	39.10
7930				#5 @ 16" E.W.		12.05	28	40.05
8100	Economy	10"	2	#4 @ 48" E.W.		16.75	21	37.75
8120				#5 @ 32" E.W.		17	21	38
8130				#5 @ 16" E.W.		17.45	22	39.45
8200	Double	10"	2	#4 @ 48" E.W.		15.50	17.20	32.70
8220				#5 @ 32" E.W.		15.75	17.35	33.10
8230				#5 @ 16" E.W.		16.20	18	34.20
8300	Fire	10"	2	#4 @ 48" E.W.		28.50	24.50	53
8320				#5 @ 32" E.W.		29	24.50	53.50
8330				#5 @ 16" E.W.		29.50	25	54.50
8400	Roman	10"	2	#4 @ 48" E.W.		17.80	25	42.80
8420				#5 @ 32" E.W.		18.05	25	43.05
8430				#5 @ 16" E.W.		18.50	25.50	44

B2010 Exterior Walls

B2010 127	Reinforced Brick Walls

	TYPE	THICKNESS (IN.)	WYTHE	REINF. & SPACING		COST PER S.F.		
						MAT.	INST.	TOTAL
8500	Norman	10"	2	#4 @ 48" E.W.		13.20	20.50	33.70
8520				#5 @ 32" E.W.		13.45	20.50	33.95
8530				#5 @ 16" E.W.		13.90	21.50	35.40
8600	Norwegian	10"	2	#4 @ 48" E.W.		12.20	18.40	30.60
8620				#5 @ 32" E.W.		12.45	18.55	31
8630				#5 @ 16" E.W.		12.90	19.20	32.10
8800	Triple	10"	2	#4 @ 48" E.W.		10.25	14.55	24.80
8820				#5 @ 32" E.W.		10.50	14.70	25.20
8830				#5 @ 16" E.W.		10.95	15.35	26.30

B2010 Exterior Walls

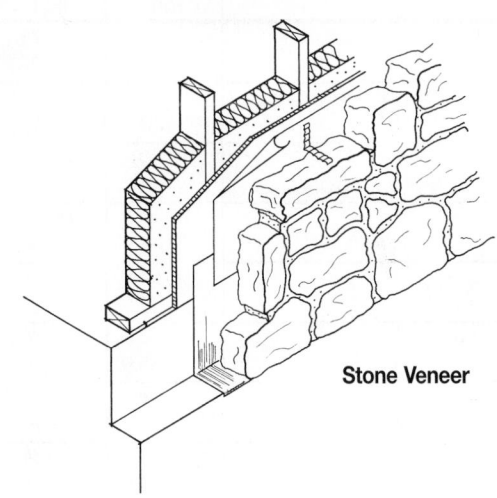

The table below lists costs per S.F. for stone veneer walls on various backup using different stone.

Stone Veneer

B2010 128	Stone Veneer	COST PER S.F.		
		MAT.	INST.	TOTAL
2000	Ashlar veneer, 4", 2" x 4" stud backup, 16" O.C., 8' high, low priced stone	15.75	24	39.75
2050	2" x 6" stud backup, 16" O.C.	17.45	28	45.45
2100	Metal stud backup, 8' high, 16" O.C.	16.55	24	40.55
2150	24" O.C.	16.20	23.50	39.70
2200	Conc. block backup, 4" thick	17.30	28.50	45.80
2300	6" thick	17.95	29	46.95
2350	8" thick	18.15	29.50	47.65
2400	10" thick	18.65	31	49.65
2500	12" thick	20	33	53
3100	High priced stone, wood stud backup, 10' high, 16" O.C.	22.50	27.50	50
3200	Metal stud backup, 10' high, 16" O.C.	23	27.50	50.50
3250	24" O.C.	23	27	50
3300	Conc. block backup, 10' high, 4" thick	24	32	56
3350	6" thick	24.50	32	56.50
3400	8" thick	25.50	34	59.50
3450	10" thick	25.50	34.50	60
3500	12" thick	26.50	36.50	63
4000	Indiana limestone 2" thk., sawn finish, wood stud backup, 10' high, 16" O.C	34.50	14.15	48.65
4100	Metal stud backup, 10' high, 16" O.C.	36.50	17.95	54.45
4150	24" O.C.	36.50	17.95	54.45
4200	Conc. block backup, 4" thick	36	18.70	54.70
4250	6" thick	36.50	19	55.50
4300	8" thick	36.50	19.65	56.15
4350	10" thick	37.50	21	58.50
4400	12" thick	38.50	23.50	62
4450	2" thick, smooth finish, wood stud backup, 8' high, 16" O.C.	34.50	14.15	48.65
4550	Metal stud backup, 8' high, 16" O.C.	35	14.25	49.25
4600	24" O.C.	35	13.75	48.75
4650	Conc. block backup, 4" thick	36	18.50	54.50
4700	6" thick	36.50	18.80	55.30
4750	8" thick	36.50	19.45	55.95
4800	10" thick	37.50	21	58.50
4850	12" thick	38.50	23.50	62
5350	4" thick, smooth finish, wood stud backup, 8' high, 16" O.C.	37	14.15	51.15
5450	Metal stud backup, 8' high, 16" O.C.	37.50	14.50	52
5500	24" O.C.	37.50	13.95	51.45
5550	Conc. block backup, 4" thick	38.50	18.70	57.20
5600	6" thick	39	19	58

B2010 Exterior Walls

B2010 128	Stone Veneer	COST PER S.F.		
		MAT.	INST.	TOTAL
5650	8" thick	43	24	67
5700	10" thick	40	21	61
5750	12" thick	41	23.50	64.50
6000	Granite, gray or pink, 2" thick, wood stud backup, 8' high, 16" O.C.	34.50	25.50	60
6100	Metal studs, 8' high, 16" O.C.	35	26	61
6150	24" O.C.	35	25.50	60.50
6200	Conc. block backup, 4" thick	36	30	66
6250	6" thick	36.50	30.50	67
6300	8" thick	36.50	31	67.50
6350	10" thick	37.50	32.50	70
6400	12" thick	38.50	34.50	73
6900	4" thick, wood stud backup, 8' high, 16" O.C.	45.50	29.50	75
7000	Metal studs, 8' high, 16" O.C.	46	29.50	75.50
7050	24" O.C.	46	29	75
7100	Conc. block backup, 4" thick	47	34	81
7150	6" thick	47.50	34.50	82
7200	8" thick	47.50	35	82.50
7250	10" thick	48.50	36.50	85
7300	12" thick	49.50	38.50	88

B2010 Exterior Walls

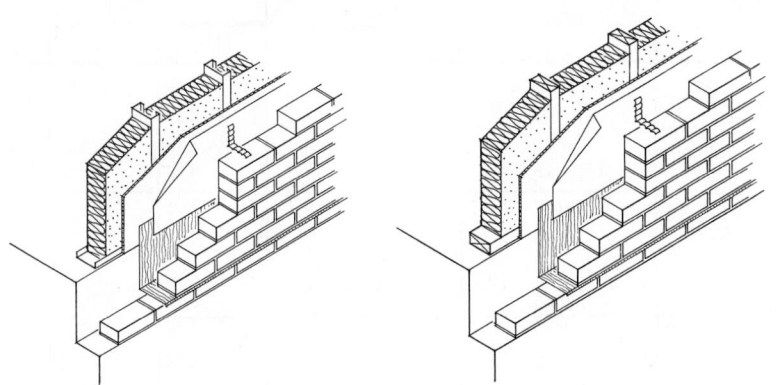

Exterior brick veneer/stud backup walls are defined in the following terms; type of brick and studs, stud spacing and bond. All systems include a brick shelf, ties to the backup and all necessary dampproofing and insulation.

B2010 129 — Brick Veneer/Wood Stud Backup

	FACE BRICK	STUD BACKUP	STUD SPACING (IN.)	BOND	FACE	COST PER S.F. MAT.	COST PER S.F. INST.	COST PER S.F. TOTAL
1100	Standard	2x4-wood	16	running		7.55	18.70	26.25
1120				common		8.55	21	29.55
1140				Flemish		9.25	24.50	33.75
1160				English		10.15	26	36.15
1400		2x6-wood	16	running		8.10	18.65	26.75
1420				common		8.85	21.50	30.35
1440				Flemish		9.55	25	34.55
1460				English		10.45	26	36.45
1700	Glazed	2x4-wood	16	running		24	19.35	43.35
1720				common		28.50	22	50.50
1740				Flemish		31.50	26	57.50
1760				English		35	27.50	62.50
2300	Engineer	2x4-wood	16	running		7.40	16.75	24.15
2320				common		8.30	18.70	27
2340				Flemish		9	22	31
2360				English		9.85	23	32.85
2900	Roman	2x4-wood	16	running		10.75	17.20	27.95
2920				common		12.45	19.35	31.80
2940				Flemish		13.55	22.50	36.05
2960				English		15.10	24	39.10
4100	Norwegian	2x4-wood	16	running		7.95	13.45	21.40
4120				common		9	14.85	23.85
4140				Flemish		9.75	17.20	26.95
4160				English		10.70	17.90	28.60

B2010 130 — Brick Veneer/Metal Stud Backup

	FACE BRICK	STUD BACKUP	STUD SPACING (IN.)	BOND	FACE	COST PER S.F. MAT.	COST PER S.F. INST.	COST PER S.F. TOTAL
5050	Standard	16 ga x 6"LB	16	running		9.15	19.60	28.75
5100		25ga.x6"NLB	24	running		7.10	18.30	25.40
5120				common		8.10	21	29.10
5140				Flemish		8.45	23.50	31.95
5160				English		9.70	25.50	35.20

363

B2010 Exterior Walls

B2010 130	Brick Veneer/Metal Stud Backup

	FACE BRICK	STUD BACKUP	STUD SPACING (IN.)	BOND	FACE	COST PER S.F.		
						MAT.	INST.	TOTAL
5200	Standard	20ga.x3-5/8"NLB	16	running		7.20	19.05	26.25
5220				common		8.20	21.50	29.70
5240				Flemish		8.90	25	33.90
5260				English		9.80	26.50	36.30
5400		16ga.x3-5/8"LB	16	running		8.05	19.30	27.35
5420				common		9.05	22	31.05
5440				Flemish		9.75	25.50	35.25
5460				English		10.65	26.50	37.15
5500			24	running		7.65	18.80	26.45
5520				common		8.70	21.50	30.20
5540				Flemish		9.40	25	34.40
5560				English		10.30	26	36.30
5700	Glazed	25ga.x6"NLB	24	running		23.50	18.95	42.45
5720				common		28	21.50	49.50
5740				Flemish		31	25.50	56.50
5760				English		34.50	27	61.50
5800		20ga.x3-5/8"NLB	24	running		23.50	19.15	42.65
5820				common		28	22	50
5840				Flemish		31	26	57
5860				English		34.50	27.50	62
6000		16ga.x3-5/8"LB	16	running		24.50	19.95	44.45
6020				common		29	22.50	51.50
6040				Flemish		32	26.50	58.50
6060				English		35.50	28	63.50
6300	Engineer	25ga.x6"NLB	24	running		6.95	16.35	23.30
6320				common		7.85	18.30	26.15
6340				Flemish		8.55	21.50	30.05
6360				English		9.40	22.50	31.90
6400		20ga.x3-5/8"NLB	16	running		7.05	17.10	24.15
6420				common		7.95	19.05	27
6440				Flemish		8.65	22.50	31.15
6460				English		9.50	23.50	33
6900	Roman	25ga.x6"NLB	24	running		10.30	16.80	27.10
6920				common		12	18.90	30.90
6940				Flemish		13.10	22	35.10
6960				English		14.65	23.50	38.15
7000		20ga.x3-5/8"NLB	16	running		10.40	17.55	27.95
7020				common		12.10	19.70	31.80
7040				Flemish		13.20	23	36.20
7060				English		14.75	24.50	39.25
7500	Norman	25ga.x6"NLB	24	running		8	14.35	22.35
7520				common		9.15	15.95	25.10
7540				Flemish		18.60	18.60	37.20
7560				English		11.05	19.60	30.65
7600		20ga.x3-5/8"NLB	24	running		8	14.55	22.55
7620				common		9.15	16.15	25.30
7640				Flemish		18.60	18.80	37.40
7660				English		11.05	19.80	30.85
8100	Norwegian	25ga.x6"NLB	24	running		7.50	13.05	20.55
8120				common		8.55	14.45	23
8140				Flemish		9.30	16.80	26.10
8160				English		10.25	17.50	27.75

B2010 Exterior Walls

B2010 130				Brick Veneer/Metal Stud Backup				
	FACE BRICK	STUD BACKUP	STUD SPACING (IN.)	BOND	FACE	COST PER S.F.		
						MAT.	INST.	TOTAL
8400	Norwegian	16ga.x3-5/8"LB	16	running		8.45	14.05	22.50
8420				common		9.50	15.45	24.95
8440				Flemish		10.25	17.80	28.05
8460				English		11.20	18.50	29.70

B2010 Exterior Walls

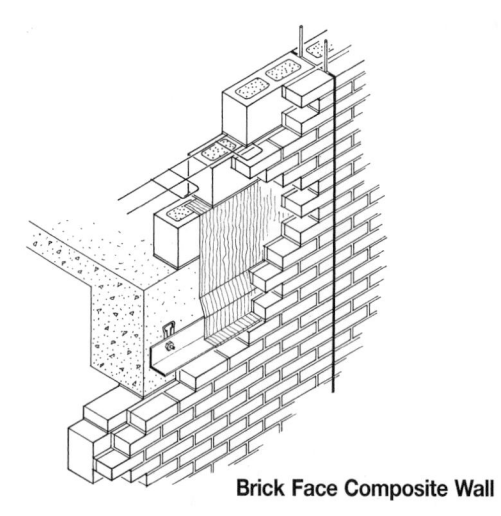

Brick Face Composite Wall

B2010 132		Brick Face Composite Wall - Double Wythe						
	FACE BRICK	BACKUP MASONRY	BACKUP THICKNESS (IN.)	BACKUP CORE FILL		COST PER S.F.		
						MAT.	INST.	TOTAL
1000	Standard	common brick	4	none		11.25	29	40.25
1040		SCR brick	6	none		13.30	25.50	38.80
1080		conc. block	4	none		8.45	23	31.45
1120			6	perlite		10.10	24	34.10
1160				styrofoam		10.40	23.50	33.90
1200			8	perlite		10.75	24.50	35.25
1240				styrofoam		10.75	24	34.75
1520		glazed block	4	none		18.65	25	43.65
1560			6	perlite		20.50	25.50	46
1600				styrofoam		21	25	46
1640			8	perlite		22	26	48
1680				styrofoam		22	25.50	47.50
1720	Standard	clay tile	4	none		12.70	22.50	35.20
1760			6	none		16.65	23	39.65
1800			8	none		19.15	23.50	42.65
1840		glazed tile	4	none		21	29.50	50.50
1880								
2000	Glazed	common brick	4	none		28	29.50	57.50
2040		SCR brick	6	none		30	26.50	56.50
2080		conc. block	4	none		25	24	49
2200			8	perlite		27.50	25.50	53
2240				styrofoam		27.50	25	52.50
2280		L.W. block	4	none		25	23.50	48.50
2400			8	perlite		28	25	53
2520		glazed block	4	none		35.50	25.50	61
2560			6	perlite		37.50	26	63.50
2600				styrofoam		37.50	26	63.50
2640			8	perlite		38.50	27	65.50
2680				styrofoam		38.50	26.50	65

B2010 Exterior Walls

B2010 132	Brick Face Composite Wall - Double Wythe

	FACE BRICK	BACKUP MASONRY	BACKUP THICKNESS (IN.)	BACKUP CORE FILL		COST PER S.F.		
						MAT.	INST.	TOTAL
2720	Glazed	clay tile	4	none		29.50	23	52.50
2760			6	none		33.50	23.50	57
2800			8	none		36	24.50	60.50
2840		glazed tile	4	none		37.50	30	67.50
3000	Engineer	common brick	4	none		11.05	27	38.05
3040		SCR brick	6	none		13.15	24	37.15
3080		conc. block	4	none		8.25	21.50	29.75
3200			8	perlite		10.55	22.50	33.05
3280		L.W. block	4	none		6.35	15.10	21.45
3320			6	perlite		10.30	22	32.30
3520		glazed block	4	none		18.50	23	41.50
3560			6	perlite		20.50	23.50	44
3600				styrofoam		21	23	44
3640			8	perlite		21.50	24.50	46
3680				styrofoam		21.50	24	45.50
3720		clay tile	4	none		12.55	20.50	33.05
3800			8	none		19	22	41
3840		glazed tile	4	none		20.50	27.50	48
4000	Roman	common brick	4	none		14.45	27.50	41.95
4040		SCR brick	6	none		16.55	24	40.55
4080		conc. block	4	none		11.65	21.50	33.15
4120			6	perlite		13.30	22.50	35.80
4200			8	perlite		13.95	23	36.95
5000	Norman	common brick	4	none		12.15	25	37.15
5040		SCR brick	6	none		14.25	22	36.25
5280		L.W. block	4	none		9.45	19.20	28.65
5320			6	perlite		11.40	20	31.40
5720		clay tile	4	none		13.65	18.40	32.05
5760			6	none		17.60	19.05	36.65
5840		glazed tile	4	none		21.50	25.50	47
6000	Norwegian	common brick	4	none		11.65	23.50	35.15
6040		SCR brick	6	none		13.70	20.50	34.20
6080		conc. block	4	none		8.85	18.05	26.90
6120			6	perlite		10.50	18.90	29.40
6160				styrofoam		10.80	18.55	29.35
6200			8	perlite		11.15	19.45	30.60
6520		glazed block	4	none		19.05	19.65	38.70
6560			6	perlite		21	20.50	41.50
7000	Utility	common brick	4	none		12.05	22.50	34.55
7040		SCR brick	6	none		14.15	19.30	33.45
7080		conc. block	4	none		9.25	16.80	26.05
7120			6	perlite		10.90	17.65	28.55
7160				styrofoam		11.20	17.30	28.50
7200			8	perlite		11.55	18.20	29.75
7240				styrofoam		11.55	17.75	29.30
7280		L.W. block	4	none		9.35	16.65	26
7320			6	perlite		11.30	17.45	28.75
7520		glazed block	4	none		19.50	18.40	37.90
7560			6	perlite		21.50	19.10	40.60
7720		clay tile	4	none		13.55	15.85	29.40
7760			6	none		17.50	16.50	34
7840		glazed tile	4	none		21.50	23	44.50

B2010 Exterior Walls

B2010 133	Brick Face Composite Wall - Triple Wythe

	FACE BRICK	MIDDLE WYTHE	INSIDE MASONRY	TOTAL THICKNESS (IN.)		COST PER S.F.		
						MAT.	INST.	TOTAL
8000	Standard	common brick	standard brick	12		16.25	41.50	57.75
8100		4" conc. brick	standard brick	12		15.65	42.50	58.15
8120		4" conc. brick	common brick	12		16.10	42	58.10
8200	Glazed	common brick	standard brick	12		33	42	75
8300		4" conc. brick	standard brick	12		32.50	43	75.50
8320		4" conc. brick	glazed brick	12		32.50	42.50	75

B2010   Exterior Walls

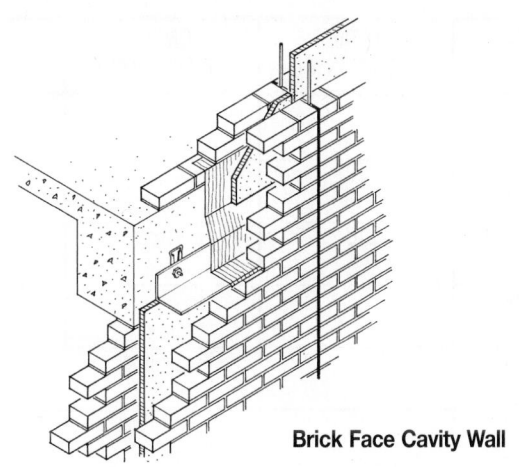

Brick Face Cavity Wall

B2010 134	Brick Face Cavity Wall

	FACE BRICK	BACKUP MASONRY	TOTAL THICKNESS (IN.)	CAVITY INSULATION		COST PER S.F.		
						MAT.	INST.	TOTAL
1000	Standard	4" common brick	10	polystyrene		11.10	29	40.10
1020				none		10.80	28	38.80
1040		6" SCR brick	12	polystyrene		13.20	26	39.20
1060				none		12.90	25	37.90
1080		4" conc. block	10	polystyrene		8.30	23.50	31.80
1100				none		8	22.50	30.50
1120		6" conc. block	12	polystyrene		8.95	24	32.95
1140				none		8.65	23	31.65
1160		4" L.W. block	10	polystyrene		8.45	23	31.45
1180				none		8.15	22.50	30.65
1200		6" L.W. block	12	polystyrene		9.35	23.50	32.85
1220				none		9.05	23	32.05
1240		4" glazed block	10	polystyrene		18.55	25	43.55
1260				none		18.25	24	42.25
1280		6" glazed block	12	polystyrene		18.55	25	43.55
1300				none		18.25	24	42.25
1320		4" clay tile	10	polystyrene		12.60	22.50	35.10
1340				none		12.30	21.50	33.80
1360		4" glazed tile	10	polystyrene		20.50	29.50	50
1380				none		20.50	29	49.50
1500	Glazed	4" common brick	10	polystyrene		27.50	29.50	57
1520				none		27.50	29	56.50
1580		4" conc. block	10	polystyrene		25	24	49
1600				none		24.50	23	47.50
1660		4" L.W. block	10	polystyrene		25	24	49
1680				none		25	23	48
1740		4" glazed block	10	polystyrene		35	25.50	60.50
1760				none		35	25	60
1820		4" clay tile	10	polystyrene		29	23	52
1840				none		29	22.50	51.50
1860		4" glazed tile	10	polystyrene		33	23.50	56.50
1880				none		33	23	56
2000	Engineer	4" common brick	10	polystyrene		10.95	27	37.95
2020				none		10.65	26.50	37.15
2080		4" conc. block	10	polystyrene		8.15	21.50	29.65
2100				none		7.85	20.50	28.35

For customer support on your Square Foot Costs with RSMeans data, call 800.448.8182.

B2010 Exterior Walls

B2010 134	Brick Face Cavity Wall

	FACE BRICK	BACKUP MASONRY	TOTAL THICKNESS (IN.)	CAVITY INSULATION		COST PER S.F.		
						MAT.	INST.	TOTAL
2160	Engineer	4" L.W. block	10	polystyrene		8.25	21.50	29.75
2180				none		7.95	20.50	28.45
2240		4" glazed block	10	polystyrene		18.40	23	41.40
2260				none		18.10	22	40.10
2320		4" clay tile	10	polystyrene		12.45	20.50	32.95
2340				none		12.15	19.70	31.85
2360		4" glazed tile	10	polystyrene		20.50	27.50	48
2380				none		20	27.	47
2500	Roman	4" common brick	10	polystyrene		14.35	27.50	41.85
2520				none		14.05	26.50	40.55
2580		4" conc. block	10	polystyrene		11.50	22	33.50
2600				none		11.20	21	32.20
2660		4" L.W. block	10	polystyrene		11.65	21.50	33.15
2680				none		11.35	21	32.35
2740		4" glazed block	10	polystyrene		22	23.50	45.50
2760				none		21.50	22.50	44
2820		4" clay tile	10	polystyrene		15.80	21	36.80
2840				none		15.50	20	35.50
2860		4" glazed tile	10	polystyrene		24	28	52
2880				none		23.50	27.50	51
3000	Norman	4" common brick	10	polystyrene		12.05	25	37.05
3020				none		11.75	24.50	36.25
3080		4" conc. block	10	polystyrene		9.20	19.40	28.60
3100				none		8.90	18.65	27.55
3160		4" L.W. block	10	polystyrene		9.35	19.25	28.60
3180				none		9.05	18.50	27.55
3240		4" glazed block	10	polystyrene		19.45	21	40.45
3260				none		19.15	20	39.15
3320		4" clay tile	10	polystyrene		13.50	18.45	31.95
3340				none		13.20	17.70	30.90
3360		4" glazed tile	10	polystyrene		21.50	25.50	47
3380				none		21.50	25	46.50
3500	Norwegian	4" common brick	10	polystyrene		11.50	24	35.50
3520				none		11.20	23	34.20
3580		4" conc. block	10	polystyrene		8.70	18.10	26.80
3600				none		8.40	17.35	25.75
3660		4" L.W. block	10	polystyrene		8.85	17.95	26.80
3680				none		8.55	17.20	25.75
3740		4" glazed block	10	polystyrene		18.95	19.70	38.65
3760				none		18.65	18.95	37.60
3820		4" clay tile	10	polystyrene		13	17.15	30.15
3840				none		12.70	16.40	29.10
3860		4" glazed tile	10	polystyrene		21	24.50	45.50
3880				none		21	23.50	44.50
4000	Utility	4" common brick	10	polystyrene		11.95	22.50	34.45
4020				none		11.65	21.50	33.15
4080		4" conc. block	10	polystyrene		9.10	16.85	25.95
4100				none		8.80	16.10	24.90
4160		4" L.W. block	10	polystyrene		9.25	16.70	25.95
4180				none		8.95	15.95	24.90
4240		4" glazed block	10	polystyrene		19.35	18.45	37.80
4260				none		19.05	17.70	36.75

B2010 Exterior Walls

| B2010 134 | Brick Face Cavity Wall | | | | | | | |

	FACE BRICK	BACKUP MASONRY	TOTAL THICKNESS (IN.)	CAVITY INSULATION		COST PER S.F.		
						MAT.	INST.	TOTAL
4320	Utility	4" clay tile	10	polystyrene		13.40	15.90	29.30
4340				none		13.10	15.15	28.25
4360		4" glazed tile	10	polystyrene		21.50	23	44.50
4380				none		21	22.50	43.50

| B2010 135 | Brick Face Cavity Wall - Insulated Backup | | | | | | | |

	FACE BRICK	BACKUP MASONRY	TOTAL THICKNESS (IN.)	BACKUP CORE FILL		COST PER S.F.		
						MAT.	INST.	TOTAL
5100	Standard	6" conc. block	10	perlite		9.65	23.50	33.15
5120				styrofoam		10	23	33
5180		6" L.W. block	10	perlite		10.05	23.50	33.55
5200				styrofoam		10.35	23	33.35
5260		6" glazed block	10	perlite		20	25	45
5280				styrofoam		20.50	24.50	45
5340		6" clay tile	10	none		16.25	22.50	38.75
5360		8" clay tile	12	none		18.75	23	41.75
5600	Glazed	6" conc. block	10	perlite		26.50	24	50.50
5620				styrofoam		26.50	23.50	50
5680		6" L.W. block	10	perlite		26.50	24	50.50
5700				styrofoam		27	23.50	50.50
5760		6" glazed block	10	perlite		37	25.50	62.50
5780				styrofoam		37	25	62
5840		6" clay tile	10	none		33	23	56
5860		8" clay tile	8	none		35.50	23.50	59
6100	Engineer	6" conc. block	10	perlite		9.50	21.50	31
6120				styrofoam		9.80	21	30.80
6180		6" L.W. block	10	perlite		9.90	21.50	31.40
6200				styrofoam		10.20	21	31.20
6260		6" glazed block	10	perlite		20	23	43
6280				styrofoam		20.50	22.50	43
6340		6" clay tile	10	none		16.10	20.50	36.60
6360		8" clay tile	12	none		18.60	21	39.60
6600	Roman	6" conc. block	10	perlite		12.85	22	34.85
6620				styrofoam		13.20	21.50	34.70
6680		6" L.W. block	10	perlite		13.25	22	35.25
6700				styrofoam		13.60	21.50	35.10
6760		6" glazed block	10	perlite		23.50	23.50	47
6780				styrofoam		24	23	47
6840		6" clay tile	10	none		19.45	21	40.45
6860		8" clay tile	12	none		22	21.50	43.50
7100	Norman	6" conc. block	10	perlite		10.55	19.50	30.05
7120				styrofoam		10.90	19.15	30.05
7180		6" L.W. block	10	perlite		10.95	19.30	30.25
7200				styrofoam		11.30	18.95	30.25
7260		6" glazed block	10	perlite		21	21	42
7280				styrofoam		21.50	20.50	42
7340		6" clay tile	10	none		17.15	18.35	35.50
7360		8" clay tile	12	none		19.70	19.15	38.85
7600	Norwegian	6" conc. block	10	perlite		10.05	18.20	28.25
7620				styrofoam		10.40	17.85	28.25
7680		6" L.W. block	10	perlite		10.45	18	28.45
7700				styrofoam		10.75	17.65	28.40

371

For customer support on your Square Foot Costs with RSMeans data, call 800.448.8182.

B2010 Exterior Walls

B2010 135	Brick Face Cavity Wall - Insulated Backup

	FACE BRICK	BACKUP MASONRY	TOTAL THICKNESS (IN.)	BACKUP CORE FILL		COST PER S.F.		
						MAT.	INST.	TOTAL
7760	Norwegian	6" glazed block	10	perlite		20.50	19.65	40.15
7780				styrofoam		21	19.30	40.30
7840		6" clay tile	10	none		16.65	17.05	33.70
7860		8" clay tile	12	none		19.15	17.85	37
8100	Utility	6" conc. block	10	perlite		10.45	16.95	27.40
8120				styrofoam		10.80	16.60	27.40
8180		6" L.W. block	10	perlite		10.85	16.75	27.60
8200				styrofoam		11.20	16.40	27.60
8220		8" L.W. block	12	perlite		12	17.30	29.30
8240				styrofoam		12.05	16.85	28.90
8260		6" glazed block	10	perlite		21	18.40	39.40
8280				styrofoam		21.50	18.05	39.55
8300		8" glazed block	12	perlite		22	19.05	41.05
8340		6" clay tile	10	none		17.05	15.80	32.85
8360		8" clay tile	12	none		19.60	16.60	36.20

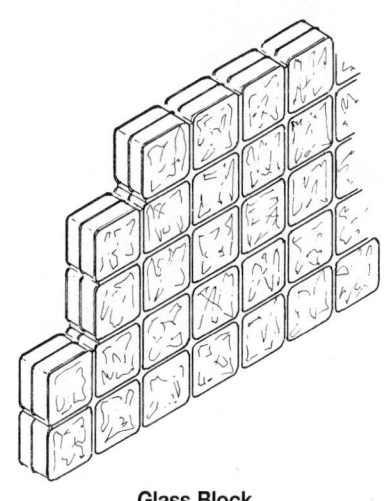

Glass Block

B2010 140	Glass Block	COST PER S.F.		
		MAT.	INST.	TOTAL
2300	Glass block 4″ thick, 6″x6″ plain, under 1,000 S.F.	29	27.50	56.50
2400	1,000 to 5,000 S.F.	28	24	52
2500	Over 5,000 S.F.	27.50	22	49.50
2600	Solar reflective, under 1,000 S.F.	41	37	78
2700	1,000 to 5,000 S.F.	39.50	32	71.50
2800	Over 5,000 S.F.	38.50	30	68.50
3500	8″x8″ plain, under 1,000 S.F.	18.25	20.50	38.75
3600	1,000 to 5,000 S.F.	17.90	17.60	35.50
3700	Over 5,000 S.F.	17.35	15.85	33.20
3800	Solar reflective, under 1,000 S.F.	25.50	27.50	53
3900	1,000 to 5,000 S.F.	25	23.50	48.50
4000	Over 5,000 S.F.	24	21	45
5000	12″x12″ plain, under 1,000 S.F.	26.50	18.85	45.35
5100	1,000 to 5,000 S.F.	26	15.85	41.85
5200	Over 5,000 S.F.	25.50	14.50	40
5300	Solar reflective, under 1,000 S.F.	37.50	25.50	63
5400	1,000 to 5,000 S.F.	36.50	21	57.50
5600	Over 5,000 S.F.	36	19.20	55.20
5800	3″ thinline, 6″x6″ plain, under 1,000 S.F.	26	27.50	53.50
5900	Over 5,000 S.F.	25	22	47
6000	Solar reflective, under 1,000 S.F.	36.50	37	73.50
6100	Over 5,000 S.F.	35	30	65
6200	8″x8″ plain, under 1,000 S.F.	14.70	20.50	35.20
6300	Over 5,000 S.F.	14.05	15.85	29.90
6400	Solar reflective, under 1,000 S.F.	20.50	27.50	48
6500	Over 5,000 S.F.	19.55	21	40.55

B2010 Exterior Walls

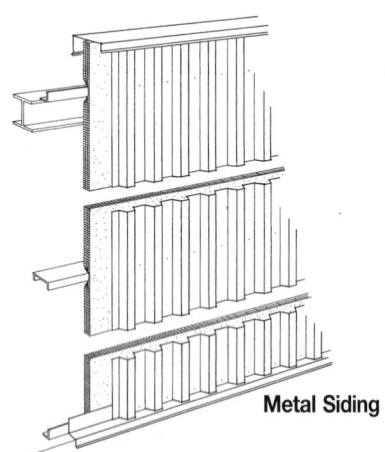

Metal Siding

The table below lists costs for metal siding of various descriptions, not including the steel frame, or the structural steel, of a building. Costs are per S.F. including all accessories and insulation.

B2010 146	Metal Siding Panel	COST PER S.F.		
		MAT.	INST.	TOTAL
1400	Metal siding aluminum panel, corrugated, .024" thick, natural	3.39	4.47	7.86
1450	Painted	3.55	4.47	8.02
1500	.032" thick, natural	3.68	4.47	8.15
1550	Painted	4.27	4.47	8.74
1600	Ribbed 4" pitch, .032" thick, natural	3.66	4.47	8.13
1650	Painted	4.32	4.47	8.79
1700	.040" thick, natural	4.22	4.47	8.69
1750	Painted	4.78	4.47	9.25
1800	.050" thick, natural	4.75	4.47	9.22
1850	Painted	5.35	4.47	9.82
1900	Ribbed 8" pitch, .032" thick, natural	3.49	4.29	7.78
1950	Painted	4.14	4.31	8.45
2000	.040" thick, natural	4.05	4.31	8.36
2050	Painted	4.60	4.35	8.95
2100	.050" thick, natural	4.58	4.33	8.91
2150	Painted	5.20	4.37	9.57
3000	Steel, corrugated or ribbed, 29 Ga., .0135" thick, galvanized	2.29	4.05	6.34
3050	Colored	3.28	4.10	7.38
3100	26 Ga., .0179" thick, galvanized	2.40	4.07	6.47
3150	Colored	3.34	4.12	7.46
3200	24 Ga., .0239" thick, galvanized	2.87	4.09	6.96
3250	Colored	3.62	4.14	7.76
3300	22 Ga., .0299" thick, galvanized	3.39	4.11	7.50
3350	Colored	4.32	4.16	8.48
3400	20 Ga., .0359" thick, galvanized	3.39	4.11	7.50
3450	Colored	4.58	4.40	8.98
4100	Sandwich panels, factory fab., 1" polystyrene, steel core, 26 Ga., galv.	5.90	6.40	12.30
4200	Colored, 1 side	7.65	6.40	14.05
4300	2 sides	9.35	6.40	15.75
4400	2" polystyrene core, 26 Ga., galvanized	10.10	6.40	16.50
4500	Colored, 1 side	8.75	6.40	15.15
4600	2 sides	10.45	6.40	16.85
4700	22 Ga., baked enamel exterior	13.50	6.75	20.25
4800	Polyvinyl chloride exterior	14.30	6.75	21.05
5100	Textured aluminum, 4' x 8' x 5/16" plywood backing, single face	4.54	3.89	8.43
5200	Double face	5.95	3.89	9.84

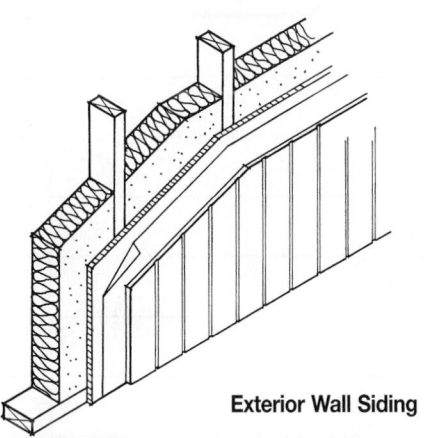

The table below lists costs per S.F. for exterior walls with wood siding. A variety of systems are presented using both wood and metal studs at 16″ and 24″ O.C.

Exterior Wall Siding

B2010 148	Panel, Shingle & Lap Siding	COST PER S.F.		
		MAT.	INST.	TOTAL
1400	Wood siding w/2″ x 4″ studs, 16″ O.C., insul. wall, 5/8″ text 1-11 fir ply.	3.57	6.30	9.87
1450	5/8″ text 1-11 cedar plywood	4.92	6.10	11.02
1500	1″ x 4″ vert T.&G. redwood	10.30	7.30	17.60
1600	1″ x 8″ vert T.&G. redwood	10.75	6.35	17.10
1650	1″ x 5″ rabbetted cedar bev. siding	6.90	6.55	13.45
1700	1″ x 6″ cedar drop siding	7	6.60	13.60
1750	1″ x 12″ rough sawn cedar	3.50	6.35	9.85
1800	1″ x 12″ sawn cedar, 1″ x 4″ battens	7.30	5.95	13.25
1850	1″ x 10″ redwood shiplap siding	7.10	6.15	13.25
1900	18″ no. 1 red cedar shingles, 5-1/2″ exposed	4.92	8.30	13.22
1950	6″ exposed	4.71	8.05	12.76
2000	6-1/2″ exposed	4.51	7.75	12.26
2100	7″ exposed	4.30	7.50	11.80
2150	7-1/2″ exposed	4.10	7.20	11.30
3000	8″ wide aluminum siding	4.53	5.65	10.18
3150	8″ plain vinyl siding	3.29	5.75	9.04
3250	8″ insulated vinyl siding	3.64	6.40	10.04
3300				
3400	2″ x 6″ studs, 16″ O.C., insul. wall, w/ 5/8″ text 1-11 fir plywood	4	6.50	10.50
3500	5/8″ text 1-11 cedar plywood	5.35	6.50	11.85
3600	1″ x 4″ vert T.&G. redwood	10.75	7.45	18.20
3700	1″ x 8″ vert T.&G. redwood	11.20	6.50	17.70
3800	1″ x 5″ rabbetted cedar bev siding	7.35	6.75	14.10
3900	1″ x 6″ cedar drop siding	7.45	6.80	14.25
4000	1″ x 12″ rough sawn cedar	3.93	6.55	10.48
4200	1″ x 12″ sawn cedar, 1″ x 4″ battens	7.75	6.10	13.85
4500	1″ x 10″ redwood shiplap siding	7.55	6.30	13.85
4550	18″ no. 1 red cedar shingles, 5-1/2″ exposed	5.35	8.50	13.85
4600	6″ exposed	5.15	8.20	13.35
4650	6-1/2″ exposed	4.94	7.95	12.89
4700	7″ exposed	4.73	7.65	12.38
4750	7-1/2″ exposed	4.53	7.40	11.93
4800	8″ wide aluminum siding	4.96	5.85	10.81
4850	8″ plain vinyl siding	4.83	10.15	14.98
4900	8″ insulated vinyl siding	5.20	10.75	15.95
4950	8″ fiber cement siding	4.73	10.95	15.68
5000	2″ x 6″ studs, 24″ O.C., insul. wall, 5/8″ text 1-11, fir plywood	3.83	6.10	9.93
5050	5/8″ text 1-11 cedar plywood	5.20	6.10	11.30
5100	1″ x 4″ vert T.&G. redwood	11.70	11.30	23
5150	1″ x 8″ vert T.&G. redwood	11	6.15	17.15

B2010 Exterior Walls

B2010 148	Panel, Shingle & Lap Siding	COST PER S.F.		
		MAT.	INST.	TOTAL
5200	1" x 5" rabbetted cedar bev siding	7.20	6.35	13.55
5250	1" x 6" cedar drop siding	7.30	6.40	13.70
5300	1" x 12" rough sawn cedar	3.76	6.15	9.91
5400	1" x 12" sawn cedar, 1" x 4" battens	7.55	5.75	13.30
5450	1" x 10" redwood shiplap siding	7.35	5.95	13.30
5500	18" no. 1 red cedar shingles, 5-1/2" exposed	5.20	8.10	13.30
5550	6" exposed	4.97	7.85	12.82
5650	7" exposed	4.56	7.30	11.86
5700	7-1/2" exposed	4.36	7	11.36
5750	8" wide aluminum siding	4.79	5.45	10.24
5800	8" plain vinyl siding	3.55	5.55	9.10
5850	8" insulated vinyl siding	3.90	6.20	10.10
5875	8" fiber cement siding	4.56	10.55	15.11
5900	3-5/8" metal studs, 16 Ga. 16" O.C. insul. wall, 5/8" text 1-11 fir plywood	4.40	6.65	11.05
5950	5/8" text 1-11 cedar plywood	5.75	6.65	12.40
6000	1" x 4" vert T.&G. redwood	11.15	7.60	18.75
6050	1" x 8" vert T.&G. redwood	11.60	6.70	18.30
6100	1" x 5" rabbetted cedar bev siding	7.75	6.90	14.65
6150	1" x 6" cedar drop siding	7.85	6.95	14.80
6200	1" x 12" rough sawn cedar	4.33	6.70	11.03
6250	1" x 12" sawn cedar, 1" x 4" battens	8.15	6.30	14.45
6300	1" x 10" redwood shiplap siding	7.95	6.45	14.40
6350	18" no. 1 red cedar shingles, 5-1/2" exposed	5.75	8.65	14.40
6500	6" exposed	5.55	8.35	13.90
6550	6-1/2" exposed	5.35	8.10	13.45
6600	7" exposed	5.15	7.80	12.95
6650	7-1/2" exposed	5.15	7.60	12.75
6700	8" wide aluminum siding	5.35	6	11.35
6750	8" plain vinyl siding	4.12	5.90	10.02
6800	8" insulated vinyl siding	4.47	6.50	10.97
7000	3-5/8" metal studs, 16 Ga. 24" O.C. insul wall, 5/8" text 1-11 fir plywood	4.03	5.90	9.93
7050	5/8" text 1-11 cedar plywood	5.40	5.90	11.30
7100	1" x 4" vert T.&G. redwood	10.75	7.10	17.85
7150	1" x 8" vert T.&G. redwood	11.20	6.15	17.35
7200	1" x 5" rabbetted cedar bev siding	7.40	6.40	13.80
7250	1" x 6" cedar drop siding	7.45	6.40	13.85
7300	1" x 12" rough sawn cedar	3.96	6.15	10.11
7350	1" x 12" sawn cedar 1" x 4" battens	7.75	5.75	13.50
7400	1" x 10" redwood shiplap siding	7.55	5.95	13.50
7450	18" no. 1 red cedar shingles, 5-1/2" exposed	5.60	8.40	14
7500	6" exposed	5.40	8.15	13.55
7550	6-1/2" exposed	5.15	7.85	13
7600	7" exposed	4.97	7.60	12.57
7650	7-1/2" exposed	4.56	7.05	11.61
7700	8" wide aluminum siding	4.99	5.50	10.49
7750	8" plain vinyl siding	3.75	5.60	9.35
7800	8" insul. vinyl siding	4.10	6.20	10.30

B2010 Exterior Walls

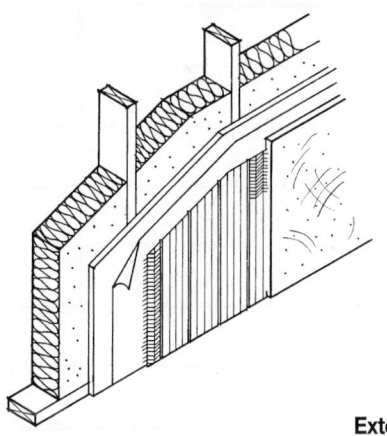

The table below lists costs for some typical stucco walls including all the components as demonstrated in the component block below. Prices are presented for backup walls using wood studs, metal studs and CMU.

Exterior Stucco Wall

B2010 151	Stucco Wall	COST PER S.F.		
		MAT.	INST.	TOTAL
2100	Cement stucco, 7/8" th., plywood sheathing, stud wall, 2" x 4", 16" O.C.	3.21	10.30	13.51
2200	24" O.C.	3.09	10	13.09
2300	2" x 6", 16" O.C.	3.64	10.45	14.09
2400	24" O.C.	3.47	10.10	13.57
2500	No sheathing, metal lath on stud wall, 2" x 4", 16" O.C.	2.31	8.95	11.26
2600	24" O.C.	2.19	8.65	10.84
2700	2" x 6", 16" O.C.	2.74	9.10	11.84
2800	24" O.C.	2.57	8.75	11.32
2900	1/2" gypsum sheathing, 3-5/8" metal studs, 16" O.C.	3.40	9.35	12.75
2950	24" O.C.	3.03	8.80	11.83
3000	Cement stucco, 5/8" th., 2 coats on std. CMU block, 8"x 16", 8" thick	3.66	12.50	16.16
3100	10" thick	5.35	12.80	18.15
3200	12" thick	5.50	14.75	20.25
3300	Std. light Wt. block 8" x 16", 8" thick	4.47	12.30	16.77
3400	10" thick	5.30	12.60	17.90
3500	12" thick	5.50	14.45	19.95
3600	3 coat stucco, self furring metal lath 3.4 Lb/SY, on 8" x 16", 8" thick	4.60	14.75	19.35
3700	10" thick	5.40	13.50	18.90
3800	12" thick	8.60	18.20	26.80
3900	Lt. Wt. block, 8" thick	4.54	13	17.54
4000	10" thick	5.35	13.30	18.65
4100	12" thick	5.60	15.15	20.75

B2010 152	E.I.F.S.	COST PER S.F.		
		MAT.	INST.	TOTAL
5100	E.I.F.S., plywood sheathing, stud wall, 2" x 4", 16" O.C., 1" EPS	3.86	10.35	14.21
5110	2" EPS	4.16	10.35	14.51
5120	3" EPS	4.46	10.35	14.81
5130	4" EPS	4.76	10.35	15.11
5140	2" x 6", 16" O.C., 1" EPS	4.29	10.50	14.79
5150	2" EPS	4.79	10.65	15.44
5160	3" EPS	4.89	10.50	15.39
5170	4" EPS	5.20	10.50	15.70
5180	Cement board sheathing, 3-5/8" metal studs, 16" O.C., 1" EPS	4.72	12.70	17.42
5190	2" EPS	5	12.70	17.70
5200	3" EPS	5.30	12.70	18
5210	4" EPS	6.45	12.70	19.15
5220	6" metal studs, 16" O.C., 1" EPS	5.30	12.75	18.05
5230	2" EPS	5.80	12.90	18.70

B2010 Exterior Walls

B2010 152	E.I.F.S.	COST PER S.F.		
		MAT.	INST.	TOTAL
5240	3" EPS	5.90	12.75	18.65
5250	4" EPS	7.05	12.75	19.80
5260	CMU block, 8" x 8" x 16", 1" EPS	5.05	14.75	19.80
5270	2" EPS	6.85	19.20	26.05
5280	3" EPS	5.65	14.75	20.40
5290	4" EPS	6.80	16.30	23.10
5300	8" x 10" x 16", 1" EPS	6.70	15.05	21.75
5310	2" EPS	7	15.05	22.05
5320	3" EPS	7.30	15.05	22.35
5330	4" EPS	7.60	15.05	22.65
5340	8" x 12" x 16", 1" EPS	6.85	17	23.85
5350	2" EPS	7.15	17	24.15
5360	3" EPS	7.45	17	24.45
5370	4" EPS	7.75	17	24.75

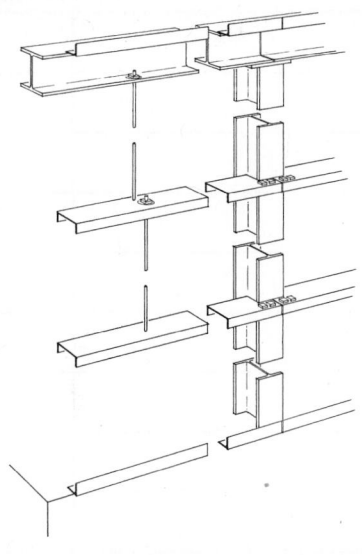

Description: The table below lists costs, $/S.F., for channel girts with sag rods and connector angles top and bottom for various column spacings, building heights and wind loads. Additive costs are shown for wind columns.

How to Use this Table: Add the cost of girts, sag rods, and angles to the framing costs of steel buildings clad in metal or composition siding. If the column spacing is in excess of the column spacing shown, use intermediate wind columns. Additive costs are shown under "Wind Columns".

B2010 154		Metal Siding Support						
	BLDG. HEIGHT (FT.)	WIND LOAD (P.S.F.)	COL. SPACING (FT.)		INTERMEDIATE COLUMNS	COST PER S.F.		
						MAT.	INST.	TOTAL
3000	18	20	20			2.46	3.93	6.39
3100					wind cols.	1.27	.30	1.57
3200		20	25			2.69	3.99	6.68
3300					wind cols.	1.01	.23	1.24
3400		20	30			2.96	4.06	7.02
3500					wind cols.	.84	.20	1.04
3600		20	35			3.26	4.13	7.39
3700					wind cols.	.72	.17	.89
3800		30	20			2.72	4	6.72
3900					wind cols.	1.27	.30	1.57
4000		30	25			2.96	4.06	7.02
4100					wind cols.	1.01	.23	1.24
4200		30	30			3.28	4.13	7.41
4300					wind cols.	1.15	.27	1.42
4600	30	20	20			2.38	2.78	5.16
4700					wind cols.	1.79	.43	2.22
4800		20	25			2.67	2.84	5.51
4900					wind cols.	1.43	.34	1.77
5000		20	30			3.01	2.93	5.94
5100					wind cols.	1.41	.34	1.75
5200		20	35			3.36	3.01	6.37
5300					wind cols.	1.40	.33	1.73
5400		30	20			2.70	2.85	5.55
5500					wind cols.	2.12	.50	2.62
5600		30	25			3	2.92	5.92
5700					wind cols.	2.02	.48	2.50
5800		30	30			3.38	3.01	6.39
5900					wind cols.	1.90	.45	2.35

B2010 Exterior Walls

B2010 160	Pole Barn Exterior Wall	COST PER S.F.		
		MAT.	INST.	TOTAL
2000	Pole barn exterior wall, pressure treated pole in concrete, 8' O.C.			
3000	Steel siding, 8' eave	4.63	10.45	15.08
3050	10' eave	4.42	9.60	14.02
3100	12' eave	4.20	8.80	13
3150	14' eave	4.07	8.30	12.37
3200	16' eave	3.98	7.95	11.93
4000	Aluminum siding, 8' eave	4.37	10.45	14.82
4050	10' eave	4.10	9.45	13.55
4100	12' eave	3.94	8.80	12.74
4150	14' eave	3.81	8.30	12.11
4200	16' eave	3.72	7.95	11.67
5000	Wood siding, 8' eave	5.10	9.20	14.30
5050	10' eave	4.83	8.20	13.03
5100	12' eave	4.67	7.50	12.17
5150	14' eave	4.54	7.05	11.59
5200	16' eave	4.45	6.65	11.10
6000	Plywood siding, 8' eave	3.88	9.15	13.03
6050	10' eave	3.61	8.15	11.76
6100	12' eave	3.45	7.45	10.90
6150	14' eave	3.32	7	10.32
6200	16' eave	3.23	6.60	9.83

B2020   Exterior Windows

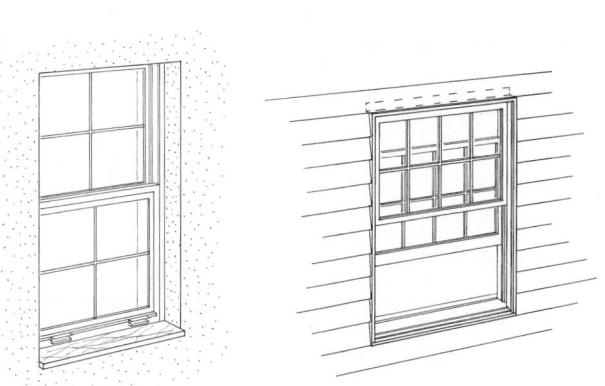

The table below lists window systems by material, type and size. Prices between sizes listed can be interpolated with reasonable accuracy. Prices include frame, hardware, and casing as illustrated in the component block below.

B2020 102				Wood Windows				
	MATERIAL	TYPE	GLAZING	SIZE	DETAIL	MAT.	INST.	TOTAL
3000	Wood	double hung	std. glass	2'-8" x 4'-6"		249	268	517
3050				3'-0" x 5'-6"		330	310	640
3100			insul. glass	2'-8" x 4'-6"		265	268	533
3150				3'-0" x 5'-6"		355	310	665
3200		sliding	std. glass	3'-4" x 2'-7"		330	223	553
3250				4'-4" x 3'-3"		345	244	589
3300				5'-4" x 6'-0"		440	294	734
3350			insul. glass	3'-4" x 2'-7"		405	261	666
3400				4'-4" x 3'-3"		420	283	703
3450				5'-4" x 6'-0"		530	335	865
3500		awning	std. glass	2'-10" x 1'-9"		231	134	365
3600				4'-4" x 2'-8"		375	130	505
3700			insul. glass	2'-10" x 1'-9"		282	155	437
3800				4'-4" x 2'-8"		460	142	602
3900		casement	std. glass	1'-10" x 3'-2"	1 lite	340	176	516
3950				4'-2" x 4'-2"	2 lite	605	234	839
4000				5'-11" x 5'-2"	3 lite	905	300	1,205
4050				7'-11" x 6'-3"	4 lite	1,275	360	1,635
4100				9'-11" x 6'-3"	5 lite	1,675	420	2,095
4150			insul. glass	1'-10" x 3'-2"	1 lite	340	176	516
4200				4'-2" x 4'-2"	2 lite	620	234	854
4250				5'-11" x 5'-2"	3 lite	955	300	1,255
4300				7'-11" x 6'-3"	4 lite	1,350	360	1,710
4350				9'-11" x 6'-3"	5 lite	1,675	420	2,095
4400		picture	std. glass	4'-6" x 4'-6"		480	300	780
4450				5'-8" x 4'-6"		535	335	870
4500		picture	insul. glass	4'-6" x 4'-6"		585	350	935
4550				5'-8" x 4'-6"		655	390	1,045
4600		fixed bay	std. glass	8' x 5'		1,650	555	2,205
4650				9'-9" x 5'-4"		1,175	785	1,960
4700			insul. glass	8' x 5'		2,250	555	2,805
4750				9'-9" x 5'-4"		1,275	785	2,060
4800		casement bay	std. glass	8' x 5'		1,725	640	2,365
4850			insul. glass	8' x 5'		1,875	640	2,515
4900		vert. bay	std. glass	8' x 5'		1,875	640	2,515
4950			insul. glass	8' x 5'		1,950	640	2,590

For customer support on your Square Foot Costs with RSMeans data, call 800.448.8182.

381

B2020 Exterior Windows

B2020 104 — Steel Windows

	MATERIAL	TYPE	GLAZING	SIZE	DETAIL	COST PER UNIT		
						MAT.	INST.	TOTAL
5000	Steel	double hung	1/4" tempered	2'-8" x 4'-6"		890	203	1,093
5050				3'-4" x 5'-6"		1,350	310	1,660
5100			insul. glass	2'-8" x 4'-6"		935	234	1,169
5150				3'-4" x 5'-6"		1,425	355	1,780
5200		horz. pivoted	std. glass	2' x 2'		274	68	342
5250				3' x 3'		615	153	768
5300				4' x 4'		1,100	271	1,371
5350				6' x 4'		1,650	405	2,055
5400			insul. glass	2' x 2'		290	78	368
5450				3' x 3'		650	176	826
5500				4' x 4'		1,150	310	1,460
5550				6' x 4'		1,750	470	2,220
5600		picture window	std. glass	3' x 3'		375	153	528
5650				6' x 4'		995	405	1,400
5700			insul. glass	3' x 3'		410	176	586
5750				6' x 4'		1,100	470	1,570
5800		industrial security	std. glass	2'-9" x 4'-1"		825	190	1,015
5850				4'-1" x 5'-5"		1,625	375	2,000
5900			insul. glass	2'-9" x 4'-1"		870	219	1,089
5950				4'-1" x 5'-5"		1,700	430	2,130
6000		comm. projected	std. glass	3'-9" x 5'-5"		1,325	345	1,670
6050				6'-9" x 4'-1"		1,800	465	2,265
6100			insul. glass	3'-9" x 5'-5"		1,400	395	1,795
6150				6'-9" x 4'-1"		1,900	535	2,435
6200		casement	std. glass	4'-2" x 4'-2"	2 lite	1,200	294	1,494
6250			insul. glass	4'-2" x 4'-2"		1,250	340	1,590
6300			std. glass	5'-11" x 5'-2"	3 lite	2,500	520	3,020
6350			insul. glass	5'-11" x 5'-2"		2,600	595	3,195

B2020 106 — Aluminum Windows

	MATERIAL	TYPE	GLAZING	SIZE	DETAIL	COST PER UNIT		
						MAT.	INST.	TOTAL
6400	Aluminum	double hung	insul. glass	3'-0" x 4'-0"		530	143	673
6450				4'-5" x 5'-3"		450	179	629
6500			insul. glass	3'-1" x 3'-2"		475	172	647
6550				4'-5" x 5'-3"		540	215	755
6600		sliding	std. glass	3' x 2'		242	143	385
6650				5' x 3'		370	159	529
6700				8' x 4'		395	238	633
6750				9' x 5'		595	355	950
6800			insul. glass	3' x 2'		260	143	403
6850				5' x 3'		430	159	589
6900				8' x 4'		635	238	873
6950				9' x 5'		935	355	1,290
7000		single hung	std. glass	2' x 3'		233	143	376
7050				2'-8" x 6'-8"		405	179	584
7100				3'-4" x 5'-0"		335	159	494
7150			insul. glass	2' x 3'		283	143	426
7200				2'-8" x 6'-8"		530	179	709
7250				3'-4" x 5'		375	159	534
7300		double hung	std. glass	2' x 3'		320	102	422
7350				2'-8" x 6'-8"		945	300	1,245

B2020 Exterior Windows

B2020 106			Aluminum Windows					

	MATERIAL	TYPE	GLAZING	SIZE	DETAIL	COST PER UNIT		
						MAT.	INST.	TOTAL
7400	Aluminum	double hung	std. glass	3'-4" x 5'		880	282	1,162
7450			insul. glass	2' x 3'		340	117	457
7500				2'-8" x 6'-8"		1,025	345	1,370
7550				3'-4" x 5'-0"		945	325	1,270
7600		casement	std. glass	3'-1" x 3'-2"		268	165	433
7650				4'-5" x 5'-3"		650	400	1,050
7700			insul. glass	3'-1" x 3'-2"		305	190	495
7750				4'-5" x 5'-3"		740	460	1,200
7800		hinged swing	std. glass	3' x 4'		580	203	783
7850				4' x 5'		970	340	1,310
7900			insul. glass	3' x 4'		630	234	864
7950				4' x 5'		1,050	390	1,440
8200		picture unit	std. glass	2'-0" x 3'-0"		174	102	276
8250				2'-8" x 6'-8"		515	300	815
8300				3'-4" x 5'-0"		485	282	767
8350			insul. glass	2'-0" x 3'-0"		197	117	314
8400				2'-8" x 6'-8"		585	345	930
8450				3'-4" x 5'-0"		550	325	875
8500		awning type	std. glass	3'-0" x 3'-0"	2 lite	475	102	577
8550				3'-0" x 4'-0"	3 lite	550	143	693
8600				3'-0" x 5'-4"	4 lite	660	143	803
8650				4'-0" x 5'-4"	4 lite	725	159	884
8700			insul. glass	3'-0" x 3'-0"	2 lite	505	102	607
8750				3'-0" x 4'-0"	3 lite	635	143	778
8800				3'-0" x 5'-4"	4 lite	780	143	923
8850				4'-0" x 5'-4"	4 lite	865	159	1,024
9051								
9052								

B2020 210	Tubular Aluminum Framing	COST/S.F. OPNG.		
		MAT.	INST.	TOTAL
1100	Alum flush tube frame, for 1/4" glass, 1-3/4"x4", 5'x6'opng, no inter horiz	15.20	12.35	27.55
1150	One intermediate horizontal	20.50	14.70	35.20
1200	Two intermediate horizontals	26	17.05	43.05
1250	5' x 20' opening, three intermediate horizontals	14	10.55	24.55
1400	1-3/4" x 4-1/2", 5' x 6' opening, no intermediate horizontals	17.35	12.35	29.70
1450	One intermediate horizontal	23	14.70	37.70
1500	Two intermediate horizontals	28.50	17.05	45.55
1550	5' x 20' opening, three intermediate horizontals	15.55	10.55	26.10
1700	For insulating glass, 2"x4-1/2", 5'x6' opening, no intermediate horizontals	16.80	13	29.80
1750	One intermediate horizontal	21.50	15.55	37.05
1800	Two intermediate horizontals	26.50	18.05	44.55
1850	5' x 20' opening, three intermediate horizontals	14.50	11.15	25.65
2000	Thermal break frame, 2-1/4"x4-1/2", 5'x6'opng, no intermediate horizontals	17.50	13.25	30.75
2050	One intermediate horizontal	23.50	16.25	39.75
2100	Two intermediate horizontals	29	19.20	48.20
2150	5' x 20' opening, three intermediate horizontals	15.85	11.70	27.55

B2020 Exterior Windows

The table below lists costs of curtain wall and spandrel panels per S.F. Costs do not include structural framing used to hang the panels.

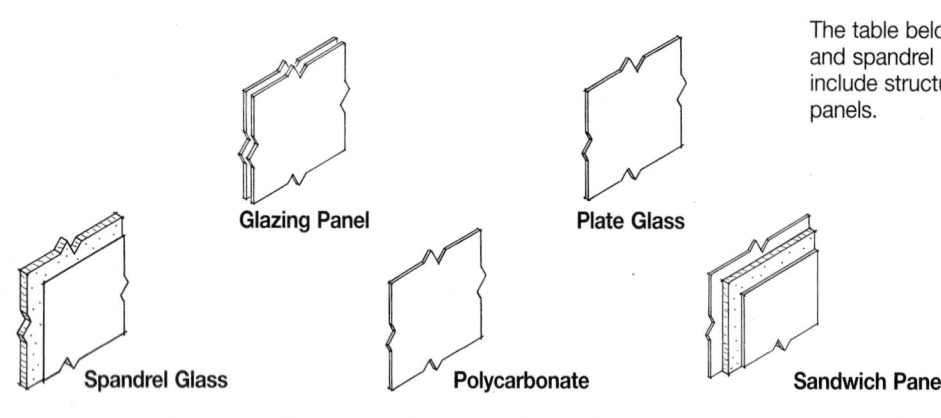

Glazing Panel

Plate Glass

Spandrel Glass

Polycarbonate

Sandwich Panel

B2020 220	Curtain Wall Panels	COST PER S.F.		
		MAT.	INST.	TOTAL
1000	Glazing panel, insulating, 1/2" thick, 2 lites 1/8" float, clear	11.40	12.35	23.75
1100	Tinted	15.95	12.35	28.30
1200	5/8" thick units, 2 lites 3/16" float, clear	15.15	13.05	28.20
1300	Tinted	15.20	13.05	28.25
1400	1" thick units, 2 lites, 1/4" float, clear	19.10	15.65	34.75
1500	Tinted	26	15.65	41.65
1600	Heat reflective film inside	30.50	13.85	44.35
1700	Light and heat reflective glass, tinted	35	13.85	48.85
2000	Plate glass, 1/4" thick, clear	10.90	9.80	20.70
2050	Tempered	7.50	9.80	17.30
2100	Tinted	10.40	9.80	20.20
2200	3/8" thick, clear	12.30	15.65	27.95
2250	Tempered	19.45	15.65	35.10
2300	Tinted	18.20	15.65	33.85
2400	1/2" thick, clear	21	21.50	42.50
2450	Tempered	29	21.50	50.50
2500	Tinted	32	21.50	53.50
2600	3/4" thick, clear	41	33.50	74.50
2650	Tempered	48	33.50	81.50
3000	Spandrel glass, panels, 1/4" plate glass insul w/fiberglass, 1" thick	18.90	9.80	28.70
3100	2" thick	23.50	9.80	33.30
3200	Galvanized steel backing, add	6.60		6.60
3300	3/8" plate glass, 1" thick	33	9.80	42.80
3400	2" thick	37.50	9.80	47.30
4000	Polycarbonate, masked, clear or colored, 1/8" thick	16.75	6.90	23.65
4100	3/16" thick	20.50	7.10	27.60
4200	1/4" thick	20	7.60	27.60
4300	3/8" thick	31.50	7.85	39.35
5000	Facing panel, textured alum., 4' x 8' x 5/16" plywood backing, sgl face	4.54	3.89	8.43
5100	Double face	5.95	3.89	9.84
5200	4' x 10' x 5/16" plywood backing, single face	4.84	3.89	8.73
5300	Double face	6.50	3.89	10.39
5400	4' x 12' x 5/16" plywood backing, single face	4.84	3.89	8.73
5500	Sandwich panel, 22 Ga. galv., both sides 2" insulation, enamel exterior	13.50	6.75	20.25
5600	Polyvinylidene fluoride exterior finish	14.30	6.75	21.05
5700	26 Ga., galv. both sides, 1" insulation, colored 1 side	7.65	6.40	14.05
5800	Colored 2 sides	9.35	6.40	15.75

B2030 Exterior Doors

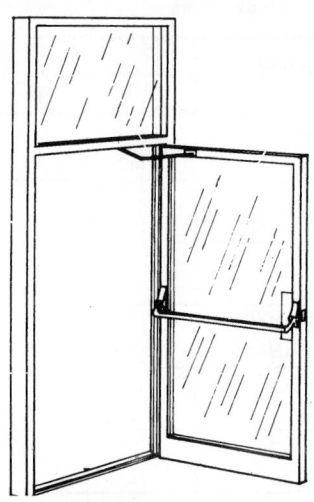

Costs are listed for exterior door systems by material, type and size. Prices between sizes listed can be interpolated with reasonable accuracy. Prices are per opening for a complete door system including frame.

B2030 110					Glazed Doors, Steel or Aluminum			
	MATERIAL	**TYPE**	**DOORS**	**SPECIFICATION**	**OPENING**	**COST PER OPNG.**		
						MAT.	**INST.**	**TOTAL**
5600	St. Stl. & glass	revolving	stock unit	manual oper.	6'-0" x 7'-0"	54,000	9,475	63,475
5650				auto Cntrls.	6'-10" x 7'-0"	71,000	10,200	81,200
5700	Bronze	revolving	stock unit	manual oper.	6'-10" x 7'-0"	55,000	19,100	74,100
5750				auto Cntrls.	6'-10" x 7'-0"	72,000	19,800	91,800
5800	St. Stl. & glass	balanced	standard	economy	3'-0" x 7'-0"	11,100	1,625	12,725
5850				premium	3'-0" x 7'-0"	18,400	2,100	20,500
6300	Alum. & glass	w/o transom	narrow stile	w/panic Hrdwre.	3'-0" x 7'-0"	2,725	1,075	3,800
6350				dbl. door, Hrdwre.	6'-0" x 7'-0"	4,575	1,775	6,350
6400			wide stile	hdwre.	3'-0" x 7'-0"	2,925	1,050	3,975
6450				dbl. door, Hdwre.	6'-0" x 7'-0"	5,975	2,125	8,100
6500			full vision	hdwre.	3'-0" x 7'-0"	3,725	1,700	5,425
6550				dbl. door, Hdwre.	6'-0" x 7'-0"	5,400	2,400	7,800
6600			non-standard	hdwre.	3'-0" x 7'-0"	3,050	1,050	4,100
6650				dbl. door, Hdwre.	6'-0" x 7'-0"	6,100	2,125	8,225
6700			bronze fin.	hdwre.	3'-0" x 7'-0"	2,500	1,050	3,550
6750				dbl. door, Hrdwre.	6'-0" x 7'-0"	5,000	2,125	7,125
6800			black fin.	hdwre.	3'-0" x 7'-0"	2,925	1,050	3,975
6850				dbl. door, Hdwre.	6'-0" x 7'-0"	5,850	2,125	7,975
6900		w/transom	narrow stile	hdwre.	3'-0" x 10'-0"	3,175	1,225	4,400
6950				dbl. door, Hdwre.	6'-0" x 10'-0"	5,050	2,150	7,200
7000			wide stile	hdwre.	3'-0" x 10'-0"	3,725	1,475	5,200
7050				dbl. door, Hdwre.	6'-0" x 10'-0"	5,750	2,550	8,300
7100			full vision	hdwre.	3'-0" x 10'-0"	4,025	1,625	5,650
7150				dbl. door, Hdwre.	6'-0" x 10'-0"	6,075	2,800	8,875
7200			non-standard	hdwre.	3'-0" x 10'-0"	3,100	1,150	4,250
7250				dbl. door, Hdwre.	6'-0" x 10'-0"	6,225	2,300	8,525
7300			bronze fin.	hdwre.	3'-0" x 10'-0"	2,575	1,150	3,725
7350				dbl. door, Hdwre.	6'-0" x 10'-0"	5,150	2,300	7,450
7400			black fin.	hdwre.	3'-0" x 10'-0"	2,975	1,150	4,125
7450				dbl. door, Hdwre.	6'-0" x 10'-0"	5,975	2,300	8,275
7500		revolving	stock design	minimum	6'-10" x 7'-0"	34,700	3,800	38,500
7550				average	6'-0" x 7'-0"	40,200	4,775	44,975

B2030 Exterior Doors

B2030 110	Glazed Doors, Steel or Aluminum

	MATERIAL	TYPE	DOORS	SPECIFICATION	OPENING	COST PER OPNG.		
						MAT.	INST.	TOTAL
7600	Alum. & glass	revolving	stock design	maximum	6'-10" x 7'-0"	48,000	6,350	54,350
7650				min., automatic	6'-10" x 7'-0"	51,500	4,500	56,000
7700				avg., automatic	6'-10" x 7'-0"	57,000	5,475	62,475
7750				max., automatic	6'-10" x 7'-0"	65,000	7,050	72,050
7800		balanced	standard	economy	3'-0" x 7'-0"	7,725	1,600	9,325
7850				premium	3'-0" x 7'-0"	9,250	2,050	11,300
7900		mall front	sliding panels	alum. fin.	16'-0" x 9'-0"	4,225	905	5,130
7950					24'-0" x 9'-0"	6,050	1,675	7,725
8000				bronze fin.	16'-0" x 9'-0"	4,925	1,050	5,975
8050					24'-0" x 9'-0"	7,050	1,950	9,000
8100			fixed panels	alum. fin.	48'-0" x 9'-0"	10,900	1,300	12,200
8150				bronze fin.	48'-0" x 9'-0"	12,700	1,525	14,225
8200		sliding entrance	5' x 7' door	electric oper.	12'-0" x 7'-6"	9,575	1,675	11,250
8250		sliding patio	temp. glass	economy	6'-0" x 7'-0"	1,850	310	2,160
8300			temp. glass	economy	12'-0" x 7'-0"	4,425	410	4,835
8350				premium	6'-0" x 7'-0"	2,775	465	3,240
8400					12'-0" x 7'-0"	6,650	615	7,265

B2030 Exterior Doors

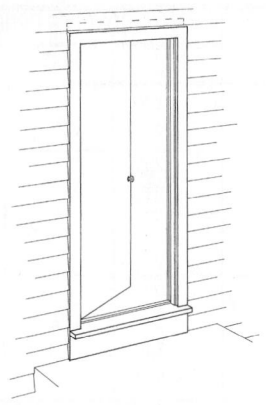

Exterior Door System

The table below lists exterior door systems by material, type and size. Prices between sizes listed can be interpolated with reasonable accuracy. Prices are per opening for a complete door system including frame and required hardware.

Wood doors in this table are designed with wood frames and metal doors with hollow metal frames. Depending upon quality the total material and installation cost of a wood door frame is about the same as a hollow metal frame.

B2030 210 — Wood Doors

	MATERIAL	TYPE	DOORS	SPECIFICATION	OPENING	COST PER OPNG. MAT.	INST.	TOTAL
2350	Birch	solid core	single door	hinged	2'-6" x 6'-8"	2,050	335	2,385
2400					2'-6" x 7'-0"	2,050	335	2,385
2450					2'-8" x 7'-0"	2,125	335	2,460
2500					3'-0" x 7'-0"	2,050	335	2,385
2550			double door		2'-6" x 6'-8"	3,925	615	4,540
2600					2'-6" x 7'-0"	3,925	615	4,540
2650					2'-8" x 7'-0"	4,100	615	4,715
2700					3'-0" x 7'-0"	3,800	535	4,335
2750	Wood	combination	storm & screen		3'-0" x 6'-8"	340	82.50	422.50
2800					3'-0" x 7'-0"	375	91	466
2850		overhead	panels, H.D.	manual oper.	8'-0" x 8'-0"	1,350	620	1,970
2900					10'-0" x 10'-0"	1,875	685	2,560
2950					12'-0" x 12'-0"	2,575	825	3,400
3000					14'-0" x 14'-0"	4,000	950	4,950
3050					20'-0" x 16'-0"	6,575	1,900	8,475
3100				electric oper.	8'-0" x 8'-0"	2,575	930	3,505
3150					10'-0" x 10'-0"	3,100	995	4,095
3200					12'-0" x 12'-0"	3,800	1,125	4,925
3250					14'-0" x 14'-0"	5,225	1,250	6,475
3300					20'-0" x 16'-0"	7,900	2,525	10,425

B2030 220 — Steel Doors

	MATERIAL	TYPE	DOORS	SPECIFICATION	OPENING	COST PER OPNG. MAT.	INST.	TOTAL
3350	Steel 18 Ga.	hollow metal	1 door w/frame	no label	2'-6" x 7'-0"	2,425	345	2,770
3400					2'-8" x 7'-0"	2,425	345	2,770
3450					3'-0" x 7'-0"	2,450	370	2,820
3500					3'-6" x 7'-0"	2,650	365	3,015
3550		hollow metal	1 door w/frame	no label	4'-0" x 8'-0"	2,875	365	3,240
3600			2 doors w/frame	no label	5'-0" x 7'-0"	4,775	640	5,415
3650					5'-4" x 7'-0"	4,775	640	5,415
3700					6'-0" x 7'-0"	4,775	640	5,415
3750					7'-0" x 7'-0"	5,225	680	5,905
3800					8'-0" x 8'-0"	5,675	680	6,355

B2030 Exterior Doors

B2030 220 — Steel Doors

	MATERIAL	TYPE	DOORS	SPECIFICATION	OPENING	COST PER OPNG.		
						MAT.	INST.	TOTAL
3850	Steel 18 Ga.	hollow metal	1 door w/frame	"A" label	2'-6" x 7'-0"	2,875	420	3,295
3900					2'-8" x 7'-0"	2,850	425	3,275
3950					3'-0" x 7'-0"	2,850	425	3,275
4000					3'-6" x 7'-0"	3,025	430	3,455
4050					4'-0" x 8'-0"	3,200	455	3,655
4100			2 doors w/frame	"A" label	5'-0" x 7'-0"	5,600	775	6,375
4150					5'-4" x 7'-0"	5,600	785	6,385
4200					6'-0" x 7'-0"	5,600	785	6,385
4250					7'-0" x 7'-0"	5,950	800	6,750
4300					8'-0" x 8'-0"	5,950	805	6,755
4350	Steel 24 Ga.	overhead	sectional	manual oper.	8'-0" x 8'-0"	1,100	620	1,720
4400					10'-0" x 10'-0"	1,425	685	2,110
4450					12'-0" x 12'-0"	1,675	825	2,500
4500					20'-0" x 14'-0"	3,950	1,775	5,725
4550				electric oper.	8'-0" x 8'-0"	2,325	930	3,255
4600					10'-0" x 10'-0"	2,650	995	3,645
4650					12'-0" x 12'-0"	2,900	1,125	4,025
4700					20'-0" x 14'-0"	5,275	2,400	7,675
4750	Steel	overhead	rolling	manual oper.	8'-0" x 8'-0"	800	895	1,695
4800					10'-0" x 10'-0"	1,175	1,025	2,200
4850					12'-0" x 12'-0"	1,325	1,200	2,525
4900					14'-0" x 14'-0"	3,300	1,775	5,075
4950					20'-0" x 12'-0"	2,300	1,600	3,900
5000					20'-0" x 16'-0"	4,025	2,375	6,400
5050				electric oper.	8'-0" x 8'-0"	2,100	1,175	3,275
5100					10'-0" x 10'-0"	2,475	1,300	3,775
5150					12'-0" x 12'-0"	2,625	1,475	4,100
5200					14'-0" x 14'-0"	4,600	2,050	6,650
5250					20'-0" x 12'-0"	3,700	1,875	5,575
5300					20'-0" x 16'-0"	5,425	2,650	8,075
5350				fire rated	10'-0" x 10'-0"	2,350	1,300	3,650
5400			rolling grille	manual oper.	10'-0" x 10'-0"	2,950	1,425	4,375
5450					15'-0" x 8'-0"	3,300	1,775	5,075
5500		vertical lift	1 door w/frame	motor operator	16'-0" x 16'-0"	24,000	5,375	29,375
5550					32'-0" x 24'-0"	55,000	3,575	58,575

B2030 230 — Aluminum Doors

	MATERIAL	TYPE	DOORS	SPECIFICATION	OPENING	COST PER OPNG.		
						MAT.	INST.	TOTAL
6000	Aluminum	combination	storm & screen	hinged	3'-0" x 6'-8"	345	88	433
6050					3'-0" x 7'-0"	380	97	477
6100		overhead	rolling grille	manual oper.	12'-0" x 12'-0"	4,750	2,525	7,275
6150				motor oper.	12'-0" x 12'-0"	6,175	2,800	8,975
6200	Alum. & Fbrgls.	overhead	heavy duty	manual oper.	12'-0" x 12'-0"	3,675	825	4,500
6250				electric oper.	12'-0" x 12'-0"	4,900	1,125	6,025

B3010   Roof Coverings

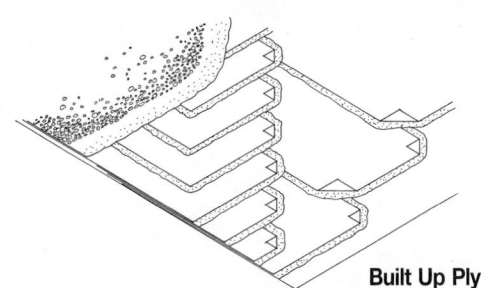

Multiple ply roofing is the most popular covering for minimum pitch roofs.

Built Up Ply

B3010 105	Built-Up	COST PER S.F.		
		MAT.	INST.	TOTAL
1200	Asphalt flood coat w/gravel; not incl. insul, flash., nailers			
1300				
1400	Asphalt base sheets & 3 plies #15 asphalt felt, mopped	1.14	2.02	3.16
1500	On nailable deck	1.18	2.11	3.29
1600	4 plies #15 asphalt felt, mopped	1.56	2.23	3.79
1700	On nailable deck	1.38	2.34	3.72
1800	Coated glass base sheet, 2 plies glass (type IV), mopped	1.22	2.02	3.24
1900	For 3 plies	1.47	2.23	3.70
2000	On nailable deck	1.38	2.34	3.72
2300	4 plies glass fiber felt (type IV), mopped	1.82	2.23	4.05
2400	On nailable deck	1.64	2.34	3.98
2500	Organic base sheet & 3 plies #15 organic felt, mopped	1.19	2.24	3.43
2600	On nailable deck	1.19	2.34	3.53
2700	4 plies #15 organic felt, mopped	1.48	2.02	3.50
2750				
2800	Asphalt flood coat, smooth surface, not incl. insul, flash., nailers			
2900	Asphalt base sheet & 3 plies #15 asphalt felt, mopped	1.21	1.85	3.06
3000	On nailable deck	1.12	1.93	3.05
3100	Coated glass fiber base sheet & 2 plies glass fiber felt, mopped	1.17	1.78	2.95
3200	On nailable deck	1.10	1.85	2.95
3300	For 3 plies, mopped	1.42	1.93	3.35
3400	On nailable deck	1.33	2.02	3.35
3700	4 plies glass fiber felt (type IV), mopped	1.67	1.93	3.60
3800	On nailable deck	1.58	2.02	3.60
3900	Organic base sheet & 3 plies #15 organic felt, mopped	1.22	1.85	3.07
4000	On nailable decks	1.13	1.93	3.06
4100	4 plies #15 organic felt, mopped	1.43	2.02	3.45
4200	Coal tar pitch with gravel surfacing			
4300	4 plies #15 tarred felt, mopped	2.19	2.11	4.30
4400	3 plies glass fiber felt (type IV), mopped	1.81	2.34	4.15
4500	Coated glass fiber base sheets 2 plies glass fiber felt, mopped	1.87	2.34	4.21
4600	On nailable decks	1.63	2.47	4.10
4800	3 plies glass fiber felt (type IV), mopped	2.51	2.11	4.62
4900	On nailable decks	2.27	2.23	4.50

B3010 Roof Coverings

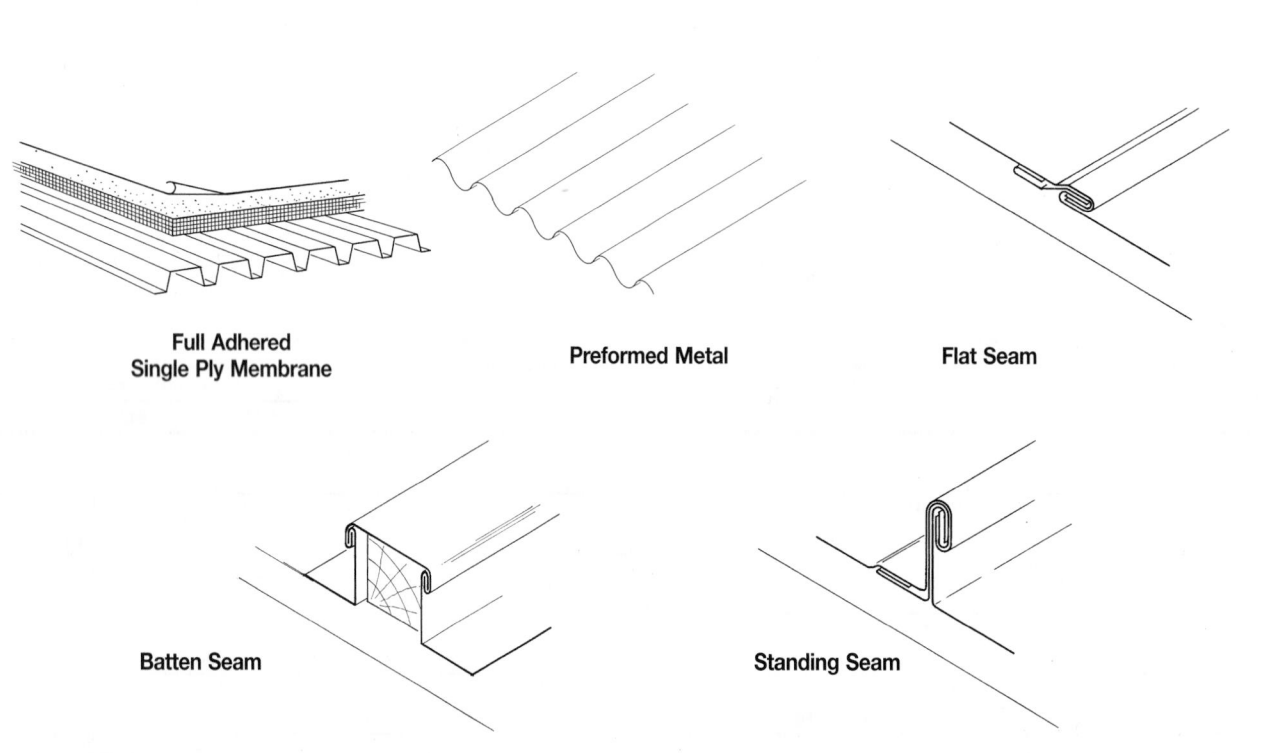

Full Adhered
Single Ply Membrane

Preformed Metal

Flat Seam

Batten Seam

Standing Seam

B3010 120	Single Ply Membrane	COST PER S.F.		
		MAT.	INST.	TOTAL
1000	CSPE (Chlorosulfonated polyethylene), 45 mils, plate attached	2.66	.83	3.49
1100	Loosely laid with stone ballast	2.66	.83	3.49
2000	EPDM (Ethylene propylene diene monomer), 45 mils, fully adhered	1.24	1.11	2.35
2100	Loosely laid with stone ballast	.94	.57	1.51
2200	Mechanically fastened with batten strips	.87	.83	1.70
3300	60 mils, fully adhered	1.40	1.11	2.51
3400	Loosely laid with stone ballast	1.11	.57	1.68
3500	Mechanically fastened with batten strips	1.03	.83	1.86
4000	Modified bit., SBS modified, granule surface cap sheet, mopped, 150 mils	1.47	2.22	3.69
4100	Smooth surface cap sheet, mopped, 145 mils	.90	2.11	3.01
4500	APP modified, granule surface cap sheet, torched, 180 mils	1.06	1.44	2.50
4600	Smooth surface cap sheet, torched, 170 mils	.85	1.37	2.22
5000	PIB (Polyisobutylene), 100 mils, fully adhered with contact cement	2.88	1.11	3.99
5100	Loosely laid with stone ballast	2.32	.57	2.89
5200	Partially adhered with adhesive	2.77	.83	3.60
5300	Hot asphalt attachment	2.65	.83	3.48
6000	Reinforced PVC, 48 mils, loose laid and ballasted with stone	1.24	.57	1.81
6100	Partially adhered with mechanical fasteners	1.17	.83	2
6200	Fully adhered with adhesive	1.68	1.11	2.79
6300	Reinforced PVC, 60 mils, loose laid and ballasted with stone	1.25	.57	1.82
6400	Partially adhered with mechanical fasteners	1.18	.83	2.01
6500	Fully adhered with adhesive	1.69	1.11	2.80

B3010 130	Preformed Metal Roofing	COST PER S.F.		
		MAT.	INST.	TOTAL
0200	Corrugated roofing, aluminum, mill finish, .0175″ thick, .272 P.S.F.	1.10	2.03	3.13
0250	.0215″ thick, .334 P.S.F.	1.36	2.03	3.39
0300	.0240″ thick, .412 P.S.F.	1.98	2.03	4.01
0350	.0320″ thick, .552 P.S.F.	3.34	2.03	5.37

B3010 Roof Coverings

B3010 130	Preformed Metal Roofing	COST PER S.F.		
		MAT.	INST.	TOTAL
0400	Painted, .0175" thick, .280 P.S.F.	1.60	2.03	3.63
0450	.0215" thick, .344 P.S.F.	1.71	2.03	3.74
0500	.0240" thick, .426 P.S.F.	2.42	2.03	4.45
0550	.0320" thick, .569 P.S.F.	3.61	2.03	5.64
0700	Fiberglass, 6 oz., .375 P.S.F.	2.72	2.43	5.15
0750	8 oz., .5 P.S.F.	4.97	2.43	7.40
0900	Steel, galvanized, 29 ga., .72 P.S.F.	1.84	2.21	4.05
0930	26 ga., .91 P.S.F.	1.58	2.32	3.90
0950	24 ga., 1.26 P.S.F.	2.22	2.43	4.65
0970	22 ga., 1.45 P.S.F.	3.30	2.56	5.86
1000	Colored, 28 ga., 1.08 P.S.F.	1.99	2.32	4.31
1050	26 ga., 1.43 P.S.F.	2.33	2.43	4.76

B3010 135	Formed Metal	COST PER S.F.		
		MAT.	INST.	TOTAL
1000	Batten seam, formed copper roofing, 3" min slope, 16 oz., 1.2 P.S.F.	14.10	6.75	20.85
1100	18 oz., 1.35 P.S.F.	15.85	7.40	23.25
2000	Zinc copper alloy, 3" min slope, .020" thick, .88 P.S.F.	14.60	6.20	20.80
3000	Flat seam, copper, 1/4" min. slope, 16 oz., 1.2 P.S.F.	10.35	6.20	16.55
3100	18 oz., 1.35 P.S.F.	11.60	6.45	18.05
5000	Standing seam, copper, 2-1/2" min. slope, 16 oz., 1.25 P.S.F.	11.35	5.70	17.05
5100	18 oz., 1.40 P.S.F.	12.60	6.20	18.80
6000	Zinc copper alloy, 2-1/2" min. slope, .020" thick, .87 P.S.F.	14.10	6.20	20.30
6100	.032" thick, 1.39 P.S.F.	20	6.75	26.75

B30 Roofing

B3010   Roof Coverings

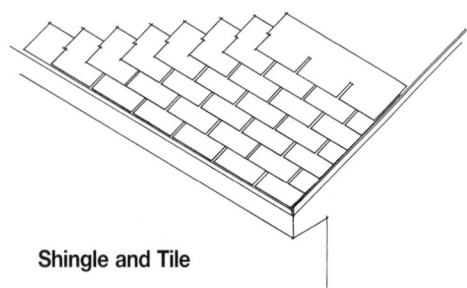

Shingle and Tile

Shingles and tiles are practical in applications where the roof slope is more than 3-1/2″ per foot of rise. Lines 1100 through 6000 list the various materials and the weight per square foot.

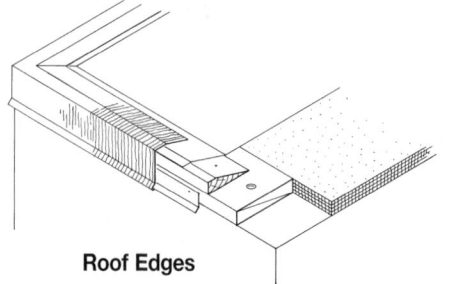

Roof Edges

The table below lists the costs for various types of roof perimeter edge treatments.

Roof edge systems include the cost per L.F. for a 2″ x 8″ treated wood nailer fastened at 4′-0″ O.C. and a diagonally cut 4″ x 6″ treated wood cant.

Roof edge and base flashing are assumed to be made from the same material.

B3010 140	Shingle & Tile	COST PER S.F.		
		MAT.	INST.	TOTAL
1095	Asphalt roofing			
1100	Strip shingles, 4″ slope, inorganic class A 210-235 lb./sq.	.89	1.25	2.14
1150	Organic, class C, 235-240 lb./sq.	1.11	1.35	2.46
1200	Premium laminated multi-layered, class A, 260-300 lb./sq.	1.64	1.86	3.50
1545	Metal roofing			
1550	Alum., shingles, colors, 3″ min slope, .019″ thick, 0.4 PSF	2.86	1.34	4.20
1850	Steel, colors, 3″ min slope, 26 gauge, 1.0 PSF	4.70	2.80	7.50
2795	Slate roofing			
2800	Slate roofing, 4″ min. slope, shingles, 3/16″ thick, 8.0 PSF	6.20	3.50	9.70
3495	Wood roofing			
3500	4″ min slope, cedar shingles, 16″ x 5″, 5″ exposure 1.6 PSF	3.37	2.66	6.03
4000	Shakes, 18″, 8-1/2″ exposure, 2.8 PSF	3.12	3.29	6.41
5095	Tile roofing			
5100	Aluminum, mission, 3″ min slope, .019″ thick, 0.65 PSF	9.25	2.57	11.82
6000	Clay tile, flat shingle, interlocking, 15″, 166 pcs/sq, fireflashed blend	5.25	3.07	8.32

B3010 320	Roof Deck Rigid Insulation	COST PER S.F.		
		MAT.	INST.	TOTAL
0100	Fiberboard low density			
0150	1″ thick R2.78	.76	.76	1.52
0300	1 1/2″ thick R4.17	1.10	.76	1.86
0350	2″ thick R5.56	1.39	.76	2.15
0370	Fiberboard high density, 1/2″ thick R1.3	.42	.64	1.06
0380	1″ thick R2.5	.74	.76	1.50
0390	1 1/2″ thick R3.8	1.03	.76	1.79
0410	Fiberglass, 3/4″ thick R2.78	.80	.64	1.44
0450	15/16″ thick R3.70	1.02	.64	1.66
0550	1-5/16″ thick R5.26	1.68	.64	2.32
0650	2 7/16″ thick R10	2.01	.76	2.77
1510	Polyisocyanurate 2#/CF density, 1″ thick	.59	.52	1.11
1550	1 1/2″ thick	.76	.55	1.31
1600	2″ thick	.93	.60	1.53
1650	2 1/2″ thick	1.21	.62	1.83
1700	3″ thick	1.33	.64	1.97
1750	3 1/2″ thick	2	.64	2.64
1800	Tapered for drainage	.69	.52	1.21

B3010 Roof Coverings

B3010 320 — Roof Deck Rigid Insulation

		COST PER S.F.		
		MAT.	INST.	TOTAL
1810	Expanded polystyrene, 1#/CF density, 3/4" thick R2.89	.34	.49	.83
1820	2" thick R7.69	.71	.55	1.26
1830	15 PSI compressive strength, 1" thick, R5	.73	.49	1.22
1835	2" thick R10	.91	.55	1.46
1840	3" thick R15	1.69	.64	2.33
2550	40 PSI compressive strength, 1" thick R5	1.08	.49	1.57
2600	2" thick R10	1.94	.55	2.49
2650	3" thick R15	2.75	.64	3.39
2700	4" thick R20	3.57	.64	4.21
2750	Tapered for drainage	1.08	.52	1.60
2810	60 PSI compressive strength, 1" thick R5	1.28	.50	1.78
2850	2" thick R10	2.31	.57	2.88
2900	Tapered for drainage	1.22	.52	1.74
3050	Composites with 2" EPS			
3060	1" Fiberboard	1.72	.64	2.36
3070	7/16" oriented strand board	1.39	.76	2.15
3080	1/2" plywood	1.70	.76	2.46
3090	1" perlite	1.41	.76	2.17
4000	Composites with 1-1/2" polyisocyanurate			
4010	1" fiberboard	1.30	.76	2.06
4020	1" perlite	1.18	.73	1.91
4030	7/16" oriented strand board	1.04	.76	1.80

B3010 420 — Roof Edges

	EDGE TYPE	DESCRIPTION	SPECIFICATION	FACE HEIGHT		COST PER L.F.		
						MAT.	INST.	TOTAL
1000	Aluminum	mill finish	.050" thick	4"		15	11.30	26.30
1100				6"		15.30	11.65	26.95
1300		duranodic	.050" thick	4"		15.95	11.30	27.25
1400				6"		16.55	11.65	28.20
1600		painted	.050" thick	4"		16.05	11.30	27.35
1700				6"		17.45	11.65	29.10
2000	Copper	plain	16 oz.	4"		35	11.30	46.30
2100				6"		45	11.65	56.65
2300			20 oz.	4"		44	12.95	56.95
2400				6"		50.50	12.45	62.95
2700	Sheet Metal	galvanized	20 Ga.	4"		18.05	13.50	31.55
2800				6"		18.10	13.50	31.60
3000			24 Ga.	4"		14.95	11.30	26.25
3100				6"		15	11.30	26.30

B3010 430 — Flashing

	MATERIAL	BACKING	SIDES	SPECIFICATION	QUANTITY	COST PER S.F.		
						MAT.	INST.	TOTAL
0040	Aluminum	none		.019"		1.54	4.08	5.62
0050				.032"		1.49	4.08	5.57
0300		fabric	2	.004"		1.76	1.79	3.55
0400		mastic		.004"		1.76	1.79	3.55
0700	Copper	none		16 oz.	<500 lbs.	8.90	5.15	14.05
0800				24 oz.	<500 lbs.	16.25	5.65	21.90
3500	PVC black	none		.010"		.29	2.08	2.37
3700				.030"		.36	2.08	2.44
4200	Neoprene			1/16"		2.86	2.08	4.94
4500	Stainless steel	none		.015"	<500 lbs.	7.30	5.15	12.45
4600	Copper clad				>2000 lbs.	7.45	3.82	11.27
5000	Plain			32 ga.		3.69	3.82	7.51

393

For customer support on your Square Foot Costs with RSMeans data, call 800.448.8182.

B30 Roofing

B3010 Roof Coverings

B3010 610 — Gutters

	SECTION	MATERIAL	THICKNESS	SIZE	FINISH	COST PER L.F. MAT.	COST PER L.F. INST.	COST PER L.F. TOTAL
0050	Box	aluminum	.027″	5″	enameled	3.28	6.05	9.33
0100					mill	3.18	6.05	9.23
0200			.032″	5″	enameled	3.87	5.85	9.72
0500		copper	16 Oz.	5″	lead coated	20.50	5.85	26.35
0600					mill	9.10	5.85	14.95
1000		steel galv.	28 Ga.	5″	enameled	2.51	5.85	8.36
1200			26 Ga.	5″	mill	2.72	5.85	8.57
1800		vinyl		4″	colors	1.47	5.35	6.82
1900				5″	colors	1.80	5.35	7.15
2300		hemlock or fir		4″x5″	treated	24	6.20	30.20
3000	Half round	copper	16 Oz.	4″	lead coated	17.10	5.85	22.95
3100					mill	9.35	5.85	15.20
3600		steel galv.	28 Ga.	5″	enameled	2.51	5.85	8.36
4102		stainless steel		5″	mill	10.80	5.85	16.65
5000		vinyl		4″	white	1.53	5.35	6.88

B3010 620 — Downspouts

	MATERIALS	SECTION	SIZE	FINISH	THICKNESS	COST PER V.L.F. MAT.	COST PER V.L.F. INST.	COST PER V.L.F. TOTAL
0100	Aluminum	rectangular	2″x3″	embossed mill	.020″	1	3.84	4.84
0150				enameled	.020″	1.50	3.84	5.34
0250			3″x4″	enameled	.024″	2.15	5.20	7.35
0300		round corrugated	3″	enameled	.020″	2.29	3.84	6.13
0350			4″	enameled	.025″	3.51	5.20	8.71
0500	Copper	rectangular corr.	2″x3″	mill	16 Oz.	9.15	3.84	12.99
0600		smooth		mill	16 Oz.	12.40	3.84	16.24
0700		rectangular corr.	3″x4″	mill	16 Oz.	10.30	5.05	15.35
1300	Steel	rectangular corr.	2″x3″	galvanized	28 Ga.	2.24	3.84	6.08
1350				epoxy coated	24 Ga.	2.64	3.84	6.48
1400		smooth		galvanized	28 Ga.	4.21	3.84	8.05
1450		rectangular corr.	3″x4″	galvanized	28 Ga.	2.42	5.05	7.47
1500				epoxy coated	24 Ga.	3.20	5.05	8.25
1550		smooth		galvanized	28 Ga.	4.75	5.05	9.80
1600		round corrugated	2″	galvanized	28 Ga.	2.28	3.84	6.12
1650			3″	galvanized	28 Ga.	2.28	3.84	6.12
1700			4″	galvanized	28 Ga.	2.29	5.05	7.34
1750			5″	galvanized	28 Ga.	3.88	5.60	9.48
2552	S.S. tubing sch.5	rectangular	3″x4″	mill		125	5.05	130.05

B3010 630 — Gravel Stop

	MATERIALS	SECTION	SIZE	FINISH	THICKNESS	COST PER L.F. MAT.	COST PER L.F. INST.	COST PER L.F. TOTAL
5100	Aluminum	extruded	4″	mill	.050″	7.20	5.05	12.25
5200			4″	duranodic	.050″	8.15	5.05	13.20
5300			8″	mill	.050″	8.55	5.85	14.40
5400			8″	duranodic	.050″	10.05	5.85	15.90
6000			12″-2 pc.	duranodic	.050″	12.10	7.30	19.40
6100	Stainless	formed	6″	mill	24 Ga.	17.40	5.40	22.80

B30 Roofing

B3020 Roof Openings

Roof Hatch

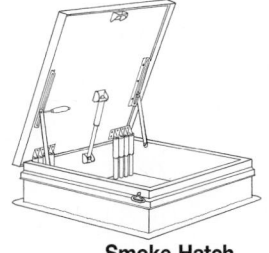
Smoke Hatch

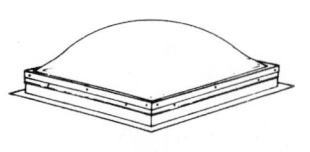
Skylight

B3020 110	Skylights	COST PER S.F.		
		MAT.	INST.	TOTAL
5100	Skylights, plastic domes, insul curbs, nom. size to 10 S.F., single glaze	33	15.20	48.20
5200	Double glazing	38	18.70	56.70
5300	10 S.F. to 20 S.F., single glazing	35	6.15	41.15
5400	Double glazing	32.50	7.70	40.20
5500	20 S.F. to 30 S.F., single glazing	26	5.25	31.25
5600	Double glazing	30	6.15	36.15
5700	30 S.F. to 65 S.F., single glazing	25	3.99	28.99
5800	Double glazing	31	5.25	36.25
6000	Sandwich panels fiberglass, 1-9/16" thick, 2 S.F. to 10 S.F.	20.50	12.15	32.65
6100	10 S.F. to 18 S.F.	18.45	9.15	27.60
6200	2-3/4" thick, 25 S.F. to 40 S.F.	29.50	8.25	37.75
6300	40 S.F. to 70 S.F.	24	7.35	31.35

B3020 210	Hatches	COST PER OPNG.		
		MAT.	INST.	TOTAL
0200	Roof hatches with curb and 1" fiberglass insulation, 2'-6"x3'-0", aluminum	975	243	1,218
0300	Galvanized steel 165 lbs.	925	243	1,168
0400	Primed steel 164 lbs.	675	243	918
0500	2'-6"x4'-6" aluminum curb and cover, 150 lbs.	1,200	270	1,470
0600	Galvanized steel 220 lbs.	1,050	270	1,320
0650	Primed steel 218 lbs.	1,100	270	1,370
0800	2'x6"x8'-0" aluminum curb and cover, 260 lbs.	2,275	370	2,645
0900	Galvanized steel, 360 lbs.	2,075	370	2,445
0950	Primed steel 358 lbs.	1,325	370	1,695
1200	For plexiglass panels, add to the above	525		525
2100	Smoke hatches, unlabeled not incl. hand winch operator, 2'-6"x3', galv	1,125	294	1,419
2200	Plain steel, 160 lbs.	815	294	1,109
2400	2'-6"x8'-0",galvanized steel, 360 lbs.	2,275	405	2,680
2500	Plain steel, 350 lbs.	1,450	405	1,855
3000	4'-0"x8'-0", double leaf low profile, aluminum cover, 359 lb.	3,100	305	3,405
3100	Galvanized steel 475 lbs.	2,575	305	2,880
3200	High profile, aluminum cover, galvanized curb, 361 lbs.	3,050	305	3,355

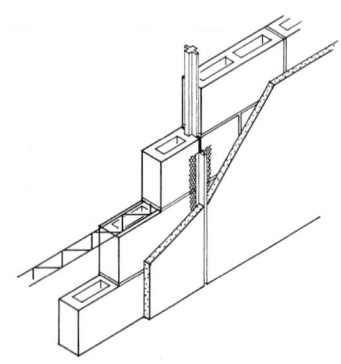

The Concrete Block Partition Systems are defined by weight and type of block, thickness, type of finish and number of sides finished. System components include joint reinforcing on alternate courses and vertical control joints.

C1010 102						COST PER S.F.		
Concrete Block Partitions - Regular Weight								
	TYPE	THICKNESS (IN.)	TYPE FINISH	SIDES FINISHED		MAT.	INST.	TOTAL
1000	Hollow	4	none	0		2.30	6.80	9.10
1010			gyp. plaster 2 coat	1		3.09	9.90	12.99
1020				2		3.88	12.95	16.83
1200			portland - 3 coat	1		2.86	10.30	13.16
1400			5/8" drywall	1		2.96	9.05	12.01
1500		6	none	0		2.91	7.30	10.21
1510			gyp. plaster 2 coat	1		3.70	10.40	14.10
1520				2		4.49	13.45	17.94
1700			portland - 3 coat	1		3.47	10.80	14.27
1900			5/8" drywall	1		3.57	9.55	13.12
1910				2		4.23	11.85	16.08
2000		8	none	0		3.10	7.80	10.90
2010			gyp. plaster 2 coat	1		3.89	10.85	14.74
2020			gyp. plaster 2 coat	2		4.68	13.95	18.63
2200			portland - 3 coat	1		3.66	11.30	14.96
2400			5/8" drywall	1		3.76	10.05	13.81
2410				2		4.42	12.35	16.77
2500		10	none	0		3.66	8.15	11.81
2510			gyp. plaster 2 coat	1		4.45	11.20	15.65
2520				2		5.25	14.30	19.55
2700			portland - 3 coat	1		4.22	11.65	15.87
2900			5/8" drywall	1		4.32	10.40	14.72
2910				2		4.98	12.70	17.68
3000	Solid	2	none	0		1.99	6.70	8.69
3010			gyp. plaster	1		2.78	9.80	12.58
3020				2		3.57	12.85	16.42
3200			portland - 3 coat	1		2.55	10.20	12.75
3400			5/8" drywall	1		2.65	8.95	11.60
3410				2		3.31	11.25	14.56
3500		4	none	0		2.58	7	9.58
3510			gyp. plaster	1		3.48	10.10	13.58
3520				2		4.16	13.15	17.31
3700			portland - 3 coat	1		3.14	10.50	13.64
3900			5/8" drywall	1		3.24	9.25	12.49
3910				2		3.90	11.55	15.45

C10 Interior Construction

C1010 Partitions

C1010 102	Concrete Block Partitions - Regular Weight

	TYPE	THICKNESS (IN.)	TYPE FINISH	SIDES FINISHED		COST PER S.F.		
						MAT.	INST.	TOTAL
4000	Solid	6	none	0		2.99	7.55	10.54
4010			gyp. plaster	1		3.78	10.65	14.43
4020				2		4.57	13.70	18.27
4200			portland - 3 coat	1		3.55	11.05	14.60
4400			5/8" drywall	1		3.65	9.80	13.45
4410				2		4.31	12.10	16.41

C1010 104	Concrete Block Partitions - Lightweight

	TYPE	THICKNESS (IN.)	TYPE FINISH	SIDES FINISHED		COST PER S.F.		
						MAT.	INST.	TOTAL
5000	Hollow	4	none	0		2.43	6.65	9.08
5010			gyp. plaster	1		3.22	9.75	12.97
5020				2		4.01	12.80	16.81
5200			portland - 3 coat	1		2.99	10.15	13.14
5400			5/8" drywall	1		3.09	8.90	11.99
5410				2		3.75	11.20	14.95
5500		6	none	0		3.30	7.10	10.40
5520			gyp. plaster	2		4.88	13.25	18.13
5700			portland - 3 coat	1		3.86	10.60	14.46
5900			5/8" drywall	1		3.96	9.35	13.31
5910				2		4.62	11.65	16.27
6000		8	none	0		3.97	7.60	11.57
6010			gyp. plaster	1		4.76	10.65	15.41
6020				2		5.55	13.75	19.30
6200			portland - 3 coat	1		4.53	11.10	15.63
6400			5/8" drywall	1		4.63	9.85	14.48
6410				2		5.30	12.15	17.45
6500		10	none	0		4.78	7.95	12.73
6510			gyp. plaster	1		5.55	11	16.55
6520				2		6.35	14.10	20.45
6700			portland - 3 coat	1		5.35	11.45	16.80
6900			5/8" drywall	1		5.45	10.20	15.65
6910				2		6.10	12.50	18.60
7000	Solid	4	none	0		2.18	6.95	9.13
7010			gyp. plaster	1		2.97	10.05	13.02
7020				2		3.76	13.10	16.86
7200			portland - 3 coat	1		2.74	10.45	13.19
7400			5/8" drywall	1		2.84	9.20	12.04
7410				2		3.50	11.50	15
7500		6	none	0		3.86	7.45	11.31
7510			gyp. plaster	1		4.79	10.60	15.39
7520				2		5.45	13.60	19.05
7700			portland - 3 coat	1		4.42	10.95	15.37
7900			5/8" drywall	1		4.52	9.70	14.22
7910				2		5.20	12	17.20
8000		8	none	0		4.32	8	12.32
8010			gyp. plaster	1		5.10	11.05	16.15
8020				2		5.90	14.15	20.05
8200			portland - 3 coat	1		4.88	11.50	16.38
8400			5/8" drywall	1		4.98	10.25	15.23
8410				2		5.65	12.55	18.20

C1010 Partitions

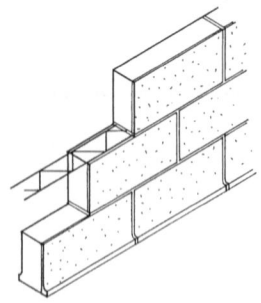

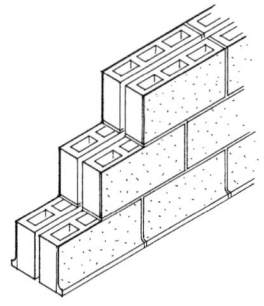

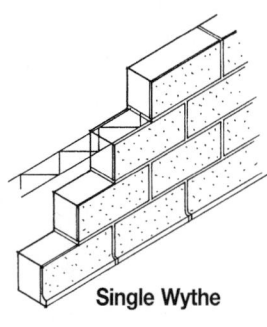
Single Wythe

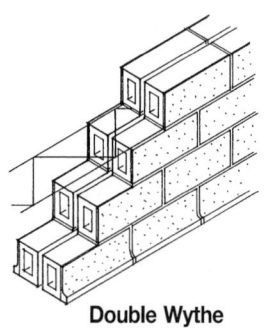
Double Wythe

C1010 120	Tile Partitions	COST PER S.F.		
		MAT.	INST.	TOTAL
1000	8W series 8"x16", 4" thick wall, reinf every 2 courses, glazed 1 side	15.25	8.15	23.40
1100	Glazed 2 sides	18.10	8.65	26.75
1200	Glazed 2 sides, using 2 wythes of 2" thick tile	22.50	15.60	38.10
1300	6" thick wall, horizontal reinf every 2 courses, glazed 1 side	31	8.50	39.50
1400	Glazed 2 sides, each face different color, 2" and 4" tile	26.50	15.95	42.45
1500	8" thick wall, glazed 2 sides using 2 wythes of 4" thick tile	30.50	16.30	46.80
1600	10" thick wall, glazed 2 sides using 1 wythe of 4" & 1 wythe of 6" tile	46.50	16.65	63.15
1700	Glazed 2 sides cavity wall, using 2 wythes of 4" thick tile	30.50	16.30	46.80
1800	12" thick wall, glazed 2 sides using 2 wythes of 6" thick tile	62	17	79
1900	Glazed 2 sides cavity wall, using 2 wythes of 4" thick tile	30.50	16.30	46.80
2100	6T series 5-1/3"x12" tile, 4" thick, non load bearing glazed one side	14.40	12.75	27.15
2200	Glazed two sides	19.15	14.40	33.55
2300	Glazed two sides, using two wythes of 2" thick tile	19	25	44
2400	6" thick, glazed one side	19.40	13.40	32.80
2500	Glazed two sides	23	15.20	38.20
2600	Glazed two sides using 2" thick tile and 4" thick tile	24	25.50	49.50
2700	8" thick, glazed one side	25.50	15.60	41.10
2800	Glazed two sides using two wythes of 4" thick tile	29	25.50	54.50
2900	Glazed two sides using 6" thick tile and 2" thick tile	29	26	55
3000	10" thick cavity wall, glazed two sides using two wythes of 4" tile	29	25.50	54.50
3100	12" thick, glazed two sides using 4" thick tile and 8" thick tile	40	28.50	68.50
3200	2" thick facing tile, glazed one side, on 6" concrete block	13.80	14.85	28.65
3300	On 8" concrete block	13.95	15.30	29.25
3400	On 10" concrete block	14.50	15.60	30.10

C1010 Partitions

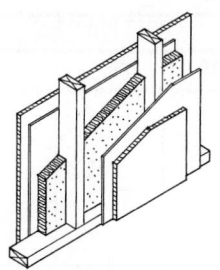

Wood Stud Framing

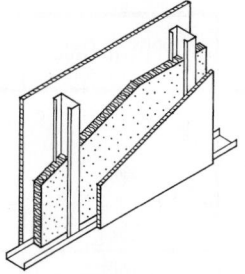

Metal Stud Framing

The Drywall Partitions/Stud Framing Systems are defined by type of drywall and number of layers, type and spacing of stud framing, and treatment on the opposite face. Components include taping and finishing.

Cost differences between regular and fire resistant drywall are negligible, and terminology is interchangeable. In some cases fiberglass insulation is included for additional sound deadening.

C1010 124 — Drywall Partitions/Wood Stud Framing

	FACE LAYER	BASE LAYER	FRAMING	OPPOSITE FACE	INSULATION	MAT.	INST.	TOTAL
1200	5/8" FR drywall	none	2 x 4, @ 16" O.C.	same	0	1.38	3.72	5.10
1250				5/8" reg. drywall	0	1.37	3.72	5.09
1300				nothing	0	.93	2.48	3.41
1400		1/4" SD gypsum	2 x 4 @ 16" O.C.	same	1-1/2" fiberglass	2.74	5.70	8.44
1425					Sound attenuation	2.97	5.80	8.77
1450				5/8" FR drywall	1-1/2" fiberglass	2.28	5.05	7.33
1475					Sound attenuation	2.51	5.10	7.61
1500				nothing	1-1/2" fiberglass	1.83	3.79	5.62
1600		resil. channels	2 x 4 @ 16", O.C.	same	1-1/2" fiberglass	2.17	7.25	9.42
1650				5/8" FR drywall	1-1/2" fiberglass	1.99	5.80	7.79
1700				nothing	1-1/2" fiberglass	1.54	4.55	6.09
1800		5/8" FR drywall	2 x 4 @ 24" O.C.	same	0	2.07	4.71	6.78
1850				5/8" FR drywall	0	1.67	4.09	5.76
1900				nothing	0	1.22	2.85	4.07
2200		5/8" FR drywall	2 rows-2 x 4	same	2" fiberglass	3.22	6.80	10.02
2250			16"O.C.	5/8" FR drywall	2" fiberglass	2.82	6.20	9.02
2300				nothing	2" fiberglass	2.37	4.95	7.32
2400	5/8" WR drywall	none	2 x 4, @ 16" O.C.	same	0	1.54	3.72	5.26
2450				5/8" FR drywall	0	1.46	3.72	5.18
2500				nothing	0	1.01	2.48	3.49
2600		5/8" FR drywall	2 x 4, @ 24" O.C.	same	0	2.23	4.71	6.94
2650				5/8" FR drywall	0	1.75	4.09	5.84
2700				nothing	0	1.30	2.85	4.15
2800	5/8" VF drywall	none	2 x 4, @ 16" O.C.	same	0	2.22	3.98	6.20
2850				5/8" FR drywall	0	1.80	3.85	5.65
2900				nothing	0	1.35	2.61	3.96
3000		5/8" FR drywall	2 x 4 , 24" O.C.	same	0	2.91	4.97	7.88
3050				5/8" FR drywall	0	2.09	4.22	6.31
3100				nothing	0	1.64	2.98	4.62
3200	1/2" reg drywall	3/8" reg drywall	2 x 4, @ 16" O.C.	same	0	2.12	4.96	7.08
3250				5/8" FR drywall	0	1.75	4.34	6.09
3300				nothing	0	1.30	3.10	4.40

C1010 126 — Drywall Partitions/Metal Stud Framing

	FACE LAYER	BASE LAYER	FRAMING	OPPOSITE FACE	INSULATION	MAT.	INST.	TOTAL
5000	5/8" FR drywall	none	1-5/8" @ 16" O.C.	same	0	1.18	3.73	4.91
5010				5/8" reg. drywall	0	1.17	3.73	4.90
5020				nothing	0	.73	2.49	3.22
5030			2-1/2"" @ 16" O.C.	same	0	1.27	3.74	5.01

C1010 Partitions

C1010 126	Drywall Partitions/Metal Stud Framing

	FACE LAYER	BASE LAYER	FRAMING	OPPOSITE FACE	INSULATION	COST PER S.F.		
						MAT.	INST.	TOTAL
5040	5/8" FR drywall	none	2-1/2" @ 16" O.C.	5/8" reg. drywall	0	1.26	3.74	5
5050				nothing	0	.82	2.50	3.32
5060			3-5/8" @ 16" O.C.	same	0	1.32	3.77	5.09
5070				5/8" reg. drywall	0	1.31	3.77	5.08
5080				nothing	0	.87	2.53	3.40
5200			1-5/8" @ 24" O.C.	same	0	1.11	3.29	4.40
5250				5/8" reg. drywall	0	1.10	3.29	4.39
5300				nothing	0	.66	2.05	2.71
5310			2-1/2" @ 24" O.C.	same	0	1.17	3.30	4.47
5320				5/8" reg. drywall	0	1.16	3.30	4.46
5330				nothing	0	.72	2.06	2.78
5400			3-5/8" @ 24" O.C.	same	0	1.21	3.31	4.52
5450				5/8" reg. drywall	0	1.20	3.31	4.51
5500				nothing	0	.76	2.07	2.83
5530		1/4" SD gypsum	1-5/8" @ 16" O.C.	same	0	2.10	5.10	7.20
5535				5/8" FR drywall	0	1.18	3.73	4.91
5540				nothing	0	1.19	3.18	4.37
5545			2-1/2" @ 16" O.C.	same	0	2.19	5.10	7.29
5550				5/8" FR drywall	0	1.27	3.74	5.01
5555				nothing	0	1.28	3.19	4.47
5560			3-5/8" @ 16" O.C.	same	0	2.24	5.15	7.39
5565				5/8" FR drywall	0	1.32	3.77	5.09
5570				nothing	0	1.33	3.22	4.55
5600			1-5/8" @ 24" O.C.	same	0	2.03	4.67	6.70
5650				5/8" FR drywall	0	1.57	3.98	5.55
5700				nothing	0	1.12	2.74	3.86
5800			2-1/2" @ 24" O.C.	same	0	2.09	4.68	6.77
5850				5/8" FR drywall	0	1.63	3.99	5.62
5900				nothing	0	1.18	2.75	3.93
5910			3-5/8" @ 24" O.C.	same	0	2.13	4.69	6.82
5920				5/8" FR drywall	0	1.67	4	5.67
5930				nothing	0	1.22	2.76	3.98
6000		5/8" FR drywall	2-1/2" @ 16" O.C.	same	0	2.15	5.30	7.45
6050				5/8" FR drywall	0	1.75	4.68	6.43
6100				nothing	0	1.30	3.44	4.74
6110			3-5/8" @ 16" O.C.	same	0	2.12	5	7.12
6120				5/8" FR drywall	0	1.72	4.39	6.11
6130				nothing	0	1.27	3.15	4.42
6170			2-1/2" @ 24" O.C.	same	0	1.97	4.54	6.51
6180				5/8" FR drywall	0	1.57	3.92	5.49
6190				nothing	0	1.12	2.68	3.80
6200			3-5/8" @ 24" O.C.	same	0	2.01	4.55	6.56
6250				5/8"FR drywall	3-1/2" fiberglass	2.13	4.39	6.52
6300				nothing	0	1.16	2.69	3.85
6310	5/8" WR drywall	none	1-5/8" @ 16" O.C.	same	0	1.34	3.73	5.07
6320				5/8" WR drywall	0	1.82	4.35	6.17
6330				nothing	0	1.29	3.11	4.40
6340			2-1/2" @ 16" O.C.	same	0	2.39	4.98	7.37
6350				5/8" WR drywall	0	1.91	4.36	6.27
6360				nothing	0	1.38	3.12	4.50
6370			3-5/8" @ 16" O.C.	same	0	2.44	5	7.44
6380				5/8" WR drywall	0	1.96	4.39	6.35
6390				nothing	0	1.43	3.15	4.58

For customer support on your Square Foot Costs with RSMeans data, call 800.448.8182.

C10 Interior Construction

C1010 Partitions

C1010 126			Drywall Partitions/Metal Stud Framing					

	FACE LAYER	BASE LAYER	FRAMING	OPPOSITE FACE	INSULATION	COST PER S.F.		
						MAT.	INST.	TOTAL
6400	5/8″ WR drywall	none	1-5/8″ @ 24″ O.C.	same	0	1.27	3.29	4.56
6450				5/8″ WR drywall	0	1.19	3.29	4.48
6500				nothing	0	.74	2.05	2.79
6510			2-1/2″ @ 24″ O.C.	same	0	2.29	4.54	6.83
6520				5/8″ WR drywall	0	1.25	3.30	4.55
6530				nothing	0	.80	2.06	2.86
6600			3-5/8″ @ 24″ O.C.	same	0	1.37	3.31	4.68
6650				5/8″ WR drywall	0	1.29	3.31	4.60
6700				nothing	0	.84	2.07	2.91
6800		5/8″ FR drywall	2-1/2″ @ 16″ O.C.	same	0	2.31	5.30	7.61
6850				5/8″ FR drywall	0	1.83	4.68	6.51
6900				nothing	0	1.38	3.44	4.82
6910			3-5/8″ @ 16″ O.C.	same	0	2.28	5	7.28
6920				5/8″ FR drywall	0	1.80	4.39	6.19
6930				nothing	0	1.35	3.15	4.50
6940			2-1/2″ @ 24″ O.C.	same	0	1.97	4.54	6.51
6950				5/8″ FR drywall	0	1.65	3.92	5.57
6960				nothing	0	1.20	2.68	3.88
7000			3-5/8″ @ 24″ O.C.	same	0	2.17	4.55	6.72
7050				5/8″FR drywall	3-1/2″ fiberglass	2.21	4.39	6.60
7100				nothing	0	1.24	2.69	3.93
7110	5/8″ VF drywall	none	1-5/8″ @ 16″ O.C.	same	0	2.02	3.99	6.01
7120				5/8″ FR drywall	0	1.66	4.35	6.01
7130				nothing	0	1.21	3.11	4.32
7140			3-5/8″ @ 16″ O.C.	same	0	2.28	5	7.28
7150				5/8″ FR drywall	0	1.80	4.39	6.19
7160				nothing	0	1.35	3.15	4.50
7200		none	1-5/8″ @ 24″ O.C.	same	0	1.95	3.55	5.50
7250				5/8″ FR drywall	0	1.53	3.42	4.95
7300				nothing	0	1.08	2.18	3.26
7400			3-5/8″ @ 24″ O.C.	same	0	2.05	3.57	5.62
7450				5/8″ FR drywall	0	1.63	3.44	5.07
7500				nothing	0	1.18	2.20	3.38

C1010 Partitions

C1010 126	Drywall Partitions/Metal Stud Framing

	FACE LAYER	BASE LAYER	FRAMING	OPPOSITE FACE	INSULATION	COST PER S.F.		
						MAT.	INST.	TOTAL
7600	5/8" FR drywall		2-1/2" @ 16" O.C.	same	0	2.99	5.55	8.54
7650				5/8" FR drywall	0	2.17	4.81	6.98
7700				nothing	0	1.72	3.57	5.29
7710			3-5/8" @ 16" O.C.	same	0	2.96	5.25	8.21
7720				5/8" FR drywall	0	2.14	4.52	6.66
7730				nothing	0	1.69	3.28	4.97
7740			2-1/2" @ 24" O.C.	same	0	2.81	4.80	7.61
7750				5/8" FR drywall	0	1.99	4.05	6.04
7760				nothing	0	1.54	2.81	4.35
7800			3-5/8" @ 24" O.C.	same	0	2.85	4.81	7.66
7850				5/8"FR drywall	3-1/2" fiberglass	2.55	4.52	7.07
7900				nothing	0	1.58	2.82	4.40

C1010 Partitions

C1010 128	Drywall Components	COST PER S.F.		
		MAT.	INST.	TOTAL
0060	Metal studs, 24" O.C. including track, load bearing, 20 gage, 2-1/2"	.71	1.16	1.87
0080	3-5/8"	.84	1.18	2.02
0100	4"	.75	1.20	1.95
0120	6"	1.12	1.23	2.35
0140	Metal studs, 24" O.C. including track, load bearing, 18 gage, 2-1/2"	.71	1.16	1.87
0160	3-5/8"	.84	1.18	2.02
0180	4"	.75	1.20	1.95
0200	6"	1.12	1.23	2.35
0220	16 gage, 2-1/2"	.82	1.32	2.14
0240	3-5/8"	.98	1.35	2.33
0260	4"	1.03	1.37	2.40
0280	6"	1.30	1.41	2.71
0300	Non load bearing, 25 gage, 1-5/8"	.21	.81	1.02
0340	3-5/8"	.31	.83	1.14
0360	4"	.34	.83	1.17
0380	6"	.41	.85	1.26
0400	20 gage, 2-1/2"	.33	1.03	1.36
0420	3-5/8"	.38	1.05	1.43
0440	4"	.46	1.05	1.51
0460	6"	.55	1.06	1.61
0540	Wood studs including blocking, shoe and double top plate, 2"x4", 12" O.C.	.60	1.55	2.15
0560	16" O.C.	.48	1.24	1.72
0580	24" O.C.	.37	.99	1.36
0600	2"x6", 12" O.C.	.92	1.77	2.69
0620	16" O.C.	.75	1.37	2.12
0640	24" O.C.	.57	1.08	1.65
0642	Furring one side only, steel channels, 3/4", 12" O.C.	.44	2.47	2.91
0644	16" O.C.	.39	2.19	2.58
0646	24" O.C.	.26	1.66	1.92
0647	1-1/2", 12" O.C.	.58	2.77	3.35
0648	16" O.C.	.52	2.42	2.94
0649	24" O.C.	.35	1.91	2.26
0650	Wood strips, 1" x 3", on wood, 12" O.C.	.47	1.12	1.59
0651	16" O.C.	.35	.84	1.19
0652	On masonry, 12" O.C.	.51	1.25	1.76
0653	16" O.C.	.38	.94	1.32
0654	On concrete, 12" O.C.	.51	2.38	2.89
0655	16" O.C.	.38	1.79	2.17
0665	Gypsum board, one face only, exterior sheathing, 1/2"	.51	1.10	1.61
0680	Interior, fire resistant, 1/2"	.41	.62	1.03
0700	5/8"	.40	.62	1.02
0720	Sound deadening board 1/4"	.46	.69	1.15
0740	Standard drywall 3/8"	.40	.62	1.02
0760	1/2"	.37	.62	.99
0780	5/8"	.39	.62	1.01
0800	Tongue & groove coreboard 1"	.85	2.57	3.42
0820	Water resistant, 1/2"	.46	.62	1.08
0840	5/8"	.48	.62	1.10
0860	Add for the following:, foil backing	.20		.20
0880	Fiberglass insulation, 3-1/2"	.52	.46	.98
0900	6"	.68	.46	1.14
0920	Rigid insulation 1"	.59	.62	1.21
0940	Resilient furring @ 16" O.C.	.23	1.94	2.17
0960	Taping and finishing	.05	.62	.67
0980	Texture spray	.04	.73	.77
1000	Thin coat plaster	.12	.77	.89
1050	Sound wall framing, 2x6 plates, 2x4 staggered studs, 12" O.C.	.73	1.52	2.25

C1010 Partitions

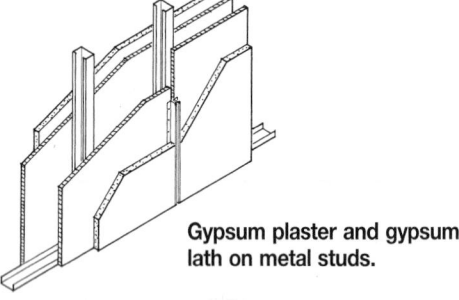

Gypsum plaster and gypsum lath on metal studs.

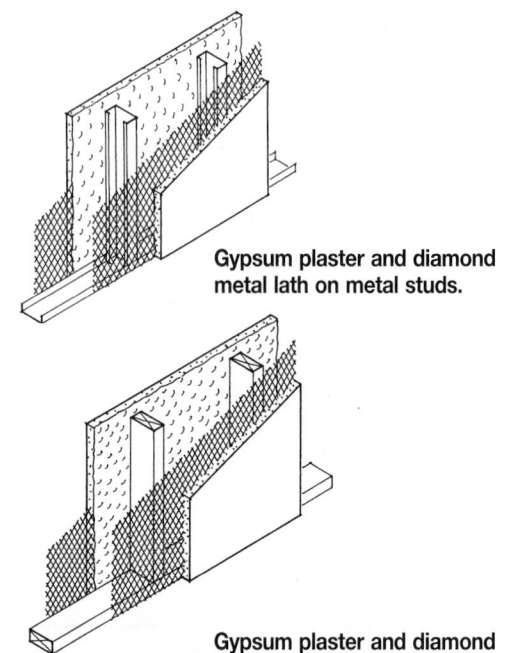

Gypsum plaster and diamond metal lath on metal studs.

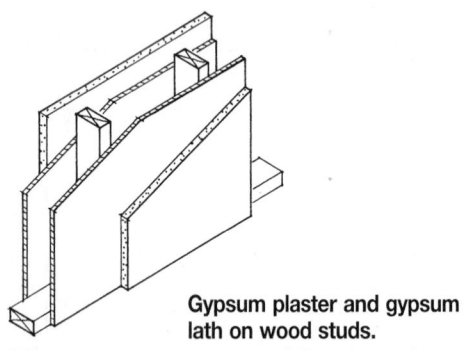

Gypsum plaster and gypsum lath on wood studs.

Gypsum plaster and diamond metal lath on wood studs.

C1010 140			Plaster Partitions/Metal Stud Framing					
	TYPE	FRAMING	LATH	OPPOSITE FACE		COST PER S.F.		
						MAT.	INST.	TOTAL
1000	2 coat gypsum	2-1/2" @ 16"O.C.	3/8" gypsum	same		3.62	9.05	12.67
1010				nothing		2.04	5.25	7.29
1100		3-1/4" @ 24"O.C.	1/2" gypsum	same		3.40	8.75	12.15
1110				nothing		1.90	4.90	6.80
1500	2 coat vermiculite	2-1/2" @ 16"O.C.	3/8" gypsum	same		2.67	9.90	12.57
1510				nothing		1.57	5.70	7.27
1600		3-1/4" @ 24"O.C.	1/2" gypsum	same		2.45	9.55	12
1610				nothing		1.43	5.35	6.78
2000	3 coat gypsum	2-1/2" @ 16"O.C.	3/8" gypsum	same		2.46	10.20	12.66
2010				nothing		1.46	5.85	7.31
2020			3.4lb. diamond	same		2.80	10.20	13
2030				nothing		1.62	5.85	7.47
2040			2.75lb. ribbed	same		2.57	10.20	12.77
2050				nothing		1.51	5.85	7.36
2100		3-1/4" @ 24"O.C.	1/2" gypsum	same		2.24	9.90	12.14
2110				nothing		1.32	5.50	6.82
2120			3.4lb. ribbed	same		2.78	9.90	12.68
2130				nothing		1.59	5.50	7.09
3500	3 coat gypsum	2-1/2" @ 16"O.C.	3/8" gypsum	same		2.94	13.05	15.99
3510	W/med. Keenes			nothing		1.70	7.25	8.95
3520			3.4lb. diamond	same		3.28	13.05	16.33
3530				nothing		1.86	7.25	9.11
3540			2.75lb. ribbed	same		3.05	13.05	16.10
3550				nothing		1.75	7.25	9
3600		3-1/4" @ 24"O.C.	1/2" gypsum	same		2.72	12.75	15.47
3610				nothing		1.56	6.90	8.46
3620			3.4lb. ribbed	same		3.26	12.75	16.01

C1010 Partitions

| C1010 140 | | Plaster Partitions/Metal Stud Framing | | | | | | |

	TYPE	FRAMING	LATH	OPPOSITE FACE		COST PER S.F.		
						MAT.	INST.	TOTAL
3630	3 coat gypsum	3/4" @ 24" O.C.	3.4lb. ribbed	nothing		1.83	6.90	8.73
4000	3 coat gypsum	2-1/2" @ 16"O.C.	3/8" gypsum	same		2.95	14.50	17.45
4010	W/hard Keenes			nothing		1.70	8	9.70
4020			3.4lb. diamond	same		3.29	14.50	17.79
4030				nothing		1.86	8	9.86
4040			2.75lb. ribbed	same		3.06	14.50	17.56
4050				nothing		1.75	8	9.75
4100		3-1/4" @ 24"O.C.	1/2" gypsum	same		2.73	14.20	16.93
4110				nothing		1.56	7.65	9.21
4120			3.4lb. ribbed	same		3.27	14.20	17.47
4130				nothing		1.83	7.65	9.48

| C1010 142 | | Plaster Partitions/Wood Stud Framing | | | | | | |

	TYPE	FRAMING	LATH	OPPOSITE FACE		COST PER S.F.		
						MAT.	INST.	TOTAL
5000	2 coat gypsum	2"x4" @ 16"O.C.	3/8" gypsum	same		3.73	8.85	12.58
5010				nothing		2.15	5.15	7.30
5100		2"x4" @ 24"O.C.	1/2" gypsum	same		3.46	8.65	12.11
5110				nothing		1.96	4.94	6.90
5500	2 coat vermiculite	2"x4" @ 16"O.C.	3/8" gypsum	same		2.78	9.65	12.43
5510				nothing		1.68	5.60	7.28
5600		2"x4" @ 24"O.C.	1/2" gypsum	same		2.51	9.50	12.01
5610				nothing		1.49	5.35	6.84
6000	3 coat gypsum	2"x4" @ 16"O.C.	3/8" gypsum	same		2.57	10	12.57
6010				nothing		1.57	5.75	7.32
6020			3.4lb. diamond	same		2.91	10.10	13.01
6030				nothing		1.73	5.80	7.53
6040			2.75lb. ribbed	same		2.65	10.15	12.80
6050				nothing		1.61	5.85	7.46
6100		2"x4" @ 24"O.C.	1/2" gypsum	same		2.30	9.85	12.15
6110				nothing		1.38	5.55	6.93
6120			3.4lb. ribbed	same		2.57	9.95	12.52
6130				nothing		1.51	5.60	7.11
7500	3 coat gypsum	2"x4" @ 16"O.C.	3/8" gypsum	same		3.05	12.85	15.90
7510	W/med Keenes			nothing		1.81	7.15	8.96
7520			3.4lb. diamond	same		3.39	12.95	16.34
7530				nothing		1.97	7.20	9.17
7540			2.75lb. ribbed	same		3.13	13	16.13
7550				nothing		1.85	7.25	9.10
7600		2"x4" @ 24"O.C.	1/2" gypsum	same		2.78	12.70	15.48
7610				nothing		1.62	6.95	8.57
7620			3.4lb. ribbed	same		3.32	12.90	16.22
7630				nothing		1.89	7.05	8.94
8000	3 coat gypsum	2"x4" @ 16"O.C.	3/8" gypsum	same		3.06	14.30	17.36
8010	W/hard Keenes			nothing		1.81	7.90	9.71
8020			3.4lb. diamond	same		3.40	14.40	17.80
8030				nothing		1.97	7.90	9.87
8040			2.75lb. ribbed	same		3.14	14.45	17.59
8050				nothing		1.85	7.95	9.80
8100		2"x4" @ 24"O.C.	1/2" gypsum	same		2.79	14.15	16.94
8110				nothing		1.62	7.65	9.27
8120			3.4lb. ribbed	same		3.33	14.35	17.68
8130				nothing		1.89	7.80	9.69

C1010 Partitions

C1010 144	Plaster Partition Components	COST PER S.F.		
		MAT.	INST.	TOTAL
0060	Metal studs, 16" O.C., including track, non load bearing, 25 gage, 1-5/8"	.37	1.26	1.63
0080	2-1/2"	.37	1.26	1.63
0100	3-1/4"	.42	1.29	1.71
0120	3-5/8"	.42	1.29	1.71
0140	4"	.47	1.30	1.77
0160	6"	.55	1.31	1.86
0180	Load bearing, 20 gage, 2-1/2"	.96	1.61	2.57
0200	3-5/8"	1.14	1.63	2.77
0220	4"	1.18	1.67	2.85
0240	6"	1.51	1.69	3.20
0260	16 gage 2-1/2"	1.13	1.82	2.95
0280	3-5/8"	1.35	1.87	3.22
0300	4"	1.41	1.90	3.31
0320	6"	1.78	1.93	3.71
0340	Wood studs, including blocking, shoe and double plate, 2"x4" , 12" O.C.	.60	1.55	2.15
0360	16" O.C.	.48	1.24	1.72
0380	24" O.C.	.37	.99	1.36
0400	2"x6" , 12" O.C.	.92	1.77	2.69
0420	16" O.C.	.75	1.37	2.12
0440	24" O.C.	.57	1.08	1.65
0460	Furring one face only, steel channels, 3/4" , 12" O.C.	.44	2.47	2.91
0480	16" O.C.	.39	2.19	2.58
0500	24" O.C.	.26	1.66	1.92
0520	1-1/2" , 12" O.C.	.58	2.77	3.35
0540	16" O.C.	.52	2.42	2.94
0560	24"O.C.	.35	1.91	2.26
0580	Wood strips 1"x3", on wood., 12" O.C.	.47	1.12	1.59
0600	16"O.C.	.35	.84	1.19
0620	On masonry, 12" O.C.	.51	1.25	1.76
0640	16" O.C.	.38	.94	1.32
0660	On concrete, 12" O.C.	.51	2.38	2.89
0680	16" O.C.	.38	1.79	2.17
0700	Gypsum lath. plain or perforated, nailed to studs, 3/8" thick	.39	.76	1.15
0720	1/2" thick	.31	.80	1.11
0740	Clipped to studs, 3/8" thick	.39	.86	1.25
0760	1/2" thick	.31	.92	1.23
0780	Metal lath, diamond painted, nailed to wood studs, 2.5 lb.	.51	.76	1.27
0800	3.4 lb.	.55	.80	1.35
0820	Screwed to steel studs, 2.5 lb.	.51	.80	1.31
0840	3.4 lb.	.53	.86	1.39
0860	Rib painted, wired to steel, 2.75 lb	.44	.86	1.30
0880	3.4 lb	.58	.92	1.50
0900	4.0 lb	.70	.99	1.69
0910				
0920	Gypsum plaster, 2 coats	.46	2.93	3.39
0940	3 coats	.65	3.52	4.17
0960	Perlite or vermiculite plaster, 2 coats	.73	3.34	4.07
0980	3 coats	.78	4.16	4.94
1000	Stucco, 3 coats, 1" thick, on wood framing	1.02	5.90	6.92
1020	On masonry	.42	4.57	4.99
1100	Metal base galvanized and painted 2-1/2" high	.71	2.42	3.13

C1010 Partitions

C1010 205	Partitions	COST PER S.F.		
		MAT.	INST.	TOTAL
0360	Folding accordion, vinyl covered, acoustical, 3 lb. S.F., 17 ft max. hgt	30	12.35	42.35
0380	5 lb. per S.F. 27 ft max height	42	13	55
0400	5.5 lb. per S.F., 17 ft. max height	49	13.70	62.70
0420	Commercial, 1.75 lb per S.F., 8 ft. max height	28.50	5.50	34
0440	2.0 Lb per S.F., 17 ft. max height	29.50	8.25	37.75
0460	Industrial, 4.0 lb. per S.F. 27 ft. max height	44	16.45	60.45
0480	Vinyl clad wood or steel, electric operation 6 psf	62.50	7.70	70.20
0500	Wood, non acoustic, birch or mahogany	33.50	4.12	37.62
0560	Folding leaf, alum framed acoustical 12' high., 5.5 lb/S.F. standard trim	46	20.50	66.50
0580	Premium trim	56	41	97
0600	6.5 lb. per S.F., standard trim	48.50	20.50	69
0620	Premium trim	60	41	101
0640	Steel acoustical, 7.5 per S.F., vinyl faced, standard trim	68.50	20.50	89
0660	Premium trim	83.50	41	124.50
0680	Wood acoustic type, vinyl faced to 18' high 6 psf, economy trim	64.50	20.50	85
0700	Standard trim	77	27.50	104.50
0720	Premium trim	99.50	41	140.50
0740	Plastic lam. or hardwood faced, standard trim	66.50	20.50	87
0760	Premium trim	71	41	112
0780	Wood, low acoustical type to 12 ft. high 4.5 psf	48	24.50	72.50
0840	Demountable, trackless wall, cork finish, semi acous, 1-5/8" th, unsealed	42.50	3.80	46.30
0860	Sealed	47	6.50	53.50
0880	Acoustic, 2" thick, unsealed	40	4.05	44.05
0900	Sealed	61.50	5.50	67
0920	In-plant modular office system, w/prehung steel door			
0940	3" thick honeycomb core panels			
0960	12' x 12', 2 wall	15.90	.77	16.67
0970	4 wall	16.60	1.02	17.62
0980	16' x 16', 2 wall	17.25	.54	17.79
0990	4 wall	11	.54	11.54
1000	Gypsum, demountable, 3" to 3-3/4" thick x 9' high, vinyl clad	7.35	2.83	10.18
1020	Fabric clad	18.30	3.11	21.41
1040	1.75 system, vinyl clad hardboard, paper honeycomb core panel			
1060	1-3/4" to 2-1/2" thick x 9' high	12.30	2.83	15.13
1080	Unitized gypsum panel system, 2" to 2-1/2" thick x 9' high			
1100	Vinyl clad gypsum	15.85	2.83	18.68
1120	Fabric clad gypsum	26	3.11	29.11
1140	Movable steel walls, modular system			
1160	Unitized panels, 48" wide x 9' high			
1180	Baked enamel, pre-finished	17.85	2.28	20.13
1200	Fabric clad	26	2.44	28.44
1300	Metal panel partition, load bearing studs, 16 gage, corrugate/ribbed panel,	4.10	4.99	9.09

C1020 Interior Doors

| Single Leaf | Sliding Entrance | Rolling Overhead |

C1020 102	Special Doors	COST PER OPNG.		
		MAT.	INST.	TOTAL
2500	Single leaf, wood, 3'-0"x7'-0"x1 3/8", birch, solid core	490	230	720
2510	Hollow core	420	220	640
2530	Hollow core, lauan	405	220	625
2540	Louvered pine	565	220	785
2550	Paneled pine	625	220	845
2600	Hollow metal, comm. quality, flush, 3'-0"x7'-0"x1-3/8"	980	242	1,222
2650	3'-0"x10'-0" openings with panel	1,400	315	1,715
2700	Metal fire, comm. quality, 3'-0"x7'-0"x1-3/8"	1,200	252	1,452
3200	Double leaf, wood, hollow core, 2 - 3'-0"x7'-0"x1-3/8"	650	420	1,070
3300	Hollow metal, comm. quality, B label, 2'-3'-0"x7'-0"x1-3/8"	2,225	530	2,755
3400	6'-0"x10'-0" opening, with panel	2,675	650	3,325
3500	Double swing door system, 12'-0"x7'-0", mill finish	9,900	3,275	13,175
3700	Black finish	10,000	3,375	13,375
3800	Sliding entrance door and system mill finish	10,800	2,975	13,775
3900	Bronze finish	11,800	3,200	15,000
4000	Black finish	12,300	3,325	15,625
4100	Sliding panel mall front, 16'x9' opening, mill finish	4,225	905	5,130
4200	Bronze finish	5,500	1,175	6,675
4300	Black finish	6,750	1,450	8,200
4400	24'x9' opening mill finish	6,050	1,675	7,725
4500	Bronze finish	7,875	2,175	10,050
4600	Black finish	9,675	2,675	12,350
4700	48'x9' opening mill finish	10,900	1,300	12,200
4800	Bronze finish	14,200	1,700	15,900
4900	Black finish	17,400	2,075	19,475
5000	Rolling overhead steel door, manual, 8' x 8' high	800	895	1,695
5100	10' x 10' high	1,175	1,025	2,200
5200	20' x 10' high	3,275	1,425	4,700
5300	12' x 12' high	1,325	1,200	2,525
5400	Motor operated, 8' x 8' high	2,100	1,175	3,275
5500	10' x 10' high	2,475	1,300	3,775
5600	20' x 10' high	4,575	1,700	6,275
5700	12' x 12' high	2,625	1,475	4,100
5800	Roll up grille, aluminum, manual, 10' x 10' high, mill finish	3,300	1,750	5,050
5900	Bronze anodized	5,100	1,750	6,850
6000	Motor operated, 10' x 10' high, mill finish	4,725	2,025	6,750
6100	Bronze anodized	6,525	2,025	8,550
6200	Steel, manual, 10' x 10' high	2,950	1,425	4,375
6300	15' x 8' high	3,300	1,775	5,075
6400	Motor operated, 10' x 10' high	4,375	1,700	6,075
6500	15' x 8' high	4,725	2,050	6,775
8970	Counter door, rolling, 6' high, 14' wide, aluminum	3,225	880	4,105

C1030   Fittings

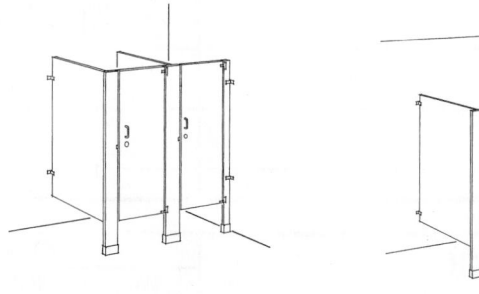
Toilet Units

Entrance Screens

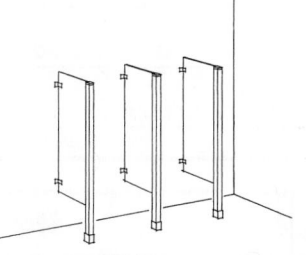
Urinal Screens

C1030 110	Toilet Partitions	COST PER UNIT		
		MAT.	INST.	TOTAL
0380	Toilet partitions, cubicles, ceiling hung, marble	1,825	600	2,425
0400	Painted metal	610	310	920
0420	Plastic laminate	605	310	915
0430	Phenolic	1,000	310	1,310
0440	Polymer plastic	1,100	310	1,410
0460	Stainless steel	1,250	310	1,560
0480	Handicap addition	515		515
0520	Floor and ceiling anchored, marble	1,925	480	2,405
0540	Painted metal	660	247	907
0560	Plastic laminate	840	247	1,087
0570	Phenolic	950	247	1,197
0580	Polymer plastic	1,100	247	1,347
0600	Stainless steel	1,375	247	1,622
0620	Handicap addition	360		360
0660	Floor mounted marble	1,200	400	1,600
0680	Painted metal	685	176	861
0700	Plastic laminate	605	176	781
0710	Phenolic	825	176	1,001
0720	Polymer plastic	975	176	1,151
0740	Stainless steel	1,350	176	1,526
0760	Handicap addition	360		360
0780	Juvenile deduction	48		48
0820	Floor mounted with headrail marble	1,325	400	1,725
0840	Painted metal	430	206	636
0860	Plastic laminate	915	206	1,121
0870	Phenolic	825	206	1,031
0880	Polymer plastic	945	206	1,151
0900	Stainless steel	1,175	206	1,381
0920	Handicap addition	360		360
0960	Wall hung, painted metal	735	176	911
1020	Stainless steel	1,900	176	2,076
1040	Handicap addition	360		360
1080	Entrance screens, floor mounted, 54" high, marble	770	134	904
1100	Painted metal	266	82.50	348.50
1140	Stainless steel	1,050	82.50	1,132.50
1150	Polymer plastic	555	206	761
1300	Urinal screens, floor mounted, 24" wide, plastic laminate	232	154	386
1320	Marble	745	183	928
1330	Polymer plastic	480	206	686
1340	Painted metal	253	154	407
1380	Stainless steel	645	154	799
1428	Wall mounted wedge type, painted metal	153	124	277

C1030 Fittings

C1030 110	Toilet Partitions	COST PER UNIT		
		MAT.	INST.	TOTAL
1460	Stainless steel	670	124	794
1500	Partitions, shower stall, single wall, painted steel, 2'-8" x 2'-8"	990	292	1,282
1510	Fiberglass, 2'-8" x 2'-8"	850	325	1,175
1520	Double wall, enameled steel, 2'-8" x 2'-8"	1,275	292	1,567
1530	Stainless steel, 2'-8" x 2'-8"	2,625	292	2,917
1560	Tub enclosure, sliding panels, tempered glass, aluminum frame	440	365	805
1570	Chrome/brass frame, clear glass	1,300	485	1,785

C1030 310	Storage Specialties, EACH	COST EACH		
		MAT.	INST.	TOTAL
0200	Lockers, steel, 1-tier, 5' to 6' high, per opng, 1 wide, knock down constr.	180	52	232
0210	3 wide	335	61	396
0220	2- tier	111	28	139
0235	Two person	365	36.50	401.50
0240	Duplex	370	36.50	406.50
0245	Welded, athletic type, ventilated, 1- tier	370	97.50	467.50
0600	Shelving, metal industrial, braced, 3' wide, 1' deep	26.50	12.25	38.75
0610	2' deep	40	13	53

C1030 510	Identifying/Visual Aid Specialties, EACH	COST EACH		
		MAT.	INST.	TOTAL
0100	Control boards, magnetic, porcelain finish, framed, 24" x 18"	219	154	373
0110	96" x 48"	1,175	247	1,422
0120	Directory boards, outdoor, black plastic, 36" x 24"	770	620	1,390
0130	36" x 36"	890	825	1,715
0140	Indoor, economy, open faced, 18" x 24"	191	176	367
0510	Street, reflective alum., dbl. face, 4 way, w/bracket	207	41	248
0520	Letters, cast aluminum, 1/2" deep, 4" high	30	34.50	64.50
0530	1" deep, 10" high	66	34.50	100.50
0540	Plaques, cast aluminum, 20" x 30"	2,050	310	2,360
0550	Cast bronze, 36" x 48"	5,375	620	5,995

C1030 520	Identifying/Visual Aid Specialties, S.F.	COST PER S.F.		
		MAT.	INST.	TOTAL
0100	Bulletin board, cork sheets, no frame, 1/4" thick	1.74	4.26	6
0120	Aluminum frame, 1/4" thick, 3' x 5'	10.30	5.15	15.45
0200	Chalkboards, wall hung, alum, frame & chalktrough	13.80	2.73	16.53
0210	Wood frame & chalktrough	12.40	2.95	15.35
0220	Sliding board, one board with back panel	65.50	2.64	68.14
0230	Two boards with back panel	97	2.64	99.64
0240	Liquid chalk type, alum. frame & chalktrough	15.35	2.73	18.08
0250	Wood frame & chalktrough	35	2.73	37.73

C1030 710	Bath and Toilet Accessories, EACH	COST EACH		
		MAT.	INST.	TOTAL
0100	Specialties, bathroom accessories, st. steel, curtain rod, 5' long, 1" diam	30	47.50	77.50
0120	Dispenser, towel, surface mounted	47	38.50	85.50
0140	Grab bar, 1-1/4" diam., 12" long	33	25.50	58.50
0160	Mirror, framed with shelf, 18" x 24"	208	31	239
0170	72" x 24"	285	103	388
0180	Toilet tissue dispenser, surface mounted, single roll	19.85	20.50	40.35
0200	Towel bar, 18" long	47	27	74
0300	Medicine cabinets, sliding mirror doors, 20" x 16" x 4-3/4", unlighted	137	88	225
0310	24" x 19" x 8-1/2", lighted	227	124	351

C1030 Fittings

C1030 730	Bath and Toilet Accessories, L.F.	COST PER L.F.		
		MAT.	INST.	TOTAL
0100	Partitions, hospital curtain, ceiling hung, polyester oxford cloth	27	6	33
0110	Designer oxford cloth	16.70	7.65	24.35

C1030 830	Fabricated Cabinets, EACH	COST EACH		
		MAT.	INST.	TOTAL
0110	Household, base, hardwood, one top drawer & one door below x 12" wide	330	50	380
0115	24" wide	455	55.50	510.50
0120	Four drawer x 24" wide	430	55.50	485.50
0130	Wall, hardwood, 30" high with one door x 12" wide	287	56	343
0140	Two doors x 48" wide	625	67	692

C1030 830	Fabricated Counters, L.F.	COST PER L.F.		
		MAT.	INST.	TOTAL
0150	Counter top-laminated plastic, stock, economy	19	20.50	39.50
0160	Custom-square edge, 7/8" thick	17.15	46	63.15
0170	School, counter, wood, 32" high	277	62	339
0180	Metal, 84" high	630	82.50	712.50

C1030 910	Other Fittings, EACH	COST EACH		
		MAT.	INST.	TOTAL
0500	Mail boxes, horizontal, rear loaded, aluminum, 5" x 6" x 15" deep	47	18.15	65.15
0510	Front loaded, aluminum, 10" x 12" x 15" deep	105	31	136
0520	Vertical, front loaded, aluminum, 15" x 5" x 6" deep	49.50	18.15	67.65
0530	Bronze, duranodic finish	53.50	18.15	71.65
0540	Letter slot, post office	132	77	209
0550	Mail counter, window, post office, with grille	620	310	930
0700	Turnstiles, one way, 4' arm, 46" diam., manual	2,050	247	2,297
0710	Electric	2,400	1,025	3,425
0720	3 arm, 5'-5" diam. & 7' high, manual	6,550	1,225	7,775
0730	Electric	8,725	2,050	10,775

C2010 Stair Construction

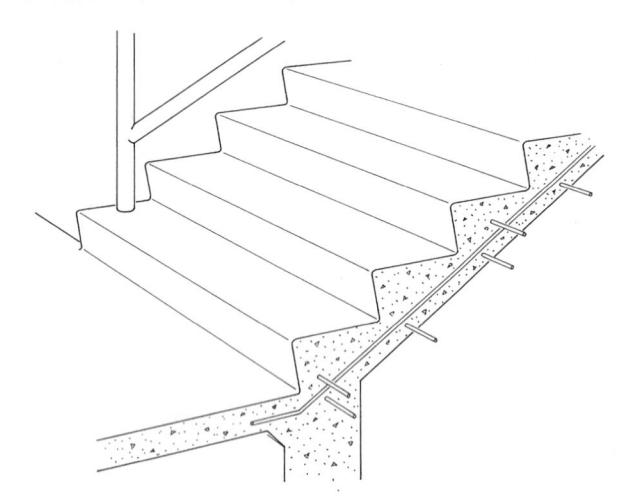

The table below lists the cost per flight for 4'-0" wide stairs. Side walls are not included. Railings are included in the prices.

C2010 110	Stairs	COST PER FLIGHT		
		MAT.	INST.	TOTAL
0470	Stairs, C.I.P. concrete, w/o landing, 12 risers, w/o nosing	1,475	2,500	3,975
0480	With nosing	2,400	2,725	5,125
0550	W/landing, 12 risers, w/o nosing	1,650	3,075	4,725
0560	With nosing	2,550	3,300	5,850
0570	16 risers, w/o nosing	2,000	3,900	5,900
0580	With nosing	3,225	4,175	7,400
0590	20 risers, w/o nosing	2,375	4,675	7,050
0600	With nosing	3,900	5,050	8,950
0610	24 risers, w/o nosing	2,750	5,475	8,225
0620	With nosing	4,575	5,900	10,475
0630	Steel, grate type w/nosing & rails, 12 risers, w/o landing	5,575	1,200	6,775
0640	With landing	7,825	1,625	9,450
0660	16 risers, with landing	9,675	2,050	11,725
0680	20 risers, with landing	11,500	2,450	13,950
0700	24 risers, with landing	13,400	2,825	16,225
0701				
0710	Metal pan stairs for concrete in-fill, picket rail, 12 risers, w/o landing	8,225	1,200	9,425
0720	With landing	10,800	1,800	12,600
0740	16 risers, with landing	13,600	2,200	15,800
0760	20 risers, with landing	16,300	2,600	18,900
0780	24 risers, with landing	19,100	2,975	22,075
0790	Cast iron tread & pipe rail, 12 risers, w/o landing	8,275	1,200	9,475
0800	With landing	10,900	1,800	12,700
1120	Wood, prefab box type, oak treads, wood rails 3'-6" wide, 14 risers	2,400	515	2,915
1150	Prefab basement type, oak treads, wood rails 3'-0" wide, 14 risers	1,150	127	1,277

C3010 Wall Finishes

C3010 230	Paint & Covering	COST PER S.F.		
		MAT.	INST.	TOTAL
0060	Painting, interior on plaster and drywall, brushwork, primer & 1 coat	.14	.79	.93
0080	Primer & 2 coats	.22	1.05	1.27
0100	Primer & 3 coats	.30	1.28	1.58
0120	Walls & ceilings, roller work, primer & 1 coat	.14	.53	.67
0140	Primer & 2 coats	.22	.67	.89
0160	Woodwork incl. puttying, brushwork, primer & 1 coat	.14	1.14	1.28
0180	Primer & 2 coats	.22	1.51	1.73
0200	Primer & 3 coats	.30	2.05	2.35
0260	Cabinets and casework, enamel, primer & 1 coat	.15	1.28	1.43
0280	Primer & 2 coats	.23	1.58	1.81
0300	Masonry or concrete, latex, brushwork, primer & 1 coat	.32	1.07	1.39
0320	Primer & 2 coats	.42	1.53	1.95
0340	Addition for block filler	.20	1.31	1.51
0380	Fireproof paints, intumescent, 1/8" thick 3/4 hour	2.43	1.05	3.48
0400	3/16" thick 1 hour	5.40	1.58	6.98
0420	7/16" thick 2 hour	7	3.66	10.66
0440	1-1/16" thick 3 hour	11.40	7.30	18.70
0500	Gratings, primer & 1 coat	.35	1.60	1.95
0600	Pipes over 12" diameter	.83	5.10	5.93
0700	Structural steel, brushwork, light framing 300-500 S.F./Ton	.10	1.81	1.91
0720	Heavy framing 50-100 S.F./Ton	.10	.91	1.01
0740	Spraywork, light framing 300-500 S.F./Ton	.11	.40	.51
0760	Heavy framing 50-100 S.F./Ton	.11	.45	.56
0800	Varnish, interior wood trim, no sanding sealer & 1 coat	.08	1.28	1.36
0820	Hardwood floor, no sanding 2 coats	.17	.27	.44
0840	Wall coatings, acrylic glazed coatings, minimum	.39	.98	1.37
0860	Maximum	.80	1.68	2.48
0880	Epoxy coatings, solvent based	.50	.98	1.48
0900	Water based	.35	3.01	3.36
0940	Exposed epoxy aggregate, troweled on, 1/16" to 1/4" aggregate, topping mix	.77	2.18	2.95
0960	Integral mix	1.65	3.94	5.59
0980	1/2" to 5/8" aggregate, topping mix	1.49	3.94	5.43
1000	Integral mix	2.59	6.40	8.99
1020	1" aggregate, topping mix	2.63	5.70	8.33
1040	Integral mix	4	9.30	13.30
1060	Sprayed on, topping mix	.63	1.74	2.37
1080	Water based	1.30	3.53	4.83
1100	High build epoxy 50 mil, solvent based	.84	1.31	2.15
1120	Water based	1.42	5.40	6.82
1140	Laminated epoxy with fiberglass solvent based	.90	1.74	2.64
1160	Water based	1.64	3.53	5.17
1180	Sprayed perlite or vermiculite 1/16" thick, solvent based	.31	.17	.48
1200	Water based	.91	.80	1.71
1260	Wall coatings, vinyl plastic, solvent based	.42	.70	1.12
1280	Water based	1.02	2.13	3.15
1300	Urethane on smooth surface, 2 coats, solvent based	.37	.45	.82
1320	Water based	.66	.77	1.43
1340	3 coats, solvent based	.47	.61	1.08
1360	Water based	.99	1.09	2.08
1380	Ceramic-like glazed coating, cementitious, solvent based	.53	1.16	1.69
1400	Water based	1.01	1.49	2.50
1420	Resin base, solvent based	.36	.80	1.16
1440	Water based	.64	1.55	2.19
1460	Wall coverings, aluminum foil	1.13	1.86	2.99
1500	Vinyl backing	6.05	2.14	8.19
1520	Cork tiles, 12"x12", light or dark, 3/16" thick	4.69	2.14	6.83
1540	5/16" thick	3.66	2.18	5.84
1560	Basketweave, 1/4" thick	3.73	2.14	5.87

C3010 Wall Finishes

C3010 230	Paint & Covering	COST PER S.F.		
		MAT.	INST.	TOTAL
1580	Natural, non-directional, 1/2" thick	7.45	2.14	9.59
1600	12"x36", granular, 3/16" thick	1.43	1.33	2.76
1620	1" thick	1.83	1.39	3.22
1640	12"x12", polyurethane coated, 3/16" thick	4.43	2.14	6.57
1660	5/16" thick	6.60	2.18	8.78
1661	Paneling, prefinished plywood, birch	1.32	2.94	4.26
1662	Mahogany, African	2.70	3.09	5.79
1663	Philippine (lauan)	.72	2.47	3.19
1664	Oak or cherry	2.20	3.09	5.29
1665	Rosewood	3.41	3.86	7.27
1666	Teak	3.43	3.09	6.52
1667	Chestnut	5.80	3.29	9.09
1668	Pecan	2.75	3.09	5.84
1669	Walnut	2.71	3.09	5.80
1670	Wood board, knotty pine, finished	2.65	5.10	7.75
1671	Rough sawn cedar	4.02	5.10	9.12
1672	Redwood	5.85	5.10	10.95
1673	Aromatic cedar	3.05	5.45	8.50
1680	Cork wallpaper, paper backed, natural	1.76	1.07	2.83
1700	Color	3.12	1.07	4.19
1720	Gypsum based, fabric backed, minimum	.85	.64	1.49
1740	Average	1.39	.71	2.10
1760	Small quantities	.75	.80	1.55
1780	Vinyl wall covering, fabric back, light weight	1.18	.80	1.98
1800	Medium weight	1	1.07	2.07
1820	Heavy weight	1.54	1.18	2.72
1840	Wall paper, double roll, solid pattern, avg. workmanship	.67	.80	1.47
1860	Basic pattern, avg. workmanship	1.31	.96	2.27
1880	Basic pattern, quality workmanship	2.32	1.18	3.50
1900	Grass cloths with lining paper, minimum	1.52	1.28	2.80
1920	Maximum	3.47	1.47	4.94
1940	Ceramic tile, thin set, 4-1/4" x 4-1/4"	2.87	5.25	8.12
1942				
1960	12" x 12"	5.15	6.20	11.35

C3010 235	Paint Trim	COST PER L.F.		
		MAT.	INST.	TOTAL
2040	Painting, wood trim, to 6" wide, enamel, primer & 1 coat	.15	.64	.79
2060	Primer & 2 coats	.23	.81	1.04
2080	Misc. metal brushwork, ladders	.71	6.40	7.11
2100	Pipes, to 4" dia.	.10	1.35	1.45
2120	6" to 8" dia.	.21	2.70	2.91
2140	10" to 12" dia.	.65	4.03	4.68
2160	Railings, 2" pipe	.25	3.20	3.45
2180	Handrail, single	.19	1.28	1.47
2185	Caulking & Sealants, Polyurethane, In place, 1 or 2 component, 1/2" X 1/4"	.38	2.14	2.52

C3020 Floor Finishes

C3020 410	Tile & Covering	COST PER S.F.		
		MAT.	INST.	TOTAL
0060	Carpet tile, nylon, fusion bonded, 18" x 18" or 24" x 24", 24 oz.	3.77	.77	4.54
0080	35 oz.	4.44	.77	5.21
0100	42 oz.	5.75	.77	6.52
0140	Carpet, tufted, nylon, roll goods, 12' wide, 26 oz.	2.69	.82	3.51
0160	36 oz.	4.42	.82	5.24
0180	Woven, wool, 36 oz.	12	.88	12.88
0200	42 oz.	13.30	.88	14.18
0220	Padding, add to above, 2.7 density	.73	.41	1.14
0240	13.0 density	.98	.41	1.39
0260	Composition flooring, acrylic, 1/4" thick	1.93	5.90	7.83
0280	3/8" thick	2.53	6.80	9.33
0300	Epoxy, 3/8" thick	3.39	4.55	7.94
0320	1/2" thick	4.88	6.25	11.13
0340	Epoxy terrazzo, granite chips	7	6.80	13.80
0360	Recycled porcelain	10.65	9.10	19.75
0380	Mastic, hot laid, 1-1/2" thick, minimum	5	4.45	9.45
0400	Maximum	6.40	5.90	12.30
0420	Neoprene 1/4" thick, minimum	4.93	5.65	10.58
0440	Maximum	6.70	7.15	13.85
0460	Polyacrylate with ground granite 1/4", granite chips	4.10	4.18	8.28
0480	Recycled porcelain	7.55	6.40	13.95
0500	Polyester with colored quartz chips 1/16", minimum	3.72	2.88	6.60
0520	Maximum	5.80	4.55	10.35
0540	Polyurethane with vinyl chips, clear	8.35	2.88	11.23
0560	Pigmented	12.15	3.58	15.73
0600	Concrete topping, granolithic concrete, 1/2" thick	.40	4.83	5.23
0620	1" thick	.80	4.96	5.76
0640	2" thick	1.60	5.70	7.30
0660	Heavy duty 3/4" thick, minimum	.54	7.50	8.04
0680	Maximum	.98	8.95	9.93
0700	For colors, add to above, minimum	.45	1.73	2.18
0720	Maximum	.75	1.90	2.65
0740	Exposed aggregate finish, minimum	.22	.90	1.12
0760	Maximum	.37	1.21	1.58
0780	Abrasives, .25 P.S.F. add to above, minimum	.61	.66	1.27
0800	Maximum	.92	.66	1.58
0820	Dust on coloring, add, minimum	.45	.43	.88
0840	Maximum	.75	.90	1.65
0860	Floor coloring using 0.6 psf powdered color, 1/2" integral, minimum	5.90	4.83	10.73
0880	Maximum	6.20	4.83	11.03
0900	Dustproofing, add, minimum	.18	.30	.48
0920	Maximum	.65	.43	1.08
0930	Paint	.42	1.53	1.95
0940	Hardeners, metallic add, minimum	.47	.66	1.13
0960	Maximum	1.42	.98	2.40
0980	Non-metallic, minimum	.18	.66	.84
1000	Maximum	.54	.98	1.52
1020	Integral topping and finish, 1:1:2 mix, 3/16" thick	.14	2.85	2.99
1040	1/2" thick	.36	3	3.36
1060	3/4" thick	.54	3.35	3.89
1080	1" thick	.72	3.81	4.53
1100	Terrazzo, minimum	3.93	18.15	22.08
1120	Maximum	7.25	22.50	29.75
1340	Cork tile, minimum	5.75	1.76	7.51
1360	Maximum	7.90	1.76	9.66
1380	Polyethylene, in rolls, minimum	4.49	2.02	6.51
1400	Maximum	7.85	2.02	9.87
1420	Polyurethane, thermoset, minimum	6.20	5.55	11.75

417

C3020 Floor Finishes

C3020 410	Tile & Covering	COST PER S.F.		
		MAT.	INST.	TOTAL
1440	Maximum	7.25	11.10	18.35
1460	Rubber, sheet goods, minimum	9.10	4.62	13.72
1480	Maximum	13.25	6.15	19.40
1500	Tile, minimum	6.35	1.39	7.74
1520	Maximum	11.45	2.02	13.47
1580	Vinyl, composition tile, minimum	1.34	1.11	2.45
1600	Maximum	1.95	1.11	3.06
1620	Vinyl tile, 3/32", minimum	4.08	1.11	5.19
1640	Maximum	3.55	1.11	4.66
1660	Sheet goods, plain pattern/colors	4.73	2.22	6.95
1680	Intricate pattern/colors	8.25	2.77	11.02
1720	Tile, ceramic natural clay	6.55	5.45	12
1730	Marble, synthetic 12"x12"x5/8"	12.20	16.60	28.80
1740	Ceramic tile, floors, porcelain type, 1 color, color group 2, 1" x 1"	7.05	5.45	12.50
1760	Ceramic flr porcelain 1 color, color group 2 2" x 2" epoxy grout	8.35	6.40	14.75
1800	Quarry tile, mud set, minimum	9.10	7.10	16.20
1820	Maximum	10.50	9.05	19.55
1840	Thin set, deduct		1.42	1.42
1850	Tile, natural stone, marble, in mortar bed, 12" x 12" x 3/8" thick	17.60	26.50	44.10
1860	Terrazzo precast, minimum	5.50	7.40	12.90
1880	Maximum	12.60	7.40	20
1900	Non-slip, minimum	27	18.30	45.30
1920	Maximum	28	24.50	52.50
1960	Stone flooring, polished marble in mortar bed	19.25	26.50	45.75
2020	Wood block, end grain factory type, natural finish, 2" thick	5.40	4.94	10.34
2040	Fir, vertical grain, 1"x4", no finish, minimum	3.58	2.42	6
2060	Maximum	3.80	2.42	6.22
2080	Prefinished white oak, prime grade, 2-1/4" wide	5.75	3.63	9.38
2100	3-1/4" wide	6.30	3.34	9.64
2120	Maple strip, sanded and finished, minimum	5.55	5.30	10.85
2140	Maximum	6.45	5.30	11.75
2160	Oak strip, sanded and finished, minimum	3.99	5.30	9.29
2180	Maximum	4.88	5.30	10.18
2200	Parquetry, sanded and finished, plain pattern	6.25	5.50	11.75
2220	Intricate pattern	11.10	7.85	18.95
2260	Add for sleepers on concrete, treated, 24" O.C., 1"x2"	1.98	3.18	5.16
2280	1"x3"	2.15	2.48	4.63
2300	2"x4"	1.13	1.25	2.38
2340	Underlayment, plywood, 3/8" thick	1.14	.82	1.96
2350	1/2" thick	1.35	.85	2.20
2360	5/8" thick	1.50	.88	2.38
2370	3/4" thick	1.62	.95	2.57
2380	Particle board, 3/8" thick	.45	.82	1.27
2390	1/2" thick	.47	.85	1.32
2400	5/8" thick	.61	.88	1.49
2410	3/4" thick	.74	.95	1.69
2420	Hardboard, 4' x 4', .215" thick	.75	.82	1.57
9200	Vinyl, composition tile, 12" x 12" x 1/8" thick, recycled content	2.73	1.11	3.84

C3020 600	Bases, Curbs & Trim	COST PER UNIT		
		MAT.	INST.	TOTAL
0050	1/8" vinyl base, 2-1/2" H, straight or cove, std. colors	.77	1.76	2.53
0055	4" H	1.31	1.76	3.07
0060	6" H	1.62	1.76	3.38
0065	Corners, 2-1/2" H	2.43	1.76	4.19
0070	4" H	2.77	1.76	4.53
0075	6" H	3.12	1.76	4.88

C3020 Floor Finishes

C3020 600	Bases, Curbs & Trim	COST PER UNIT		
		MAT.	INST.	TOTAL
0080	1/8" rubber base, 2-1/2" H, straight or cove, std. colors	1.21	1.76	2.97
0085	4" H	1.41	1.76	3.17
0090	6" H	2.09	1.76	3.85
0095	Corners, 2-1/2" H	2.71	1.76	4.47
0100	4" H	2.79	1.76	4.55
0105	6" H	3.42	1.76	5.18

C3030 Ceiling Finishes

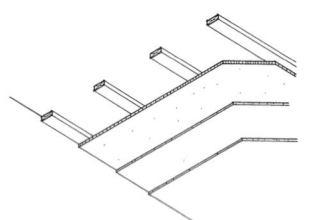

2 Coats of Plaster on Gypsum
Lath on Wood Furring

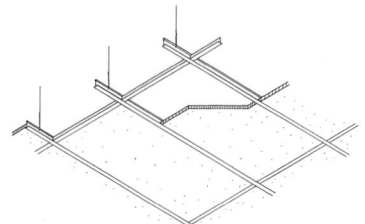

Fiberglass Board on
Exposed Suspended Grid System

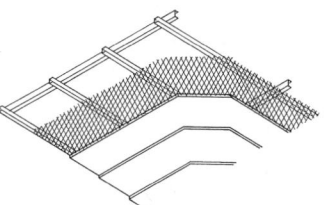

Plaster and Metal Lath
on Metal Furring

C3030 105 — Plaster Ceilings

	TYPE	LATH	FURRING	SUPPORT		MAT.	INST.	TOTAL
						COST PER S.F.		
2400	2 coat gypsum	3/8" gypsum	1"x3" wood, 16" O.C.	wood		1.34	6.25	7.59
2500	Painted			masonry		1.37	6.40	7.77
2600				concrete		1.37	7.15	8.52
2700	3 coat gypsum	3.4# metal	1"x3" wood, 16" O.C.	wood		1.69	6.70	8.39
2800	Painted			masonry		1.72	6.80	8.52
2900				concrete		1.72	7.60	9.32
3000	2 coat perlite	3/8" gypsum	1"x3" wood, 16" O.C.	wood		1.61	6.55	8.16
3100	Painted			masonry		1.64	6.65	8.29
3200				concrete		1.64	7.40	9.04
3300	3 coat perlite	3.4# metal	1"x3" wood, 16" O.C.	wood		1.77	6.70	8.47
3400	Painted			masonry		1.80	6.80	8.60
3500				concrete		1.80	7.60	9.40
3600	2 coat gypsum	3/8" gypsum	3/4" CRC, 12" O.C.	1-1/2" CRC, 48"O.C.		1.43	7.70	9.13
3700	Painted		3/4" CRC, 16" O.C.	1-1/2" CRC, 48"O.C.		1.38	6.95	8.33
3800			3/4" CRC, 24" O.C.	1-1/2" CRC, 48"O.C.		1.25	6.35	7.60
3900	2 coat perlite	3/8" gypsum	3/4" CRC, 12" O.C.	1-1/2" CRC, 48"O.C		1.70	8.30	10
4000	Painted		3/4" CRC, 16" O.C.	1-1/2" CRC, 48"O.C.		1.65	7.50	9.15
4100			3/4" CRC, 24" O.C.	1-1/2" CRC, 48"O.C.		1.52	6.90	8.42
4200	3 coat gypsum	3.4# metal	3/4" CRC, 12" O.C.	1-1/2" CRC, 36" O.C.		1.95	10	11.95
4300	Painted		3/4" CRC, 16" O.C.	1-1/2" CRC, 36" O.C.		1.90	9.20	11.10
4400			3/4" CRC, 24" O.C.	1-1/2" CRC, 36" O.C.		1.77	8.60	10.37
4500	3 coat perlite	3.4# metal	3/4" CRC, 12" O.C.	1-1/2" CRC,36" O.C.		2.08	10.90	12.98
4600	Painted		3/4" CRC, 16" O.C.	1-1/2" CRC, 36" O.C.		2.03	10.15	12.18
4700			3/4" CRC, 24" O.C.	1-1/2" CRC, 36" O.C.		1.90	9.55	11.45

C3030 110 — Drywall Ceilings

	TYPE	FINISH	FURRING	SUPPORT		MAT.	INST.	TOTAL
						COST PER S.F.		
4800	1/2" F.R. drywall	painted and textured	1"x3" wood, 16" O.C.	wood		.99	3.86	4.85
4900				masonry		1.02	3.99	5.01
5000				concrete		1.02	4.75	5.77
5100	5/8" F.R. drywall	painted and textured	1"x3" wood, 16" O.C.	wood		.98	3.86	4.84
5200				masonry		1.01	3.99	5
5300				concrete		1.01	4.75	5.76

C3030 Ceiling Finishes

C3030 110	Drywall Ceilings					COST PER S.F.		
	TYPE	FINISH	FURRING	SUPPORT		MAT.	INST.	TOTAL
5400	1/2″ F.R. drywall	painted and textured	7/8″ resil. channels	24″ O.C.		.79	3.75	4.54
5500			1″x2″ wood	stud clips		.99	3.63	4.62
5602		painted	1-5/8″ metal studs	24″ O.C.		.85	3.35	4.20
5700	5/8″ F.R. drywall	painted and textured	1-5/8″metal studs	24″ O.C.		.84	3.35	4.19
5702								

C3030 140	Plaster Ceiling Components	COST PER S.F.		
		MAT.	INST.	TOTAL
0060	Plaster, gypsum incl. finish			
0080	3 coats	.65	3.93	4.58
0100	Perlite, incl. finish, 2 coats	.73	3.93	4.66
0120	3 coats	.78	4.87	5.65
0140	Thin coat on drywall	.12	.77	.89
0200	Lath, gypsum, 3/8″ thick	.39	1.06	1.45
0220	1/2″ thick	.31	1.10	1.41
0240	5/8″ thick	.35	1.29	1.64
0260	Metal, diamond, 2.5 lb.	.51	.86	1.37
0280	3.4 lb.	.55	.92	1.47
0300	Flat rib, 2.75 lb.	.44	.86	1.30
0320	3.4 lb.	.58	.92	1.50
0440	Furring, steel channels, 3/4″ galvanized , 12″ O.C.	.44	2.77	3.21
0460	16″ O.C.	.39	2	2.39
0480	24″ O.C.	.26	1.38	1.64
0500	1-1/2″ galvanized , 12″ O.C.	.58	3.06	3.64
0520	16″ O.C.	.52	2.24	2.76
0540	24″ O.C.	.35	1.49	1.84
0560	Wood strips, 1″x3″, on wood, 12″ O.C.	.47	1.76	2.23
0580	16″ O.C.	.35	1.32	1.67
0600	24″ O.C.	.24	.88	1.12
0620	On masonry, 12″ O.C.	.51	1.93	2.44
0640	16″ O.C.	.38	1.45	1.83
0660	24″ O.C.	.26	.97	1.23
0680	On concrete, 12″ O.C.	.51	2.94	3.45
0700	16″ O.C.	.38	2.21	2.59
0720	24″ O.C.	.26	1.47	1.73
0940	Paint on plaster or drywall, roller work, primer + 1 coat	.14	.53	.67
0960	Primer + 2 coats	.22	.67	.89

C3030 210	Acoustical Ceilings					COST PER S.F.		
	TYPE	TILE	GRID	SUPPORT		MAT.	INST.	TOTAL
5800	5/8″ fiberglass board	24″ x 48″	tee	suspended		2.50	1.85	4.35
5900		24″ x 24″	tee	suspended		2.75	2.03	4.78
6000	3/4″ fiberglass board	24″ x 48″	tee	suspended		4.41	1.89	6.30
6100		24″ x 24″	tee	suspended		4.66	2.07	6.73
6500	5/8″ mineral fiber	12″ x 12″	1″x3″ wood, 12″ O.C.	wood		3.04	3.82	6.86
6600				masonry		3.08	3.99	7.07
6700				concrete		3.08	5	8.08
6800	3/4″ mineral fiber	12″ x 12″	1″x3″ wood, 12″ O.C.	wood		3.83	3.82	7.65
6900				masonry		3.83	3.82	7.65
7000				concrete		3.83	3.82	7.65
7100	3/4″mineral fiber on	12″ x 12″	25 ga. channels	runners		4.26	4.77	9.03
7102	5/8″ F.R. drywall							

421

C3030 Ceiling Finishes

| C3030 210 | Acoustical Ceilings | | | | | COST PER S.F. | | |

	TYPE	TILE	GRID	SUPPORT		MAT.	INST.	TOTAL
7200 7201 7202	5/8" plastic coated Mineral fiber	12" x 12"	25 ga. channels	adhesive backed		3.01	2.06	5.07
7300 7301 7302	3/4" plastic coated Mineral fiber	12" x 12"		adhesive backed		3.80	2.06	5.86
7400 7401 7402	3/4" mineral fiber	12" x 12"	conceal 2" bar & channels	suspended		2.95	4.65	7.60

C3030 240	Acoustical Ceiling Components	COST PER S.F.		
		MAT.	INST.	TOTAL
2480	Ceiling boards, eggcrate, acrylic, 1/2" x 1/2" x 1/2" cubes	2.02	1.24	3.26
2500	Polystyrene, 3/8" x 3/8" x 1/2" cubes	1.77	1.21	2.98
2520	1/2" x 1/2" x 1/2" cubes	2.04	1.24	3.28
2540	Fiberglass boards, plain, 5/8" thick	1.39	.99	2.38
2560	3/4" thick	3.30	1.03	4.33
2580	Grass cloth faced, 3/4" thick	3.29	1.24	4.53
2600	1" thick	3.97	1.27	5.24
2620	Luminous panels, prismatic, acrylic	3.22	1.54	4.76
2640	Polystyrene	1.85	1.54	3.39
2660	Flat or ribbed, acrylic	4.62	1.54	6.16
2680	Polystyrene	2.60	1.54	4.14
2700	Drop pan, white, acrylic	6.05	1.54	7.59
2720	Polystyrene	4.94	1.54	6.48
2740	Mineral fiber boards, 5/8" thick, standard	.96	.91	1.87
2760	Plastic faced	2.88	1.54	4.42
2780	2 hour rating	1.41	.91	2.32
2800	Perforated aluminum sheets, .024 thick, corrugated painted	2.97	1.26	4.23
2820	Plain	5.35	1.24	6.59
3080	Mineral fiber, plastic coated, 12" x 12" or 12" x 24", 5/8" thick	2.57	2.06	4.63
3100	3/4" thick	3.36	2.06	5.42
3120	Fire rated, 3/4" thick, plain faced	1.56	2.06	3.62
3140	Mylar faced	2.31	2.06	4.37
3160	Add for ceiling primer	.13		.13
3180	Add for ceiling cement	.44		.44
3240	Suspension system, furring, 1" x 3" wood 12" O.C.	.47	1.76	2.23
3260	T bar suspension system, 2' x 4' grid	.87	.77	1.64
3280	2' x 2' grid	1.12	.95	2.07
3300	Concealed Z bar suspension system 12" module	1.02	1.19	2.21
3320	Add to above for 1-1/2" carrier channels 4' O.C.	.13	1.31	1.44
3340	Add to above for carrier channels for recessed lighting	.24	1.34	1.58

D10 Conveying

D1010 Elevators and Lifts

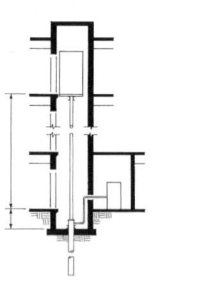

Hydraulic

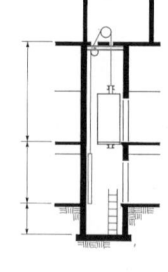

Traction Geared

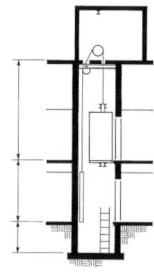

Traction Gearless

D1010 110	Hydraulic	COST EACH		
		MAT.	INST.	TOTAL
1300	Pass. elev., 1500 lb., 2 Floors, 100 FPM	53,000	20,100	73,100
1400	5 Floors, 100 FPM	90,500	53,500	144,000
1600	2000 lb., 2 Floors, 100 FPM	54,000	20,100	74,100
1700	5 floors, 100 FPM	91,500	53,500	145,000
1900	2500 lb., 2 Floors, 100 FPM	56,500	20,100	76,600
2000	5 floors, 100 FPM	94,000	53,500	147,500
2200	3000 lb., 2 Floors, 100 FPM	58,000	20,100	78,100
2300	5 floors, 100 FPM	95,500	53,500	149,000
2500	3500 lb., 2 Floors, 100 FPM	61,500	20,100	81,600
2600	5 floors, 100 FPM	99,000	53,500	152,500
2800	4000 lb., 2 Floors, 100 FPM	63,000	20,100	83,100
2900	5 floors, 100 FPM	100,500	53,500	154,000
3100	4500 lb., 2 Floors, 100 FPM	66,500	20,100	86,600
3200	5 floors, 100 FPM	104,000	53,500	157,500
4000	Hospital elevators, 3500 lb., 2 Floors, 100 FPM	88,000	20,100	108,100
4100	5 floors, 100 FPM	133,000	53,500	186,500
4300	4000 lb., 2 Floors, 100 FPM	88,000	20,100	108,100
4400	5 floors, 100 FPM	133,000	53,500	186,500
4600	4500 lb., 2 Floors, 100 FPM	96,000	20,100	116,100
4800	5 floors, 100 FPM	141,000	53,500	194,500
4900	5000 lb., 2 Floors, 100 FPM	100,000	20,100	120,100
5000	5 floors, 100 FPM	145,000	53,500	198,500
6700	Freight elevators (Class "B"), 3000 lb., 2 Floors, 50 FPM	119,500	25,600	145,100
6800	5 floors, 100 FPM	188,000	67,000	255,000
7000	4000 lb., 2 Floors, 50 FPM	124,500	25,600	150,100
7100	5 floors, 100 FPM	193,500	67,000	260,500
7500	10,000 lb., 2 Floors, 50 FPM	149,500	25,600	175,100
7600	5 floors, 100 FPM	218,000	67,000	285,000
8100	20,000 lb., 2 Floors, 50 FPM	178,500	25,600	204,100
8200	5 Floors, 100 FPM	247,000	67,000	314,000

D1010 140	Traction Geared Elevators	COST EACH		
		MAT.	INST.	TOTAL
1300	Passenger, 2000 Lb., 5 floors, 200 FPM	135,500	50,500	186,000
1500	15 floors, 350 FPM	272,000	152,500	424,500
1600	2500 Lb., 5 floors, 200 FPM	140,000	50,500	190,500
1800	15 floors, 350 FPM	276,500	152,500	429,000
2200	3500 Lb., 5 floors, 200 FPM	142,500	50,500	193,000
2400	15 floors, 350 FPM	278,500	152,500	431,000
2500	4000 Lb., 5 floors, 200 FPM	143,500	50,500	194,000
2700	15 floors, 350 FPM	280,000	152,500	432,500

D1010 Elevators and Lifts

D1010 140	Traction Geared Elevators	COST EACH		
		MAT.	INST.	TOTAL
2800	4500 Lb., 5 floors, 200 FPM	146,500	50,500	197,000
3000	15 floors, 350 FPM	283,000	152,500	435,500
3100	5000 Lb., 5 floors, 200 FPM	149,500	50,500	200,000
3300	15 floors, 350 FPM	286,000	152,500	438,500
4000	Hospital, 3500 Lb., 5 floors, 200 FPM	139,000	50,500	189,500
4200	15 floors, 350 FPM	326,500	152,500	479,000
4300	4000 Lb., 5 floors, 200 FPM	139,000	50,500	189,500
4500	15 floors, 350 FPM	326,500	152,500	479,000
4600	4500 Lb., 5 floors, 200 FPM	146,000	50,500	196,500
4800	15 floors, 350 FPM	333,000	152,500	485,500
4900	5000 Lb., 5 floors, 200 FPM	148,000	50,500	198,500
5100	15 floors, 350 FPM	335,500	152,500	488,000
6000	Freight, 4000 Lb., 5 floors, 50 FPM class 'B'	160,500	52,000	212,500
6200	15 floors, 200 FPM class 'B'	350,000	181,500	531,500
6300	8000 Lb., 5 floors, 50 FPM class 'B'	187,000	52,000	239,000
6500	15 floors, 200 FPM class 'B'	377,000	181,500	558,500
7000	10,000 Lb., 5 floors, 50 FPM class 'B'	218,500	52,000	270,500
7200	15 floors, 200 FPM class 'B'	618,500	181,500	800,000
8000	20,000 Lb., 5 floors, 50 FPM class 'B'	242,500	52,000	294,500
8200	15 floors, 200 FPM class 'B'	642,500	181,500	824,000

D1010 150	Traction Gearless Elevators	COST EACH		
		MAT.	INST.	TOTAL
1700	Passenger, 2500 Lb., 10 floors, 200 FPM	291,500	131,500	423,000
1900	30 floors, 600 FPM	600,000	335,000	935,000
2000	3000 Lb., 10 floors, 200 FPM	292,000	131,500	423,500
2200	30 floors, 600 FPM	600,500	335,000	935,500
2300	3500 Lb., 10 floors, 200 FPM	294,000	131,500	425,500
2500	30 floors, 600 FPM	602,500	335,000	937,500
2700	50 floors, 800 FPM	869,500	539,000	1,408,500
2800	4000 Lb., 10 floors, 200 FPM	295,000	131,500	426,500
3000	30 floors, 600 FPM	603,500	335,000	938,500
3200	50 floors, 800 FPM	870,500	539,000	1,409,500
3300	4500 Lb., 10 floors, 200 FPM	298,000	131,500	429,500
3500	30 floors, 600 FPM	606,500	335,000	941,500
3700	50 floors, 800 FPM	873,500	539,000	1,412,500
3800	5000 Lb., 10 floors, 200 FPM	301,000	131,500	432,500
4000	30 floors, 600 FPM	609,500	335,000	944,500
4200	50 floors, 800 FPM	876,500	539,000	1,415,500
6000	Hospital, 3500 Lb., 10 floors, 200 FPM	316,000	131,500	447,500
6200	30 floors, 600 FPM	730,000	335,000	1,065,000
6400	4000 Lb., 10 floors, 200 FPM	316,000	131,500	447,500
6600	30 floors, 600 FPM	730,000	335,000	1,065,000
6800	4500 Lb., 10 floors, 200 FPM	322,500	131,500	454,000
7000	30 floors, 600 FPM	736,500	335,000	1,071,500
7200	5000 Lb., 10 floors, 200 FPM	324,500	131,500	456,000
7400	30 floors, 600 FPM	738,500	335,000	1,073,500

For customer support on your Square Foot Costs with RSMeans data, call 800.448.8182.

425

D2010 Plumbing Fixtures

One Piece Wall Hung Water Closet

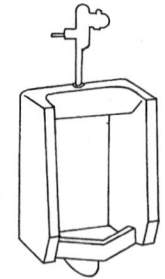

Wall Hung Urinal

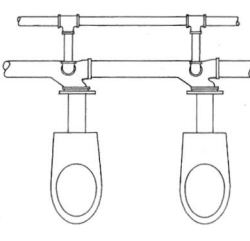

Side By Side Water Closet Group

Floor Mount Water Closet

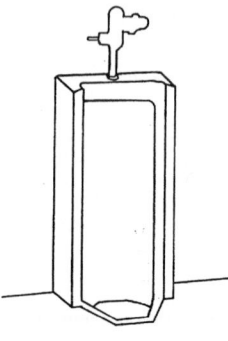

Stall Type Urinal

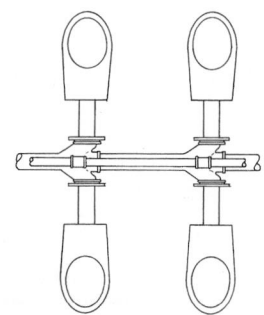

Back to Back Water Closet Group

D2010 110	Water Closet Systems	COST EACH		
		MAT.	INST.	TOTAL
1800	Water closet, vitreous china			
1840	Tank type, wall hung			
1880	Close coupled two piece	1,725	795	2,520
1920	Floor mount, one piece	1,300	845	2,145
1960	One piece low profile	1,275	845	2,120
2000	Two piece close coupled	725	845	1,570
2040	Bowl only with flush valve			
2080	Wall hung	2,500	900	3,400
2120	Floor mount	870	860	1,730
2122				
2160	Floor mount, ADA compliant with 18" high bowl	890	880	1,770

D2010 120	Water Closets, Group	COST EACH		
		MAT.	INST.	TOTAL
1760	Water closets, battery mount, wall hung, side by side, first closet	2,625	930	3,555
1800	Each additional water closet, add	2,525	875	3,400
3000	Back to back, first pair of closets	4,425	1,225	5,650
3100	Each additional pair of closets, back to back	4,350	1,200	5,550
9000	Back to back, first pair of closets, auto sensor flush valve, 1.28 gpf	4,900	1,325	6,225
9100	Ea additional pair of cls, back to back, auto sensor flush valve, 1.28 gpf	4,725	1,250	5,975

D2010 210	Urinal Systems	COST EACH		
		MAT.	INST.	TOTAL
2000	Urinal, vitreous china, wall hung	590	890	1,480
2040	Stall type	1,325	1,075	2,400

D2010   Plumbing Fixtures

Systems are complete with trim and rough-in (supply, waste and vent) to connect to supply branches and waste mains.

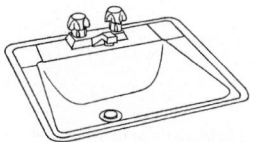
Vanity Top

Wall Hung

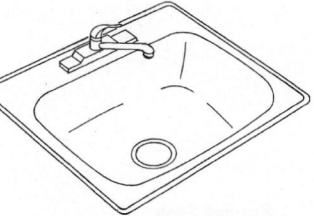
Counter Top Single Bowl

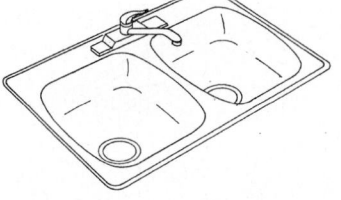
Counter Top Double Bowl

D2010 310	Lavatory Systems	COST EACH		
		MAT.	INST.	TOTAL
1560	Lavatory w/trim, vanity top, PE on CI, 20" x 18", Vanity top by others.	595	795	1,390
1600	19" x 16" oval	395	795	1,190
1640	18" round	695	795	1,490
1680	Cultured marble, 19" x 17"	405	795	1,200
1720	25" x 19"	430	795	1,225
1760	Stainless, self-rimming, 25" x 22"	615	795	1,410
1800	17" x 22"	605	795	1,400
1840	Steel enameled, 20" x 17"	405	820	1,225
1880	19" round	435	820	1,255
1920	Vitreous china, 20" x 16"	500	835	1,335
1960	19" x 16"	505	835	1,340
2000	22" x 13"	505	835	1,340
2040	Wall hung, PE on CI, 18" x 15"	905	880	1,785
2080	19" x 17"	955	880	1,835
2120	20" x 18"	785	880	1,665
2160	Vitreous china, 18" x 15"	695	905	1,600
2200	19" x 17"	650	905	1,555
2240	24" x 20"	785	905	1,690
2300	20" x 27", handicap	1,550	975	2,525

D2010 410	Kitchen Sink Systems	COST EACH		
		MAT.	INST.	TOTAL
1720	Kitchen sink w/trim, countertop, PE on CI, 24"x21", single bowl	640	875	1,515
1760	30" x 21" single bowl	1,025	875	1,900
1800	32" x 21" double bowl	745	940	1,685
1880	Stainless steel, 19" x 18" single bowl	980	875	1,855
1920	25" x 22" single bowl	1,050	875	1,925
1960	33" x 22" double bowl	1,400	940	2,340
2000	43" x 22" double bowl	1,600	955	2,555
2040	44" x 22" triple bowl	1,600	990	2,590
2080	44" x 24" corner double bowl	1,125	955	2,080
2120	Steel, enameled, 24" x 21" single bowl	885	875	1,760
2160	32" x 21" double bowl	925	940	1,865
2240	Raised deck, PE on CI, 32" x 21", dual level, double bowl	830	1,200	2,030
2280	42" x 21" dual level, triple bowl	1,400	1,300	2,700

D2010 Plumbing Fixtures

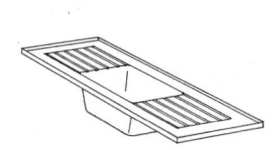

Laboratory Sink

Service Sink

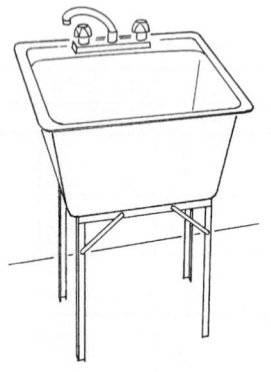

Single Compartment Sink

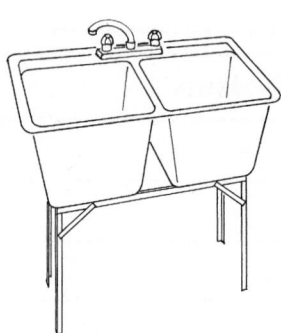

Double Compartment Sink

Systems are complete with trim and rough-in (supply, waste and vent) to connect to supply branches and waste mains.

D2010 420	Laundry Sink Systems	COST EACH		
		MAT.	INST.	TOTAL
1740	Laundry sink w/trim, PE on CI, black iron frame			
1760	24" x 20", single compartment	955	855	1,810
1840	48" x 21" double compartment	1,200	930	2,130
1920	Molded stone, on wall, 22" x 21" single compartment	505	855	1,360
1960	45"x 21" double compartment	675	930	1,605
2040	Plastic, on wall or legs, 18" x 23" single compartment	465	840	1,305
2080	20" x 24" single compartment	485	840	1,325
2120	36" x 23" double compartment	565	905	1,470
2160	40" x 24" double compartment	645	905	1,550

D2010 430	Laboratory Sink Systems	COST EACH		
		MAT.	INST.	TOTAL
1580	Laboratory sink w/trim,			
1590	Stainless steel, single bowl,			
1600	Double drainboard, 54" x 24" O.D.	1,450	1,125	2,575
1640	Single drainboard, 47" x 24"O.D.	1,150	1,125	2,275
1670	Stainless steel, double bowl,			
1680	70" x 24" O.D.	1,575	1,125	2,700
1750	Polyethylene, single bowl,			
1760	Flanged, 14-1/2" x 14-1/2" O.D.	535	1,000	1,535
1800	18-1/2" x 18-1/2" O.D.	655	1,000	1,655
1840	23-1/2" x 20-1/2" O.D.	675	1,000	1,675
1920	Polypropylene, cup sink, oval, 7" x 4" O.D.	420	885	1,305
1960	10" x 4-1/2" O.D.	450	885	1,335
1961				

For customer support on your Square Foot Costs with RSMeans data, call 800.448.8182.

D2010 Plumbing Fixtures

D2010 440	Service Sink Systems	COST EACH		
		MAT.	INST.	TOTAL
4260	Service sink w/trim, PE on CI, corner floor, 28" x 28", w/rim guard	2,150	1,125	3,275
4300	Wall hung w/rim guard, 22" x 18"	2,475	1,300	3,775
4340	24" x 20"	2,575	1,300	3,875
4380	Vitreous china, wall hung 22" x 20"	2,475	1,300	3,775
4383				

D2010 Plumbing Fixtures

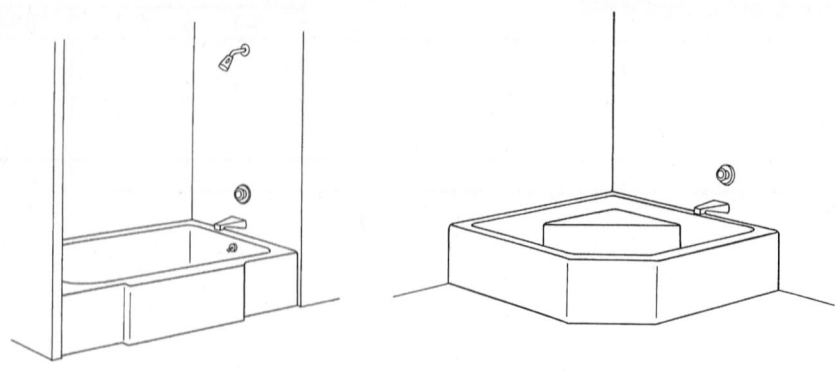

Recessed Bathtub

Corner Bathtub

Systems are complete with trim and rough-in (supply, waste and vent) to connect to supply branches and waste mains.

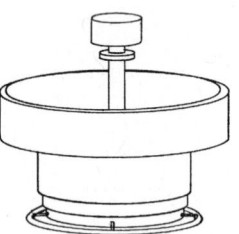

Circular Wash Fountain

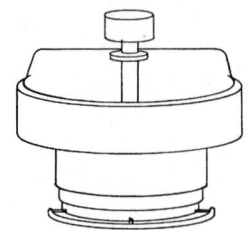

Semi-Circular Wash Fountain

D2010 510	Bathtub Systems	COST EACH		
		MAT.	INST.	TOTAL
2000	Bathtub, recessed, P.E. on Cl., 48" x 42"	3,575	985	4,560
2040	72" x 36"	3,650	1,100	4,750
2080	Mat bottom, 5' long	1,800	955	2,755
2120	5'-6" long	2,425	985	3,410
2160	Corner, 48" x 42"	3,575	955	4,530
2200	Formed steel, enameled, 4'-6" long	1,025	885	1,910

D2010 610	Group Wash Fountain Systems	COST EACH		
		MAT.	INST.	TOTAL
1740	Group wash fountain, precast terrazzo			
1760	Circular, 36" diameter	8,650	1,450	10,100
1800	54" diameter	10,700	1,575	12,275
1840	Semi-circular, 36" diameter	7,650	1,450	9,100
1880	54" diameter	10,100	1,575	11,675
1960	Stainless steel, circular, 36" diameter	7,850	1,350	9,200
2000	54" diameter	9,525	1,500	11,025
2040	Semi-circular, 36" diameter	6,150	1,350	7,500
2080	54" diameter	8,375	1,500	9,875
2160	Thermoplastic, circular, 36" diameter	5,500	1,100	6,600
2200	54" diameter	6,325	1,275	7,600
2240	Semi-circular, 36" diameter	5,100	1,100	6,200
2280	54" diameter	6,125	1,275	7,400

D2010 Plumbing Fixtures

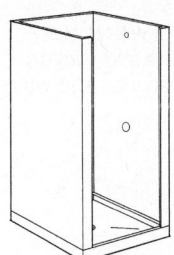

Square Shower Stall

Corner Angle Shower Stall

Systems are complete with trim and rough-in (supply, waste and vent) to connect to supply branches and waste mains.

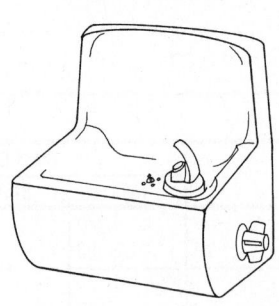

Wall Mounted, Low Back

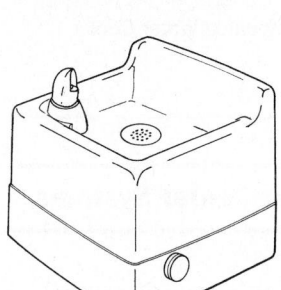

Wall Mounted, No Back

D2010 710	Shower Systems	COST EACH		
		MAT.	INST.	TOTAL
1560	Shower, stall, baked enamel, molded stone receptor, 30″ square	1,600	915	2,515
1600	32″ square	1,600	930	2,530
1640	Terrazzo receptor, 32″ square	1,800	930	2,730
1680	36″ square	1,950	940	2,890
1720	36″ corner angle	2,075	385	2,460
1800	Fiberglass one piece, three walls, 32″ square	705	905	1,610
1840	36″ square	765	905	1,670
1880	Polypropylene, molded stone receptor, 30″ square	1,025	1,325	2,350
1920	32″ square	1,050	1,325	2,375
1960	Built-in head, arm, bypass, stops and handles	122	345	467
2050	Shower, stainless steel panels, handicap			
2100	w/fixed and handheld head, control valves, grab bar, and seat	4,175	4,125	8,300
2500	Shower, group with six heads, thermostatic mix valves & balancing valve	12,600	1,000	13,600
2520	Five heads	9,050	920	9,970

D2010 810	Drinking Fountain Systems	COST EACH		
		MAT.	INST.	TOTAL
1740	Drinking fountain, one bubbler, wall mounted			
1760	Non recessed			
1800	Bronze, no back	1,325	525	1,850
1840	Cast iron, enameled, low back	1,300	525	1,825
1880	Fiberglass, 12″ back	2,450	525	2,975
1920	Stainless steel, no back	1,325	525	1,850
1960	Semi-recessed, poly marble	1,275	525	1,800
2040	Stainless steel	1,675	525	2,200
2080	Vitreous china	1,200	525	1,725
2120	Full recessed, poly marble	2,050	525	2,575
2200	Stainless steel	1,900	525	2,425
2240	Floor mounted, pedestal type, aluminum	2,850	715	3,565
2320	Bronze	2,475	715	3,190
2360	Stainless steel	2,375	715	3,090

D2010 Plumbing Fixtures

Wall Hung Water Cooler

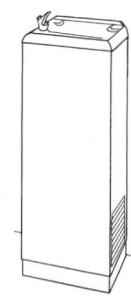

Floor Mounted Water Cooler

Systems are complete with trim and rough-in (supply, waste and vent) to connect to supply branches and waste mains.

D2010 820	Water Cooler Systems	COST EACH		
		MAT.	INST.	TOTAL
1840	Water cooler, electric, wall hung, 8.2 G.P.H.	1,225	675	1,900
1880	Dual height, 14.3 G.P.H.	1,850	695	2,545
1920	Wheelchair type, 7.5 G.P.H.	1,325	675	2,000
1960	Semi recessed, 8.1 G.P.H.	1,125	675	1,800
2000	Full recessed, 8 G.P.H.	2,675	725	3,400
2040	Floor mounted, 14.3 G.P.H.	1,275	590	1,865
2080	Dual height, 14.3 G.P.H.	1,600	715	2,315
2120	Refrigerated compartment type, 1.5 G.P.H.	1,900	590	2,490

D2010 Plumbing Fixtures

Two Fixture   Three Fixture

*Common wall is with adjacent bathrom

D2010 920	Two Fixture Bathroom, Two Wall Plumbing	COST EACH		
		MAT.	INST.	TOTAL
1180	Bathroom, lavatory & water closet, 2 wall plumbing, stand alone	1,900	2,200	4,100
1200	Share common plumbing wall*	1,725	1,875	3,600

D2010 922	Two Fixture Bathroom, One Wall Plumbing	COST EACH		
		MAT.	INST.	TOTAL
2220	Bathroom, lavatory & water closet, one wall plumbing, stand alone	1,800	1,975	3,775
2240	Share common plumbing wall*	1,425	1,675	3,100

D2010 924	Three Fixture Bathroom, One Wall Plumbing	COST EACH		
		MAT.	INST.	TOTAL
1150	Bathroom, three fixture, one wall plumbing			
1160	Lavatory, water closet & bathtub			
1170	Stand alone	3,150	2,550	5,700
1180	Share common plumbing wall *	2,700	1,850	4,550

D2010 926	Three Fixture Bathroom, Two Wall Plumbing	COST EACH		
		MAT.	INST.	TOTAL
2130	Bathroom, three fixture, two wall plumbing			
2140	Lavatory, water closet & bathtub			
2160	Stand alone	3,175	2,600	5,775
2180	Long plumbing wall common *	2,850	2,075	4,925
3610	Lavatory, bathtub & water closet			
3620	Stand alone	3,450	2,950	6,400
3640	Long plumbing wall common *	3,225	2,675	5,900
4660	Water closet, corner bathtub & lavatory			
4680	Stand alone	4,975	2,625	7,600
4700	Long plumbing wall common *	4,500	1,975	6,475
6100	Water closet, stall shower & lavatory			
6120	Stand alone	3,525	2,950	6,475
6140	Long plumbing wall common *	3,300	2,725	6,025
7060	Lavatory, corner stall shower & water closet			
7080	Stand alone	3,775	2,600	6,375
7100	Short plumbing wall common *	3,225	1,750	4,975

D2010 Plumbing Fixtures

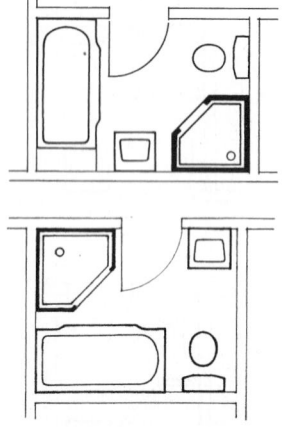

Four Fixture Bathroom Systems consisting of a lavatory, water closet, bathtub, shower and rough-in service piping.

• Prices for plumbing and fixtures only.

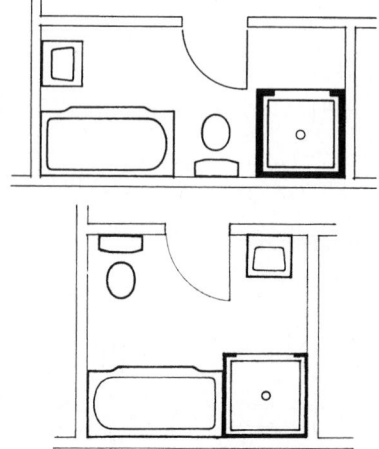

*Common wall is with an adjacent bathroom.

D2010 928	Four Fixture Bathroom, Two Wall Plumbing	COST EACH		
		MAT.	INST.	TOTAL
1140	Bathroom, four fixture, two wall plumbing			
1150	Bathtub, water closet, stall shower & lavatory			
1160	Stand alone	4,425	2,825	7,250
1180	Long plumbing wall common *	3,950	2,175	6,125
2260	Bathtub, lavatory, corner stall shower & water closet			
2280	Stand alone	5,075	2,825	7,900
2320	Long plumbing wall common *	4,600	2,175	6,775
3620	Bathtub, stall shower, lavatory & water closet			
3640	Stand alone	4,900	3,550	8,450
3660	Long plumbing wall (opp. door) common *	4,425	2,900	7,325

D2010 930	Four Fixture Bathroom, Three Wall Plumbing	COST EACH		
		MAT.	INST.	TOTAL
4680	Bathroom, four fixture, three wall plumbing			
4700	Bathtub, stall shower, lavatory & water closet			
4720	Stand alone	5,750	3,900	9,650
4760	Long plumbing wall (opposite door) common *	5,575	3,600	9,175

D2010 Plumbing Fixtures

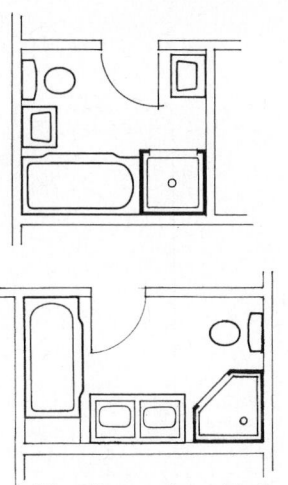

Five Fixture Bathroom Systems consisting of two lavatories, a water closet, bathtub, shower and rough-in service piping.

- Prices for plumbing and fixtures only.

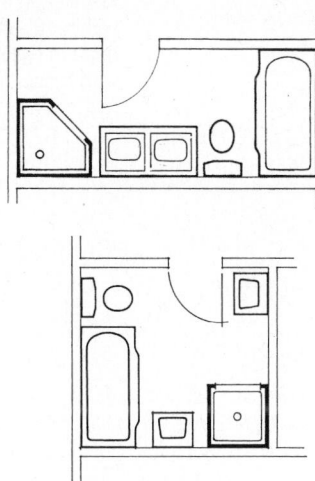

*Common wall is with an adjacent bathroom.

D2010 932	Five Fixture Bathroom, Two Wall Plumbing	COST EACH		
		MAT.	INST.	TOTAL
1320	Bathroom, five fixture, two wall plumbing			
1340	Bathtub, water closet, stall shower & two lavatories			
1360	Stand alone	5,675	4,525	10,200
1400	One short plumbing wall common *	5,200	3,875	9,075

D2010 934	Five Fixture Bathroom, Three Wall Plumbing	COST EACH		
		MAT.	INST.	TOTAL
2360	Bathroom, five fixture, three wall plumbing			
2380	Water closet, bathtub, two lavatories & stall shower			
2400	Stand alone	6,325	4,550	10,875
2440	One short plumbing wall common *	5,850	3,900	9,750

D2010 936	Five Fixture Bathroom, One Wall Plumbing	COST EACH		
		MAT.	INST.	TOTAL
4080	Bathroom, five fixture, one wall plumbing			
4100	Bathtub, two lavatories, corner stall shower & water closet			
4120	Stand alone	6,025	4,050	10,075
4160	Share common wall *	5,025	2,625	7,650

For customer support on your Square Foot Costs with RSMeans data, call 800.448.8182.

D2020 Domestic Water Distribution

Installation includes piping and fittings within 10' of heater. Gas and oil fired heaters require vent piping (not included with these units).

Gas Fired

Oil Fired

D2020 220	Gas Fired Water Heaters - Residential Systems	COST EACH		
		MAT.	INST.	TOTAL
2200	Gas fired water heater, residential, 100°F rise			
2260	30 gallon tank, 32 GPH	2,650	1,550	4,200
2300	40 gallon tank, 32 GPH	2,850	1,750	4,600
2340	50 gallon tank, 63 GPH	3,025	1,750	4,775
2380	75 gallon tank, 63 GPH	4,250	1,950	6,200
2420	100 gallon tank, 63 GPH	4,425	2,050	6,475
2422				

D2020 230	Oil Fired Water Heaters - Residential Systems	COST EACH		
		MAT.	INST.	TOTAL
2200	Oil fired water heater, residential, 100°F rise			
2220	30 gallon tank, 103 GPH	2,075	1,450	3,525
2260	50 gallon tank, 145 GPH	2,600	1,600	4,200
2300	70 gallon tank, 164 GPH	3,675	1,825	5,500
2340	85 gallon tank, 181 GPH	12,000	1,850	13,850

D2020   Domestic Water Distribution

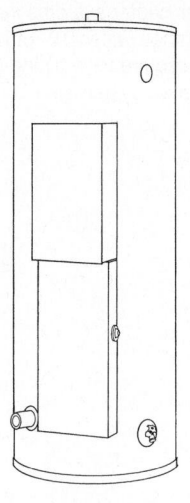

Systems below include piping and fittings within 10' of heater. Electric water heaters do not require venting. Gas fired heaters require vent piping (not included in these prices).

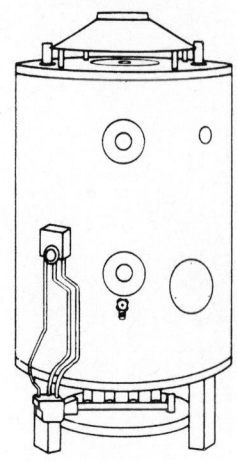

Electric

Gas Fired

D2020 240	Electric Water Heaters - Commercial Systems	COST EACH		
		MAT.	INST.	TOTAL
1800	Electric water heater, commercial, 100°F rise			
1820	50 gallon tank, 9 KW 37 GPH	8,625	1,300	9,925
1860	80 gal, 12 KW 49 GPH	11,000	1,600	12,600
1900	36 KW 147 GPH	14,900	1,750	16,650
1940	120 gal, 36 KW 147 GPH	16,100	1,875	17,975
1980	150 gal, 120 KW 490 GPH	50,000	2,000	52,000
2020	200 gal, 120 KW 490 GPH	53,000	2,075	55,075
2060	250 gal, 150 KW 615 GPH	58,500	2,425	60,925
2100	300 gal, 180 KW 738 GPH	63,500	2,525	66,025
2140	350 gal, 30 KW 123 GPH	47,200	2,750	49,950
2180	180 KW 738 GPH	65,500	2,750	68,250
2220	500 gal, 30 KW 123 GPH	61,500	3,200	64,700
2260	240 KW 984 GPH	99,500	3,200	102,700
2300	700 gal, 30 KW 123 GPH	74,500	3,675	78,175
2340	300 KW 1230 GPH	110,500	3,675	114,175
2380	1000 gal, 60 KW 245 GPH	90,000	5,125	95,125
2420	480 KW 1970 GPH	146,500	5,125	151,625
2460	1500 gal, 60 KW 245 GPH	131,500	6,325	137,825
2500	480 KW 1970 GPH	181,000	6,325	187,325

D2020 250	Gas Fired Water Heaters - Commercial Systems	COST EACH		
		MAT.	INST.	TOTAL
1760	Gas fired water heater, commercial, 100°F rise			
1780	75.5 MBH input, 63 GPH	5,450	2,000	7,450
1820	95 MBH input, 86 GPH	9,525	2,000	11,525
1860	100 MBH input, 91 GPH	9,775	2,100	11,875
1900	115 MBH input, 110 GPH	9,850	2,150	12,000
1980	155 MBH input, 150 GPH	12,300	2,425	14,725
2020	175 MBH input, 168 GPH	12,600	2,600	15,200
2060	200 MBH input, 192 GPH	12,800	2,925	15,725
2100	240 MBH input, 230 GPH	13,600	3,175	16,775
2140	300 MBH input, 278 GPH	15,000	3,625	18,625
2180	390 MBH input, 374 GPH	17,800	3,675	21,475
2220	500 MBH input, 480 GPH	23,900	3,975	27,875
2260	600 MBH input, 576 GPH	27,600	4,300	31,900

For customer support on your Square Foot Costs with RSMeans data, call 800.448.8182.

D2020 Domestic Water Distribution

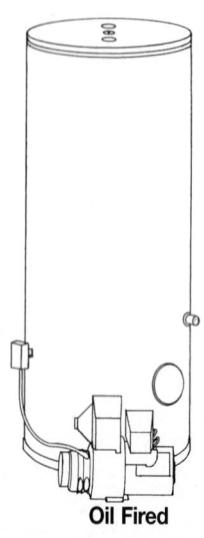

Oil fired water heater systems include piping and fittings within 10' of heater. Oil fired heaters require vent piping (not included in these systems).

Oil Fired

D2020 260	Oil Fired Water Heaters - Commercial Systems	COST EACH		
		MAT.	INST.	TOTAL
1800	Oil fired water heater, commercial, 100°F rise			
1820	140 gal., 140 MBH input, 134 GPH	29,300	1,700	31,000
1900	140 gal., 255 MBH input, 247 GPH	31,600	2,150	33,750
1940	140 gal., 270 MBH input, 259 GPH	38,900	2,450	41,350
1980	140 gal., 400 MBH input, 384 GPH	40,100	2,825	42,925
2060	140 gal., 720 MBH input, 691 GPH	42,600	2,950	45,550
2100	221 gal., 300 MBH input, 288 GPH	56,500	3,225	59,725
2140	221 gal., 600 MBH input, 576 GPH	62,500	3,275	65,775
2180	221 gal., 800 MBH input, 768 GPH	63,500	3,400	66,900
2220	201 gal., 1000 MBH input, 960 GPH	65,000	3,425	68,425
2260	201 gal., 1250 MBH input, 1200 GPH	67,000	3,500	70,500
2300	201 gal., 1500 MBH input, 1441 GPH	72,000	3,575	75,575
2340	411 gal., 600 MBH input, 576 GPH	73,000	3,675	76,675
2380	411 gal., 800 MBH input, 768 GPH	75,500	3,775	79,275
2420	411 gal., 1000 MBH input, 960 GPH	78,500	4,350	82,850
2460	411 gal., 1250 MBH input, 1200 GPH	80,000	4,475	84,475
2500	397 gal., 1500 MBH input, 1441 GPH	85,000	4,600	89,600
2540	397 gal., 1750 MBH input, 1681 GPH	87,000	4,775	91,775

D2040 Rain Water Drainage

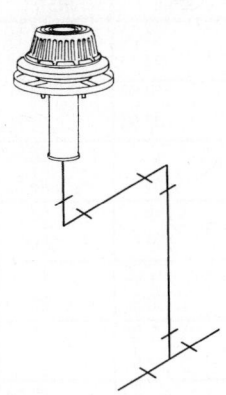

Design Assumptions: Vertical conductor size is based on a maximum rate of rainfall of 4" per hour. To convert roof area to other rates multiply "Max. S.F. Roof Area" shown by four and divide the result by desired local rate. The answer is the local roof area that may be handled by the indicated pipe diameter.

Basic cost is for roof drain, 10' of vertical leader and 10' of horizontal, plus connection to the main.

Pipe Dia.	Max. S.F. Roof Area	Gallons per Min.
2"	544	23
3"	1610	67
4"	3460	144
5"	6280	261
6"	10,200	424
8"	22,000	913

D2040 210	Roof Drain Systems	COST EACH		
		MAT.	INST.	TOTAL
1880	Roof drain, DWV PVC, 2" diam., piping, 10' high	370	755	1,125
1920	For each additional foot add	9.50	23	32.50
1960	3" diam., 10' high	395	880	1,275
2000	For each additional foot add	9.40	25.50	34.90
2040	4" diam., 10' high	470	985	1,455
2080	For each additional foot add	11.20	28	39.20
2120	5" diam., 10' high	1,775	1,150	2,925
2160	For each additional foot add	42.50	31.50	74
2200	6" diam., 10' high	1,525	1,250	2,775
2240	For each additional foot add	19.10	34.50	53.60
2280	8" diam., 10' high	3,225	2,125	5,350
2320	For each additional foot add	38.50	43.50	82
3940	C.I., soil, single hub, service wt., 2" diam. piping, 10' high	805	825	1,630
3980	For each additional foot add	18.60	21.50	40.10
4120	3" diam., 10' high	915	890	1,805
4160	For each additional foot add	22	22.50	44.50
4200	4" diam., 10' high	1,425	970	2,395
4240	For each additional foot add	38.50	24.50	63
4280	5" diam., 10' high	1,900	1,075	2,975
4320	For each additional foot add	47.50	27.50	75
4360	6" diam., 10' high	2,100	1,150	3,250
4400	For each additional foot add	46	28.50	74.50
4440	8" diam., 10' high	4,025	2,350	6,375
4480	For each additional foot add	70	48.50	118.50
6040	Steel galv. sch 40 threaded, 2" diam. piping, 10' high	835	795	1,630
6080	For each additional foot add	14.25	21	35.25
6120	3" diam., 10' high	1,375	1,150	2,525
6160	For each additional foot add	23	31.50	54.50
6200	4" diam., 10' high	2,025	1,475	3,500
6240	For each additional foot add	26.50	37.50	64
6280	5" diam., 10' high	2,075	1,225	3,300
6320	For each additional foot add	34.50	36.50	71
6360	6" diam, 10' high	2,475	1,575	4,050
6400	For each additional foot add	36.50	50	86.50
6440	8" diam., 10' high	4,575	2,275	6,850
6480	For each additional foot add	55.50	56.50	112

D2090 Other Plumbing Systems

D2090 810	Piping - Installed - Unit Costs	COST PER L.F.		
		MAT.	INST.	TOTAL
0840	Cast iron, soil, B & S, service weight, 2" diameter	18.60	21.50	40.10
0860	3" diameter	22	22.50	44.50
0880	4" diameter	38.50	24.50	63
0900	5" diameter	47.50	27.50	75
0920	6" diameter	46	28.50	74.50
0940	8" diameter	70	48.50	118.50
0960	10" diameter	111	53	164
0980	12" diameter	159	59.50	218.50
1040	No hub, 1-1/2" diameter	17.60	19	36.60
1060	2" diameter	18.80	20	38.80
1080	3" diameter	21.50	21	42.50
1100	4" diameter	37.50	23	60.50
1120	5" diameter	47.50	25.50	73
1140	6" diameter	45.50	26.50	72
1160	8" diameter	76	41.50	117.50
1180	10" diameter	124	46.50	170.50
1220	Copper tubing, hard temper, solder, type K, 1/2" diameter	5.75	9.60	15.35
1260	3/4" diameter	9.20	10.10	19.30
1280	1" diameter	13.50	11.35	24.85
1300	1-1/4" diameter	16.25	13.35	29.60
1320	1-1/2" diameter	19.25	15	34.25
1340	2" diameter	28.50	18.70	47.20
1360	2-1/2" diameter	43.50	22.50	66
1380	3" diameter	58.50	25	83.50
1400	4" diameter	105	35.50	140.50
1480	5" diameter	143	42	185
1500	6" diameter	211	55	266
1520	8" diameter	360	61.50	421.50
1560	Type L, 1/2" diameter	3.68	9.25	12.93
1600	3/4" diameter	4.94	9.85	14.79
1620	1" diameter	7.60	11	18.60
1640	1-1/4" diameter	12.60	12.90	25.50
1660	1-1/2" diameter	11.35	14.40	25.75
1680	2" diameter	19.70	17.85	37.55
1700	2-1/2" diameter	28	21.50	49.50
1720	3" diameter	47	24	71
1740	4" diameter	68	34.50	102.50
1760	5" diameter	158	39.50	197.50
1780	6" diameter	159	52.50	211.50
1800	8" diameter	266	58	324
1840	Type M, 1/2" diameter	4.07	8.90	12.97
1880	3/4" diameter	5.75	9.60	15.35
1900	1" diameter	9	10.70	19.70
1920	1-1/4" diameter	11.85	12.50	24.35
1940	1-1/2" diameter	15.25	13.85	29.10
1960	2" diameter	22.50	17	39.50
1980	2-1/2" diameter	33.50	21	54.50
2000	3" diameter	41	23	64
2020	4" diameter	80.50	33.50	114
2060	6" diameter	213	50	263
2080	8" diameter	355	55	410
2120	Type DWV, 1-1/4" diameter	12.75	12.50	25.25
2160	1-1/2" diameter	12.35	13.85	26.20
2180	2" diameter	19.85	17	36.85
2200	3" diameter	27	23	50
2220	4" diameter	60.50	33.50	94
2240	5" diameter	107	37.50	144.50
2260	6" diameter	157	50	207

D2090 Other Plumbing Systems

D2090 810	Piping - Installed - Unit Costs	COST PER L.F.		
		MAT.	INST.	TOTAL
2280	8" diameter	480	55	535
2800	Plastic,PVC, DWV, schedule 40, 1-1/4" diameter	9.35	17.85	27.20
2820	1-1/2" diameter	8.50	21	29.50
2830	2" diameter	9.50	23	32.50
2840	3" diameter	9.40	25.50	34.90
2850	4" diameter	11.20	28	39.20
2890	6" diameter	19.10	34.50	53.60
3010	Pressure pipe 200 PSI, 1/2" diameter	5.45	13.85	19.30
3030	3/4" diameter	5.90	14.70	20.60
3040	1" diameter	9.20	16.30	25.50
3050	1-1/4" diameter	10.15	17.85	28
3060	1-1/2" diameter	10.75	21	31.75
3070	2" diameter	11	23	34
3080	2-1/2" diameter	12.25	24	36.25
3090	3" diameter	13.25	25.50	38.75
3100	4" diameter	35	28	63
3110	6" diameter	31.50	34.50	66
3120	8" diameter	49	43.50	92.50
4000	Steel, schedule 40, black, threaded, 1/2" diameter	4.20	11.90	16.10
4020	3/4" diameter	4.59	12.30	16.89
4030	1" diameter	4.82	14.15	18.97
4040	1-1/4" diameter	5.65	15.15	20.80
4050	1-1/2" diameter	6.25	16.85	23.10
4060	2" diameter	13.15	21	34.15
4070	2-1/2" diameter	17.65	27	44.65
4080	3" diameter	20.50	31.50	52
4090	4" diameter	23.50	37.50	61
4100	Grooved, 5" diameter	47.50	36.50	84
4110	6" diameter	53.50	50	103.50
4120	8" diameter	89	56.50	145.50
4130	10" diameter	119	67.50	186.50
4140	12" diameter	133	77.50	210.50
4200	Galvanized, threaded, 1/2" diameter	4.24	11.90	16.14
4220	3/4" diameter	4.87	12.30	17.17
4230	1" diameter	5.35	14.15	19.50
4240	1-1/4" diameter	6.20	15.15	21.35
4250	1-1/2" diameter	7.05	16.85	23.90
4260	2" diameter	14.25	21	35.25
4270	2-1/2" diameter	19.55	27	46.55
4280	3" diameter	23	31.50	54.50
4290	4" diameter	26.50	37.50	64
4300	Grooved, 5" diameter	34.50	36.50	71
4310	6" diameter	36.50	50	86.50
4320	8" diameter	55.50	56.50	112
4330	10" diameter	113	67.50	180.50
4340	12" diameter	137	77.50	214.50
5010	Flanged, black, 1" diameter	11.35	20	31.35
5020	1-1/4" diameter	12.40	22	34.40
5030	1-1/2" diameter	12.85	24	36.85
5040	2" diameter	15.05	31.50	46.55
5050	2-1/2" diameter	18.80	39.50	58.30
5060	3" diameter	21.50	44	65.50
5070	4" diameter	34	54.50	88.50
5080	5" diameter	60	67	127
5090	6" diameter	65	86.50	151.50
5100	8" diameter	107	113	220
5110	10" diameter	151	135	286
5120	12" diameter	182	155	337

441

D2090 Other Plumbing Systems

D2090 810	Piping - Installed - Unit Costs	COST PER L.F.		
		MAT.	INST.	TOTAL
5720	Grooved joints, black, 3/4" diameter	6.65	10.55	17.20
5730	1" diameter	6.45	11.90	18.35
5740	1-1/4" diameter	7.65	12.90	20.55
5750	1-1/2" diameter	8.35	14.70	23.05
5760	2" diameter	9.60	18.70	28.30
5770	2-1/2" diameter	12.60	23.50	36.10
5900	3" diameter	15.25	27	42.25
5910	4" diameter	26.50	30	56.50
5920	5" diameter	47.50	36.50	84
5930	6" diameter	53.50	50	103.50
5940	8" diameter	89	56.50	145.50
5950	10" diameter	119	67.50	186.50
5960	12" diameter	133	77.50	210.50

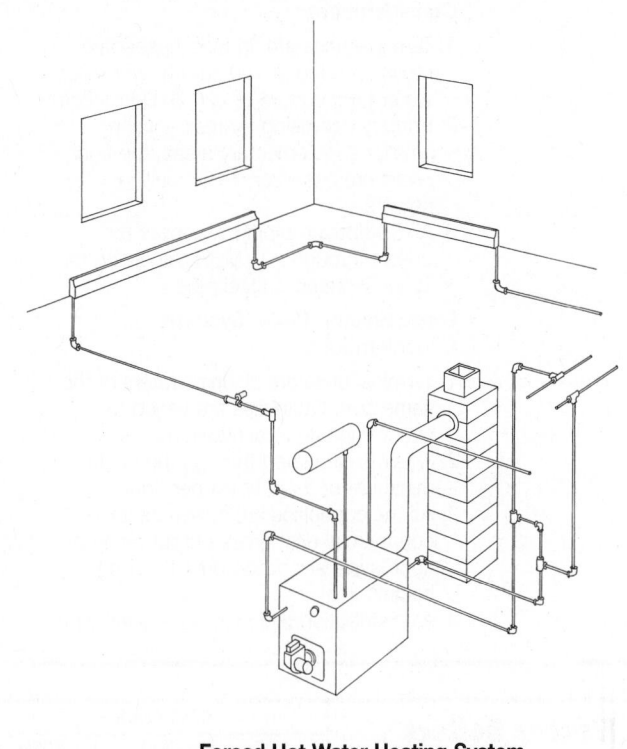

Forced Hot Water Heating System

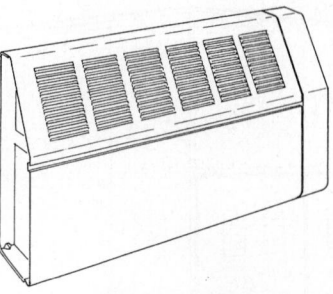

Fin Tube Radiation

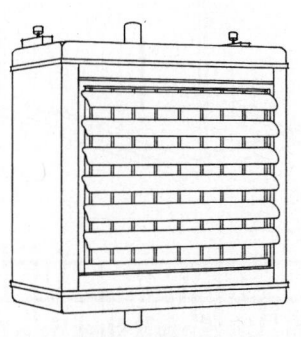

Terminal Unit Heater

D3010 510	Apartment Building Heating - Fin Tube Radiation	COST PER S.F.		
		MAT.	INST.	TOTAL
1740	Heating systems, fin tube radiation, forced hot water			
1760	1,000 S.F. area, 10,000 C.F. volume	7.43	7.23	14.66
1800	10,000 S.F. area, 100,000 C.F. volume	3.30	4.36	7.66
1840	20,000 S.F. area, 200,000 C.F. volume	3.57	4.90	8.47
1880	30,000 S.F. area, 300,000 C.F. volume	3.48	4.76	8.24

D3010 520	Commercial Building Heating - Fin Tube Radiation	COST PER S.F.		
		MAT.	INST.	TOTAL
1940	Heating systems, fin tube radiation, forced hot water			
1960	1,000 S.F. bldg, one floor	18.90	16.40	35.30
2000	10,000 S.F., 100,000 C.F., total two floors	4.69	6.20	10.89
2040	100,000 S.F., 1,000,000 C.F., total three floors	2.17	2.77	4.94
2080	1,000,000 S.F., 10,000,000 C.F., total five floors	1.13	1.45	2.58

D3010 530	Commercial Bldg. Heating - Terminal Unit Heaters	COST PER S.F.		
		MAT.	INST.	TOTAL
1860	Heating systems, terminal unit heaters, forced hot water			
1880	1,000 S.F. bldg., one floor	18.50	15	33.50
1920	10,000 S.F. bldg., 100,000 C.F. total two floors	4.18	5.10	9.28
1960	100,000 S.F. bldg., 1,000,000 C.F. total three floors	2.19	2.54	4.73
2000	1,000,000 S.F. bldg., 10,000,000 C.F. total five floors	1.43	1.63	3.06

443

For customer support on your Square Foot Costs with RSMeans data, call 800.448.8182.

D3020 Heat Generating Systems

Small Electric Boiler Systems Considerations:

1. Terminal units are fin tube baseboard radiation rated at 720 BTU/hr with 200° water temperature or 820 BTU/hr steam.
2. Primary use being for residential or smaller supplementary areas, the floor levels are based on 7-1/2" ceiling heights.
3. All distribution piping is copper for boilers through 205 MBH. All piping for larger systems is steel pipe.

Large Electric Boiler Systems Considerations:

1. Terminal units are all unit heaters of the same size. Quantities are varied to accommodate total requirements.
2. All air is circulated through the heaters a minimum of three times per hour.
3. As the capacities are adequate for commercial use, gross output rating by floor levels are based on 10' ceiling height.
4. All distribution piping is black steel pipe.

D3020 102	Small Heating Systems, Hydronic, Electric Boilers	COST PER S.F.		
		MAT.	INST.	TOTAL
1100	Small heating systems, hydronic, electric boilers			
1120	Steam, 1 floor, 1480 S.F., 61 M.B.H.	7.70	9.32	17.02
1160	3,000 S.F., 123 M.B.H.	5.95	8.20	14.15
1200	5,000 S.F., 205 M.B.H.	4.85	7.55	12.40
1240	2 floors, 12,400 S.F., 512 M.B.H.	3.94	7.50	11.44
1280	3 floors, 24,800 S.F., 1023 M.B.H.	4.57	7.40	11.97
1320	34,750 S.F., 1,433 M.B.H.	4.19	7.20	11.39
1360	Hot water, 1 floor, 1,000 S.F., 41 M.B.H.	13.15	5.15	18.30
1400	2,500 S.F., 103 M.B.H.	8.90	9.30	18.20
1440	2 floors, 4,850 S.F., 205 M.B.H.	8.20	11.20	19.40
1480	3 floors, 9,700 S.F., 410 M.B.H.	8.80	11.60	20.40

D3020 104	Large Heating Systems, Hydronic, Electric Boilers	COST PER S.F.		
		MAT.	INST.	TOTAL
1230	Large heating systems, hydronic, electric boilers			
1240	9,280 S.F., 135 K.W., 461 M.B.H., 1 floor	4.45	3.47	7.92
1280	14,900 S.F., 240 K.W., 820 M.B.H., 2 floors	5.85	5.65	11.50
1320	18,600 S.F., 296 K.W., 1,010 M.B.H., 3 floors	5.80	6.10	11.90
1360	26,100 S.F., 420 K.W., 1,432 M.B.H., 4 floors	5.70	6	11.70
1400	39,100 S.F., 666 K.W., 2,273 M.B.H., 4 floors	4.87	5	9.87
1440	57,700 S.F., 900 K.W., 3,071 M.B.H., 5 floors	4.63	4.96	9.59
1480	111,700 S.F., 1,800 K.W., 6,148 M.B.H., 6 floors	4.10	4.26	8.36
1520	149,000 S.F., 2,400 K.W., 8,191 M.B.H., 8 floors	4.07	4.25	8.32
1560	223,300 S.F., 3,600 K.W., 12,283 M.B.H., 14 floors	4.34	4.83	9.17

D3020 Heat Generating Systems

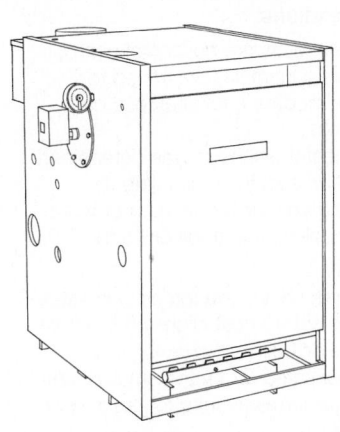

Boiler Selection: The maximum allowable working pressures are limited by ASME "Code for Heating Boilers" to 15 PSI for steam and 160 PSI for hot water heating boilers, with a maximum temperature limitation of 250° F. Hot water boilers are generally rated for a working pressure of 30 PSI. High pressure boilers are governed by the ASME "Code for Power Boilers" which is used almost universally for boilers operating over 15 PSIG. High pressure boilers used for a combination of heating/process loads are usually designed for 150 PSIG.

Boiler ratings are usually indicated as either Gross or Net Output. The Gross Load is equal to the Net Load plus a piping and pickup allowance. When this allowance cannot be determined, divide the gross output rating by 1.25 for a value equal to or greater than the next heat loss requirement of the building.

Table below lists installed cost per boiler and includes insulating jacket, standard controls, burner and safety controls. Costs do not include piping or boiler base pad. Outputs are Gross.

D3020 106	Boilers, Hot Water & Steam	COST EACH		
		MAT.	INST.	TOTAL
0600	Boiler, electric, steel, hot water, 12 K.W., 41 M.B.H.	5,650	1,575	7,225
0620	30 K.W., 103 M.B.H.	6,175	1,725	7,900
0640	60 K.W., 205 M.B.H.	6,725	1,875	8,600
0660	120 K.W., 410 M.B.H.	7,275	2,300	9,575
0680	210 K.W., 716 M.B.H.	8,700	3,425	12,125
0700	510 K.W., 1,739 M.B.H.	24,600	6,425	31,025
0720	720 K.W., 2,452 M.B.H.	30,500	7,225	37,725
0740	1,200 K.W., 4,095 M.B.H.	38,800	8,300	47,100
0760	2,100 K.W., 7,167 M.B.H.	70,000	10,500	80,500
0780	3,600 K.W., 12,283 M.B.H.	103,500	17,600	121,100
0820	Steam, 6 K.W., 20.5 M.B.H.	4,675	1,725	6,400
0840	24 K.W., 81.8 M.B.H.	5,500	1,875	7,375
0860	60 K.W., 205 M.B.H.	7,600	2,050	9,650
0880	150 K.W., 512 M.B.H.	11,000	3,175	14,175
0900	510 K.W., 1,740 M.B.H.	35,900	7,850	43,750
0920	1,080 K.W., 3,685 M.B.H.	44,700	11,300	56,000
0940	2,340 K.W., 7,984 M.B.H.	95,000	17,600	112,600
0980	Gas, cast iron, hot water, 80 M.B.H.	2,125	1,975	4,100
1000	100 M.B.H.	2,700	2,150	4,850
1020	163 M.B.H.	3,325	2,900	6,225
1040	280 M.B.H.	4,600	3,200	7,800
1060	544 M.B.H.	10,400	5,700	16,100
1080	1,088 M.B.H.	14,300	7,225	21,525
1100	2,000 M.B.H.	22,600	11,300	33,900
1120	2,856 M.B.H.	32,600	14,500	47,100
1140	4,720 M.B.H.	78,500	19,900	98,400
1160	6,970 M.B.H.	139,500	32,500	172,000
1180	For steam systems under 2,856 M.B.H., add 8%			
1520	Oil, cast iron, hot water, 109 M.B.H.	2,450	2,400	4,850
1540	173 M.B.H.	3,125	2,900	6,025
1560	236 M.B.H.	3,975	3,400	7,375
1580	1,084 M.B.H.	11,600	7,700	19,300
1600	1,600 M.B.H.	17,000	11,000	28,000
1620	2,480 M.B.H.	21,800	14,100	35,900
1640	3,550 M.B.H.	29,000	16,900	45,900
1660	Steam systems same price as hot water			

445

D3020 Heat Generating Systems

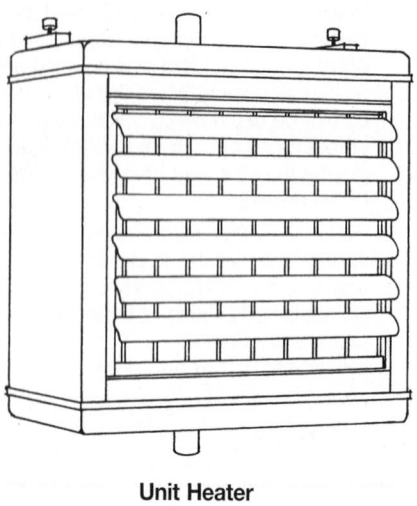

Unit Heater

Fossil Fuel Boiler Systems Considerations:

1. Terminal units are horizontal unit heaters. Quantities are varied to accommodate total heat loss per building.
2. Unit heater selection was determined by their capacity to circulate the building volume a minimum of three times per hour in addition to the BTU output.
3. Systems shown are forced hot water. Steam boilers cost slightly more than hot water boilers. However, this is compensated for by the smaller size or fewer terminal units required with steam.
4. Floor levels are based on 10′ story heights.
5. MBH requirements are gross boiler output.

D3020 108	Heating Systems, Unit Heaters	COST PER S.F.		
		MAT.	INST.	TOTAL
1260	Heating systems, hydronic, fossil fuel, terminal unit heaters,			
1280	Cast iron boiler, gas, 80 M.B.H., 1,070 S.F. bldg.	12.06	11.12	23.18
1320	163 M.B.H., 2,140 S.F. bldg.	8	7.60	15.60
1360	544 M.B.H., 7,250 S.F. bldg.	5.85	5.35	11.20
1400	1,088 M.B.H., 14,500 S.F. bldg.	4.82	5.05	9.87
1440	3,264 M.B.H., 43,500 S.F. bldg.	3.91	3.84	7.75
1480	5,032 M.B.H., 67,100 S.F. bldg.	4.15	3.92	8.07
1520	Oil, 109 M.B.H., 1,420 S.F. bldg.	13.35	9.90	23.25
1560	235 M.B.H., 3,150 S.F. bldg.	7.60	7.05	14.65
1600	940 M.B.H., 12,500 S.F. bldg.	5.20	4.65	9.85
1640	1,600 M.B.H., 21,300 S.F. bldg.	4.97	4.45	9.42
1680	2,480 M.B.H., 33,100 S.F. bldg.	4.72	4.01	8.73
1720	3,350 M.B.H., 44,500 S.F. bldg.	4.46	4.09	8.55
1760	Coal, 148 M.B.H., 1,975 S.F. bldg.	96	6.50	102.50
1800	300 M.B.H., 4,000 S.F. bldg.	52.50	5.10	57.60
1840	2,360 M.B.H., 31,500 S.F. bldg.	11.75	4.07	15.82

D3020 Heat Generating Systems

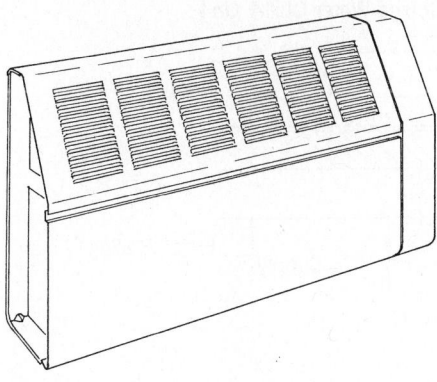

Fossil Fuel Boiler System Considerations:

1. Terminal units are commercial steel fin tube radiation. Quantities are varied to accommodate total heat loss per building.
2. Systems shown are forced hot water. Steam boilers cost slightly more than hot water boilers. However, this is compensated for by the smaller size or fewer terminal units required with steam.
3. Floor levels are based on 10' story heights.
4. MBH requirements are gross boiler output.

Fin Tube Radiator

D3020 110	Heating System, Fin Tube Radiation	COST PER S.F.		
		MAT.	INST.	TOTAL
3230	Heating systems, hydronic, fossil fuel, fin tube radiation			
3240	Cast iron boiler, gas, 80 MBH, 1,070 S.F. bldg.	14.77	16.73	31.50
3280	169 M.B.H., 2,140 S.F. bldg.	9.25	10.60	19.85
3320	544 M.B.H., 7,250 S.F. bldg.	7.80	9.05	16.85
3360	1,088 M.B.H., 14,500 S.F. bldg.	6.90	8.90	15.80
3400	3,264 M.B.H., 43,500 S.F. bldg.	6.15	7.85	14
3440	5,032 M.B.H., 67,100 S.F. bldg.	6.40	7.95	14.35
3480	Oil, 109 M.B.H., 1,420 S.F. bldg.	18	18.25	36.25
3520	235 M.B.H., 3,150 S.F. bldg.	9.50	10.70	20.20
3560	940 M.B.H., 12,500 S.F. bldg.	7.30	8.55	15.85
3600	1,600 M.B.H., 21,300 S.F. bldg.	7.20	8.45	15.65
3640	2,480 M.B.H., 33,100 S.F. bldg.	6.95	8.05	15
3680	3,350 M.B.H., 44,500 S.F. bldg.	6.70	8.10	14.80
3720	Coal, 148 M.B.H., 1,975 S.F. bldg.	98	10.10	108.10
3760	300 M.B.H., 4,000 S.F. bldg.	54.50	8.70	63.20
3800	2,360 M.B.H., 31,500 S.F. bldg.	13.90	8.05	21.95
4080	Steel boiler, oil, 97 M.B.H., 1,300 S.F. bldg.	12.75	15	27.75
4120	315 M.B.H., 4,550 S.F. bldg.	6.70	7.65	14.35
4160	525 M.B.H., 7,000 S.F. bldg.	8.60	8.75	17.35
4200	1,050 M.B.H., 14,000 S.F. bldg.	8.35	8.75	17.10
4240	2,310 M.B.H., 30,800 S.F. bldg.	7.35	8.05	15.40
4280	3,150 M.B.H., 42,000 S.F. bldg.	7.15	8.20	15.35

D3030 Cooling Generating Systems

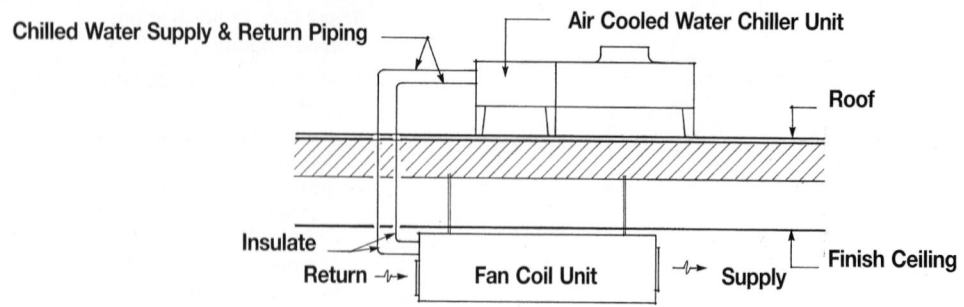

*Cooling requirements would lead to a choice of multiple chillers.

D3030 110	Chilled Water, Air Cooled Condenser Systems	COST PER S.F.		
		MAT.	INST.	TOTAL
1180	Packaged chiller, air cooled, with fan coil unit			
1200	Apartment corridors, 3,000 S.F., 5.50 ton	7.77	9.27	17.04
1240	6,000 S.F., 11.00 ton	6.20	7.35	13.55
1280	10,000 S.F., 18.33 ton	5.25	5.55	10.80
1320	20,000 S.F., 36.66 ton	4.05	4.02	8.07
1360	40,000 S.F., 73.33 ton	4.96	4.03	8.99
1440	Banks and libraries, 3,000 S.F., 12.50 ton	11.05	10.75	21.80
1480	6,000 S.F., 25.00 ton	9.95	8.70	18.65
1520	10,000 S.F., 41.66 ton	7.70	6.05	13.75
1560	20,000 S.F., 83.33 ton	8.25	5.45	13.70
1600	40,000 S.F., 167 ton*			
1680	Bars and taverns, 3,000 S.F., 33.25 ton	18.60	12.60	31.20
1720	6,000 S.F., 66.50 ton	16.35	9.50	25.85
1760	10,000 S.F., 110.83 ton	12.45	2.86	15.31
1800	20,000 S.F., 220 ton*			
1840	40,000 S.F., 440 ton*			
1920	Bowling alleys, 3,000 S.F., 17.00 ton	13.75	12.05	25.80
1960	6,000 S.F., 34.00 ton	10.45	8.40	18.85
2000	10,000 S.F., 56.66 ton	8.85	6	14.85
2040	20,000 S.F., 113.33 ton	9.15	5.50	14.65
2080	40,000 S.F., 227 ton*			
2160	Department stores, 3,000 S.F., 8.75 ton	10.65	10.30	20.95
2200	6,000 S.F., 17.50 ton	7.75	7.90	15.65
2240	10,000 S.F., 29.17 ton	6.25	5.65	11.90
2280	20,000 S.F., 58.33 ton	5.15	4.15	9.30
2320	40,000 S.F., 116.66 ton	6	4.15	10.15
2400	Drug stores, 3,000 S.F., 20.00 ton	15.50	12.45	27.95
2440	6,000 S.F., 40.00 ton	12.40	9.20	21.60
2480	10,000 S.F., 66.66 ton	12.65	7.75	20.40
2520	20,000 S.F., 133.33 ton	10.85	5.75	16.60
2560	40,000 S.F., 267 ton*			
2640	Factories, 2,000 S.F., 10.00 ton	9.85	10.35	20.20
2680	6,000 S.F., 20.00 ton	8.75	8.30	17.05
2720	10,000 S.F., 33.33 ton	6.75	5.80	12.55
2760	20,000 S.F., 66.66 ton	7.30	5.20	12.50
2800	40,000 S.F., 133.33 ton	6.50	4.25	10.75
2880	Food supermarkets, 3,000 S.F., 8.50 ton	10.45	10.25	20.70
2920	6,000 S.F., 17.00 ton	7.65	7.85	15.50
2960	10,000 S.F., 28.33 ton	5.95	5.50	11.45
3000	20,000 S.F., 56.66 ton	4.93	4.05	8.98

D30 HVAC

D3030 Cooling Generating Systems

D3030 110	Chilled Water, Air Cooled Condenser Systems	COST PER S.F.		
		MAT.	INST.	TOTAL
3040	40,000 S.F., 113.33 ton	5.85	4.16	10.01
3120	Medical centers, 3,000 S.F., 7.00 ton	9.50	10	19.50
3160	6,000 S.F., 14.00 ton	6.90	7.60	14.50
3200	10,000 S.F., 23.33 ton	6.05	5.80	11.85
3240	20,000 S.F., 46.66 ton	4.76	4.23	8.99
3280	40,000 S.F., 93.33 ton	5.40	4.09	9.49
3360	Offices, 3,000 S.F., 9.50 ton	9.50	10.20	19.70
3400	6,000 S.F., 19.00 ton	8.50	8.25	16.75
3440	10,000 S.F., 31.66 ton	6.55	5.75	12.30
3480	20,000 S.F., 63.33 ton	7.35	5.30	12.65
3520	40,000 S.F., 126.66 ton	6.45	4.30	10.75
3600	Restaurants, 3,000 S.F., 15.00 ton	12.30	11.15	23.45
3640	6,000 S.F., 30.00 ton	10.10	8.45	18.55
3680	10,000 S.F., 50.00 ton	8.60	6.15	14.75
3720	20,000 S.F., 100.00 ton	9.45	5.80	15.25
3760	40,000 S.F., 200 ton*			
3840	Schools and colleges, 3,000 S.F., 11.50 ton	10.60	10.60	21.20
3880	6,000 S.F., 23.00 ton	9.50	8.55	18.05
3920	10,000 S.F., 38.33 ton	7.30	5.95	13.25
3960	20,000 S.F., 76.66 ton	8.15	5.50	13.65
4000	40,000 S.F., 153 ton*			

449

D3030 Cooling Generating Systems

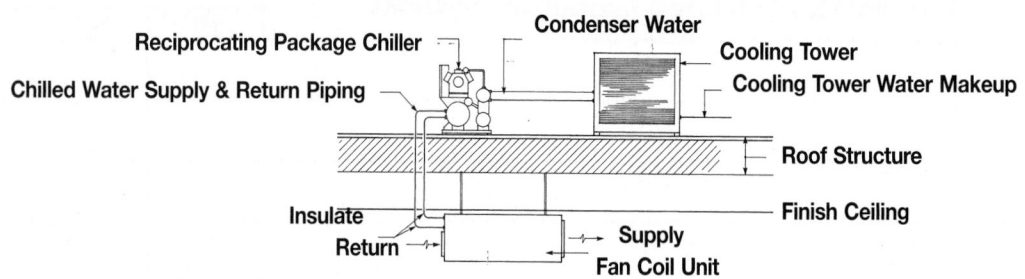

Reciprocating Package Chiller

Chilled Water Supply & Return Piping

Condenser Water

Cooling Tower

Cooling Tower Water Makeup

Roof Structure

Finish Ceiling

Insulate

Return

Supply

Fan Coil Unit

*Cooling requirements would lead to a choice of multiple chillers

D3030 115	Chilled Water, Cooling Tower Systems	COST PER S.F.		
		MAT.	INST.	TOTAL
1300	Packaged chiller, water cooled, with fan coil unit			
1320	Apartment corridors, 4,000 S.F., 7.33 ton	7.45	9.05	16.50
1360	6,000 S.F., 11.00 ton	6.05	7.75	13.80
1400	10,000 S.F., 18.33 ton	6	5.80	11.80
1440	20,000 S.F., 26.66 ton	4.53	4.19	8.72
1480	40,000 S.F., 73.33 ton	5.45	4.18	9.63
1520	60,000 S.F., 110.00 ton	5.30	4.43	9.73
1600	Banks and libraries, 4,000 S.F., 16.66 ton	12.90	9.85	22.75
1640	6,000 S.F., 25.00 ton	10.45	8.60	19.05
1680	10,000 S.F., 41.66 ton	8.35	6.35	14.70
1720	20,000 S.F., 83.33 ton	9.30	5.75	15.05
1760	40,000 S.F., 166.66 ton	8.70	7.40	16.10
1800	60,000 S.F., 250.00 ton	8.90	7.85	16.75
1880	Bars and taverns, 4,000 S.F., 44.33 ton	19.40	12.20	31.60
1920	6,000 S.F., 66.50 ton	21	12.65	33.65
1960	10,000 S.F., 110.83 ton	18.65	9.65	28.30
2000	20,000 S.F., 221.66 ton	19.55	10.05	29.60
2040	40,000 S.F., 440 ton*			
2080	60,000 S.F., 660 ton*			
2160	Bowling alleys, 4,000 S.F., 22.66 ton	14.75	11	25.75
2200	6,000 S.F., 34.00 ton	12.30	9.25	21.55
2240	10,000 S.F., 56.66 ton	10.10	6.90	17
2280	20,000 S.F., 113.33 ton	10.50	6.25	16.75
2320	40,000 S.F., 226.66 ton	11.30	7.40	18.70
2360	60,000 S.F., 340 ton			
2440	Department stores, 4,000 S.F., 11.66 ton	8.10	9.75	17.85
2480	6,000 S.F., 17.50 ton	9.60	8	17.60
2520	10,000 S.F., 29.17 ton	6.75	5.85	12.60
2560	20,000 S.F., 58.33 ton	5.35	4.42	9.77
2600	40,000 S.F., 116.66 ton	6.35	4.49	10.84
2640	60,000 S.F., 175.00 ton	7.75	7.20	14.95
2720	Drug stores, 4,000 S.F., 26.66 ton	15.35	11.05	26.40
2760	6,000 S.F., 40.00 ton	13.20	9.40	22.60
2800	10,000 S.F., 66.66 ton	13.45	8.55	22
2840	20,000 S.F., 133.33 ton	11.20	6.65	17.85
2880	40,000 S.F., 266.67 ton	11.85	8.35	20.20
2920	60,000 S.F., 400 ton*			
3000	Factories, 4,000 S.F., 13.33 ton	11.20	9.40	20.60
3040	6,000 S.F., 20.00 ton	9.45	8.25	17.70
3080	10,000 S.F., 33.33 ton	7.35	6	13.35
3120	20,000 S.F., 66.66 ton	7.55	5.55	13.10
3160	40,000 S.F., 133.33 ton	6.65	4.72	11.37

D3030 Cooling Generating Systems

D3030 115	Chilled Water, Cooling Tower Systems	COST PER S.F.		
		MAT.	INST.	TOTAL
3200	60,000 S.F., 200.00 ton	8.10	7.55	15.65
3280	Food supermarkets, 4,000 S.F., 11.33 ton	8	9.70	17.70
3320	6,000 S.F., 17.00 ton	8.45	8	16.45
3360	10,000 S.F., 28.33 ton	6.65	5.80	12.45
3400	20,000 S.F., 56.66 ton	5.45	4.43	9.88
3440	40,000 S.F., 113.33 ton	6.35	4.51	10.86
3480	60,000 S.F., 170.00 ton	7.70	7.20	14.90
3560	Medical centers, 4.000 S.F., 9.33 ton	6.95	8.90	15.85
3600	6,000 S.F., 14.00 ton	8.35	7.70	16.05
3640	10,000 S.F., 23.33 ton	6.35	5.70	12.05
3680	20,000 S.F., 46.66 ton	4.95	4.32	9.27
3720	40,000 S.F., 93.33 ton	5.95	4.38	10.33
3760	60,000 S.F., 140.00 ton	6.70	7.30	14
3840	Offices, 4,000 S.F., 12.66 ton	10.85	9.30	20.15
3880	6,000 S.F., 19.00 ton	9.45	8.40	17.85
3920	10,000 S.F., 31.66 ton	7.35	6.10	13.45
3960	20,000 S.F., 63.33 ton	7.50	5.55	13.05
4000	40,000 S.F., 126.66 ton	7.65	7.05	14.70
4040	60,000 S.F., 190.00 ton	7.85	7.45	15.30
4120	Restaurants, 4,000 S.F., 20.00 ton	13.20	10.25	23.45
4160	6,000 S.F., 30.00 ton	11.10	8.70	19.80
4200	10,000 S.F., 50.00 ton	9.35	6.65	16
4240	20,000 S.F., 100.00 ton	10.35	6.20	16.55
4280	40,000 S.F., 200.00 ton	9.10	7.15	16.25
4320	60,000 S.F., 300.00 ton	9.90	8.10	18
4400	Schools and colleges, 4,000 S.F., 15.33 ton	12.25	9.65	21.90
4440	6,000 S.F., 23.00 ton	9.95	8.45	18.40
4480	10,000 S.F., 38.33 ton	7.95	6.25	14.20
4520	20,000 S.F., 76.66 ton	9	5.70	14.70
4560	40,000 S.F., 153.33 ton	8.20	7.25	15.45
4600	60,000 S.F., 230.00 ton	8.25	7.55	15.80

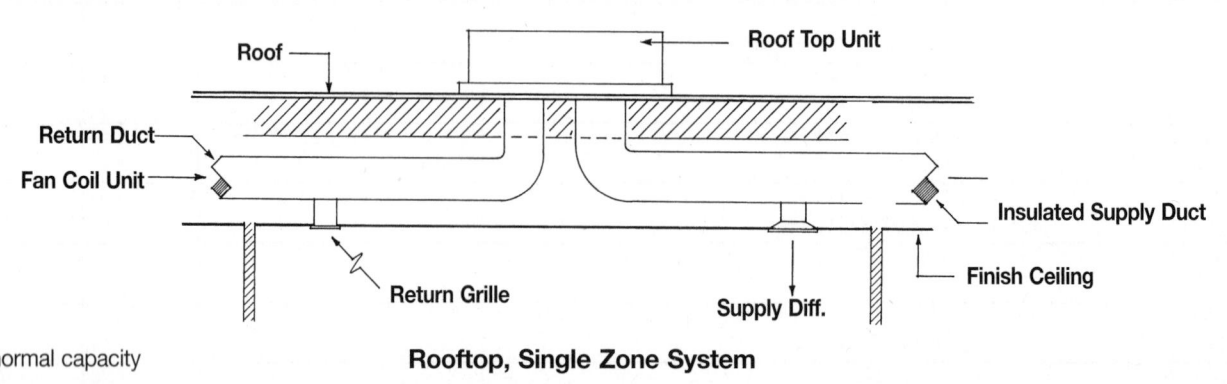

Rooftop, Single Zone System

*Above normal capacity

D3050 150	Rooftop Single Zone Unit Systems	COST PER S.F.		
		MAT.	INST.	TOTAL
1260	Rooftop, single zone, air conditioner			
1280	Apartment corridors, 500 S.F., .92 ton	3.25	4.30	7.55
1320	1,000 S.F., 1.83 ton	3.22	4.30	7.52
1360	1500 S.F., 2.75 ton	2.08	3.55	5.63
1400	3,000 S.F., 5.50 ton	2.17	3.41	5.58
1440	5,000 S.F., 9.17 ton	2.26	3.10	5.36
1480	10,000 S.F., 18.33 ton	3.39	2.92	6.31
1560	Banks or libraries, 500 S.F., 2.08 ton	7.35	9.80	17.15
1600	1,000 S.F., 4.17 ton	4.73	8.05	12.78
1640	1,500 S.F., 6.25 ton	4.93	7.75	12.68
1680	3,000 S.F., 12.50 ton	5.15	7.05	12.20
1720	5,000 S.F., 20.80 ton	7.70	6.60	14.30
1760	10,000 S.F., 41.67 ton	5.95	6.60	12.55
1840	Bars and taverns, 500 S.F. 5.54 ton	11.90	13	24.90
1880	1,000 S.F., 11.08 ton	12.45	11.15	23.60
1920	1,500 S.F., 16.62 ton	11.90	10.35	22.25
1960	3,000 S.F., 33.25 ton	15.75	9.90	25.65
2000	5,000 S.F., 55.42 ton	13.45	9.90	23.35
2040	10,000 S.F., 110.83 ton*			
2080	Bowling alleys, 500 S.F., 2.83 ton	6.45	10.95	17.40
2120	1,000 S.F., 5.67 ton	6.70	10.55	17.25
2160	1,500 S.F., 8.50 ton	7	9.60	16.60
2200	3,000 S.F., 17.00 ton	6.70	9.20	15.90
2240	5,000 S.F., 28.33 ton	8.70	8.95	17.65
2280	10,000 S.F., 56.67 ton	7.50	8.95	16.45
2360	Department stores, 500 S.F., 1.46 ton	5.15	6.85	12
2400	1,000 S.F., 2.92 ton	3.31	5.65	8.96
2480	3,000 S.F., 8.75 ton	3.60	4.94	8.54
2520	5,000 S.F., 14.58 ton	3.45	4.73	8.18
2560	10,000 S.F., 29.17 ton	4.47	4.61	9.08
2640	Drug stores, 500 S.F., 3.33 ton	7.55	12.90	20.45
2680	1,000 S.F., 6.67 ton	7.90	12.40	20.30
2720	1,500 S.F., 10.00 ton	8.20	11.30	19.50
2760	3,000 S.F., 20.00 ton	12.30	10.60	22.90
2800	5,000 S.F., 33.33 ton	10.20	10.55	20.75
2840	10,000 S.F., 66.67 ton	8.80	10.55	19.35
2920	Factories, 500 S.F., 1.67 ton	5.85	7.85	13.70
3000	1,500 S.F., 5.00 ton	3.94	6.20	10.14
3040	3,000 S.F., 10.00 ton	4.11	5.65	9.76
3080	5,000 S.F., 16.67 ton	3.94	5.40	9.34
3120	10,000 S.F., 33.33 ton	5.10	5.25	10.35
3200	Food supermarkets, 500 S.F., 1.42 ton	4.98	6.65	11.63

D3050 Terminal & Package Units

D3050 150	Rooftop Single Zone Unit Systems	COST PER S.F.		
		MAT.	INST.	TOTAL
3240	1,000 S.F., 2.83 ton	3.19	5.45	8.64
3280	1,500 S.F., 4.25 ton	3.35	5.25	8.60
3320	3,000 S.F., 8.50 ton	3.49	4.80	8.29
3360	5,000 S.F., 14.17 ton	3.35	4.60	7.95
3400	10,000 S.F., 28.33 ton	4.34	4.48	8.82
3480	Medical centers, 500 S.F., 1.17 ton	4.11	5.50	9.61
3520	1,000 S.F., 2.33 ton	4.11	5.45	9.56
3560	1,500 S.F., 3.50 ton	2.65	4.52	7.17
3640	5,000 S.F., 11.67 ton	2.88	3.95	6.83
3680	10,000 S.F., 23.33 ton	4.31	3.71	8.02
3760	Offices, 500 S.F., 1.58 ton	5.55	7.45	13
3800	1,000 S.F., 3.17 ton	3.60	6.15	9.75
3840	1,500 S.F., 4.75 ton	3.75	5.90	9.65
3880	3,000 S.F., 9.50 ton	3.91	5.35	9.26
3920	5,000 S.F., 15.83 ton	3.75	5.15	8.90
3960	10,000 S.F., 31.67 ton	4.86	5	9.86
4000	Restaurants, 500 S.F., 2.50 ton	8.80	11.75	20.55
4040	1,000 S.F., 5.00 ton	5.90	9.30	15.20
4080	1,500 S.F., 7.50 ton	6.15	8.45	14.60
4120	3,000 S.F., 15.00 ton	5.90	8.10	14
4160	5,000 S.F., 25.00 ton	9.25	7.95	17.20
4200	10,000 S.F., 50.00 ton	6.60	7.90	14.50
4240	Schools and colleges, 500 S.F., 1.92 ton	6.75	9	15.75
4280	1,000 S.F., 3.83 ton	4.35	7.45	11.80
4360	3,000 S.F., 11.50 ton	4.72	6.50	11.22
4400	5,000 S.F., 19.17 ton	7.10	6.10	13.20

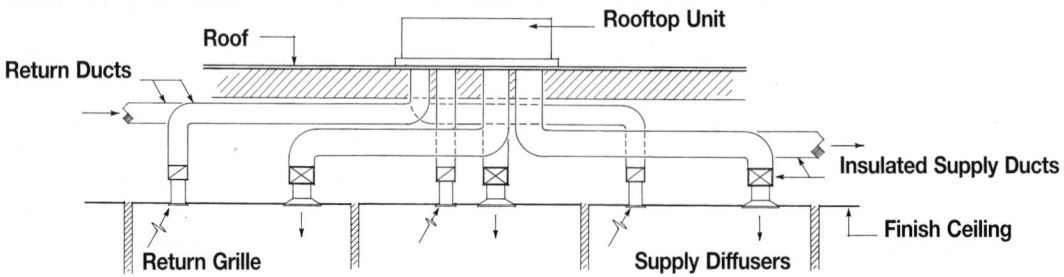

*Note A: Small single zone unit recommended.

*Note B: A combination of multizone units recommended.

D3050 155	Rooftop Multizone Unit Systems	COST PER S.F.		
		MAT.	INST.	TOTAL
1240	Rooftop, multizone, air conditioner			
1260	Apartment corridors, 1,500 S.F., 2.75 ton. See Note A.			
1280	3,000 S.F., 5.50 ton	11.13	5.50	16.63
1320	10,000 S.F., 18.30 ton	7.70	5.30	13
1360	15,000 S.F., 27.50 ton	7.75	5.30	13.05
1400	20,000 S.F., 36.70 ton	7.90	5.30	13.20
1440	25,000 S.F., 45.80 ton	7.35	5.30	12.65
1520	Banks or libraries, 1,500 S.F., 6.25 ton	25.50	12.50	38
1560	3,000 S.F., 12.50 ton	21.50	12.25	33.75
1600	10,000 S.F., 41.67 ton	16.75	12.10	28.85
1640	15,000 S.F., 62.50 ton	12.25	12	24.25
1680	20,000 S.F., 83.33 ton	12.30	12	24.30
1720	25,000 S.F., 104.00 ton	10.80	11.95	22.75
1800	Bars and taverns, 1,500 S.F., 16.62 ton	55	18.05	73.05
1840	3,000 S.F., 33.24 ton	46	17.40	63.40
1880	10,000 S.F., 110.83 ton	27	17.30	44.30
1920	15,000 S.F., 165 ton, See Note B.			
1960	20,000 S.F., 220 ton, See Note B.			
2000	25,000 S.F., 275 ton, See Note B.			
2080	Bowling alleys, 1,500 S.F., 8.50 ton	34.50	17.05	51.55
2120	3,000 S.F., 17.00 ton	29	16.65	45.65
2160	10,000 S.F., 56.70 ton	23	16.45	39.45
2200	15,000 S.F., 85.00 ton	16.70	16.35	33.05
2240	20,000 S.F., 113.00 ton	14.70	16.25	30.95
2280	25,000 S.F., 140.00 ton see Note B.			
2360	Department stores, 1,500 S.F., 4.37 ton, See Note A.			
2400	3,000 S.F., 8.75 ton	17.70	8.75	26.45
2440	10,000 S.F., 29.17 ton	12.60	8.40	21
2520	20,000 S.F., 58.33 ton	8.60	8.40	17
2560	25,000 S.F., 72.92 ton	8.60	8.40	17
2640	Drug stores, 1,500 S.F., 10.00 ton	40.50	20	60.50
2680	3,000 S.F., 20.00 ton	28	19.30	47.30
2720	10,000 S.F., 66.66 ton	19.60	19.20	38.80
2760	15,000 S.F., 100.00 ton	17.35	19.15	36.50
2800	20,000 S.F., 135 ton, See Note B.			
2840	25,000 S.F., 165 ton, See Note B.			
2920	Factories, 1,500 S.F., 5 ton, See Note A.			
3000	10,000 S.F., 33.33 ton	14.40	9.60	24

D3050 Terminal & Package Units

D3050 155	Rooftop Multizone Unit Systems	COST PER S.F.		
		MAT.	INST.	TOTAL
3040	15,000 S.F., 50.00 ton	13.40	9.70	23.10
3080	20,000 S.F., 66.66 ton	9.80	9.60	19.40
3120	25,000 S.F., 83.33 ton	9.80	9.60	19.40
3200	Food supermarkets, 1,500 S.F., 4.25 ton, See Note A.			
3240	3,000 S.F., 8.50 ton	17.20	8.50	25.70
3280	10,000 S.F., 28.33 ton	11.95	8.15	20.10
3320	15,000 S.F., 42.50 ton	11.40	8.25	19.65
3360	20,000 S.F., 56.67 ton	8.35	8.15	16.50
3400	25,000 S.F., 70.83 ton	8.35	8.15	16.50
3480	Medical centers, 1,500 S.F., 3.5 ton, See Note A.			
3520	3,000 S.F., 7.00 ton	14.20	7	21.20
3560	10,000 S.F., 23.33 ton	9.55	6.75	16.30
3600	15,000 S.F., 35.00 ton	10.05	6.75	16.80
3640	20,000 S.F., 46.66 ton	9.35	6.75	16.10
3680	25,000 S.F., 58.33 ton	6.85	6.70	13.55
3760	Offices, 1,500 S.F., 4.75 ton, See Note A.			
3800	3,000 S.F., 9.50 ton	19.25	9.50	28.75
3840	10,000 S.F., 31.66 ton	13.65	9.15	22.80
3920	20,000 S.F., 63.33 ton	9.35	9.10	18.45
3960	25,000 S.F., 79.16 ton	9.35	9.10	18.45
4000	Restaurants, 1,500 S.F., 7.50 ton	30.50	15.05	45.55
4040	3,000 S.F., 15.00 ton	25.50	14.70	40.20
4080	10,000 S.F., 50.00 ton	20	14.50	34.50
4120	15,000 S.F., 75.00 ton	14.70	14.40	29.10
4160	20,000 S.F., 100.00 ton	13	14.40	27.40
4200	25,000 S.F., 125 ton, See Note B.			
4240	Schools and colleges, 1,500 S.F., 5.75 ton	23.50	11.50	35
4320	10,000 S.F., 38.33 ton	15.40	11.15	26.55
4360	15,000 S.F., 57.50 ton	11.30	11.05	22.35
4400	20,000 S.F.,76.66 ton	11.30	11.05	22.35
4440	25,000 S.F., 95.83 ton	10.40	11.05	21.45

D3050 Terminal & Package Units

Self-Contained Water Cooled System

D3050 160	Self-contained, Water Cooled Unit Systems	COST PER S.F.		
		MAT.	INST.	TOTAL
1280	Self-contained, water cooled unit	3.45	3.50	6.95
1300	Apartment corridors, 500 S.F., .92 ton	3.45	3.50	6.95
1320	1,000 S.F., 1.83 ton	3.43	3.49	6.92
1360	3,000 S.F., 5.50 ton	2.71	2.93	5.64
1400	5,000 S.F., 9.17 ton	2.60	2.71	5.31
1440	10,000 S.F., 18.33 ton	3.78	2.43	6.21
1520	Banks or libraries, 500 S.F., 2.08 ton	7.55	3.57	11.12
1560	1,000 S.F., 4.17 ton	6.15	6.65	12.80
1600	3,000 S.F., 12.50 ton	5.90	6.20	12.10
1640	5,000 S.F., 20.80 ton	8.60	5.55	14.15
1680	10,000 S.F., 41.66 ton	6.95	5.55	12.50
1760	Bars and taverns, 500 S.F., 5.54 ton	15.55	6	21.55
1800	1,000 S.F., 11.08 ton	15.25	10.55	25.80
1840	3,000 S.F., 33.25 ton	20	8.35	28.35
1880	5,000 S.F., 55.42 ton	17.10	8.70	25.80
1920	10,000 S.F., 110.00 ton	16.60	8.60	25.20
2000	Bowling alleys, 500 S.F., 2.83 ton	10.25	4.86	15.11
2040	1,000 S.F., 5.66 ton	8.40	9.05	17.45
2080	3,000 S.F., 17.00 ton	11.70	7.55	19.25
2120	5,000 S.F., 28.33 ton	10.55	7.25	17.80
2160	10,000 S.F., 56.66 ton	8.95	7.40	16.35
2200	Department stores, 500 S.F., 1.46 ton	5.30	2.50	7.80
2240	1,000 S.F., 2.92 ton	4.31	4.65	8.96
2280	3,000 S.F., 8.75 ton	4.14	4.32	8.46
2320	5,000 S.F., 14.58 ton	6	3.87	9.87
2360	10,000 S.F., 29.17 ton	5.45	3.73	9.18
2440	Drug stores, 500 S.F., 3.33 ton	12.10	5.70	17.80
2480	1,000 S.F., 6.66 ton	9.85	10.65	20.50
2520	3,000 S.F., 20.00 ton	13.75	8.85	22.60
2560	5,000 S.F., 33.33 ton	14.40	8.75	23.15
2600	10,000 S.F., 66.66 ton	10.50	8.75	19.25
2680	Factories, 500 S.F., 1.66 ton	6	2.86	8.86
2720	1,000 S.F. 3.37 ton	4.98	5.40	10.38
2760	3,000 S.F., 10.00 ton	4.72	4.94	9.66
2800	5,000 S.F., 16.66 ton	6.85	4.43	11.28
2840	10,000 S.F., 33.33 ton	6.20	4.27	10.47

D3050 Terminal & Package Units

D3050 160	Self-contained, Water Cooled Unit Systems	COST PER S.F.		
		MAT.	INST.	TOTAL
2920	Food supermarkets, 500 S.F., 1.42 ton	5.15	2.44	7.59
2960	1,000 S.F., 2.83 ton	5.30	5.40	10.70
3000	3,000 S.F., 8.50 ton	4.19	4.53	8.72
3040	5,000 S.F., 14.17 ton	4.02	4.20	8.22
3080	10,000 S.F., 28.33 ton	5.30	3.63	8.93
3160	Medical centers, 500 S.F., 1.17 ton	4.22	1.99	6.21
3200	1,000 S.F., 2.33 ton	4.38	4.45	8.83
3240	3,000 S.F., 7.00 ton	3.45	3.72	7.17
3280	5,000 S.F., 11.66 ton	3.31	3.45	6.76
3320	10,000 S.F., 23.33 ton	4.81	3.10	7.91
3400	Offices, 500 S.F., 1.58 ton	5.75	2.72	8.47
3440	1,000 S.F., 3.17 ton	5.95	6.05	12
3480	3,000 S.F., 9.50 ton	4.49	4.69	9.18
3520	5,000 S.F., 15.83 ton	6.55	4.20	10.75
3560	10,000 S.F., 31.67 ton	5.90	4.05	9.95
3640	Restaurants, 500 S.F., 2.50 ton	9.05	4.29	13.34
3680	1,000 S.F., 5.00 ton	7.40	8	15.40
3720	3,000 S.F., 15.00 ton	7.10	7.40	14.50
3760	5,000 S.F., 25.00 ton	10.30	6.65	16.95
3800	10,000 S.F., 50.00 ton	6.35	6.10	12.45
3880	Schools and colleges, 500 S.F., 1.92 ton	6.95	3.28	10.23
3920	1,000 S.F., 3.83 ton	5.65	6.10	11.75
3960	3,000 S.F., 11.50 ton	5.45	5.70	11.15
4000	5,000 S.F., 19.17 ton	7.90	5.10	13
4040	10,000 S.F., 38.33 ton	6.40	5.10	11.50

D3050 Terminal & Package Units

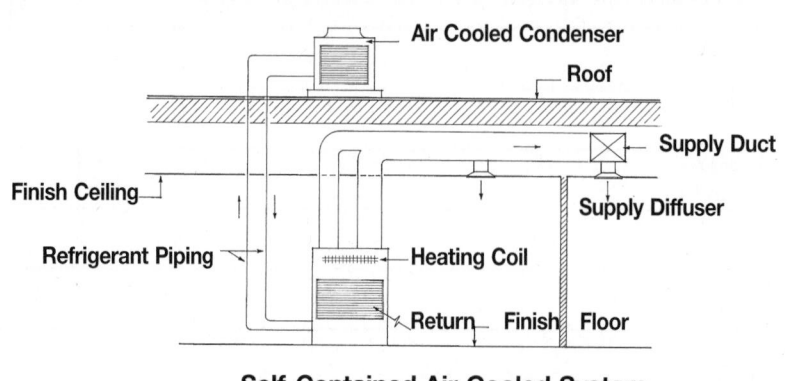

Self-Contained Air Cooled System

D3050 165	Self-contained, Air Cooled Unit Systems	COST PER S.F.		
		MAT.	INST.	TOTAL
1300	Self-contained, air cooled unit			
1320	Apartment corridors, 500 S.F., .92 ton	5.30	4.50	9.80
1360	1,000 S.F., 1.83 ton	5.25	4.44	9.69
1400	3,000 S.F., 5.50 ton	4.80	4.16	8.96
1440	5,000 S.F., 9.17 ton	3.93	3.95	7.88
1480	10,000 S.F., 18.33 ton	3.29	3.62	6.91
1560	Banks or libraries, 500 S.F., 2.08 ton	11.55	5.70	17.25
1600	1,000 S.F., 4.17 ton	10.85	9.40	20.25
1640	3,000 S.F., 12.50 ton	8.95	8.95	17.90
1680	5,000 S.F., 20.80 ton	7.60	8.25	15.85
1720	10,000 S.F., 41.66 ton	8.55	8.05	16.60
1800	Bars and taverns, 500 S.F., 5.54 ton	26.50	12.45	38.95
1840	1,000 S.F., 11.08 ton	23.50	18.05	41.55
1880	3,000 S.F., 33.25 ton	22.50	15.60	38.10
1920	5,000 S.F., 55.42 ton	23	15.60	38.60
1960	10,000 S.F., 110.00 ton	23.50	15.55	39.05
2040	Bowling alleys, 500 S.F., 2.83 ton	15.85	7.80	23.65
2080	1,000 S.F., 5.66 ton	14.85	12.80	27.65
2120	3,000 S.F., 17.00 ton	10.25	11.20	21.45
2160	5,000 S.F., 28.33 ton	11.80	10.95	22.75
2200	10,000 S.F., 56.66 ton	11.90	10.95	22.85
2240	Department stores, 500 S.F., 1.46 ton	8.15	4.02	12.17
2280	1,000 S.F., 2.92 ton	7.65	6.60	14.25
2320	3,000 S.F., 8.75 ton	6.25	6.25	12.50
2360	5,000 S.F., 14.58 ton	6.25	6.25	12.50
2400	10,000 S.F., 29.17 ton	6.05	5.65	11.70
2480	Drug stores, 500 S.F., 3.33 ton	18.65	9.20	27.85
2520	1,000 S.F., 6.66 ton	17.40	15.05	32.45
2560	3,000 S.F., 20.00 ton	12.20	13.20	25.40
2600	5,000 S.F., 33.33 ton	13.85	12.90	26.75
2640	10,000 S.F., 66.66 ton	14.10	12.90	27
2720	Factories, 500 S.F., 1.66 ton	9.45	4.64	14.09
2760	1,000 S.F., 3.33 ton	8.75	7.55	16.30
2800	3,000 S.F., 10.00 ton	7.15	7.15	14.30
2840	5,000 S.F., 16.66 ton	6	6.60	12.60
2880	10,000 S.F., 33.33 ton	6.95	6.45	13.40
2960	Food supermarkets, 500 S.F., 1.42 ton	7.90	3.91	11.81
3000	1,000 S.F., 2.83 ton	8.05	6.85	14.90
3040	3,000 S.F., 8.50 ton	7.40	6.40	13.80
3080	5,000 S.F., 14.17 ton	6.10	6.10	12.20

D3050 Terminal & Package Units

D3050 165	Self-contained, Air Cooled Unit Systems	COST PER S.F.		
		MAT.	INST.	TOTAL
3120	10,000 S.F., 28.33 ton	5.90	5.45	11.35
3200	Medical centers, 500 S.F., 1.17 ton	6.55	3.22	9.77
3240	1,000 S.F., 2.33 ton	6.70	5.65	12.35
3280	3,000 S.F., 7.00 ton	6.10	5.30	11.40
3320	5,000 S.F., 16.66 ton	5.05	5.05	10.10
3360	10,000 S.F., 23.33 ton	4.21	4.61	8.82
3440	Offices, 500 S.F., 1.58 ton	8.90	4.37	13.27
3480	1,000 S.F., 3.16 ton	9.15	7.70	16.85
3520	3,000 S.F., 9.50 ton	6.80	6.85	13.65
3560	5,000 S.F., 15.83 ton	5.80	6.25	12.05
3600	10,000 S.F., 31.66 ton	6.60	6.10	12.70
3680	Restaurants, 500 S.F., 2.50 ton	13.95	6.90	20.85
3720	1,000 S.F., 5.00 ton	13.10	11.30	24.40
3760	3,000 S.F., 15.00 ton	10.75	10.75	21.50
3800	5,000 S.F., 25.00 ton	10.35	9.65	20
3840	10,000 S.F., 50.00 ton	11	9.65	20.65
3920	Schools and colleges, 500 S.F., 1.92 ton	10.70	5.30	16
3960	1,000 S.F., 3.83 ton	10.05	8.70	18.75
4000	3,000 S.F., 11.50 ton	8.25	8.25	16.50
4040	5,000 S.F., 19.17 ton	6.90	7.60	14.50
4080	10,000 S.F., 38.33 ton	7.85	7.40	15.25

For customer support on your Square Foot Costs with RSMeans data, call 800.448.8182.

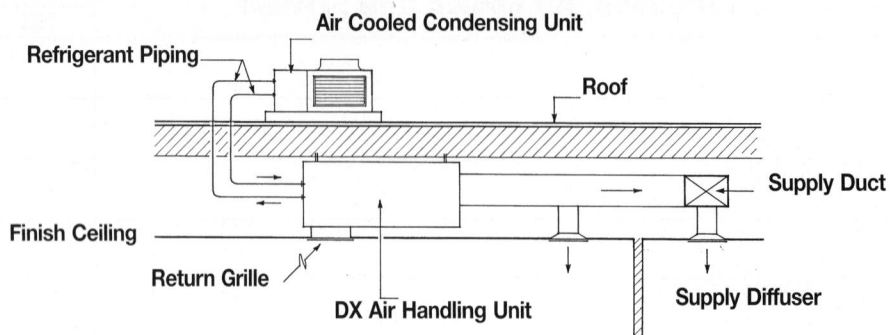

*Cooling requirements would lead to more than one system.

D3050 170	Split Systems With Air Cooled Condensing Units	COST PER S.F.		
		MAT.	INST.	TOTAL
1260	Split system, air cooled condensing unit			
1280	Apartment corridors, 1,000 S.F., 1.83 ton	2.23	2.78	5.01
1320	2,000 S.F., 3.66 ton	1.76	2.82	4.58
1360	5,000 S.F., 9.17 ton	1.64	3.51	5.15
1400	10,000 S.F., 18.33 ton	1.77	3.85	5.62
1440	20,000 S.F., 36.66 ton	2.10	3.91	6.01
1520	Banks and libraries, 1,000 S.F., 4.17 ton	4	6.40	10.40
1560	2,000 S.F., 8.33 ton	3.74	8	11.74
1600	5,000 S.F., 20.80 ton	4.03	8.75	12.78
1640	10,000 S.F., 41.66 ton	4.79	8.90	13.69
1680	20,000 S.F., 83.32 ton	5.75	9.25	15
1760	Bars and taverns, 1,000 S.F., 11.08 ton	8.45	12.60	21.05
1800	2,000 S.F., 22.16 ton	10.90	15.10	26
1840	5,000 S.F., 55.42 ton	10.50	14.10	24.60
1880	10,000 S.F., 110.84 ton	13.10	14.70	27.80
1920	20,000 S.F., 220 ton*			
2000	Bowling alleys, 1,000 S.F., 5.66 ton	5.55	11.95	17.50
2040	2,000 S.F., 11.33 ton	5.10	10.85	15.95
2080	5,000 S.F., 28.33 ton	5.50	11.90	17.40
2120	10,000 S.F., 56.66 ton	6.50	12.10	18.60
2160	20,000 S.F., 113.32 ton	8.90	13.10	22
2320	Department stores, 1,000 S.F., 2.92 ton	2.81	4.40	7.21
2360	2,000 S.F., 5.83 ton	2.86	6.15	9.01
2400	5,000 S.F., 14.58 ton	2.61	5.60	8.21
2440	10,000 S.F., 29.17 ton	2.83	6.10	8.93
2480	20,000 S.F., 58.33 ton	3.35	6.20	9.55
2560	Drug stores, 1,000 S.F., 6.66 ton	6.55	14.05	20.60
2600	2,000 S.F., 13.32 ton	5.95	12.80	18.75
2640	5,000 S.F., 33.33 ton	7.65	14.20	21.85
2680	10,000 S.F., 66.66 ton	7.90	14.85	22.75
2720	20,000 S.F., 133.32 ton*			
2800	Factories, 1,000 S.F., 3.33 ton	3.21	5	8.21
2840	2,000 S.F., 6.66 ton	3.26	7.05	10.31
2880	5,000 S.F., 16.66 ton	3.23	7	10.23
2920	10,000 S.F., 33.33 ton	3.82	7.10	10.92
2960	20,000 S.F., 66.66 ton	3.93	7.40	11.33
3040	Food supermarkets, 1,000 S.F., 2.83 ton	2.73	4.27	7
3080	2,000 S.F., 5.66 ton	2.78	6	8.78
3120	5,000 S.F., 14.66 ton	2.53	5.45	7.98
3160	10,000 S.F., 28.33 ton	2.75	5.95	8.70
3200	20,000 S.F., 56.66 ton	3.26	6.05	9.31
3280	Medical centers, 1,000 S.F., 2.33 ton	2.50	3.45	5.95

D3050 Terminal & Package Units

D3050 170	Split Systems With Air Cooled Condensing Units	COST PER S.F.		
		MAT.	INST.	TOTAL
3320	2,000 S.F., 4.66 ton	2.29	4.92	7.21
3360	5,000 S.F., 11.66 ton	2.09	4.48	6.57
3400	10,000 S.F., 23.33 ton	2.26	4.89	7.15
3440	20,000 S.F., 46.66 ton	2.68	4.97	7.65
3520	Offices, 1,000 S.F., 3.17 ton	3.05	4.78	7.83
3560	2,000 S.F., 6.33 ton	3.09	6.70	9.79
3600	5,000 S.F., 15.83 ton	2.84	6.05	8.89
3640	10,000 S.F., 31.66 ton	3.07	6.65	9.72
3680	20,000 S.F., 63.32 ton	3.74	7.05	10.79
3760	Restaurants, 1,000 S.F., 5.00 ton	4.90	10.55	15.45
3800	2,000 S.F., 10.00 ton	4.48	9.60	14.08
3840	5,000 S.F., 25.00 ton	4.85	10.50	15.35
3880	10,000 S.F., 50.00 ton	5.75	10.65	16.40
3920	20,000 S.F., 100.00 ton	7.85	11.55	19.40
4000	Schools and colleges, 1,000 S.F., 3.83 ton	3.68	5.90	9.58
4040	2,000 S.F., 7.66 ton	3.43	7.35	10.78
4080	5,000 S.F., 19.17 ton	3.71	8.05	11.76
4120	10,000 S.F., 38.33 ton	4.40	8.15	12.55

For customer support on your Square Foot Costs with RSMeans data, call 800.448.8182.

D3090   Other HVAC Systems/Equip

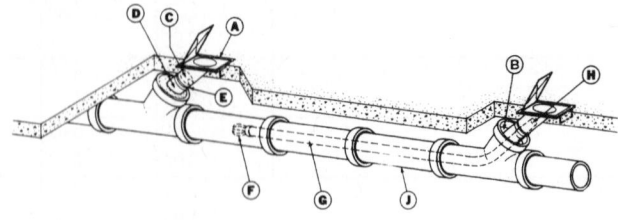

Cast Iron Garage Exhaust System

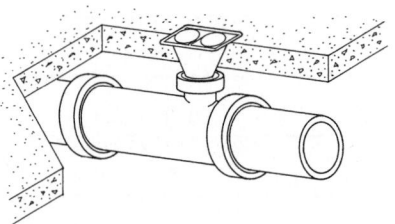

Dual Exhaust System

D3090 320	Garage Exhaust Systems	COST PER BAY		
		MAT.	INST.	TOTAL
1040	Garage, single 3″ exhaust outlet, cars & light trucks, one bay	4,500	2,050	6,550
1060	Additional bays up to seven bays	1,175	555	1,730
1500	4″ outlet, trucks, one bay	4,525	2,050	6,575
1520	Additional bays up to six bays	1,200	555	1,755
1600	5″ outlet, diesel trucks, one bay	4,800	2,050	6,850
1650	Additional single bays up to six bays	1,625	655	2,280
1700	Two adjoining bays	4,800	2,050	6,850
2000	Dual exhaust, 3″ outlets, pair of adjoining bays	5,725	2,575	8,300
2100	Additional pairs of adjoining bays	1,800	655	2,455

D4010 Sprinklers

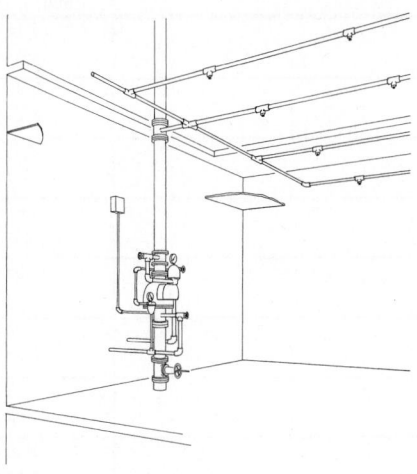

Dry Pipe System: A system employing automatic sprinklers attached to a piping system containing air under pressure, the release of which from the opening of sprinklers permits the water pressure to open a valve known as a "dry pipe valve". The water then flows into the piping system and out the opened sprinklers.

All areas are assumed to be open.

D4010 310	Dry Pipe Sprinkler Systems	COST PER S.F.		
		MAT.	INST.	TOTAL
0520	Dry pipe sprinkler systems, steel, black, sch. 40 pipe			
0530	Light hazard, one floor, 500 S.F.	10.95	6.60	17.55
0560	1000 S.F.	6	3.85	9.85
0580	2000 S.F.	5.70	3.90	9.60
0600	5000 S.F.	2.98	2.67	5.65
0620	10,000 S.F.	2.11	2.20	4.31
0640	50,000 S.F.	1.52	1.93	3.45
0660	Each additional floor, 500 S.F.	2.21	3.22	5.43
0680	1000 S.F.	1.84	2.64	4.48
0700	2000 S.F.	1.83	2.45	4.28
0720	5000 S.F.	1.53	2.12	3.65
0740	10,000 S.F.	1.41	1.95	3.36
0760	50,000 S.F.	1.27	1.72	2.99
1000	Ordinary hazard, one floor, 500 S.F.	11.15	6.65	17.80
1020	1000 S.F.	6.25	3.90	10.15
1040	2000 S.F.	5.85	4.06	9.91
1060	5000 S.F.	3.49	2.86	6.35
1080	10,000 S.F.	2.82	2.89	5.71
1100	50,000 S.F.	2.49	2.69	5.18
1140	Each additional floor, 500 S.F.	2.43	3.30	5.73
1160	1000 S.F.	2.25	2.97	5.22
1180	2000 S.F.	2.22	2.70	4.92
1200	5000 S.F.	2.11	2.35	4.46
1220	10,000 S.F.	1.90	2.28	4.18
1240	50,000 S.F.	1.86	1.96	3.82
1500	Extra hazard, one floor, 500 S.F.	15	8.30	23.30
1520	1000 S.F.	9.15	6	15.15
1540	2000 S.F.	6.45	5.20	11.65
1560	5000 S.F.	3.97	3.96	7.93
1580	10,000 S.F.	4.26	3.74	8
1600	50,000 S.F.	4.60	3.66	8.26
1660	Each additional floor, 500 S.F.	3.27	4.11	7.38
1680	1000 S.F.	3.20	3.91	7.11
1700	2000 S.F.	2.99	3.90	6.89
1720	5000 S.F.	2.61	3.40	6.01
1740	10,000 S.F.	3.21	3.09	6.30
1760	50,000 S.F.	3.29	2.99	6.28
2020	Grooved steel, black, sch. 40 pipe, light hazard, one floor, 2000 S.F.	5.55	3.35	8.90

463

D4010 Sprinklers

D4010 310	Dry Pipe Sprinkler Systems	COST PER S.F.		
		MAT.	INST.	TOTAL
2060	10,000 S.F.	2.16	1.90	4.06
2100	Each additional floor, 2000 S.F.	1.93	1.99	3.92
2150	10,000 S.F.	1.46	1.65	3.11
2200	Ordinary hazard, one floor, 2000 S.F.	5.85	3.54	9.39
2250	10,000 S.F.	2.65	2.46	5.11
2300	Each additional floor, 2000 S.F.	2.23	2.18	4.41
2350	10,000 S.F.	1.96	2.21	4.17
2400	Extra hazard, one floor, 2000 S.F.	6.65	4.46	11.11
2450	10,000 S.F.	3.83	3.14	6.97
2500	Each additional floor, 2000 S.F.	3.19	3.20	6.39
2550	10,000 S.F.	2.94	2.74	5.68
3050	Grooved steel, black, sch. 10 pipe, light hazard, one floor, 2000 S.F.	5.50	3.31	8.81
3100	10,000 S.F.	2.09	1.87	3.96
3150	Each additional floor, 2000 S.F.	1.86	1.95	3.81
3200	10,000 S.F.	1.39	1.62	3.01
3250	Ordinary hazard, one floor, 2000 S.F.	5.80	3.51	9.31
3300	10,000 S.F.	2.53	2.40	4.93
3350	Each additional floor, 2000 S.F.	2.14	2.15	4.29
3400	10,000 S.F.	1.84	2.15	3.99
3450	Extra hazard, one floor, 2000 S.F.	6.60	4.44	11.04
3500	10,000 S.F.	3.58	3.09	6.67
3550	Each additional floor, 2000 S.F.	3.11	3.18	6.29
3600	10,000 S.F.	2.82	2.70	5.52
4050	Copper tubing, type M, light hazard, one floor, 2000 S.F.	6	3.31	9.31
4100	10,000 S.F.	2.67	1.92	4.59
4150	Each additional floor, 2000 S.F.	2.38	1.99	4.37
4200	10,000 S.F.	1.98	1.68	3.66
4250	Ordinary hazard, one floor, 2000 S.F.	6.45	3.68	10.13
4300	10,000 S.F.	3.30	2.24	5.54
4350	Each additional floor, 2000 S.F.	3.17	2.28	5.45
4400	10,000 S.F.	2.55	1.95	4.50
4450	Extra hazard, one floor, 2000 S.F.	7.40	4.49	11.89
4500	10,000 S.F.	5.70	3.40	9.10
4550	Each additional floor, 2000 S.F.	3.92	3.23	7.15
4600	10,000 S.F.	4.15	2.98	7.13
5050	Copper tubing, type M, T-drill system, light hazard, one floor			
5060	2000 S.F.	6	3.09	9.09
5100	10,000 S.F.	2.55	1.59	4.14
5150	Each additional floor, 2000 S.F.	2.38	1.77	4.15
5200	10,000 S.F.	1.86	1.35	3.21
5250	Ordinary hazard, one floor, 2000 S.F.	6.25	3.15	9.40
5300	10,000 S.F.	3.19	2	5.19
5350	Each additional floor, 2000 S.F.	2.59	1.79	4.38
5400	10,000 S.F.	2.39	1.66	4.05
5450	Extra hazard, one floor, 2000 S.F.	6.85	3.71	10.56
5500	10,000 S.F.	4.84	2.50	7.34
5550	Each additional floor, 2000 S.F.	3.36	2.45	5.81
5600	10,000 S.F.	3.29	2.08	5.37

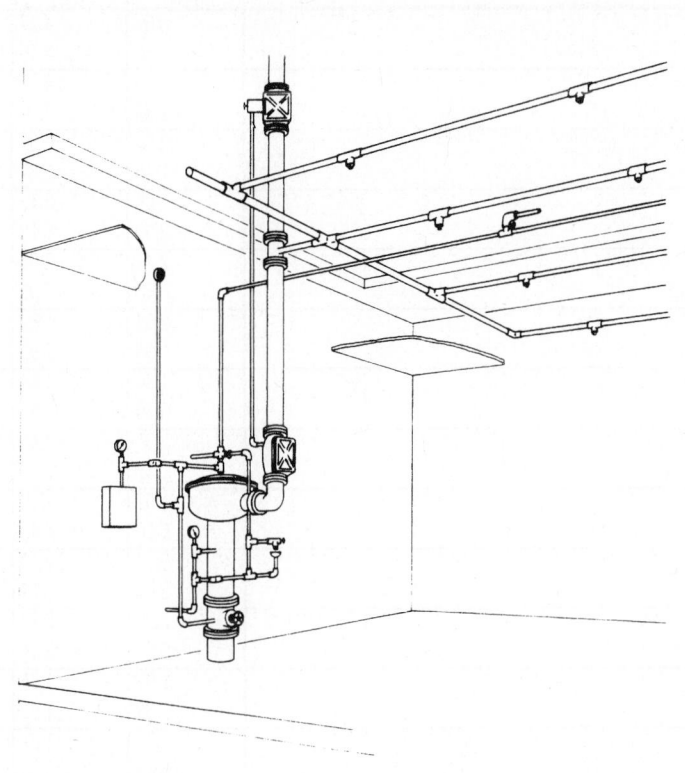

Preaction System: A system employing automatic sprinklers attached to a piping system containing air that may or may not be under pressure, with a supplemental heat responsive system of generally more sensitive characteristics than the automatic sprinklers themselves, installed in the same areas as the sprinklers. Actuation of the heat responsive system, as from a fire, opens a valve which permits water to flow into the sprinkler piping system and to be discharged from those sprinklers which were opened by heat from the fire.

All areas are assumed to be opened.

D4010 350	Preaction Sprinkler Systems	COST PER S.F.		
		MAT.	INST.	TOTAL
0520	Preaction sprinkler systems, steel, black, sch. 40 pipe			
0530	Light hazard, one floor, 500 S.F.	10.65	5.25	15.90
0560	1000 S.F.	6.05	3.95	10
0580	2000 S.F.	5.55	3.89	9.44
0600	5000 S.F.	2.85	2.66	5.51
0620	10,000 S.F.	1.99	2.19	4.18
0640	50,000 S.F.	1.40	1.93	3.33
0660	Each additional floor, 500 S.F.	2.41	2.86	5.27
0680	1000 S.F.	1.87	2.64	4.51
0700	2000 S.F.	1.86	2.45	4.31
0720	5000 S.F.	1.40	2.11	3.51
0740	10,000 S.F.	1.29	1.94	3.23
0760	50,000 S.F.	1.20	1.77	2.97
1000	Ordinary hazard, one floor, 500 S.F.	11.10	5.70	16.80
1020	1000 S.F.	6.10	3.89	9.99
1040	2000 S.F.	6	4.06	10.06
1060	5000 S.F.	3.21	2.84	6.05
1080	10,000 S.F.	2.48	2.87	5.35
1100	50,000 S.F.	2.15	2.67	4.82
1140	Each additional floor, 500 S.F.	2.87	3.31	6.18
1160	1000 S.F.	1.92	2.68	4.60
1180	2000 S.F.	1.76	2.67	4.43
1200	5000 S.F.	1.95	2.53	4.48
1220	10,000 S.F.	1.79	2.62	4.41
1240	50,000 S.F.	1.63	2.31	3.94
1500	Extra hazard, one floor, 500 S.F.	14.60	7.30	21.90
1520	1000 S.F.	8.35	5.50	13.85
1540	2000 S.F.	5.95	5.20	11.15

D4010 Sprinklers

D4010 350	Preaction Sprinkler Systems	COST PER S.F.		
		MAT.	INST.	TOTAL
1560	5000 S.F.	3.67	4.27	7.94
1580	10,000 S.F.	3.69	4.15	7.84
1600	50,000 S.F.	3.95	4.07	8.02
1660	Each additional floor, 500 S.F.	3.33	4.10	7.43
1680	1000 S.F.	2.65	3.89	6.54
1700	2000 S.F.	2.45	3.87	6.32
1720	5000 S.F.	2.05	3.42	5.47
1740	10,000 S.F.	2.50	3.13	5.63
1760	50,000 S.F.	2.47	2.95	5.42
2020	Grooved steel, black, sch. 40 pipe, light hazard, one floor, 2000 S.F.	5.60	3.35	8.95
2060	10,000 S.F.	2.04	1.89	3.93
2100	Each additional floor of 2000 S.F.	1.96	1.99	3.95
2150	10,000 S.F.	1.34	1.64	2.98
2200	Ordinary hazard, one floor, 2000 S.F.	5.70	3.53	9.23
2250	10,000 S.F.	2.31	2.44	4.75
2300	Each additional floor, 2000 S.F.	2.07	2.17	4.24
2350	10,000 S.F.	1.62	2.19	3.81
2400	Extra hazard, one floor, 2000 S.F.	6.10	4.43	10.53
2450	10,000 S.F.	3.13	3.10	6.23
2500	Each additional floor, 2000 S.F.	2.65	3.17	5.82
2550	10,000 S.F.	2.20	2.70	4.90
3050	Grooved steel, black, sch. 10 pipe light hazard, one floor, 2000 S.F.	5.55	3.31	8.86
3100	10,000 S.F.	1.97	1.86	3.83
3150	Each additional floor, 2000 S.F.	1.89	1.95	3.84
3200	10,000 S.F.	1.27	1.61	2.88
3250	Ordinary hazard, one floor, 2000 S.F.	5.55	3.28	8.83
3300	10,000 S.F.	1.89	2.37	4.26
3350	Each additional floor, 2000 S.F.	1.98	2.14	4.12
3400	10,000 S.F.	1.50	2.13	3.63
3450	Extra hazard, one floor, 2000 S.F.	6.05	4.41	10.46
3500	10,000 S.F.	2.84	3.05	5.89
3550	Each additional floor, 2000 S.F.	2.57	3.15	5.72
3600	10,000 S.F.	2.08	2.66	4.74
4050	Copper tubing, type M, light hazard, one floor, 2000 S.F.	6.05	3.31	9.36
4100	10,000 S.F.	2.55	1.91	4.46
4150	Each additional floor, 2000 S.F.	2.42	1.99	4.41
4200	10,000 S.F.	1.56	1.66	3.22
4250	Ordinary hazard, one floor, 2000 S.F.	6.30	3.67	9.97
4300	10,000 S.F.	2.96	2.22	5.18
4350	Each additional floor, 2000 S.F.	2.40	2.02	4.42
4400	10,000 S.F.	2.01	1.77	3.78
4450	Extra hazard, one floor, 2000 S.F.	6.85	4.46	11.31
4500	10,000 S.F.	4.92	3.36	8.28
4550	Each additional floor, 2000 S.F.	3.38	3.20	6.58
4600	10,000 S.F.	3.41	2.94	6.35
5050	Copper tubing, type M, T-drill system, light hazard, one floor			
5060	2000 S.F.	6	3.09	9.09
5100	10,000 S.F.	2.43	1.58	4.01
5150	Each additional floor, 2000 S.F.	2.41	1.77	4.18
5200	10,000 S.F.	1.74	1.34	3.08
5250	Ordinary hazard, one floor, 2000 S.F.	6.05	3.14	9.19
5300	10,000 S.F.	2.85	1.98	4.83
5350	Each additional floor, 2000 S.F.	2.43	1.80	4.23
5400	10,000 S.F.	2.16	1.73	3.89
5450	Extra hazard, one floor, 2000 S.F.	6.30	3.68	9.98
5500	10,000 S.F.	4.06	2.46	6.52
5550	Each additional floor, 2000 S.F.	2.82	2.42	5.24
5600	10,000 S.F.	2.55	2.04	4.59

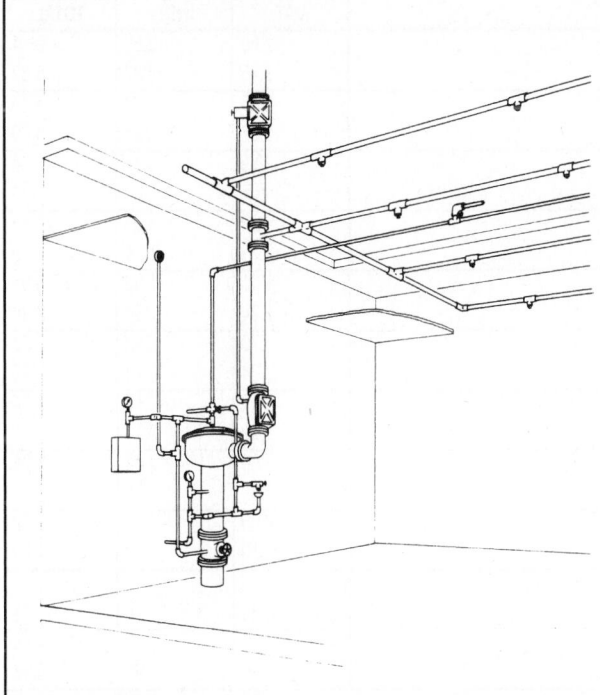

Deluge System: A system employing open sprinklers attached to a piping system connected to a water supply through a valve which is opened by the operation of a heat responsive system installed in the same areas as the sprinklers. When this valve opens, water flows into the piping system and discharges from all sprinklers attached thereto.

D4010 370	Deluge Sprinkler Systems	COST PER S.F.		
		MAT.	INST.	TOTAL
0520	Deluge sprinkler systems, steel, black, sch. 40 pipe			
0530	Light hazard, one floor, 500 S.F.	24	5.30	29.30
0560	1000 S.F.	12.65	3.79	16.44
0580	2000 S.F.	9.20	3.90	13.10
0600	5000 S.F.	4.32	2.67	6.99
0620	10,000 S.F.	2.72	2.19	4.91
0640	50,000 S.F.	1.54	1.93	3.47
0660	Each additional floor, 500 S.F.	2.41	2.86	5.27
0680	1000 S.F.	1.87	2.64	4.51
0700	2000 S.F.	1.86	2.45	4.31
0720	5000 S.F.	1.40	2.11	3.51
0740	10,000 S.F.	1.29	1.94	3.23
0760	50,000 S.F.	1.20	1.77	2.97
1000	Ordinary hazard, one floor, 500 S.F.	25.50	6.10	31.60
1020	1000 S.F.	12.75	3.92	16.67
1040	2000 S.F.	9.70	4.07	13.77
1060	5000 S.F.	4.68	2.85	7.53
1080	10,000 S.F.	3.21	2.87	6.08
1100	50,000 S.F.	2.35	2.71	5.06
1140	Each additional floor, 500 S.F.	2.87	3.31	6.18
1160	1000 S.F.	1.92	2.68	4.60
1180	2000 S.F.	1.76	2.67	4.43
1200	5000 S.F.	1.83	2.33	4.16
1220	10,000 S.F.	1.72	2.31	4.03
1240	50,000 S.F.	1.56	2.12	3.68
1500	Extra hazard, one floor, 500 S.F.	28	7.35	35.35
1520	1000 S.F.	15.55	5.70	21.25
1540	2000 S.F.	9.60	5.20	14.80
1560	5000 S.F.	4.87	3.94	8.81
1580	10,000 S.F.	4.31	3.78	8.09
1600	50,000 S.F.	4.43	3.73	8.16
1660	Each additional floor, 500 S.F.	3.33	4.10	7.43

D4010 Sprinklers

D4010 370	Deluge Sprinkler Systems	COST PER S.F.		
		MAT.	INST.	TOTAL
1680	1000 S.F.	2.65	3.89	6.54
1700	2000 S.F.	2.45	3.87	6.32
1720	5000 S.F.	2.05	3.42	5.47
1740	10,000 S.F.	2.55	3.25	5.80
1760	50,000 S.F.	2.59	3.15	5.74
2000	Grooved steel, black, sch. 40 pipe, light hazard, one floor			
2020	2000 S.F.	9.30	3.36	12.66
2060	10,000 S.F.	2.79	1.91	4.70
2100	Each additional floor, 2,000 S.F.	1.96	1.99	3.95
2150	10,000 S.F.	1.34	1.64	2.98
2200	Ordinary hazard, one floor, 2000 S.F.	5.70	3.53	9.23
2250	10,000 S.F.	3.04	2.44	5.48
2300	Each additional floor, 2000 S.F.	2.07	2.17	4.24
2350	10,000 S.F.	1.62	2.19	3.81
2400	Extra hazard, one floor, 2000 S.F.	9.80	4.44	14.24
2450	10,000 S.F.	3.89	3.11	7
2500	Each additional floor, 2000 S.F.	2.65	3.17	5.82
2550	10,000 S.F.	2.20	2.70	4.90
3000	Grooved steel, black, sch. 10 pipe, light hazard, one floor			
3050	2000 S.F.	8.55	3.18	11.73
3100	10,000 S.F.	2.70	1.86	4.56
3150	Each additional floor, 2000 S.F.	1.89	1.95	3.84
3200	10,000 S.F.	1.27	1.61	2.88
3250	Ordinary hazard, one floor, 2000 S.F.	9.30	3.51	12.81
3300	10,000 S.F.	2.62	2.37	4.99
3350	Each additional floor, 2000 S.F.	1.98	2.14	4.12
3400	10,000 S.F.	1.50	2.13	3.63
3450	Extra hazard, one floor, 2000 S.F.	9.70	4.42	14.12
3500	10,000 S.F.	3.57	3.05	6.62
3550	Each additional floor, 2000 S.F.	2.57	3.15	5.72
3600	10,000 S.F.	2.08	2.66	4.74
4000	Copper tubing, type M, light hazard, one floor			
4050	2000 S.F.	9.70	3.32	13.02
4100	10,000 S.F.	3.28	1.91	5.19
4150	Each additional floor, 2000 S.F.	2.41	1.99	4.40
4200	10,000 S.F.	1.56	1.66	3.22
4250	Ordinary hazard, one floor, 2000 S.F.	9.95	3.68	13.63
4300	10,000 S.F.	3.69	2.22	5.91
4350	Each additional floor, 2000 S.F.	2.40	2.02	4.42
4400	10,000 S.F.	2.01	1.77	3.78
4450	Extra hazard, one floor, 2000 S.F.	10.55	4.47	15.02
4500	10,000 S.F.	5.70	3.37	9.07
4550	Each additional floor, 2000 S.F.	3.38	3.20	6.58
4600	10,000 S.F.	3.41	2.94	6.35
5000	Copper tubing, type M, T-drill system, light hazard, one floor			
5050	2000 S.F.	9.70	3.10	12.80
5100	10,000 S.F.	3.16	1.58	4.74
5150	Each additional floor, 2000 S.F.	2.40	1.80	4.20
5200	10,000 S.F.	1.74	1.34	3.08
5250	Ordinary hazard, one floor, 2000 S.F.	9.75	3.15	12.90
5300	10,000 S.F.	3.58	1.98	5.56
5350	Each additional floor, 2000 S.F.	2.43	1.78	4.21
5400	10,000 S.F.	2.16	1.73	3.89
5450	Extra hazard, one floor, 2000 S.F.	9.95	3.69	13.64
5500	10,000 S.F.	4.79	2.46	7.25
5550	Each additional floor, 2000 S.F.	2.82	2.42	5.24
5600	10,000 S.F.	2.55	2.04	4.59

D4010 Sprinklers

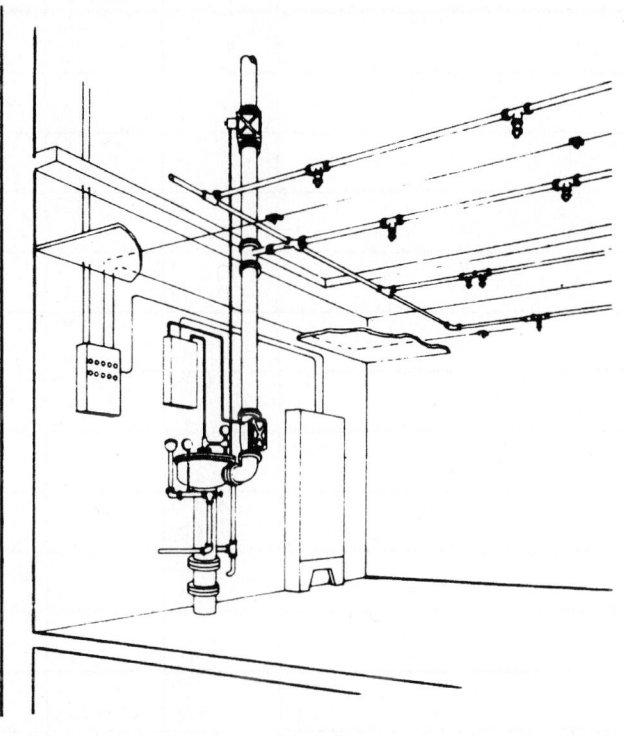

On-off multicycle is a fixed fire protection sprinkler system utilizing water as its extinguishing agent. It is a time delayed, recycling, preaction type which automatically shuts the water off when heat is reduced below the detector operating temperature and turns the water back on when that temperature is exceeded.

The system senses a fire condition through a closed circuit electrical detector system which controls water flow to the fire automatically. Batteries supply up to 90 hours emergency power supply for system operation. The piping system is dry (until water is required) and is monitored with pressurized air. Should any leak in the system piping occur, an alarm will sound, but water will not enter the system until heat is sensed by a Firecycle detector.

D4010 390	On-off multicycle Sprinkler Systems	COST PER S.F.		
		MAT.	INST.	TOTAL
0520	On-off multicycle sprinkler systems, steel, black, sch. 40 pipe			
0530	Light hazard, one floor, 500 S.F.	48.50	10.25	58.75
0560	1000 S.F.	25	6.45	31.45
0580	2000 S.F.	14.90	4.76	19.66
0600	5000 S.F.	6.60	3.01	9.61
0620	10,000 S.F.	3.91	2.38	6.29
0640	50,000 S.F.	1.82	1.96	3.78
0660	Each additional floor of 500 S.F.	2.41	2.88	5.29
0680	1000 S.F.	1.87	2.64	4.51
0700	2000 S.F.	1.56	2.44	4
0720	5000 S.F.	1.40	2.12	3.52
0740	10,000 S.F.	1.35	1.95	3.30
0760	50,000 S.F.	1.24	1.76	3
1000	Ordinary hazard, one floor, 500 S.F.	49	10.70	59.70
1020	1000 S.F.	25	6.40	31.40
1040	2000 S.F.	15.05	4.92	19.97
1060	5000 S.F.	6.95	3.19	10.14
1080	10,000 S.F.	4.40	3.06	7.46
1100	50,000 S.F.	2.85	3.03	5.88
1140	Each additional floor, 500 S.F.	2.87	3.33	6.20
1160	1000 S.F.	1.92	2.68	4.60
1180	2000 S.F.	1.98	2.46	4.44
1200	5000 S.F.	1.83	2.34	4.17
1220	10,000 S.F.	1.62	2.27	3.89
1240	50,000 S.F.	1.56	2	3.56
1500	Extra hazard, one floor, 500 S.F.	52.50	12.35	64.85
1520	1000 S.F.	27	8	35
1540	2000 S.F.	15.25	6.05	21.30
1560	5000 S.F.	7.15	4.28	11.43
1580	10,000 S.F.	5.55	4.32	9.87

469

D4010 390	On-off multicycle Sprinkler Systems	COST PER S.F.		
		MAT.	INST.	TOTAL
1600	50,000 S.F.	4.78	4.81	9.59
1660	Each additional floor, 500 S.F.	3.33	4.12	7.45
1680	1000 S.F.	2.65	3.89	6.54
1700	2000 S.F.	2.45	3.87	6.32
1720	5000 S.F.	2.05	3.43	5.48
1740	10,000 S.F.	2.56	3.14	5.70
1760	50,000 S.F.	2.59	3.04	5.63
2020	Grooved steel, black, sch. 40 pipe, light hazard, one floor			
2030	2000 S.F.	14.95	4.22	19.17
2060	10,000 S.F.	4.29	2.86	7.15
2100	Each additional floor, 2000 S.F.	1.96	1.99	3.95
2150	10,000 S.F.	1.40	1.65	3.05
2200	Ordinary hazard, one floor, 2000 S.F.	15.05	4.40	19.45
2250	10,000 S.F.	4.59	2.77	7.36
2300	Each additional floor, 2000 S.F.	2.07	2.17	4.24
2350	10,000 S.F.	1.68	2.20	3.88
2400	Extra hazard, one floor, 2000 S.F.	15.45	5.30	20.75
2450	10,000 S.F.	4.98	3.27	8.25
2500	Each additional floor, 2000 S.F.	2.65	3.17	5.82
2550	10,000 S.F.	2.26	2.71	4.97
3050	Grooved steel, black, sch. 10 pipe, light hazard, one floor,			
3060	2000 S.F.	14.85	4.18	19.03
3100	10,000 S.F.	3.89	2.05	5.94
3150	Each additional floor, 2000 S.F.	1.89	1.95	3.84
3200	10,000 S.F.	1.33	1.62	2.95
3250	Ordinary hazard, one floor, 2000 S.F.	14.95	4.37	19.32
3300	10,000 S.F.	4.11	2.57	6.68
3350	Each additional floor, 2000 S.F.	1.98	2.14	4.12
3400	10,000 S.F.	1.56	2.14	3.70
3450	Extra hazard, one floor, 2000 S.F.	15.40	5.30	20.70
3500	10,000 S.F.	4.73	3.22	7.95
3550	Each additional floor, 2000 S.F.	2.57	3.15	5.72
3600	10,000 S.F.	2.14	2.67	4.81
4060	Copper tubing, type M, light hazard, one floor, 2000 S.F.	15.35	4.18	19.53
4100	10,000 S.F.	4.47	2.10	6.57
4150	Each additional floor, 2000 S.F.	2.41	1.99	4.40
4200	10,000 S.F.	1.92	1.68	3.60
4250	Ordinary hazard, one floor, 2000 S.F.	15.60	4.54	20.14
4300	10,000 S.F.	4.88	2.41	7.29
4350	Each additional floor, 2000 S.F.	2.40	2.02	4.42
4400	10,000 S.F.	2.04	1.76	3.80
4450	Extra hazard, one floor, 2000 S.F.	16.20	5.35	21.55
4500	10,000 S.F.	6.90	3.58	10.48
4550	Each additional floor, 2000 S.F.	3.38	3.20	6.58
4600	10,000 S.F.	3.47	2.95	6.42
5060	Copper tubing, type M, T-drill system, light hazard, one floor 2000 S.F.	15.35	3.96	19.31
5100	10,000 S.F.	4.35	1.77	6.12
5150	Each additional floor, 2000 S.F.	2.58	1.87	4.45
5200	10,000 S.F.	1.80	1.35	3.15
5250	Ordinary hazard, one floor, 2000 S.F.	15.40	4.01	19.41
5300	10,000 S.F.	4.77	2.17	6.94
5350	Each additional floor, 2000 S.F.	2.43	1.78	4.21
5400	10,000 S.F.	2.22	1.74	3.96
5450	Extra hazard, one floor, 2000 S.F.	15.65	4.55	20.20
5500	10,000 S.F.	5.95	2.63	8.58
5550	Each additional floor, 2000 S.F.	2.82	2.42	5.24
5600	10,000 S.F.	2.61	2.05	4.66

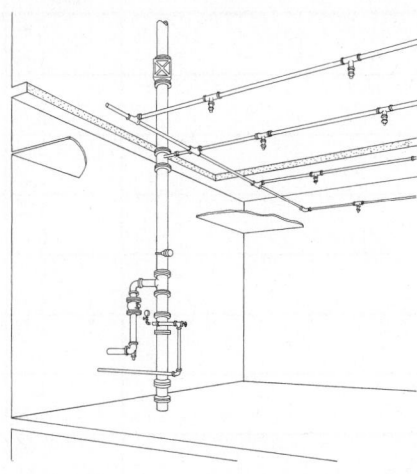

Wet Pipe System. A system employing automatic sprinklers attached to a piping system containing water and connected to a water supply so that water discharges immediately from sprinklers opened by heat from a fire.

All areas are assumed to be open.

D4010 410	Wet Pipe Sprinkler Systems	COST PER S.F.		
		MAT.	INST.	TOTAL
0520	Wet pipe sprinkler systems, steel, black, sch. 40 pipe			
0530	Light hazard, one floor, 500 S.F.	2.97	3.39	6.36
0560	1000 S.F.	6.10	3.50	9.60
0580	2000 S.F.	5.55	3.53	9.08
0600	5000 S.F.	2.61	2.52	5.13
0620	10,000 S.F.	1.69	2.10	3.79
0640	50,000 S.F.	1.04	1.89	2.93
0660	Each additional floor, 500 S.F.	1.24	2.88	4.12
0680	1000 S.F.	1.30	2.68	3.98
0700	2000 S.F.	1.26	2.42	3.68
0720	5000 S.F.	.92	2.09	3.01
0740	10,000 S.F.	.87	1.92	2.79
0760	50,000 S.F.	.70	1.49	2.19
1000	Ordinary hazard, one floor, 500 S.F.	3.33	3.63	6.96
1020	1000 S.F.	6.15	3.44	9.59
1040	2000 S.F.	5.70	3.69	9.39
1060	5000 S.F.	2.97	2.70	5.67
1080	10,000 S.F.	2.18	2.78	4.96
1100	50,000 S.F.	1.75	2.60	4.35
1140	Each additional floor, 500 S.F.	1.67	3.26	4.93
1160	1000 S.F.	1.32	2.65	3.97
1180	2000 S.F.	1.46	2.66	4.12
1200	5000 S.F.	1.50	2.54	4.04
1220	10,000 S.F.	1.37	2.60	3.97
1240	50,000 S.F.	1.25	2.29	3.54
1500	Extra hazard, one floor, 500 S.F.	11.75	5.60	17.35
1520	1000 S.F.	7.75	4.89	12.64
1540	2000 S.F.	5.95	4.98	10.93
1560	5000 S.F.	3.63	4.37	8
1580	10,000 S.F.	3.34	4.10	7.44
1600	50,000 S.F.	3.78	4.01	7.79
1660	Each additional floor, 500 S.F.	2.13	4.05	6.18
1680	1000 S.F.	2.05	3.86	5.91
1700	2000 S.F.	1.85	3.84	5.69
1720	5000 S.F.	1.57	3.40	4.97
1740	10,000 S.F.	2.08	3.11	5.19
1760	50,000 S.F.	2.10	2.98	5.08
2020	Grooved steel, black sch. 40 pipe, light hazard, one floor, 2000 S.F.	5.60	2.99	8.59

471

D4010 Sprinklers

D4010 410	Wet Pipe Sprinkler Systems	COST PER S.F.		
		MAT.	INST.	TOTAL
2060	10,000 S.F.	2.21	1.88	4.09
2100	Each additional floor, 2000 S.F.	1.36	1.96	3.32
2150	10,000 S.F.	.92	1.62	2.54
2200	Ordinary hazard, one floor, 2000 S.F.	5.70	3.17	8.87
2250	10,000 S.F.	2.01	2.35	4.36
2300	Each additional floor, 2000 S.F.	1.47	2.14	3.61
2350	10,000 S.F.	1.20	2.17	3.37
2400	Extra hazard, one floor, 2000 S.F.	6.10	4.07	10.17
2450	10,000 S.F.	2.79	3.01	5.80
2500	Each additional floor, 2000 S.F.	2.05	3.14	5.19
2550	10,000 S.F.	1.78	2.68	4.46
3050	Grooved steel, black sch. 10 pipe, light hazard, one floor, 2000 S.F.	5.55	2.95	8.50
3100	10,000 S.F.	1.67	1.77	3.44
3150	Each additional floor, 2000 S.F.	1.29	1.92	3.21
3200	10,000 S.F.	.85	1.59	2.44
3250	Ordinary hazard, one floor, 2000 S.F.	5.60	3.14	8.74
3300	10,000 S.F.	1.89	2.29	4.18
3350	Each additional floor, 2000 S.F.	1.38	2.11	3.49
3400	10,000 S.F.	1.08	2.11	3.19
3450	Extra hazard, one floor, 2000 S.F.	6.05	4.05	10.10
3500	10,000 S.F.	2.54	2.96	5.50
3550	Each additional floor, 2000 S.F.	1.97	3.12	5.09
3600	10,000 S.F.	1.66	2.64	4.30
4050	Copper tubing, type M, light hazard, one floor, 2000 S.F.	6	2.95	8.95
4100	10,000 S.F.	2.25	1.82	4.07
4150	Each additional floor, 2000 S.F.	1.81	1.96	3.77
4200	10,000 S.F.	1.44	1.65	3.09
4250	Ordinary hazard, one floor, 2000 S.F.	6.30	3.31	9.61
4300	10,000 S.F.	2.66	2.13	4.79
4350	Each additional floor, 2000 S.F.	2.07	2.16	4.23
4400	10,000 S.F.	1.79	1.91	3.70
4450	Extra hazard, one floor, 2000 S.F.	6.85	4.10	10.95
4500	10,000 S.F.	4.62	3.27	7.89
4550	Each additional floor, 2000 S.F.	2.78	3.17	5.95
4600	10,000 S.F.	2.99	2.92	5.91
5050	Copper tubing, type M, T-drill system, light hazard, one floor			
5060	2000 S.F.	6	2.73	8.73
5100	10,000 S.F.	2.13	1.49	3.62
5150	Each additional floor, 2000 S.F.	1.81	1.74	3.55
5200	10,000 S.F.	1.32	1.32	2.64
5250	Ordinary hazard, one floor, 2000 S.F.	6.05	2.78	8.83
5300	10,000 S.F.	2.55	1.89	4.44
5350	Each additional floor, 2000 S.F.	1.83	1.75	3.58
5400	10,000 S.F.	1.74	1.71	3.45
5450	Extra hazard, one floor, 2000 S.F.	6.30	3.32	9.62
5500	10,000 S.F.	3.76	2.37	6.13
5550	Each additional floor, 2000 S.F.	2.31	2.44	4.75
5600	10,000 S.F.	2.13	2.02	4.15

D40 Fire Protection

D4020 Standpipes

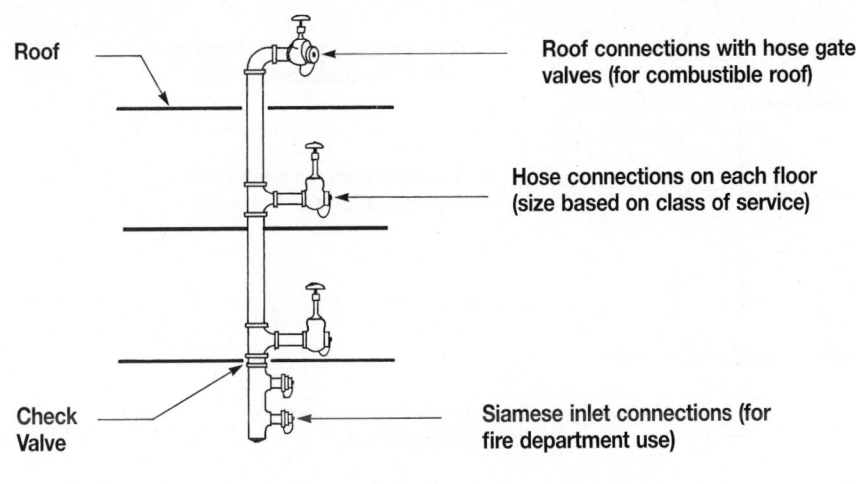

Roof

Roof connections with hose gate valves (for combustible roof)

Hose connections on each floor (size based on class of service)

Check Valve

Siamese inlet connections (for fire department use)

D4020 310	Wet Standpipe Risers, Class I	COST PER FLOOR		
		MAT.	INST.	TOTAL
0550	Wet standpipe risers, Class I, steel, black, sch. 40, 10' height			
0560	4" diameter pipe, one floor	5,700	3,700	9,400
0580	Additional floors	1,400	1,150	2,550
0600	6" diameter pipe, one floor	9,450	6,475	15,925
0620	Additional floors	2,475	1,825	4,300
0640	8" diameter pipe, one floor	14,100	7,825	21,925
0660	Additional floors	3,600	2,200	5,800

D4020 310	Wet Standpipe Risers, Class II	COST PER FLOOR		
		MAT.	INST.	TOTAL
1030	Wet standpipe risers, Class II, steel, black sch. 40, 10' height			
1040	2" diameter pipe, one floor	2,475	1,325	3,800
1060	Additional floors	925	515	1,440
1080	2-1/2" diameter pipe, one floor	3,450	1,950	5,400
1100	Additional floors	995	600	1,595

D4020 310	Wet Standpipe Risers, Class III	COST PER FLOOR		
		MAT.	INST.	TOTAL
1530	Wet standpipe risers, Class III, steel, black, sch. 40, 10' height			
1540	4" diameter pipe, one floor	5,825	3,700	9,525
1560	Additional floors	1,225	960	2,185
1580	6" diameter pipe, one floor	9,575	6,475	16,050
1600	Additional floors	2,550	1,825	4,375
1620	8" diameter pipe, one floor	14,300	7,825	22,125
1640	Additional floors	3,650	2,200	5,850

D4020 Standpipes

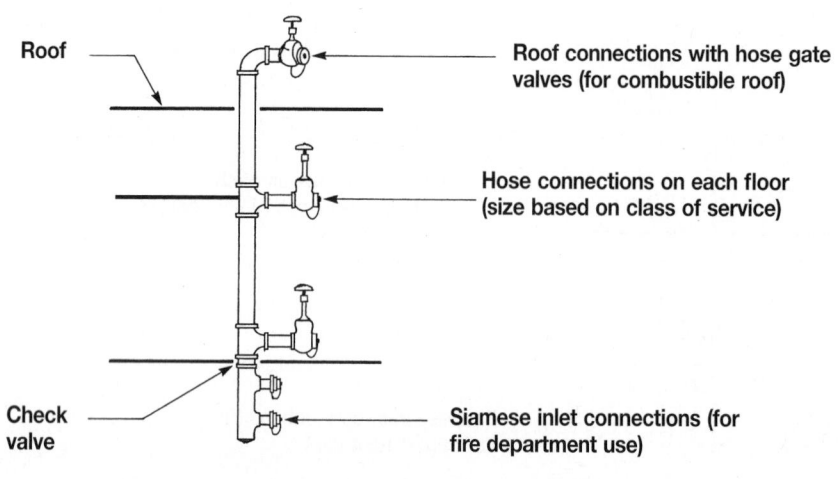

Roof

Roof connections with hose gate valves (for combustible roof)

Hose connections on each floor (size based on class of service)

Check valve

Siamese inlet connections (for fire department use)

D4020 330	Dry Standpipe Risers, Class I	COST PER FLOOR		
		MAT.	INST.	TOTAL
0530	Dry standpipe riser, Class I, steel, black, sch. 40, 10' height			
0540	4" diameter pipe, one floor	3,550	2,975	6,525
0560	Additional floors	1,275	1,075	2,350
0580	6" diameter pipe, one floor	7,200	5,125	12,325
0600	Additional floors	2,350	1,750	4,100
0620	8" diameter pipe, one floor	11,000	6,225	17,225
0640	Additional floors	3,475	2,125	5,600

D4020 330	Dry Standpipe Risers, Class II	COST PER FLOOR		
		MAT.	INST.	TOTAL
1030	Dry standpipe risers, Class II, steel, black, sch. 40, 10' height			
1040	2" diameter pipe, one floor	2,200	1,375	3,575
1060	Additional floors	800	455	1,255
1080	2-1/2" diameter pipe, one floor	2,875	1,625	4,500
1100	Additional floors	870	540	1,410

D4020 330	Dry Standpipe Risers, Class III	COST PER FLOOR		
		MAT.	INST.	TOTAL
1530	Dry standpipe risers, Class III, steel, black, sch. 40, 10' height			
1540	4" diameter pipe, one floor	3,600	2,925	6,525
1560	Additional floors	1,100	980	2,080
1580	6" diameter pipe, one floor	7,250	5,125	12,375
1600	Additional floors	2,425	1,750	4,175
1620	8" diameter pipe, one floor	11,100	6,225	17,325
1640	Additional floors	3,525	2,125	5,650

D4020 Standpipes

D4020 410	Fire Hose Equipment	COST EACH		
		MAT.	INST.	TOTAL
0100	Adapters, reducing, 1 piece, FxM, hexagon, cast brass, 2-1/2" x 1-1/2"	72.50		72.50
0200	Pin lug, 1-1/2" x 1"	52.50		52.50
0250	3" x 2-1/2"	158		158
0300	For polished chrome, add 75% mat.			
0400	Cabinets, D.S. glass in door, recessed, steel box, not equipped			
0500	Single extinguisher, steel door & frame	141	163	304
0550	Stainless steel door & frame	249	163	412
0600	Valve, 2-1/2" angle, steel door & frame	196	109	305
0650	Aluminum door & frame	238	109	347
0700	Stainless steel door & frame	320	109	429
0750	Hose rack assy, 2-1/2" x 1-1/2" valve & 100' hose, steel door & frame	365	217	582
0800	Aluminum door & frame	540	217	757
0850	Stainless steel door & frame	720	217	937
0900	Hose rack assy & extinguisher,2-1/2"x1-1/2" valve & hose,steel door & frame	320	260	580
0950	Aluminum	690	260	950
1000	Stainless steel	700	260	960
1550	Compressor, air, dry pipe system, automatic, 200 gal., 3/4 H.P.	1,250	555	1,805
1600	520 gal., 1 H.P.	1,550	555	2,105
1650	Alarm, electric pressure switch (circuit closer)	116	28	144
2500	Couplings, hose, rocker lug, cast brass, 1-1/2"	71.50		71.50
2550	2-1/2"	56		56
3000	Escutcheon plate, for angle valves, polished brass, 1-1/2"	16.70		16.70
3050	2-1/2"	27.50		27.50
3500	Fire pump, electric, w/controller, fittings, relief valve			
3550	4" pump, 30 H.P., 500 G.P.M.	17,000	4,050	21,050
3600	5" pump, 40 H.P., 1000 G.P.M.	19,100	4,600	23,700
3650	5" pump, 100 H.P., 1000 G.P.M.	25,900	5,100	31,000
3700	For jockey pump system, add	3,150	650	3,800
5000	Hose, per linear foot, synthetic jacket, lined,			
5100	300 lb. test, 1-1/2" diameter	3.30	.50	3.80
5150	2-1/2" diameter	6.10	.59	6.69
5200	500 lb. test, 1-1/2" diameter	2.64	.50	3.14
5250	2-1/2" diameter	5.90	.59	6.49
5500	Nozzle, plain stream, polished brass, 1-1/2" x 10"	65.50		65.50
5550	2-1/2" x 15" x 13/16" or 1-1/2"	117		117
5600	Heavy duty combination adjustable fog and straight stream w/handle 1-1/2"	460		460
5650	2-1/2" direct connection	515		515
6000	Rack, for 1-1/2" diameter hose 100 ft. long, steel	99	65	164
6050	Brass	153	65	218
6500	Reel, steel, for 50 ft. long 1-1/2" diameter hose	156	93	249
6550	For 75 ft. long 2-1/2" diameter hose	340	93	433
7050	Siamese, w/plugs & chains, polished brass, sidewalk, 4" x 2-1/2" x 2-1/2"	805	520	1,325
7100	6" x 2-1/2" x 2-1/2"	820	650	1,470
7200	Wall type, flush, 4" x 2-1/2" x 2-1/2"	835	260	1,095
7250	6" x 2-1/2" x 2-1/2"	980	283	1,263
7300	Projecting, 4" x 2-1/2" x 2-1/2"	585	260	845
7350	6" x 2-1/2" x 2-1/2"	980	283	1,263
7400	For chrome plate, add 15% mat.			
8000	Valves, angle, wheel handle, 300 Lb., rough brass, 1-1/2"	125	60.50	185.50
8050	2-1/2"	227	103	330
8100	Combination pressure restricting, 1-1/2"	106	60.50	166.50
8150	2-1/2"	229	103	332
8200	Pressure restricting, adjustable, satin brass, 1-1/2"	360	60.50	420.50
8250	2-1/2"	425	103	528
8300	Hydrolator, vent and drain, rough brass, 1-1/2"	125	60.50	185.50
8350	2-1/2"	125	60.50	185.50
8400	Cabinet assy, incls. adapter, rack, hose, and nozzle	940	395	1,335

D4090 Other Fire Protection Systems

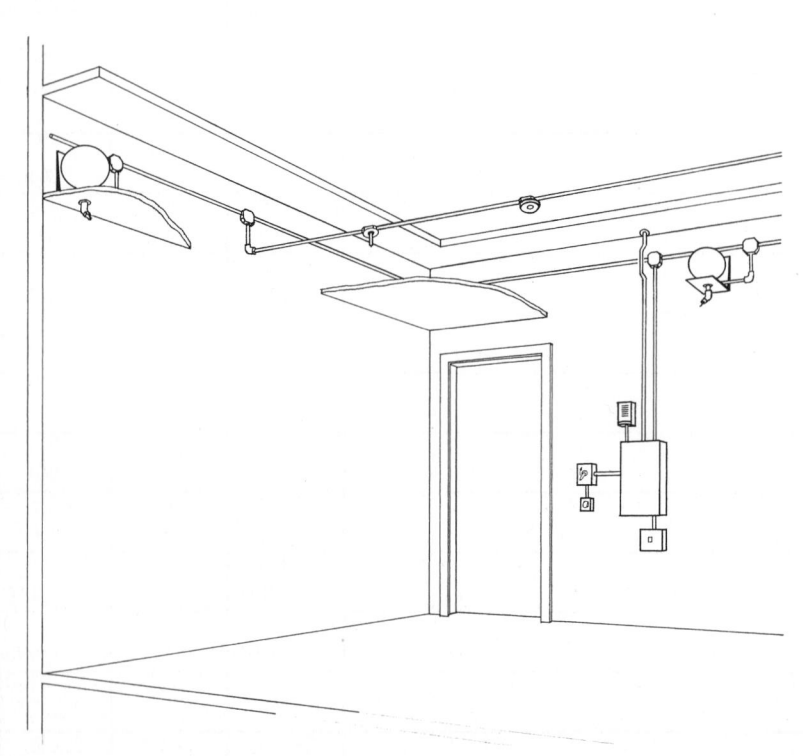

General: Automatic fire protection (suppression) systems other than water sprinklers may be desired for special environments, high risk areas, isolated locations or unusual hazards. Some typical applications would include:

1. Paint dip tanks
2. Securities vaults
3. Electronic data processing
4. Tape and data storage
5. Transformer rooms
6. Spray booths
7. Petroleum storage
8. High rack storage

Piping and wiring costs are highly variable and are not included.

D4090 910	Fire Suppression Unit Components	COST EACH		
		MAT.	INST.	TOTAL
0020	Detectors with brackets			
0040	Fixed temperature heat detector	56.50	99.50	156
0060	Rate of temperature rise detector	56.50	87	143.50
0080	Ion detector (smoke) detector	132	112	244
0200	Extinguisher agent			
0240	200 lb FM200, container	7,275	335	7,610
0280	75 lb carbon dioxide cylinder	1,450	225	1,675
0320	Dispersion nozzle			
0340	FM200 1-1/2" dispersion nozzle	200	53.50	253.50
0380	Carbon dioxide 3" x 5" dispersion nozzle	158	41.50	199.50
0420	Control station			
0440	Single zone control station with batteries	1,025	695	1,720
0470	Multizone (4) control station with batteries	3,050	1,400	4,450
0500	Electric mechanical release	1,025	360	1,385
0550	Manual pull station	87	125	212
0640	Battery standby power 10" x 10" x 17"	455	174	629
0740	Bell signalling device	137	87	224
0900	Standard low-rise sprinkler accessory package, 3 story	3,675	1,925	5,600
1000	Standard mid-rise sprinkler accessory package, 8 story	7,700	3,700	11,400
1100	Standard high-rise sprinkler accessory package, 16 story	18,400	7,250	25,650

D4090 920	FM200 Systems	COST PER C.F.		
		MAT.	INST.	TOTAL
0820	Average FM200 system, minimum			1.94
0840	Maximum			3.85

476

D5010 Electrical Service/Distribution

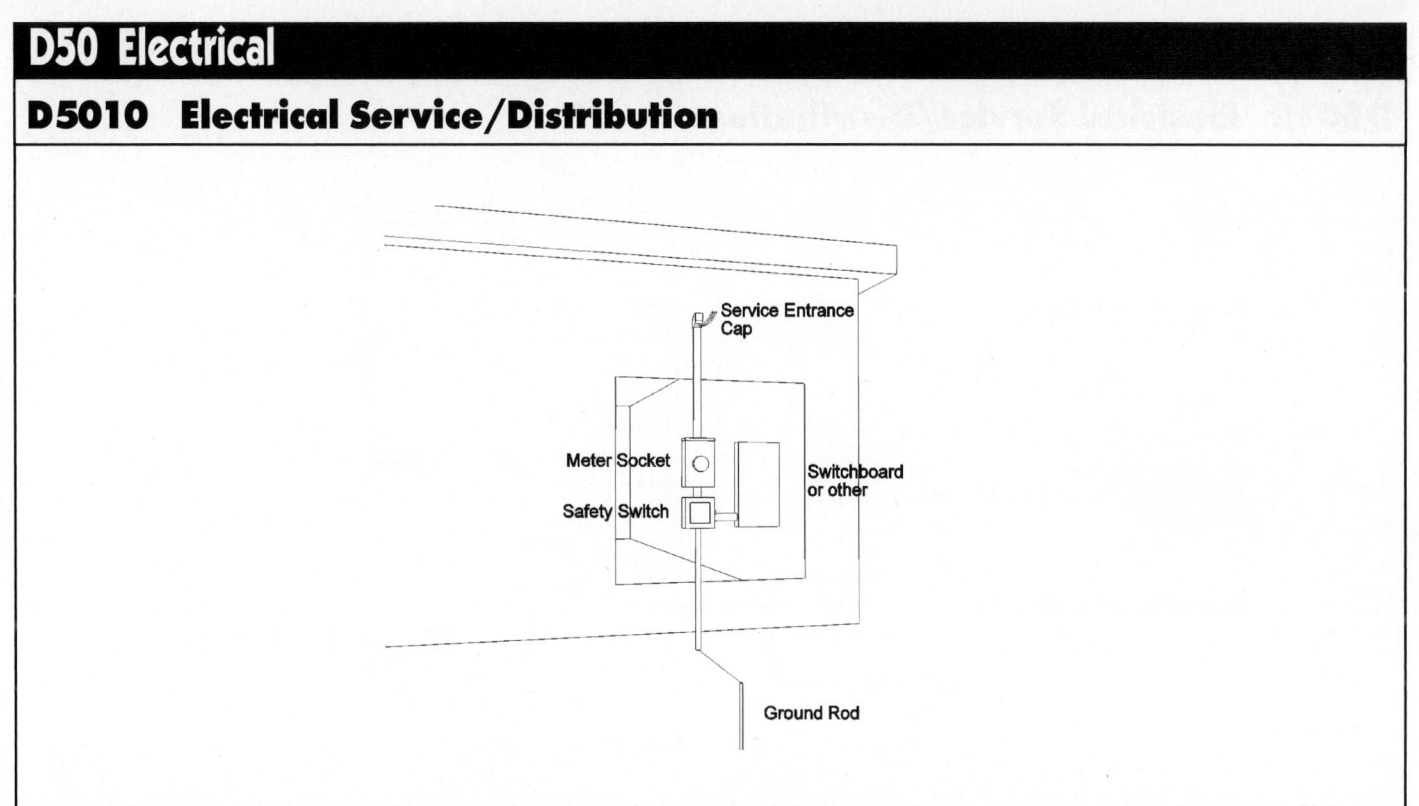

D5010 120	Overhead Electric Service, 3 Phase - 4 Wire	COST EACH		
		MAT.	INST.	TOTAL
0200	Service installation, includes breakers, metering, 20' conduit & wire			
0220	3 phase, 4 wire, 120/208 volts, 60 A	500	1,150	1,650
0240	100 A	675	1,275	1,950
0245	100 A w/circuit breaker	1,475	1,600	3,075
0280	200 A	1,175	1,775	2,950
0285	200 A w/circuit breaker	2,850	2,225	5,075
0320	400 A	2,350	3,525	5,875
0325	400 A w/circuit breaker	5,525	4,400	9,925
0360	600 A	3,850	5,200	9,050
0365	600 A, w/switchboard	8,250	6,600	14,850
0400	800 A	6,100	6,250	12,350
0405	800 A, w/switchboard	10,500	7,825	18,325
0440	1000 A	7,425	7,700	15,125
0445	1000 A, w/switchboard	12,700	9,450	22,150
0480	1200 A	9,700	8,775	18,475
0485	1200 A, w/groundfault switchboard	32,500	10,900	43,400
0520	1600 A	12,400	11,500	23,900
0525	1600 A, w/groundfault switchboard	37,100	13,700	50,800
0560	2000 A	16,000	13,900	29,900
0565	2000 A, w/groundfault switchboard	43,800	17,200	61,000
0610	1 phase, 3 wire, 120/240 volts, 100 A (no safety switch)	160	630	790
0615	100 A w/load center	330	1,200	1,530
0620	200 A	360	860	1,220
0625	200 A w/load center	660	1,825	2,485

For customer support on your Square Foot Costs with RSMeans data, call 800.448.8182.

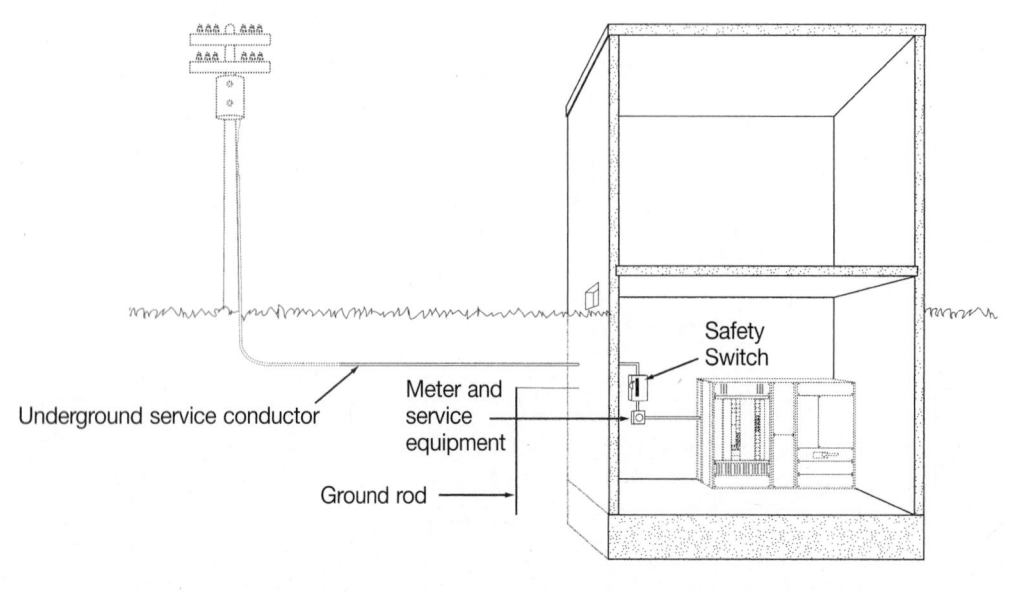

D5010 130	Underground Electric Service	COST EACH		
		MAT.	INST.	TOTAL
0950	Underground electric service including excavation, backfill and compaction			
1000	3 phase, 4 wire, 277/480 volts, 2000 A	29,700	22,200	51,900
1050	2000 A w/groundfault switchboard	56,500	24,700	81,200
1100	1600 A	23,300	17,900	41,200
1150	1600 A w/groundfault switchboard	48,000	20,200	68,200
1200	1200 A	18,100	14,300	32,400
1250	1200 A w/groundfault switchboard	40,900	16,400	57,300
1400	800 A	12,800	11,800	24,600
1450	800 A w/switchboard	17,200	13,400	30,600
1500	600 A	8,350	10,800	19,150
1550	600 A w/switchboard	12,700	12,200	24,900
1600	1 phase, 3 wire, 120/240 volts, 200 A	2,950	3,675	6,625
1650	200 A w/load center	3,250	4,625	7,875
1700	100 A	2,350	2,975	5,325
1750	100 A w/load center	2,650	3,950	6,600

D5010   Electrical Service/Distribution

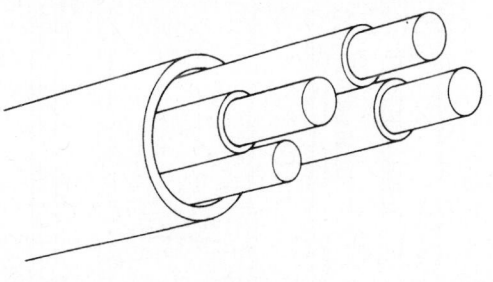

D5010 230	Feeder Installation	COST PER L.F.		
		MAT.	INST.	TOTAL
0200	Feeder installation 600 V, including RGS conduit and XHHW wire, 60 A	7.10	12.80	19.90
0240	100 A	8.80	16.65	25.45
0280	200 A	22.50	25	47.50
0320	400 A	45	50	95
0360	600 A	72.50	82	154.50
0400	800 A	94	97	191
0440	1000 A	117	128	245
0480	1200 A	145	164	309
0520	1600 A	188	194	382
0560	2000 A	235	256	491
1200	Branch installation 600 V, including EMT conduit and THW wire, 15 A	1.25	6.25	7.50
1240	20 A	1.45	6.65	8.10
1280	30 A	1.79	6.85	8.64
1320	50 A	3.53	9.55	13.08
1360	65 A	3.85	10.15	14
1400	85 A	5.95	11.75	17.70
1440	100 A	7.10	12	19.10
1480	130 A	8.05	14.05	22.10
1520	150 A	11.85	16.10	27.95
1560	200 A	12.75	18.35	31.10

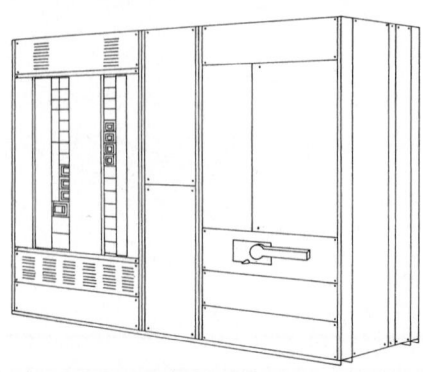

D5010 240	Switchgear	COST PER EACH		
		MAT.	INST.	TOTAL
0190	Switchgear installation, including switchboard, panels, & circuit breaker			
0200	120/208 V, 3 phase, 400 A	8,950	2,975	11,925
0240	600 A	11,700	3,325	15,025
0280	800 A	15,000	3,800	18,800
0300	1000 A	18,600	4,250	22,850
0320	1200 A	19,400	4,775	24,175
0360	1600 A	31,700	5,500	37,200
0400	2000 A	38,200	6,025	44,225
0500	277/480 V, 3 phase, 400 A	12,200	5,525	17,725
0520	600 A	16,700	6,350	23,050
0540	800 A	19,200	6,950	26,150
0560	1000 A	24,500	7,750	32,250
0580	1200 A	26,300	8,625	34,925
0600	1600 A	39,000	9,500	48,500
0620	2000 A	49,200	10,500	59,700

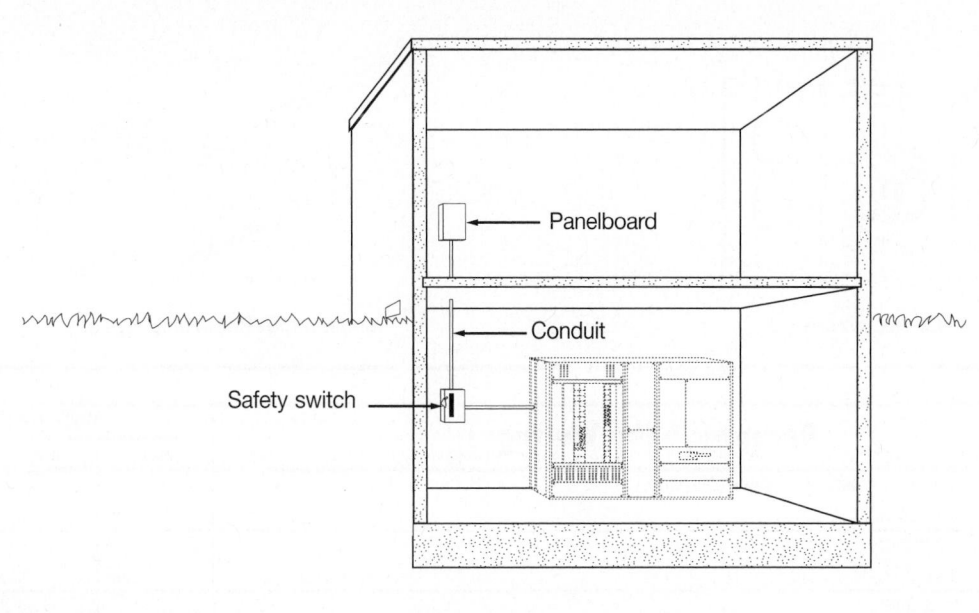

D5010 250	Panelboard	COST EACH		
		MAT.	INST.	TOTAL
0900	Panelboards, NQOD, 4 wire, 120/208 volts w/conductor & conduit			
1000	100 A, 0 stories, 0' horizontal	2,100	1,475	3,575
1020	1 stories, 25' horizontal	2,475	2,100	4,575
1040	5 stories, 50' horizontal	3,175	3,375	6,550
1060	10 stories, 75' horizontal	4,000	4,825	8,825
1080	225A, 0 stories, 0' horizontal	3,800	1,925	5,725
2000	1 stories, 25' horizontal	4,850	3,075	7,925
2020	5 stories, 50' horizontal	6,925	5,325	12,250
2040	10 stories, 75' horizontal	9,325	7,950	17,275
2060	400A, 0 stories, 0' horizontal	5,600	2,900	8,500
2080	1 stories, 25' horizontal	7,050	4,750	11,800
3000	5 stories, 50' horizontal	9,875	8,425	18,300
3020	10 stories, 75' horizontal	16,500	16,500	33,000
3040	600 A, 0 stories, 0' horizontal	8,250	3,475	11,725
3060	1 stories, 25' horizontal	10,600	6,525	17,125
3080	5 stories, 50' horizontal	15,200	12,500	27,700
4000	10 stories, 75' horizontal	26,900	24,600	51,500
4010	Panelboards, NEHB, 4 wire, 277/480 volts w/conductor, conduit, & safety switch			
4020	100 A, 0 stories, 0' horizontal, includes safety switch	3,150	2,025	5,175
4040	1 stories, 25' horizontal	3,525	2,650	6,175
4060	5 stories, 50' horizontal	4,225	3,900	8,125
4080	10 stories, 75' horizontal	5,050	5,350	10,400
5000	225 A, 0 stories, 0' horizontal	4,700	2,450	7,150
5020	1 stories, 25' horizontal	5,750	3,600	9,350
5040	5 stories, 50' horizontal	7,825	5,850	13,675
5060	10 stories, 75' horizontal	10,200	8,475	18,675
5080	400 A, 0 stories, 0' horizontal	7,425	3,800	11,225
6000	1 stories, 25' horizontal	8,875	5,650	14,525
6020	5 stories, 50' horizontal	11,700	9,325	21,025
6040	10 stories, 75' horizontal	18,300	17,400	35,700
6060	600 A, 0 stories, 0' horizontal	10,600	4,825	15,425
6080	1 stories, 25' horizontal	12,900	7,875	20,775
7000	5 stories, 50' horizontal	17,500	13,900	31,400
7020	10 stories, 75' horizontal	29,300	26,000	55,300

D5020 Lighting and Branch Wiring

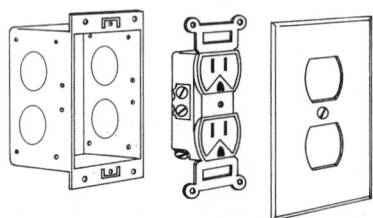

Duplex Wall Receptacle

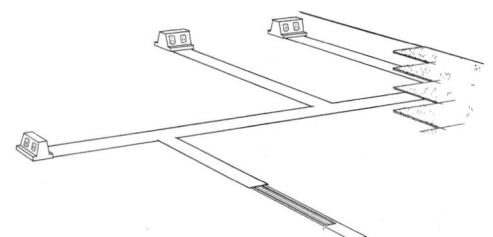

Undercarpet Receptacle System

D5020 110	Receptacle (by Wattage)	COST PER S.F.		
		MAT.	INST.	TOTAL
0190	Receptacles include plate, box, conduit, wire & transformer when required			
0200	2.5 per 1000 S.F., .3 watts per S.F.	.36	1.55	1.91
0240	With transformer	.45	1.63	2.08
0280	4 per 1000 S.F., .5 watts per S.F.	.42	1.80	2.22
0320	With transformer	.57	1.92	2.49
0360	5 per 1000 S.F., .6 watts per S.F.	.14	.71	.85
0400	With transformer	.69	2.29	2.98
0440	8 per 1000 S.F., .9 watts per S.F.	.52	2.38	2.90
0480	With transformer	.82	2.63	3.45
0520	10 per 1000 S.F., 1.2 watts per S.F.	.56	2.57	3.13
0560	With transformer	.93	2.88	3.81
0600	16.5 per 1000 S.F., 2.0 watts per S.F.	.67	3.21	3.88
0640	With transformer	1.29	3.72	5.01
0680	20 per 1000 S.F., 2.4 watts per S.F.	.70	3.50	4.20
0720	With transformer	1.45	4.12	5.57

D5020 115	Receptacles, Floor	COST PER S.F.		
		MAT.	INST.	TOTAL
0200	Receptacle systems, underfloor duct, 5' on center, low density	7.50	3.43	10.93
0240	High density	8.10	4.41	12.51
0280	7' on center, low density	5.95	2.95	8.90
0320	High density	6.50	3.93	10.43
0400	Poke thru fittings, low density	1.32	1.67	2.99
0440	High density	2.62	3.32	5.94
0520	Telepoles, using Romex, low density	1.27	1.03	2.30
0560	High density	2.53	2.06	4.59
0600	Using EMT, low density	1.34	1.37	2.71
0640	High density	2.70	2.72	5.42
0720	Conduit system with floor boxes, low density	1.18	1.17	2.35
0760	High density	2.37	2.34	4.71
0840	Undercarpet power system, 3 conductor with 5 conductor feeder, low density	1.53	.41	1.94
0880	High density	3.03	.85	3.88

D5020 Lighting and Branch Wiring

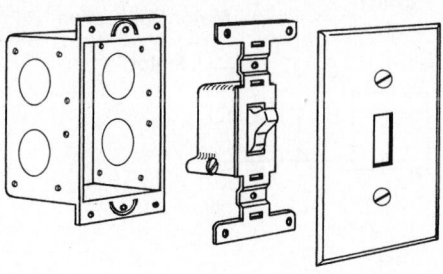

Description: Table D5020 130 includes the cost for switch, plate, box, conduit in slab or EMT exposed and copper wire. Add 20% for exposed conduit.

No power required for switches.

Federal energy guidelines recommend the maximum lighting area controlled per switch shall not exceed 1000 S.F. and that areas over 500 S.F. shall be so controlled that total illumination can be reduced by at least 50%.

D5020 130	Wall Switch by Sq. Ft.	COST PER S.F.		
		MAT.	INST.	TOTAL
0200	Wall switches, 1.0 per 1000 S.F.	.05	.26	.31
0240	1.2 per 1000 S.F.	.05	.27	.32
0280	2.0 per 1000 S.F.	.08	.40	.48
0320	2.5 per 1000 S.F.	.09	.50	.59
0360	5.0 per 1000 S.F.	.22	1.08	1.30
0400	10.0 per 1000 S.F.	.46	2.19	2.65

D5020 135	Miscellaneous Power	COST PER S.F.		
		MAT.	INST.	TOTAL
0200	Miscellaneous power, to .5 watts	.03	.12	.15
0240	.8 watts	.04	.18	.22
0280	1 watt	.06	.24	.30
0320	1.2 watts	.07	.28	.35
0360	1.5 watts	.08	.33	.41
0400	1.8 watts	.10	.38	.48
0440	2 watts	.11	.45	.56
0480	2.5 watts	.13	.55	.68
0520	3 watts	.17	.65	.82

D5020 140	Central A. C. Power (by Wattage)	COST PER S.F.		
		MAT.	INST.	TOTAL
0200	Central air conditioning power, 1 watt	.07	.27	.34
0220	2 watts	.08	.31	.39
0240	3 watts	.15	.47	.62
0280	4 watts	.16	.47	.63
0320	6 watts	.24	.63	.87
0360	8 watts	.34	.67	1.01
0400	10 watts	.48	.79	1.27

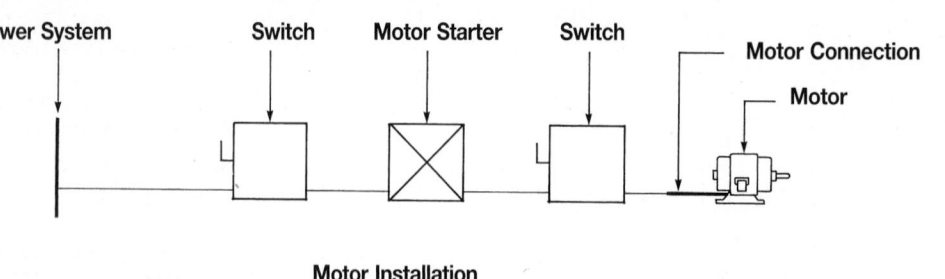

Motor Installation

D5020 145	Motor Installation	COST EACH		
		MAT.	INST.	TOTAL
0200	Motor installation, single phase, 115V, 1/3 HP motor size	470	1,050	1,520
0240	1 HP motor size	490	1,050	1,540
0280	2 HP motor size	540	1,125	1,665
0320	3 HP motor size	620	1,150	1,770
0360	230V, 1 HP motor size	470	1,075	1,545
0400	2 HP motor size	510	1,075	1,585
0440	3 HP motor size	570	1,150	1,720
0520	Three phase, 200V, 1-1/2 HP motor size	545	1,175	1,720
0560	3 HP motor size	655	1,275	1,930
0600	5 HP motor size	660	1,425	2,085
0640	7-1/2 HP motor size	685	1,450	2,135
0680	10 HP motor size	1,050	1,825	2,875
0720	15 HP motor size	1,475	2,050	3,525
0760	20 HP motor size	1,775	2,350	4,125
0800	25 HP motor size	1,850	2,350	4,200
0840	30 HP motor size	2,975	2,750	5,725
0880	40 HP motor size	3,625	3,275	6,900
0920	50 HP motor size	6,600	3,825	10,425
0960	60 HP motor size	6,850	4,025	10,875
1000	75 HP motor size	8,350	4,625	12,975
1040	100 HP motor size	24,700	5,425	30,125
1080	125 HP motor size	25,400	5,975	31,375
1120	150 HP motor size	28,500	7,025	35,525
1160	200 HP motor size	29,300	8,250	37,550
1240	230V, 1-1/2 HP motor size	525	1,175	1,700
1280	3 HP motor size	635	1,275	1,910
1320	5 HP motor size	635	1,425	2,060
1360	7-1/2 HP motor size	635	1,425	2,060
1400	10 HP motor size	990	1,725	2,715
1440	15 HP motor size	1,125	1,900	3,025
1480	20 HP motor size	1,675	2,275	3,950
1520	25 HP motor size	1,775	2,350	4,125
1560	30 HP motor size	1,825	2,350	4,175
1600	40 HP motor size	3,475	3,175	6,650
1640	50 HP motor size	3,700	3,375	7,075
1680	60 HP motor size	6,600	3,825	10,425
1720	75 HP motor size	7,775	4,350	12,125
1760	100 HP motor size	8,450	4,850	13,300
1800	125 HP motor size	25,600	5,600	31,200
1840	150 HP motor size	26,400	6,350	32,750
1880	200 HP motor size	27,000	7,075	34,075
1960	460V, 2 HP motor size	650	1,175	1,825
2000	5 HP motor size	760	1,275	2,035
2040	10 HP motor size	730	1,425	2,155
2080	15 HP motor size	995	1,625	2,620
2120	20 HP motor size	1,050	1,750	2,800

D5020 Lighting and Branch Wiring

D5020 145	Motor Installation	COST EACH		
		MAT.	INST.	TOTAL
2160	25 HP motor size	1,125	1,825	2,950
2200	30 HP motor size	1,475	1,975	3,450
2240	40 HP motor size	1,825	2,125	3,950
2280	50 HP motor size	2,000	2,350	4,350
2320	60 HP motor size	3,150	2,725	5,875
2360	75 HP motor size	3,650	3,025	6,675
2400	100 HP motor size	3,975	3,375	7,350
2440	125 HP motor size	6,900	3,850	10,750
2480	150 HP motor size	8,575	4,300	12,875
2520	200 HP motor size	9,300	4,850	14,150
2600	575V, 2 HP motor size	650	1,175	1,825
2640	5 HP motor size	760	1,275	2,035
2680	10 HP motor size	730	1,425	2,155
2720	20 HP motor size	995	1,625	2,620
2760	25 HP motor size	1,050	1,750	2,800
2800	30 HP motor size	1,475	1,975	3,450
2840	50 HP motor size	1,575	2,050	3,625
2880	60 HP motor size	3,100	2,725	5,825
2920	75 HP motor size	3,150	2,725	5,875
2960	100 HP motor size	3,650	3,025	6,675
3000	125 HP motor size	6,650	3,775	10,425
3040	150 HP motor size	6,900	3,850	10,750
3080	200 HP motor size	8,625	4,375	13,000

D5020 155	Motor Feeder	COST PER L.F.		
		MAT.	INST.	TOTAL
0200	Motor feeder systems, single phase, feed up to 115V 1HP or 230V 2 HP	2.06	8.20	10.26
0240	115V 2HP, 230V 3HP	2.16	8.35	10.51
0280	115V 3HP	2.47	8.70	11.17
0360	Three phase, feed to 200V 3HP, 230V 5HP, 460V 10HP, 575V 10HP	2.17	8.85	11.02
0440	200V 5HP, 230V 7.5HP, 460V 15HP, 575V 20HP	2.32	9.05	11.37
0520	200V 10HP, 230V 10HP, 460V 30HP, 575V 30HP	2.78	9.55	12.33
0600	200V 15HP, 230V 15HP, 460V 40HP, 575V 50HP	3.84	10.95	14.79
0680	200V 20HP, 230V 25HP, 460V 50HP, 575V 60HP	5.10	13.90	19
0760	200V 25HP, 230V 30HP, 460V 60HP, 575V 75HP	5.95	14.10	20.05
0840	200V 30HP	6.70	14.60	21.30
0920	230V 40HP, 460V 75HP, 575V 100HP	8.90	15.90	24.80
1000	200V 40HP	10.40	17.05	27.45
1080	230V 50HP, 460V 100HP, 575V 125HP	12.55	18.80	31.35
1160	200V 50HP, 230V 60HP, 460V 125HP, 575V 150HP	14.55	19.95	34.50
1240	200V 60HP, 460V 150HP	18.30	23.50	41.80
1320	230V 75HP, 575V 200HP	21	24.50	45.50
1400	200V 75HP	23.50	25	48.50
1480	230V 100HP, 460V 200HP	29.50	29	58.50
1560	200V 100HP	41	36	77
1640	230V 125HP	41	36	77
1720	200V 125HP, 230V 150HP	46.50	45	91.50
1800	200V 150HP	57.50	49	106.50
1880	200V 200HP	74	52	126
1960	230V 200HP	68.50	54	122.50

485

D50 Electrical

D5020 Lighting and Branch Wiring

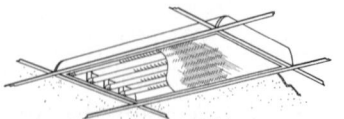

Fluorescent Fixture

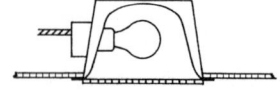

Incandescent Fixture

D5020 210	Fluorescent Fixtures (by Wattage)	COST PER S.F.		
		MAT.	INST.	TOTAL
0190	Fluorescent fixtures recess mounted in ceiling			
0195	T12, standard 40 watt lamps			
0200	1 watt per S.F., 20 FC, 5 fixtures @40 watts per 1000 S.F.	.71	2.04	2.75
0240	2 watt per S.F., 40 FC, 10 fixtures @40 watt per 1000 S.F.	1.41	4.02	5.43
0280	3 watt per S.F., 60 FC, 15 fixtures @40 watt per 1000 S.F	2.12	6.05	8.17
0320	4 watt per S.F., 80 FC, 20 fixtures @40 watt per 1000 S.F.	2.82	8.05	10.87
0400	5 watt per S.F., 100 FC, 25 fixtures @40 watt per 1000 S.F.	3.52	10.10	13.62
0450	T8, energy saver 32 watt lamps			
0500	0.8 watt per S.F., 20 FC, 5 fixtures @32 watt per 1000 S.F.	.78	2.04	2.82
0520	1.6 watt per S.F., 40 FC, 10 fixtures @32 watt per 1000 S.F.	1.54	4.02	5.56
0540	2.4 watt per S.F., 60 FC, 15 fixtures @ 32 watt per 1000 S.F	2.31	6.05	8.36
0560	3.2 watt per S.F., 80 FC, 20 fixtures @32 watt per 1000 S.F.	3.08	8.05	11.13
0580	4 watt per S.F., 100 FC, 25 fixtures @32 watt per 1000 S.F.	3.85	10.10	13.95

D5020 216	Incandescent Fixture (by Wattage)	COST PER S.F.		
		MAT.	INST.	TOTAL
0190	Incandescent fixture recess mounted, type A			
0200	1 watt per S.F., 8 FC, 6 fixtures per 1000 S.F.	.96	1.64	2.60
0240	2 watt per S.F., 16 FC, 12 fixtures per 1000 S.F.	1.92	3.28	5.20
0280	3 watt per S.F., 24 FC, 18 fixtures, per 1000 S.F.	2.85	4.88	7.73
0320	4 watt per S.F., 32 FC, 24 fixtures per 1000 S.F.	3.80	6.50	10.30
0400	5 watt per S.F., 40 FC, 30 fixtures per 1000 S.F.	4.76	8.15	12.91

D5020 Lighting and Branch Wiring

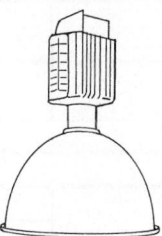

High Bay Fixture

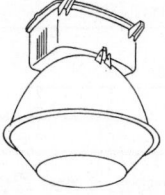

Low Bay Fixture

D5020 226	H.I.D. Fixture, High Bay, 16' (by Wattage)	COST PER S.F.		
		MAT.	INST.	TOTAL
0190	High intensity discharge fixture, 16' above work plane			
0240	1 watt/S.F., type E, 42 FC, 1 fixture/1000 S.F.	1.03	1.72	2.75
0280	Type G, 52 FC, 1 fixture/1000 S.F.	1.03	1.72	2.75
0320	Type C, 54 FC, 2 fixture/1000 S.F.	1.20	1.92	3.12
0440	2 watt/S.F., type E, 84 FC, 2 fixture/1000 S.F.	2.05	3.50	5.55
0480	Type G, 105 FC, 2 fixture/1000 S.F.	2.05	3.50	5.55
0520	Type C, 108 FC, 4 fixture/1000 S.F.	2.39	3.84	6.23
0640	3 watt/S.F., type E, 126 FC, 3 fixture/1000 S.F.	3.08	5.25	8.33
0680	Type G, 157 FC, 3 fixture/1000 S.F.	3.08	5.25	8.33
0720	Type C, 162 FC, 6 fixture/1000 S.F.	3.59	5.75	9.34
0840	4 watt/S.F., type E, 168 FC, 4 fixture/1000 S.F.	4.10	7	11.10
0880	Type G, 210 FC, 4 fixture/1000 S.F.	4.10	7	11.10
0920	Type C, 243 FC, 9 fixture/1000 S.F.	5.25	8.05	13.30
1040	5 watt/S.F., type E, 210 FC, 5 fixture/1000 S.F.	5.15	8.70	13.85
1080	Type G, 262 FC, 5 fixture/1000 S.F.	5.15	8.70	13.85
1120	Type C, 297 FC, 11 fixture/1000 S.F.	6.45	9.95	16.40

D5020 234	H.I.D. Fixture, High Bay, 30' (by Wattage)	COST PER S.F.		
		MAT.	INST.	TOTAL
0190	High intensity discharge fixture, 30' above work plane			
0240	1 watt/S.F., type E, 37 FC, 1 fixture/1000 S.F.	1.13	2.18	3.31
0280	Type G, 45 FC., 1 fixture/1000 S.F.	1.13	2.18	3.31
0320	Type F, 50 FC, 1 fixture/1000 S.F.	.96	1.71	2.67
0440	2 watt/S.F., type E, 74 FC, 2 fixtures/1000 S.F.	2.27	4.34	6.61
0480	Type G, 92 FC, 2 fixtures/1000 S.F.	2.27	4.34	6.61
0520	Type F, 100 FC, 2 fixtures/1000 S.F.	1.94	3.48	5.42
0640	3 watt/S.F., type E, 110 FC, 3 fixtures/1000 S.F.	3.42	6.60	10.02
0680	Type G, 138 FC, 3 fixtures/1000 S.F.	3.42	6.60	10.02
0720	Type F, 150 FC, 3 fixtures/1000 S.F.	2.90	5.20	8.10
0840	4 watt/S.F., type E, 148 FC, 4 fixtures/1000 S.F.	4.54	8.75	13.29
0880	Type G, 185 FC, 4 fixtures/1000 S.F.	4.54	8.75	13.29
0920	Type F, 200 FC, 4 fixtures/1000 S.F.	3.88	6.95	10.83
1040	5 watt/S.F., type E, 185 FC, 5 fixtures/1000 S.F.	5.70	10.95	16.65
1080	Type G, 230 FC, 5 fixtures/1000 S.F.	5.70	10.95	16.65
1120	Type F, 250 FC, 5 fixtures/1000 S.F.	4.85	8.65	13.50

D5020 238	H.I.D. Fixture, Low Bay, 8'-10' (by Wattage)	COST PER S.F.		
		MAT.	INST.	TOTAL
0190	High intensity discharge fixture, 8'-10' above work plane			
0240	1 watt/S.F., type J, 30 FC, 4 fixtures/1000 S.F.	2.33	3.55	5.88
0280	Type K, 29 FC, 5 fixtures/1000 S.F.	2.38	3.19	5.57
0400	2 watt/S.F., type J, 52 FC, 7 fixtures/1000 S.F.	4.14	6.50	10.64
0440	Type K, 63 FC, 11 fixtures/1000 S.F.	5.15	6.70	11.85
0560	3 watt/S.F., type J, 81 FC, 11 fixtures/1000 S.F.	6.40	9.85	16.25

D5020 Lighting and Branch Wiring

D5020 238	H.I.D. Fixture, Low Bay, 8'-10' (by Wattage)	COST PER S.F.		
		MAT.	INST.	TOTAL
0600	Type K, 92 FC, 16 fixtures/1000 S.F.	7.55	9.90	17.45
0720	4 watt/S.F., type J, 103 FC, 14 fixtures/1000 S.F.	8.25	12.95	21.20
0760	Type K, 127 FC, 22 fixtures/1000 S.F.	10.30	13.40	23.70
0880	5 watt/S.F., type J, 133 FC, 18 fixtures/1000 S.F.	10.55	16.30	26.85
0920	Type K, 155 FC, 27 fixtures/1000 S.F.	12.70	16.60	29.30

D5020 242	H.I.D. Fixture, Low Bay, 16' (by Wattage)	COST PER S.F.		
		MAT.	INST.	TOTAL
0190	High intensity discharge fixture, mounted 16' above work plane			
0240	1 watt/S.F., type J, 28 FC, 4 fixt./1000 S.F.	2.45	4.02	6.47
0280	Type K, 27 FC, 5 fixt./1000 S.F.	2.72	4.50	7.22
0400	2 watt/S.F., type J, 48 FC, 7 fixt/1000 S.F.	4.46	7.80	12.26
0440	Type K, 58 FC, 11 fixt/1000 S.F.	5.80	9.20	15
0560	3 watt/S.F., type J, 75 FC, 11 fixt/1000 S.F.	6.90	11.80	18.70
0600	Type K, 85 FC, 16 fixt/1000 S.F.	8.50	13.70	22.20
0720	4 watt/S.F., type J, 95 FC, 14 fixt/1000 S.F.	8.95	15.55	24.50
0760	Type K, 117 FC, 22 fixt/1000 S.F.	11.60	18.40	30
0880	5 watt/S.F., type J, 122 FC, 18 fixt/1000 S.F.	11.35	19.60	30.95
0920	Type K, 143 FC, 27 fixt/1000 S.F.	14.30	23	37.30

D50 Electrical

D5030 Communications and Security

D5030 910	Communication & Alarm Systems	COST EACH		
		MAT.	INST.	TOTAL
0200	Communication & alarm systems, includes outlets, boxes, conduit & wire			
0210	Sound system, 6 outlets	5,650	8,800	14,450
0220	12 outlets	8,050	14,100	22,150
0240	30 outlets	14,100	26,600	40,700
0280	100 outlets	42,100	89,000	131,100
0320	Fire detection systems, non-addressable, 12 detectors	3,350	7,300	10,650
0360	25 detectors	5,900	12,300	18,200
0400	50 detectors	11,600	24,400	36,000
0440	100 detectors	21,700	44,400	66,100
0450	Addressable type, 12 detectors	4,900	7,350	12,250
0452	25 detectors	8,950	12,400	21,350
0454	50 detectors	17,200	24,400	41,600
0456	100 detectors	33,500	44,900	78,400
0458	Fire alarm control panel, 8 zone, excluding wire and conduit	1,050	1,400	2,450
0459	12 zone	2,300	2,075	4,375
0460	Fire alarm command center, addressable without voice, excl. wire & conduit	4,975	1,225	6,200
0462	Addressable with voice	10,500	1,925	12,425
0480	Intercom systems, 6 stations	4,000	6,025	10,025
0560	25 stations	13,200	23,000	36,200
0640	100 stations	50,500	84,500	135,000
0680	Master clock systems, 6 rooms	4,650	9,800	14,450
0720	12 rooms	7,450	16,700	24,150
0760	20 rooms	10,500	23,700	34,200
0800	30 rooms	17,200	43,600	60,800
0840	50 rooms	28,000	73,500	101,500
0920	Master TV antenna systems, 6 outlets	2,250	6,200	8,450
0960	12 outlets	4,200	11,600	15,800
1000	30 outlets	14,300	26,800	41,100
1040	100 outlets	52,500	87,500	140,000

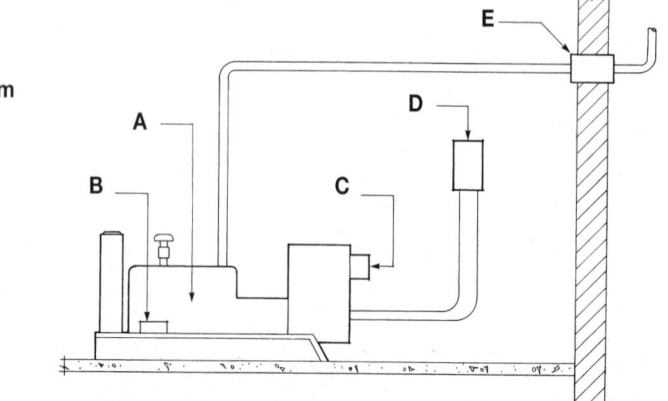

Generator System

A: Engine
B: Battery
C: Charger
D: Transfer Switch
E: Muffler

D5090 210	Generators (by kW)	COST PER kW		
		MAT.	INST.	TOTAL
0190	Generator sets, include battery, charger, muffler & transfer switch			
0200	Gas/gasoline operated, 3 phase, 4 wire, 277/480 volt, 7.5 kW	1,250	305	1,555
0240	11.5 kW	1,150	231	1,381
0280	20 kW	785	149	934
0320	35 kW	535	97.50	632.50
0360	80 kW	385	119	504
0400	100 kW	335	115	450
0440	125 kW	550	109	659
0480	185 kW	490	82.50	572.50
0560	Diesel engine with fuel tank, 30 kW	405	114	519
0600	50 kW	440	89.50	529.50
0640	75 kW	340	71.50	411.50
0680	100 kW	305	60.50	365.50
0760	150 kW	276	48	324
0840	200 kW	249	39	288
0880	250 kW	214	32.50	246.50
0960	350 kW	184	27	211
1040	500 kW	201	21	222

E1010 Commercial Equipment

E1010 110	Security/Vault, EACH	COST EACH		
		MAT.	INST.	TOTAL
0100	Bank equipment, drive up window, drawer & mike, no glazing, economy	8,100	1,425	9,525
0110	Deluxe	10,400	2,850	13,250
0120	Night depository, economy	8,575	1,425	10,000
0130	Deluxe	12,300	2,850	15,150
0140	Pneumatic tube systems, 2 station, standard	28,900	5,325	34,225
0150	Teller, automated, 24 hour, single unit	49,200	5,325	54,525
0160	Teller window, bullet proof glazing, 44" x 60"	4,875	980	5,855
0170	Pass through, painted steel, 72" x 40"	4,875	1,775	6,650
0300	Safe, office type, 1 hr. rating, 34" x 20" x 20"	2,475		2,475
0310	4 hr. rating, 62" x 33" x 20"	10,500		10,500
0320	Data storage, 4 hr. rating, 63" x 44" x 16"	15,800		15,800
0330	Jewelers, 63" x 44" x 16"	15,600		15,600
0340	Money, "B" label, 9" x 14" x 14"	620		620
0350	Tool and torch resistive, 24" x 24" x 20"	9,200	243	9,443
0500	Security gates-scissors type, painted steel, single, 6' high, 5-1/2' wide	248	355	603
0510	Double gate, 7-1/2' high, 14' wide	680	715	1,395

E1010 510	Mercantile Equipment, EACH	COST EACH		
		MAT.	INST.	TOTAL
0015	Barber equipment, chair, hydraulic, economy	655	25.50	680.50
0020	Deluxe	4,150	38.50	4,188.50
0100	Checkout counter, single belt	3,750	97	3,847
0110	Double belt, power take-away	5,375	108	5,483
0200	Display cases, freestanding, glass and aluminum, 3'-6" x 3' x 1'-0" deep	1,575	154	1,729
0220	Wall mounted, glass and aluminum, 3' x 4' x 1'-4" deep	2,550	247	2,797
0320	Frozen food, chest type, 12 ft. long	8,600	415	9,015

E1010 610	Laundry/Dry Cleaning, EACH	COST EACH		
		MAT.	INST.	TOTAL
0100	Laundry equipment, dryers, gas fired, residential, 16 lb. capacity	785	250	1,035
0110	Commercial, 30 lb. capacity, single	3,850	250	4,100
0120	Dry cleaners, electric, 20 lb. capacity	37,100	7,225	44,325
0130	30 lb. capacity	59,500	9,625	69,125
0140	Ironers, commercial, 120" with canopy, 8 roll	180,500	20,600	201,100
0150	Institutional, 110", single roll	37,400	3,475	40,875
0160	Washers, residential, 4 cycle	1,075	250	1,325
0170	Commercial, coin operated, deluxe	3,875	250	4,125

E1020 Institutional Equipment

E1020 110 — Ecclesiastical Equipment, EACH

		COST EACH		
		MAT.	INST.	TOTAL
0090	Church equipment, altar, wood, custom, plain	2,850	440	3,290
0100	Granite, custom, deluxe	40,100	6,000	46,100
0110	Baptistry, fiberglass, economy	6,100	1,600	7,700
0120	Bells & carillons, keyboard operation	20,200	8,950	29,150
0130	Confessional, wood, single, economy	3,525	1,025	4,550
0140	Double, deluxe	20,700	3,100	23,800
0150	Steeples, translucent fiberglas, 30" square, 15' high	10,200	1,850	12,050

E1020 130 — Ecclesiastical Equipment, L.F.

		COST PER L.F.		
		MAT.	INST.	TOTAL
0100	Arch. equip., church equip. pews, bench type, hardwood, economy	110	31	141
0110	Deluxe	192	41	233

E1020 210 — Library Equipment, EACH

		COST EACH		
		MAT.	INST.	TOTAL
0110	Library equipment, carrels, metal, economy	295	124	419
0120	Hardwood, deluxe	1,775	154	1,929

E1020 230 — Library Equipment, L.F.

		COST PER L.F.		
		MAT.	INST.	TOTAL
0100	Library equipment, book shelf, metal, single face, 90" high x 10" shelf	129	51.50	180.50
0110	Double face, 90" high x 10" shelf	430	111	541
0120	Charging desk, built-in, with counter, plastic laminate	510	88	598

E1020 310 — Theater and Stage Equipment, EACH

		COST EACH		
		MAT.	INST.	TOTAL
0200	Movie equipment, changeover, economy	565		565
0210	Film transport, incl. platters and autowind, economy	5,975		5,975
0220	Lamphouses, incl. rectifiers, xenon, 1000W	7,900	350	8,250
0230	4000W	12,600	465	13,065
0240	Projector mechanisms, 35 mm, economy	13,600		13,600
0250	Deluxe	18,600		18,600
0260	Sound systems, incl. amplifier, single, economy	4,100	775	4,875
0270	Dual, Dolby/super sound	20,600	1,750	22,350
0280	Projection screens, wall hung, manual operation, 50 S.F., economy	355	124	479
0290	Electric operation, 100 S.F., deluxe	2,925	620	3,545

E1020 320 — Theater and Stage Equipment, S.F.

		COST PER S.F.		
		MAT.	INST.	TOTAL
0090	Movie equipment, projection screens, rigid in wall, acrylic, 1/4" thick	51	6.05	57.05
0100	1/2" thick	59.50	9.05	68.55
0110	Stage equipment, curtains, velour, medium weight	9.45	2.06	11.51
0120	Silica based yarn, fireproof	17.65	24.50	42.15
0130	Stages, portable with steps, folding legs, 8" high	45.50		45.50
0140	Telescoping platforms, aluminum, deluxe	58	32	90

E1020 330 — Theater and Stage Equipment, L.F.

		COST PER L.F.		
		MAT.	INST.	TOTAL
0100	Stage equipment, curtain track, heavy duty	72.50	68.50	141
0110	Lights, border, quartz, colored	200	35	235

E1020 610 — Detention Equipment, EACH

		COST PER EACH		
		MAT.	INST.	TOTAL
0110	Detention equipment, cell front rolling door, 7/8" bars, 5' x 7' high	5,975	1,500	7,475
0120	Cells, prefab., including front, 5' x 7' x 7' deep	10,700	2,000	12,700

E1020 Institutional Equipment

E1020 610	Detention Equipment, EACH	COST PER EACH		
		MAT.	INST.	TOTAL
0130	Doors and frames, 3' x 7', single plate	5,450	745	6,195
0140	Double plate	6,625	745	7,370
0150	Toilet apparatus, incl wash basin	3,750	1,075	4,825
0160	Visitor cubicle, vision panel, no intercom	3,700	1,500	5,200

E1020 710	Laboratory Equipment, EACH	COST PER EACH		
		MAT.	INST.	TOTAL
0110	Laboratory equipment, glassware washer, distilled water, economy	7,350	805	8,155
0120	Deluxe	16,000	1,450	17,450
0140	Radio isotope	19,000		19,000

E1020 720	Laboratory Equipment, S.F.	COST PER S.F.		
		MAT.	INST.	TOTAL
0100	Arch. equip., lab equip., counter tops, acid proof, economy	52	15.05	67.05
0110	Stainless steel	181	15.05	196.05

E1020 730	Laboratory Equipment, L.F.	COST PER L.F.		
		MAT.	INST.	TOTAL
0110	Laboratory equipment, cabinets, wall, open	232	62	294
0120	Base, drawer units	620	68.50	688.50
0130	Fume hoods, not incl. HVAC, economy	615	229	844
0140	Deluxe incl. fixtures	1,150	515	1,665

E1020 810	Medical Equipment, EACH	COST EACH		
		MAT.	INST.	TOTAL
0100	Dental equipment, central suction system, economy	1,175	625	1,800
0110	Compressor-air, deluxe	9,625	1,275	10,900
0120	Chair, hydraulic, economy	2,700	1,275	3,975
0130	Deluxe	4,675	2,575	7,250
0140	Drill console with accessories, economy	2,700	400	3,100
0150	Deluxe	5,700	400	6,100
0160	X-ray unit, portable	2,700	160	2,860
0170	Panoramic unit	18,700	1,075	19,775
0300	Medical equipment, autopsy table, standard	10,800	750	11,550
0310	Deluxe	18,100	1,250	19,350
0320	Incubators, economy	3,675		3,675
0330	Deluxe	15,200		15,200
0700	Station, scrub-surgical, single, economy	5,900	250	6,150
0710	Dietary, medium, with ice	23,300		23,300
0720	Sterilizers, general purpose, single door, 20" x 20" x 28"	13,700		13,700
0730	Floor loading, double door, 28" x 67" x 52"	246,000		246,000
0740	Surgery tables, standard	16,400	1,025	17,425
0750	Deluxe	27,600	1,425	29,025
0770	Tables, standard, with base cabinets, economy	1,175	410	1,585
0780	Deluxe	5,900	620	6,520
0790	X-ray, mobile, economy	20,800		20,800
0800	Stationary, deluxe	303,500		303,500

E1030 Vehicular Equipment

E1030 110	Vehicular Service Equipment, EACH	COST EACH		
		MAT.	INST.	TOTAL
0110	Automotive equipment, compressors, electric, 1-1/2 H.P., std. controls	530	1,175	1,705
0120	5 H.P., dual controls	3,550	1,750	5,300
0130	Hoists, single post, 4 ton capacity, swivel arms	7,300	4,375	11,675
0140	Dual post, 12 ton capacity, adjustable frame	12,500	920	13,420
0150	Lube equipment, 3 reel type, with pumps	11,200	3,500	14,700
0160	Product dispenser, 6 nozzles, w/vapor recovery, not incl. piping, installed	27,900		27,900
0800	Scales, dial type, built in floor, 5 ton capacity, 8' x 6' platform	10,100	3,700	13,800
0810	10 ton capacity, 9' x 7' platform	9,725	5,300	15,025
0820	Truck (including weigh bridge), 20 ton capacity, 24' x 10'	13,900	6,175	20,075

E1030 210	Parking Control Equipment, EACH	COST EACH		
		MAT.	INST.	TOTAL
0110	Parking equipment, automatic gates, 8 ft. arm, one way	3,625	1,275	4,900
0120	Traffic detectors, single treadle	2,200	580	2,780
0130	Booth for attendant, economy	7,800		7,800
0140	Deluxe	28,700		28,700
0150	Ticket printer/dispenser, rate computing	9,175	995	10,170
0160	Key station on pedestal	680	340	1,020

E1030 310	Loading Dock Equipment, EACH	COST EACH		
		MAT.	INST.	TOTAL
0110	Dock bumpers, rubber blocks, 4-1/2" thick, 10" high, 14" long	60	24	84
0120	6" thick, 20" high, 11" long	137	47.50	184.50
0130	Dock boards, H.D., 5' x 5', aluminum, 5000 lb. capacity	1,550		1,550
0140	16,000 lb. capacity	1,400		1,400
0150	Dock levelers, hydraulic, 7' x 8', 10 ton capacity	6,325	1,450	7,775
0160	Dock lifters, platform, 6' x 6', portable, 3000 lb. capacity	10,500		10,500
0170	Dock shelters, truck, scissor arms, economy	2,325	620	2,945
0180	Deluxe	2,875	1,225	4,100

E1090   Other Equipment

E1090 210	Solid Waste Handling Equipment, EACH	COST EACH		
		MAT.	INST.	TOTAL
0110	Waste handling, compactors, single bag, 250 lbs./hr., hand fed	17,600	730	18,330
0120	Heavy duty industrial, 5 C.Y. capacity	38,000	3,500	41,500
0130	Incinerator, electric, 100 lbs./hr., economy	74,500	3,375	77,875
0140	Gas, 2000 lbs./hr., deluxe	436,000	28,500	464,500
0150	Shredder, no baling, 35 tons/hr.	350,000		350,000
0160	Incl. baling, 50 tons/day	698,500		698,500

E1090 350	Food Service Equipment, EACH	COST EACH		
		MAT.	INST.	TOTAL
0110	Kitchen equipment, bake oven, single deck	6,300	168	6,468
0120	Broiler, without oven	4,250	168	4,418
0130	Commercial dish washer, semiautomatic, 50 racks/hr.	7,675	1,025	8,700
0140	Automatic, 275 racks/hr.	28,800	4,375	33,175
0150	Cooler, beverage, reach-in, 6 ft. long	3,525	225	3,750
0160	Food warmer, counter, 1.65 kw	790		790
0170	Fryers, with submerger, single	1,525	193	1,718
0180	Double	2,825	270	3,095
0185	Ice maker, 1000 lb. per day, with bin	5,675	1,350	7,025
0190	Kettles, steam jacketed, 20 gallons	9,575	296	9,871
0200	Range, restaurant type, burners, 2 ovens and 24" griddle	5,875	225	6,100
0210	Range hood, incl. carbon dioxide system, elect. stove	2,175	450	2,625
0220	Gas stove	2,575	450	3,025

E1090 360	Food Service Equipment, S.F.	COST PER S.F.		
		MAT.	INST.	TOTAL
0110	Refrigerators, prefab, walk-in, 7'-6" high, 6' x 6'	122	22.50	144.50
0120	12' x 20'	126	11.25	137.25

E1090 410	Residential Equipment, EACH	COST EACH		
		MAT.	INST.	TOTAL
0110	Arch. equip., appliances, range, cook top, 4 burner, economy	360	116	476
0120	Built in, single oven 30" wide, economy	880	116	996
0130	Standing, single oven-21" wide, economy	520	97	617
0135	Free standing, 30" wide, 1 oven, average	1,025	194	1,219
0140	Double oven-30" wide, deluxe	3,625	97	3,722
0150	Compactor, residential, economy	760	124	884
0160	Deluxe	1,275	206	1,481
0170	Dish washer, built-in, 2 cycles, economy	325	360	685
0180	4 or more cycles, deluxe	1,300	720	2,020
0190	Garbage disposer, sink type, economy	116	144	260
0200	Deluxe	224	144	368
0210	Refrigerator, no frost, 10 to 12 C.F., economy	485	97	582
0220	21 to 29 C.F., deluxe	2,575	325	2,900
0300	Washing machine, automatic	1,475	750	2,225

E1090 610	School Equipment, EACH	COST EACH		
		MAT.	INST.	TOTAL
0110	School equipment, basketball backstops, wall mounted, wood, fixed	1,750	1,075	2,825
0120	Suspended type, electrically operated	7,525	2,100	9,625
0130	Bleachers-telescoping, manual operation, 15 tier, economy (per seat)	125	38.50	163.50
0140	Power operation, 30 tier, deluxe (per seat)	490	68	558
0150	Weight lifting gym, universal, economy	310	970	1,280
0160	Deluxe	16,500	1,950	18,450
0170	Scoreboards, basketball, 1 side, economy	2,725	910	3,635
0180	4 sides, deluxe	17,800	12,600	30,400
0800	Vocational shop equipment, benches, metal	495	247	742
0810	Wood	790	247	1,037

E10 Equipment

E1090 Other Equipment

E1090 610	School Equipment, EACH	COST EACH		
		MAT.	INST.	TOTAL
0820	Dust collector, not incl. ductwork, 6' diam.	5,775	665	6,440
0830	Planer, 13" x 6"	1,300	310	1,610

E1090 620	School Equipment, S.F.	COST PER S.F.		
		MAT.	INST.	TOTAL
0110	School equipment, gym mats, naugahyde cover, 2" thick	5.70		5.70
0120	Wrestling, 1" thick, heavy duty	5.50		5.50

E1090 810	Athletic, Recreational, and Therapeutic Equipment, EACH	COST EACH		
		MAT.	INST.	TOTAL
0050	Bowling alley with gutters	54,500	13,000	67,500
0060	Combo. table and ball rack	1,475		1,475
0070	Bowling alley automatic scorer	11,200		11,200
0110	Sauna, prefabricated, incl. heater and controls, 7' high, 6' x 4'	5,650	940	6,590
0120	10' x 12'	13,400	2,075	15,475
0130	Heaters, wall mounted, to 200 C.F.	855		855
0140	Floor standing, to 1000 C.F., 12500 W	3,975	232	4,207
0610	Shooting range incl. bullet traps, controls, separators, ceilings, economy	47,700	7,900	55,600
0620	Deluxe	67,000	13,300	80,300
0650	Sport court, squash, regulation, in existing building, economy			44,400
0660	Deluxe			49,400
0670	Racketball, regulation, in existing building, economy	48,000	11,700	59,700
0680	Deluxe	52,000	23,400	75,400
0700	Swimming pool equipment, diving stand, stainless steel, 1 meter	11,100	455	11,555
0710	3 meter	18,400	3,100	21,500
0720	Diving boards, 16 ft. long, aluminum	4,750	455	5,205
0730	Fiberglass	3,700	455	4,155
0740	Filter system, sand, incl. pump, 6000 gal./hr.	2,350	830	3,180
0750	Lights, underwater, 12 volt with transformer, 300W	395	695	1,090
0760	Slides, fiberglass with aluminum handrails & ladder, 6' high, straight	4,050	770	4,820
0780	12' high, straight with platform	16,900	1,025	17,925

E1090 820	Athletic, Recreational, and Therapeutic Equipment, S.F.	COST PER S.F.		
		MAT.	INST.	TOTAL
0110	Swimming pools, residential, vinyl liner, metal sides	24	8.35	32.35
0120	Concrete sides	29	15.35	44.35
0130	Gunite shell, plaster finish, 350 S.F.	53.50	31.50	85
0140	800 S.F.	43	18.40	61.40
0150	Motel, gunite shell, plaster finish	66	40	106
0160	Municipal, gunite shell, tile finish, formed gutters	254	69.50	323.50

497

E2010 Fixed Furnishings

E2010 310	Window Treatment, EACH	COST EACH		
		MAT.	INST.	TOTAL
0110	Furnishings, blinds, exterior, aluminum, louvered, 1'-4" wide x 3'-0" long	211	62	273
0120	1'-4" wide x 6'-8" long	375	68.50	443.50
0130	Hemlock, solid raised, 1'-4" wide x 3'-0" long	90	62	152
0140	1'-4" wide x 6'-9" long	158	68.50	226.50
0150	Polystyrene, louvered, 1'-3" wide x 3'-3" long	42	62	104
0160	1'-3" wide x 6'-8" long	76	68.50	144.50
0200	Interior, wood folding panels, louvered, 7" x 20" (per pair)	101	36.50	137.50
0210	18" x 40" (per pair)	152	36.50	188.50

E2010 320	Window Treatment, S.F.	COST PER S.F.		
		MAT.	INST.	TOTAL
0110	Furnishings, blinds-interior, venetian-aluminum, stock, 2" slats	5.35	1.05	6.40
0120	Custom, 1" slats, deluxe	6.50	1.05	7.55
0130	Vertical, PVC or cloth, T&B track, economy	9.50	1.34	10.84
0140	Deluxe	17.60	1.54	19.14
0150	Draperies, unlined, economy	33		33
0160	Lightproof, deluxe	64.50		64.50
0510	Shades, mylar, wood roller, single layer, non-reflective	3.30	.90	4.20
0520	Metal roller, triple layer, heat reflective	8.85	.90	9.75
0530	Vinyl, light weight, 4 ga.	.89	.90	1.79
0540	Heavyweight, 6 ga.	2.73	.90	3.63
0550	Vinyl coated cotton, lightproof decorator shades	6.15	.90	7.05
0560	Woven aluminum, 3/8" thick, light and fireproof	8.55	1.76	10.31

E2010 420	Fixed Floor Grilles and Mats, S.F.	COST PER S.F.		
		MAT.	INST.	TOTAL
0110	Floor mats, recessed, inlaid black rubber, 3/8" thick, solid	28.50	3.13	31.63
0120	Colors, 1/2" thick, perforated	40	3.13	43.13
0130	Link-including nosings, steel-galvanized, 3/8" thick	30	3.13	33.13
0140	Vinyl, in colors	29	3.13	32.13

E2010 510	Fixed Multiple Seating, EACH	COST EACH		
		MAT.	INST.	TOTAL
0110	Seating, painted steel, upholstered, economy	167	35.50	202.50
0120	Deluxe	550	44	594
0400	Seating, lecture hall, pedestal type, economy	292	56	348
0410	Deluxe	565	85	650
0500	Auditorium chair, veneer construction	300	56	356
0510	Fully upholstered, spring seat	279	56	335

For customer support on your Square Foot Costs with RSMeans data, call 800.448.8182.

E2020 Moveable Furnishings

E2020 210	Furnishings/EACH	COST EACH		
		MAT.	INST.	TOTAL
0200	Hospital furniture, beds, manual, economy	940		940
0210	Deluxe	3,100		3,100
0220	All electric, economy	2,075		2,075
0230	Deluxe	4,100		4,100
0240	Patient wall systems, no utilities, economy, per room	1,550		1,550
0250	Deluxe, per room	2,200		2,200
0300	Hotel furnishings, standard room set, economy, per room	2,800		2,800
0310	Deluxe, per room	9,375		9,375
0500	Office furniture, standard employee set, economy, per person	640		640
0510	Deluxe, per person	2,725		2,725
0550	Posts, portable, pedestrian traffic control, economy	158		158
0560	Deluxe	256		256
0700	Restaurant furniture, booth, molded plastic, stub wall and 2 seats, economy	375	310	685
0710	Deluxe	1,600	410	2,010
0720	Upholstered seats, foursome, single-economy	880	124	1,004
0730	Foursome, double-deluxe	2,175	206	2,381

E2020 220	Furniture and Accessories, L.F.	COST PER L.F.		
		MAT.	INST.	TOTAL
0210	Dormitory furniture, desk top (built-in),laminated plastc, 24"deep, economy	53.50	24.50	78
0220	30" deep, deluxe	298	31	329
0230	Dressing unit, built-in, economy	226	103	329
0240	Deluxe	680	154	834
0310	Furnishings, cabinets, hospital, base, laminated plastic	405	124	529
0320	Stainless steel	745	124	869
0330	Countertop, laminated plastic, no backsplash	56.50	31	87.50
0340	Stainless steel	182	31	213
0350	Nurses station, door type, laminated plastic	470	124	594
0360	Stainless steel	900	124	1,024
0710	Restaurant furniture, bars, built-in, back bar	237	124	361
0720	Front bar	325	124	449
0910	Wardrobes & coatracks, standing, steel, single pedestal, 30" x 18" x 63"	152		152
0920	Double face rack, 39" x 26" x 70"	137		137
0930	Wall mounted rack, steel frame & shelves, 12" x 15" x 26"	70	8.80	78.80
0940	12" x 15" x 50"	42.50	4.59	47.09

499

For customer support on your Square Foot Costs with RSMeans data, call 800.448.8182.

F1010 Special Structures

F1010 120	Air-Supported Structures, S.F.	COST PER S.F.		
		MAT.	INST.	TOTAL
0110	Air supported struc., polyester vinyl fabric, 24oz., warehouse, 5000 S.F.	30	.39	30.39
0120	50,000 S.F.	14.05	.31	14.36
0130	Tennis, 7,200 S.F.	26.50	.32	26.82
0140	24,000 S.F.	18.70	.32	19.02
0150	Woven polyethylene, 6 oz., shelter, 3,000 S.F.	17.95	.65	18.60
0160	24,000 S.F.	13.45	.32	13.77
0170	Teflon coated fiberglass, stadium cover, economy	65	.17	65.17
0180	Deluxe	77.50	.23	77.73
0190	Air supported storage tank covers, reinf. vinyl fabric, 12 oz., 400 S.F.	26.50	.55	27.05
0200	18,000 S.F.	9.65	.49	10.14

F1010 210	Pre-Engineered Structures, EACH	COST EACH		
		MAT.	INST.	TOTAL
0600	Radio towers, guyed, 40 lb. section, 50' high, 70 MPH basic wind speed	3,050	1,425	4,475
0610	90 lb. section, 400' high, wind load 70 MPH basic wind speed	42,000	16,100	58,100
0620	Self supporting, 60' high, 70 MPH basic wind speed	4,950	2,825	7,775
0630	190' high, wind load 90 MPH basic wind speed	31,100	11,300	42,400
0700	Shelters, aluminum frame, acrylic glazing, 8' high, 3' x 9'	3,475	1,250	4,725
0710	9' x 12'	8,200	1,950	10,150

F1010 320	Other Special Structures, S.F.	COST PER S.F.		
		MAT.	INST.	TOTAL
0110	Swimming pool enclosure, transluscent, freestanding, economy	56	6.20	62.20
0120	Deluxe	102	17.65	119.65
0510	Tension structures, steel frame, polyester vinyl fabric, 12,000 S.F.	18.40	2.86	21.26
0520	20,800 S.F.	18.10	2.58	20.68

F1010 330	Special Structures, EACH	COST EACH		
		MAT.	INST.	TOTAL
0110	Kiosks, round, 5' diam., 7' high, aluminum wall, illuminated	26,000		26,000
0120	Rectangular, 5' x 9', 1" insulated dbl. wall fiberglass, 7'-6" high	26,500		26,500
0220	Silos, steel prefab, 30,000 gal., painted, economy	24,700	5,650	30,350
0230	Epoxy-lined, deluxe	51,500	11,300	62,800

F1010 340	Special Structures, S.F.	COST PER S.F.		
		MAT.	INST.	TOTAL
0110	Comfort stations, prefab, mobile on steel frame, economy	211		211
0120	Permanent on concrete slab, deluxe	206	49.50	255.50
0210	Domes, bulk storage, wood framing, wood decking, 50' diam.	75.50	1.84	77.34
0220	116' diam.	36.50	2.14	38.64
0230	Steel framing, metal decking, 150' diam.	37.50	12.20	49.70
0240	400' diam.	30.50	9.35	39.85
0250	Geodesic, wood framing, wood panels, 30' diam.	38.50	2.21	40.71
0260	60' diam.	24.50	1.31	25.81
0270	Aluminum framing, acrylic panels, 40' diam.			71
0280	Aluminum panels, 400' diam.			22.50
0310	Garden house, prefab, wood, shell only, 48 S.F.	56	6.20	62.20
0320	200 S.F.	34	25.50	59.50
0410	Greenhouse, shell-stock, residential, lean-to, 8'-6" long x 3'-10" wide	48	36.50	84.50
0420	Freestanding, 8'-6" long x 13'-6" wide	50.50	11.45	61.95
0430	Commercial-truss frame, under 2000 S.F., deluxe			14.75
0440	Over 5,000 S.F., economy			13.55
0450	Institutional-rigid frame, under 500 S.F., deluxe			33
0460	Over 2,000 S.F., economy			12.35
0510	Hangar, prefab, galv. roof and walls, bottom rolling doors, economy	15.25	4.85	20.10
0520	Electric bifolding doors, deluxe	14	6.30	20.30

F1020 Integrated Construction

F1020 110	Integrated Construction, EACH	COST EACH		
		MAT.	INST.	TOTAL
0110	Integrated ceilings, radiant electric, 2' x 4' panel, manila finish	305	28	333
0120	ABS plastic finish	122	53.50	175.50

F1020 120	Integrated Construction, S.F.	COST PER S.F.		
		MAT.	INST.	TOTAL
0110	Integrated ceilings, Luminaire, suspended, 5' x 5' modules, 50% lighted	5.55	14.80	20.35
0120	100% lighted	7.20	26.50	33.70
0130	Dimensionaire, 2' x 4' module tile system, no air bar	2.68	2.47	5.15
0140	With air bar, deluxe	3.93	2.47	6.40
0220	Pedestal access floor pkg., inc. stl. pnls, peds. & stringers, w/vinyl cov.	29	3.29	32.29
0230	With high pressure laminate covering	27	3.29	30.29
0240	With carpet covering	28.50	3.29	31.79
0250	Aluminum panels, no stringers, no covering	36.50	2.47	38.97

F1020 250	Special Purpose Room, EACH	COST EACH		
		MAT.	INST.	TOTAL
0110	Portable booth, acoustical, 27 db 1000 hz., 15 S.F. floor	4,225		4,225
0120	55 S.F. flr.	8,700		8,700

F1020 260	Special Purpose Room, S.F.	COST PER S.F.		
		MAT.	INST.	TOTAL
0110	Anechoic chambers, 7' high, 100 cps cutoff, 25 S.F.			2,975
0120	200 cps cutoff, 100 S.F.			1,550
0130	Audiometric rooms, under 500 S.F.	66	25	91
0140	Over 500 S.F.	64.50	20.50	85
0300	Darkrooms, shell, not including door, 240 S.F., 8' high	31	10.30	41.30
0310	64 S.F., 12' high	79.50	19.30	98.80
0510	Music practice room, modular, perforated steel, under 500 S.F.	42.50	17.65	60.15
0520	Over 500 S.F.	36	15.45	51.45

F1020 330	Special Construction, L.F.	COST PER L.F.		
		MAT.	INST.	TOTAL
0110	Spec. const., air curtains, shipping & receiving, 8'high x 5'wide, economy	815	146	961
0120	20' high x 8' wide, heated, deluxe	3,700	365	4,065
0130	Customer entrance, 10' high x 5' wide, economy	1,175	146	1,321
0140	12' high x 4' wide, heated, deluxe	2,525	183	2,708

503

For customer support on your Square Foot Costs with RSMeans data, call 800.448.8182.

F1030 Special Construction Systems

F1030 120	Sound, Vibration, and Seismic Construction, S.F.	COST PER S.F.		
		MAT.	INST.	TOTAL
0020	Special construction, acoustical, enclosure, 4" thick, 8 psf panels	34.50	25.50	60
0030	Reverb chamber, 4" thick, parallel walls	49	31	80
0110	Sound absorbing panels, 2'-6" x 8', painted metal	13	8.60	21.60
0120	Vinyl faced	10.15	7.70	17.85
0130	Flexible transparent curtain, clear	7.85	10.20	18.05
0140	With absorbing foam, 75% coverage	11	10.20	21.20
0150	Strip entrance, 2/3 overlap	8.40	16.20	24.60
0160	Full overlap	10.60	19.05	29.65
0200	Audio masking system, plenum mounted, over 10,000 S.F.	.77	.32	1.09
0210	Ceiling mounted, under 5,000 S.F.	1.33	.58	1.91

F1030 210	Radiation Protection, EACH	COST EACH		
		MAT.	INST.	TOTAL
0110	Shielding, lead x-ray protection, radiography room, 1/16" lead, economy	12,300	4,650	16,950
0120	Deluxe	14,700	7,750	22,450
0210	Deep therapy x-ray, 1/4" lead, economy	34,200	14,500	48,700
0220	Deluxe	42,200	19,400	61,600

F1030 220	Radiation Protection, S.F.	COST PER S.F.		
		MAT.	INST.	TOTAL
0110	Shielding, lead, gypsum board, 5/8" thick, 1/16" lead	13.30	7.35	20.65
0120	1/8" lead	27	8.40	35.40
0130	Lath, 1/16" thick	12	8.60	20.60
0140	1/8" thick	24.50	9.70	34.20
0150	Radio frequency, galvanized steel, prefab type, economy	5.80	3.29	9.09
0160	Radio frequency, door, copper/wood laminate, 4" x 7'	9.95	8.80	18.75

F1030 910	Other Special Construction Systems, EACH	COST EACH		
		MAT.	INST.	TOTAL
0110	Disappearing stairways, folding, pine, 8'-6" ceiling	195	154	349
0220	Automatic electric, wood, 8' to 9' ceiling	10,300	1,225	11,525
0300	Fireplace prefabricated, freestanding or wall hung, painted	1,775	475	2,250
0310	Stainless steel	3,575	685	4,260
0320	Woodburning stoves, cast iron, economy, less than 1500 S.F. htg. area	1,475	950	2,425
0330	Greater than 2000 S.F. htg. area	3,075	1,550	4,625

504

For customer support on your Square Foot Costs with RSMeans data, call 800.448.8182.

F10 Special Construction

F1040 Special Facilities

F1040 210	Ice Rinks, EACH	COAT EACH		
		MAT.	INST.	TOTAL
0100	Ice skating rink, 85' x 200', 55° system, 5 mos., 100 ton			664,000
0110	90° system, 12 mos., 135 ton			751,000
0120	Dash boards, acrylic screens, polyethylene coated plywood	156,000	41,600	197,600
0130	Fiberglass and aluminum construction	179,000	41,600	220,600

F1040 510	Liquid & Gas Storage Tanks, EACH	COST EACH		
		MAT.	INST.	TOTAL
0100	Tanks, steel, ground level, 100,000 gal.			231,000
0110	10,000,000 gal.			5,255,000
0120	Elevated water, 50,000 gal.		189,500	440,000
0130	1,000,000 gal.		884,500	1,951,500
0150	Cypress wood, ground level, 3,000 gal.	10,500	12,300	22,800
0160	Redwood, ground level, 45,000 gal.	82,000	33,400	115,400

For customer support on your Square Foot Costs with RSMeans data, call 800.448.8182.

F10 Special Construction

F1040 Special Facilities

F1040 910	Special Construction, EACH	COST EACH		
		MAT.	INST.	TOTAL
0110	Special construction, bowling alley incl. pinsetter, scorer etc., economy	44,600	12,400	57,000
0120	Deluxe	62,500	13,700	76,200
0130	For automatic scorer, economy, add	6,750		6,750
0140	Deluxe, add	11,200		11,200
0300	Control tower, modular, 12' x 10', incl. instrumentation, economy			882,000
0310	Deluxe			1,386,000
0400	Garage costs, residential, prefab, wood, single car economy	6,850	1,225	8,075
0410	Two car deluxe	17,200	2,475	19,675
0500	Hangars, prefab, galv. steel, bottom rolling doors, economy (per plane)	18,800	5,375	24,175
0510	Electrical bi-folding doors, deluxe (per plane)	14,000	7,400	21,400

For customer support on your Square Foot Costs with RSMeans data, call 800.448.8182.

G1030 Site Earthwork

G1030 805	Trenching Common Earth	COST PER L.F.		
		MAT.	INST.	TOTAL
1310	Trenching, common earth, no slope, 2' wide, 2' deep, 3/8 C.Y. bucket		2.66	2.66
1330	4' deep, 3/8 C.Y. bucket		5.25	5.25
1360	10' deep, 1 C.Y. bucket		11.90	11.90
1400	4' wide, 2' deep, 3/8 C.Y. bucket		5.46	5.46
1420	4' deep, 1/2 C.Y. bucket		9.23	9.23
1450	10' deep, 1 C.Y. bucket		25.65	25.65
1480	18' deep, 2-1/2 C.Y. bucket		40.20	40.20
1520	6' wide, 6' deep, 5/8 C.Y. bucket w/trench box		24.90	24.90
1540	10' deep, 1 C.Y. bucket		33.25	33.25
1570	20' deep, 3-1/2 C.Y. bucket		56.50	56.50
1640	8' wide, 12' deep, 1-1/2 C.Y. bucket w/trench box		46.15	46.15
1680	24' deep, 3-1/2 C.Y. bucket		88.50	88.50
1730	10' wide, 20' deep, 3-1/2 C.Y. bucket w/trench box		79.50	79.50
1740	24' deep, 3-1/2 C.Y. bucket		103	103
3500	1 to 1 slope, 2' wide, 2' deep, 3/8 C.Y. bucket		5.26	5.26
3540	4' deep, 3/8 C.Y. bucket		15.62	15.62
3600	10' deep, 1 C.Y. bucket		71.50	71.50
3800	4' wide, 2' deep, 3/8 C.Y. bucket		8.06	8.06
3840	4' deep, 1/2 C.Y. bucket		18.39	18.39
3900	10' deep, 1 C.Y. bucket		90.50	90.50
4030	6' wide, 6' deep, 5/8 C.Y. bucket w/trench box		49.30	49.30
4050	10' deep, 1 C.Y. bucket		79.50	79.50
4080	20' deep, 3-1/2 C.Y. bucket		240	240
4500	8' wide, 12' deep, 1-1/2 C.Y. bucket w/trench box		118.50	118.50
4650	24' deep, 3-1/2 C.Y. bucket		362	362
4800	10' wide, 20' deep, 3-1/2 C.Y. bucket w/trench box		247	247
4850	24' deep, 3-1/2 C.Y. bucket		384	384

G1030 806	Trenching Loam & Sandy Clay	COST PER L.F.		
		MAT.	INST.	TOTAL
1310	Trenching, loam & sandy clay, no slope, 2' wide, 2' deep, 3/8 C.Y. bucket		2.61	2.61

G1030 807	Trenching Sand & Gravel	COST PER L.F.		
		MAT.	INST.	TOTAL
1310	Trenching, sand & gravel, no slope, 2' wide, 2' deep, 3/8 C.Y. bucket		2.45	2.45
1320	3' deep, 3/8 C.Y. bucket		4.07	4.07
1330	4' deep, 3/8 C.Y. bucket		4.86	4.86
1340	6' deep, 3/8 C.Y. bucket		6.20	6.20
1350	8' deep, 1/2 C.Y. bucket		8.20	8.20
1360	10' deep, 1 C.Y. bucket		8.75	8.75
1400	4' wide, 2' deep, 3/8 C.Y. bucket		5	5
1410	3' deep, 3/8 C.Y. bucket		7.40	7.40
1420	4' deep, 1/2 C.Y. bucket		8.55	8.55
1430	6' deep, 1/2 C.Y. bucket		13.45	13.45
1440	8' deep, 1/2 C.Y. bucket		19.35	19.35
1450	10' deep, 1 C.Y. bucket		19.30	19.30
1460	12' deep, 1 C.Y. bucket		24	24
1470	15' deep, 1-1/2 C.Y. bucket		28	28
1480	18' deep, 2-1/2 C.Y. bucket		31.50	31.50
1520	6' wide, 6' deep, 5/8 C.Y. bucket w/trench box		23	23
1530	8' deep, 3/4 C.Y. bucket		30	30
1540	10' deep, 1 C.Y. bucket		30	30
1550	12' deep, 1-1/2 C.Y. bucket		33	33
1560	16' deep, 2 C.Y. bucket		43	43
1570	20' deep, 3-1/2 C.Y. bucket		51.50	51.50
1580	24' deep, 3-1/2 C.Y. bucket		63.50	63.50

G1030 Site Earthwork

G1030 807	Trenching Sand & Gravel	COST PER L.F.		
		MAT.	INST.	TOTAL
1640	8' wide, 12' deep, 1-1/2 C.Y. bucket w/trench box		43.50	43.50
1650	15' deep, 1-1/2 C.Y. bucket		54.50	54.50
1660	18' deep, 2-1/2 C.Y. bucket		62.50	62.50
1680	24' deep, 3-1/2 C.Y. bucket		83	83
1730	10' wide, 20' deep, 3-1/2 C.Y. bucket w/trench box		83	83
1740	24' deep, 3-1/2 C.Y. bucket		104	104
1780	12' wide, 20' deep, 3-1/2 C.Y. bucket w/trench box		99.50	99.50
1790	25' deep, 3-1/2 C.Y. bucket		129	129
1800	1/2:1 slope, 2' wide, 2' deep, 3/8 C.Y. bucket		3.66	3.66
1810	3' deep, 3/8 C.Y. bucket		6.35	6.35
1820	4' deep, 3/8 C.Y. bucket		9.65	9.65
1840	6' deep, 3/8 C.Y. bucket		15.40	15.40
1860	8' deep, 1/2 C.Y. bucket		24.50	24.50
1880	10' deep, 1 C.Y. bucket		30.50	30.50
2300	4' wide, 2' deep, 3/8 C.Y. bucket		6.20	6.20
2310	3' deep, 3/8 C.Y. bucket		10.15	10.15
2320	4' deep, 1/2 C.Y. bucket		12.80	12.80
2340	6' deep, 1/2 C.Y. bucket		23.50	23.50
2360	8' deep, 1/2 C.Y. bucket		39	39
2380	10' deep, 1 C.Y. bucket		43.50	43.50
2400	12' deep, 1 C.Y. bucket		73	73
2430	15' deep, 1-1/2 C.Y. bucket		81.50	81.50
2460	18' deep, 2-1/2 C.Y. bucket		128	128
2840	6' wide, 6' deep, 5/8 C.Y. bucket w/trench box		34.50	34.50
2860	8' deep, 3/4 C.Y. bucket		50.50	50.50
2880	10' deep, 1 C.Y. bucket		55	55
2900	12' deep, 1-1/2 C.Y. bucket		67	67
2940	16' deep, 2 C.Y. bucket		102	102
2980	20' deep, 3-1/2 C.Y. bucket		139	139
3020	24' deep, 3-1/2 C.Y. bucket		193	193
3100	8' wide, 12' deep, 1-1/4 C.Y. bucket w/trench box		78	78
3120	15' deep, 1-1/2 C.Y. bucket		108	108
3140	18' deep, 2-1/2 C.Y. bucket		137	137
3180	24' deep, 3-1/2 C.Y. bucket		213	213
3270	10' wide, 20' deep, 3-1/2 C.Y. bucket w/trench box		171	171
3280	24' deep, 3-1/2 C.Y. bucket		233	233
3370	12' wide, 20' deep, 3-1/2 C.Y. bucket w/trench box		191	191
3380	25' deep, 3-1/2 C.Y. bucket		269	269
3500	1:1 slope, 2' wide, 2' deep, 3/8 C.Y. bucket		6.50	6.50
3520	3' deep, 3/8 C.Y. bucket		9.05	9.05
3540	4' deep, 3/8 C.Y. bucket		14.50	14.50
3560	6' deep, 3/8 C.Y. bucket		15.40	15.40
3580	8' deep, 1/2 C.Y. bucket		41	41
3600	10' deep, 1 C.Y. bucket		52.50	52.50
3800	4' wide, 2' deep, 3/8 C.Y. bucket		7.40	7.40
3820	3' deep, 3/8 C.Y. bucket		12.85	12.85
3840	4' deep, 1/2 C.Y. bucket		17.05	17.05
3860	6' deep, 1/2 C.Y. bucket		34	34
3880	8' deep, 1/2 C.Y. bucket		58	58
3900	10' deep, 1 C.Y. bucket		68.50	68.50
3920	12' deep, 1 C.Y. bucket		97.50	97.50
3940	15' deep, 1-1/2 C.Y. bucket		135	135
3960	18' deep, 2-1/2 C.Y. bucket		175	175
4030	6' wide, 6' deep, 5/8 C.Y. bucket w/trench box		46	46
4040	8' deep, 3/4 C.Y. bucket		71	71
4050	10' deep, 1 C.Y. bucket		80.50	80.50
4060	12' deep, 1-1/2 C.Y. bucket		101	101
4070	16' deep, 2 C.Y. bucket		161	161

G1030 Site Earthwork

G1030 807	Trenching Sand & Gravel	COST PER L.F.		
		MAT.	INST.	TOTAL
4080	20' deep, 3-1/2 C.Y. bucket		227	227
4090	24' deep, 3-1/2 C.Y. bucket		320	320
4500	8' wide, 12' deep, 1-1/2 C.Y. bucket w/trench box		112	112
4550	15' deep, 1-1/2 C.Y. bucket		161	161
4600	18' deep, 2-1/2 C.Y. bucket		211	211
4650	24' deep, 3-1/2 C.Y. bucket		340	340
4800	10' wide, 20' deep, 3-1/2 C.Y. bucket w/trench box		258	258
4850	24' deep, 3-1/2 C.Y. bucket		360	360
4950	12' wide, 20' deep, 3-1/2 C.Y. bucket w/trench box		274	274
4980	25' deep, 3-1/2 C.Y. bucket		410	410

G1030 815	Pipe Bedding	COST PER L.F.		
		MAT.	INST.	TOTAL
1440	Pipe bedding, side slope 0 to 1, 1' wide, pipe size 6" diameter	1.71	.87	2.58
1460	2' wide, pipe size 8" diameter	3.70	1.90	5.60
1500	Pipe size 12" diameter	3.86	1.97	5.83
1600	4' wide, pipe size 20" diameter	9.25	4.74	13.99
1660	Pipe size 30" diameter	9.70	4.98	14.68
1680	6' wide, pipe size 32" diameter	16.60	8.50	25.10
1740	8' wide, pipe size 60" diameter	33	16.85	49.85
1780	12' wide, pipe size 84" diameter	60.50	31	91.50

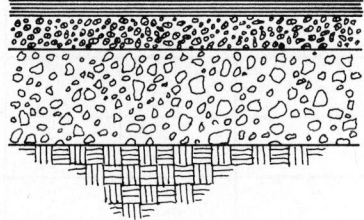

The Bituminous Roadway Systems are listed for pavement thicknesses between 3-l/2″ and 7″ and crushed stone bases from 3″ to 22″ in depth. Systems costs are expressed per linear foot for varying widths of two and multi-lane roads. Earth moving is not included. Granite curbs and line painting are added as required system components.

G2010 232	Bituminous Roadways Crushed Stone	COST PER L.F.		
		MAT.	INST.	TOTAL
1050	Bitum. roadway, two lanes, 3-1/2″ th. pvmt., 3″ th. crushed stone, 24′ wide	79.50	54	133.50
1100	28′ wide	86.50	54.50	141
1150	32′ wide	94	63	157
1200	4″ thick crushed stone, 24′ wide	82	52	134
1210	28′ wide	89.50	55.50	145
1220	32′ wide	97.50	59	156.50
1222	36′ wide	105	63	168
1224	40′ wide	113	66.50	179.50
1230	6″ thick crushed stone, 24′ wide	87	53.50	140.50
1240	28′ wide	96	58	154
1250	32′ wide	105	61.50	166.50
1252	36′ wide	113	65.50	178.50
1254	40′ wide	122	70	192
1256	8″ thick crushed stone, 24′ wide	92.50	55	147.50
1258	28′ wide	102	59.50	161.50
1260	32′ wide	112	64	176
1262	36′ wide	121	68	189
1264	40′ wide	131	72.50	203.50
1300	9″ thick crushed stone, 24′ wide	95.50	56.50	152
1350	28′ wide	106	61.50	167.50
1400	32′ wide	116	65.50	181.50
1410	10″ thick crushed stone, 24′ wide	98	57	155
1412	28′ wide	108	61.50	169.50
1414	32′ wide	119	66	185
1416	36′ wide	129	70.50	199.50
1418	40′ wide	140	75	215
1420	12″ thick crushed stone, 24′ wide	103	58.50	161.50
1422	28′ wide	114	63	177
1424	32′ wide	126	68.50	194.50
1426	36′ wide	137	73	210
1428	40′ wide	148	78	226
1430	15″ thick crushed stone, 24′ wide	111	61	172
1432	28′ wide	124	66.50	190.50
1434	32′ wide	137	72	209
1436	36′ wide	149	76.50	225.50
1438	40′ wide	162	82	244
1440	18″ thick crushed stone, 24′ wide	119	63.50	182.50
1442	28′ wide	133	69	202
1444	32′ wide	147	74.50	221.50
1446	36′ wide	161	80.50	241.50
1448	40′ wide	175	86	261
1550	4″ thick pavement., 4″ thick crushed stone, 24′ wide	87	52.50	139.50

511

G2010 Roadways

G2010 232	Bituminous Roadways Crushed Stone	COST PER L.F.		
		MAT.	INST.	TOTAL
1600	28' wide	96	56.50	152.50
1650	32' wide	105	60.50	165.50
1652	36' wide	113	64	177
1654	40' wide	122	68	190
1700	6" thick crushed stone , 24' wide	92.50	54.50	147
1710	28' wide	102	58.50	160.50
1720	32' wide	111	62.50	173.50
1722	36' wide	121	66.50	187.50
1724	40' wide	131	70.50	201.50
1726	8" thick crushed stone , 24' wide	97.50	56	153.50
1728	28' wide	108	60	168
1730	32' wide	119	64.50	183.50
1732	36' wide	129	69	198
1734	40' wide	139	73	212
1800	10" thick crushed stone, 24' wide	103	58	161
1850	28' wide	115	61	176
1900	32' wide	126	67	193
1902	36' wide	137	71.50	208.50
1904	40' wide	142	74.50	216.50

G20 Site Improvements

G2020 Parking Lots

G2020 210	Parking Lots Gravel Base	COST PER CAR		
		MAT.	INST.	TOTAL
1500	Parking lot, 90° angle parking, 3" bituminous paving, 6" gravel base	780	445	1,225
1540	10" gravel base	875	515	1,390
1560	4" bituminous paving, 6" gravel base	990	475	1,465
1600	10" gravel base	1,075	540	1,615
1620	6" bituminous paving, 6" gravel base	1,325	505	1,830
1660	10" gravel base	1,425	570	1,995
1800	60° angle parking, 3" bituminous paving, 6" gravel base	780	445	1,225
1840	10" gravel base	875	515	1,390
1860	4" bituminous paving, 6" gravel base	990	475	1,465
1900	10" gravel base	1,075	540	1,615
1920	6" bituminous paving, 6" gravel base	1,325	505	1,830
1960	10" gravel base	1,425	570	1,995
2200	45° angle parking, 3" bituminous paving, 6" gravel base	795	455	1,250
2240	10" gravel base	900	525	1,425
2260	4" bituminous paving, 6" gravel base	1,000	480	1,480
2300	10" gravel base	1,125	555	1,680
2320	6" bituminous paving, 6" gravel base	1,350	520	1,870
2360	10" gravel base	1,450	585	2,035

G20 Site Improvements

G2040 Site Development

G2040 810	Flagpoles	COST EACH		
		MAT.	INST.	TOTAL
0110	Flagpoles, on grade, aluminum, tapered, 20' high	1,200	705	1,905
0120	70' high	9,850	1,775	11,625
0130	Fiberglass, tapered, 23' high	745	705	1,450
0140	59' high	5,725	1,575	7,300
0150	Concrete, internal halyard, 20' high	1,525	565	2,090
0160	100' high	19,900	1,400	21,300

G2040 950	Other Site Development, EACH	COST EACH		
		MAT.	INST.	TOTAL
0110	Grandstands, permanent, closed deck, steel, economy (per seat)			29.50
0120	Deluxe (per seat)			33.50
0130	Composite design, economy (per seat)			41
0140	Deluxe (per seat)			100

For customer support on your Square Foot Costs with RSMeans data, call 800.448.8182.

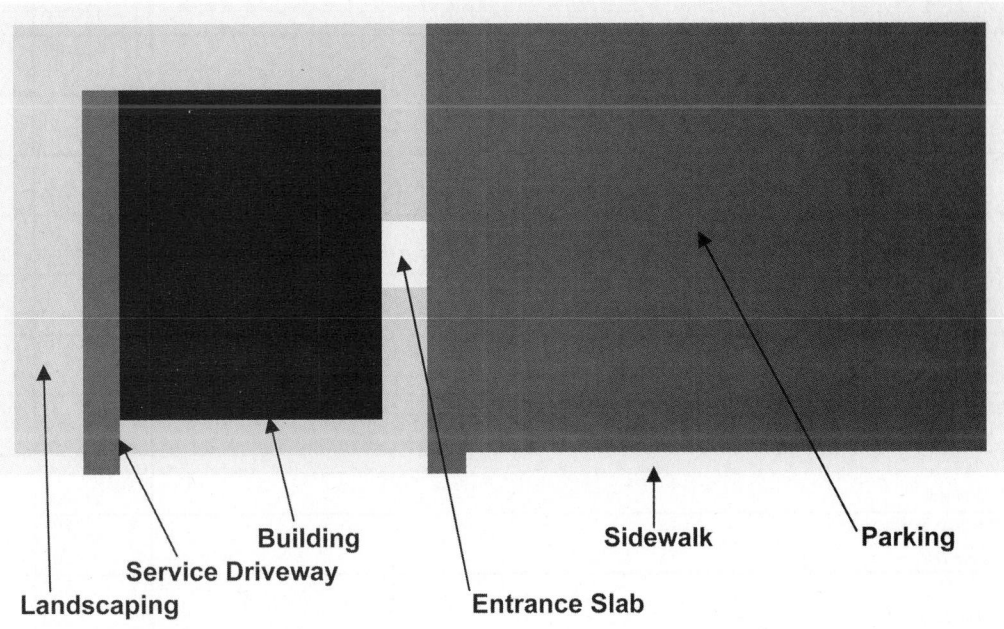

Building
Service Driveway
Landscaping
Sidewalk
Parking
Entrance Slab

Hidden Features
Utilities and Lighting
(Features do not represent actual sizes)

G2040 990	Site Development Components for Buildings	COST EACH		
		MAT.	INST.	TOTAL
0970	Assume minimal and balanced cut & fill, no rock, no demolition, no haz mat.			
0975	Lines can be adjusted linearly +/- 20% within same use & number of floors.			
0980	60,000 S.F. 2-Story Office Bldg on 3.3 Acres w/20% Green Space			
1000	Site Preparation	410	48,000	48,410
1002	Utilities	103,000	108,500	211,500
1004	Pavement	308,000	193,500	501,500
1006	Stormwater Management	270,500	62,000	332,500
1008	Sidewalks	15,800	19,900	35,700
1010	Exterior lighting	54,000	40,900	94,900
1014	Landscaping	77,000	53,000	130,000
1028	80,000 S.F. 4-Story Office Bldg on 3.5 Acres w/20% Green Space			
1030	Site preparation	420	51,500	51,920
1032	Utilities	95,000	99,000	194,000
1034	Pavement	369,000	231,500	600,500
1036	Stormwater Managmement	285,000	65,000	350,000
1038	Sidewalks	12,900	16,300	29,200
1040	Lighting	54,000	43,100	97,100
1042	Landscaping	79,000	55,000	134,000
1058	16,000 S.F. 1-Story Medical Office Bldg on 1.7 Acres w/25% Green Space			
1060	Site preparation	295	37,000	37,295
1062	Utilities	73,000	79,500	152,500
1064	Pavement	143,000	91,500	234,500
1066	Stormwater Management	132,000	30,200	162,200
1068	Sidewalks	11,500	14,600	26,100
1070	Lighting	33,100	23,700	56,800
1072	Landscaping	52,000	32,600	84,600
1098	36,000 S.F. 3-Story Apartment Bldg on 1.6 Acres w/25% Green Space			
1100	Site preparation	285	36,400	36,685

515

G2040 Site Development

G2040 990	Site Development Components for Buildings	COST EACH		
		MAT.	INST.	TOTAL
1102	Utilities	68,000	72,500	140,500
1104	Pavement	145,000	92,000	237,000
1106	Stormwater Management	123,000	28,200	151,200
1108	Sidewalks	9,975	12,600	22,575
1110	Lighting	30,900	22,500	53,400
1112	Landscaping	50,000	31,000	81,000
1128	130,000 S.F. 3-Story Hospital Bldg on 5.9 Acres w/17% Green Space			
1130	Site preparation	550	67,500	68,050
1132	Utilities	137,500	142,500	280,000
1134	Pavement	628,000	392,500	1,020,500
1136	Stormwater Management	506,500	116,000	622,500
1138	Sidewalks	18,900	23,900	42,800
1140	Lighting	88,500	72,500	161,000
1142	Landscaping	110,500	84,000	194,500
1158	60,000 S.F. 1-Story Light Manufacturing Bldg on 3.9 Acres w/18% Green Space			
1160	Site preparation	450	51,000	51,450
1162	Utilities	123,500	135,500	259,000
1164	Pavement	303,000	168,000	471,000
1166	Stormwater Management	333,500	76,500	410,000
1168	Sidewalks	22,300	28,200	50,500
1170	Lighting	65,000	46,700	111,700
1172	Landscaping	83,500	59,500	143,000
1198	8,000 S.F. 1-Story Restaurant Bldg on 1.4 Acres w/26% Green Space			
1200	Site preparation	269	35,600	35,869
1202	Utilities	61,000	63,500	124,500
1204	Pavement	137,000	88,000	225,000
1206	Stormwater Management	108,500	24,800	133,300
1208	Sidewalks	8,125	10,300	18,425
1210	Lighting	27,600	20,800	48,400
1212	Landscaping	46,900	28,500	75,400
1228	20,000 S.F. 1-Story Retail Store Bldg on 2.1 Acres w/23% Green Space			
1230	Site preparation	330	42,700	43,030
1232	Utilities	81,500	88,500	170,000
1234	Pavement	186,000	118,000	304,000
1236	Stormwater Management	169,000	38,700	207,700
1238	Sidewalks	12,900	16,300	29,200
1240	Lighting	38,000	27,400	65,400
1242	Landscaping	59,500	38,500	98,000
1258	60,000 S.F. 1-Story Warehouse on 3.1 Acres w/19% Green Space			
1260	Site preparation	400	47,200	47,600
1262	Utilities	111,500	126,500	238,000
1264	Pavement	188,000	95,000	283,000
1266	Stormwater Management	259,500	59,500	319,000
1268	Sidewalks	22,300	28,200	50,500
1270	Lighting	54,500	36,300	90,800
1272	Landscaping	72,000	49,100	121,100

G3030 Storm Sewer

G3030 210	Manholes & Catch Basins	COST PER EACH		
		MAT.	INST.	TOTAL
1920	Manhole/catch basin, brick, 4' I.D. riser, 4' deep	1,400	1,900	3,300
1980	10' deep	3,000	4,475	7,475
3200	Block, 4' I.D. riser, 4' deep	1,200	1,525	2,725
3260	10' deep	2,525	3,700	6,225
4620	Concrete, cast-in-place, 4' I.D. riser, 4' deep	1,350	2,625	3,975
4680	10' deep	3,150	6,250	9,400
5820	Concrete, precast, 4' I.D. riser, 4' deep	1,675	1,375	3,050
5880	10' deep	3,250	3,200	6,450
6200	6' I.D. riser, 4' deep	3,725	2,025	5,750
6260	10' deep	7,175	4,750	11,925

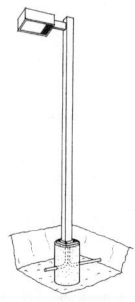

Light Pole

G4020 210	Light Pole (Installed)	COST EACH		
		MAT.	INST.	TOTAL
0200	Light pole, aluminum, 20' high, 1 arm bracket	1,400	1,250	2,650
0240	2 arm brackets	1,550	1,250	2,800
0280	3 arm brackets	1,700	1,300	3,000
0320	4 arm brackets	1,850	1,300	3,150
0360	30' high, 1 arm bracket	2,450	1,575	4,025
0400	2 arm brackets	2,600	1,575	4,175
0440	3 arm brackets	2,750	1,625	4,375
0480	4 arm brackets	2,900	1,625	4,525
0680	40' high, 1 arm bracket	3,000	2,100	5,100
0720	2 arm brackets	3,150	2,100	5,250
0760	3 arm brackets	3,275	2,150	5,425
0800	4 arm brackets	3,425	2,175	5,600
0840	Steel, 20' high, 1 arm bracket	1,500	1,325	2,825
0880	2 arm brackets	1,600	1,325	2,925
0920	3 arm brackets	1,525	1,375	2,900
0960	4 arm brackets	1,625	1,375	3,000
1000	30' high, 1 arm bracket	1,950	1,700	3,650
1040	2 arm brackets	2,050	1,700	3,750
1080	3 arm brackets	1,975	1,725	3,700
1120	4 arm brackets	2,075	1,725	3,800
1320	40' high, 1 arm bracket	2,550	2,275	4,825
1360	2 arm brackets	2,650	2,275	4,925
1400	3 arm brackets	2,575	2,300	4,875
1440	4 arm brackets	2,675	2,300	4,975

Reference Section

All the reference information is in one section making it easy to find what you need to know ... and easy to use the data set on a daily basis. This section is visually identified by a vertical black bar on the edge of pages.

In this Reference Section, we've included General Conditions; Location Factors for adjusting costs to the region you are in; Historical Cost Indexes for cost comparisons over time; a Glossary; an explanation of all Abbreviations used in the data set; and sample Estimating Forms.

Table of Contents

General Conditions, Overhead & Profit

The total building costs in the Commercial/Industrial/Institutional section include a 10% allowance for general conditions and a 15% allowance for the general contractor's overhead and profit and contingencies.

General contractor overhead includes indirect costs such as permits, Workers' Compensation, insurances, supervision and bonding fees. Overhead will vary with the size of project, the contractor's operating procedures and location. Profits will vary with economic activity and local conditions.

Contingencies provide for unforeseen construction difficulties which include material shortages and weather. In all situations, the appraiser should give consideration to possible adjustment of the 25% factor used in developing the Commercial/Industrial/Institutional models.

Architectural Fees

Tabulated below are typical percentage fees by project size, for good professional architectural service. Fees may vary from those listed depending upon degree of design difficulty and economic conditions in any particular area.

Rates can be interpolated horizontally and vertically. Various portions of the same project requiring different rates should be adjusted proportionately. For alterations, add 50% to the fee for the first $500,000 of project cost and add 25% to the fee for project cost over $500,000.

Architectural fees tabulated below include Engineering fees.

Insurance Exclusions

Many insurance companies exclude from coverage such items as architect's fees, excavation, foundations below grade, underground piping and site preparation. Since exclusions vary among insurance companies, it is recommended that for greatest accuracy each exclusion be priced separately using the unit-in-place section.

As a rule of thumb, exclusions can be calculated at 9% of total building cost plus the appropriate allowance for architect's fees.

Building Types	Total Project Size in Thousands of Dollars						
	100	250	500	1,000	5,000	10,000	50,000
Factories, garages, warehouses, repetitive housing	9.0%	8.0%	7.0%	6.2%	5.3%	4.9%	4.5%
Apartments, banks, schools, libraries, offices, municipal buildings	12.2	12.3	9.2	8.0	7.0	6.6	6.2
Churches, hospitals, homes, laboratories, museums, research	15.0	13.6	12.7	11.9	9.5	8.8	8.0
Memorials, monumental work, decorative furnishings	—	16.0	14.5	13.1	10.0	9.0	8.3

520

For customer support on your Square Foot Costs with RSMeans data, call 800.448.8182.

Location Factors - Residential/Commercial

Costs shown in *Square Foot Costs with RSMeans data* are based on national averages for materials and installation. To adjust these costs to a specific location, simply multiply the base cost by the factor for that city. The data is arranged alphabetically by state and postal zip code numbers. For a city not listed, use the factor for a nearby city with similar economic characteristics.

STATE/ZIP	CITY	Residential	Commercial
ALABAMA			
350-352	Birmingham	.83	.85
354	Tuscaloosa	.84	.86
355	Jasper	.83	.86
356	Decatur	.83	.85
357-358	Huntsville	.83	.85
359	Gadsden	.82	.85
360-361	Montgomery	.82	.85
362	Anniston	.81	.83
363	Dothan	.84	.85
364	Evergreen	.81	.85
365-366	Mobile	.83	.84
367	Selma	.82	.85
368	Phenix City	.83	.85
369	Butler	.81	.85
ALASKA			
995-996	Anchorage	1.22	1.17
997	Fairbanks	1.25	1.18
998	Juneau	1.23	1.17
999	Ketchikan	1.26	1.24
ARIZONA			
850,853	Phoenix	.87	.88
851,852	Mesa/Tempe	.86	.87
855	Globe	.85	.87
856-857	Tucson	.85	.86
859	Show Low	.86	.88
860	Flagstaff	.87	.89
863	Prescott	.86	.88
864	Kingman	.85	.87
865	Chambers	.87	.88
ARKANSAS			
716	Pine Bluff	.78	.83
717	Camden	.76	.80
718	Texarkana	.78	.80
719	Hot Springs	.75	.80
720-722	Little Rock	.80	.83
723	West Memphis	.78	.82
724	Jonesboro	.77	.81
725	Batesville	.75	.79
726	Harrison	.76	.79
727	Fayetteville	.74	.79
728	Russellville	.75	.78
729	Fort Smith	.81	.81
CALIFORNIA			
900-902	Los Angeles	1.15	1.13
903-905	Inglewood	1.12	1.09
906-908	Long Beach	1.12	1.10
910-912	Pasadena	1.11	1.09
913-916	Van Nuys	1.14	1.11
917-918	Alhambra	1.15	1.10
919-921	San Diego	1.10	1.09
922	Palm Springs	1.11	1.09
923-924	San Bernardino	1.14	1.08
925	Riverside	1.14	1.11
926-927	Santa Ana	1.13	1.08
928	Anaheim	1.14	1.11
930	Oxnard	1.14	1.10
931	Santa Barbara	1.14	1.10
932-933	Bakersfield	1.11	1.08
934	San Luis Obispo	1.15	1.10
935	Mojave	1.12	1.07
936-938	Fresno	1.18	1.11
939	Salinas	1.21	1.15
940-941	San Francisco	1.33	1.29
942,956-958	Sacramento	1.19	1.14
943	Palo Alto	1.27	1.20
944	San Mateo	1.30	1.21
945	Vallejo	1.23	1.17
946	Oakland	1.29	1.23
947	Berkeley	1.32	1.23
948	Richmond	1.31	1.20
949	San Rafael	1.30	1.23
950	Santa Cruz	1.25	1.18
951	San Jose	1.30	1.23
952	Stockton	1.19	1.12
953	Modesto	1.18	1.12

STATE/ZIP	CITY	Residential	Commercial
CALIFORNIA (CONT'D)			
954	Santa Rosa	1.27	1.21
955	Eureka	1.23	1.14
959	Marysville	1.20	1.14
960	Redding	1.24	1.18
961	Susanville	1.21	1.17
COLORADO			
800-802	Denver	.88	.91
803	Boulder	.90	.89
804	Golden	.85	.89
805	Fort Collins	.88	.90
806	Greeley	.88	.89
807	Fort Morgan	.85	.87
808-809	Colorado Springs	.84	.89
810	Pueblo	.84	.87
811	Alamosa	.84	.88
812	Salida	.82	.87
813	Durango	.86	.87
814	Montrose	.81	.87
815	Grand Junction	.92	.91
816	Glenwood Springs	.81	.87
CONNECTICUT			
060	New Britain	1.11	1.06
061	Hartford	1.09	1.07
062	Willimantic	1.11	1.07
063	New London	1.10	1.05
064	Meriden	1.10	1.06
065	New Haven	1.11	1.08
066	Bridgeport	1.12	1.07
067	Waterbury	1.11	1.07
068	Norwalk	1.11	1.07
069	Stamford	1.12	1.10
D.C.			
200-205	Washington	.92	.95
DELAWARE			
197	Newark	1.02	1.04
198	Wilmington	1.01	1.04
199	Dover	1.02	1.04
FLORIDA			
320,322	Jacksonville	.80	.82
321	Daytona Beach	.83	.84
323	Tallahassee	.80	.83
324	Panama City	.81	.83
325	Pensacola	.84	.85
326,344	Gainesville	.80	.83
327-328,347	Orlando	.81	.84
329	Melbourne	.83	.86
330-332,340	Miami	.79	.82
333	Fort Lauderdale	.80	.82
334,349	West Palm Beach	.80	.81
335-336,346	Tampa	.82	.84
337	St. Petersburg	.81	.85
338	Lakeland	.79	.84
339,341	Fort Myers	.79	.82
342	Sarasota	.83	.85
GEORGIA			
300-303,399	Atlanta	.88	.88
304	Statesboro	.79	.85
305	Gainesville	.82	.84
306	Athens	.81	.85
307	Dalton	.84	.87
308-309	Augusta	.87	.87
310-312	Macon	.84	.86
313-314	Savannah	.85	.86
315	Waycross	.81	.84
316	Valdosta	.75	.83
317,398	Albany	.84	.86
318-319	Columbus	.84	.86
HAWAII			
967	Hilo	1.19	1.17
968	Honolulu	1.21	1.19

For customer support on your Square Foot Costs with RSMeans data, call 800.448.8182.

Location Factors - Residential/Commercial

STATE/ZIP	CITY	Residential	Commercial
STATES & POSS.			
969	Guam	.95	1.02
IDAHO			
832	Pocatello	.89	.92
833	Twin Falls	.89	.92
834	Idaho Falls	.87	.91
835	Lewiston	.98	.99
836-837	Boise	.90	.91
838	Coeur d'Alene	.98	.98
ILLINOIS			
600-603	North Suburban	1.22	1.17
604	Joliet	1.24	1.18
605	South Suburban	1.22	1.17
606-608	Chicago	1.26	1.21
609	Kankakee	1.16	1.12
610-611	Rockford	1.13	1.11
612	Rock Island	.98	.99
613	La Salle	1.12	1.10
614	Galesburg	1.04	1.03
615-616	Peoria	1.07	1.05
617	Bloomington	1.04	1.04
618-619	Champaign	1.05	1.05
620-622	East St. Louis	1.01	1.01
623	Quincy	1.03	1.00
624	Effingham	1.04	1.02
625	Decatur	1.03	1.02
626-627	Springfield	1.04	1.03
628	Centralia	1.01	1.00
629	Carbondale	1.00	1.00
INDIANA			
460	Anderson	.90	.90
461-462	Indianapolis	.92	.92
463-464	Gary	1.04	1.03
465-466	South Bend	.90	.91
467-468	Fort Wayne	.88	.88
469	Kokomo	.90	.88
470	Lawrenceburg	.87	.87
471	New Albany	.87	.87
472	Columbus	.91	.90
473	Muncie	.91	.90
474	Bloomington	.93	.91
475	Washington	.91	.91
476-477	Evansville	.90	.91
478	Terre Haute	.90	.92
479	Lafayette	.92	.90
IOWA			
500-503,509	Des Moines	.92	.94
504	Mason City	.85	.87
505	Fort Dodge	.84	.86
506-507	Waterloo	.88	.90
508	Creston	.88	.90
510-511	Sioux City	.89	.90
512	Sibley	.75	.81
513	Spencer	.76	.83
514	Carroll	.87	.89
515	Council Bluffs	.87	.91
516	Shenandoah	.81	.88
520	Dubuque	.89	.91
521	Decorah	.87	.87
522-524	Cedar Rapids	.94	.93
525	Ottumwa	.87	.88
526	Burlington	.90	.90
527-528	Davenport	.99	.98
KANSAS			
660-662	Kansas City	1.00	.99
664-666	Topeka	.87	.90
667	Fort Scott	.91	.89
668	Emporia	.84	.88
669	Belleville	.82	.86
670-672	Wichita	.82	.87
673	Independence	.90	.89
674	Salina	.81	.87
675	Hutchinson	.80	.84
676	Hays	.83	.87
677	Colby	.86	.88
678	Dodge City	.83	.89
679	Liberal	.82	.87
KENTUCKY			
400-402	Louisville	.87	.88
403-405	Lexington	.87	.89

STATE/ZIP	CITY	Residential	Commercial
KENTUCKY (CONT'D)			
406	Frankfort	.86	.88
407-409	Corbin	.82	.86
410	Covington	.85	.89
411-412	Ashland	.90	.94
413-414	Campton	.87	.88
415-416	Pikeville	.86	.90
417-418	Hazard	.86	.87
420	Paducah	.88	.89
421-422	Bowling Green	.87	.88
423	Owensboro	.88	.90
424	Henderson	.88	.88
425-426	Somerset	.84	.86
427	Elizabethtown	.84	.85
LOUISIANA			
700-701	New Orleans	.86	.86
703	Thibodaux	.82	.83
704	Hammond	.78	.81
705	Lafayette	.86	.84
706	Lake Charles	.87	.85
707-708	Baton Rouge	.86	.85
710-711	Shreveport	.83	.84
712	Monroe	.81	.84
713-714	Alexandria	.82	.84
MAINE			
039	Kittery	.88	.90
040-041	Portland	.92	.93
042	Lewiston	.92	.92
043	Augusta	.88	.90
044	Bangor	.91	.91
045	Bath	.88	.89
046	Machias	.87	.88
047	Houlton	.88	.88
048	Rockland	.87	.87
049	Waterville	.87	.88
MARYLAND			
206	Waldorf	.90	.92
207-208	College Park	.89	.93
209	Silver Spring	.90	.92
210-212	Baltimore	.92	.94
214	Annapolis	.90	.92
215	Cumberland	.91	.92
216	Easton	.87	.89
217	Hagerstown	.91	.93
218	Salisbury	.85	.86
219	Elkton	.93	.92
MASSACHUSETTS			
010-011	Springfield	1.05	1.03
012	Pittsfield	1.04	1.01
013	Greenfield	1.03	1.02
014	Fitchburg	1.11	1.06
015-016	Worcester	1.12	1.08
017	Framingham	1.13	1.09
018	Lowell	1.14	1.10
019	Lawrence	1.14	1.11
020-022, 024	Boston	1.19	1.15
023	Brockton	1.12	1.08
025	Buzzards Bay	1.10	1.05
026	Hyannis	1.09	1.06
027	New Bedford	1.11	1.07
MICHIGAN			
480,483	Royal Oak	.99	.97
481	Ann Arbor	1.00	.99
482	Detroit	1.02	1.01
484-485	Flint	.93	.94
486	Saginaw	.90	.93
487	Bay City	.90	.93
488-489	Lansing	.90	.94
490	Battle Creek	.88	.90
491	Kalamazoo	.88	.89
492	Jackson	.90	.93
493,495	Grand Rapids	.88	.91
494	Muskegon	.86	.88
496	Traverse City	.82	.87
497	Gaylord	.86	.88
498-499	Iron Mountain	.86	.90
MINNESOTA			
550-551	Saint Paul	1.06	1.06
553-555	Minneapolis	1.07	1.06
556-558	Duluth	.99	1.01

For customer support on your Square Foot Costs with RSMeans data, call 800.448.8182.

Location Factors - Residential/Commercial

STATE/ZIP	CITY	Residential	Commercial
MINNESOTA (CONT'D)			
559	Rochester	.99	1.00
560	Mankato	.96	.97
561	Windom	.91	.94
562	Willmar	.93	.96
563	St. Cloud	1.01	1.02
564	Brainerd	.92	.97
565	Detroit Lakes	.90	.95
566	Bemidji	.91	.97
567	Thief River Falls	.90	.93
MISSISSIPPI			
386	Clarksdale	.73	.77
387	Greenville	.81	.84
388	Tupelo	.74	.79
389	Greenwood	.77	.77
390-392	Jackson	.82	.83
393	Meridian	.80	.82
394	Laurel	.78	.79
395	Biloxi	.84	.83
396	McComb	.74	.77
397	Columbus	.75	.79
MISSOURI			
630-631	St. Louis	1.03	1.03
633	Bowling Green	.97	.97
634	Hannibal	.95	.96
635	Kirksville	.91	.96
636	Flat River	.97	.98
637	Cape Girardeau	.91	.96
638	Sikeston	.89	.94
639	Poplar Bluff	.89	.94
640-641	Kansas City	1.02	1.01
644-645	St. Joseph	.97	.96
646	Chillicothe	.96	.96
647	Harrisonville	.99	.99
648	Joplin	.90	.90
650-651	Jefferson City	.93	.96
652	Columbia	.91	.96
653	Sedalia	.90	.96
654-655	Rolla	.96	.97
656-658	Springfield	.88	.92
MONTANA			
590-591	Billings	.87	.90
592	Wolf Point	.86	.90
593	Miles City	.87	.89
594	Great Falls	.87	.90
595	Havre	.83	.89
596	Helena	.87	.89
597	Butte	.84	.89
598	Missoula	.84	.87
599	Kalispell	.84	.87
NEBRASKA			
680-681	Omaha	.90	.91
683-685	Lincoln	.90	.90
686	Columbus	.89	.90
687	Norfolk	.89	.90
688	Grand Island	.87	.90
689	Hastings	.89	.90
690	McCook	.83	.86
691	North Platte	.86	.88
692	Valentine	.83	.87
693	Alliance	.82	.88
NEVADA			
889-891	Las Vegas	1.02	1.04
893	Ely	1.00	.99
894-895	Reno	.90	.95
897	Carson City	.91	.94
898	Elko	.93	.92
NEW HAMPSHIRE			
030	Nashua	.98	.96
031	Manchester	.97	.96
032-033	Concord	.97	.96
034	Keene	.91	.88
035	Littleton	.94	.88
036	Charleston	.89	.86
037	Claremont	.89	.86
038	Portsmouth	.95	.94

STATE/ZIP	CITY	Residential	Commercial
NEW JERSEY			
070-071	Newark	1.20	1.16
072	Elizabeth	1.22	1.15
073	Jersey City	1.18	1.14
074-075	Paterson	1.20	1.15
076	Hackensack	1.19	1.14
077	Long Branch	1.19	1.13
078	Dover	1.19	1.14
079	Summit	1.20	1.14
080,083	Vineland	1.18	1.12
081	Camden	1.19	1.13
082,084	Atlantic City	1.22	1.13
085-086	Trenton	1.19	1.14
087	Point Pleasant	1.19	1.14
088-089	New Brunswick	1.22	1.15
NEW MEXICO			
870-872	Albuquerque	.83	.88
873	Gallup	.82	.89
874	Farmington	.83	.89
875	Santa Fe	.84	.89
877	Las Vegas	.82	.88
878	Socorro	.82	.87
879	Truth/Consequences	.81	.86
880	Las Cruces	.82	.86
881	Clovis	.83	.88
882	Roswell	.83	.89
883	Carrizozo	.83	.90
884	Tucumcari	.84	.89
NEW YORK			
100-102	New York	1.40	1.35
103	Staten Island	1.32	1.30
104	Bronx	1.36	1.29
105	Mount Vernon	1.17	1.17
106	White Plains	1.20	1.17
107	Yonkers	1.24	1.22
108	New Rochelle	1.21	1.17
109	Suffern	1.19	1.13
110	Queens	1.37	1.33
111	Long Island City	1.40	1.34
112	Brooklyn	1.38	1.33
113	Flushing	1.39	1.34
114	Jamaica	1.38	1.33
115,117,118	Hicksville	1.26	1.26
116	Far Rockaway	1.38	1.34
119	Riverhead	1.28	1.26
120-122	Albany	1.02	1.03
123	Schenectady	1.04	1.03
124	Kingston	1.13	1.14
125-126	Poughkeepsie	1.25	1.18
127	Monticello	1.14	1.13
128	Glens Falls	.96	.98
129	Plattsburgh	1.01	.99
130-132	Syracuse	.99	1.00
133-135	Utica	.97	.98
136	Watertown	.95	.99
137-139	Binghamton	.99	.99
140-142	Buffalo	1.09	1.06
143	Niagara Falls	1.05	1.04
144-146	Rochester	1.01	1.01
147	Jamestown	.93	.95
148-149	Elmira	.96	.99
NORTH CAROLINA			
270,272-274	Greensboro	.94	.86
271	Winston-Salem	.94	.85
275-276	Raleigh	.93	.84
277	Durham	.94	.87
278	Rocky Mount	.92	.84
279	Elizabeth City	.91	.86
280	Gastonia	.96	.85
281-282	Charlotte	.94	.85
283	Fayetteville	.94	.87
284	Wilmington	.93	.84
285	Kinston	.92	.84
286	Hickory	.93	.84
287-288	Asheville	.94	.85
289	Murphy	.95	.84
NORTH DAKOTA			
580-581	Fargo	.90	.92
582	Grand Forks	.88	.91
583	Devils Lake	.90	.91
584	Jamestown	.88	.90
585	Bismarck	.89	.92

523

Location Factors - Residential/Commercial

STATE/ZIP	CITY	Residential	Commercial
NORTH DAKOTA (CONT'D)			
586	Dickinson	.88	.92
587	Minot	.87	.92
588	Williston	.88	.91
OHIO			
430-432	Columbus	.91	.92
433	Marion	.90	.89
434-436	Toledo	.96	.96
437-438	Zanesville	.88	.90
439	Steubenville	.93	.94
440	Lorain	.94	.94
441	Cleveland	.97	.97
442-443	Akron	.96	.95
444-445	Youngstown	.93	.92
446-447	Canton	.92	.91
448-449	Mansfield	.90	.91
450	Hamilton	.89	.89
451-452	Cincinnati	.91	.90
453-454	Dayton	.89	.89
455	Springfield	.90	.90
456	Chillicothe	.93	.92
457	Athens	.91	.92
458	Lima	.89	.91
OKLAHOMA			
730-731	Oklahoma City	.81	.84
734	Ardmore	.82	.84
735	Lawton	.83	.85
736	Clinton	.82	.84
737	Enid	.83	.85
738	Woodward	.78	.82
739	Guymon	.82	.83
740-741	Tulsa	.82	.83
743	Miami	.80	.81
744	Muskogee	.80	.82
745	McAlester	.74	.79
746	Ponca City	.80	.81
747	Durant	.79	.81
748	Shawnee	.79	.82
749	Poteau	.79	.81
OREGON			
970-972	Portland	1.00	1.00
973	Salem	.99	1.00
974	Eugene	1.00	.99
975	Medford	.98	.99
976	Klamath Falls	.99	.99
977	Bend	1.01	.99
978	Pendleton	.99	.98
979	Vale	.97	.91
PENNSYLVANIA			
150-152	Pittsburgh	1.01	1.02
153	Washington	.96	.99
154	Uniontown	.93	.98
155	Bedford	.90	.96
156	Greensburg	.96	.99
157	Indiana	.94	.98
158	Dubois	.91	.97
159	Johnstown	.91	.97
160	Butler	.92	.96
161	New Castle	.93	.96
162	Kittanning	.93	.96
163	Oil City	.93	.95
164-165	Erie	.94	.95
166	Altoona	.90	.94
167	Bradford	.96	.97
168	State College	.92	.95
169	Wellsboro	.92	.95
170-171	Harrisburg	.95	.96
172	Chambersburg	.89	.93
173-174	York	.91	.94
175-176	Lancaster	.94	.95
177	Williamsport	.90	.92
178	Sunbury	.92	.94
179	Pottsville	.92	.95
180	Lehigh Valley	1.06	1.05
181	Allentown	1.03	1.01
182	Hazleton	.91	.95
183	Stroudsburg	.96	1.01
184-185	Scranton	.97	.98
186-187	Wilkes-Barre	.94	.96
188	Montrose	.93	.96
189	Doylestown	1.07	1.07

STATE/ZIP	CITY	Residential	Commercial
PENNSYLVANIA (CONT'D)			
190-191	Philadelphia	1.17	1.15
193	Westchester	1.09	1.08
194	Norristown	1.09	1.09
195-196	Reading	.98	1.00
PUERTO RICO			
009	San Juan	.77	.80
RHODE ISLAND			
028	Newport	1.07	1.03
029	Providence	1.08	1.04
SOUTH CAROLINA			
290-292	Columbia	.95	.85
293	Spartanburg	.95	.85
294	Charleston	.96	.86
295	Florence	.92	.85
296	Greenville	.94	.85
297	Rock Hill	.94	.85
298	Aiken	.95	.85
299	Beaufort	.85	.80
SOUTH DAKOTA			
570-571	Sioux Falls	.87	.89
572	Watertown	.76	.81
573	Mitchell	.78	.82
574	Aberdeen	.79	.84
575	Pierre	.83	.87
576	Mobridge	.78	.84
577	Rapid City	.86	.89
TENNESSEE			
370-372	Nashville	.83	.86
373-374	Chattanooga	.83	.86
375,380-381	Memphis	.84	.86
376	Johnson City	.76	.82
377-379	Knoxville	.80	.83
382	McKenzie	.72	.79
383	Jackson	.75	.81
384	Columbia	.79	.83
385	Cookeville	.70	.79
TEXAS			
750	McKinney	.80	.84
751	Waxahachie	.80	.84
752-753	Dallas	.83	.85
754	Greenville	.80	.84
755	Texarkana	.82	.83
756	Longview	.80	.83
757	Tyler	.82	.84
758	Palestine	.78	.81
759	Lufkin	.79	.82
760-761	Fort Worth	.82	.83
762	Denton	.84	.83
763	Wichita Falls	.84	.81
764	Eastland	.82	.80
765	Temple	.80	.78
766-767	Waco	.84	.81
768	Brownwood	.80	.81
769	San Angelo	.78	.81
770-772	Houston	.83	.85
773	Huntsville	.80	.83
774	Wharton	.80	.84
775	Galveston	.82	.84
776-777	Beaumont	.85	.84
778	Bryan	.78	.82
779	Victoria	.81	.84
780	Laredo	.80	.82
781-782	San Antonio	.80	.83
783-784	Corpus Christi	.83	.83
785	McAllen	.84	.82
786-787	Austin	.79	.82
788	Del Rio	.81	.83
789	Giddings	.80	.81
790-791	Amarillo	.79	.82
792	Childress	.81	.82
793-794	Lubbock	.80	.84
795-796	Abilene	.82	.82
797	Midland	.84	.85
798-799,885	El Paso	.79	.82
UTAH			
840-841	Salt Lake City	.84	.90
842,844	Ogden	.82	.87
843	Logan	.83	.88

Location Factors - Residential/Commercial

STATE/ZIP	CITY	Residential	Commercial
UTAH (CONT'D)			
845	Price	.84	.88
846-847	Provo	.83	.89
VERMONT			
050	White River Jct.	.96	.91
051	Bellows Falls	1.01	.96
052	Bennington	1.00	.94
053	Brattleboro	1.01	.96
054	Burlington	.93	.92
056	Montpelier	.99	.94
057	Rutland	.92	.91
058	St. Johnsbury	.97	.91
059	Guildhall	.97	.90
VIRGINIA			
220-221	Fairfax	1.01	.93
222	Arlington	1.03	.94
223	Alexandria	1.04	.93
224-225	Fredericksburg	1.00	.92
226	Winchester	1.01	.92
227	Culpeper	1.00	.91
228	Harrisonburg	.86	.85
229	Charlottesville	.88	.87
230-232	Richmond	1.03	.90
233-235	Norfolk	.95	.87
236	Newport News	.95	.86
237	Portsmouth	.87	.83
238	Petersburg	.95	.88
239	Farmville	.85	.80
240-241	Roanoke	1.00	.88
242	Bristol	.91	.82
243	Pulaski	.83	.83
244	Staunton	.86	.85
245	Lynchburg	.98	.89
246	Grundy	.81	.80
WASHINGTON			
980-981,987	Seattle	1.05	1.05
982	Everett	1.06	1.02
983-984	Tacoma	1.03	1.02
985	Olympia	1.01	1.01
986	Vancouver	.98	1.01
988	Wenatchee	.95	.97
989	Yakima	1.00	1.01
990-992	Spokane	.97	.94
993	Richland	.97	.97
994	Clarkston	.94	.93
WEST VIRGINIA			
247-248	Bluefield	.93	.94
249	Lewisburg	.92	.95
250-253	Charleston	.95	.96
254	Martinsburg	.89	.93
255-257	Huntington	.97	.97
258-259	Beckley	.93	.95
260	Wheeling	.93	.97
261	Parkersburg	.91	.95
262	Buckhannon	.92	.96
263-264	Clarksburg	.92	.96
265	Morgantown	.92	.96
266	Gassaway	.92	.95
267	Romney	.89	.93
268	Petersburg	.89	.93
WISCONSIN			
530,532	Milwaukee	1.05	1.03
531	Kenosha	1.05	1.01
534	Racine	1.04	1.02
535	Beloit	.98	.98
537	Madison	.98	.98
538	Lancaster	.97	.95
539	Portage	.94	.95
540	New Richmond	.95	.94
541-543	Green Bay	1.03	.98
544	Wausau	.96	.95
545	Rhinelander	.93	.95
546	La Crosse	.97	.96
547	Eau Claire	.99	.98
548	Superior	.95	.96
549	Oshkosh	.93	.93
WYOMING			
820	Cheyenne	.86	.89
821	Yellowstone Nat. Pk.	.83	.88
822	Wheatland	.79	.87

STATE/ZIP	CITY	Residential	Commercial
WYOMING (CONT'D)			
823	Rawlins	.86	.90
824	Worland	.83	.88
825	Riverton	.82	.88
826	Casper	.83	.88
827	Newcastle	.85	.88
828	Sheridan	.86	.89
829-831	Rock Springs	.88	.90
CANADIAN FACTORS (reflect Canadian currency)			
ALBERTA			
	Calgary	1.07	1.10
	Edmonton	1.07	1.11
	Fort McMurray	1.10	1.10
	Lethbridge	1.07	1.07
	Lloydminster	1.02	1.02
	Medicine Hat	1.03	1.02
	Red Deer	1.03	1.02
BRITISH COLUMBIA			
	Kamloops	1.00	1.02
	Prince George	.99	1.03
	Vancouver	1.01	1.06
	Victoria	1.01	1.02
MANITOBA			
	Brandon	1.07	1.02
	Portage la Prairie	.97	.95
	Winnipeg	.95	1.00
NEW BRUNSWICK			
	Bathurst	.89	.91
	Dalhousie	.90	.91
	Fredericton	.94	.98
	Moncton	.91	.93
	Newcastle	.90	.91
	St. John	.99	.98
NEWFOUNDLAND			
	Corner Brook	1.06	1.03
	St. Johns	1.04	1.07
NORTHWEST TERRITORIES			
	Yellowknife	1.12	1.13
NOVA SCOTIA			
	Bridgewater	.93	.95
	Dartmouth	1.02	1.02
	Halifax	1.01	1.04
	New Glasgow	1.01	1.01
	Sydney	1.00	1.00
	Truro	.93	.95
	Yarmouth	1.01	1.01
ONTARIO			
	Barrie	1.10	1.06
	Brantford	1.09	1.06
	Cornwall	1.08	1.04
	Hamilton	1.06	1.09
	Kingston	1.08	1.05
	Kitchener	1.03	1.03
	London	1.05	1.08
	North Bay	1.16	1.10
	Oshawa	1.06	1.04
	Ottawa	1.06	1.09
	Owen Sound	1.09	1.05
	Peterborough	1.07	1.04
	Sarnia	1.10	1.06
	Sault Ste. Marie	1.04	1.02
	St. Catharines	1.04	1.02
	Sudbury	1.02	1.02
	Thunder Bay	1.07	1.03
	Timmins	1.06	1.03
	Toronto	1.08	1.11
	Windsor	1.06	1.02
PRINCE EDWARD ISLAND			
	Charlottetown	.89	.96
	Summerside	.95	.97
QUEBEC			
	Cap-de-la-Madeleine	1.07	.99
	Charlesbourg	1.07	.99
	Chicoutimi	1.10	1.01
	Gatineau	1.06	.99

STATE/ZIP	CITY	Residential	Commercial
QUEBEC (CONT'D)			
	Granby	1.06	
	Hull	1.06	.99
	Joliette	1.07	1.00
	Laval	1.06	.99
	Montreal	1.05	1.07
	Quebec City	1.06	1.07
	Rimouski	1.10	1.02
	Rouyn-Noranda	1.06	.99
	Saint-Hyacinthe	1.06	.99
	Sherbrooke	1.06	.99
	Sorel	1.07	1.00
	Saint-Jerome	1.06	.99
	Trois-Rivieres	1.17	1.06
SASKATCHEWAN			
	Moose Jaw	.89	.91
	Prince Albert	.88	.90
	Regina	1.08	1.11
	Saskatoon	1.04	1.03
YUKON			
	Whitehorse	1.03	1.08

Historical Cost Indexes

The following tables are the estimated Historical Cost Indexes based on a 30-city national average with a base of 100 on January 1, 1993.

The indexes may be used to:

1. Estimate and compare construction costs for different years in the same city.
2. Estimate and compare construction costs in different cities for the same year.
3. Estimate and compare construction costs in different cities for different years.
4. Compare construction trends in any city with the national average.

EXAMPLES

1. Estimate and compare construction costs for different years in the same city.

   A. To estimate the construction cost of a building in Lexington, KY in 1970, knowing that it cost $915,000 in 2018.

   Index Lexington, KY in 1970 = 26.9
   Index Lexington, KY in 2018 = 188.1

   $$\frac{\text{Index 1970}}{\text{Index 2018}} \quad x \quad \text{Cost 2018} \quad = \quad \text{Cost 1970}$$

   $$\frac{26.9}{188.1} \quad x \quad \$915,000 \quad = \quad \$130,853$$

   Construction Cost in Lexington, KY in 1970 = $130,853

   B. To estimate the current construction cost of a building in Boston, MA that was built in 1980 for $900,000.

   Index Boston, MA in 1980 = 64.0
   Index Boston, MA in 2018 = 242.4

   $$\frac{\text{Index 2018}}{\text{Index 1980}} \quad x \quad \text{Cost 1980} \quad = \quad \text{Cost 2018}$$

   $$\frac{242.4}{64.0} \quad x \quad \$900,000 \quad = \quad \$3,408,750$$

   Construction Cost in Boston in 2018 = $3,408,750

2. Estimate and compare construction costs in different cities for the same year.

   To compare the construction cost of a building in Topeka, KS in 2018 with the known cost of $800,000 in Baltimore, MD in 2018

   Index Topeka, KS in 2018   = 190.0
   Index Baltimore, MD in 2018 = 199.7

   $$\frac{\text{Index Topeka}}{\text{Index Baltimore}} \quad x \quad \text{Cost Baltimore} = \text{Cost Topeka}$$

   $$\frac{190.0}{199.7} \quad x \quad \$800,000 \quad = \$761,142$$

   Construction Cost in Topeka in 2016 = $761,142

3. Estimate and compare construction costs in different cities for different years.

   To compare the construction cost of a building in Detroit, MI in 2018 with the known construction cost of $5,000,000 for the same building in San Francisco, CA in 1980.

   Index Detroit, MI in 2018   = 214.2
   Index San Francisco, CA in 1980 = 75.2

   $$\frac{\text{Index Detroit 2018}}{\text{Index San Francisco 1980}} \quad x \text{ Cost San Francisco 1980} = \text{Cost Detroit 2018}$$

   $$\frac{214.2}{75.2} \quad x \quad \$5,000,000 \quad = \quad \$14,242,021$$

   Construction Cost in Detroit in 2018   = $14,242,021

4. Compare construction trends in any city with the national average.

   To compare the construction cost in Las Vegas, NV from 1975 to 2018 with the increase in the National Average during the same time period.

   Index Las Vegas, NV for 1975   =   42.8   For 2018 =   222.3
   Index 30 City Average for 1975 =   43.7   For 2018 =   215.8

   A. National Average escalation = $\frac{\text{Index} - \text{30 City 2018}}{\text{Index} - \text{30 City 1975}}$
      From 1975 to 2018

   $$= \frac{215.8}{43.7}$$

   National Average escalation
   From 1975 to 2018   =   4.94 or increased by 494%

   B. Escalation for Las Vegas, NV = $\frac{\text{Index Las Vegas, NV 2018}}{\text{Index Las Vegas, NV 1975}}$
      From 1975 to 2018

   $$= \frac{222.3}{42.8}$$

   Las Vegas escalation
   From 1975 to 2018   =   5.19 or increased by 519%

   Conclusion: Construction costs in Las Vegas are higher than National Average costs and increased at a greater rate from 1975 to 2018 than the National Average.

Historical Cost Indexes

Year	National 30 City Average	Alabama Birmingham	Huntsville	Mobile	Montgomery	Tuscaloosa	Alaska Anchorage	Arizona Phoenix	Tuscon	Arkansas Fort Smith	Little Rock	California Anaheim	Bakersfield	Fresno	Los Angeles	Oxnard
Jan 2018	215.8	180.4E	180.1E	179.7E	180.3E	181.4E	248.7E	185.7E	183.1E	172.3E	174.6E	236E	231.5E	234.4E	240.2E	234.2E
2017	209.4	178.6	178.3	177.9	178.5	179.6	246.2	183.8	181.3	170.6	172.8	233.6	229.2	232.0	237.8	231.9
2016	207.7	185.6	184.2	184.1	183.5	186.3	246.3	181.7	178.1	169.8	174.6	223.6	221.3	224.3	226.2	222.5
2015	204.0	184.8	182.1	185.4	182.7	184.5	244.7	181.0	177.9	167.6	171.1	217.3	214.6	218.8	218.5	215.7
2014	203.0	180.3	175.8	170.6	163.1	165.9	241.0	180.0	177.0	165.1	168.4	214.2	215.2	217.0	217.3	214.8
2013	196.9	173.3	168.7	165.7	158.7	161.6	235.3	174.1	169.2	161.2	163.2	207.0	205.6	210.3	210.7	207.8
2012	194.0	169.1	162.7	163.2	153.8	154.6	232.5	172.2	166.4	158.7	161.0	204.1	202.9	207.3	207.2	204.7
2011	185.7	163.0	156.3	156.9	147.6	148.3	225.1	163.8	159.5	152.0	154.1	196.1	194.6	199.8	199.2	197.0
2010	181.6	159.6	152.9	153.1	144.4	144.9	218.3	160.7	157.3	150.3	152.5	192.8	190.8	195.0	194.9	192.8
2009	182.5	162.7	157.2	155.6	148.8	149.4	222.9	161.3	156.8	149.2	156.3	194.5	192.3	195.0	196.6	194.5
2008	171.0	150.3	146.9	143.7	138.4	138.8	210.8	152.2	148.4	138.6	145.4	182.7	179.8	183.2	184.7	182.6
2007	165.0	146.9	143.3	140.3	134.8	135.4	206.8	147.7	143.1	134.8	141.7	175.1	173.7	176.7	177.5	176.2
2006	156.2	135.7	133.6	126.7	124.0	122.2	196.4	137.9	134.8	123.2	127.2	166.8	164.9	169.8	167.3	167.4
2005	146.7	127.9	125.4	119.3	116.6	114.6	185.6	128.5	124.1	115.4	119.4	156.5	153.0	157.9	157.1	156.4
2004	132.8	115.9	112.9	107.0	104.9	102.9	167.0	116.5	113.6	103.8	108.6	142.1	139.5	143.8	142.0	140.9
2003	129.7	113.1	110.7	104.8	102.6	100.8	163.5	113.9	110.6	101.8	105.8	139.4	137.2	142.1	139.6	138.7
2002	126.7	110.0	103.4	103.6	101.7	99.7	159.1	113.3	110.2	100.4	102.1	136.5	133.4	136.0	136.4	136.3
2001	122.2	106.0	100.5	100.6	98.3	95.9	152.6	109.0	106.4	97.3	98.7	132.5	129.3	132.8	132.4	132.4
2000	118.9	104.1	98.9	99.1	94.2	94.6	148.3	106.9	104.9	94.4	95.7	129.4	125.2	129.4	129.9	129.7
1999	116.6	101.2	97.4	97.7	92.5	92.8	145.9	105.5	103.3	93.1	94.4	127.9	123.6	126.9	128.7	128.2
1998	113.6	96.2	94.0	94.8	90.3	89.8	143.8	102.1	101.0	90.6	91.4	125.2	120.3	123.9	125.8	124.7
1995	105.6	87.8	88.0	88.8	84.5	83.5	138.0	96.1	95.5	85.0	85.6	120.1	115.7	117.5	120.9	120.0
1990	93.2	79.4	77.6	82.7	78.4	75.2	125.8	86.4	87.0	77.1	77.9	107.7	102.7	103.3	107.5	107.4
1985	81.8	71.1	70.5	72.7	70.9	67.9	116.0	78.1	77.7	69.6	71.0	95.4	92.2	92.6	94.6	96.6
1980	60.7	55.2	54.1	56.8	56.8	54.0	91.4	63.7	62.4	53.1	55.8	68.7	69.4	68.7	67.4	69.9
1975	43.7	40.0	40.9	41.8	40.1	37.8	57.3	44.5	45.2	39.0	38.7	47.6	46.6	47.7	48.3	47.0
1970	27.8	24.1	25.1	25.8	25.4	24.2	43.0	27.2	28.5	24.3	22.3	31.0	30.8	31.1	29.0	31.0
1965	21.5	19.6	19.3	19.6	19.5	18.7	34.9	21.8	22.0	18.7	18.5	23.9	23.7	24.0	22.7	23.9
1960	19.5	17.9	17.5	17.8	17.7	16.9	31.7	19.9	20.0	17.0	16.8	21.7	21.5	21.8	20.6	21.7
1955	16.3	14.8	14.7	14.9	14.9	14.2	26.6	16.7	16.7	14.3	14.5	18.2	18.1	18.3	17.3	18.2
1950	13.5	12.2	12.1	12.3	12.3	11.7	21.9	13.8	13.8	11.8	11.6	15.1	15.0	15.1	14.3	15.1

Year	National 30 City Average	California Riverside	Sacramento	San Diego	San Francisco	Santa Barbara	Stockton	Vallejo	Colorado Colorado Springs	Denver	Pueblo	Connecticut Bridgeport	Bristol	Hartford	New Britain	New Haven
Jan 2018	215.8	235.6E	240.3E	231.6E	273.5E	233.4E	238.6E	249.3E	189.2E	189.7E	185E	226.7E	225.2E	226.2E	224.8E	227.3E
2017	209.4	233.2	237.9	229.3	270.8	231.1	236.2	246.8	187.3	187.8	183.1	224.4	222.9	223.9	222.5	225.0
2016	207.7	223.3	227.3	218.3	256.6	221.9	227.9	236.0	189.4	191.0	186.2	226.4	225.6	226.1	225.0	226.9
2015	204.0	217.1	221.0	213.1	250.2	215.7	222.8	230.2	185.1	186.6	183.5	224.1	223.3	224.2	222.8	225.0
2014	203.0	214.1	221.5	211.4	248.0	214.7	218.4	227.9	187.8	189.1	184.7	223.8	222.8	223.9	222.4	224.1
2013	196.9	207.0	215.0	203.2	241.0	207.9	211.8	222.2	180.7	183.4	180.0	217.8	217.2	218.4	216.8	218.5
2012	194.0	204.1	211.9	198.9	237.6	204.8	208.8	218.9	179.4	182.2	177.9	213.9	213.3	214.4	212.9	214.7
2011	185.7	195.8	203.1	192.3	227.9	196.2	201.3	210.3	171.7	174.1	170.7	205.2	204.2	205.1	203.9	205.8
2010	181.6	192.7	196.7	188.2	223.0	192.9	195.9	204.6	170.5	172.6	168.1	198.8	199.1	199.8	198.8	200.7
2009	182.5	193.1	198.1	191.6	224.9	193.6	194.5	202.7	168.4	172.0	166.7	199.0	197.7	198.7	197.4	199.3
2008	171.0	181.3	186.0	179.9	210.6	181.5	183.2	191.6	158.8	161.9	157.9	185.8	184.8	185.7	184.5	186.1
2007	165.0	174.7	179.1	173.6	201.1	175.2	178.2	185.9	152.6	155.9	152.3	179.6	178.3	179.6	178.0	179.5
2006	156.2	166.0	172.2	164.0	191.2	166.4	171.0	177.9	146.1	149.2	145.0	169.5	168.0	169.5	167.7	169.8
2005	146.7	155.4	161.1	153.8	179.7	155.7	160.0	167.2	138.3	141.1	136.4	160.8	159.4	159.6	159.1	160.8
2004	132.8	141.0	147.2	139.0	163.6	141.4	144.2	148.6	125.6	127.2	123.5	143.6	142.9	142.4	142.7	144.3
2003	129.7	138.7	144.9	136.6	162.1	139.2	141.8	146.4	123.1	123.7	120.7	141.3	140.2	140.3	140.0	141.6
2002	126.7	135.3	138.4	134.2	157.9	135.9	136.9	143.8	118.7	121.3	116.7	133.8	133.6	133.1	133.3	134.8
2001	122.2	131.4	135.5	129.7	151.8	131.5	134.1	140.2	113.3	117.2	113.1	128.6	128.5	128.5	128.3	128.6
2000	118.9	128.0	131.5	127.1	146.9	128.7	130.7	137.1	109.8	111.8	108.8	122.7	122.6	122.9	122.4	122.7
1999	116.6	126.5	129.4	124.9	145.1	127.2	127.9	135.3	107.0	109.1	107.1	121.1	121.3	121.2	121.1	121.3
1998	113.6	123.7	125.9	121.3	141.9	123.7	124.7	132.5	103.3	106.5	103.7	119.1	119.4	120.0	119.7	120.0
1995	105.6	119.2	119.5	115.4	133.8	119.0	119.0	122.5	96.1	98.9	96.8	116.0	116.5	116.9	116.3	116.5
1990	93.2	107.0	104.9	105.6	121.8	106.4	105.4	111.4	86.9	88.8	88.8	96.3	95.9	96.6	95.9	96.5
1985	81.8	94.8	92.2	94.2	106.2	93.2	93.7	97.0	79.6	81.0	80.4	86.1	86.2	87.2	86.1	86.3
1980	60.7	68.7	71.3	68.1	75.2	71.1	71.2	71.9	60.7	60.9	59.5	61.5	60.7	61.9	60.7	61.4
1975	43.7	47.3	49.1	47.7	49.8	46.1	47.6	46.5	43.1	42.7	42.5	45.2	45.5	46.0	45.2	46.0
1970	27.8	31.0	32.1	30.7	31.6	31.3	31.8	31.8	27.6	26.1	27.3	29.2	28.2	29.6	28.2	29.3
1965	21.5	23.9	24.8	24.0	23.7	24.1	24.5	24.5	21.3	20.9	21.0	22.4	21.7	22.6	21.7	23.2
1960	19.5	21.7	22.5	21.7	21.5	21.9	22.3	22.2	19.3	19.0	19.1	20.0	19.8	20.0	19.8	20.0
1955	16.3	18.2	18.9	18.2	18.0	18.4	18.7	18.6	16.2	15.9	16.0	16.8	16.6	16.8	16.6	16.8
1950	13.5	15.0	15.6	15.0	14.9	15.2	15.4	15.4	13.4	13.2	13.2	13.9	13.7	13.8	13.7	13.9

Historical Cost Indexes

Year	National 30 City Average	Connecticut Norwalk	Connecticut Stamford	Connecticut Waterbury	Delaware Wilmington	D.C. Washington	Florida Fort Lauderdale	Florida Jacksonville	Florida Miami	Florida Orlando	Florida Tallahassee	Florida Tampa	Georgia Albany	Georgia Atlanta	Georgia Columbus	Georgia Macon
Jan 2018	215.8	226.5E	233.1E	226.5E	219.9E	201.1E	175.5E	174.4E	175.2E	178.3E	175.9E	178.8E	181.2E	187E	1819.3E	181.2E
2017	209.4	224.2	230.8	224.2	217.7	199.1	173.7	172.7	173.4	176.5	174.1	177.0	179.4	185.1	1801.1	179.4
2016	207.7	232.7	233.3	225.7	215.7	201.3	177.7	176.3	179.1	180.4	173.4	180.6	176.3	185.0	173.6	172.4
2015	204.0	230.2	230.3	223.8	211.7	197.3	176.1	174.2	178.1	178.3	172.5	178.9	168.6	178.3	171.3	170.0
2014	203.0	230.3	230.5	223.2	210.2	197.0	176.5	171.0	178.4	176.5	164.4	183.2	166.6	177.5	170.2	168.2
2013	196.9	223.1	224.2	217.3	203.6	191.7	173.2	168.4	175.8	174.8	160.8	180.2	163.2	173.3	166.7	164.9
2012	194.0	219.8	220.0	213.4	201.0	189.6	170.9	165.8	173.7	172.8	158.5	177.8	158.3	170.7	160.6	159.3
2011	185.7	211.0	211.2	204.6	193.3	182.3	164.4	159.4	167.0	166.1	151.7	171.4	151.8	164.0	154.3	152.7
2010	181.6	198.3	203.8	199.4	187.7	179.0	160.3	155.4	163.3	162.5	148.1	167.5	148.3	161.0	150.9	149.4
2009	182.5	198.8	204.3	198.4	188.9	181.4	160.7	152.0	165.2	163.9	145.1	165.1	152.0	164.5	155.6	153.5
2008	171.0	185.2	189.2	185.3	177.5	170.4	149.6	142.5	152.9	153.0	135.3	155.6	140.4	153.2	143.9	142.5
2007	165.0	178.6	183.3	179.1	173.4	163.1	145.6	139.0	148.9	148.1	131.7	152.0	135.8	148.0	140.1	138.1
2006	156.7	169.7	173.7	169.1	158.3	153.0	136.0	126.5	136.7	134.3	118.4	136.6	123.6	139.9	126.3	124.3
2005	146.7	160.9	164.1	160.5	149.6	142.7	126.0	119.2	127.1	126.0	110.7	127.7	115.7	131.7	118.9	116.1
2004	132.8	143.6	146.5	143.2	136.1	126.8	114.8	107.8	115.6	113.2	100.6	116.6	102.8	119.6	102.0	105.0
2003	129.7	140.6	145.1	140.8	133.0	124.6	109.0	105.6	110.5	107.8	98.0	103.5	100.5	115.8	99.1	102.7
2002	126.7	133.5	136.2	133.9	129.4	120.3	107.1	104.6	107.4	106.7	97.3	102.8	99.7	113.0	98.3	101.9
2001	122.2	128.2	131.5	128.8	124.8	115.9	104.3	100.8	104.6	103.8	94.8	100.3	96.4	109.1	95.5	99.0
2000	118.9	121.5	126.4	123.2	117.4	113.8	102.4	99.0	101.8	101.2	93.6	98.9	94.9	106.1	94.1	97.0
1999	116.6	120.0	122.0	121.2	116.6	111.5	101.4	98.0	100.8	100.1	92.5	97.9	93.5	102.9	92.8	95.8
1998	113.6	118.8	120.8	120.1	112.3	109.6	99.4	96.2	99.0	98.4	90.9	96.3	91.4	100.8	90.4	93.3
1995	105.6	115.5	117.9	117.3	106.1	102.3	94.0	90.9	93.7	93.5	86.5	92.2	85.0	92.0	82.6	86.8
1990	93.2	96.3	98.9	95.1	92.5	90.4	83.9	81.1	84.0	82.9	78.4	85.0	75.0	80.4	74.0	76.9
1985	81.8	85.3	86.8	85.6	81.1	78.8	76.7	73.4	78.3	73.9	70.6	77.3	66.9	70.3	66.1	68.1
1980	60.7	60.7	60.9	62.3	58.5	59.6	55.3	55.8	56.5	56.7	53.5	57.2	51.7	54.0	51.1	51.3
1975	43.7	44.7	45.0	46.3	42.9	43.7	42.1	40.3	43.2	41.5	38.1	41.3	37.5	38.4	36.2	36.5
1970	27.8	28.1	28.2	28.9	27.0	26.3	25.7	22.8	27.0	26.2	24.5	24.2	23.8	25.2	22.8	23.4
1965	21.5	21.6	21.7	22.2	20.9	21.8	19.8	17.4	19.3	20.2	18.9	18.6	18.3	19.8	17.6	18.0
1960	19.5	19.7	19.7	20.2	18.9	19.4	18.0	15.8	17.6	18.3	17.2	16.9	16.7	17.1	16.0	16.4
1955	16.3	16.5	16.5	17.0	15.9	16.3	15.1	13.2	14.7	15.4	14.4	14.1	14.0	14.4	13.4	13.7
1950	13.5	13.6	13.7	14.0	13.1	13.4	12.5	11.0	12.2	12.7	11.9	11.7	11.5	11.9	11.0	11.3

Year	National 30 City Average	Georgia Savannah	Hawaii Honolulu	Idaho Boise	Idaho Pocatello	Illinois Chicago	Illinois Decatur	Illinois Joliet	Illinois Peoria	Illinois Rockford	Illinois Springfield	Indiana Anderson	Indiana Evansville	Indiana Fort Wayne	Indiana Gary	Indiana Indianapolis
Jan 2018	215.8	182.3E	252.2E	194.5E	195.4E	254.2E	217.5E	250.4E	222.6E	235.5E	218.2E	191.7E	194.2E	187.1E	219E	193.7E
2017	209.4	180.5	249.7	192.5	193.4	251.7	215.3	247.9	220.4	233.1	216.0	189.8	192.3	185.2	216.8	191.8
2016	207.7	175.1	251.7	191.1	190.2	244.9	212.9	244.3	217.5	229.6	215.1	190.5	192.3	184.4	216.8	192.5
2015	204.0	170.5	250.5	185.4	185.3	238.9	209.0	238.7	213.7	225.3	210.5	186.4	190.1	183.0	212.4	189.6
2014	203.0	167.2	239.7	184.1	184.4	238.1	206.0	236.8	212.5	224.4	208.2	184.5	188.1	180.8	210.1	188.4
2013	196.9	163.5	231.5	179.6	179.9	231.1	200.8	228.7	207.2	217.2	203.2	177.8	182.5	175.9	203.6	182.7
2012	194.0	158.9	227.9	170.4	171.5	226.2	195.7	222.7	199.7	211.8	197.2	175.2	177.8	171.8	197.9	180.6
2011	185.7	151.7	219.0	162.8	163.6	217.6	187.9	216.8	193.4	205.8	189.4	168.4	169.7	165.0	191.0	173.0
2010	181.6	147.9	214.6	161.7	162.4	210.6	181.2	208.3	186.2	197.6	182.2	162.8	165.9	160.0	183.7	168.1
2009	182.5	152.8	218.6	162.6	163.4	209.0	182.2	205.9	186.2	196.5	184.0	164.9	166.9	163.0	185.2	170.3
2008	171.0	141.1	205.5	153.2	152.8	195.7	168.7	185.5	172.2	180.2	168.5	152.4	155.8	151.0	169.0	159.2
2007	165.0	136.9	200.4	148.0	148.1	186.5	160.5	177.7	166.0	175.0	159.2	149.0	152.6	147.7	165.4	154.1
2006	156.2	124.0	191.9	141.6	141.2	177.8	153.8	168.7	155.2	163.3	152.8	139.8	144.4	139.1	154.5	144.6
2005	146.7	116.9	181.2	133.8	132.5	164.4	145.0	158.3	146.5	151.2	145.4	131.4	136.2	131.0	146.4	137.4
2004	132.8	105.6	161.2	122.0	120.5	149.3	129.3	145.1	133.9	138.1	130.3	120.9	123.5	120.1	132.2	125.0
2003	129.7	103.0	159.4	120.2	118.7	146.2	127.4	143.5	132.4	137.3	128.4	119.7	121.4	118.2	131.4	122.5
2002	126.7	102.0	157.2	118.3	117.1	141.2	123.8	138.5	129.6	132.9	125.3	117.5	119.0	116.4	129.1	120.5
2001	122.2	99.0	150.0	114.3	113.4	135.8	120.1	133.7	124.3	127.8	119.8	113.4	115.6	112.1	123.4	116.4
2000	118.9	97.5	144.8	112.9	112.1	131.2	115.1	124.6	119.0	122.2	116.2	109.8	111.5	108.4	117.8	113.2
1999	116.6	96.0	143.0	110.2	109.6	129.6	113.0	122.6	116.4	120.7	113.8	107.1	109.3	106.8	112.8	110.6
1998	113.6	93.7	140.4	107.4	107.0	125.2	110.1	119.8	113.8	115.5	111.1	105.0	107.2	104.5	111.1	108.0
1995	105.6	87.4	130.3	99.5	98.2	114.2	98.5	110.5	102.3	103.6	98.1	96.4	97.2	95.0	100.7	100.1
1990	93.2	77.9	104.7	88.2	88.1	98.4	90.9	98.4	93.7	94.0	90.1	84.6	89.3	83.4	88.4	87.1
1985	81.8	68.9	94.7	78.0	78.0	82.4	81.9	83.4	83.7	83.0	81.5	75.2	79.8	75.0	77.8	77.1
1980	60.7	52.2	68.9	60.3	59.5	62.8	62.3	63.4	64.5	61.6	61.1	56.5	59.0	56.7	59.8	57.9
1975	43.7	36.9	44.6	40.8	40.5	45.7	43.1	44.5	44.7	42.7	42.4	39.5	41.7	39.9	41.9	40.6
1970	27.8	21.0	30.4	26.7	26.6	29.1	28.0	28.6	29.0	27.5	27.6	25.4	26.4	25.5	27.1	26.2
1965	21.5	16.4	21.8	20.6	20.5	22.7	21.5	22.1	22.4	21.2	21.3	19.5	20.4	19.7	20.8	20.7
1960	19.5	14.9	19.8	18.7	18.6	20.2	19.6	20.0	20.3	19.2	19.3	17.7	18.7	17.9	18.9	18.4
1955	16.3	12.5	16.6	15.7	15.6	16.9	16.4	16.8	17.0	16.1	16.2	14.9	15.7	15.0	15.9	15.5
1950	13.5	10.3	13.7	13.0	12.9	14.0	13.6	13.9	14.0	13.3	13.4	12.3	12.9	12.4	13.1	12.8

Year	National 30 City Average	Indiana Muncie	South Bend	Terre Haute	Iowa Cedar Rapids	Davenport	Des Moines	Sioux City	Waterloo	Kansas Topeka	Wichita	Kentucky Lexington	Louisville	Louisiana Baton Rouge	Lake Charles	New Orleans
Jan 2018	215.8	191.4E	194.3E	195.1E	196.8E	207.8E	197E	190.1E	190.4E	190E	184.9E	188.1E	186.9E	181E	180E	181.1E
2017	209.4	189.5	192.4	193.1	194.8	205.7	195.0	188.2	188.5	188.1	183.0	186.2	185.0	179.2	178.2	179.3
2016	207.7	191.0	190.7	192.8	191.4	199.9	190.0	184.0	181.0	181.0	191.0	186.0	188.1	180.1	176.7	180.0
2015	204.0	186.0	188.6	190.6	188.8	196.9	189.0	179.1	177.8	175.7	175.1	184.4	185.9	176.6	175.4	177.6
2014	203.0	185.0	185.5	189.2	188.3	195.9	188.5	178.9	177.4	173.6	173.7	183.7	186.1	171.3	170.9	177.5
2013	196.9	179.3	179.3	182.6	183.4	190.8	181.8	173.2	171.3	168.8	167.3	179.1	182.4	166.3	166.2	173.1
2012	194.0	175.8	174.3	178.9	180.3	184.6	179.2	168.8	169.2	162.3	161.4	174.8	177.5	164.4	163.8	171.7
2011	185.7	169.2	168.1	171.3	173.8	178.1	172.7	162.3	162.8	156.5	155.3	167.8	170.8	156.9	156.5	162.1
2010	181.6	163.1	163.5	167.8	170.4	175.6	170.2	159.0	160.2	152.8	151.3	158.7	166.9	153.7	153.4	160.3
2009	182.5	164.5	167.1	168.0	165.9	172.0	162.1	156.3	147.6	154.5	152.4	160.7	167.5	156.9	154.4	161.9
2008	171.0	153.0	153.1	155.9	157.3	163.4	152.4	147.8	139.2	144.9	143.1	152.0	156.8	144.0	141.5	149.3
2007	165.0	149.7	149.5	152.7	152.4	157.6	148.9	144.0	135.0	141.0	138.7	147.6	150.9	139.4	138.0	145.2
2006	156.2	141.2	141.3	144.0	145.8	151.1	142.9	137.3	128.4	134.7	132.9	129.0	143.0	129.4	129.8	135.8
2005	146.7	131.0	133.8	135.2	135.7	142.1	133.5	129.4	120.0	125.9	125.6	121.5	135.1	120.8	117.7	126.2
2004	132.8	120.0	120.6	123.0	122.7	128.6	121.6	116.2	108.3	112.6	113.4	110.1	120.3	105.6	109.0	115.3
2003	129.7	118.7	119.2	121.7	120.9	126.4	120.2	114.6	105.9	109.4	111.3	107.8	118.7	103.3	106.1	112.4
2002	126.7	116.7	117.1	119.9	116.0	121.1	116.7	111.4	103.7	107.8	109.6	106.2	116.4	102.5	105.0	110.6
2001	122.2	112.8	111.6	115.5	112.5	117.5	113.2	107.7	100.6	104.3	105.2	103.3	112.7	99.4	101.3	104.6
2000	118.9	109.1	106.9	110.8	108.8	112.6	108.9	99.0	98.1	101.0	101.1	101.4	109.3	97.8	99.7	102.1
1999	116.6	105.7	103.7	107.9	104.1	109.2	107.6	96.7	96.4	98.7	99.3	99.7	106.6	96.1	97.6	99.5
1998	113.6	103.6	102.3	106.0	102.5	106.6	103.8	95.1	94.8	97.4	97.4	97.4	101.5	94.7	96.1	97.7
1995	105.6	95.6	94.6	97.2	95.4	96.6	94.4	88.2	88.9	91.1	91.0	91.6	94.6	89.6	92.7	91.6
1990	93.2	83.9	85.1	89.1	87.1	86.5	86.9	80.4	80.3	83.2	82.6	82.8	82.6	82.0	84.1	84.5
1985	81.8	75.2	76.0	79.2	75.9	77.6	77.1	72.1	72.3	75.0	74.7	75.5	74.5	74.9	78.5	78.2
1980	60.7	56.1	58.3	59.7	62.7	59.6	61.8	59.3	57.7	58.9	58.0	59.3	59.8	59.1	60.0	57.2
1975	43.7	39.2	40.3	41.9	43.1	41.0	43.1	41.0	40.0	42.0	43.1	42.7	42.5	40.4	40.6	41.5
1970	27.8	25.3	26.2	27.0	28.2	26.9	27.6	26.6	25.9	27.0	25.5	26.9	25.9	25.0	26.7	27.2
1965	21.5	19.5	20.2	20.9	21.7	20.8	21.7	20.5	20.0	20.8	19.6	20.7	20.3	19.4	20.6	20.4
1960	19.5	17.8	18.3	18.9	19.7	18.8	19.5	18.6	18.2	18.9	17.8	18.8	18.4	17.6	18.7	20.4
1955	16.3	14.9	15.4	15.9	16.6	15.8	16.3	15.6	15.2	15.8	15.0	15.8	15.4	14.8	15.7	18.5
1950	13.5	12.3	12.7	13.1	13.7	13.1	13.5	12.9	12.6	13.1	12.4	13.0	12.8	12.2	13.0	12.8

Year	National 30 City Average	Louisiana Shreve-port	Maine Lewis-ton	Portland	Maryland Balti-more	Massachusetts Boston	Brockton	Fall River	Law-rence	Lowell	New Bedford	Pitts-field	Spring-field	Wor-cester	Michigan Ann Arbor	Dear-born
Jan 2018	215.8	179.7E	193.9E	196.7E	199.7E	242.4E	228.7E	227.3E	236.3E	233.6E	226.7E	213.8E	215.7E	228.4E	210.8E	213.2E
2017	209.4	177.9	192.0	194.7	197.7	240.0	226.4	225.0	233.9	231.3	224.4	211.7	213.5	226.1	208.7	211.1
2016	207.7	173.2	197.2	200.0	193.8	245.0	230.5	229.0	236.9	235.5	228.2	216.1	218.3	230.1	211.3	212.5
2015	204.0	170.9	193.7	196.9	189.2	240.7	228.3	228.2	234.5	233.2	227.4	212.8	214.6	226.8	209.8	211.5
2014	203.0	167.7	192.4	195.4	187.6	239.1	229.1	229.3	234.8	233.9	228.5	213.3	214.8	227.3	206.1	208.4
2013	196.9	159.0	187.9	189.5	183.2	232.9	221.6	221.8	225.3	223.4	221.1	205.9	206.5	219.3	201.2	202.4
2012	194.0	153.2	180.7	182.4	180.8	229.3	217.2	217.8	221.6	219.3	217.0	200.2	201.9	215.2	196.1	199.2
2011	185.7	146.7	167.4	169.1	173.6	219.3	208.2	206.0	211.6	210.1	205.2	191.2	194.5	206.6	186.7	189.4
2010	181.6	145.1	162.1	163.7	168.3	214.3	202.7	200.7	204.9	204.7	200.0	185.9	189.4	201.1	183.3	186.8
2009	182.5	147.6	159.4	161.8	169.1	211.8	199.2	197.6	201.9	201.9	196.9	185.1	187.1	198.2	179.2	186.9
2008	171.0	135.9	149.1	150.9	157.7	198.6	186.4	185.1	188.7	189.1	184.5	171.3	173.7	184.5	169.6	176.4
2007	165.0	132.4	146.4	148.2	152.5	191.8	180.4	179.7	182.0	181.7	179.1	166.8	169.1	180.4	166.6	172.5
2006	156.2	125.2	140.2	139.8	144.9	180.4	171.6	170.4	172.3	173.5	170.4	157.3	159.7	171.6	161.1	165.5
2005	146.7	117.4	131.7	131.4	135.8	169.6	161.3	159.6	162.2	162.3	159.6	147.6	150.7	158.8	147.5	155.8
2004	132.8	105.7	119.8	119.4	121.4	154.1	144.4	143.6	146.4	146.7	143.6	131.9	136.5	143.2	135.9	142.1
2003	129.7	103.9	117.7	117.3	118.1	150.2	139.6	140.6	143.3	143.6	140.5	128.9	133.8	139.0	134.1	139.0
2002	126.7	102.1	117.5	117.1	115.6	145.6	136.0	135.0	136.9	137.3	134.9	124.7	128.3	134.9	131.4	134.3
2001	122.2	98.3	114.6	114.3	111.7	140.9	132.1	131.4	133.1	132.9	131.3	120.3	124.5	130.1	126.8	129.8
2000	118.9	96.2	105.7	105.4	107.7	138.9	129.4	128.4	129.7	130.3	128.3	116.7	121.0	127.3	124.3	125.8
1999	116.6	94.8	105.0	104.7	106.4	136.2	126.7	126.3	127.3	127.4	126.2	115.0	118.8	123.8	117.5	122.7
1998	113.6	92.0	102.8	102.5	104.1	132.8	125.0	124.7	125.0	125.3	124.6	114.2	117.1	122.6	116.0	119.7
1995	105.6	86.7	96.8	96.5	96.1	128.6	119.6	117.7	120.5	119.6	117.2	110.7	112.7	114.8	106.1	110.5
1990	93.2	80.3	88.5	88.6	85.6	110.9	103.7	102.9	105.2	102.8	102.6	98.7	101.4	103.2	93.2	96.5
1985	81.8	73.6	76.7	77.0	72.7	92.8	88.8	88.7	89.4	88.2	88.6	85.0	85.6	86.7	80.1	82.7
1980	60.7	58.7	57.3	58.5	53.6	64.0	63.7	64.1	63.4	62.7	63.1	61.8	62.0	62.3	62.9	64.0
1975	43.7	40.5	41.9	42.1	39.8	46.6	45.8	45.7	46.2	45.7	46.1	45.5	45.8	46.0	44.1	44.9
1970	27.8	26.4	26.5	25.8	25.1	29.2	29.1	29.2	29.1	28.9	29.0	28.6	28.5	28.7	28.5	28.9
1965	21.5	20.3	20.4	19.4	20.2	23.0	22.5	22.5	22.5	22.2	22.4	22.0	22.4	22.1	22.0	22.3
1960	19.5	18.5	18.6	17.6	17.5	20.5	20.4	20.4	20.4	20.2	20.3	20.0	20.1	20.1	20.0	20.2
1955	16.3	15.5	15.6	14.7	14.7	17.2	17.1	17.1	17.1	16.9	17.1	16.8	16.9	16.8	16.7	16.9
1950	13.5	12.8	12.9	12.2	12.1	14.2	14.1	14.1	14.1	14.0	14.1	13.8	14.0	13.9	13.8	14.0

530

Historical Cost Indexes

Year	National 30 City Average	Michigan						Minnesota			Mississippi		Missouri			
		Detroit	Flint	Grand Rapids	Kala-mazoo	Lansing	Sagi-naw	Duluth	Minne-apolis	Roches-ter	Biloxi	Jackson	Kansas City	St. Joseph	St. Louis	Spring-field
Jan 2018	215.8	214.2E	201.2E	192.8E	191.5E	200.1E	197.3E	214.2E	223.4E	209.7E	177.1E	177.4E	215.1E	204.9E	215.3E	194.9E
2017	209.4	212.1	199.2	190.9	189.6	198.1	195.3	212.1	221.2	207.6	175.3	175.6	212.9	202.8	213.1	192.9
2016	207.7	212.8	200.2	191.0	189.7	200.9	195.9	217.6	227.2	212.4	173.3	177.7	213.2	205.5	212.1	194.6
2015	204.0	211.8	198.9	191.2	190.4	199.6	194.1	212.9	222.0	209.1	167.3	174.2	210.0	202.3	210.1	191.8
2014	203.0	208.7	196.0	187.9	188.4	195.7	191.7	211.1	220.7	207.5	164.6	169.8	210.8	199.3	208.6	189.5
2013	196.9	203.1	190.6	182.4	182.4	190.2	186.5	205.1	216.3	201.4	160.9	164.3	204.7	193.1	202.4	181.6
2012	194.0	200.2	186.3	176.4	178.9	185.9	182.8	203.1	214.7	199.7	157.6	160.5	200.8	189.0	198.0	177.7
2011	185.7	190.3	178.5	169.9	171.9	178.0	174.6	195.7	208.1	193.8	151.4	154.3	191.1	178.8	190.6	168.3
2010	181.6	187.5	176.2	160.6	167.3	175.8	171.9	193.1	203.8	188.9	148.0	151.0	186.1	174.2	185.9	164.3
2009	182.5	187.8	176.0	156.7	166.3	173.7	168.6	191.8	203.1	188.0	152.8	156.6	185.6	172.3	187.2	163.7
2008	171.0	177.0	164.4	140.2	155.7	161.6	158.7	177.4	190.6	175.0	141.8	147.1	175.5	164.2	176.2	152.9
2007	165.0	172.9	161.7	137.2	152.6	158.7	156.1	174.2	184.5	171.4	137.7	131.2	169.0	157.8	170.6	145.3
2006	156.2	165.8	153.8	131.1	146.0	152.5	150.6	166.2	173.9	161.9	123.3	117.1	162.0	152.4	159.3	139.4
2005	146.7	156.2	145.1	123.1	134.9	142.0	140.6	157.3	164.6	152.8	116.0	109.9	151.3	143.1	149.4	131.7
2004	132.8	141.9	129.6	112.8	122.5	129.7	127.6	139.1	150.1	137.0	105.3	99.3	135.1	126.9	135.5	116.3
2003	129.7	138.7	127.9	110.9	120.8	127.7	126.0	136.7	146.5	134.5	101.4	96.7	131.9	124.8	133.3	113.9
2002	126.7	134.2	126.9	107.4	119.9	125.3	124.4	131.9	139.5	130.0	101.0	96.3	128.4	122.7	129.6	112.5
2001	122.2	129.4	122.4	104.3	115.2	120.1	119.7	131.4	136.1	124.9	98.7	93.9	121.8	114.7	125.5	106.7
2000	118.9	125.3	119.8	102.7	111.6	117.6	115.9	124.1	131.1	120.1	97.2	92.9	118.2	111.9	122.4	104.3
1999	116.6	122.6	113.7	100.9	106.1	111.5	110.1	120.3	126.5	117.1	95.7	91.8	114.9	106.1	119.8	101.1
1998	113.6	119.5	112.3	99.7	104.9	110.1	108.7	117.7	124.6	115.5	92.1	89.5	108.2	104.2	115.9	98.9
1995	105.6	110.1	104.1	91.1	96.4	98.3	102.3	100.3	111.9	102.5	84.2	83.5	99.7	96.1	106.3	89.5
1990	93.2	96.0	91.7	84.0	87.4	90.2	88.9	94.3	100.5	95.6	76.4	75.6	89.5	86.8	94.0	82.6
1985	81.8	81.6	80.7	75.8	79.4	80.0	80.5	85.2	87.9	86.5	69.4	68.5	78.2	77.3	80.8	73.2
1980	60.7	62.5	63.1	59.8	61.7	60.3	63.3	63.4	64.1	64.8	54.3	54.3	59.1	60.4	59.7	56.7
1975	43.7	45.8	44.9	41.8	43.7	44.1	44.8	45.2	46.0	45.3	37.9	37.7	42.7	45.4	44.8	41.1
1970	27.8	29.7	28.5	26.7	28.1	27.8	28.7	28.9	29.5	28.9	24.4	21.3	25.3	28.3	28.4	26.0
1965	21.5	22.1	21.9	20.6	21.7	21.5	22.2	22.3	23.4	22.3	18.8	16.4	20.6	21.8	21.8	20.0
1960	19.5	20.1	20.0	18.7	19.7	19.5	20.1	20.3	20.5	20.2	17.1	14.9	19.0	19.8	19.5	18.2
1955	16.3	16.8	16.7	15.7	16.5	16.3	16.9	17.0	17.2	17.0	14.3	12.5	16.0	16.6	16.3	15.2
1950	13.5	13.9	13.8	13.0	13.6	13.5	13.9	14.0	14.2	14.0	11.8	10.3	13.2	13.7	13.5	12.6

Year	National 30 City Average	Montana		Nebraska		Nevada		New Hampshire		New Jersey					NM	NY
		Billings	Great Falls	Lincoln	Omaha	Las Vegas	Reno	Man-chester	Nashua	Camden	Jersey City	Newark	Pater-son	Trenton	Albu-querque	Albany
Jan 2018	215.8	192.3E	192.9E	191.5E	192E	222.3E	201.8E	202.7E	202E	238.7E	241.1E	245.2E	243.2E	241.6E	187.8E	215.7E
2017	209.4	190.4	191.0	189.6	190.1	220.1	199.8	200.7	200.0	236.3	238.7	242.7	240.8	239.2	185.9	213.5
2016	207.7	191.8	191.1	186.8	187.6	221.1	198.8	200.9	200.5	228.8	229.6	232.9	231.3	229.3	181.6	209.6
2015	204.0	188.1	188.4	183.7	184.7	215.5	194.9	198.7	197.8	224.5	226.1	230.1	228.5	227.6	178.3	208.1
2014	203.0	185.4	186.3	181.1	183.7	210.9	195.7	198.8	197.3	224.6	226.6	230.6	227.8	226.6	178.0	205.1
2013	196.9	180.5	181.1	175.5	180.6	206.8	190.4	193.8	192.5	217.7	219.3	223.6	221.2	218.7	173.6	195.3
2012	194.0	178.6	179.5	169.3	177.1	202.6	186.8	189.2	188.3	213.6	215.2	219.2	217.5	213.7	171.0	191.2
2011	185.7	168.8	170.8	162.8	169.6	195.7	179.8	176.8	176.1	205.5	207.0	210.3	209.2	206.4	163.3	180.9
2010	181.6	166.5	168.6	159.3	165.8	193.7	175.6	172.5	172.0	201.2	203.3	206.3	205.3	200.8	162.5	177.8
2009	182.5	165.3	166.0	161.7	165.0	191.5	176.5	174.0	173.5	196.8	200.5	203.6	202.0	198.6	163.0	178.1
2008	171.0	153.1	154.6	152.0	154.8	176.2	167.0	162.5	161.9	184.0	187.3	189.6	188.7	185.6	152.6	166.0
2007	165.0	147.7	147.8	147.7	150.3	166.7	161.4	159.3	158.7	179.3	183.3	185.2	184.6	181.7	146.7	159.4
2006	156.2	140.5	140.8	132.4	140.7	160.0	154.5	146.7	146.4	167.7	171.0	173.6	172.2	169.9	140.1	151.6
2005	146.7	131.9	132.4	124.1	132.8	149.5	145.4	136.8	136.5	159.0	162.4	163.9	163.0	161.5	130.4	142.2
2004	132.8	118.2	117.9	112.1	119.4	137.1	130.5	124.2	123.9	143.4	145.4	147.2	146.6	146.0	118.3	128.6
2003	129.7	115.8	115.8	110.2	117.3	133.8	128.3	122.4	122.2	142.0	144.3	145.7	145.0	143.0	116.4	126.8
2002	126.7	114.6	114.5	108.2	115.0	131.9	126.4	119.9	119.6	135.9	138.5	140.2	140.1	137.6	114.6	122.6
2001	122.2	117.5	117.7	101.3	111.6	127.8	122.5	116.2	116.2	133.4	136.7	136.9	136.8	136.0	111.4	119.2
2000	118.9	113.7	113.9	98.8	107.0	125.8	118.2	111.9	111.9	128.4	130.5	132.6	132.4	130.6	109.0	116.5
1999	116.6	112.1	112.7	96.6	104.8	121.9	114.4	109.6	109.6	125.3	128.8	131.6	129.5	129.6	106.7	114.6
1998	113.6	109.7	109.1	94.9	101.2	118.1	111.4	110.1	110.0	124.3	127.7	128.9	128.5	127.9	103.8	113.0
1995	105.6	104.7	104.9	85.6	93.4	108.5	105.0	100.9	100.8	107.0	112.2	111.9	112.1	111.2	96.3	103.6
1990	93.2	92.9	93.8	78.8	85.0	96.3	94.5	86.3	86.3	93.0	93.5	93.8	97.2	94.5	84.9	93.2
1985	81.8	83.9	84.3	71.5	77.5	87.6	85.0	78.1	78.1	81.3	83.3	83.5	84.5	82.8	76.7	79.5
1980	60.7	63.9	64.6	58.5	63.5	64.6	63.0	56.6	56.0	58.6	60.6	60.1	60.0	58.9	59.0	59.5
1975	43.7	43.1	43.8	40.9	43.2	42.8	41.9	41.3	40.8	42.3	43.4	44.2	43.9	43.8	40.3	43.9
1970	27.8	28.5	28.9	26.4	26.8	29.4	28.0	26.2	25.6	27.2	27.8	29.0	27.8	27.4	26.4	28.3
1965	21.5	22.0	22.3	20.3	20.6	22.4	21.6	20.6	19.7	20.9	21.4	23.8	21.4	21.3	20.6	22.3
1960	19.5	20.0	20.3	18.5	18.7	20.2	19.6	18.0	17.9	19.0	19.4	19.4	19.4	19.2	18.5	19.3
1955	16.3	16.7	17.0	15.5	15.7	16.9	16.4	15.1	15.0	16.0	16.3	16.3	16.3	16.1	15.6	16.2
1950	13.5	13.9	14.0	12.8	13.0	14.0	13.6	12.5	12.4	13.2	13.5	13.5	13.5	13.3	12.9	13.4

531

Historical Cost Indexes

Year	National 30 City Average	New York Binghamton	Buffalo	New York	Rochester	Schenectady	Syracuse	Utica	Yonkers	North Carolina Charlotte	Durham	Greensboro	Raleigh	Winston-Salem	N. Dakota Fargo	Ohio Akron
Jan 2018	215.8	210.5E	223E	285.8E	215.2E	216E	211E	208E	259.4E	178.8E	182.1E	180E	177.6E	179.8E	194.8E	201.4E
2017	209.4	208.4	220.8	282.9	213.0	213.8	208.9	205.9	256.8	177.0	180.3	178.2	175.8	178.0	192.8	199.4
2016	207.7	202.6	213.7	270.5	208.4	209.8	203.6	200.8	247.9	180.6	182.0	179.8	174.6	179.6	185.5	202.4
2015	204.0	202.4	209.6	268.0	203.5	207.4	200.9	198.1	243.4	173.1	172.3	170.5	167.6	170.6	181.0	199.1
2014	203.0	202.4	207.8	267.7	202.2	204.2	198.8	196.4	245.6	165.5	165.2	163.9	161.6	164.3	177.1	199.2
2013	196.9	195.5	200.7	259.4	194.3	195.5	193.6	188.5	230.0	159.6	158.4	158.7	157.5	159.1	169.6	191.8
2012	194.0	192.9	198.1	256.3	191.2	191.4	189.6	185.2	227.1	156.4	155.6	156.3	155.4	157.2	167.3	187.7
2011	185.7	176.1	188.1	245.6	181.9	181.5	180.8	173.5	218.9	142.6	143.8	142.2	142.8	140.9	160.3	180.8
2010	181.6	173.8	184.0	241.4	179.6	179.0	176.7	170.5	217.1	140.5	141.6	140.0	140.6	138.8	156.2	174.5
2009	182.5	172.9	184.4	239.9	179.6	178.8	176.9	171.3	217.5	145.1	145.8	144.0	145.1	142.7	154.3	175.7
2008	171.0	160.7	173.9	226.8	168.0	167.2	165.9	161.6	202.1	135.8	136.6	134.6	135.3	133.8	145.0	165.4
2007	165.0	155.7	168.6	215.2	163.5	159.8	159.8	155.0	196.5	132.8	133.7	131.7	132.1	131.0	139.8	159.0
2006	156.2	147.5	159.2	204.5	155.2	151.1	150.9	146.4	186.0	125.1	124.6	123.8	124.2	123.0	133.2	152.8
2005	146.7	137.5	149.4	194.0	147.3	142.2	141.9	136.9	176.1	110.5	112.1	112.1	112.1	111.1	125.3	144.4
2004	132.8	123.5	136.1	177.7	132.0	128.4	127.3	124.4	161.8	98.9	100.0	100.1	100.3	99.4	113.0	131.9
2003	129.7	121.8	132.8	173.4	130.3	126.7	124.7	121.7	160.0	96.2	97.5	97.5	97.8	96.8	110.6	129.6
2002	126.7	119.0	128.5	170.1	127.1	123.0	121.8	118.4	154.8	94.8	96.1	96.1	96.3	95.5	106.3	127.2
2001	122.2	116.0	125.2	164.4	123.1	120.1	118.6	115.3	151.4	91.5	92.9	92.9	93.3	92.3	103.1	123.5
2000	118.9	112.4	122.3	159.2	120.0	117.5	115.1	112.4	144.8	90.3	91.6	91.6	91.9	91.0	97.9	117.8
1999	116.6	108.6	120.2	155.9	116.8	114.7	113.7	108.5	140.6	89.3	90.2	90.3	90.5	90.1	96.8	116.0
1998	113.6	109.0	119.1	154.4	117.2	114.2	113.6	108.6	141.4	88.2	89.1	89.2	89.4	89.0	95.2	113.1
1995	105.6	99.3	110.1	140.7	106.6	104.6	104.0	97.7	129.4	81.8	82.6	82.6	82.7	82.6	88.4	103.4
1990	93.2	87.0	94.1	118.1	94.6	93.9	91.0	85.6	111.4	74.8	75.4	75.5	75.5	75.3	81.3	94.6
1985	81.8	77.5	83.2	94.9	81.6	80.1	81.0	77.0	92.8	66.9	67.6	67.7	67.6	67.5	73.4	86.8
1980	60.7	58.0	60.6	66.0	60.5	60.3	61.6	58.5	65.9	51.1	52.2	52.5	51.7	50.8	57.4	62.3
1975	43.7	42.5	45.1	49.5	44.8	43.7	44.8	42.7	47.1	36.1	37.0	37.0	37.3	36.1	39.3	44.7
1970	27.8	27.0	28.9	32.8	29.2	27.8	28.5	26.9	30.0	20.9	23.7	23.6	23.4	23.0	25.8	28.3
1965	21.5	20.8	22.2	25.5	22.8	21.4	21.9	20.7	23.1	16.0	18.2	18.2	18.0	17.7	19.9	21.8
1960	19.5	18.9	19.9	21.5	19.7	19.5	19.9	18.8	21.0	14.4	16.6	16.6	16.4	16.1	18.1	19.9
1955	16.3	15.8	16.7	18.1	16.5	16.3	16.7	15.8	17.6	12.1	13.9	13.9	13.7	13.5	15.2	16.6
1950	13.5	13.1	13.8	14.9	13.6	13.5	13.8	13.0	14.5	10.0	11.5	11.5	11.4	11.2	12.5	13.7

Year	National 30 City Average	Ohio Canton	Cincinnati	Cleveland	Columbus	Dayton	Lorain	Springfield	Toledo	Youngstown	Oklahoma Lawton	Oklahoma City	Tulsa	Oregon Eugene	Portland	PA Allentown
Jan 2018	215.8	194.1E	188E	204.7E	193.5E	188E	198.9E	189.7E	203.4E	196.2E	179.8E	178.8E	176.6E	209.4E	211.6E	215.9E
2017	209.4	192.2	186.1	202.6	191.6	186.1	196.9	187.8	201.4	194.2	178.0	177.0	174.8	207.3	209.5	213.7
2016	207.7	195.0	190.1	205.9	195.2	189.3	199.9	190.6	206.1	197.9	177.7	180.1	172.0	209.1	209.2	214.8
2015	204.0	191.1	188.1	203.4	192.4	187.9	196.7	188.1	201.0	193.3	175.1	177.1	170.0	203.1	204.3	209.8
2014	203.0	190.8	188.1	203.1	192.4	185.8	196.7	187.2	199.7	193.8	169.3	170.9	166.7	200.4	201.4	209.0
2013	196.9	183.9	181.9	195.5	187.1	180.7	187.6	181.5	193.5	186.5	162.7	165.5	161.7	194.7	195.6	203.2
2012	194.0	180.2	178.6	191.4	183.5	177.6	184.1	178.0	190.4	183.3	158.1	158.0	151.3	190.1	191.8	199.3
2011	185.7	172.1	171.3	184.0	175.7	168.4	176.8	168.4	183.1	176.5	151.7	151.6	144.8	184.9	186.6	192.2
2010	181.6	167.5	166.5	180.2	170.3	163.4	173.0	164.6	176.5	171.7	149.5	149.4	143.1	181.5	182.9	187.9
2009	182.5	168.4	168.1	181.6	171.2	165.3	174.0	166.0	178.5	172.8	152.5	153.3	146.8	181.3	183.6	188.3
2008	171.0	158.4	159.0	170.6	160.2	157.5	163.9	157.5	168.2	161.5	141.3	140.6	136.2	172.3	174.7	175.7
2007	165.0	153.2	152.4	164.9	155.1	149.4	158.1	150.3	161.6	156.7	136.2	135.6	132.5	168.5	169.5	169.4
2006	156.2	147.0	144.9	156.9	146.9	142.9	151.8	143.4	154.5	150.3	129.5	129.5	125.7	160.5	161.7	160.1
2005	146.7	137.7	136.8	147.9	138.0	134.5	143.3	134.8	145.5	141.1	121.6	121.6	117.8	150.9	152.2	150.2
2004	132.8	125.6	124.3	135.6	126.2	121.1	131.5	122.0	132.8	129.3	109.3	109.4	107.3	136.8	137.8	133.9
2003	129.7	123.7	121.3	132.7	123.7	119.4	126.1	120.0	130.4	126.2	107.5	107.8	105.0	134.2	135.9	130.3
2002	126.7	120.9	119.2	130.0	120.9	117.5	124.3	118.0	128.4	123.6	106.3	106.0	104.1	132.2	133.9	128.1
2001	122.2	117.7	115.9	125.9	117.1	114.0	120.4	114.7	123.8	119.9	102.0	102.1	99.7	129.8	131.2	123.5
2000	118.9	112.7	110.1	121.3	112.5	109.0	115.0	109.2	115.7	114.1	98.7	98.9	97.5	126.3	127.4	119.9
1999	116.6	110.6	107.9	118.7	109.5	107.1	112.0	106.4	113.6	111.9	97.5	97.4	96.2	120.9	124.3	117.8
1998	113.6	108.5	105.0	114.8	106.8	104.5	109.4	104.1	111.2	109.0	94.3	94.5	94.5	120.0	122.2	115.7
1995	105.6	98.8	97.1	106.4	99.1	94.6	97.1	92.0	100.6	100.1	85.3	88.0	89.0	112.2	114.3	108.3
1990	93.2	92.1	86.7	95.1	88.1	85.9	89.4	83.4	92.9	91.8	77.1	79.9	80.0	98.0	99.4	95.0
1985	81.8	83.7	78.2	86.2	77.7	76.3	80.7	74.3	84.7	82.7	71.2	73.9	73.8	88.7	90.1	81.9
1980	60.7	60.6	59.8	61.0	58.6	56.6	60.8	56.2	64.0	61.5	52.2	55.5	57.2	68.9	68.4	58.7
1975	43.7	44.0	43.7	44.6	42.2	40.5	42.7	40.0	44.9	44.5	38.6	38.6	39.2	44.6	44.9	42.3
1970	27.8	27.8	28.2	29.6	26.7	27.0	27.5	25.4	29.1	29.6	24.1	23.4	24.1	30.4	30.0	27.3
1965	21.5	21.5	20.7	21.1	20.2	20.0	21.1	19.6	21.3	21.0	18.5	17.8	20.6	23.4	22.6	21.0
1960	19.5	19.5	19.3	19.6	18.7	18.1	19.2	17.8	19.4	19.1	16.9	16.1	18.1	21.2	21.1	19.1
1955	16.3	16.3	16.2	16.5	15.7	15.2	16.1	14.9	16.2	16.0	14.1	13.5	15.1	17.8	17.7	16.0
1950	13.5	13.5	13.3	13.6	13.0	12.5	13.3	12.3	13.4	13.2	11.7	11.2	12.5	14.7	14.6	13.2

Historical Cost Indexes

Year	National 30 City Average	Pennsylvania Erie	Harris-burg	Phila-delphia	Pitts-burgh	Reading	Scranton	RI Provi-dence	South Carolina Charles-ton	Colum-bia	South Dakota Rapid City	Sioux Falls	Tennessee Chatta-nooga	Knox-ville	Memphis	Nash-ville
Jan 2018	215.8	201.5E	204.5E	242.2E	215.8E	212.8E	209.5E	220.8E	180.2E	178.3E	187.2E	186.7E	183.7E	175.1E	182.7E	182.3E
2017	209.4	199.5	202.4	239.8	213.6	210.7	207.4	218.6	178.4	176.5	185.3	184.8	181.9	173.3	180.9	180.5
2016	207.7	199.7	202.4	238.2	212.2	207.6	207.0	221.6	176.8	176.6	174.7	175.2	177.4	173.9	181.9	181.8
2015	204.0	194.5	200.3	234.7	209.5	204.2	201.7	219.9	171.3	171.9	167.7	168.0	174.3	171.0	177.5	179.9
2014	203.0	191.4	200.3	232.1	207.1	201.4	201.4	220.3	171.0	162.5	168.6	166.8	174.1	170.9	176.5	178.2
2013	196.9	187.2	192.5	223.6	201.4	196.1	196.0	213.4	166.3	157.9	159.6	161.2	167.4	159.3	169.3	172.8
2012	194.0	182.0	187.9	221.8	196.8	190.8	191.5	209.3	155.3	148.1	156.7	158.8	163.8	155.8	166.3	167.5
2011	185.7	176.0	180.9	212.8	187.7	184.3	185.3	196.8	149.4	142.3	149.5	151.3	157.3	149.3	158.4	160.7
2010	181.6	171.0	175.2	209.3	182.3	179.8	180.8	192.8	147.4	140.2	147.3	149.0	152.4	144.7	154.2	156.7
2009	182.5	171.9	176.5	209.5	181.7	180.7	181.3	192.9	150.8	144.3	148.2	151.1	155.7	148.0	157.2	159.8
2008	171.0	161.0	165.7	196.2	169.2	164.9	167.9	178.2	142.1	134.8	138.4	141.3	136.7	133.4	146.0	147.5
2007	165.0	156.0	159.4	188.4	163.2	164.9	163.1	174.4	139.3	131.7	129.5	133.0	133.7	130.6	143.1	144.4
2006	156.2	148.7	150.7	177.8	155.5	155.0	153.9	163.6	122.4	118.7	122.8	126.6	127.0	123.9	136.5	135.5
2005	146.7	140.6	141.5	166.4	146.6	145.5	142.9	155.6	110.2	109.6	115.2	118.5	116.0	115.1	128.6	127.9
2004	132.8	127.0	126.2	148.4	133.1	128.7	129.0	138.8	98.9	98.4	103.9	107.3	105.0	104.5	115.6	115.8
2003	129.7	124.5	123.8	145.5	130.5	126.3	126.0	135.9	96.1	95.7	101.3	104.0	102.8	102.3	111.1	110.9
2002	126.7	122.1	121.1	142.0	127.8	123.4	124.2	131.1	94.8	93.9	100.2	103.2	102.2	101.2	107.6	109.2
2001	122.2	118.9	118.4	136.9	124.1	120.4	121.1	127.8	92.1	91.3	96.8	99.7	98.2	96.5	102.6	104.3
2000	118.9	114.6	114.0	132.1	120.9	115.9	117.7	122.8	89.9	89.1	94.2	98.0	96.9	95.0	100.8	100.8
1999	116.6	113.8	112.8	129.8	119.6	114.8	116.3	121.6	89.0	88.1	92.4	95.8	95.9	93.4	99.7	98.6
1998	113.6	109.8	110.5	126.6	117.1	112.5	113.7	120.3	88.0	87.2	90.7	93.6	94.9	92.4	97.2	96.8
1995	105.6	99.8	100.6	117.1	106.3	103.6	103.8	111.1	82.7	82.2	84.0	84.7	89.2	86.0	91.2	87.6
1990	93.2	88.5	89.5	98.5	91.2	90.8	91.3	94.1	73.7	74.5	77.2	78.4	79.9	75.1	81.3	77.1
1985	81.8	79.6	77.2	82.2	81.5	77.7	80.0	83.0	65.9	66.9	69.7	71.2	72.5	67.7	74.3	66.7
1980	60.7	59.4	56.9	58.7	61.3	57.7	58.4	59.2	50.6	51.1	54.8	57.1	55.0	51.6	55.9	53.1
1975	43.7	43.4	42.2	44.5	44.5	43.6	41.9	42.7	35.0	36.4	37.7	39.6	39.1	37.3	40.7	37.0
1970	27.8	27.5	25.8	27.5	28.7	27.1	27.0	27.3	22.8	22.9	24.8	25.8	24.9	22.0	23.0	22.8
1965	21.5	21.1	20.1	21.7	22.4	20.9	20.8	21.9	17.6	17.6	19.1	19.9	19.2	17.1	18.3	17.4
1960	19.5	19.1	18.6	19.4	19.7	19.0	18.9	19.1	16.0	16.0	17.3	18.1	17.4	15.5	16.6	15.8
1955	16.3	16.1	15.6	16.3	16.5	15.9	15.9	16.0	13.4	13.4	14.5	15.1	14.6	13.0	13.9	13.3
1950	13.5	13.3	12.9	13.5	13.6	13.2	13.1	13.2	11.1	11.1	12.0	12.5	12.1	10.7	11.5	10.9

Year	National 30 City Average	Texas Abi-lene	Ama-rillo	Austin	Beau-mont	Corpus Christi	Dallas	El Paso	Fort Worth	Houston	Lubbock	Odessa	San Antonio	Waco	Wichita Falls	Utah Ogden
Jan 2018	215.8	175.8E	174.6E	174.3E	179.8E	180.7E	181.8E	175.3E	176.7E	181E	178.4E	177.4E	177.3E	172.4E	172.1E	185.2E
2017	209.4	174.0	172.8	172.6	178.0	178.9	180.0	173.5	174.9	179.2	176.6	175.6	175.5	170.7	170.4	183.3
2016	207.7	173.7	171.0	173.3	178.3	175.5	177.9	172.8	174.1	180.5	175.2	172.6	173.1	170.3	170.1	179.4
2015	204.0	170.8	170.7	172.3	174.0	172.8	174.4	168.8	172.5	176.4	173.3	171.5	171.8	167.3	169.0	177.1
2014	203.0	160.9	167.2	167.5	165.2	167.3	172.4	156.2	169.5	176.0	166.9	159.2	169.5	161.7	160.6	174.2
2013	196.9	155.2	161.7	158.7	160.2	157.7	167.4	151.2	161.7	169.4	159.3	151.5	164.3	156.1	155.9	167.7
2012	194.0	152.5	158.6	154.3	158.2	151.8	165.2	149.1	159.4	167.8	156.6	149.1	160.7	153.8	152.9	165.6
2011	185.7	145.9	151.9	147.6	152.4	145.0	157.9	142.5	152.8	160.8	150.0	142.5	152.6	147.3	146.6	158.7
2010	181.6	144.2	150.1	144.7	150.5	142.1	155.1	140.8	149.6	157.3	148.1	140.7	147.9	145.8	145.0	157.5
2009	182.5	143.5	147.3	146.4	149.8	141.7	155.5	141.1	150.1	160.9	144.6	139.1	150.8	146.7	146.1	155.1
2008	171.0	132.4	137.0	137.5	140.3	133.2	144.1	130.7	138.3	149.1	134.7	128.9	141.0	136.1	136.1	144.8
2007	165.0	128.5	132.2	131.6	137.1	128.8	138.6	126.2	134.8	146.0	130.2	125.0	136.4	132.1	132.1	140.0
2006	156.2	121.6	125.7	125.5	130.1	122.0	131.1	120.3	127.6	138.2	123.2	118.3	129.8	125.2	125.3	133.4
2005	146.7	113.8	117.3	117.9	121.3	114.4	123.7	112.5	119.4	129.0	115.5	110.5	121.3	116.1	117.3	126.0
2004	132.8	102.9	106.3	105.7	108.5	102.9	112.0	101.5	108.4	115.9	104.7	99.9	108.4	104.8	105.1	115.2
2003	129.7	100.5	104.4	103.9	106.1	100.4	109.3	99.5	105.5	113.4	102.3	97.6	104.8	102.7	103.0	112.7
2002	126.7	99.9	101.9	102.4	104.5	98.9	107.9	98.7	104.6	111.5	100.5	95.9	103.7	100.5	101.4	110.8
2001	122.2	93.4	98.4	99.8	102.3	96.6	103.8	95.5	100.9	107.8	97.6	93.4	100.5	97.8	97.3	107.7
2000	118.9	93.4	98.1	99.1	101.6	96.9	102.7	92.4	99.9	106.0	97.6	93.4	99.4	97.3	96.9	104.6
1999	116.6	91.8	94.5	96.0	99.7	94.0	101.0	90.7	97.6	104.6	96.0	92.1	98.0	94.8	95.5	103.3
1998	113.6	89.8	92.7	94.2	97.9	91.8	97.9	88.4	94.5	101.3	93.3	90.1	94.8	92.7	92.9	98.5
1995	105.6	85.2	87.4	89.3	93.7	87.4	91.4	85.2	89.5	95.9	88.4	85.6	88.9	86.4	86.8	92.2
1990	93.2	78.0	80.1	81.3	86.5	79.3	84.5	76.7	82.1	85.4	81.5	78.6	80.7	79.6	80.3	83.4
1985	81.8	71.1	72.5	74.5	79.3	72.3	77.6	69.4	75.1	79.6	74.0	71.2	73.9	71.7	73.3	75.2
1980	60.7	53.4	55.2	54.5	57.6	54.5	57.9	53.1	57.0	59.4	55.6	57.2	55.0	54.9	55.4	62.2
1975	43.7	37.6	39.0	39.0	39.6	38.1	40.7	38.0	40.4	41.2	38.9	37.9	39.0	38.6	38.0	40.0
1970	27.8	24.5	24.9	24.9	25.7	24.5	25.5	23.7	25.9	25.4	25.1	24.6	23.3	24.8	24.5	26.8
1965	21.5	18.9	19.2	19.2	19.9	18.9	19.9	19.0	19.9	20.0	19.4	19.0	18.5	19.2	18.9	20.6
1960	19.5	17.1	17.4	17.4	18.1	17.1	18.2	17.0	18.1	18.2	17.6	17.3	16.8	17.4	17.2	18.8
1955	16.3	14.4	14.6	14.6	15.1	14.4	15.3	14.3	15.2	15.2	14.8	14.5	14.1	14.6	14.4	15.7
1950	13.5	11.9	12.1	12.1	12.5	11.9	12.6	11.8	12.5	12.6	12.2	12.0	11.6	12.1	11.9	13.0

533

Historical Cost Indexes

Year	National 30 City Average	Utah Salt Lake City	Vermont Bur-lington	Vermont Rutland	Virginia Alex-andria	Virginia Newport News	Virginia Norfolk	Virginia Rich-mond	Virginia Roanoke	Washington Seattle	Washington Spokane	Washington Tacoma	West Virginia Charles-ton	West Virginia Hunt-ington	Wisconsin Green Bay	Wisconsin Kenosha
Jan 2018	215.8	190.7E	194.8E	191.4E	197.6E	182.5E	182.3E	184.7E	187E	221.5E	200.8E	218E	203.5E	206.6E	209E	215.8E
2017	209.4	188.8	192.8	189.5	195.6	180.7	180.5	182.8	185.1	219.3	198.8	215.8	201.5	204.5	206.9	213.6
2016	207.7	184.7	193.6	190.2	195.6	178.3	178.2	178.9	178.0	213.4	195.1	209.7	204.2	206.5	207.4	212.8
2015	204.0	182.1	191.6	188.5	192.5	175.8	177.4	177.2	175.3	209.9	193.1	207.2	199.9	201.3	200.8	205.4
2014	203.0	177.1	191.3	188.6	190.5	174.4	175.7	176.4	173.5	209.8	190.8	205.2	197.0	199.5	198.9	205.2
2013	196.9	171.1	178.7	176.6	185.0	170.0	171.5	170.7	168.3	203.4	184.3	199.1	187.2	193.2	193.9	201.2
2012	194.0	168.8	172.4	170.5	182.4	168.3	169.7	169.4	166.4	201.9	181.6	194.3	183.4	186.4	190.0	195.5
2011	185.7	161.8	158.8	157.1	174.5	159.8	161.1	159.0	154.2	194.1	175.5	187.6	177.5	180.3	181.4	187.6
2010	181.6	160.9	156.3	154.6	170.9	156.7	158.0	156.6	151.8	191.8	171.5	186.0	171.6	176.1	179.0	185.1
2009	182.5	160.4	158.3	156.7	172.5	160.0	161.9	160.7	155.6	188.5	173.0	186.3	174.6	177.4	175.5	182.5
2008	171.0	149.7	147.2	145.8	161.8	150.8	150.8	150.9	145.9	176.9	162.4	174.9	162.8	165.7	165.6	172.2
2007	165.0	144.6	144.4	143.2	155.0	146.2	146.1	147.3	142.9	171.4	156.7	168.7	158.7	160.1	159.4	164.8
2006	156.2	137.7	131.7	130.7	146.2	133.7	134.6	134.8	129.7	162.9	150.1	160.7	149.4	151.4	152.7	157.1
2005	146.7	129.4	124.6	123.8	136.5	123.5	124.4	125.4	112.2	153.9	141.9	151.5	140.8	141.0	144.4	148.3
2004	132.8	117.8	113.0	112.3	121.5	108.9	110.2	110.9	99.7	138.0	127.6	134.5	124.8	125.6	128.9	132.4
2003	129.7	116.0	110.7	110.1	119.5	104.8	106.2	108.6	97.0	134.9	125.9	133.0	123.3	123.6	127.4	130.9
2002	126.7	113.7	109.0	108.4	115.1	102.9	104.1	106.6	95.2	132.7	123.9	131.4	121.2	120.9	123.0	127.3
2001	122.2	109.1	105.7	105.2	110.8	99.8	100.3	102.9	92.1	127.9	120.3	125.7	114.6	117.5	119.1	123.6
2000	118.9	106.5	98.9	98.3	108.1	96.5	97.6	100.2	90.7	124.6	118.3	122.9	111.5	114.4	114.6	119.1
1999	116.6	104.5	98.2	97.7	106.1	95.6	96.5	98.8	89.8	123.3	116.7	121.6	110.6	113.4	112.1	115.8
1998	113.6	99.5	97.8	97.3	104.1	93.7	93.9	97.0	88.3	119.4	114.3	118.3	106.7	109.0	109.5	112.9
1995	105.6	93.1	91.1	90.8	96.3	86.0	86.4	87.8	82.8	113.7	107.4	112.8	95.8	97.2	97.6	97.9
1990	93.2	84.3	83.0	82.9	86.1	76.3	76.7	77.6	76.1	100.1	98.5	100.5	86.1	86.8	86.7	87.8
1985	81.8	75.9	74.8	74.9	75.1	68.7	68.8	69.5	67.2	88.3	89.0	91.2	77.7	77.7	76.7	77.4
1980	60.7	57.0	55.3	58.3	57.3	52.5	52.4	54.3	51.3	67.9	66.3	66.7	57.7	58.3	58.6	58.3
1975	43.7	40.1	41.8	43.9	41.7	37.2	36.9	37.1	37.1	44.9	44.4	44.5	41.0	40.0	40.9	40.5
1970	27.8	26.1	25.4	26.8	26.2	23.9	21.5	22.0	23.7	28.8	29.3	29.6	26.1	25.8	26.4	26.5
1965	21.5	20.0	19.8	20.6	20.2	18.4	17.1	17.2	18.3	22.4	22.5	22.8	20.1	19.9	20.3	20.4
1960	19.5	18.4	18.0	18.8	18.4	16.7	15.4	15.6	16.6	20.4	20.8	20.8	18.3	18.1	18.4	18.6
1955	16.3	15.4	15.1	15.7	15.4	14.0	12.9	13.1	13.9	17.1	17.4	17.4	15.4	15.2	15.5	15.6
1950	13.5	12.7	12.4	13.0	12.7	11.6	10.7	10.8	11.5	14.1	14.4	14.4	12.7	12.5	12.8	12.9

Year	National 30 City Average	Wisconsin Mad-ison	Wisconsin Mil-waukee	Wisconsin Racine	Wyoming Chey-enne	Canada Calgary	Canada Edmon-ton	Canada Ham-ilton	Canada London	Canada Montreal	Canada Ottawa	Canada Quebec	Canada Tor-onto	Canada Van-couver	Canada Win-nipeg
Jan 2018	215.8	209.5E	217.4E	216E	186.9E	233.7E	234.7E	230.1E	227.3E	224E	230E	225.2E	233.7E	223.7E	209.8E
2017	209.4	207.4	215.2	213.8	185.0	231.4	232.3	227.8	225.0	221.8	227.7	222.9	231.4	221.5	207.7
2016	207.7	207.4	214.4	212.4	178.6	229.1	229.4	221.7	221.3	218.5	223.4	218.9	228.6	219.3	205.3
2015	204.0	202.0	209.4	205.1	175.4	226.6	226.3	221.2	217.0	220.4	220.4	220.4	225.9	217.1	204.1
2014	203.0	202.4	210.2	205.3	173.6	228.2	229.1	222.1	218.9	219.8	214.8	218.3	227.4	218.8	202.9
2013	196.9	197.8	204.6	200.5	167.6	222.7	223.3	216.0	213.7	214.5	214.1	212.6	220.9	216.2	200.2
2012	194.0	191.2	202.2	194.9	165.0	219.1	219.6	211.1	208.2	209.6	209.7	207.5	216.6	214.0	200.7
2011	185.7	183.4	191.9	187.0	155.3	212.6	212.4	204.2	200.1	202.1	202.0	200.6	209.6	207.1	192.6
2010	181.6	181.3	187.1	184.4	154.4	200.6	200.7	197.7	193.7	193.9	195.6	192.7	200.7	192.7	182.9
2009	182.5	180.0	187.3	182.8	155.2	205.4	206.6	203.0	198.9	198.8	200.1	197.4	205.6	199.6	188.5
2008	171.0	168.4	176.3	173.0	146.5	190.2	191.0	194.3	188.6	186.7	188.0	186.8	194.6	184.9	174.4
2007	165.0	160.7	168.9	164.7	141.3	183.1	184.4	186.8	182.4	181.7	182.4	182.0	187.7	180.6	170.1
2006	156.2	152.8	158.8	157.2	128.0	163.6	164.8	169.0	164.9	158.6	163.7	159.0	170.6	169.3	155.9
2005	146.7	145.2	148.4	147.9	118.9	154.4	155.7	160.0	156.2	149.2	155.1	150.1	162.6	159.4	146.9
2004	132.8	129.7	134.2	132.7	104.7	138.8	139.4	142.6	140.4	134.5	141.0	135.7	146.0	141.7	129.7
2003	129.7	128.4	131.1	131.5	102.7	133.6	134.2	139.1	137.0	130.2	137.4	131.4	142.3	137.0	124.4
2002	126.7	124.5	128.2	126.5	101.7	122.5	122.2	136.3	134.1	127.2	134.9	128.4	139.8	134.6	121.4
2001	122.2	120.6	123.9	123.3	99.0	117.5	117.4	131.5	129.1	124.4	130.2	125.5	134.7	130.2	117.2
2000	118.9	116.6	120.5	118.7	98.1	115.9	115.8	130.0	127.5	122.8	128.8	124.1	133.1	128.4	115.6
1999	116.6	115.9	117.4	115.5	96.9	115.3	115.2	128.1	125.6	120.8	126.8	121.9	131.2	127.1	115.2
1998	113.6	110.8	113.4	112.7	95.4	112.5	112.4	126.3	123.9	119.0	124.7	119.6	128.5	123.8	113.7
1995	105.6	96.5	103.9	97.8	87.6	107.4	107.4	119.9	117.5	110.8	118.2	111.5	121.6	116.2	107.6
1990	93.2	84.3	88.9	87.3	79.1	98.0	97.1	103.8	101.4	99.0	102.7	96.8	104.6	103.2	95.1
1985	81.8	74.3	77.4	77.0	72.3	90.2	89.1	85.8	84.9	82.4	83.9	79.9	86.5	89.8	83.4
1980	60.7	56.8	58.8	58.1	56.9	64.9	63.3	63.7	61.9	59.2	60.9	59.3	60.9	65.0	61.7
1975	43.7	40.7	43.3	40.7	40.6	42.2	41.6	42.9	41.6	39.7	41.5	39.0	42.2	42.4	39.2
1970	27.8	26.5	29.4	26.5	26.0	28.9	28.6	28.5	27.8	25.6	27.6	26.0	25.6	26.0	23.1
1965	21.5	20.6	21.8	20.4	20.0	22.3	22.0	22.0	21.4	18.7	21.2	20.1	19.4	20.5	17.5
1960	19.5	18.1	19.0	18.6	18.2	20.2	20.0	20.0	19.5	17.0	19.3	18.2	17.6	18.6	15.8
1955	16.3	15.2	15.9	15.6	15.2	17.0	16.8	16.7	16.3	14.3	16.2	15.3	14.8	15.5	13.3
1950	13.5	12.5	13.2	12.9	12.6	14.0	13.9	13.8	13.5	11.8	13.4	12.6	12.2	12.8	10.9

534

Back by customer demand!

You asked and we listened. For customer convenience and estimating ease, we have made the 2018 Project Costs available for download at **www.RSMeans.com/2018books**. You will also find sample estimates, an RSMeans data overview video and book registration form to receive quarterly data updates throughout 2018.

Estimating Tips

- The cost figures available in the download were derived from hundreds of projects contained in the RSMeans database of completed construction projects. They include the contractor's overhead and profit. The figures have been adjusted to January of the current year.

- These projects were located throughout the U.S. and reflect a tremendous variation in square foot (S.F.) costs. This is due to differences, not only in labor and material costs, but also in individual owners' requirements. For instance, a bank in a large city would have different features than one in a rural area. This is true of all the different types of buildings analyzed. Therefore, caution should be exercised when using these Project Costs. For example, for courthouses, costs in the database are local courthouse costs and will not apply to the larger, more elaborate federal courthouses.

- None of the figures "go with" any others. All individual cost items were computed and tabulated separately. Thus, the sum of the median figures for plumbing, HVAC, and electrical will not normally total up to the total mechanical and electrical costs arrived at by separate analysis and tabulation of the projects.

- Each building was analyzed as to total and component costs and percentages. The figures were arranged in ascending order with the results tabulated as shown. The 1/4 column shows that 25% of the projects had lower costs and 75% had higher. The 3/4 column shows that 75% of the projects had lower costs and 25% had higher. The median column shows that 50% of the projects had lower costs and 50% had higher.

- Project Costs are useful in the conceptual stage when no details are available. As soon as details become available in the project design, the square foot approach should be discontinued and the project priced as to its particular components. When more precision is required, or for estimating the replacement cost of specific buildings, the current edition of *RSMeans Square Foot Costs* should be used.

- In using the figures in this section, it is recommended that the median column be used for preliminary figures if no additional information is available. The median figures, when multiplied by the total city construction cost index figures (see City Cost Indexes) and then multiplied by the project size modifier at the end of this section, should present a fairly accurate base figure, which would then have to be adjusted in view of the estimator's experience, local economic conditions, code requirements, and the owner's particular requirements. There is no need to factor the percentage figures, as these should remain constant from city to city.

- The editors of this data would greatly appreciate receiving cost figures on one or more of your recent projects, which would then be included in the averages for next year. All cost figures received will be kept confidential, except that they will be averaged with other similar projects to arrive at square foot cost figures for next year.

See the website above for details and the discount available for submitting one or more of your projects.

Did you know?

RSMeans data is available through our online application with 24/7 access:

- Search for unit prices by keyword
- Leverage the most up-to-date data
- Build and export estimates

Try it free for 30 days!
www.rsmeans.com/2018freetrial

50 17 00 | Project Costs

			UNIT	UNIT COSTS			% OF TOTAL		
				1/4	MEDIAN	3/4	1/4	MEDIAN	3/4
01	0000	**Auto Sales with Repair**	S.F.						
	0100	Architectural		98.50	110	119	58%	64%	67%
	0200	Plumbing		8.25	8.65	11.55	4.84%	5.20%	6.80%
	0300	Mechanical		11.05	14.80	16.35	6.40%	8.70%	10.15%
	0400	Electrical		17	21	26.50	9.05%	11.70%	15.90%
	0500	Total Project Costs		165	173	177			
02	0000	**Banking Institutions**	S.F.						
	0100	Architectural		149	183	222	59%	65%	69%
	0200	Plumbing		6	8.35	11.60	2.12%	3.39%	4.19%
	0300	Mechanical		11.90	16.45	19.40	4.41%	5.10%	10.75%
	0400	Electrical		29	35	54	10.45%	13.05%	15.90%
	0500	Total Project Costs		247	278	340			
03	0000	**Court House**	S.F.						
	0100	Architectural		78.50	78.50	78.50	54.50%	54.50%	54.50%
	0200	Plumbing		2.96	2.96	2.96	2.07%	2.07%	2.07%
	0300	Mechanical		18.55	18.55	18.55	12.95%	12.95%	12.95%
	0400	Electrical		24	24	24	16.60%	16.60%	16.60%
	0500	Total Project Costs		143	143	143			
04	0000	**Data Centers**	S.F.						
	0100	Architectural		177	177	177	68%	68%	68%
	0200	Plumbing		9.70	9.70	9.70	3.71%	3.71%	3.71%
	0300	Mechanical		24.50	24.50	24.50	9.45%	9.45%	9.45%
	0400	Electrical		23.50	23.50	23.50	9%	9%	9%
	0500	Total Project Costs		261	261	261			
05	0000	**Detention Centers**	S.F.						
	0100	Architectural		164	174	184	52%	53%	60.50%
	0200	Plumbing		17.35	21	25.50	5.15%	7.10%	7.25%
	0300	Mechanical		22	31.50	37.50	7.55%	9.50%	13.80%
	0400	Electrical		36	42.50	55.50	10.90%	14.85%	17.95%
	0500	Total Project Costs		278	293	345			
06	0000	**Fire Stations**	S.F.						
	0100	Architectural		97	120	174	49%	55.50%	63%
	0200	Plumbing		9.45	13.30	15.75	4.86%	5.80%	6.30%
	0300	Mechanical		12.80	18.65	24	5.45%	8.25%	9.70%
	0400	Electrical		21.50	27.50	32.50	8.35%	12.80%	14.70%
	0500	Total Project Costs		195	222	294			
07	0000	**Gymnasium**	S.F.						
	0100	Architectural		82	108	108	57%	64.50%	64.50%
	0200	Plumbing		2.01	6.60	6.60	1.58%	3.48%	3.48%
	0300	Mechanical		3.09	28	28	2.42%	14.65%	14.65%
	0400	Electrical		10.10	19.60	19.60	7.95%	10.35%	10.35%
	0500	Total Project Costs		128	189	189			
08	0000	**Hospitals**	S.F.						
	0100	Architectural		99.50	164	178	43%	47.50%	48%
	0200	Plumbing		7.30	13.95	30	6%	7.45%	7.65%
	0300	Mechanical		48.50	54.50	71	14.20%	17.95%	23.50%
	0400	Electrical		22	44	57.50	10.95%	13.75%	16.85%
	0500	Total Project Costs		233	345	375			
09	0000	**Industrial Buildings**	S.F.						
	0100	Architectural		42	66.50	216	46%	54%	56.50%
	0200	Plumbing		1.62	6.10	12.30	2%	3.06%	6.30%
	0300	Mechanical		4.47	8.50	40.50	4.77%	5.55%	14.80%
	0400	Electrical		6.80	7.80	65	7.85%	13.55%	16.20%
	0500	Total Project Costs		74.50	97	400			
10	0000	**Medical Clinics & Offices**	S.F.						
	0100	Architectural		79.50	118	152	48.50%	55.50%	62.50%
	0200	Plumbing		8.10	12.20	19.65	4.44%	6.40%	8.05%
	0300	Mechanical		13.70	25	42.50	8.30%	11.05%	17.25%
	0400	Electrical		18.60	25	36	8.55%	12.10%	14.65%
	0500	Total Project Costs		156	208	272			

For customer support on your Square Foot Costs with RSMeans data, call 800.448.8182.

		50 17 00 | Project Costs	UNIT	UNIT COSTS			% OF TOTAL		
				1/4	MEDIAN	3/4	1/4	MEDIAN	3/4
11	0000	**Mixed Use**	S.F.						
	0100	Architectural		85	120	182	45.50%	52.50%	70%
	0200	Plumbing		5.70	8.70	11	2.84%	3.44%	4.18%
	0300	Mechanical		14.05	36.50	65	8.05%	13.75%	20%
	0400	Electrical		14.85	36.50	49	8.30%	12.95%	15.80%
	0500	Total Project Costs	↓	179	201	320			
12	0000	**Multi-Family Housing**	S.F.						
	0100	Architectural		70.50	106	158	56.50%	61.50%	66.50%
	0200	Plumbing		5.90	10.20	13.90	4.81%	6.30%	7.40%
	0300	Mechanical		6.60	8.80	35.50	4.92%	6.90%	11.20%
	0400	Electrical		9.20	14.35	17.40	6.20%	8.45%	10.25%
	0500	Total Project Costs	↓	104	194	233			
13	0000	**Nursing Home & Assisted Living**	S.F.						
	0100	Architectural		66.50	87	116	51.50%	56.50%	69%
	0200	Plumbing		6.75	10.45	11.70	6.05%	7.40%	8.80%
	0300	Mechanical		5.90	7.55	17.05	4.04%	6.70%	9.55%
	0400	Electrical		9.75	12.80	21.50	7%	10.05%	12.20%
	0500	Total Project Costs	↓	114	135	179			
14	0000	**Office Buildings**	S.F.						
	0100	Architectural		90.50	120	168	56.50%	62%	71.50%
	0200	Plumbing		4.73	7.05	11.45	2.62%	3.48%	5.85%
	0300	Mechanical		10.25	15.80	23.50	5.50%	8.20%	11.10%
	0400	Electrical		11.85	20	31.50	7.75%	10%	12.70%
	0500	Total Project Costs	↓	150	186	262			
15	0000	**Parking Garage**	S.F.						
	0100	Architectural		29.50	36	37.50	70%	79%	88%
	0200	Plumbing		.97	1.01	1.90	2.05%	2.70%	2.83%
	0300	Mechanical		.75	1.16	4.39	2.11%	3.62%	3.81%
	0400	Electrical		2.58	2.83	5.90	5.30%	6.35%	7.95%
	0500	Total Project Costs	↓	36	43.50	47			
16	0000	**Parking Garage/Mixed Use**	S.F.						
	0100	Architectural		95.50	104	106	61%	62%	65.50%
	0200	Plumbing		3.06	4.01	6.15	2.47%	2.72%	3.66%
	0300	Mechanical		13.10	14.75	21	7.80%	13.10%	13.60%
	0400	Electrical		13.70	19.70	20.50	8.20%	12.65%	18.15%
	0500	Total Project Costs	↓	156	162	168			
17	0000	**Police Stations**	S.F.						
	0100	Architectural		108	121	152	49%	56.50%	61%
	0200	Plumbing		14.25	17.10	17.20	5.05%	5.55%	9.05%
	0300	Mechanical		32.50	45	46.50	13%	14.55%	16.55%
	0400	Electrical		24.50	26.50	28	9.15%	12.10%	14%
	0500	Total Project Costs	↓	202	248	282			
18	0000	**Police/Fire**	S.F.						
	0100	Architectural		105	105	320	55.50%	66%	68%
	0200	Plumbing		8.40	8.70	32	5.45%	5.50%	5.55%
	0300	Mechanical		12.85	20.50	73.50	8.35%	12.70%	12.80%
	0400	Electrical		14.65	18.70	84	9.50%	11.75%	14.55%
	0500	Total Project Costs	↓	154	159	580			
19	0000	**Public Assembly Buildings**	S.F.						
	0100	Architectural		119	148	200	58%	61.50%	64%
	0200	Plumbing		5.65	7.40	11.45	2.60%	3.32%	4.01%
	0300	Mechanical		14.95	23	33	6.90%	9.15%	12.75%
	0400	Electrical		19.70	24.50	38.50	8.65%	10.75%	13%
	0500	Total Project Costs	↓	188	242	335			
20	0000	**Recreational**	S.F.						
	0100	Architectural		103	162	219	53.50%	59%	66%
	0200	Plumbing		8.40	15.30	20	3.12%	4.76%	7.40%
	0300	Mechanical		12.65	19.25	30.50	5.35%	7.90%	12.75%
	0400	Electrical		14.15	24.50	33	7.35%	8.50%	10.65%
	0500	Total Project Costs	↓	183	272	375			

50 17 00 | Project Costs

			UNIT	UNIT COSTS			% OF TOTAL		
				1/4	MEDIAN	3/4	1/4	MEDIAN	3/4
21	0000	**Restaurants**	S.F.						
	0100	Architectural		50.50	179	231	57.50%	60.50%	78%
	0200	Plumbing		9.45	29.50	37	7.35%	7.75%	8.95%
	0300	Mechanical		7.25	18.30	44.50	6.50%	10.80%	18.30%
	0400	Electrical		11.65	21.50	48.50	6.75%	8.45%	12.45%
	0500	Total Project Costs		65	283	400			
22	0000	**Retail**	S.F.						
	0100	Architectural		52	57.50	104	60%	60%	64.50%
	0200	Plumbing		5.35	8.05	10.15	5.05%	6.70%	9%
	0300	Mechanical		4.88	7	8.60	5.05%	6.15%	6.60%
	0400	Electrical		6.85	10.95	17.60	7.90%	10.15%	12.25%
	0500	Total Project Costs		87	106	173			
23	0000	**Schools**	S.F.						
	0100	Architectural		88.50	114	152	52.50%	57%	61.50%
	0200	Plumbing		7.10	9.85	14.40	3.75%	4.82%	7%
	0300	Mechanical		16.85	23	35.50	9.50%	12.25%	14.55%
	0400	Electrical		16.65	23	29	9.45%	11.40%	13.05%
	0500	Total Project Costs		151	206	272			
24	0000	**University, College & Private School Classroom & Admin Buildings**	S.F.						
	0100	Architectural		115	142	179	50.50%	55%	59.50%
	0200	Plumbing		6.55	10.15	14.35	2.74%	4.30%	6.35%
	0300	Mechanical		24.50	35.50	43	10.10%	12.15%	14.70%
	0400	Electrical		18.55	26	31.50	7.65%	9.50%	11.55%
	0500	Total Project Costs		191	264	350			
25	0000	**University, College & Private School Dormitories**	S.F.						
	0100	Architectural		75	132	140	54.50%	65%	68.50%
	0200	Plumbing		9.95	14.05	21	6.45%	7.30%	9.15%
	0300	Mechanical		4.46	18.95	30	4.13%	9%	12.05%
	0400	Electrical		5.30	18.40	28	4.75%	7.35%	12.30%
	0500	Total Project Costs		111	211	249			
26	0000	**University, College & Private School Science, Eng. & Lab Buildings**	S.F.						
	0100	Architectural		129	137	178	50.50%	56.50%	58%
	0200	Plumbing		8.90	9.30	24.50	3.29%	3.95%	8.40%
	0300	Mechanical		40.50	64	64	13.20%	21.50%	23.50%
	0400	Electrical		26	30	35.50	8.95%	9.70%	12.85%
	0500	Total Project Costs		265	271	295			
27	0000	**University, College & Private School Student Union Buildings**	S.F.						
	0100	Architectural		69.50	102	102	51%	59.50%	59.50%
	0200	Plumbing		5	23	23	4.27%	11.45%	11.45%
	0300	Mechanical		11.25	29.50	29.50	9.60%	14.55%	14.55%
	0400	Electrical		15.40	25.50	25.50	12.80%	13.15%	13.15%
	0500	Total Project Costs		117	201	201			
28	0000	**Warehouses**	S.F.						
	0100	Architectural		44	68.50	162	61.50%	67.50%	72%
	0200	Plumbing		2.28	4.88	9.40	2.82%	3.72%	5%
	0300	Mechanical		2.70	15.40	24	4.56%	8.20%	10.70%
	0400	Electrical		4.91	18.45	30.50	7.50%	10.10%	18.30%
	0500	Total Project Costs		65.50	116	226			

Square Foot Project Size Modifier

One factor that affects the S.F. cost of a particular building is the size. In general, for buildings built to the same specifications in the same locality, the larger building will have the lower S.F. cost. This is due mainly to the decreasing contribution of the exterior walls plus the economy of scale usually achievable in larger buildings. The Area Conversion Scale shown below will give a factor to convert costs for the typical size building to an adjusted cost for the particular project.

The Square Foot Base Size lists the median costs, most typical project size in our accumulated data, and the range in size of the projects.

The Size Factor for your project is determined by dividing your project area in S.F. by the typical project size for the particular Building Type. With this factor, enter the Area Conversion Scale at the appropriate Size Factor and determine the appropriate cost multiplier for your building size.

Example: Determine the cost per S.F. for a 152,600 S.F. Multi-family housing.

$$\frac{\text{Proposed building area} = 152{,}600 \text{ S.F.}}{\text{Typical size from below} = 76{,}300 \text{ S.F.}} = 2.00$$

Enter Area Conversion scale at 2.0, intersect curve, read horizontally the appropriate cost multiplier of .94. Size adjusted cost becomes .94 x $194.00 = $182.36 based on national average costs.

Note: For Size Factors less than .50, the Cost Multiplier is 1.1
For Size Factors greater than 3.5, the Cost Multiplier is .90

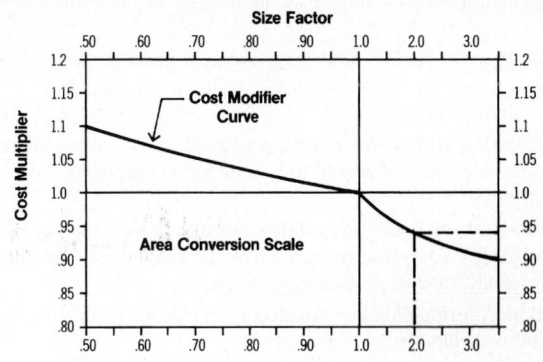

System	Median Cost (Total Project Costs)	Typical Size Gross S.F. (Median of Projects)	Typical Range (Low – High) (Projects)
Auto Sales with Repair	$173.00	25,300	8,200 – 28,700
Banking Institutions	278.00	26,300	3,300 – 28,700
Detention Centers	293.00	42,000	12,300 – 183,300
Fire Stations	222.00	14,600	6,300 – 29,600
Hospitals	345.00	137,500	54,700 – 410,300
Industrial Buildings	$97.00	16,900	5,100 – 200,600
Medical Clinics & Offices	208.00	5,500	2,600 – 327,000
Mixed Use	201.00	49,900	14,400 – 49,900
Multi-Family Housing	194.00	76,300	12,500 – 1,161,500
Nursing Home & Assisted Living	135.00	16,200	1,500 – 242,600
Office Buildings	186.00	10,000	1,100 – 930,000
Parking Garage	43.50	174,600	99,900 – 287,000
Parking Garage/Mixed Use	162.00	5,300	5,300 – 318,000
Police Stations	248.00	15,400	15,400 – 31,600
Public Assembly Buildings	242.00	30,500	2,200 – 235,300
Recreational	272.00	2,300	1,500 – 223,800
Restaurants	283.00	6,100	5,500 – 42,000
Retail	106.00	28,700	5,800 – 61,000
Schools	206.00	30,000	5,500 – 410,800
University, College & Private School Classroom & Admin Buildings	264.00	89,200	9,400 – 196,200
University, College & Private School Dormitories	211.00	50,800	1,500 – 126,900
University, College & Private School Science, Eng. & Lab Buildings	271.00	39,800	36,000 – 117,600
Warehouses	116.00	2,100	600 – 303,800

539

Glossary

Accent lighting–Fixtures or directional beams of light arranged to bring attention to an object or area.

Acoustical material–A material fabricated for the sole purpose of absorbing sound.

Addition–An expansion to an existing structure generally in the form of a room, floor(s) or wing(s). An increase in the floor area or volume of a structure.

Aggregate–Materials such as sand, gravel, stone, vermiculite, perlite and fly ash (slag) which are essential components in the production of concrete, mortar or plaster.

Air conditioning system–An air treatment to control the temperature, humidity and cleanliness of air and to provide for its distribution throughout the structure.

Air curtain–A stream of air directed downward to prevent the loss of hot or cool air from the structure and inhibit the entrance of dust and insects. Generally installed at loading platforms.

Anodized aluminum–Aluminum treated by an electrolytic process to produce an oxide film that is corrosion resistant.

Arcade–A covered passageway between buildings often with shops and offices on one or both sides.

Ashlar–A square-cut building stone or a wall constructed of cut stone.

Asphalt paper–A sheet paper material either coated or saturated with asphalt to increase its strength and resistance to water.

Asphalt shingles–Shingles manufactured from saturated roofing felts, coated with asphalt and topped with mineral granules to prevent weathering.

Assessment ratio–The ratio between the market value and assessed valuation of a property.

Back-up–That portion of a masonry wall behind the exterior facing-usually load bearing.

Balcony–A platform projecting from a building either supported from below or cantilevered. Usually protected by railing.

Bay–Structural component consisting of beams and columns occurring consistently throughout the structure.

Beam–A horizontal structural framing member that transfers loads to vertical members (columns) or bearing walls.

Bid–An offer to perform work described in a contract at a specified price.

Board foot–A unit or measure equal in volume to a board one foot long, one foot wide and one inch thick.

Booster pump–A supplemental pump installed to increase or maintain adequate pressure in the system.

Bowstring roof–A roof supported by trusses fabricated in the shape of a bow and tied together by a straight member.

Brick veneer–A facing of brick laid against, but not bonded to, a wall.

Bridging–A method of bracing joists for stiffness, stability and load distribution.

Broom finish–A method of finishing concrete by lightly dragging a broom over freshly placed concrete.

Built-up roofing–Installed on flat or extremely low-pitched roofs, a roof covering composed of plies or laminations of saturated roofing felts alternated with layers of coal tar pitch or asphalt and surfaced with a layer of gravel or slag in a thick coat of asphalt and finished with a capping sheet.

Caisson–A watertight chamber to expedite work on foundations or structure below water level.

Cantilever–A structural member supported only at one end.

Carport–An automobile shelter having one or more sides open to the weather.

Casement window–A vertical opening window having one side fixed.

Caulking–A resilient compound generally having a silicone or rubber base used to prevent infiltration of water or outside air.

Cavity wall–An exterior wall usually of masonry having an inner and outer wall separated by a continuous air space for thermal insulation.

Cellular concrete–A lightweight concrete consisting of cement mixed with gas-producing materials which, in turn, forms bubbles resulting in good insulating qualities.

Chattel mortgage–A secured interest in a property as collateral for payment of a note.

Clapboard–A wood siding used as exterior covering in frame construction. It is applied horizontally with grain running lengthwise. The thickest section of the board is on the bottom.

Clean room–An environmentally controlled room usually found in medical facilities or precision manufacturing spaces where it is essential to eliminate dust, lint or pathogens.

Close studding–A method of construction whereby studs are spaced close together and the intervening spaces are plastered.

Cluster housing–A closely grouped series of houses resulting in a high density land use.

Cofferdam–A watertight enclosure used for foundation construction in waterfront areas. Water is pumped out of the cofferdam allowing free access to the work area.

Collar joint–Joint between the collar beam and roof rafters.

Column–A vertical structural member supporting horizontal members (beams) along the direction of its longitudinal axis.

Combination door (windows)–Door (windows) having interchangeable screens and glass for seasonal use.

Common area–Spaces either inside or outside the building designated for use by the occupant of the building but not the general public.

Common wall–A wall used jointly by two dwelling units.

Compound wall–A wall constructed of more than one material.

Concrete–A composite material consisting of sand, coarse aggregate (gravel, stone or slag), cement and water that when mixed and allowed to harden forms a hard stone-like material.

Concrete floor hardener–A mixture of chemicals applied to the surface of concrete to produce a dense, wear-resistant bearing surface.

Conduit–A tube or pipe used to protect electric wiring.

Coping–The top cover or capping on a wall.

Craneway–Steel or concrete column and beam supports and rails on which a crane travels.

Curtain wall–A non-bearing exterior wall not supported by beams or girders of the steel frame.

Dampproofing–Coating of a surface to prevent the passage of water or moisture.

Dead load–Total weight of all structural components plus permanently attached fixtures and equipment.

Decibel–Unit of acoustical measurement.

Decorative block–A concrete masonry unit having a special treatment on its face.

Deed restriction–A restriction on the use of a property as set forth in the deed.

Depreciation–A loss in property value caused by physical aging, functional or economic obsolescence.

Direct heating–Heating of spaces by means of exposed heated surfaces (stove, fire, radiators, etc.).

Distributed load–A load that acts evenly over a structural member.

Dock bumper–A resilient material attached to a loading dock to absorb the impact of trucks backing in.

Dome–A curved roof shape spanning an area.

Dormer–A structure projecting from a sloping roof.

Double-hung window–A window having two vertical sliding sashes-one covering the upper section and one covering the lower.

Downspout–A vertical pipe for carrying rain water from the roof to the ground.

Glossary

Drain tile–A hollow tile used to drain water-soaked soil.

Dressed size–In lumber about 3/8″ - 1/2″ less than nominal size after planing and sawing.

Drop panel–The depressed surface on the bottom side of a flat concrete slab which surrounds a column.

Dry pipe sprinkler system–A sprinkler system that is activated with water only in case of fire. Used in areas susceptible to freezing or to avoid the hazards of leaking pipes.

Drywall–An interior wall constructed of gypsum board, plywood or wood paneling. No water is required for application.

Duct–Pipe for transmitting warm or cold air. A pipe containing electrical cable or wires.

Dwelling–A structure designed as living quarters for one or more families.

Easement–A right of way or free access over land owned by another.

Eave–The lower edge of a sloping roof that overhangs the side wall of the structure.

Economic rent–That rent on a property sufficient to pay all operating costs exclusive of services and utilities.

Economic life–The term during which a structure is expected to be profitable. Generally shorter than the physical life of the structure.

Economic obsolescence–Loss in value due to unfavorable economic influences occurring from outside the structure itself.

Effective age–The age a structure appears to be based on observed physical condition determined by degree of maintenance and repair.

Elevator–A mechanism for vertical transport of personnel or freight equipped with car or platform.

Ell–A secondary wing or addition to a structure at right angles to the main structure.

Eminent domain–The right of the state to take private property for public use.

Envelope–The shape of a building indicating volume.

Equity–Value of an owner's interest calculated by subtracting outstanding mortgages and expenses from the value of the property.

Escheat–The assumption of ownership by the state, of property whose owner cannot be determined.

Facade–The exterior face of a building sometimes decorated with elaborate detail.

Face brick–Brick manufactured to present an attractive appearance.

Facing block–A concrete masonry unit having a decorative exterior finish.

Feasibility study–A detailed evaluation of a proposed project to determine its financial potential.

Felt paper–Paper sheathing used on walls as an insulator and to prevent infiltration of moisture.

Fenestration–Pertaining to the density and arrangement of windows in a structure.

Fiberboard–A building material composed of wood fiber compressed with a binding agent, produced in sheet form.

Fiberglass–Fine spun filaments of glass processed into various densities to produce thermal and acoustical insulation.

Field house–A long structure used for indoor athletic events.

Finish floor–The top or wearing surface of the floor system.

Fireproofing–The use of fire-resistant materials for the protection of structural members to ensure structural integrity in the event of fire.

Fire stop–A material or member used to seal an opening to prevent the spread of fire.

Flashing–A thin, impervious material such as copper or sheet metal used to prevent air or water penetration.

Float finish–A concrete finish accomplished by using a flat tool with a handle on the back.

Floor drain–An opening installed in a floor for removing excess water into a plumbing system.

Floor load–The live load the floor system has been designed for and which may be applied safely.

Flue–A heat-resistant enclosed passage in a chimney to remove gaseous products of combustion from a fireplace or boiler.

Footing–The lowest portion of the foundation wall that transmits the load directly to the soil.

Foundation–Below grade wall system that supports the structure.

Foyer–A portion of the structure that serves as a transitional space between the interior and exterior.

Frontage–That portion of a lot that is placed facing a street, public way or body of water.

Frost action–The effects of freezing and thawing on materials and the resultant structural damage.

Functional obsolescence–An inadequacy caused by outmoded design, dated construction materials or over or undersized areas, all of which cause excessive operating costs.

Furring strips–Wood or metal channels used for attaching gypsum or metal lath to masonry walls as a finish or for leveling purposes.

Gambrel roof (mansard)–A roof system having two pitches on each side.

Garden apartment–Two or three story apartment building with common outside areas.

General contractor–The prime contractor responsible for work on the construction site.

Girder–A principal horizontal supporting member.

Girt–A horizontal framing member for adding rigid support to columns which also acts as a support for sheathing or siding.

Grade beam–That portion of the foundation system that directly supports the exterior walls of the building.

Gross area–The total enclosed floor area of a building.

Ground area–The area computed by the exterior dimensions of the structure.

Ground floor–The floor of a building in closest proximity to the ground.

Grout–A mortar containing a high water content used to fill joints and cavities in masonry work.

Gutter–A wood or sheet metal conduit set along the building eaves used to channel rainwater to leaders or downspouts.

Gunite–A concrete mix placed pneumatically.

Gypsum–Hydrated calcium sulphate used as both a retarder in Portland cement and as a major ingredient in plaster of Paris. Used in sheets as a substitute for plaster.

Hall–A passageway providing access to various parts of a building-a large room for entertainment or assembly.

Hangar–A structure for the storage or repair of aircraft.

Head room–The distance between the top of the finished floor to the bottom of the finished ceiling.

Hearth–The floor of a fireplace and the adjacent area of fireproof material.

Heat pump–A mechanical device for providing either heating or air conditioning.

Hip roof–A roof whose four sides meet at a common point with no gabled ends.

I-beam–A structural member having a cross-section resembling the letter "I".

Insulation–Material used to reduce the effects of heat, cold or sound.

Jack rafter–An unusually short rafter generally found in hip roofs.

Jalousie–Adjustable glass louvers which pivot simultaneously in a common frame.

Jamb–The vertical member on either side of a door frame or window frame.

For customer support on your Square Foot Costs with RSMeans data, call 800.448.8182.

541

Joist–Parallel beams of timber, concrete or steel used to support floor and ceiling.

Junction box–A box that protects splices or joints in electrical wiring.

Kalamein door–Solid core wood doors clad with galvanized sheet metal.

Kip–A unit of weight equal to 1,000 pounds.

Lally column–A concrete-filled steel pipe used as a vertical support.

Laminated beam–A beam built up by gluing together several pieces of timber.

Landing–A platform between flights of stairs.

Lath–Strips of wood or metal used as a base for plaster.

Lead-lined door, sheetrock–Doors or sheetrock internally lined with sheet lead to provide protection from X-ray radiation.

Lean-to–A small shed or building addition with a single pitched roof attached to the exterior wall of the main building.

Lintel–A horizontal framing member used to carry a load over a wall opening.

Live load–The moving or movable load on a structure composed of furnishings, equipment or personnel weight.

Load bearing partition–A partition that supports a load in addition to its own weight.

Loading dock leveler–An adjustable platform for handling off-on loading of trucks.

Loft building–A commercial/industrial type building containing large, open unpartitioned floor areas.

Louver window–A window composed of a series of sloping, overlapping blades or slats that may be adjusted to admit varying degrees of air or light.

Main beam–The principal load bearing beam used to transmit loads directly to columns.

Mansard roof–A roof having a double pitch on all four sides, the lower level having the steeper pitch.

Mansion–An extremely large and imposing residence.

Masonry–The utilization of brick, stone, concrete or block for walls and other building components.

Mastic–A sealant or adhesive compound generally used as either a binding agent or floor finish.

Membrane fireproofing–A coating of metal lath and plaster to provide resistance to fire and heat.

Mesh reinforcement–An arrangement of wire tied or welded at their intersection to provide strength and resistance to cracking.

Metal lath–A diamond-shaped metallic base for plaster.

Mezzanine–A low story situated between two main floors.

Mill construction–A heavy timber construction that achieves fire resistance by using large wood structural members, noncombustible bearing and non-bearing walls and omitting concealed spaces under floors and roof.

Mixed occupancy–Two or more classes of occupancy in a single structure.

Molded brick–A specially shaped brick used for decorative purposes.

Monolithic concrete–Concrete poured in a continuous process so there are no joints.

Movable partition–A non-load bearing demountable partition that can be relocated and can be either ceiling height or partial height.

Moving walk–A continually moving horizontal passenger carrying device.

Multi-zone system–An air conditioning system that is capable of handling several individual zones simultaneously.

Net floor area–The usable, occupied area of a structure excluding stairwells, elevator shafts and wall thicknesses.

Non-combustible construction–Construction in which the walls, partitions and the structural members are of non-combustible materials.

Nurses call system–An electrically operated system for use by patients or personnel for summoning a nurse.

Occupancy rate–The number of persons per room, per dwelling unit, etc.

One-way joist connection–A framing system for floors and roofs in a concrete building consisting of a series of parallel joists supported by girders between columns.

Open web joist–A lightweight prefabricated metal chord truss.

Parapet–A low guarding wall at the point of a drop. That portion of the exterior wall that extends above the roof line.

Parging–A thin coat of mortar applied to a masonry wall used primarily for waterproofing purposes.

Parquet–Inlaid wood flooring set in a simple design.

Peaked roof–A roof of two or more slopes that rises to a peak.

Penthouse–A structure occupying usually half the roof area and used to house elevator, HVAC equipment, or other mechanical or electrical systems.

Pier–A column designed to support a concentrated load.

Pilaster–A column usually formed of the same material and integral with, but projecting from, a wall.

Pile–A concrete, wood or steel column usually less than 2′ in diameter which is driven or otherwise introduced into the soil to carry a vertical load or provide vertical support.

Pitched roof–A roof having one or more surfaces with a pitch greater than 10°.

Plenum–The space between the suspended ceiling and the floor above.

Plywood–Structural wood made of three or more layers of veneer and bonded together with glue.

Porch–A structure attached to a building to provide shelter for an entranceway or to serve as a semi-enclosed space.

Post and beam framing–A structural framing system where beams rest on posts rather than bearing walls.

Post-tensioned concrete–Concrete that has the reinforcing tendons tensioned after the concrete has set.

Pre-cast concrete–Concrete structural components fabricated at a location other than in-place.

Pre-stressed concrete–Concrete that has the reinforcing tendons tensioned prior to the concrete setting.

Purlins–A horizontal structural member supporting the roof deck and resting on the trusses, girders, beams or rafters.

Rafters–Structural members supporting the roof deck and covering.

Resilient flooring–A manufactured interior floor covering material in either sheet or tile form, that is resilient.

Ridge–The horizontal line at the junction of two sloping roof edges.

Rigid frame–A structural framing system in which all columns and beams are rigidly connected; there are no hinged joints.

Roof drain–A drain designed to accept rainwater and discharge it into a leader or downspout.

Rough floor–A layer of boards or plywood nailed to the floor joists that serves as a base for the finished floor.

Sanitary sewer–A sewer line designed to carry only liquid or waterborne waste from the structure to a central treatment plant.

Sawtooth roof–Roof shape found primarily in industrial roofs that creates the appearance of the teeth of a saw.

Semi-rigid frame–A structural system wherein the columns and beams are so attached that there is some flexibility at the joints.

Septic tank–A covered tank in which waste matter is decomposed by natural bacterial action.

Service elevator–A combination passenger and freight elevator.

Sheathing–The first covering of exterior studs by boards, plywood or particle board.

Shoring–Temporary bracing for structural components during construction.

Span–The horizontal distance between supports.

Specification–A detailed list of materials and requirements for construction of a building.

Storm sewer–A system of pipes used to carry rainwater or surface waters.

Stucco–A cement plaster used to cover exterior wall surfaces usually applied over a wood or metal lath base.

Stud–A vertical wooden structural component.

Subfloor–A system of boards, plywood or particle board laid over the floor joists to form a base.

Superstructure–That portion of the structure above the foundation or ground level.

Terrazzo–A durable floor finish made of small chips of colored stone or marble, embedded in concrete, then polished to a high sheen.

Tile–A thin piece of fired clay, stone or concrete used for floor, roof or wall finishes.

Unit cost–The cost per unit of measurement.

Vault–A room especially designed for storage.

Veneer–A thin surface layer covering a base of common material.

Vent–An opening serving as an outlet for air.

Wainscot–The lower portion of an interior wall whose surface differs from that of the upper wall.

Wall bearing construction–A structural system where the weight of the floors and roof are carried directly by the masonry walls rather than the structural framing system.

Waterproofing–Any of a number of materials applied to various surfaces to prevent the infiltration of water.

Weatherstrip–A thin strip of metal, wood or felt used to cover the joint between the door or window sash and the jamb, casing or sill to keep out air, dust, rain, etc.

Wing–A building section or addition projecting from the main structure.

X-ray protection–Lead encased in sheetrock or plaster to prevent the escape of radiation.

Abbreviation	Meaning
A	Area Square Feet; Ampere
AAFES	Army and Air Force Exchange Service
ABS	Acrylonitrile Butadiene Stryrene; Asbestos Bonded Steel
A.C., AC	Alternating Current; Air-Conditioning; Asbestos Cement; Plywood Grade A & C
ACI	American Concrete Institute
ACR	Air Conditioning Refrigeration
ADA	Americans with Disabilities Act
AD	Plywood, Grade A & D
Addit.	Additional
Adh.	Adhesive
Adj.	Adjustable
af	Audio-frequency
AFFF	Aqueous Film Forming Foam
AFUE	Annual Fuel Utilization Efficiency
AGA	American Gas Association
Agg.	Aggregate
A.H., Ah	Ampere Hours
A hr.	Ampere-hour
A.H.U., AHU	Air Handling Unit
A.I.A.	American Institute of Architects
AIC	Ampere Interrupting Capacity
Allow.	Allowance
alt., alt	Alternate
Alum.	Aluminum
a.m.	Ante Meridiem
Amp.	Ampere
Anod.	Anodized
ANSI	American National Standards Institute
APA	American Plywood Association
Approx.	Approximate
Apt.	Apartment
Asb.	Asbestos
A.S.B.C.	American Standard Building Code
Asbe.	Asbestos Worker
ASCE	American Society of Civil Engineers
A.S.H.R.A.E.	American Society of Heating, Refrig. & AC Engineers
ASME	American Society of Mechanical Engineers
ASTM	American Society for Testing and Materials
Attchmt.	Attachment
Avg., Ave.	Average
AWG	American Wire Gauge
AWWA	American Water Works Assoc.
Bbl.	Barrel
B&B, BB	Grade B and Better; Balled & Burlapped
B&S	Bell and Spigot
B.&W.	Black and White
b.c.c.	Body-centered Cubic
B.C.Y.	Bank Cubic Yards
BE	Bevel End
B.F.	Board Feet
Bg. cem.	Bag of Cement
BHP	Boiler Horsepower; Brake Horsepower
B.I.	Black Iron
bidir.	bidirectional
Bit., Bitum.	Bituminous
Bit., Conc.	Bituminous Concrete
Bk.	Backed
Bkrs.	Breakers
Bldg., bldg	Building
Blk.	Block
Bm.	Beam
Boil.	Boilermaker
bpm	Blows per Minute
BR	Bedroom
Brg., brng.	Bearing
Brhe.	Bricklayer Helper
Bric.	Bricklayer
Brk., brk	Brick
brkt	Bracket
Brs.	Brass
Brz.	Bronze
Bsn.	Basin
Btr.	Better
BTU	British Thermal Unit
BTUH	BTU per Hour
Bu.	Bushels
BUR	Built-up Roofing
BX	Interlocked Armored Cable
°C	Degree Centigrade
c	Conductivity, Copper Sweat
C	Hundred; Centigrade
C/C	Center to Center, Cedar on Cedar
C-C	Center to Center
Cab	Cabinet
Cair.	Air Tool Laborer
Cal.	Caliper
Calc	Calculated
Cap.	Capacity
Carp.	Carpenter
C.B.	Circuit Breaker
C.C.A.	Chromate Copper Arsenate
C.C.F.	Hundred Cubic Feet
cd	Candela
cd/sf	Candela per Square Foot
CD	Grade of Plywood Face & Back
CDX	Plywood, Grade C & D, exterior glue
Cefi.	Cement Finisher
Cem.	Cement
CF	Hundred Feet
C.F.	Cubic Feet
CFM	Cubic Feet per Minute
CFRP	Carbon Fiber Reinforced Plastic
c.g.	Center of Gravity
CHW	Chilled Water; Commercial Hot Water
C.I., CI	Cast Iron
C.I.P., CIP	Cast in Place
Circ.	Circuit
C.L.	Carload Lot
CL	Chain Link
Clab.	Common Laborer
Clam	Common Maintenance Laborer
C.L.F.	Hundred Linear Feet
CLF	Current Limiting Fuse
CLP	Cross Linked Polyethylene
cm	Centimeter
CMP	Corr. Metal Pipe
CMU	Concrete Masonry Unit
CN	Change Notice
Col.	Column
CO_2	Carbon Dioxide
Comb.	Combination
comm.	Commercial, Communication
Compr.	Compressor
Conc.	Concrete
Cont., cont	Continuous; Continued, Container
Corkbd.	Cork Board
Corr.	Corrugated
Cos	Cosine
Cot	Cotangent
Cov.	Cover
C/P	Cedar on Paneling
CPA	Control Point Adjustment
Cplg.	Coupling
CPM	Critical Path Method
CPVC	Chlorinated Polyvinyl Chloride
C.Pr.	Hundred Pair
CRC	Cold Rolled Channel
Creos.	Creosote
Crpt.	Carpet & Linoleum Layer
CRT	Cathode-ray Tube
CS	Carbon Steel, Constant Shear Bar Joist
Csc	Cosecant
C.S.F.	Hundred Square Feet
CSI	Construction Specifications Institute
CT	Current Transformer
CTS	Copper Tube Size
Cu	Copper, Cubic
Cu. Ft.	Cubic Foot
cw	Continuous Wave
C.W.	Cool White; Cold Water
Cwt.	100 Pounds
C.W.X.	Cool White Deluxe
C.Y.	Cubic Yard (27 cubic feet)
C.Y./Hr.	Cubic Yard per Hour
Cyl.	Cylinder
d	Penny (nail size)
D	Deep; Depth; Discharge
Dis., Disch.	Discharge
Db	Decibel
Dbl.	Double
DC	Direct Current
DDC	Direct Digital Control
Demob.	Demobilization
d.f.t.	Dry Film Thickness
d.f.u.	Drainage Fixture Units
D.H.	Double Hung
DHW	Domestic Hot Water
DI	Ductile Iron
Diag.	Diagonal
Diam., Dia	Diameter
Distrib.	Distribution
Div.	Division
Dk.	Deck
D.L.	Dead Load; Diesel
DLH	Deep Long Span Bar Joist
dlx	Deluxe
Do.	Ditto
DOP	Dioctyl Phthalate Penetration Test (Air Filters)
Dp., dp	Depth
D.P.S.T.	Double Pole, Single Throw
Dr.	Drive
DR	Dimension Ratio
Drink.	Drinking
D.S.	Double Strength
D.S.A.	Double Strength A Grade
D.S.B.	Double Strength B Grade
Dty.	Duty
DWV	Drain Waste Vent
DX	Deluxe White, Direct Expansion
dyn	Dyne
e	Eccentricity
E	Equipment Only; East; Emissivity
Ea.	Each
EB	Encased Burial
Econ.	Economy
E.C.Y	Embankment Cubic Yards
EDP	Electronic Data Processing
EIFS	Exterior Insulation Finish System
E.D.R.	Equiv. Direct Radiation
Eq.	Equation
EL	Elevation
Elec.	Electrician; Electrical
Elev.	Elevator; Elevating
EMT	Electrical Metallic Conduit; Thin Wall Conduit
Eng.	Engine, Engineered
EPDM	Ethylene Propylene Diene Monomer
EPS	Expanded Polystyrene
Eqhv.	Equip. Oper., Heavy
Eqlt.	Equip. Oper., Light
Eqmd.	Equip. Oper., Medium
Eqmm.	Equip. Oper., Master Mechanic
Eqol.	Equip. Oper., Oilers
Equip.	Equipment
ERW	Electric Resistance Welded

Abbreviations

E.S.	Energy Saver	H	High Henry	Lath.	Lather
Est.	Estimated	HC	High Capacity	Lav.	Lavatory
esu	Electrostatic Units	H.D., HD	Heavy Duty; High Density	lb.; #	Pound
E.W.	Each Way	H.D.O.	High Density Overlaid	L.B., LB	Load Bearing; L Conduit Body
EWT	Entering Water Temperature	HDPE	High Density Polyethylene Plastic	L. & E.	Labor & Equipment
Excav.	Excavation	Hdr.	Header	lb./hr.	Pounds per Hour
excl	Excluding	Hdwe.	Hardware	lb./L.F.	Pounds per Linear Foot
Exp., exp	Expansion, Exposure	H.I.D., HID	High Intensity Discharge	lbf/sq.in.	Pound-force per Square Inch
Ext., ext	Exterior; Extension	Help.	Helper Average	L.C.L.	Less than Carload Lot
Extru.	Extrusion	HEPA	High Efficiency Particulate Air	L.C.Y.	Loose Cubic Yard
f.	Fiber Stress		Filter	Ld.	Load
F	Fahrenheit; Female; Fill	Hg	Mercury	LE	Lead Equivalent
Fab., fab	Fabricated; Fabric	HIC	High Interrupting Capacity	LED	Light Emitting Diode
FBGS	Fiberglass	HM	Hollow Metal	L.F.	Linear Foot
F.C.	Footcandles	HMWPE	High Molecular Weight	L.F. Hdr	Linear Feet of Header
f.c.c.	Face-centered Cubic		Polyethylene	L.F. Nose	Linear Foot of Stair Nosing
f'c.	Compressive Stress in Concrete;	HO	High Output	L.F. Rsr	Linear Foot of Stair Riser
	Extreme Compressive Stress	Horiz.	Horizontal	Lg.	Long; Length; Large
F.E.	Front End	H.P., HP	Horsepower; High Pressure	L & H	Light and Heat
FEP	Fluorinated Ethylene Propylene	H.P.F.	High Power Factor	LH	Long Span Bar Joist
	(Teflon)	Hr.	Hour	L.H.	Labor Hours
F.G.	Flat Grain	Hrs./Day	Hours per Day	L.L., LL	Live Load
F.H.A.	Federal Housing Administration	HSC	High Short Circuit	L.L.D.	Lamp Lumen Depreciation
Fig.	Figure	Ht.	Height	lm	Lumen
Fin.	Finished	Htg.	Heating	lm/sf	Lumen per Square Foot
FIPS	Female Iron Pipe Size	Htrs.	Heaters	lm/W	Lumen per Watt
Fixt.	Fixture	HVAC	Heating, Ventilation & Air-	LOA	Length Over All
FJP	Finger jointed and primed		Conditioning	log	Logarithm
Fl. Oz.	Fluid Ounces	Hvy.	Heavy	L-O-L	Lateralolet
Flr.	Floor	HW	Hot Water	long.	Longitude
Flrs.	Floors	Hyd.; Hydr.	Hydraulic	L.P., LP	Liquefied Petroleum; Low Pressure
FM	Frequency Modulation;	Hz	Hertz (cycles)	L.P.F.	Low Power Factor
	Factory Mutual	I.	Moment of Inertia	LR	Long Radius
Fmg.	Framing	IBC	International Building Code	L.S.	Lump Sum
FM/UL	Factory Mutual/Underwriters Labs	I.C.	Interrupting Capacity	Lt.	Light
Fdn.	Foundation	ID	Inside Diameter	Lt. Ga.	Light Gauge
FNPT	Female National Pipe Thread	I.D.	Inside Dimension; Identification	L.T.L.	Less than Truckload Lot
Fori.	Foreman, Inside	I.F.	Inside Frosted	Lt. Wt.	Lightweight
Foro.	Foreman, Outside	I.M.C.	Intermediate Metal Conduit	L.V.	Low Voltage
Fount.	Fountain	In.	Inch	M	Thousand; Material; Male;
fpm	Feet per Minute	Incan.	Incandescent		Light Wall Copper Tubing
FPT	Female Pipe Thread	Incl.	Included; Including	M²CA	Meters Squared Contact Area
Fr	Frame	Int.	Interior	m/hr.; M.H.	Man-hour
F.R.	Fire Rating	Inst.	Installation	mA	Milliampere
FRK	Foil Reinforced Kraft	Insul., insul	Insulation/Insulated	Mach.	Machine
FSK	Foil/Scrim/Kraft	I.P.	Iron Pipe	Mag. Str.	Magnetic Starter
FRP	Fiberglass Reinforced Plastic	I.P.S., IPS	Iron Pipe Size	Maint.	Maintenance
FS	Forged Steel	IPT	Iron Pipe Threaded	Marb.	Marble Setter
FSC	Cast Body; Cast Switch Box	I.W.	Indirect Waste	Mat; Mat'l.	Material
Ft., ft	Foot; Feet	J	Joule	Max.	Maximum
Ftng.	Fitting	J.I.C.	Joint Industrial Council	MBF	Thousand Board Feet
Ftg.	Footing	K	Thousand; Thousand Pounds;	MBH	Thousand BTU's per hr.
Ft lb.	Foot Pound		Heavy Wall Copper Tubing, Kelvin	MC	Metal Clad Cable
Furn.	Furniture	K.A.H.	Thousand Amp. Hours	MCC	Motor Control Center
FVNR	Full Voltage Non-Reversing	kcmil	Thousand Circular Mils	M.C.F.	Thousand Cubic Feet
FVR	Full Voltage Reversing	KD	Knock Down	MCFM	Thousand Cubic Feet per Minute
FXM	Female by Male	K.D.A.T.	Kiln Dried After Treatment	M.C.M.	Thousand Circular Mils
Fy.	Minimum Yield Stress of Steel	kg	Kilogram	MCP	Motor Circuit Protector
g	Gram	kG	Kilogauss	MD	Medium Duty
G	Gauss	kgf	Kilogram Force	MDF	Medium-density fibreboard
Ga.	Gauge	kHz	Kilohertz	M.D.O.	Medium Density Overlaid
Gal., gal.	Gallon	Kip	1000 Pounds	Med.	Medium
Galv., galv	Galvanized	KJ	Kilojoule	MF	Thousand Feet
GC/MS	Gas Chromatograph/Mass	K.L.	Effective Length Factor	M.F.B.M.	Thousand Feet Board Measure
	Spectrometer	K.L.F.	Kips per Linear Foot	Mfg.	Manufacturing
Gen.	General	Km	Kilometer	Mfrs.	Manufacturers
GFI	Ground Fault Interrupter	KO	Knock Out	mg	Milligram
GFRC	Glass Fiber Reinforced Concrete	K.S.F.	Kips per Square Foot	MGD	Million Gallons per Day
Glaz.	Glazier	K.S.I.	Kips per Square Inch	MGPH	Million Gallons per Hour
GPD	Gallons per Day	kV	Kilovolt	MH, M.H.	Manhole; Metal Halide; Man-Hour
gpf	Gallon per Flush	kVA	Kilovolt Ampere	MHz	Megahertz
GPH	Gallons per Hour	kVAR	Kilovar (Reactance)	Mi.	Mile
gpm, GPM	Gallons per Minute	KW	Kilowatt	MI	Malleable Iron; Mineral Insulated
GR	Grade	KWh	Kilowatt-hour	MIPS	Male Iron Pipe Size
Gran.	Granular	L	Labor Only; Length; Long;	mj	Mechanical Joint
Grnd.	Ground		Medium Wall Copper Tubing	m	Meter
GVW	Gross Vehicle Weight	Lab.	Labor	mm	Millimeter
GWB	Gypsum Wall Board	lat	Latitude	Mill.	Millwright
				Min., min.	Minimum, Minute

545

Misc.	Miscellaneous	PCM	Phase Contrast Microscopy	SBS	Styrene Butadiere Styrene
ml	Milliliter, Mainline	PDCA	Painting and Decorating	SC	Screw Cover
M.L.F.	Thousand Linear Feet		Contractors of America	SCFM	Standard Cubic Feet per Minute
Mo.	Month	P.E., PE	Professional Engineer;	Scaf.	Scaffold
Mobil.	Mobilization		Porcelain Enamel;	Sch., Sched.	Schedule
Mog.	Mogul Base		Polyethylene; Plain End	S.C.R.	Modular Brick
MPH	Miles per Hour	P.E.C.I.	Porcelain Enamel on Cast Iron	S.D.	Sound Deadening
MPT	Male Pipe Thread	Perf.	Perforated	SDR	Standard Dimension Ratio
MRGWB	Moisture Resistant Gypsum	PEX	Cross Linked Polyethylene	S.E.	Surfaced Edge
	Wallboard	Ph.	Phase	Sel.	Select
MRT	Mile Round Trip	P.I.	Pressure Injected	SER, SEU	Service Entrance Cable
ms	Millisecond	Pile.	Pile Driver	S.F.	Square Foot
M.S.F.	Thousand Square Feet	Pkg.	Package	S.F.C.A.	Square Foot Contact Area
Mstz.	Mosaic & Terrazzo Worker	Pl.	Plate	S.F. Flr.	Square Foot of Floor
M.S.Y.	Thousand Square Yards	Plah.	Plasterer Helper	S.F.G.	Square Foot of Ground
Mtd., mtd., mtd	Mounted	Plas.	Plasterer	S.F. Hor.	Square Foot Horizontal
Mthe.	Mosaic & Terrazzo Helper	plf	Pounds Per Linear Foot	SFR	Square Feet of Radiation
Mtng.	Mounting	Pluh.	Plumber Helper	S.F. Shlf.	Square Foot of Shelf
Mult.	Multi; Multiply	Plum.	Plumber	S4S	Surface 4 Sides
MUTCD	Manual on Uniform Traffic Control	Ply.	Plywood	Shee.	Sheet Metal Worker
	Devices	p.m.	Post Meridiem	Sin.	Sine
M.V.A.	Million Volt Amperes	Pntd.	Painted	Skwk.	Skilled Worker
M.V.A.R.	Million Volt Amperes Reactance	Pord.	Painter, Ordinary	SL	Saran Lined
MV	Megavolt	pp	Pages	S.L.	Slimline
MW	Megawatt	PP, PPL	Polypropylene	Sldr.	Solder
MXM	Male by Male	P.P.M.	Parts per Million	SLH	Super Long Span Bar Joist
MYD	Thousand Yards	Pr.	Pair	S.N.	Solid Neutral
N	Natural; North	P.E.S.B.	Pre-engineered Steel Building	SO	Stranded with oil resistant inside
nA	Nanoampere	Prefab.	Prefabricated		insulation
NA	Not Available; Not Applicable	Prefin.	Prefinished	S-O-L	Socketolet
N.B.C.	National Building Code	Prop.	Propelled	sp	Standpipe
NC	Normally Closed	PSF, psf	Pounds per Square Foot	S.P.	Static Pressure; Single Pole; Self-
NEMA	National Electrical Manufacturers	PSI, psi	Pounds per Square Inch		Propelled
	Assoc.	PSIG	Pounds per Square Inch Gauge	Spri.	Sprinkler Installer
NEHB	Bolted Circuit Breaker to 600V.	PSP	Plastic Sewer Pipe	spwg	Static Pressure Water Gauge
NFPA	National Fire Protection Association	Pspr.	Painter, Spray	S.P.D.T.	Single Pole, Double Throw
NLB	Non-Load-Bearing	Psst.	Painter, Structural Steel	SPF	Spruce Pine Fir; Sprayed
NM	Non-Metallic Cable	P.T.	Potential Transformer		Polyurethane Foam
nm	Nanometer	P. & T.	Pressure & Temperature	S.P.S.T.	Single Pole, Single Throw
No.	Number	Ptd.	Painted	SPT	Standard Pipe Thread
NO	Normally Open	Ptns.	Partitions	Sq.	Square; 100 Square Feet
N.O.C.	Not Otherwise Classified	Pu	Ultimate Load	Sq. Hd.	Square Head
Nose.	Nosing	PVC	Polyvinyl Chloride	Sq. In.	Square Inch
NPT	National Pipe Thread	Pvmt.	Pavement	S.S.	Single Strength; Stainless Steel
NQOD	Combination Plug-on/Bolt on	PRV	Pressure Relief Valve	S.S.B.	Single Strength B Grade
	Circuit Breaker to 240V.	Pwr.	Power	sst, ss	Stainless Steel
N.R.C., NRC	Noise Reduction Coefficient/	Q	Quantity Heat Flow	Sswk.	Structural Steel Worker
	Nuclear Regulator Commission	Qt.	Quart	Sswl.	Structural Steel Welder
N.R.S.	Non Rising Stem	Quan., Qty.	Quantity	St.; Stl.	Steel
ns	Nanosecond	Q.C.	Quick Coupling	STC	Sound Transmission Coefficient
NTP	Notice to Proceed	r	Radius of Gyration	Std.	Standard
nW	Nanowatt	R	Resistance	Stg.	Staging
OB	Opposing Blade	R.C.P.	Reinforced Concrete Pipe	STK	Select Tight Knot
OC	On Center	Rect.	Rectangle	STP	Standard Temperature & Pressure
OD	Outside Diameter	recpt.	Receptacle	Stpi.	Steamfitter, Pipefitter
O.D.	Outside Dimension	Reg.	Regular	Str.	Strength; Starter; Straight
ODS	Overhead Distribution System	Reinf.	Reinforced	Strd.	Stranded
O.G.	Ogee	Req'd.	Required	Struct.	Structural
O.H.	Overhead	Res.	Resistant	Sty.	Story
O&P	Overhead and Profit	Resi.	Residential	Subj.	Subject
Oper.	Operator	RF	Radio Frequency	Subs.	Subcontractors
Opng.	Opening	RFID	Radio-frequency Identification	Surf.	Surface
Orna.	Ornamental	Rgh.	Rough	Sw.	Switch
OSB	Oriented Strand Board	RGS	Rigid Galvanized Steel	Swbd.	Switchboard
OS&Y	Outside Screw and Yoke	RHW	Rubber, Heat & Water Resistant;	S.Y.	Square Yard
OSHA	Occupational Safety and Health		Residential Hot Water	Syn.	Synthetic
	Act	rms	Root Mean Square	S.Y.P.	Southern Yellow Pine
Ovhd.	Overhead	Rnd.	Round	Sys.	System
OWG	Oil, Water or Gas	Rodm.	Rodman	t.	Thickness
Oz.	Ounce	Rofc.	Roofer, Composition	T	Temperature; Ton
P.	Pole; Applied Load; Projection	Rofp.	Roofer, Precast	Tan	Tangent
p.	Page	Rohe.	Roofer Helpers (Composition)	T.C.	Terra Cotta
Pape.	Paperhanger	Rots.	Roofer, Tile & Slate	T & C	Threaded and Coupled
P.A.P.R.	Powered Air Purifying Respirator	R.O.W.	Right of Way	T.D.	Temperature Difference
PAR	Parabolic Reflector	RPM	Revolutions per Minute	TDD	Telecommunications Device for
P.B., PB	Push Button	R.S.	Rapid Start		the Deaf
Pc., Pcs.	Piece, Pieces	Rsr	Riser	T.E.M.	Transmission Electron Microscopy
P.C.	Portland Cement; Power Connector	RT	Round Trip	temp	Temperature, Tempered, Temporary
P.C.F.	Pounds per Cubic Foot	S.	Suction; Single Entrance; South	TFFN	Nylon Jacketed Wire

Abbreviations

TFE	Tetrafluoroethylene (Teflon)	U.L., UL	Underwriters Laboratory	w/	With
T. & G.	Tongue & Groove;	Uld.	Unloading	W.C., WC	Water Column; Water Closet
	Tar & Gravel	Unfin.	Unfinished	W.F.	Wide Flange
Th., Thk.	Thick	UPS	Uninterruptible Power Supply	W.G.	Water Gauge
Thn.	Thin	URD	Underground Residential	Wldg.	Welding
Thrded	Threaded		Distribution	W. Mile	Wire Mile
Tilf.	Tile Layer, Floor	US	United States	W-O-L	Weldolet
Tilh.	Tile Layer, Helper	USGBC	U.S. Green Building Council	W.R.	Water Resistant
THHN	Nylon Jacketed Wire	USP	United States Primed	Wrck.	Wrecker
THW.	Insulated Strand Wire	UTMCD	Uniform Traffic Manual For Control	WSFU	Water Supply Fixture Unit
THWN	Nylon Jacketed Wire		Devices	W.S.P.	Water, Steam, Petroleum
T.L., TL	Truckload	UTP	Unshielded Twisted Pair	WT., Wt.	Weight
T.M.	Track Mounted	V	Volt	WWF	Welded Wire Fabric
Tot.	Total	VA	Volt Amperes	XFER	Transfer
T-O-L	Threadolet	VAT	Vinyl Asbestos Tile	XFMR	Transformer
tmpd	Tempered	V.C.T.	Vinyl Composition Tile	XHD	Extra Heavy Duty
TPO	Thermoplastic Polyolefin	VAV	Variable Air Volume	XHHW	Cross-Linked Polyethylene Wire
T.S.	Trigger Start	VC	Veneer Core	XLPE	Insulation
Tr.	Trade	VDC	Volts Direct Current	XLP	Cross-linked Polyethylene
Transf.	Transformer	Vent.	Ventilation	Xport	Transport
Trhv.	Truck Driver, Heavy	Vert.	Vertical	Y	Wye
Trlr	Trailer	V.F.	Vinyl Faced	yd	Yard
Trlt.	Truck Driver, Light	V.G.	Vertical Grain	yr	Year
TTY	Teletypewriter	VHF	Very High Frequency	Δ	Delta
TV	Television	VHO	Very High Output	%	Percent
T.W.	Thermoplastic Water Resistant	Vib.	Vibrating	~	Approximately
	Wire	VLF	Vertical Linear Foot	Ø	Phase; diameter
UCI	Uniform Construction Index	VOC	Volatile Organic Compound	@	At
UF	Underground Feeder	Vol.	Volume	#	Pound; Number
UGND	Underground Feeder	VRP	Vinyl Reinforced Polyester	<	Less Than
UHF	Ultra High Frequency	W	Wire; Watt; Wide; West	>	Greater Than
U.I.	United Inch			Z	Zone

RESIDENTIAL
COST ESTIMATE

OWNER'S NAME: _____   APPRAISER: _____

RESIDENCE ADDRESS: _____   PROJECT: _____

CITY, STATE, ZIP CODE: _____   DATE: _____

CLASS OF CONSTRUCTION	RESIDENCE TYPE	CONFIGURATION	EXTERIOR WALL SYSTEM
☐ ECONOMY	☐ 1 STORY	☐ DETACHED	☐ WOOD SIDING—WOOD FRAME
☐ AVERAGE	☐ 1-1/2 STORY	☐ TOWN/ROW HOUSE	☐ BRICK VENEER—WOOD FRAME
☐ CUSTOM	☐ 2 STORY	☐ SEMI-DETACHED	☐ STUCCO ON WOOD FRAME
☐ LUXURY	☐ 2-1/2 STORY		☐ PAINTED CONCRETE BLOCK
	☐ 3 STORY	OCCUPANCY	☐ SOLID MASONRY (AVERAGE & CUSTOM)
	☐ BI-LEVEL	☐ ONE STORY	☐ STONE VENEER—WOOD FRAME
	☐ TRI-LEVEL	☐ TWO FAMILY	☐ SOLID BRICK (LUXURY)
		☐ THREE FAMILY	☐ SOLID STONE (LUXURY)
		☐ OTHER _____	

*LIVING AREA (Main Building)	*LIVING AREA (Wing or Ell) ( )	*LIVING AREA (Wing or Ell) ( )
First Level _____ S.F.	First Level _____ S.F.	First Level _____ S.F.
Second Level _____ S.F.	Second Level _____ S.F.	Second Level _____ S.F.
Third Level _____ S.F.	Third Level _____ S.F.	Third Level _____ S.F.
Total _____ S.F.	Total _____ S.F.	Total _____ S.F.

*Basement Area is not part of living area.

MAIN BUILDING	COSTS PER S.F. LIVING AREA
Cost per Square Foot of Living Area, from Page _____	$
Basement Addition: _____ % Finished, _____ % Unfinished	+
Roof Cover Adjustment: _____ Type, Page _____ (Add or Deduct)	( )
Central Air Conditioning: ☐ Separate Ducts ☐ Heating Ducts, Page _____	+
Heating System Adjustment: _____ Type, Page _____ (Add or Deduct)	( )
Main Building: Adjusted Cost per S.F. of Living Area	$

MAIN BUILDING TOTAL COST	$ _____ /S.F.	x	_____ S.F.	x	_____	=	$ _____
	Cost per S.F. Living Area		Living Area		Town/Row House Multiplier (Use 1 for Detached)		TOTAL COST

WING OR ELL ( ) _____ STORY	COSTS PER S.F. LIVING AREA
Cost per Square Foot of Living Area, from Page _____	$
Basement Addition: _____ % Finished, _____ % Unfinished	+
Roof Cover Adjustment: _____ Type, Page _____ (Add or Deduct)	( )
Central Air Conditioning: ☐ Separate Ducts ☐ Heating Ducts, Page _____	+
Heating System Adjustment: _____ Type, Page _____ (Add or Deduct)	( )
Wing or Ell ( ): Adjusted Cost per S.F. of Living Area	$

WING OR ELL ( ) TOTAL COST	$ _____ /S.F.	x	_____ S.F.	=	$ _____
	Cost per S.F. Living Area		Living Area		TOTAL COST

WING OR ELL ( ) _____ STORY	COSTS PER S.F. LIVING AREA
Cost per Square Foot of Living Area, from Page _____	$
Basement Addition: _____ % Finished, _____ % Unfinished	+
Roof Cover Adjustment: _____ Type, Page _____ (Add or Deduct)	( )
Central Air Conditioning: ☐ Separate Ducts ☐ Heating Ducts, Page _____	+
Heating System Adjustment: _____ Type, Page _____ (Add or Deduct)	( )
Wing or Ell ( ): Adjusted Cost per S.F. of Living Area	$

WING OR ELL ( ) TOTAL COST	$ _____ /S.F.	x	_____ S.F.	=	$ _____
	Cost per S.F. Living Area		Living Area		TOTAL COST

TOTAL THIS PAGE [_____]

Page 1 of 2

For customer support on your Square Foot Costs with RSMeans data, call 800.448.8182.

RESIDENTIAL
COST ESTIMATE

Total Page 1			$	
	QUANTITY	UNIT COST		
Additional Bathrooms: _____ Full, _____ Half				
Finished Attic: _____ Ft. x _____ Ft.	S.F.		+	
Breezeway: ☐ Open ☐ Enclosed _____ Ft. x _____ Ft.	S.F.		+	
Covered Porch: ☐ Open ☐ Enclosed _____ Ft. x _____ Ft.	S.F.		+	
Fireplace: ☐ Interior Chimney ☐ Exterior Chimney ☐ No. of Flues ☐ Additional Fireplaces			+	
Appliances:			+	
Kitchen Cabinets Adjustment: (+/−)				
☐ Garage ☐ Carport: _____ Car(s) Description _____ (+/−)				
Miscellaneous:			+	

ADJUSTED TOTAL BUILDING COST $ _____

REPLACEMENT COST	
ADJUSTED TOTAL BUILDING COST	$ _____
Site Improvements	
(A) Paving & Sidewalks	$ _____
(B) Landscaping	$ _____
(C) Fences	$ _____
(D) Swimming Pool	$ _____
(E) Miscellaneous	$ _____
TOTAL	$ _____
Location Factor	x _____
Location Replacement Cost	$ _____
Depreciation	−$ _____
LOCAL DEPRECIATED COST	$ _____

INSURANCE COST	
ADJUSTED TOTAL BUILDING COST	$ _____
Insurance Exclusions	
(A) Footings, Site Work, Underground Piping	−$ _____
(B) Architects' Fees	−$ _____
Total Building Cost Less Exclusion	$ _____
Location Factor	x _____
LOCAL INSURABLE REPLACEMENT COST	$ _____

SKETCH AND ADDITIONAL CALCULATIONS

CII APPRAISAL

5. DATE: _____

1. SUBJECT PROPERTY: _____
6. APPRAISER: _____

2. BUILDING: _____

3. ADDRESS: _____

4. BUILDING USE: _____
7. YEAR BUILT: _____

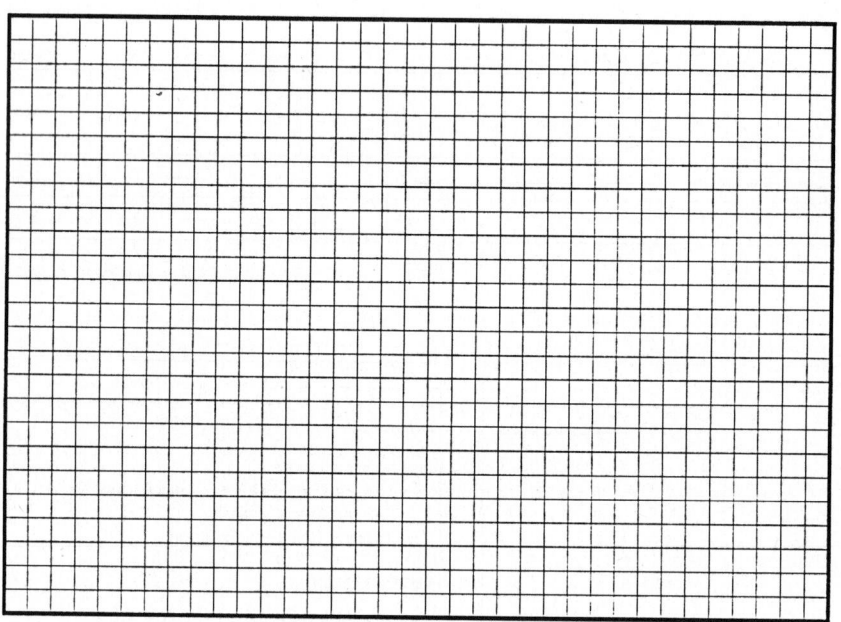

8. EXTERIOR WALL CONSTRUCTION: _____

9. FRAME: _____

10. GROUND FLOOR AREA: _____ S.F.   11. GROSS FLOOR AREA (EXCL. BASEMENT): _____ S.F.

12. NUMBER OF STORIES: _____   13. STORY HEIGHT: _____

14. PERIMETER: _____ L.F.   15. BASEMENT AREA: _____ S.F.

16. GENERAL COMMENTS: _____

_____

_____

_____

_____

_____

_____

_____

NO.		DESCRIPTION		UNIT	UNIT COST	NEW SF COST	MODEL SF COST	+/- CHANGE
A SUBSTRUCTURE								
A	1010	Standard Foundations	Bay size:	S.F. Gnd.				
A	1030	Slab on Grade	Material: Thickness:	S.F. Slab				
A	2010	Basement Excavation Depth:	Area:	S.F. Gnd.				
A	2020	Basement Walls		L.F. Walls				
B SHELL								
		B10 Superstructure						
B	1010	Floor Construction	Elevated floors:	S.F. Floor				
				S.F. Floor				
B	1020	Roof Construction		S.F. Roof				
		B20 Exterior Enclosure						
B	2010	Exterior walls Material:	Thickness: % of wall	S.F. Walls				
		Material:	Thickness: % of wall	S.F. Walls				
B	2020	Exterior Windows Type:	% of wall	S.F. Wind.				
		Type:	% of wall Each	S.F. Wind.				
B	2030	Exterior Doors Type:	Number:	Each				
		Type:	Number:	Each				
		B30 Roofing						
B	3010	Roof Coverings Material:		S.F. Roof				
		Material:		S.F. Roof				
B	3020	Roof Openings		S.F. Opng.				
C INTERIORS								
C	1010	Partitions: Material:	Density:	S.F. Part.				
		Material:	Density:					
C	1020	Interior Doors Type:	Number:	Each				
C	1030	Fittings		Each				
C	2010	Stair Construction		Flight				
C	3010	Wall Finishes Material:	% of Wall	S.F. Walls				
		Material:		S.F. Walls				
C	3020	Floor Finishes Material:		S.F. Floor				
		Material:		S.F. Floor				
		Material:		S.F. Floor				
C	3030	Ceiling Finishes Material:		S.F. Ceil.				

For customer support on your Square Foot Costs with RSMeans data, call 800.448.8182.

551

	NO.	SYSTEM/COMPONENT	DESCRIPTION	UNIT	UNIT COST	NEW S.F. COST	MODEL S.F. COST	+/- CHANGE
D		**SERVICES**						
		D10 Conveying						
D	1010	Elevators & Lifts	Type: Capacity: Stops:	Each				
D	1020	Escalators & Moving Walks	Type:	Each				
		D20 Plumbing						
D	2010	Plumbing		Each				
D	2020	Domestic Water Distribution		S.F. Floor				
D	2040	Rain Water Drainage		S.F. Roof				
		D30 HVAC						
D	3010	Energy Supply		Each				
D	3020	Heat Generating Systems	Type:	S.F. Floor				
D	3030	Cooling Generating Systems	Type:	S.F. Floor				
D	3090	Other HVAC Sys. & Equipment		Each				
		D40 Fire Protection						
D	4010	Sprinklers		S.F. Floor				
D	4020	Standpipes		S.F. Floor				
		D50 Electrical						
D	5010	Electrical Service/Distribution		S.F. Floor				
D	5020	Lighting & Branch Wiring		S.F. Floor				
D	5030	Communications & Security		S.F. Floor				
D	5090	Other Electrical Systems		S.F. Floor				
E		**EQUIPMENT & FURNISHINGS**						
E	1010	Commercial Equipment		Each				
E	1020	Institutional Equipment		Each				
E	1030	Vehicular Equipment		Each				
E	1090	Other Equipment		Each				
F		**SPECIAL CONSTRUCTION**						
F	1020	Integrated Construction		S.F.				
F	1040	Special Facilities		S.F.				
G		**BUILDING SITEWORK**						

ITEM		Total Change
17	Model 020 total ____ $_____	
18	Adjusted S.F. cost ____ $_____ item 17 +/- changes	
19	Building area - from item 11 _____ S.F. x adjusted S.F. cost......................................$...
20	Basement area - from item 15 _____ S.F. x S.F. cost $ ....$..
21	Base building sub-total - item 20 + item 19 ....$.....
22	Miscellaneous addition (quality, etc.) ....$......
23	Sub-total - item 22 + 21 ....$......
24	General conditions -25 % of item 23 ....$......
25	Sub-total - item 24 + item 23..$......
26	Architects fees _____ % of item 25 ....$
27	Sub-total - item 26 + item 27 ....$......
28	Location modifier ....x....
29	Local replacement cost - item 28 x item 27 ....$......
30	Depreciation _____ % of item 29 ....$
31	Depreciated local replacement cost - item 29 less item 30 ....$.....
32	Exclusions ....$.....
33	Net depreciated replacement cost - item 31 less item 32 ....$....

Base Cost per square foot floor area: *(from square foot table)*

Specify source:

Page: _____ , Model # _____ , Area _____ S.F.,
Exterior wall _____ , Frame _____

Adjustments for exterior wall variation:

Size Adjustment:
(Interpolate)

Height Adjustment: _____ + _____ = _____

Adjusted Base Cost per square foot:

Building Cost $ _____ x _____ = _____
Adjusted Base Cost   Floor Area
per square foot

Basement Cost $ _____ x _____ = _____
Basement Cost   Basement Area

Lump Sum Additions

TOTAL BUILDING COST *(Sum of above costs)* _____

Modifications: *(complexity, workmanship, size)* +/- _____ % _____

Location Modifier: City _____ Date _____ x _____

Local cost of replacement _____

Less depreciation: _____ - _____

Local cost of replacement less depreciation $ _____

For customer support on your Square Foot Costs with RSMeans data, call 800.448.8182.

553

...st Data Selection Guide

...following table provides definitive information on the content of each cost data publication. The number of lines of data provided in each unit price or ...emblies division, as well as the number of crews, is listed for each data set. The presence of other elements such as reference tables, square foot models, ...ipment rental costs, historical cost indexes, and city cost indexes, is also indicated. You can use the table to help select the RSMeans data set that has ...quantity and type of information you most need in your work.

Cost Divisions	Building Construction	Mechanical	Electrical	Commercial Renovation	Square Foot	Site Work Landsc.	Green Building	Interior	Concrete Masonry	Open Shop	Heavy Construction	Light Commercial	Facilities Construction	Plumbing	Residential
1	590	411	428	530	0	524	200	331	473	589	527	273	1065	421	178
2	777	280	86	733	0	991	207	399	214	776	733	481	1220	287	274
3	1688	340	230	1081	0	1480	986	354	2034	1688	1690	482	1788	316	389
4	960	21	0	920	0	725	180	615	1158	928	615	533	1175	0	447
5	1901	158	155	1093	0	852	1799	1106	729	1901	1037	979	1918	204	746
6	2453	18	18	2111	0	110	589	1528	281	2449	123	2141	2125	22	2661
7	1596	215	128	1634	0	580	763	532	523	1593	26	1329	1697	227	1049
8	2140	80	3	2733	0	255	1140	1813	105	2142	0	2328	2966	0	1552
9	2107	86	45	1931	0	309	455	2193	412	2050	15	1756	2356	54	1521
10	1090	17	10	685	0	234	32	899	136	1090	34	589	1181	237	224
11	1097	201	166	541	0	135	56	925	29	1064	0	231	1117	164	110
12	548	0	2	299	0	219	147	1551	14	515	0	273	1574	23	217
13	744	149	158	253	0	366	125	254	78	720	267	109	760	115	104
14	273	36	0	223	0	0	0	257	0	273	0	12	293	16	6
21	130	0	41	37	0	0	0	296	0	130	0	121	668	688	259
22	1165	7557	160	1226	0	1572	1063	849	20	1154	1681	875	7505	9414	719
23	1194	6995	581	938	0	157	898	787	38	1177	110	886	5235	1917	480
25	0	0	14	14	0	0	0	0	0	0	0	0	0	0	0
26	1512	491	10456	1293	0	811	644	1159	55	1438	600	1360	10237	399	636
27	94	0	447	101	0	0	0	71	0	94	39	67	388	0	56
28	143	79	223	124	0	0	28	97	0	127	0	70	209	57	41
31	1510	733	610	806	0	3265	288	7	1217	1455	3276	604	1569	660	613
32	836	49	8	905	0	4472	353	405	314	808	1889	440	1751	142	487
33	1246	1076	534	252	0	3021	38	0	237	523	3058	128	1698	2085	154
34	107	0	47	4	0	190	0	0	31	62	212	0	136	0	0
35	18	0	0	0	0	327	0	0	0	18	442	0	84	0	0
41	62	0	0	33	0	8	0	22	0	61	31	0	68	14	0
44	75	79	0	0	0	0	0	0	0	0	0	0	75	75	0
46	23	16	0	0	0	274	261	0	0	23	264	0	33	33	0
48	10	0	38	2	0	0	23	0	22	10	17	10	23	0	10
Totals	26089	19087	14588	20502	0	20877	10275	16450	8098	24858	16686	16077	50914	17570	12933

Asem Div	Building Construction	Mechanical	Electrical	Commercial Renovation	Square Foot	Site Work Landscape	Assemblies	Green Building	Interior	Concrete Masonry	Heavy Construction	Light Commercial	Facilities Construction	Plumbing	Asm Div	Residential
A		15	0	188	164	577	598	0	0	536	571	154	24	0	1	378
B		0	0	848	2554	0	5661	56	329	1976	368	2094	174	0	2	211
C		0	0	647	954	0	1334	0	1642	146	0	844	255	0	3	588
D		1057	941	712	1859	72	2538	330	825	0	0	1345	1105	1088	4	851
E		0	0	86	261	0	301	0	5	0	0	258	5	0	5	391
F		0	0	0	114	0	143	0	0	0	0	114	0	0	6	357
G		527	447	318	312	3377	792	0	0	534	1349	205	293	677	7	307
															8	760
															9	80
															10	0
															11	0
															12	0
Totals		1599	1388	2799	6218	4026	11367	386	2801	3192	2288	5014	1856	1765		3923

Reference Section	Building Construction Costs	Mechanical	Electrical	Commercial Renovation	Square Foot	Site Work Landscape	Assem.	Green Building	Interior	Concrete Masonry	Open Shop	Heavy Construction	Light Commercial	Facilities Construction	Plumbing	Resi.
Reference Tables	yes	yes	yes	yes	no	yes	yes	yes	yes	yes	yes	yes	yes	yes	yes	yes
Models					111			25					50			28
Crews	578	578	578	556		578		578	578	578	555	578	555	556	578	555
Equipment Rental Costs	yes	yes	yes	yes		yes		yes	yes	yes	yes	yes	yes	yes	yes	yes
Historical Cost Indexes	yes	yes	yes	yes	yes	yes	yes	yes	yes	yes	yes	yes	yes	yes	yes	no
City Cost Indexes	yes	yes	yes	yes	yes	yes	yes	yes	yes	yes	yes	yes	yes	yes	yes	yes

Our Online Estimating Solution
Competitive Cost Estimates Made Easy

Our online estimating solution is a web-based service that provides accurate and up-to-date cost information to help you build competitive estimates or budgets in less time.

Quick, intuitive, easy to use and automatically updated, you'll gain instant access to hundreds of thousands of material, labor, and equipment costs from RSMeans' comprehensive database, delivering the information you need to build budgets and competitive estimates every time.

With our online estimating solutions, you can perform quick searches to locate specific costs and adjust costs to reflect prices in your geographic area. Tag and store your favorites for fast access to your frequently used line items and assemblies and clone estimates to save time. System notifications will alert you as updated data becomes available. This data is automatically updated throughout the year.

Our visual, interactive estimating features help you create, manage, save and share estimates with ease! You'll enjoy increased flexibility with customizable advanced reports. Easily edit custom report templates and import your company logo onto your estimates.

	Core	Advanced	Complete
Unit Prices	✓	✓	✓
Assemblies	⊘	✓	✓
Sq Foot Models	⊘	⊘	✓
Editable Sq Foot Models	⊘	⊘	✓
Editable Assembly Components	⊘	✓	✓
Custom Cost Data	⊘	✓	✓
User Defined Components	⊘	✓	✓
Advanced Reporting & Customization	⊘	✓	✓
Union Labor Type	✓	✓	✓

Continue to check our website at www.RSMeans.com for more product offerings.

Estimate with Precision
Find everything you need to develop complete, accurate estimates.
- Verified costs for construction materials
- Equipment rental costs
- Crew sizing, labor hours and labor rates
- Localized costs for U.S. and Canada

Save Time & Increase Efficiency
Make cost estimating and calculating faster and easier than ever with secure, online estimating tools.
- Quickly locate costs in the searchable database
- Create estimates in minutes with RSMeans cost lines
- Tag and store favorites for fast access to frequently used items

Improve Planning & Decision-Making
Back your estimates with complete, accurate and up-to date cost data for informed business decisions.
- Verify construction costs from third parties
- Check validity of subcontractor proposals
- Evaluate material and assembly alternatives

Increase Profits
Use our online estimating solution to estimate projects quickly and accurately, so you can gain an edge over your competition.
- Create accurate and competitive bids
- Minimize the risk of cost overruns
- Reduce variability

Access the data online

Search for unit prices by keyword   Leverage the most up-to-date data   Build and export estimates

Try it free for 30 days   www.rsmeans.com/2018freetrial

2018 Seminar Schedule   📞 877-620-6245

Note: call for exact dates, locations, and details as some cities are subject to change.

Location	Dates	Location	Dates
Seattle, WA	January and August	San Francisco, CA	June
Dallas/Ft. Worth, TX	January	Bethesda, MD	June
Austin, TX	February	El Segundo, CA	August
Anchorage, AK	March and September	Dallas, TX	September
Las Vegas, NV	March	Raleigh, NC	October
New Orleans, LA	March	Salt Lake City, UT	October
Washington, DC	April and September	Baltimore, MD	November
Phoenix, AZ	April	Orlando, FL	November
Toronto	May	San Diego, CA	December
Denver, CO	May	San Antonio, TX	December

Gordian also offers a suite of online RSMeans data self-paced offerings. Check our website www.RSMeans.com for more information.

Self-Paced Professional Development Courses

Training on how to use RSMeans data and estimating tools, as well as Professional Development courses on industry topics are now offered in a convenient self-paced format. These courses are on-demand and allow more flexibility to learn around your busy schedule, while saving the cost of travel and time.

Current course offerings include:

Facilities Construction Estimating—our best-selling live class now available as an on-demand training course! Let the subject matter experts of construction estimating—the RSMeans Engineering Staff—walk you through the basics and much more of estimating for renovation and facilities construction.

RSMeansOnline.com Training—learn the ins and outs of the flagship delivery method of RSMeans data!

The Construction Process—how much do you and your team really know about the ins and outs of the "contract-side" of a construction project? This self-paced course will clarify best practices for items such as schedules, change orders, and project closeout.

These self-paced training courses can be completed over the course of 45 days and are comprised of multiple lessons with documentation, video presentation, software simulation, assessment quizzes and certificate of completion.

Site Work Estimating with RSMeans data

This new one-day program focuses directly on site work costs, a unique portion of most construction projects that often is the wild card in determining whether you have developed a good estimate or not. Accurately scoping, quantifying, and pricing site preparation, underground utility work, and improvements to exterior site elements are often the most difficult estimating tasks on any project. The program takes the participant from preparing a never-developed site through underground utility installation, pad preparation, paving and sidewalks, and landscaping. Attendees will use the full array of site work cost data through the RSMeans online program and participate in exercises to strengthen their estimating skills.

Some of what you'll learn:

- Evaluation of site work and understanding site scope of work.
- Site work estimating topics including: site clearing, grading, excavation, disposal and trucking of materials, erosion control devices, backfill and compaction, underground utilities, paving, sidewalks, fences & gates, and seeding & planting.
- Unit price site work estimates—Correct use of RSMeans site work cost data to develop a cost estimate.
- Using and modifying assemblies—Save valuable time when estimating site work activities using custom assemblies.

Who should attend: Engineers, contractors, estimators, project managers, owner's representatives, and others who are concerned with the proper preparation and/or evaluation of site work estimates.

Please bring a laptop with ability to access the internet.

Training for our Online Estimating Solution

Construction estimating is vital to the decision-making process at each state of every project. Our online solution works the way you do. It's systematic, flexible and intuitive. In this one-day class you will see how you can estimate any phase of any project faster and better.

Some of what you'll learn:
- Customizing our online estimating solution
- Making the most of RSMeans "Circle Reference" numbers
- How to integrate your cost data
- Generating reports, exporting estimates to MS Excel, sharing, collaborating and more

Also offered as a self-paced or on-site training program!

Facilities Construction Estimating

In this two-day course, professionals working in facilities management can get help with their daily challenges to establish budgets for all phases of a project.

Some of what you'll learn:
- Determining the full scope of a project
- Identifying the scope of risks and opportunities
- Creative solutions to estimating issues
- Organizing estimates for presentation and discussion
- Special techniques for repair/remodel and maintenance projects
- Negotiating project change orders

Who should attend: facility managers, engineers, contractors, facility tradespeople, planners, and project managers.

Construction Cost Estimating: Concepts and Practice

This one-day introductory course to improve estimating skills and effectiveness starts with the details of interpreting bid documents and ends with the summary of the estimate and bid submission.

Some of what you'll learn:
- Using the plans and specifications to create estimates
- The takeoff process—deriving all tasks with correct quantities
- Developing pricing using various sources; how subcontractor pricing fits in
- Summarizing the estimate to arrive at the final number
- Formulas for area and cubic measure, adding waste and adjusting productivity to specific projects
- Evaluating subcontractors' proposals and prices
- Adding insurance and bonds
- Understanding how labor costs are calculated
- Submitting bids and proposals

Who should attend: project managers, architects, engineers, owners' representatives, contractors, and anyone who's responsible for budgeting or estimating construction projects.

Maintenance & Repair Estimating for Facilities

This two-day course teaches attendees how to plan, budget, and estimate the cost of ongoing and preventive maintenance and repair for existing buildings and grounds.

Some of what you'll learn:
- The most financially favorable maintenance, repair, and replacement scheduling and estimating
- Auditing and value engineering facilities
- Preventive planning and facilities upgrading
- Determining both in-house and contract-out service costs
- Annual, asset-protecting M&R plan

Who should attend: facility managers, maintenance supervisors, buildings and grounds superintendents, plant managers, planners, estimators, and others involved in facilities planning and budgeting.

Practical Project Management for Construction Professionals

In this two-day course, acquire the essential knowledge and develop the skills to effectively and efficiently execute the day-to-day responsibilities of the construction project manager.

Some of what you'll learn:
- General conditions of the construction contract
- Contract modifications: change orders and construction change directives
- Negotiations with subcontractors and vendors
- Effective writing: notification and communications
- Dispute resolution: claims and liens

Who should attend: architects, engineers, owners' representatives, and project managers.

Mechanical & Electrical Estimating

This two-day course teaches attendees how to prepare more accurate and complete mechanical/electrical estimates, avoid the pitfalls of omission and double-counting, and understand the composition and rationale within the RSMeans mechanical/electrical database.

Some of what you'll learn:
- The unique way mechanical and electrical systems are interrelated
- M&E estimates—conceptual, planning, budgeting, and bidding stages
- Order of magnitude, square foot, assemblies, and unit price estimating
- Comparative cost analysis of equipment and design alternatives

Who should attend: architects, engineers, facilities managers, mechanical and electrical contractors, and others who need a highly reliable method for developing, understanding, and evaluating mechanical and electrical contracts.

Visit RSMeans.com/O

Unit Price Estimating

This interactive two-day seminar teaches attendees how to interpret project information and process it into final, detailed estimates with the greatest accuracy level.

The most important credential an estimator can take to the job is the ability to visualize construction and estimate accurately.

Some of what you'll learn:
- Interpreting the design in terms of cost
- The most detailed, time-tested methodology for accurate pricing
- Key cost drivers—material, labor, equipment, staging, and subcontracts
- Understanding direct and indirect costs for accurate job cost accounting and change order management

Who should attend: corporate and government estimators and purchasers, architects, engineers, and others who need to produce accurate project estimates.

Training for our CD Estimating Solution

This one-day course helps users become more familiar with the functionality of the CD. Each menu, icon, screen, and function found in the program is explained in depth. Time is devoted to hands-on estimating exercises.

Some of what you'll learn:
- Searching the database using all navigation methods
- Exporting RSMeans data to your preferred spreadsheet format
- Viewing crews, assembly components, and much more
- Automatically regionalizing the database

This training session requires you to bring a laptop computer to class.

When you register for this course you will receive an outline for your laptop requirements.

Also offered as a self-paced or on-site training program!

Facilities Estimating Using the CD

This two-day class combines hands-on skill-building with best estimating practices and real-life problems. You will learn key concepts, tips, pointers, and guidelines to save time and avoid cost oversights and errors.

Some of what you'll learn:
- Estimating process concepts
- Customizing and adapting RSMeans cost data
- Establishing scope of work to account for all known variables
- Budget estimating: when, why, and how
- Site visits: what to look for and what you can't afford to overlook
- How to estimate repair and remodeling variables

This training session requires you to bring a laptop computer to class.

Who should attend: facility managers, architects, engineers, contractors, facility tradespeople, planners, project managers, and anyone involved with JOC, SABRE, or IDIQ.

Life Cycle Cost Estimating for Facility Asset Managers

Life Cycle Cost Estimating will take the attendee through choosing the correct RSMeans database to use and then correctly applying RSMeans data to their specific life cycle application. Conceptual estimating through RSMeans new building models, conceptual estimating of major existing building projects through RSMeans renovation models, pricing specific renovation elements, estimating repair, replacement and preventive maintenance costs today and forward up to 30 years will be covered.

Some of what you'll learn:
- Cost implications of managing assets
- Planning projects and initial & life cycle costs
- How to use RSMeans data online

Who should attend: facilities owners and managers and anyone involved in the financial side of the decision making process in the planning, design, procurement, and operation of facility real assets.

Please bring a laptop with ability to access the internet.

Assessing Scope of Work for Facilities Construction Estimating

This two-day practical training program addresses the vital importance of understanding the scope of projects in order to produce accurate cost estimates for facility repair and remodeling.

Some of what you'll learn:
- Discussions of site visits, plans/specs, record drawings of facilities, and site-specific lists
- Review of CSI divisions, including means, methods, materials, and the challenges of scoping each topic
- Exercises in scope identification and scope writing for accurate estimating of projects
- Hands-on exercises that require scope, take-off, and pricing

Who should attend: corporate and government estimators, planners, facility managers, and others who need to produce accurate project estimates.

Building Systems and the Construction Process

This one-day course was written to assist novices and those outside the industry in obtaining a solid understanding of the construction process - from both a building systems and construction administration approach.

Some of what you'll learn:
- Various systems used and how components come together to create a building
- Start with foundation and end with the physical systems of the structure such as HVAC and Electrical
- Focus on the process from start of design through project closeout

This training session requires you to bring a laptop computer to class.

Who should attend: building professionals or novices to help make the crossover to the construction industry; suited for anyone responsible for providing high level oversight on construction projects.

Professional Development

569

Registration Information

Register early and save up to $100!

Register 30 days before the start date of a seminar and save $100 off your total fee. Note: This discount can be applied only once per order. It cannot be applied to team discount registrations or any other special offer.

How to register

By Phone
Register by phone at 877-620-6245

Online
Register online at www.RSMeans.com/products/seminars.aspx

Note: Purchase Orders or Credits Cards are required to register.

Two-day seminar registration fee - $1,045.

One-Day Construction Cost Estimating or Building Systems and the Construction Process - $630.

Government pricing

All federal government employees save off the regular seminar price. Other promotional discounts cannot be combined with the government discount.

Team discount program

For over five attendee registrations. Call for pricing: 781-422-5115

Refund policy

Cancellations will be accepted up to ten business days prior to the seminar start. There are no refunds for cancellations received later than ten working days prior to the first day of the seminar. A $150 processing fee will be applied for all cancellations. Written notice of the cancellation is required. Substitutions can be made at any time before the session starts. No-shows are subject to the full seminar fee.

Note: Pricing subject to change.

AACE approved courses

Many seminars described and offered here have been approved for 14 hours (1.4 recertification credits) of credit by the AACE International Certification Board toward meeting the continuing education requirements for recertification as a Certified Cost Engineer/Certified Cost Consultant.

AIA Continuing Education

We are registered with the AIA Continuing Education System (AIA/CES) and are committed to developing qualit learning activities in accordance with the CES criteria. Mar seminars meet the AIA/CES criteria for Quality Level 2. AIA members may receive 14 learning units (LUs) for each two day RSMeans course.

Daily course schedule

The first day of each seminar session begins at 8:30 a.m. and ends at 4:30 p.m. The second day begins at 8:00 a.m and ends at 4:00 p.m. Participants are urged to bring a hand-held calculator since many actual problems will be worked out in each session.

Continental breakfast

Your registration includes the cost of a continental breakfast and a morning and afternoon refreshment break These informal segments allow you to discuss topics of mutual interest with other seminar attendees. (You are fre to make your own lunch and dinner arrangements.)

Hotel/transportation arrangements

We arrange to hold a block of rooms at most host hotels. To take advantage of special group rates when making you reservation, be sure to mention that you are attending the RSMeans Institute data seminar. You are, of course, free to stay at the lodging place of your choice. (Hotel reservatio and transportation arrangements should be made directly by seminar attendees.)

Important

Class sizes are limited, so please register as soon as possible.